D0903495

NELSON

Chemistry 11

NELSON

Chemistry 11

Authors

Dr. Frank Jenkins
University of Alberta

Hans van Kessel
St. Albert Protestant Schools

Lucille Davies
Limestone District School Board

Dr. Oliver Lantz
University of Alberta

Patricia Thomas
Ottawa-Carleton District School Board

Dick Tompkins
Edmonton Public School Board

Program Consultant

Maurice Di Giuseppe
Toronto Catholic District School Board

NELSON
THOMSON LEARNING™

Australia • Canada • Mexico • Singapore • Spain • United Kingdom • United States

Nelson Chemistry 11

Dr. Frank Jenkins
Hans van Kessel
Lucille Davies
Dr. Oliver Lantz
Patricia Thomas
Dick Tompkins

Director of Publishing
David Steele

Publisher
Kevin Martindale

Program Manager
Colin Bisset

Developmental Editors
Julia Lee
Betty Robinson

Editorial Assistant
Matthew Roberts

Senior Managing Editor
Nicola Balfour

Senior Production Editor
Rosalyn Steiner

Copy Editor
Ruth Peckover

Proofreader
Laura Edlund

Production Coordinator
Sharon Latta Paterson

Creative Director
Angela Cluer

Art Management
Suzanne Peden

Design and Composition
Marnie Benedict
Anne Bradley
Susan Calverley
Angela Cluer
Zenaida Diores
Krista Donnelly
Erich Falkenberg
Tammy Gay
Julie Greener
Alicja Jamorski
Linda Neale
Peter Papayanakis
Suzanne Peden
Ken Phipps
June Reynolds
Peggy Rhodes
Katherine Strain
Janet Zanette

Cover Design
Katherine Strain

Cover Image
© Daryl Benson/Masterfile

Photo Research and Permissions
Hamish Robertson

Printer
Transcontinental Printing Inc.

Printed and bound in Canada
1 2 3 4 04 03 02 01

For more information contact Nelson Thomson Learning, 1120 Birchmount Road, Toronto, Ontario, M1K 5G4. Or you can visit our Internet site at http://www.nelson.com

National Library of Canada Cataloguing in Publication Data

Main entry under title:
Nelson chemistry 11

Includes index.
ISBN 0-17-612101-3

1. Chemistry.
I. Jenkins, Frank, 1944- .

QD33.N44 2001 540
C2001-930492-7

Acknowledgments
Nelson Thomson Learning and the authors of *Nelson Chemistry 11* thank the staff and students of Mary Ward Catholic Secondary School for the use of their facilities, and for the grace and generosity of their help.

Reviewers

NELSON SENIOR SCIENCE ADVISORY PANEL

Oksana Baczynsky
Toronto District School Board, ON

Terry Brown
Thames Valley District School Board, ON

Doug De La Matter
Renfrew District School Board, ON

Xavier Fazio
Halton Catholic District School Board, ON

Ted Gibb
Thames Valley District School Board, ON

Diana Hall
Ottawa-Carleton District School Board, ON

William Konrad
Formerly of Lambton Kent District School Board, ON

Clifford Sampson
Appleby College, ON

ASSESSMENT CONSULTANT

Damian Cooper
Halton District School Board, ON

SKILLS HANDBOOK CONSULTANT

Barry LeDrew
Formerly of Newfoundland Department of Education

ACCURACY REVIEWERS

Karen A. Henderson
Physical Sciences
University of Toronto at Scarborough

Dr. David Stone
Department of Chemistry
University of Toronto

Carl Twiddy
Formerly of York Region District School Board, ON

SAFETY REVIEWERS

Ian Mackellar
STAO Safety Committee, ON

TEACHER REVIEWERS

Thomas S.H. Baxter
Lakehead District School Board, ON

Tom Bruin
Trillium-Lakelands District School Board, ON

Luis Bustos
Peterborough/Victoria/Northumberland and Clarington Catholic District School Board, ON

Richard Christensen
Peel District School Board, ON

John A. Dmytrasz
Dufferin-Peel Catholic District School Board, ON

Brenda Forbes
York Region District School Board, ON

Ann Harrison
Niagara Catholic District School Board, ON

Paul Inwood
Superior-Greenstone District School Board

Lucy Kisway
Hamilton-Wentworth District School Board

Lisa Maclachlan
Dufferin-Peel Catholic District School Board

Robert Nalepa
Halifax Regional School Board, NS

Al Orlando
Nipissing–Parry Sound Catholic District School Board

Mirella Palmisano
Dufferin-Peel Catholic District School Board

Mike Penrose
Peel District School Board

Randy Ralph
Avalon East School Board, NF

Marc James Robillard
Lambton Kent District School Board

Milan Sanader
Dufferin-Peel Catholic District School Board

Ann Senechal
School District 14, NB

Frank Villella
Hamilton-Wentworth Catholic District School Board

Contents

Unit 1

Matter and Chemical Bonding

Unit 1 Are You Ready? 4

Chapter 1 The Nature of Matter 8

1.1 Elements and the Periodic Table 10
INVESTIGATION 1.1.1: Element or Compound? 12
1.2 Developing a Model of the Atom 22
ACTIVITY 1.2.1: Developing a Model of a Black Box 23
1.3 Understanding Atomic Mass 26
ACTIVITY 1.3.1: Modelling Half-Life 31
Careers in Nuclear Science 33
1.4 Toward a Modern Atomic Theory 37
INVESTIGATION 1.4.1: Atomic Spectra 40
ACTIVITY 1.4.1: Creating a Flame Test Key 42
INVESTIGATION 1.4.2: Identifying an Unknown Metal in a Compound with a Flame Test 44
1.5 Trends in the Periodic Table 48
LAB EXERCISE 1.5.1: Reactivity of Alkali Metals 49
ACTIVITY 1.5.1: Graphing First Ionization Energy 54
ACTIVITY 1.5.2: Graphing Electronegativity 57
Chapter 1 Summary 61
Chapter 1 Review 62

Chapter 2 Chemical Bonding 64

2.1 Classifying Compounds 66
INVESTIGATION 2.1.1: Comparing Ionic and Molecular Compounds 66
2.2 Ionic Bonding 68
2.3 Covalent Bonding 75
2.4 Electronegativity, Polar Bonds, and Polar Molecules 82
2.5 The Names and Formulas of Compounds 89
Chapter 2 Summary 102
Chapter 2 Review 103

Chapter 3 Chemical Reactions 106

3.1 Recognizing and Understanding Chemical Changes 108
ACTIVITY 3.1.1: Understanding Chemical Reactions 112
3.2 Combustion, Synthesis, and Decomposition Reactions 114
INVESTIGATION 3.2.1: The Combustion of Butane 116
3.3 Single Displacement Reactions 123
INVESTIGATION 3.3.1: A Single Displacement Reaction 124
INVESTIGATION 3.3.2: Developing an Activity Series 125
INVESTIGATION 3.3.3: Testing the Activity Series 131
LAB EXERCISE 3.3.1: Corrosion in Seawater 132
Careers in Metallurgy 135
3.4 Double Displacement Reactions 136
INVESTIGATION 3.4.1: A Double Displacement Reaction 139
INVESTIGATION 3.4.2: Identifying an Unknown 141
Chapter 3 Summary 145
Chapter 3 Review 146

Unit 1 Performance Task: New Elements 148
Unit 1 Review 150

Unit 2

Quantities in Chemical Reactions

Unit 2 Are You Ready? 156

Chapter 4 Quantities in Chemical Formulas 158

4.1 Proportions in Compounds 160
4.2 Relative Atomic Mass and Isotopic Abundance 163
LAB EXERCISE 4.2.1: Determination of Relative Atomic Mass 166
4.3 The Mole and Molar Mass 167
4.4 Calculations Involving the Mole Concept 171
4.5 Percentage Composition 178
INVESTIGATION 4.5.1: Percentage Composition by Mass of Magnesium Oxide 180
4.6 Empirical and Molecular Formulas 185
4.7 Calculating Chemical Formulas 187

INVESTIGATION 4.7.1: Determining the Formula of an Unknown Hydrate 195
Chapter 4 Summary 198
Chapter 4 Review 199

Chapter 5 Quantities in Chemical Equations 202

5.1 Quantitative Analysis 204
INVESTIGATION 5.1.1: Quantitative Analysis of Sodium Carbonate Solution 206
Quantitative Careers 208
5.2 Balancing Chemical Equations 210
5.3 Balancing Nuclear Equations 216
5.4 Calculating Masses of Reactants and Products 223
LAB EXERCISE 5.4.1: Testing Gravimetric Stoichiometry 227
5.5 Calculating Limiting and Excess Reagents 230
LAB EXERCISE 5.5.1: Designing and Testing Gravimetric Stoichiometry: Calculating an Excess Reagent 232
INVESTIGATION 5.5.1: Which Reagent Is Limiting and How Much Precipitate Is Formed? 236
5.6 The Yield of a Chemical Reaction 238
INVESTIGATION 5.6.1: Determining Percentage Yield in a Chemical Reaction 241
5.7 Chemistry in Technology 245
Chapter 5 Summary 251
Chapter 5 Review 252

Unit 2 Performance Task: Quantitative Analysis of a Reaction 254
Unit 2 Review 256

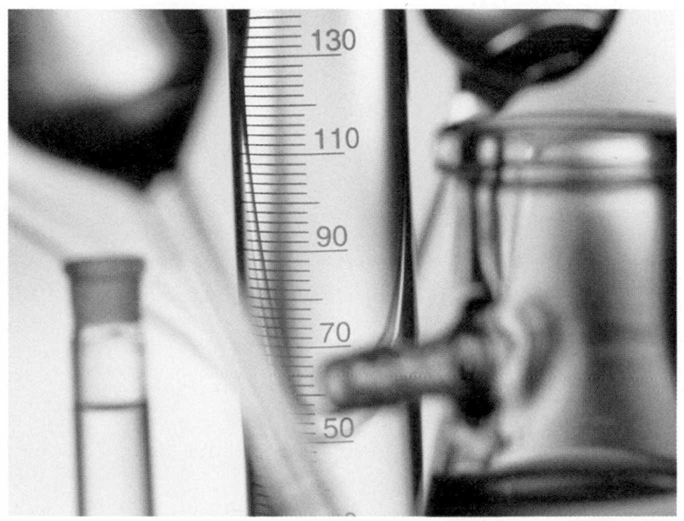

Unit 3
Solutions and Solubility

Unit 3 Are You Ready? 262

Chapter 6 The Nature and Properties of Solutions 264

6.1 Defining a Solution 266
INVESTIGATION 6.1.1: Qualitative Chemical Analysis 268
LAB EXERCISE 6.1.1: Identification of Solutions 270
6.2 Explaining Solutions 272
LAB EXERCISE 6.2.1: Testing a Hypothesis on Dissolving 272
INVESTIGATION 6.2.1: Polar and Nonpolar Solutes 274
LAB EXERCISE 6.2.2: Predicting High and Low Solubilities 275
6.3 Solution Concentration 281
6.4 Drinking Water 291
6.5 Solution Preparation 300
ACTIVITY 6.5.1: A Standard Solution from a Solid 301
ACTIVITY 6.5.2: A Standard Solution by Dilution 305
Chapter 6 Summary 308
Chapter 6 Review 309

Chapter 7 Solubility and Reactions 312

7.1 Solubility 314
INVESTIGATION 7.1.1: Solubility Curve of a Solid 314
LAB EXERCISE 7.1.1: Solubility of a Gas 318
INVESTIGATION 7.1.2: The Solubility of Sodium Chloride in Water 322
7.2 Hard Water Treatment 328
7.3 Reactions in Solution 331
INVESTIGATION 7.3.1: Precipitation Reactions in Solution 331
7.4 Waste Water Treatment 337
7.5 Qualitative Chemical Analysis 341
INVESTIGATION 7.5.1: Sequential Chemical Analysis in Solution 345
7.6 Quantitative Analysis 347
LAB EXERCISE 7.6.1: Quantitative Analysis in Solution 349
INVESTIGATION 7.6.1: Percentage Yield of Barium Sulfate 354
Chapter 7 Summary 357
Chapter 7 Review 358

Chapter 8 Acids and Bases 360

8.1 Understanding Acids and Bases 362
8.2 pH of a Solution 368
INVESTIGATION 8.2.1: Dilution and pH 372
8.3 Working with Solutions 376
Career Solutions 377
8.4 Acid–Base Theories 378
INVESTIGATION 8.4.1: Testing Arrhenius' Acid–Base Definitions 381
8.5 Acid–Base Reactions 393
INVESTIGATION 8.5.1: Titration Analysis of Vinegar 398
Chapter 8 Summary 402
Chapter 8 Review 403

Unit 3 Performance Task: Analysis of ASA 406
Unit 3 Review 408

Unit 4

Gases and Atmospheric Chemistry

Unit 4 Are You Ready? 414

Chapter 9 The Gas State 416

9.1 States of Matter 418
ACTIVITY 9.1.1: Molecular Motion 421
9.2 Gas Laws 423
INVESTIGATION 9.2.1: Pressure and Volume of a Gas 426
INVESTIGATION 9.2.2: Temperature and Volume of a Gas 429
9.3 Compressed Gases 441
9.4 The Ideal Gas Law 443
INVESTIGATION 9.4.1: Determining the Molar Mass of a Gas 446
9.5 Air Quality 449
Chapter 9 Summary 455
Chapter 9 Review 456

Chapter 10 Gas Mixtures and Reactions 458

10.1 Mixtures of Gases 460
10.2 Reactions of Gases 466
INVESTIGATION 10.2.1: Molar Volume of a Gas 472
10.3 The Ozone Layer 475
10.4 Gas Stoichiometry 480
INVESTIGATION 10.4.1: Magnesium and Hydrochloric Acid: Testing the Gas Stoichiometry Method 484
10.5 Applications of Gases 487
Careers with Gases 490
Chapter 10 Summary 491
Chapter 10 Review 492

Unit 4 Performance Task: A Study of a Technological System 494
Unit 4 Review 496

Unit 5

Hydrocarbons and Energy

Unit 5 Are You Ready? 502

Chapter 11 Hydrocarbons 504

11.1 Organic Compounds 506
11.2 Refining Petroleum 512
LAB EXERCISE 11.2.1: Fractional Distillation 515
INVESTIGATION 11.2.1: Destructive Distillation 518
11.3 Combustion of Hydrocarbons 519
INVESTIGATION 11.3.1: Combustion of a Hydrocarbon 524
11.4 Alkanes and Cycloalkanes 528
11.5 Alkenes and Alkynes 543
INVESTIGATION 11.5.1: Evidence for Multiple Bonds 550
ACTIVITY 11.5.1: Structures and Properties of Isomers 552
INVESTIGATION 11.5.2: Preparation and Properties of Ethyne (Acetylene) 553
Careers in the Petrochemical Industry 557
Chapter 11 Summary 560
Chapter 11 Review 561

Chapter 12 Energy from Hydrocarbons 564

12.1 Classifying Energy Changes 566
INVESTIGATION 12.1.1: Building a Water Heater 572
12.2 Calorimetry 573
INVESTIGATION 12.2.1: Hot and Cold Packs 575
INVESTIGATION 12.2.2: Specific Heat of Combustion 580
12.3 Heats of Reaction 582
INVESTIGATION 12.3.1: Combustion of Octane 588
12.4 Our Use of Fossil Fuels 591
Chapter 12 Summary 595
Chapter 12 Review 596

Unit 5 Performance Task: A Study of Gasoline 598
Unit 5 Review 600

Appendixes 604

Appendix A Skills Handbook 606
Appendix B Safety Skills 624
Appendix C Reference 630
Appendix D Answers 638

Glossary 640
Index 647
Credits 652
Periodic Table 654

Unit 1

Matter and Chemical Bonding

"Chemistry is a field of science that quickly embraces new technologies, so it was a natural area to capture my attention. I have always been fascinated by new, high-tech equipment. In my work, studying reactions, I get to use two remarkable devices: two ultrafast pulsed lasers that fill two laboratories. They produce very powerful bursts of light that last for time periods as short as 100 fs. (A femtosecond, 10^{-15} s, is an extremely short period of time. There are more femtoseconds in one second than there have been days since the beginning of the universe!) We use these lasers to study reactions in which bonds form and break very quickly. We are trying to learn how these reactions work. We do this fundamental research because it is challenging and exciting, and because when scientists and technologists understand more about chemical processes, we will be able to design cleaner and more efficient processes for industry, which will benefit every aspect of our lives."

Frances L. Cozens, Associate Professor, Dalhousie University, Halifax

Overall Expectations

In this unit, you will be able to

- show that you understand the relationship among periodic trends, types of chemical bonding, and the properties of ionic and molecular compounds;
- carry out laboratory studies of chemical reactions, and analyze chemical reactions in terms of the type of reaction and the reactivity of starting materials;
- use appropriate symbols and formulas to represent the structure and bonding of chemical substances; and
- describe how an understanding of matter and its properties can lead to the production of useful substances and new technologies.

Unit

1

Matter and Chemical Bonding

Are You Ready?

Safety and Technical Skills

1. Draw a floor plan of the laboratory where you will be working. On your plan indicate the location of the following:
   (a) entrances (exits), including the fire exit
   (b) storage for aprons and eye protection
   (c) eyewash station
   (d) first-aid kit
   (e) fire extinguisher(s)
   (f) fire blanket and/or shower
   (g) container for broken glass
2. (a) What should you do if your clothing catches fire?
   (b) What should you do if someone else's clothing catches fire?
3. Examine **Figure 1**. What safety rules are the students breaking?

Figure 1
What is unsafe in this picture?

4. (a) Identify the WHMIS symbols in **Figure 2**.
   (b) What should you do immediately if any chemical comes in contact with your skin?

Figure 2
The Workplace Hazardous Materials Information System (WHMIS) provides information regarding hazardous products.

5. Describe the procedure for lighting a burner by giving the correct sequence for the photographs in **Figure 3**.

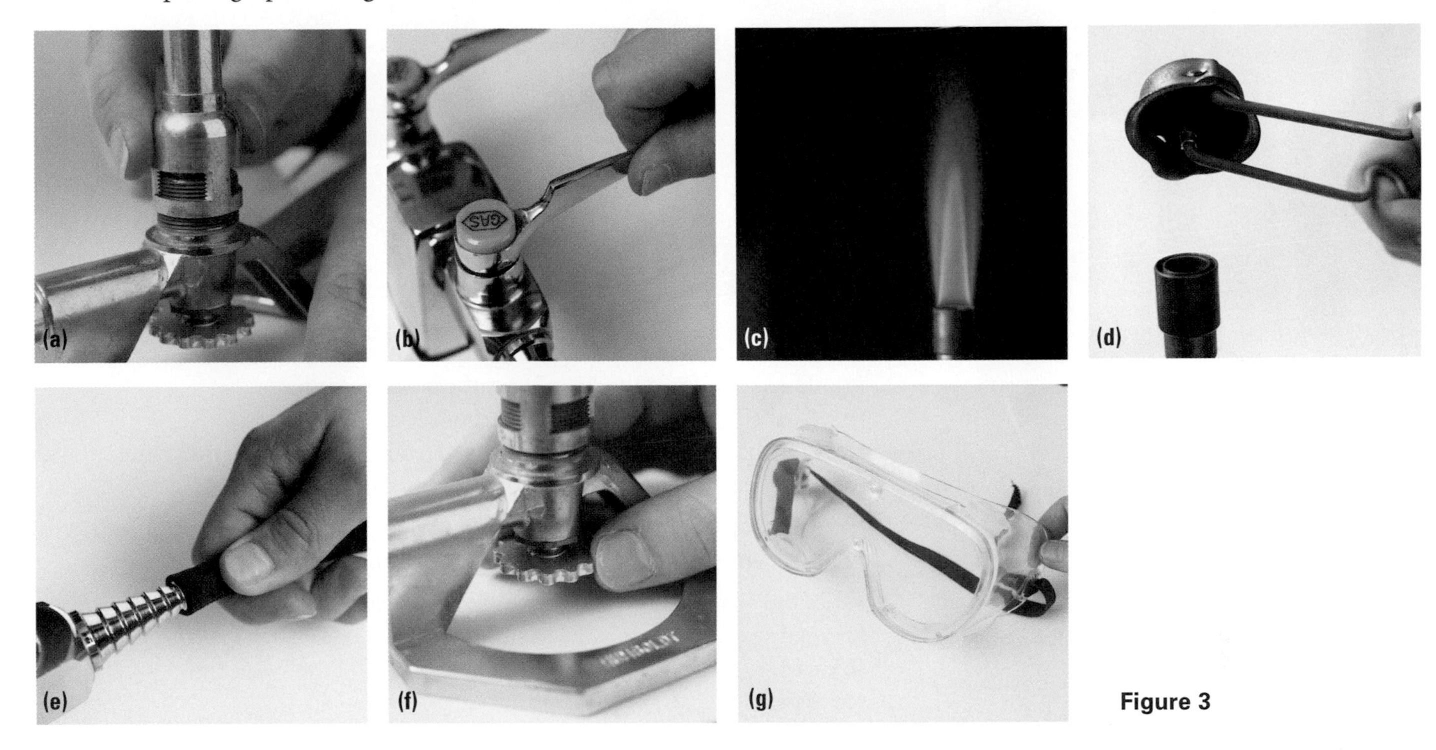

Figure 3

Inquiry and Communication Skills

6. (a) Design an experiment to answer the following scientific question: "How does altitude affect the boiling point of pure water?"
   (b) Identify dependent, independent, and controlled variables in your design.
7. Describe a standard test for each of the following substances:
   (a) oxygen gas
   (b) hydrogen gas
   (c) carbon dioxide gas

Knowledge and Understanding

8. Classify the following statements as empirical (based on measurements or observations) or theoretical (explanations or models).
   (a) Carbon burns with a yellow flame.
   (b) Carbon atoms react with oxygen molecules to produce carbon dioxide molecules.
   (c) Carbon burns faster if you blow on it.
   (d) Global warming results when carbon dioxide molecules absorb infrared radiation.
9. For each of (**a**), (**b**), and (**c**) in **Figure 4**, describe the evidence that might lead you to believe that a chemical change has taken place.

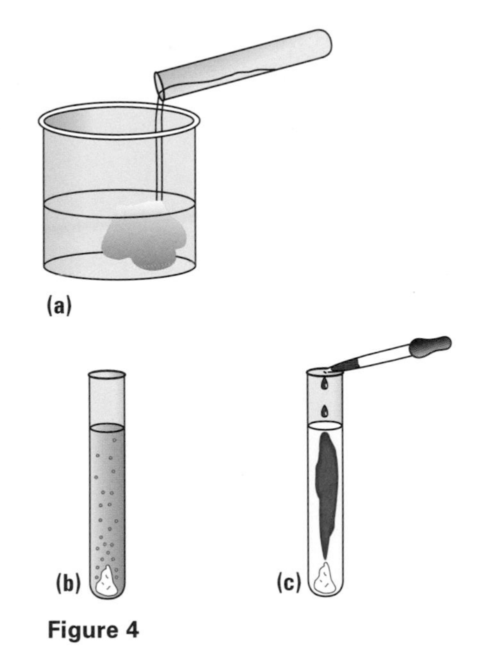

Figure 4

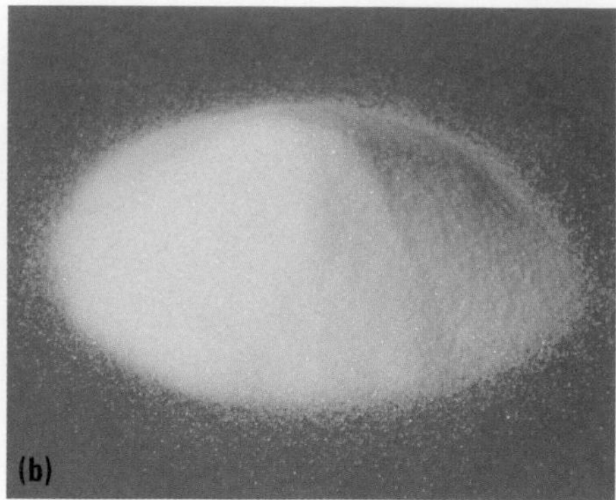

Figure 5
(a) Salad dressing
(b) Sucrose (sugar)
(c) Aluminum foil
(d) Flavoured drink

10. Classify the following statements as qualitative or quantitative.
    (a) The flame from the burning carbon was 4 cm high.
    (b) Coal is a primary source of carbon.
    (c) Coal has a higher carbon-to-hydrogen ratio than with other fuels.
    (d) Carbon is a black solid at standard conditions.
11. Identify each of the substances in **Figure 5** as an element, a compound, a homogeneous mixture, or a heterogeneous mixture.
12. Write the symbol for each of the following elements:
    (a) hydrogen
    (b) carbon
    (c) oxygen
    (d) sodium
    (e) aluminum
    (f) chlorine
    (g) calcium
    (h) iodine
    (i) lithium
    (j) gold
    (k) potassium
13. (a) Using the periodic table at the end of this text, identify two physical properties of the element zinc.
    (b) From your own experience, describe two chemical properties of the element oxygen.
14. Use the periodic table to answer the following questions concerning calcium.
    (a) In what period is it located?
    (b) In what group is it located?
    (c) What is its atomic number?
    (d) What is its atomic mass?
    (e) In what state would it be found at room temperature?

15. Identify the elements represented by the Bohr-Rutherford diagrams in **Figure 6**.

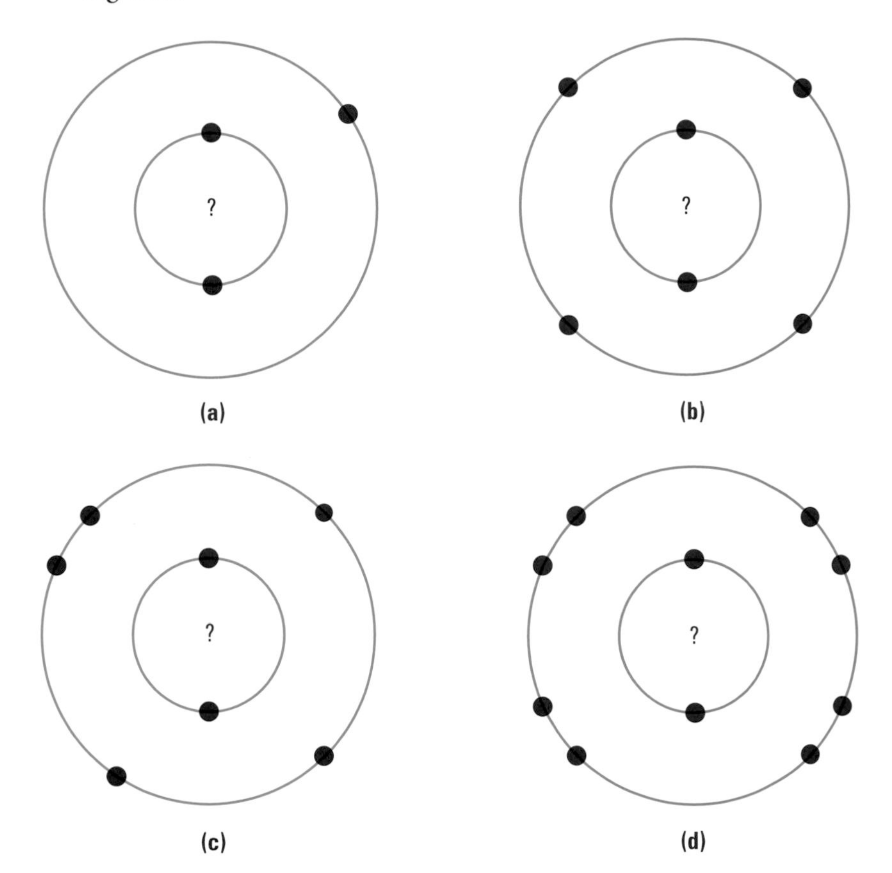

Figure 6
What elements do these diagrams represent?

16. Balance each of the following reaction equations:
   (a) $KClO_{3(s)} \rightarrow KCl_{(s)} + O_{2(g)}$
   (b) $Fe_{(s)} + O_{2(g)} \rightarrow Fe_2O_{3(s)}$
   (c) $Mg_{(s)} + HCl_{(aq)} \rightarrow MgCl_{2(aq)} + H_{2(g)}$
   (d) $N_{2(g)} + H_{2(g)} \rightarrow NH_{3(g)}$
17. Write word equations for each of the reactions in question 16.

Mathematical Skills

18. A student attempting to identify a pure substance from its density obtained the evidence shown in **Table 1**.
   (a) Construct a mass–volume graph from the evidence in **Table 1**.
   (b) From the graph, what mass of the substance has a volume of 12.7 mL?
   (c) Predict the volume of a sample of the same substance that has a mass of 8.0 g.

Table 1: Mass and Volume of an Unknown Solid

Mass (g)	Volume (mL)
1.2	3.6
1.8	5.5
2.3	6.9
3.1	9.2
6.9	20.7

Chapter

1

The Nature of Matter

In this chapter, you will be able to

- identify chemical substances and reactions in everyday use or of environmental significance;
- define atomic number, mass number, and atomic mass, and explain how these concepts relate to isotopes and radioisotopes;
- demonstrate an understanding of the periodic law, and describe how electron arrangements and forces in atoms can explain trends in the periodic table;
- compare and explain the reactivity of a series of elements, and relate this reactivity to their position in the periodic table;
- construct, analyze, and interpret graphs of properties that display trends within the periodic table.

It is possible to argue that chemistry has been responsible for some of the hazards of modern life. We are, after all, seeing environmental damage resulting from resource extraction, we are discovering the toxic effects of some products, and we are experiencing increasing difficulty disposing of our garbage, most of which is artificially produced, thanks to chemistry. However, to argue that way would be to ignore the underlying truth: Chemistry has been fundamental to the development of society as we know it. We now have cleaner fuel, more durable and safer paints, easy-care clothing, inexpensive fertilizers, life-saving pharmaceuticals, corrosion-resistant tools and machinery, and unusual new materials that we are using in interesting new ways (**Figure 1**). All of this innovation has made our lives better to some degree.

Chemistry is just another way to say "the understanding of the nature of matter." In order to further their understanding, chemists through the ages have relied upon scientific inquiry, carrying out investigations and making careful observations. The periodic table is an elegant way of summing up many of those investigations and our knowledge about matter and its constituents, the elements. The observations that went into the creation of the periodic table also helped to create the modern atomic theory. In turn, we can explain many of the patterns in the properties of the elements in terms of atomic theory. In this chapter, we will discuss the patterns used to classify the elements, and consider how these patterns are explained by atomic theory.

Reflect on your Learning

1. (a) Based on your current understanding of the atom, what are the constituents of an atom, and how are they arranged? According to this model, how do the atoms of the various elements differ from each other?
   (b) Based on your model of the atom, explain the organization of the elements in the periodic table.
2. What patterns in properties are you aware of among the elements of the periodic table? Suggest some explanations for these patterns, using your model of the atom.
3. (a) In what ways has the periodic table been useful to you in previous chemistry courses?
   (b) Examine the periodic table at the end of this text. Identify some differences between this table and the ones you used in previous grades. Speculate about how these differences might help you in this course.

Try This **Activity**

Observing and Classifying Matter

Every day we look at things and make decisions about what they are and what they might be used for. We know that a muffin is food, but a similarly sized and coloured rock is not. A cup of milk is liquid; a cup of flour is not. Making observations and classifying them are often the first steps toward forming an explanation of the nature of matter. This activity will encourage you to exercise both of these skills.

Materials: small samples of elements, such as carbon, sulfur, aluminum, silicon, magnesium, oxygen, iron, copper, tin, and lead

(a) Write a description of each sample.

- Divide the samples into two or more groups.

(b) Provide a rationale for your classification.

- Divide the samples a different way, again into two or more groups.

(c) Provide a rationale for your second classification.

- Divide the samples a third way, again into two or more groups.

(d) Provide a rationale for your third classification.
(e) Revisit your answer to question (a) and add to your descriptions.
(f) You were probably able to describe the samples in more detail at the end of the activity than you were at the beginning. Why?
(g) Are there other ways to classify the substances? Suggest some procedures you could use that could help you in further classification.

Wash your hands thoroughly after completing this activity.

Figure 1
The board, the clothes, and perhaps the fun, are products of chemistry.

1.1 Elements and the Periodic Table

element: a pure substance that cannot be broken down into simpler substances by chemical means (empirical definition); a substance composed entirely of only one kind of atom (theoretical definition)

compound: a pure substance that can be broken down by chemical means to produce two or more pure substances (empirical definition); a substance containing atoms of more than one element combined in fixed proportions (theoretical definition)

metal: an element that is a conductor of electricity, malleable, ductile, and lustrous

SATP (standard ambient temperature and pressure): exactly 25°C and 100 kPa

STP (standard temperature and pressure): exactly 0°C and 101.325 kPa

empirical definition: a statement that defines an object or process in terms of observable properties

Long before recorded history, humans used **elements** for many purposes. Copper, silver, and gold were put to many uses, both decorative and practical, in China and western Asia. At the dawn of recorded history, ancient peoples discovered that another element, tin, could be combined with copper to make a much harder material (bronze) from which they made stronger cutting tools, more effective weapons, and mirrors (**Figure 1**). About 2500 B.C., residents of what is now Turkey learned to extract iron from iron ore. The Egyptians used cobalt to make blue glass and antimony in cosmetics. In the first few centuries A.D., the Romans discovered how to use lead to make water pipes and eating utensils.

Many modern scientists study elements in great detail. As a result, we have developed thousands of new uses for elements and theories to explain their properties. We even have technologies that can create images of the atoms that make up elements. Elements are pure substances that cannot be broken down into simpler substances by chemical means. However, elements can be chemically combined to form more complex pure substances known as **compounds**.

The majority of the known elements are **metals**. All metals except mercury are solids at **SATP (standard ambient temperature and pressure)**, which is defined as exactly 25°C and 100 kPa. When you work with gases in Chapter 9, you will find that there is another set of standard conditions, **STP (standard temperature and pressure)**, which is exactly 0°C and 101.325 kPa.

From many observations of the properties of elements, scientists have developed an **empirical definition** for metals: They are malleable, ductile, and conductors of electricity. Metals are also described as *lustrous*, or shiny. You may be very familiar with some metals (e.g., iron, copper, calcium) but less so with others (e.g., vanadium, rhodium, osmium).

Figure 1
Bronze was put to use by artists from the moment of its discovery. This piece was created by an artist in ancient China.

The remaining known elements are mostly nonmetals. **Nonmetals** are generally nonconductors of electricity in their solid form. At SATP, the nonmetals are mostly gases or solids (**Figure 2**). Solid nonmetals are brittle and lack the lustre of metals. Some of the more familiar nonmetals include oxygen, chlorine, sulfur, and neon.

Some elements clearly do not fit the empirical definition for either metals or nonmetals. These elements are members of a small class known as **metalloids**. These elements are found near the blue "staircase line" that divides metals from nonmetals on the periodic table (**Figure 3**). Boron, silicon, and antimony are all metalloids.

nonmetal: an element that is generally a nonconductor of electricity and is brittle

metalloid: an element located near the "staircase line" on the periodic table; having some metallic and some nonmetallic properties

IUPAC: the International Union of Pure and Applied Chemistry; the international body that approves chemical names, symbols, and units

Naming Elements

As there are over 100 known elements; memorizing all of their names is a formidable task. Communicating across language barriers can be even more daunting. This issue was addressed by Swedish chemist Jöns Jakob Berzelius (1779–1848) in the early 19th century. In 1814, he suggested using a code of letters as symbols for elements. In this system, which is still used today, the symbol for each element consists of either a single uppercase letter or an uppercase letter followed by a lowercase letter. Because Latin was a common language of communication in Berzelius's day, many of the symbols were derived from the Latin names for the elements (**Table 1**). Today, although the names of elements are different in different languages, the same symbols are used in all languages. Scientific communication throughout the world depends on this language of symbols, which is international, precise, logical, and simple. The International Union of Pure and Applied Chemistry (**IUPAC**) specifies rules for chemical names and symbols.

Table 1: Selected Symbols and Names of Elements

International symbol	Latin	English	French	German
Ag	argentum	silver	argent	Silber
Au	aurum	gold	or	Gold
Cu	cuprum	copper	cuivre	Kupfer
Fe	ferrum	iron	fer	Eisen
Hg	hydrargyrum	mercury	mercure	Quecksilber
K	kalium	potassium	potassium	Kalium
Na	natrium	sodium	sodium	Natrium
Pb	plumbum	lead	plomb	Blei
Sb	stibium	antimony	antimonie	Antimon
Sn	stannum	tin	étain	Zinn

These rules are summarized in many scientific references, such as the *Handbook of Chemistry and Physics*, and used all over the world.

For elements, the first letter (only) of the symbol is always an uppercase letter. For example, the symbol for calcium is Ca, *not* CA, ca, or cA. New elements are still being synthesized. There are rules for naming these new elements. The names of new metallic elements end in "ium." The new elements first get a temporary name, with symbols consisting of three letters. Later permanent names and symbols are given, by a vote of IUPAC representatives from each country. The permanent name might reflect the country in which the element was discovered, or pay tribute to a notable scientist.

Practice

Understanding Concepts

1. Referring to the periodic table, classify each of the following elements as metals, metalloids, or nonmetals.
   (a) iron
   (b) aluminum
   (c) gallium
   (d) carbon
   (e) silver
   (f) oxygen
   (g) silicon
2. (a) What does the acronym IUPAC represent?
   (b) In a paragraph, explain why this organization is necessary.
3. State three sources of names for elements.

Making Connections

4. Choose five household products. List the ingredients from their product labels. Classify the ingredients as either elements or compounds. Further classify each of the elements, including elements in compounds, as metal, metalloid, or nonmetal.
5. Aluminum is a metal that has two important technological advantages: It has a low density and is easily cast, for example, into cooking pots and components for engines. Use the Internet to research the possible link between aluminum and Alzheimer's disease. Do you consider aluminum to be a significant environmental hazard?

   Follow the links for Nelson Chemistry 11, 1.1.

GO TO www.science.nelson.com

DID YOU KNOW?

Why Set a Standard?

SATP, standard ambient temperature and pressure, was established by international agreement of scientists to approximate normal conditions in the laboratory. For convenience and comparison, scientists need to communicate the value of a constant, such as density or solubility, under the same conditions as other scientists. Conditions such as these are called standard conditions. It is often easier to determine and more convenient to use values for SATP than for STP. It's not much fun working at 0°C, and the water keeps freezing!

Figure 2
Nonmetallic elements are rarely seen. The solids, such as these piles of sulfur, are generally used as raw materials to produce other substances.

12	13	14	15	16	17
	5 B	6 C	7 N	8 O	9 F
	13 Al	14 Si	15 P	16 S	17 Cl
30 Zn	31 Ga	32 Ge	33 As	34 Se	35 Br
48 Cd	49 In	50 Sn	51 Sb	52 Te	53 I
80 Hg	81 Tl	82 Pb	83 Bi	84 Po	85 At

Figure 3
The properties of some elements, which are like both metals and nonmetals, have led to the creation of the metalloids category of elements.

6. Choose two elements from the periodic table—one metal and one nonmetal. Research the discovery of these two elements. Include in your report how, when, where, and by whom they were discovered. What are some common industrial or technological applications of these elements?

   Follow the links for Nelson Chemistry 11, 1.1.

   GO TO www.science.nelson.com

Reflecting

7. Why would it be unwise for each country to choose its own names and symbols for elements?
8. Classification is not restricted to science. To make the world easier to understand, we classify music, food, vehicles, and people. Give an example of a useful classification system that you have encountered in your life. In what way is it useful? Describe a harmful example. Why do you think it is harmful?

DID YOU KNOW ?

Properties of Gold

Gold is easily identified as a metal because of its properties. It is an excellent conductor of electricity and so ductile that it can be drawn into the very fine wires needed in tiny electronic devices. It is so malleable that it can be hammered into gold leaf, or rolled into layers only a few atoms thick. Using this property, it would be possible to roll 1 cm^3 of gold into a layer so thin and so flat that it could cover the roof of your school.

Investigation 1.1.1

Element or Compound?

INQUIRY SKILLS

- ○ Questioning
- ○ Hypothesizing
- ● Predicting
- ○ Planning
- ● Conducting
- ● Recording
- ● Analyzing
- ● Evaluating
- ● Communicating

Before 1800, scientists distinguished elements from compounds by heating the substances to find out if they decomposed. If the products they obtained after cooling had different properties from the starting materials, then the experimenters concluded that decomposition had occurred, so the original substance was a compound, rather than an element. This experimental design was the only one known at that time.

In this investigation, you will test this experimental design by heating some samples and classifying the substances as either elements or compounds. Carry out the **Procedure** and complete the **Prediction**, **Analysis,** and **Evaluation** sections of the lab report.

You will use cobalt(II) chloride paper in a **diagnostic test** for water. If cobalt(II) chloride paper is exposed to a liquid or vapour, and the paper turns from blue to pink, then water is likely present. The presence of water above a solid sample may indicate that decomposition has taken place.

diagnostic test: an empirical test to detect the presence of a chemical

Question

Are water, bluestone, malachite, table salt, and sugar elements or compounds?

Prediction

(a) Referring to their chemical formulas, and using the definition for element and compound, predict the answer to the Question.

Experimental Design

A sample of each substance is heated, and any evidence of chemical decomposition is recorded. Decomposition is taken as evidence that the substance is a compound.

Materials

lab apron
distilled water, $H_2O_{(l)}$
malachite, $Cu(OH)_2 \cdot CuCO_{3(s)}$
sugar (sucrose), $C_{12}H_{22}O_{11(s)}$
250-mL Erlenmeyer flask
laboratory burner and striker
crucible
hot plate
utility clamp and stirring rod
piece of aluminum foil
eye protection
bluestone, $CuSO_4 \cdot 5H_2O_{(s)}$
table salt, $NaCl_{(s)}$
cobalt(II) chloride paper
laboratory scoop
ring stand and wire gauze
clay triangle
large (150-mL) test tube
medicine dropper

Wear eye protection and an apron.

Bluestone and malachite are harmful if swallowed.

Tie back hair and any articles of loose clothing.

Procedure

Part 1: Testing Water for Decomposition

1. Test some cobalt(II) chloride paper by placing a few drops of distilled water on the paper and noting any change in colour.
2. Pour distilled water into an Erlenmeyer flask until the water is about 1 cm deep. Set up the apparatus as shown in **Figure 4**(a).
3. Dry the inside of the top of the Erlenmeyer flask. Place a piece of cobalt(II) chloride paper across the mouth of the flask.
4. Carefully boil the water. Record your observations.

(a)

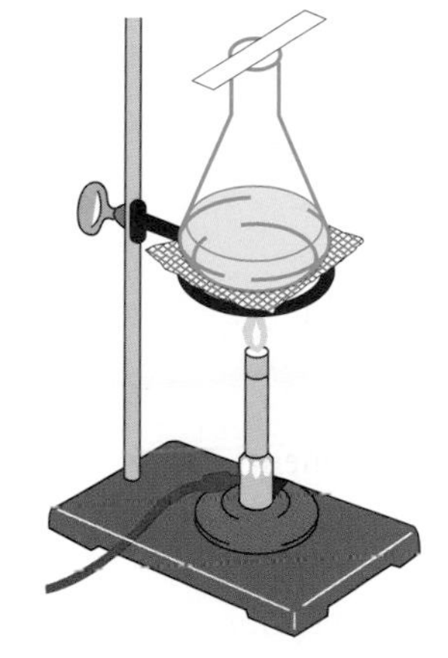

(b)

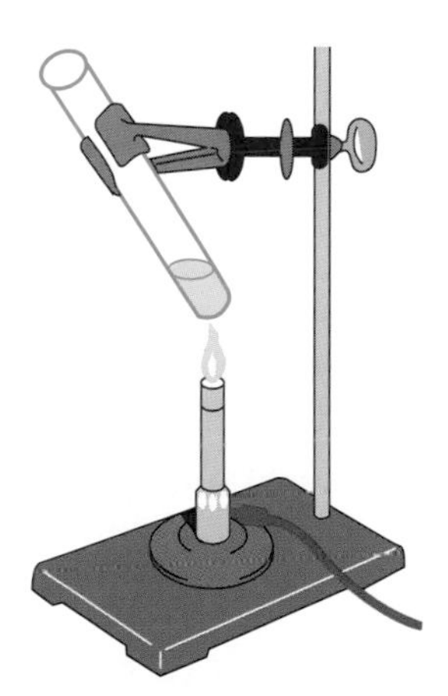

(c)

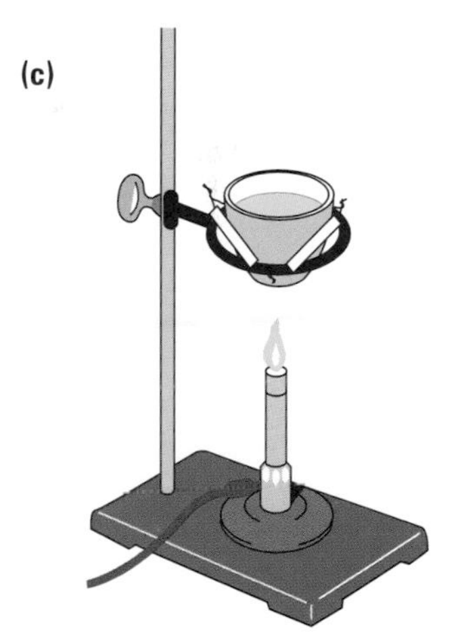

Figure 4
Methods of heating:
(a) An Erlenmeyer flask is used to funnel vapours.
(b) A test tube is used when heating small quantities of a chemical.
(c) A crucible is required when a substance must be heated strongly.

Part 2: Heating To Test for Decomposition

5. Place some bluestone to a depth of about 0.5 cm in a clean, dry test tube. Set up the apparatus as shown in **Figure 4(b)**.
6. Heat the sample carefully. Use cobalt(II) chloride paper to test for water vapour. Record your observations.
7. Set a crucible in the clay triangle on the iron ring as shown in **Figure 4(c)**. Add only enough malachite to cover the bottom of the crucible with a thin layer.
8. Heat the malachite sample slowly, with a uniform, almost invisible flame. Use cobalt(II) chloride paper to test for water vapour. Record your observations.
9. Heat the malachite sample strongly with a two-part flame. Use cobalt(II) chloride paper to test for water vapour. Record your observations.
10. Place a few grains of table salt and a few grains of sugar in two separate locations on a piece of aluminum foil. Place the foil on a hot plate.
11. Set the hot plate to maximum heat and record your observations.
12. Dispose of waste materials as instructed by your teacher.

Analysis

(b) Which substances decomposed upon heating? How do your observations support your answer?

(c) Which substances did not decompose upon heating? How do your observations support your answer?

(d) According to your observations, and the concept that heating will decompose compounds, answer the Question: Are water, bluestone, malachite, table salt, and sugar elements or compounds?

Evaluation

(e) Compare the answer you obtained in your Analysis to your Prediction. Assuming your Prediction is valid, what does this suggest about the Experimental Design?

(f) What are some limitations of using heating to determine whether a substance is an element or a compound?

Synthesis

(g) Why might you draw different conclusions from your Evidence than earlier experimenters did from their similar results?

(h) Write an alternative Experimental Design to better answer the Question.

Organizing the Elements

In 1800, 31 elements were known. By 1865, the number of identified elements had more than doubled to 63. With the discovery of more and more elements, scientists searched for a systematic way to organize their knowledge by classifying the elements. Scientists were able to make more accurate and precise measurements of mass, volume, and pressure in the course of their investigations. By studying the reactions of various elements with oxygen and using the quantitative relationships that emerged, scientists eventually determined the relative atomic mass of each element. For example, atoms of carbon were found to have a mass 12 times the mass of a hydrogen atom; and an oxygen atom has a mass 16 times that of a hydrogen atom. Atoms of hydrogen appeared to be the lightest, so

DID YOU KNOW?

Solid, Liquid, or Gas?

When representing an element or compound by its chemical formula, we can indicate its state with a subscript: (s) means solid; (l) means liquid; (aq) means "aqueous," dissolved in water; and (g) means gaseous. All of these states are at SATP—25°C and 100kPa—unless stated otherwise.

a scale was devised in which hydrogen had an atomic mass of 1 unit. The relative atomic masses of some common elements are shown in **Table 2**.

Johann Döbereiner (1780–1849) was among the first scientists to consider the idea of trends among the properties of the elements. By 1829 he had noted a similarity among the physical and chemical properties of several groups of three elements. In each case, the middle element had an atomic mass about halfway between the atomic masses of the other two. Lithium, potassium, and sodium make up one such **triad.**

Döbereiner's discovery is often referred to as the law of triads. However, for a statement to be accepted as a law, evidence must first be collected from several examples and replicated by many scientists. Laws must accurately describe and explain current observations and predict future events in a simple manner. A law is a statement of accepted knowledge. In this regard, Döbereiner's "law of triads" was never really a law. In any event, at the time, Döbereiner's idea was dismissed as coincidence.

In 1864, the English chemist John Alexander Newlands (1837–1898) arranged all of the known elements in order of increasing atomic mass. He noticed that similar physical and chemical properties appeared for every eighth element. For example, the elements lithium, sodium, and potassium are all soft, silvery-white metals. They are all highly reactive, and they form similar compounds with chlorine. There is a strong "family" resemblance among them, although the degree of reactivity increases as the atomic mass increases within the family. Newlands noticed that rubidium and cesium, although they did not follow the "eighth element" pattern, share properties with sodium, potassium, and lithium. He therefore decided to include them in the same family.

The elements that follow (in atomic mass) each of these five in Newlands's arrangement—beryllium, magnesium, calcium, strontium, and barium—also exhibit a strong family resemblance. Newlands, having an interest in opera and music, drew a parallel between the repeating properties of every eighth chemical element and the octave scale in music. Newlands called his discovery "the law of octaves." However, Newlands's law of octaves seemed to be true only for elements up to calcium. As a result, the idea was not generally accepted and drew criticism and even ridicule from members of the scientific community.

At around the same time that Newlands announced his findings, a German chemist, Julius Lothar Meyer (1830–1895), also arranged the elements in order of atomic mass. Lothar Meyer thought he found a repeating pattern in the relative volumes of the individual atoms of known elements. Unlike Newlands, Lothar Meyer observed a change in length of that repeating pattern. By 1868, Lothar Meyer had developed a table of the elements that closely resembles the modern periodic table. However, his work was not published until after the work of Dmitri Mendeleev (1834–1907), the scientist who is generally credited with the development of the modern periodic table.

Table 2: Relative Atomic Masses of Selected Elements

Element	Relative atomic mass
hydrogen	1
carbon	12
oxygen	16
sodium	23
sulfur	32
chlorine	35.5
copper	63.5
silver	108
lead	207

triad: a group of three elements with similar properties

Figure 5
Dmitri Mendeleev (1834–1907) was born in Siberia, the youngest of 17 children. Whilst employed as a chemistry professor, he explored a wide range of interests including fossil fuels, meteorology, and hot-air balloons.

Mendeleev's Periodic Table

While Döbereiner may have initiated the study of periodic (repeating) relationships among the elements, it was the Russian chemist Dmitri Mendeleev who was responsible for publishing the first **periodic law.** Mendeleev's periodic law stated that elements arranged in order of increasing atomic mass show a periodic recurrence of properties at regular intervals.

In 1869, Mendeleev (**Figure 5**) reported observing repeated patterns in the properties of elements, similar to the interpretations of Newlands and Meyer. Mendeleev first created a table listing the elements in order of atomic mass in

periodic law (according to Mendeleev): The properties of the elements are a periodic (regularly repeating) function of their atomic masses.

GROUP	I	II	III	IV	V	VI	VII	VIII
Formula of Compounds	R_2O —	RO —	R_2O_3 —	RO_2 H_4R	R_2O_5 H_3R	RO_3 H_2R	R_2O_7 HR	RO_4 —
Periods 1	H (1)							
2	Li (7)	Be (9.4)	B (11)	C (12)	N (14)	O (16)	F (19)	
3	Na (23)	Mg (24)	Al (27.3)	Si (28)	P (31)	S (32)	Cl (35.5)	
4	K (39)	Ca (40)	– (44)	Ti (48)	V (51)	Cr (52)	Mn (55)	Fe (56), Co (59) Ni (59), Cu (63)
5	[Cu (63)]	Zn (65)	– (68)	– (72)	As(75)	Se (78)	Br (80)	
6	Rb (85)	Sr (87)	?Yt (88)	Zr (90)	Nb (94)	Mo (96)	– (100)	Ru (104), Rh (104) Pd (105), Ag (108)
7	[Ag(108)]	Cd (112)	In (113)	Sn (118)	Sb (122)	Te (125)	I (127)	
8	Cs (133)	Ba (137)	?Di (138)	?Ce (140)	—	—	—	
9	—	—	—	—	—	—	—	
10	—	—	?Er (178)	?La (180)	Ta (182)	W (184)	—	Os (195), Ir (197) Pt(198), Au (199)
11	[Au (199)]	Hg (200)	Tl (204)	Pb (207)	Bi (208)	—	—	
12	—	—	—	Th (231)	—	U (240)		

Figure 6
Mendeleev's revised periodic table of 1872. Later, scientists rearranged the purple boxes to form the middle section of the modern periodic table. (For the formulas shown, "R" is used as the symbol for any element in that family.) Note that some of the symbols of the elements have changed since 1872, and modern values of atomic mass may differ.

trend or **periodic trend:** a gradual and consistent change in properties within periods or groups of the periodic table

vertical columns. Each column ended when the chemical properties of the elements started to repeat themselves, at which point a new column was started.

Mendeleev later published a revised periodic table of the elements. In this table he listed all the elements known at that time in horizontal rows, in order of atomic mass. The table is organized in such a way that elements with similar properties appear in the same column or group (**Figure 6**). This table made it very clear that there were **trends** (sometimes called **periodic trends**) among the elements: similar but gradually changing properties, such as melting and boiling points.

Mendeleev's table contained some blank spaces where no known elements appeared to fit. However, he had such confidence in his hypothesis that he proposed that those elements had not yet been discovered. For example, in the periodic table there is a gap between silicon, Si (28), and tin, Sn (118). Mendeleev predicted that an element, which he called "eka-silicon" (after silicon), would eventually be discovered and that this element would have properties related to those of silicon and tin. He made detailed predictions of the properties of this new element, using his knowledge of periodic trends. Sixteen years later, a new element was discovered in Germany. The properties of this new element, germanium, are listed in **Table 3**, beside the properties that Mendeleev had predicted for eka-silicon. The boldness of Mendeleev's quantitative predictions and their eventual success made him and his periodic table famous.

Table 3: Germanium Fulfills Mendeleev's Prediction

Property	Predicted for eka-silicon (1871)	Observed for germanium (1887)
atomic mass	72 (average of Si and Sn)	72.5
specific gravity	5.5 (average of Si and Sn)	5.35
reaction with water	none (based on none for Si and Sn)	none
reaction with acids	slight (based on Si—none; Sn—rapid)	none
oxide formula	XO_2 (based on SiO_2 and SnO_2)	GeO_2
oxide specific gravity	4.6 (average of SiO_2 and SnO_2)	4.1
chloride formula	XCl_4 (based on $SiCl_4$ and $SnCl_4$)	$GeCl_4$
chloride boiling point	86°C (average of $SiCl_4$ and $SnCl_4$)	83°C

No one in the scientific community at the time could explain *why* Mendeleev's predictions were correct—no acceptable theory of periodicity was proposed until the early 1900s. Mendeleev was working on empirical evidence alone. This mystery must have made the accuracy of his predictions even more astounding.

Practice

Understanding Concepts

9. (a) State the periodic law according to Mendeleev.
   (b) What are the limitations of the periodic law? Do you think it could be used to predict all of the properties of a new element?
10. Chlorine, fluorine, and bromine are a "triad" with increasing atomic mass. The atomic mass of fluorine is 19 and of bromine is 80. According to the law of triads, predict the atomic mass of chlorine. Compare your prediction to the accepted value in the periodic table.
11. Name three scientists who contributed to the development of the periodic table, and briefly describe the contributions of each.
12. Sulfur is a yellow solid and oxygen is a colourless gas, yet both elements are placed in the same column or family. What properties might have led Mendeleev to place them in the same family?
13. In the 1890s, an entirely new family of elements was discovered. This family consisted of unreactive gases called noble gases. Did this discovery support Mendeleev's periodic table? Explain briefly.

Reflecting

14. Science is considered by many people to be completely objective. However, the history of science shows that this is not always the case. Research the historical contributions of 19th-century scientists, such as Newlands and Lothar Meyer, to the development of the modern periodic law. Using your findings as evidence, write a brief commentary on the objectivity of science and its social and personal dimensions.

The Modern Periodic Table

Mendeleev developed his periodic table when chemists knew nothing about the internal structure of atoms. However, the beginning of the 20th century witnessed profound developments in theories about subatomic particles. In 1911, a Dutch physicist, A. van den Broek (1856–1917), suggested a rearrangement of the periodic table according to atomic number. This led to a revision of the **periodic law**. Figure 7 shows the modern periodic table. In this table every element is in

periodic law: (modern definition) When the elements are arranged in order of increasing atomic number, their properties show a periodic recurrence and gradual change.

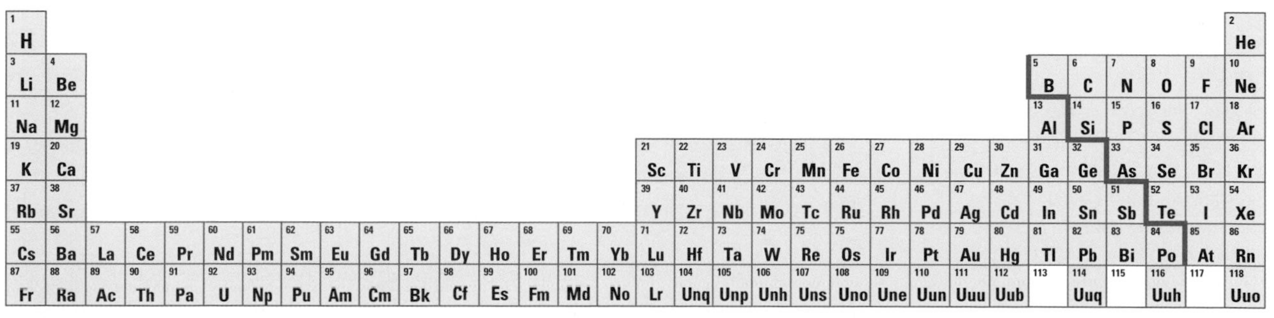

Figure 7
Because of its inconvenient shape, this extended form of the periodic table is rarely used.

sequence, but the shape of the table makes it difficult to print on a page while still including useful descriptions of each element.

The periodic table is usually printed in the form shown in **Figure 8**, with two separate rows at the bottom. Note the following important features:

group: a column of elements in the periodic table; sometimes referred to as a family

period: a row in the periodic table

- a **group**—elements with similar chemical properties in a vertical column in the main part of the table;
- a **period**—elements, arranged in a horizontal row, whose properties change from metallic on the left to nonmetallic on the right;
- the "staircase line"—a zigzag line that separates metals (to the left) from nonmetals (to the right);
- the physical state of each element at SATP (in this case red for gases, blue for liquids; the remainder are solids);
- group numbers 1 to 18.

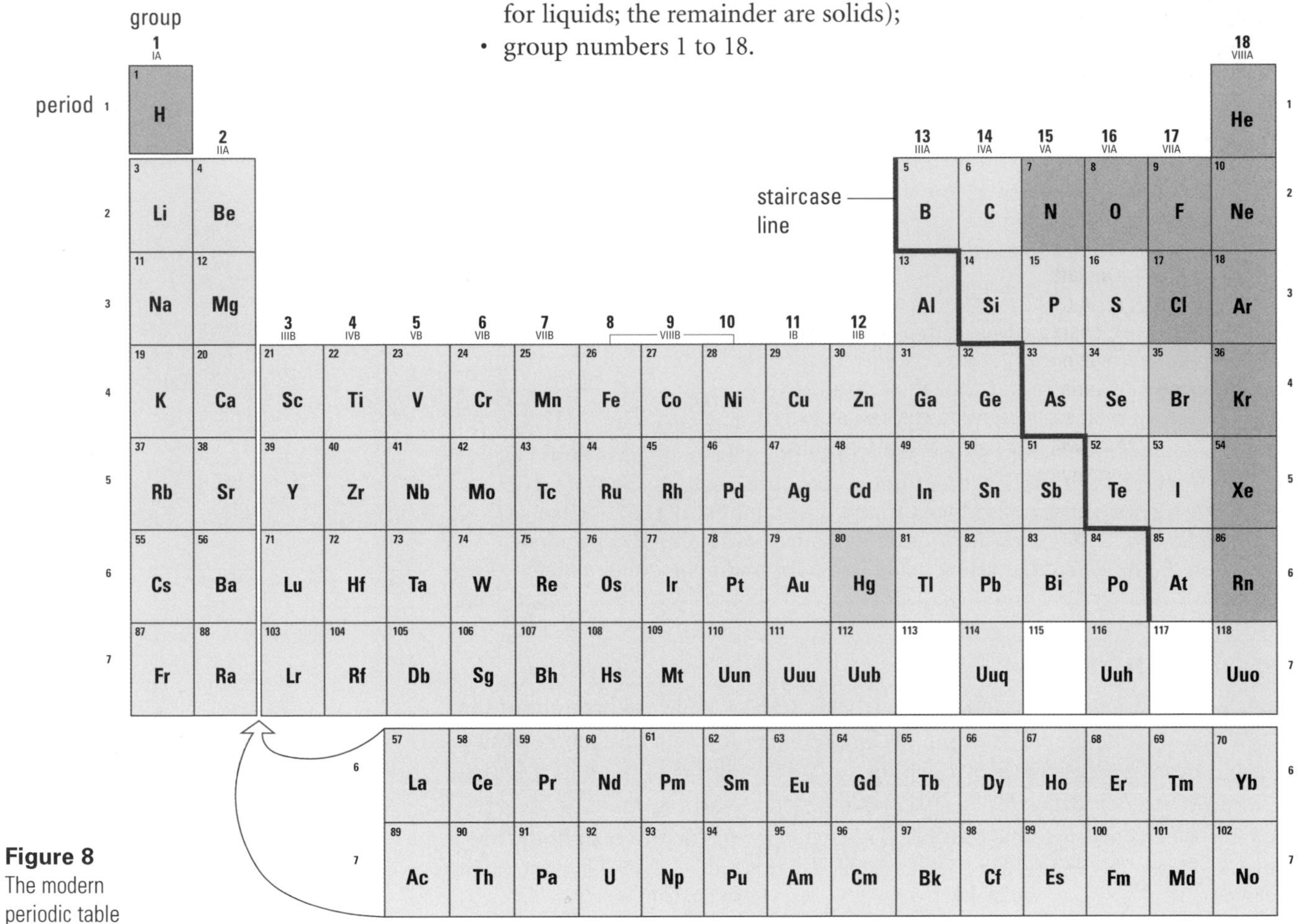

Figure 8
The modern periodic table

Names of Groups and Series of Elements

Some groups of elements and the two series of elements (those in the two horizontal rows at the bottom of the periodic table) have traditional names that are commonly used in scientific communication (**Figure 9**).

The position of hydrogen in the periodic table is a problem: It sometimes behaves like a member of the alkali metals, sometimes like halogens, and sometimes in its own unique way. Hydrogen is sufficiently different to be in a class by itself.

Although the periodic table has some limitations, scientists generally agree that the evidence supporting its general principles, as well as its usefulness in chemistry, make it an essential part of the study of chemistry.

DID YOU KNOW?

Not Completely Inert

Although the noble gases are so stable and unreactive that they remained undiscovered until the late 1800s, they can be made, under certain circumstances, to form compounds. The first known reaction involving a noble gas was a Canadian achievement. In 1962 Neil Bartlett (b. 1932), a scientist at the University of British Columbia, created a compound from xenon, platinum, and fluorine.

alkali metals
soft, silver-coloured elements; solids at SATP; exhibit metallic properties; react violently with water to form basic solutions and liberate hydrogen gas; react with halogens to form compounds similar to sodium chloride, $NaCl_{(s)}$; stored under oil or in a vacuum to prevent reaction with air

alkaline earth metals
light, very reactive metals; solids at SATP; exhibit metallic properties; form oxide coatings when exposed to air; react with oxygen to form oxides with the general chemical formula, $MO_{(s)}$; all except beryllium will react with hydrogen to form hydrides with the general chemical formula XH_2; react with water to liberate hydrogen

noble gases
gases at SATP; low melting and boiling points; extremely unreactive, making them especially interesting to chemists; krypton, xenon, and radon reluctantly form compounds with fluorine; radon is radioactive

halogens
may be solids, liquid, or gases at SATP; exhibit nonmetallic properties—not lustrous and nonconductors of electricity; extremely reactive, with fluorine being the most reactive; react readily with hydrogen and metals

representative elements
includes both metals and nonmetals from Groups 1, 2, and 13 through 17; may be solids, liquids, or gases at SATP; called representative because they most closely follow the periodic law; many form colourful compounds

transition metals
exhibit a wide range of chemical and physical properties; characteristically strong, hard metals with high melting points; good conductors of electricity; variable reactivity; form ions with variable charges; many react with oxygen to form oxides; some will react with solutions of strong acids to form hydrogen gas

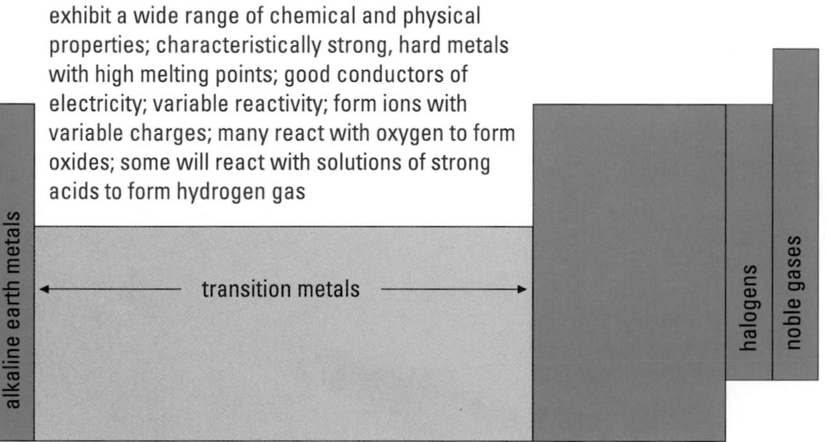

lanthanides (rare earths)

actinides including transuranic elements

lanthanides
(rare earth elements) elements with atomic numbers 57 to 70

actinides
elements with atomic numbers 89 to 102

transuranic elements
synthetic (not naturally occurring) elements with atomic numbers 93 or greater (beyond uranium)

Figure 9
The various parts of the periodic table are given specific names.

alkali metal: an element in Group 1 of the periodic table

alkaline earth metal: an element in Group 2 of the periodic table

noble gas: a element in Group 18 of the periodic table

halogen: an element in Group 17 of the periodic table

representative element: an element in any of Groups 1, 2, and 13 through 18

transition metal: an element in Groups 3 through 12 of the periodic table

lanthanide: lanthanum and the 13 elements that follow it in the 6th row of the periodic table; elements 57 to 70

actinide: actinium and the 13 elements that follow it in the 7th row of the periodic table; elements 89 to 102

transuranic elements: elements that follow uranium in the periodic table; elements 93+

Practice

Understanding Concepts

15. Create a table with five columns. At the top of the columns write the following headings: Element name; Atomic symbol; Atomic number; Group number; and State at SATP. In the left-hand column, write the names of all the elements in the second period, then complete the table.

DID YOU KNOW?

Another Word for ...

Alkali is another word for base. Alkali metals were so named because they react with water to produce a basic solution.

16. Compare the numbers of metals, nonmetals, and metalloids in the periodic table.
17. The representative elements include elements from eight groups in the periodic table. Pick an element from each of the eight groups and list its properties.
18. List two physical and two chemical properties of the
    (a) alkali metals
    (b) halogens
    (c) noble gases
19. Nitrogen and hydrogen form the well-known compound $NH_{3(g)}$, ammonia, which is used in vast quantities to make chemical fertilizer. According to the position of phosphorus in the periodic table, predict the most likely chemical formula for a compound of phosphorus and hydrogen that is also in demand in the chemical industry.

Making Connections

20. Research and report upon the properties of silicon and its use in communications technology. Use your findings to write a paragraph commenting on the validity of the statement: Future historians will define the times we live in as the Age of Silicon.

    Follow the links for Nelson Chemistry 11, 1.1.

GO TO **www.science.nelson.com**

Section 1.1 Questions

Understanding Concepts

1. Use examples of elements from the periodic table to show the effect of the periodic law.
2. The elements Li, Na, and K are arranged in descending order in a vertical column of the periodic table. Their melting points are 181°C, 97.8°C, and 63.3°C, respectively.
    (a) Using the periodic law, rather than reading the table, predict the melting point of Rb, the element immediately below K in the group.
    (b) Based on their melting points, would you classify these elements as metals or nonmetals? Give reasons.
    (c) What other physical properties would you expect these elements to share?
3. An unknown element, X, is a shiny, grey solid at SATP. When it is strongly heated in the presence of oxygen, a white, powdery solid forms on its surface.
    (a) Describe two ways in which the substance can be classified. Justify each classification.
    (b) If heated to high temperatures, the white powder is stable—it can be used to make firebrick and furnace linings. If heating of the grey solid and of the white powder had been observed by a scientist prior to 1800, what would the scientist have concluded about which was a compound and which was an element? Defend your answer.

4. Canada is rich in mineral deposits containing a variety of elements. **Table 4** lists a few examples from across Canada. In your notebook, copy and complete the table.

Table 4: Elements and Mineral Resources

Mineral resource or use	Element name	Atomic number	Element symbol	Group number	Period number	SATP state
(a) high-quality ores at Great Bear Lake, NT	radium					
(b) rich ore deposits at Bernic Lake, MB				1	6	
(c) potash deposits in Saskatchewan		19				
(d) large deposits in New Brunswick	antimony					
(e) extracted from Alberta sour natural gas			S			
(f) radiation source for cancer treatment				9	4	
(g) large ore deposits in Nova Scotia	barium					
(h) world-scale production in Sudbury, ON		28				
(i) fuel in CANDU nuclear reactors from Saskatchewan			U			
(j) fluorspar deposits in Newfoundland				17	2	
(k) large smelter in Trail, BC		30				

Applying Inquiry Skills

5. When an unknown element is added to water, it reacts violently to liberate hydrogen gas. Based on this reaction, to what groups might the element belong? What additional property could you investigate to narrow your choice of groups? Outline a possible Experimental Design.

(continued)

Making Connections

6. Select one of the elements listed in **Table 4**.
   (a) Where does it come from, how is it extracted, and what is it used for?
   (b) Does it pose any health or environmental hazards? How should it be handled?
   (c) Do the advantages of having this element available outweigh the drawbacks? To help you decide, separate your findings into two categories: advantages and drawbacks. Write a paragraph summarizing your findings and explaining your position.

   Follow the links for Nelson Chemistry 11, 1.1.

GO TO www.science.nelson.com

1.2 Developing a Model of the Atom

empirical knowledge: knowledge coming directly from observations

theoretical knowledge: knowledge based on ideas created to explain observations

model: a mental or physical representation of a theoretical concept

analogy: a comparison of a situation, object, or event with more familiar ideas, objects, or events

theory: a comprehensive set of ideas that explains a law or a large number of related observations

In all aspects of our lives, we can achieve understanding through observations (experience). The same is true in science: Understanding comes from observing the natural world and trying to make sense of those observations. All scientific knowledge can be classified as either empirical (observable) or theoretical (non-observable). Generally, **empirical knowledge** comes first, and can be as simple as a description or as complex as a powerful scientific law. For example, the physical and chemical properties of some elements were known empirically for thousands of years before we had a theory to explain these properties. This is a common occurrence: Empirical knowledge is usually well developed before any explanation is generally accepted within the scientific community. Although scientific laws are important statements summarizing considerable empirical knowledge, they contain no explanation. For an explanation—an answer to the question "Why?"—a theory is required.

So far in this chapter, you have encountered only empirical knowledge of elements, based on what has been observed. But why do the properties of elements vary across the periodic table? Why are groups of elements similar in their physical and chemical properties? Can we explain the chemical formulas of compounds formed from elements? An answer to these and other questions about elements requires a theory about what makes up elements.

Curiosity leads scientists to try to explain nature in terms of what cannot be observed. This step—formulating ideas to explain observations—is the essence of **theoretical knowledge** in science. Albert Einstein referred to theoretical knowledge as "free creations of the human mind."

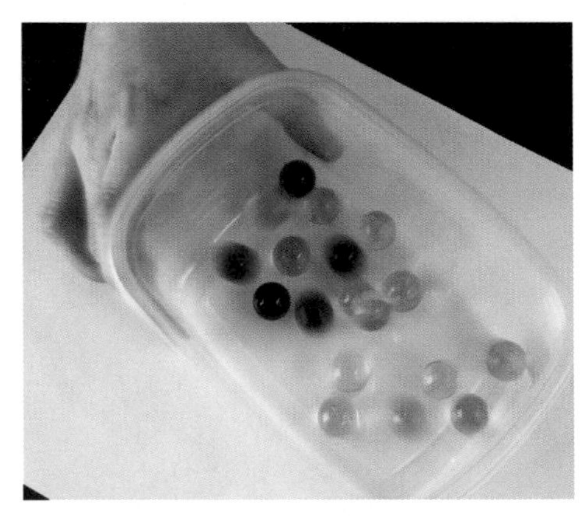

Figure 1
A physical model to represent the motion of particles described by the kinetic molecular theory of gases could be a vibrating box containing marbles. In what ways is this model useful in describing air in a sealed container? In what ways is it deficient?

It is more challenging to communicate theoretical knowledge than empirical knowledge because ideas are, by definition, abstract and cannot be seen. Theoretical knowledge can be communicated in a variety of ways such as words, symbols, **models**, and **analogies**. The difference between an analogy and a model is not always obvious. Models are representations (**Figure 1**). Analogies are comparisons. For example, some properties of a liquid can be explained using the analogy of a crowd of people in a confined space.

Theories are dynamic; they continually undergo refinement and change. To be acceptable to the scientific community, theories must

- *describe* observations in terms of non-observable ideas;
- *explain* observations by means of ideas or models;
- successfully *predict* results of future experiments; and
- be as *simple* as possible in concept and application.

Activity 1.2.1

Developing a Model of a Black Box

Have you ever tried to figure out the contents of a package you weren't allowed to open? Such an object is often referred to as a "black box." The atom is a good example of a black box. Although, by definition, we can never open a black box, we can perform operations on the box that will help us to learn more about what goes on inside it. In this activity, you will be provided with a sealed container which, together with its contents, represents a black box. You are expected to investigate your black box and provide a detailed description of it. Your description should help you to develop a model of your black box.

Materials

black box (sealed box containing an object)

Procedure

1. Obtain a "black box."
2. Manipulate the box in a variety of different ways, without opening or damaging the box. Record the results as you proceed.
3. Write a preliminary description of the object inside the box.
4. Test your model and try to improve it by doing further manipulations.

Analysis

(a) Write a detailed description of the object in the box.
(b) Which manipulations were most useful in developing your model?
(c) Can you definitely say what your object is *not* like? Can you definitely say what it *is* like?
(d) What would you require to further develop your model? Give your reasoning.

Synthesis

(e) How is investigating the contents of a black box different from investigating the nature of the atom?
(f) Everyday life provides us with many examples of black boxes—things or relationships that we cannot break open to look inside. What is a black box for some may not be for others. Give three examples of black boxes that you have encountered.

> **DID YOU KNOW ?**
>
> **Theoretical Knowledge**
>
> There is a range of types of theoretical knowledge, just as there is for empirical knowledge.
>
> - Theoretical descriptions are specific statements based on theories or models. For example, "a molecule of water is composed of two hydrogen atoms and one oxygen atom."
> - Theoretical hypotheses are ideas that are untested or tentative. For example, "protons are composed of quarks that may themselves be composed of smaller particles."
> - Theoretical definitions are general statements that characterize the nature of a substance or a process. For example, a solid is theoretically defined as "a closely packed arrangement of particles, each vibrating about a fixed location in the substance."
> - Theories are comprehensive sets of ideas that explain a large number of observations. For example, the concept that matter is composed of atoms is part of atomic theory; atomic theory explains many of the properties of materials.

Early Greek Theories of Matter

The Greek philosopher Democritus (460?–370? B.C.) first proposed an atomic theory of matter in the 5th century B.C. According to Democritus, all matter could be divided into smaller and smaller pieces until a single indivisible particle was reached. He called this particle an **atom**. He believed that different atoms are of different sizes, have regular geometric shapes, and are in constant motion. He also believed that there is empty space between atoms.

atom: the smallest particle of an element that has all the properties of that element (theoretical definition)

Aristotle (384–322 B.C.) severely criticized Democritus's theory, arguing that the idea of atoms in continuous motion in a void is illogical. The concept of a void was very controversial at the time, and of course there was no evidence that a completely empty space could exist. Instead, Aristotle supported the four-element theory of matter. Proposed by Empedocles a century earlier (c. 495–435 B.C.), the theory was based on the idea that all matter is made up of four basic substances: earth, water, air, and fire. Aristotle and his followers believed that each of these basic substances had different combinations of four specific qualities: dry, moist, cold, and hot (**Figure 2**). Aristotle's theory of the structure of matter was the prevailing model for almost 2000 years, including the period of **alchemy** in the Middle Ages (**Figure 3**). The demise of Aristotle's model followed the scientific revolution in physics and the new emphasis, in the 18th century, on **quantitative** studies. Careful observations showed that too many of the explanations and predictions using Aristotle's theory were false. Scientists and philosophers needed a new theory with a different model of matter. This led to the revival of the atom concept.

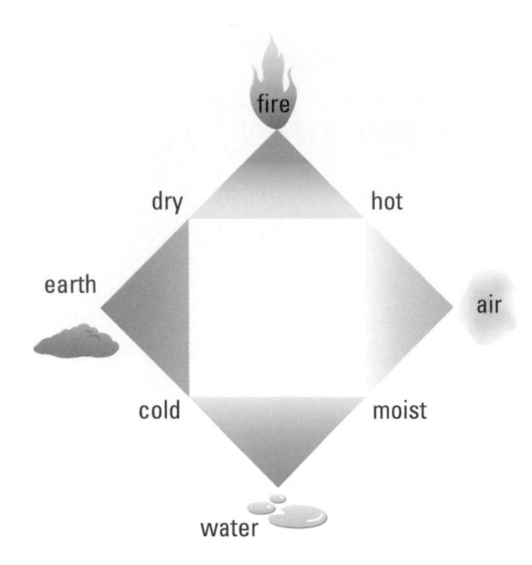

Figure 2
In Aristotle's model of matter, each basic substance, or element, possesses two of four essential qualities. For example, earth is dry and cold; fire is dry and hot. This model was based on logical thinking, but not on experimentation.

alchemy: a medieval chemical philosophy or practice, the principal goals of which were to transmute elements (e.g., lead to gold), to cure all illnesses, and to manufacture an essence that would allow long life

quantitative: involving measurements related to number or quantity

law of conservation of mass: the law stating that during a chemical reaction matter is neither created nor destroyed

law of constant composition: the law stating that compounds always have the same percentage composition by mass

Dalton's Atomic Theory

John Dalton (1766–1844), an English scientist and schoolteacher, proposed explanations for many of the known laws describing the behaviour of matter. Dalton expanded upon the atomic theory proposed by Democritus. This expanded theory was first introduced in 1803, at last replacing Aristotle's model of matter. Dalton's theory consisted of the following statements.

- All matter is composed of tiny, indivisible particles called atoms.
- All atoms of an element have identical properties.
- Atoms of different elements have different properties.
- Atoms of two or more elements can combine in constant ratios to form new substances.
- In chemical reactions, atoms join together or separate from each other but are not destroyed.

According to Dalton's theory, atoms are neither created nor destroyed in a chemical reaction. Since atoms are indivisible, and are only rearranged during chemical reactions, you must end up with the same number and kinds of atoms after a chemical reaction as you had at the beginning. Therefore, there will be no change in mass during chemical reactions. This explains the **law of conservation of mass.** Dalton's theory also suggested that atoms combine to form molecules in a fixed ratio in a given chemical reaction. Since the atoms of an element have identical properties, such as mass, and combine in constant ratios, every compound must have a fixed, definite composition. This explains the **law of constant composition.**

This introduction of atomic theory to the scientific world was followed by a period of intense investigation into the nature of matter. As a result of this investigation, Dalton's statements suggesting that the atom is indivisible and that all atoms of a given element have identical properties are no longer considered valid. Nonetheless, Dalton's theory proved very successful in explaining the laws of conservation of mass and constant composition.

Figure 3
Alchemists sought a method for transforming other metals into gold. Although they failed in their quest, they discovered new elements and compounds and developed many experimental procedures that are still used today.

Development of Atomic Theory from 1803 to 1920

By the late 1800s, several experimental results conflicted with Dalton's atomic theory. Technological advances made possible the construction of evacuated glass tubes fitted with an electrode at either end. The apparatus eventually became known as a cathode ray tube, named for the electrical discharge of particles assumed to be travelling from the negative electrode (or cathode) to the positive electrode (or anode). Several scientists used this apparatus in attempts to determine the composition of the atom. In one such experiment, conducted in 1897, J. J. Thomson (1856–1940) used a modified cathode ray tube to measure

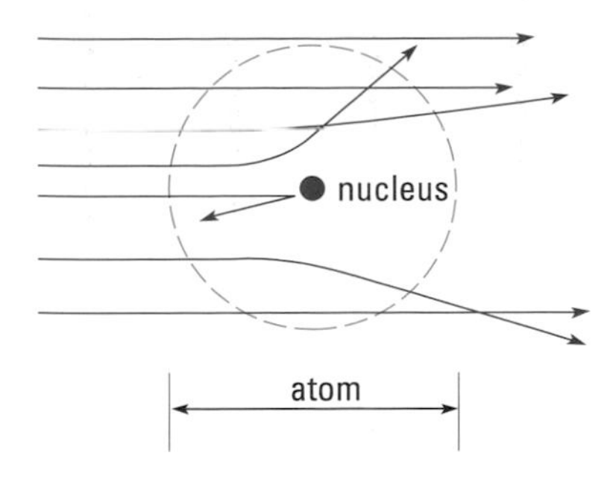

Figure 6
To explain the results of his experiment, Rutherford suggested that an atom consisted mostly of empty space, and that most of the alpha particles passed nearly straight through the gold foil because these particles did not pass close to a nucleus.

Table 1: Relative Masses and Charges of Subatomic Particles

Particle	Relative mass	Relative charge
electron	1	1–
proton	1836.12	1+
neutron	1838.65	0

Practice

Understanding Concepts

1. What is the difference between
   (a) a theory and a law?
   (b) empirical knowledge and theoretical knowledge?
2. Why is it useful for scientists to develop models of their ideas?
3. When wood is burned in a fireplace, its mass decreases. Does this observation contradict the law of conservation of mass? Justify your answer.
4. Draw a series of at least four diagrams to represent changing models of the atom, from the time of Democritus to 1932.
5. By 1932, Chadwick had modified Rutherford's model of the atom to include neutrons. According to this modified model, define each of the following:
   (a) nucleus
   (b) proton
   (c) electron
   (d) neutron

Making Connections

6. Research and describe some current technologies that have developed from or are related to cathode ray tubes.

DID YOU KNOW ?

Tiny but Massive

The word "massive" is used in everyday language to mean "very big," "huge," "enormous." In science it has a slightly different meaning: "having a relatively large mass for its size." So, a tiny nucleus can be massive, but an enormous hot-air balloon cannot!

1.3 Understanding Atomic Mass

atomic number (*Z*): the number of protons present in the nucleus of an atom of a given element

The number of protons in the nucleus determines the identity of an element and is referred to as that element's **atomic number** (*Z*). The concept of atomic number was developed by H.G.J. Moseley (1887–1915), an English physicist, subsequent to the results of Rutherford's alpha particle scattering experiments. Moseley's research work with X rays showed that the nucleus of each element has its own, unique positive charge. This positive charge increases by one as we progress, element by element, through Mendeleev's periodic table. Moseley was the first to recognize the relationship between atomic number and nuclear charge: They are equal. This discovery provided new insight into the periodic table and a rationale for listing the elements in order of the number of protons in the nucleus. Since atoms are electrically neutral, the atomic number also represents the number of electrons in an atom of an element.

By 1932, scientists had determined that the nucleus consists of protons (which are positively charged) and neutrons (which are neutral). Since both neutrons and protons are much more massive than electrons, and both reside in the nucleus, the mass of the atom is related to the number of nuclear particles (pro-the mass of the particle and its electric charge. From his analysis of the results, which included calculating the mass-to-charge ratio for the mysterious particles, Thomson made a bold suggestion. He proposed that the cathode rays were subatomic particles: a subdivision of the matter from which all elements are built. Thomson had hypothesized the existence of **electrons**.

With this new idea, Thomson developed a new model of the atom: He suggested that negatively charged electrons are distributed inside the atom, which is a positively charged sphere consisting mostly of empty space (**Figure 4(a)**). In 1904, Japanese scientist Hantaro Nagaoka (1865–1950) represented the atom as a large, positively charged sphere surrounded by a ring of negative electrons (**Figure 4(b)**). Until 1911, there was no evidence to contradict either of these models.

In 1911, an experiment was performed that tested the existing atomic models. The experiment was designed by Ernest Rutherford (1871–1937) and involved shooting alpha particles (small, positively charged particles produced by radioactive decay) through very thin pieces of gold foil. Based on J. J. Thomson's model of the atom and his belief that the atom was composed mostly of empty space, Rutherford predicted that all the alpha particles would travel through the foil largely unaffected by the atoms of gold. Although most of the alpha particles did pass easily through the foil, a small percentage of particles was deflected at large angles (**Figure 5**). Based on this evidence, Rutherford hypothesized that an atom must contain a positively charged core, the **nucleus**, which is surrounded by a predominantly empty space containing negative electrons (**Figure 6**, page 26). By finding the percentage of deflected alpha particles and the deflection angles, Rutherford determined that only a very small portion of the total volume of an atom could be attributed to the nucleus.

Despite its tiny volume, most of the mass of an atom is believed to be concentrated in the nucleus. Consequently, the nucleus is often described as a dense central core that is massive compared to the electrons. In 1914, Rutherford coined the word **proton** for the smallest unit of positive charge in the nucleus.

In 1932, James Chadwick's experiments led him to modify Rutherford's atomic model. Chadwick (1891–1974) demonstrated that atomic nuclei must contain heavy neutral particles as well as positive particles. These neutral subatomic particles are called **neutrons**. Most chemical and physical properties of elements can be explained in terms of these three subatomic particles—electrons, protons, and neutrons (**Table 1**, page 26). An atom is composed of a nucleus, containing protons and neutrons, and a number of electrons equal to the number of protons; an atom is electrically neutral.

electron: a negatively charged subatomic particle

nucleus: the small, positively charged centre of the atom

proton: a positively charged subatomic particle in the nucleus of an atom

neutron: an uncharged subatomic particle in the nucleus of an atom

Note that these are all theoretical definitions.

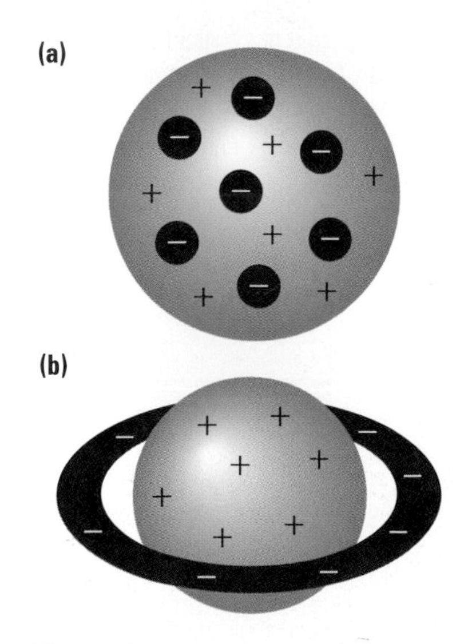

Figure 4
(a) In Thomson's model, the atom is a positive sphere with embedded electrons. This can be compared to a raisin bun in which the raisins represent the negative electrons and the bun represents the region of positive charge.
(b) In Nagaoka's model, the atom can be compared to the planet Saturn, where the planet represents the positively charged part of the atom, and the rings represent the negatively charged electrons.

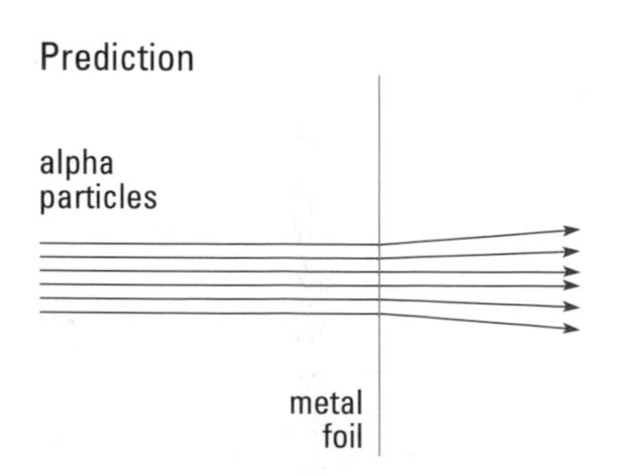

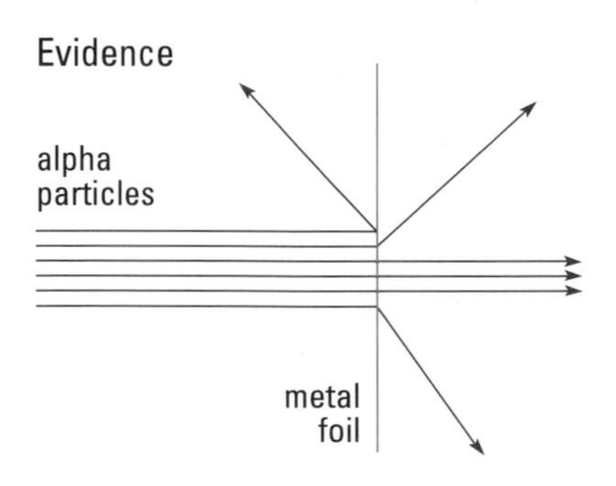

Figure 5
Rutherford's experimental observations were dramatically different from what he had expected.

tons and neutrons). The sum of the number of nuclear particles in an atom is known as the **mass number** (A) of the atom. The number of neutrons in an atom (neutron number, N) of a given element can be calculated as follows:

mass number (*A*): the sum of the number of protons and neutrons present in the nucleus of an atom

$$\text{number of neutrons} = \text{mass number} - \text{atomic number}$$

$$N = A - Z$$

According to **Figure 1**, an atom of fluorine may be represented as $^{19}_{9}\text{F}$. While the value for Z can be read directly from the periodic table, A (the mass number) is a whole-number approximation of the average atomic mass as given by the periodic table. Using this information, we can easily find the number of protons, electrons, and neutrons in an atom of any given element. In the case of fluorine, the average atom contains 9 protons and 9 electrons. Subtracting 9 from 19, we find that the atom of fluorine also contains 10 neutrons. Similarly, an atom of sodium could be represented as $^{23}_{11}\text{Na}$ and would consist of 11 protons, 11 electrons, and 12 neutrons.

$^{23}_{11}\text{Na}$ can also be written as Na-23 or sodium-23, where 23 is the mass number. This is acceptable and unambiguous because sodium always has 11 protons, so we understand that its atomic number is 11. You should be familiar with all three methods of notation.

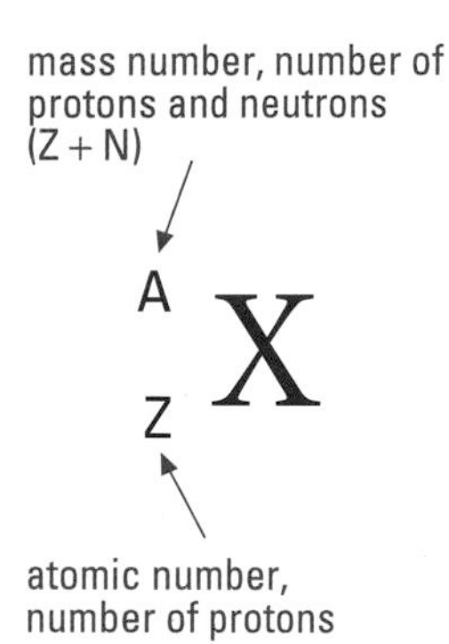

Figure 1
Symbolism representing an individual atom of an element

Sample Problem 1

The nucleus of an atom of potassium contains 19 protons and 20 neutrons.
(a) Determine the atomic number (Z) and mass number (A) of this potassium atom.
(b) Write the symbol for this potassium atom three different ways.

Solution

(a) $Z = 19$

$N = 20$

$A = Z + N$

$= 19 + 20$

$A = 39$

(b) $^{39}_{19}\text{K}$; K-39; potassium-39

Isotopes

Frederick Soddy (1877–1956), a colleague of Rutherford's, proposed that the number of neutrons can vary from atom to atom within the same element. Soddy called these variations **isotopes.** An isotope is a form of an element in which the atoms have the same number of protons as all other forms of that element, but a different number of neutrons. In other words, isotopes will have the same atomic number (Z) but different neutron numbers (N) and so different mass numbers (A). The two most abundant isotopes of chlorine are found to be $^{35}_{17}\text{Cl}$ and $^{37}_{17}\text{Cl}$. Atoms of each isotope contain 17 protons and 17 electrons. However, an atom of chlorine-35 has 18 neutrons while an atom of chlorine-37 has 20.

isotope: Atoms of an element that have the same number of protons and neutrons; there may be several isotopes of the same element that differ from each other only in the number of neutrons in their nuclei (theoretical definition).

Hydrogen has three different isotopes. More than 99% of all hydrogen atoms contain a single proton and no neutrons. The rest of the atoms represent the isotopes deuterium and tritium. An atom of deuterium contains one proton and

DID YOU KNOW ?

Exceptional Hydrogen

Hydrogen seems to be an exception to every rule in chemistry. Normally, the chemical symbol of an element does not vary, even for different isotopes. Once again, hydrogen is the exception: Deuterium is represented by the symbol D and tritium is represented by the symbol T. D is actually $^{2}_{1}H$ or H–2, and T is actually $^{3}_{1}H$ or H–3.

one neutron while an atom of tritium contains a proton and two neutrons in its nucleus.

It is the protons and electrons in atoms that are largely responsible for determining the element's chemical behaviour. Consequently, isotopes of the same element share chemical properties despite their slight difference in mass. However, their physical properties can vary considerably. For example, water containing hydrogen in the form of deuterium is known as heavy water and is represented by the chemical formula D_2O. Its use in CANDU nuclear reactors will be discussed later.

Since atoms are so small, their masses are very low. It is therefore very difficult, if not impossible, to measure the mass of an atom by conventional means. Instead, we compare atomic masses to a standard in order to determine a relative scale. At one time hydrogen, being the smallest and lightest of the elements, was used as the standard. However, since the 1960s, scientists have compared the relative **atomic mass (A_r)** of the elements to that of an isotope of carbon, carbon-12 (**Figure 2**). By convention, an atom of carbon-12 has a mass of 12 atomic mass units. Therefore, we define the **unified atomic mass unit (u)** as 1/12 of the mass of a carbon-12 atom. On this scale the proton and the neutron both have a mass close to 1 u while the electron has a mass of 0.000 55 u.

atomic mass (A_r): the relative mass of an atom on a scale on which the mass of one atom of carbon-12 is exactly 12 u

unified atomic mass unit (u): a unit of mass for atoms; 1/12 of the mass of a carbon-12 atom (theoretical definition)

mass spectrometer: a sophisticated instrument used for studying the structures of elements and compounds. One application is to determine precisely the mass and abundance of isotopes.

While some elements have only one isotopic form, most elements exist naturally as a mixture of several isotopes. A **mass spectrometer** (**Figure 3**) is a device that can be used to determine the atomic mass as well as the relative abundance of each isotope present in an element. In a mass spectrometer, gaseous atoms or molecules are accelerated by an electric field and bombarded by high-energy electrons. These speeding electrons knock electrons away from some of the particles being tested, leaving them as positively charged ions. The ions pass through a magnetic or electric field, which bends (deflects) the path they follow. The degree of deflection depends on their mass-to-charge ratio. The lower the mass of the ion, the more it is deflected. By this means, scientists have found that the average sample of the element chlorine consists of 75.77% chlorine-35 and 24.23% chlorine-37. According to the periodic table, the atomic mass of chlorine is 35.45 u. This value represents a weighted average of the atomic masses of each of the isotopes of the element. Understandably, given the greater abundance of the chlorine-35 isotope, this value is much closer to 35 u than to 37 u.

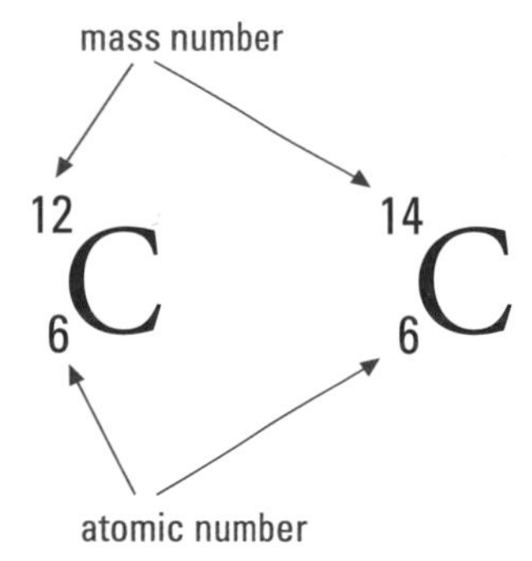

Figure 2
Two isotopes of carbon. Carbon-12 is stable, but carbon-14 is radioactive.

Figure 3
Mass spectrometers are used to compare the masses of atoms in a sample.

Practice

Understanding Concepts

1. Create a table with four columns. At the top of the columns, write the following headings: Subatomic particle; Relative atomic mass (u); Charge; Location. In the first column, write the names of the three subatomic particles, and complete the table.
2. What two particles are responsible for most of the mass of an atom?
3. Compare the mass number of an atom to its relative atomic mass. Why might they be different?
4. An atom has 14 protons and 13 neutrons. What is its mass number?
5. An atom has 15 protons and has a mass number of 31.
   (a) What is its atomic number?
   (b) How many neutrons does it have?
   (c) What element is it?
6. How many neutrons can be found in the nucleus of an atom of chlorine-37?
7. Two atoms respectively have $Z = 15$, $A = 30$ and $Z = 14$, $A = 30$. Are they isotopes of each other? Explain.
8. **Figure 4** shows a graph produced by a mass spectrometer.
   (a) What are the atomic masses of the three isotopes of magnesium?
   (b) How many neutrons would each isotope possess?
   (c) What is the relative abundance (as a percentage) of each of the three isotopes?

DID YOU KNOW ?

Weighted Averages

A good example of a weighted average is the calculation of your final mark in a course. If you did really well in the majority of your assessed work, but very poorly in one assignment, your overall mark would still be fairly high.

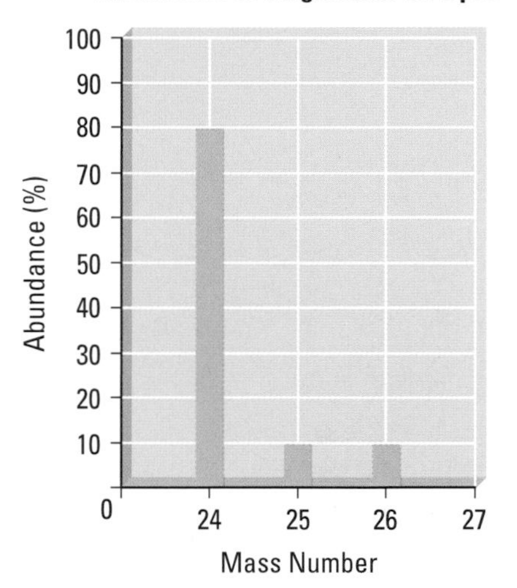

Figure 4
Mass spectrum of magnesium showing the relative abundance of its isotopes in a naturally occurring sample

Radioisotopes

Many elements have one or more isotopes that are unstable. Atoms of unstable isotopes decay, emitting radiation as their nucleus changes. Depending on the isotope, these nuclear changes might happen very quickly or extremely slowly. And the radiation they emit could be fairly harmless or very dangerous to living cells. Isotopes that can decay in this way are known as **radioisotopes** and are said to be **radioactive**. Numerous experiments have shown that radioisotopes give off three types of radiation: alpha particles, beta particles, and gamma rays (**Table 1**).

radioisotope: a radioactive isotope of an element, occurring naturally or produced artificially

radioactive: capable of spontaneously emitting radiation in the form of particles and/or gamma rays

Table 1: Characteristics of Nuclear Radiation

Radiation	Approximate speed	Penetration in air	Effective barrier
alpha (α, ${}^{4}_{2}He^{2+}$, ${}^{4}_{2}He$)	variable, but relatively slow	a few centimetres	a sheet of paper
beta (β, e^-, ${}^{0}_{-1}e$)	variable, but relatively fast	a few metres	1-2 mm of metal
gamma (γ)	very fast (speed of light)	unlimited	1 m of lead or concrete

An alpha particle is composed of two protons and two neutrons, which is equivalent to a ${}^{4}_{2}He$ nucleus. (Because it carries no electrons it has a charge of 2+. Its charge is usually omitted in nuclear equations.) The radioactive decay of uranium-238, which results in the production of an alpha particle, alters the composition of its nucleus, producing thorium-234:

$${}^{238}_{92}U \rightarrow {}^{234}_{90}Th + {}^{4}_{2}He$$

The alpha particle emitted in this radioactive decay travels at a fairly high speed (approximately 1.6×10^4 km/s).

Beta particles are high-energy electrons (and therefore carry a charge of 1–) travelling at much higher speeds than an alpha particle (approximately 1.3×10^5 km/s). When tritium (^3_1H) decays, a beta particle is produced:

$$^3_1H \rightarrow {}^3_2He + {}^{\ 0}_{-1}e$$

Evidence suggests that the beta particle, or high-energy electron, results from the conversion of a neutron into a proton and an electron. Therefore, helium-3 is produced as the number of neutrons in the tritium nucleus decreases by 1 and the number of protons increases.

Gamma radiation refers to high-energy electromagnetic waves. The emission of a gamma ray alone does not result in any changes to the mass number or atomic number of an isotope. However, other types of radiation (which do result in a change in either the atomic number or the mass number) often occur along with gamma radiation.

The discovery of radioactive decay, in a series of investigations by Harriet Brooks (1876–1933), a Canadian scientist, was a real breakthrough in developing atomic theory: Here was evidence that an atom of one element could spontaneously change into another element. Dalton's atomic theory of "tiny indivisible particles" that were unchanging was completely put to rest.

half-life: the time it takes for one-half the nuclei in a radioactive sample to decay

Every radioisotope has a characteristic property called its **half-life**. The half-life of a radioactive substance is the time taken for half of the original number of radioactive atoms to decay. The half-lives of radioisotopes vary considerably. For example, cesium-142 has a half-life of 5×10^{15} a (a means years; the SI symbol is based on "annum"—Latin for "year"), while polonium-216 has a half-life of only 0.16 s. Consider the example of a 1000-g sample of radium-226. This radioisotope has a half-life of 1590 a. After 1590 a, only 500 g of the original sample would be radium-226. The other 500 g would have decayed to form other elements. After an additional 1590 a, 250 g of the remaining 500 g of radium-226 would have decayed, leaving only 250 g of of the original radium-226.

Sample Problem 2

The half-life of cesium-137 is 30 a. What mass of cesium-137 would remain from a 12-g sample after 30 a? After 60 a?

Solution

After 30 a (one half-life), 6.0 g of cesium-137 would remain.
After 60 a (two half-lives), 3.0 g of cesium-137 would remain.

Carbon-14 has a half-life of 5730 a. It emits a beta particle as it decays. Small amounts of this isotope occur naturally in the atmosphere, where it reacts with oxygen to form radioactive carbon dioxide. Carbon-14 is absorbed by plants when a mixture of radioactive and nonradioactive carbon dioxide is taken in during photosynthesis. Carbon-14 then finds its way into other living organisms through the food chain. When a living organism dies, it stops taking in material, including carbon. Radioactive decay will gradually reduce the amount of carbon-14 present in its tissues. As a result, the ratio of radioactive carbon to nonradioactive carbon present in the organism will gradually decrease. Comparing this ratio to the normal ratio present in a living organism provides a measure of the time elapsed since the organism's death. The practice of measuring the carbon-14:carbon-12 ratio is known as **carbon-14 dating**.

carbon-14 dating: a technique that uses radioactive carbon-14 to identify the date of death of once-living material

Nonliving materials, such as rocks, can be dated using potasssium-40. Potassium-40 has a half-life of 1.3×10^9 a, which allows it to be used to date objects too old to be dated by the carbon-14 method. (As the amount of carbon-14 left in a sample becomes very small, the accuracy of the dating becomes increasingly less reliable.) Potassium-40 dating allows the investigator to set the date when the rock solidified and stopped exchanging chemicals with its surroundings.

As you can see by these examples, naturally occuring radiation is all around us. It comes from space, soil, and even food.

The radiation produced by radioactive elements can be harnessed for a wide variety of uses. In foods, radiation can be used to kill bacteria and prevent spoilage. Radioisotopes are also widely used in medicine: Cancer patients are treated with cancer-killing radiation from cobalt-60 and radium-226.

The ability of radiation to destroy cells, however, is not always an advantage. It can pose great danger, causing normal cells to mutate or even die. Acute exposure to radiation can cause severe skin burns; long-term, chronic exposure can result in various forms of cancer, birth defects, and sterility in all animals. While radiation adds immensely to our standard of living, many risks are associated with it. We must constantly evaluate the balance between risk and benefit in our application of radioisotopes.

Activity 1.3.1

Modelling Half-Life

In this activity you will use a model of a sample of radioactive material. By removing the "decayed nuclei" from the sample after each "half-life," and counting the remaining nuclei, you will collect quantitative data that can be used to plot a graph of half-life. This graph can then be used for prediction purposes.

Materials

at least 30 disks that have different faces (e.g., Othello disks, coins, or cardboard disks with a mark on one side)
a box with a lid (e.g., a shoebox)

Procedure

1. Create a table like **Table 2** to record your observations.
2. Decide which face of the disks will represent the original radioactive isotope (e.g., white, or heads).
3. Place all the disks in the box, "original" side up.
4. Put the lid on securely, and shake the box.
5. Open the box and remove all the disks that do not show the "original" side up. Record in your table the number of disks that remain after one half-life.
6. Repeat steps 4 and 5 for at least 5 more half-lives.

Table 2

Half-life	Number of disks remaining
0	30
1	?
2	?
3	?
...	

Analysis

(a) Create a graph with the number of disks remaining on the *y*-axis and the number of half-lives on the *x*-axis.

(b) Draw a best-fit curve. Describe the shape of your graph.
(c) If the units on the y-axis were "mass of radioactive sample (g)" and each half-life represented 2 a, predict the mass of radioactive material that would remain after 7 a.
(d) Does the curve of your graph ever reach the x-axis? If it doesn't, do you think it would if you had taken the activity through another few half-lives?
(e) What do the removed disks represent? Compare this model of radioactive decay with what you know of the real situation. In what ways is this a good model? How is it less than perfect?

Practice

Understanding Concepts

9. Distinguish between an isotope and a radioisotope.
10. Copy and complete **Table 3** in your notebook.
11. Write a paragraph defining the term half-life, including the terms isotope, radioisotope, radiation, and decay.
12. Radon-222 has a half-life of 4.0 d (abbreviation for "diem"—Latin for "day"). If the initial mass of the sample of this isotope is 6.8 g, calculate the mass of radon-222 remaining in the sample after
    (a) 8.0 d
    (b) 16.0 d
    (c) 32.0 d
13. Iodine-131 has a medical application: It is used to treat disorders of the thyroid gland. It has a half-life of 8.0 d. Plot a graph showing mass of iodine-131 against time over a period of 64 d, starting with a 2.0-kg sample of the radioisotope.
14. Does radioactive decay support or contradict the law of conservation of mass? Justify your answer.

Answers

12. (a) 1.7 g
    (b) 0.62 g
    (c) 0.026 g

Table 3: Emission Particles

	Alpha particle	Beta particle
(a) Another name for this particle	?	?
(b) The symbol for this particle	?	?
(c) How the nucleus of a radioisotope is altered by emission of this particle	?	?
(d) The penetrating ability of this type of radiation	?	?

Making Connections

15. Use the Internet to find a project where scientists are using or have used carbon-14 dating to find the age of artifacts. Report on the scientists' work and its implications.
    Follow the links for Nelson Chemistry 11, 1.3.
    GO TO www.science.nelson.com
16. Each of the following radioisotopes is used for beneficial purposes: chromium-51, iron-59, arsenic-74, iodine-131, phosphorus-32, sodium-24, and irridium-192. Choose one of the radioisotopes and research and report on its properties and applications. Include at least one argument in support of its use, and one argument against.
    Follow the links for Nelson Chemistry 11, 1.3.
    GO TO www.science.nelson.com
17. A Geiger counter is a device used to detect the presence of radiation from radioisotopes. Suggest some circumstances in which a Geiger counter would be useful.
18. Tritium is a radioisotope of hydrogen. Use the Internet to research the production, storage, and practical uses of tritium. What precautions must be taken when working with this radioactive substance?
    Follow the links for Nelson Chemistry 11, 1.3.
    GO TO www.science.nelson.com

Careers in Nuclear Science

The branch of physics that involves the study or use of changes within atomic nuclei is called nuclear physics. There are many careers in this field, involving many different applications of radioisotopes.

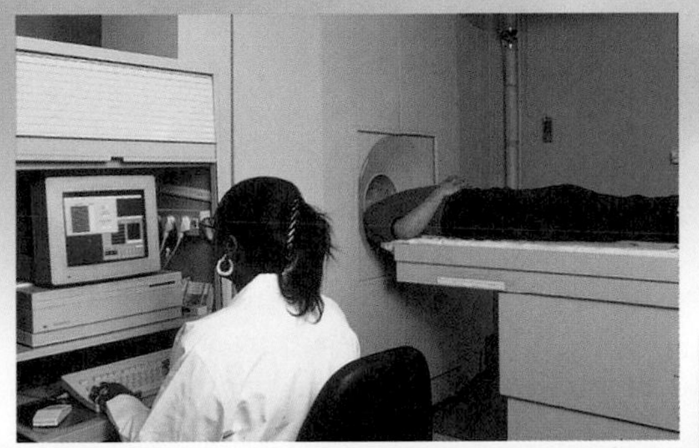

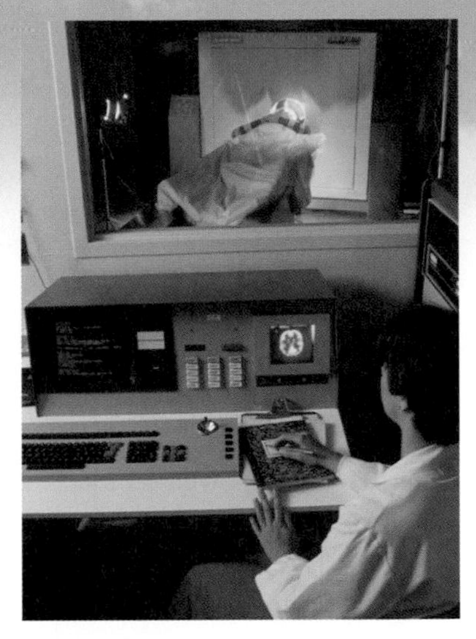

Diagnostic Radiographer

Radiographers operate computer-assisted imaging equipment to create X-ray or other images of a patient's body to diagnose injury and disease. These technologists prepare patients for diagnostic procedures, calculate an appropriate exposure time, process the results, and keep patients' records. Radiographers work under the supervision of a physician.

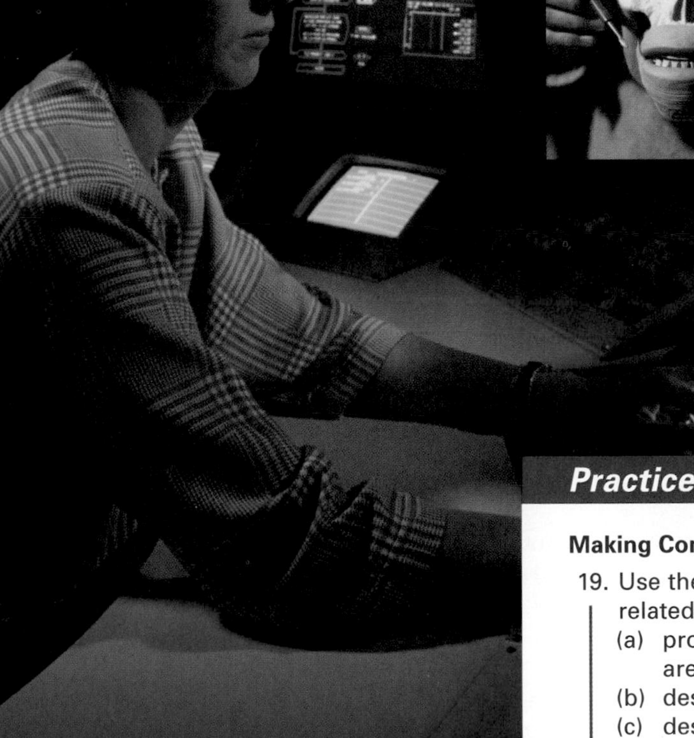

Radiologist

A radiologist is a medical professional who uses radioisotopes either for diagnosing health problems or for treating them. There have been many advances in medical radiology. There are now new, high-tech diagnostic devices and techniques, and the use of radioisotopes is becoming increasingly refined. They are now used to treat illnesses such as cancer and those caused by the overproduction of various enzymes.

Forensic Anthropologist

This fascinating career is mostly concerned with discovering when and how a person died, using the deceased person's body itself as evidence. Carbon-14 dating may be used to determine how much time has passed since death.

Nuclear Reactor Technician

Nuclear reactors provide some of Canada's electricity. The reactors must be constantly monitored, checked, and controlled. Highly trained technicians oversee the safety and output of the reactors with the assistance of state-of-the-art sensing and robotic technology.

Practice

Making Connections

19. Use the Internet to research one of the careers listed (or another related career that interests you). Write a report that
    (a) provides a general description of the work and how radioisotopes are involved;
    (b) describes the current working conditions and a typical salary;
    (c) describes the education required to work in this field;
    (d) forecasts employment trends for this field.

    Follow the links for Nelson Chemistry 11, 1.3.

GO TO www.science.nelson.com

Nuclear Power and Nuclear Waste

A Canadian-designed nuclear reactor system known as CANDU makes use of a naturally occurring radioisotope of uranium: uranium-235. The fuel in a CANDU reactor is in the form of pellets that contain uranium oxide, prepared from uranium ore. About 0.7% of the uranium in the pellets is U-235. The pellets are assembled into a fuel bundle that is placed into the calandria, or reaction vessel (**Figure 5**). A U-235 atom undergoes nuclear **fission** when a "slow" neutron collides with its nucleus, resulting in the formation of two lighter nuclei. This reaction also produces more neutrons and releases a considerable amount of energy (**Figure 6**). When these new neutrons are released, they are travelling very quickly—too quickly to cause further fission reactions. To be useful (continue the chain reaction by colliding with other U-235 nuclei), the neutrons must be slowed down. The substance used to slow the neutrons is called the moderator. In CANDU reactors the moderator is heavy water (water that contains deuterium instead of hydrogen). Heavy water is also used to cool the fuel bundles in a CANDU reactor. The heavy water is pressurized to prevent it from boiling. The hot heavy water is used to heat ordinary water, producing the steam necessary to turn turbines connected to electric generators.

fission: the splitting of a large nucleus into small nuclei

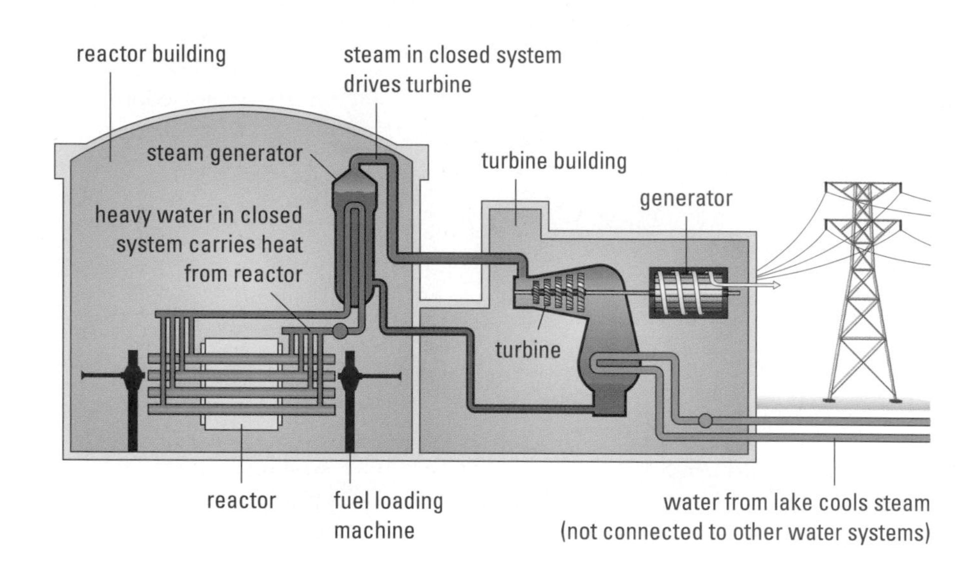

Figure 5
CANDU reactor and electricity generating station. CANDU is an acronym for Canadian deuterium and uranium.

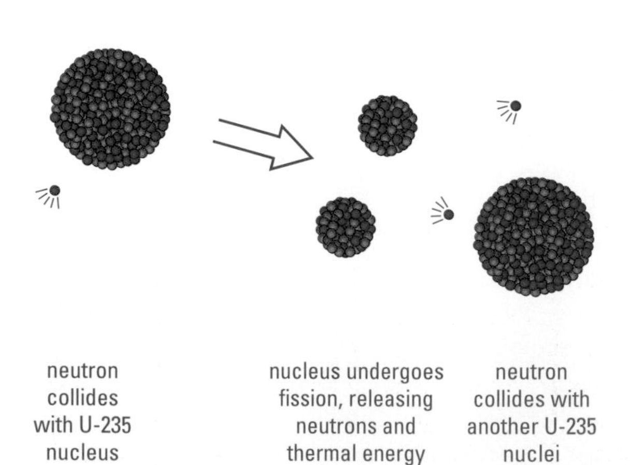

Figure 6
Nuclear fission of a uranium-235 atom

Once the fuel rods have been used, they are removed from the calandria. At the moment, the spent fuel rods are being stored in huge indoor pools of water at the nuclear power stations. On removal, the fuel rods are much more radioactive than they first were. They now contain U-238 and Pu-239 (formed when U-238 absorbs a neutron and decays by double beta emissions.) Pu-239 has a half-life of 2.44×10^4 a.

The use of nuclear energy to supply electricity is not associated with immediate environmental pollution such as acid rain, greenhouse gases, or the emission of toxic gases, all of which result from the burning of fossil fuels. Because of this, nuclear energy plants can be built close to where the power is needed. Since fresh water is required to circulate through the reactors' condensers, CANDU reactors must be constructed near a large body of water. The largest CANDU reactors in Canada are located on the shores of the Great Lakes at Bruce, Pickering, and Darlington.

Some aquatic organisms are affected by the waste heat carried by the water that cycles back into the lakes. Ground water contamination from radioactive tailings at uranium ore mines and the disposal of radioactive wastes are additional problems associated with the use of nuclear power. Plutonium, a waste material from CANDU nuclear reactors, can be used in the construction of nuclear weapons.

In April 1986, a partial meltdown of a nuclear reactor core occurred at Chernobyl in what is now Ukraine (**Figure** 7). More than 100 000 people were exposed to high levels of radiation when a steam explosion occurred in a nuclear power plant. Significant amounts of radioactive materials were released into the environment and hundreds of people who lived near the plant were treated for acute radiation sickness. Since the incident at Chernobyl, medical investigators have been tracking an increase in the incidence of thyroid cancer and birth defects in the area. The reactor at Chernobyl was not a CANDU reactor, but the accident there raised concerns about the safety of all types of reactors.

What to do with spent nuclear fuel is a thorny issue. Many suggestions have been made. The most popular at the moment is storage: keeping it where it is, in the pools within the nuclear facilities. This requires constant maintenance, and the storage sites are necessarily right next to large bodies of water, which are used for drinking water.

A longer-term alternative is disposal: for example, sealing the spent fuel in a glasslike compound and placing it in disused mines deep underground in stable rock. The concern with using underground sites for storage is that rock formations can shift and become unstable with the normal slow movement of Earth's crust. This movement would probably not be significant in our lifetimes, but the spent fuel will be radioactive for thousands of years. Engineers have identified some sites they consider to be particularly stable. One underground storage site being used by U.S. nuclear plants is in salt deposits under the Chihuahuan Desert of New Mexico. Although disposal sites have been identified in Canada, currently none has been approved.

Figure 7
Chernobyl's nuclear reactor
(a) Following a partial meltdown of the reactor core in 1986
(b) After the accident, the reactor was encased in concrete to contain the radioactivity. The concrete, which was applied in a great hurry and at the cost of several lives, is now crumbling, threatening to release still more radioactivity.

Practice

Understanding Concepts

20. Describe at least two advantages and two disadvantages associated with the use of nuclear power.
21. What isotope of uranium is used as fuel in a CANDU reactor?
22. (a) What is heavy water?
    (b) What purposes does heavy water serve in the function of a CANDU nuclear reactor?

Reflecting

23. There are risks associated with every technology, even simple ones (you can crush or cut the quick under your nails with a nail clipper). Do you, personally, feel that the risks associated with the use of nuclear power are justified by our need for energy? How do you decide which risks are acceptable, and which are not?

DECISION-MAKING SKILLS

- Define the Issue
- Identify Alternatives
- Research
- Analyze the Issue
- Defend a Decision
- Evaluate

Explore an **Issue**

Debate: Disposing of Nuclear Waste

In the case of nuclear waste, disposal can be defined as "permanent housing, without intention of retrieval." Storage is best defined as "keeping, unaltered, for an extended period." Consider these two possibilities. What is the best long-term solution to the problem of spent nuclear fuel from Ontario's CANDU reactors?

(a) In a class discussion, agree upon a resolution for your debate.

(b) Separate into small groups and research the various alternatives.

(c) As a group, decide on a position and do further research for evidence to support your position.

(d) Present and defend your group's position in a debate.

(e) Hold a postmortem on the debate. Could your group have presented or argued its position more effectively?

(f) Has your own position changed as a result of the debate? Explain.

Follow the links for Nelson Chemistry 11, 1.3.

GO TO **www.science.nelson.com**

Practice

Making Connections

24. Neutron bombardment is a technique used by forensic scientists. Use the Internet to research how this technique detects small or trace amounts of poisons in human tissue. Present your findings to the class in a way you feel is appropriate.

    Follow the links for Nelson Chemistry 11, 1.3.

GO TO **www.science.nelson.com**

25. Many radioisotopes used for medical diagnosis and therapy (such as cobalt-60 and iodine-131) or for industrial and research work are produced within the core of a nuclear reactor. Use the Internet to research and report on the use and handling precautions of one of these nuclear byproducts.

    Follow the links for Nelson Chemistry 11, 1.3.

GO TO **www.science.nelson.com**

26. Harriet Brooks, a Canadian, worked with New Zealander Ernest Rutherford at McGill University in Montreal. Reasearch and report on the significance of their contribution to atomic theory.

    Follow the links for Nelson Chemistry 11, 1.3.

GO TO **www.science.nelson.com**

Sections 1.2–1.3 Questions

Understanding Concepts

1. (a) What are the relationships among the number of protons, number of neutrons, and number of electrons in an atom?
   (b) Which of the numbers in (a) is/are related to the atomic number?
   (c) Which of the numbers in (a) is/are related to the mass number?
   (d) If the atomic mass of an element is not the same as the mass number of one atom of the element, what conclusions can you draw about that element?
2. Construct a graphic organizer to indicate the relationships among atomic number, mass number, atomic mass, isotopes, and radioisotopes.
3. Two atoms respectively have $Z = 12$, $A = 26$ and $Z = 14$, $A = 26$.
   (a) Can these two atoms be classified as isotopes of the same element? Give reasons for your answer.
   (b) Suggest an alternative classification for these atoms. Justify your choice.
4. Isotopes are classified as radioisotopes because they demonstrate the property of radioactivity. How does this property differ from other properties that we have used to classify elements?

Applying Inquiry Skills

5. When a sample of uranium-238 decays, alpha particles are emitted and the uranium nuclei are converted to thorium nuclei. Thorium-234 has a half-life of 24.10 d.
   (a) Plot a graph to predict the radioactive decay of 700.0 g of thorium-234.
   (b) Use your graph to determine how much time must pass before only 24.0 g of the sample will remain as thorium-234.

Making Connections

6. (a) Give at least three examples of ways in which radioisotopes are useful.
   (b) Suggest at least two safety precautions that may be used when handling radioisotopes.

1.4 Toward a Modern Atomic Theory

Rutherford's model of the atom raised some difficult questions. If the nucleus of an atom contained several positive protons that repelled each other, how did it stay together? Why didn't the negatively charged electrons rush toward and crash into the positively charged nucleus? In response to the first question, Rutherford suggested the idea of a nuclear force—an attractive force within the nucleus—that was much stronger than the electrostatic force of repulsion. The answer to the second question—why the electrons did not fall in to be "captured" by the nucleus—required a bold new theory created by a young Danish physicist named Niels Bohr (**Figure 1**).

Figure 1
Niels Bohr (1885–1962) developed a new theory of atomic structure and communicated this theory with an innovative model of the atom. For this work he won the Nobel Prize for Physics in 1922.

energy level: a state with definite and fixed energy in which an electron is allowed to move

orbit: a circular (spherical) path in which the electron can move around the nucleus

transition: movement of an electron from one energy level to another

ground state: the lowest energy level that an electron can occupy

Note that all of these definitions are based on Bohr's theory, and so are theoretical.

According to Bohr's theory, electrons within an atom can possess only certain discrete energies, called **energy levels**. Because of this, electron energy is said to be quantized. Each energy level, Bohr proposed, is associated with a fixed distance from the nucleus. (Later evidence showed that this is not entirely accurate.) In the Bohr model, an electron with a particular energy travels along a three-dimensional pathway called a shell or **orbit** (**Figure 2**).

These shells are designated by the principal quantum number, n, which can be any positive integer from 1 to infinity. Historically, these shells have also been designated by the symbols K, L, M, N, ... (**Figure 3**).

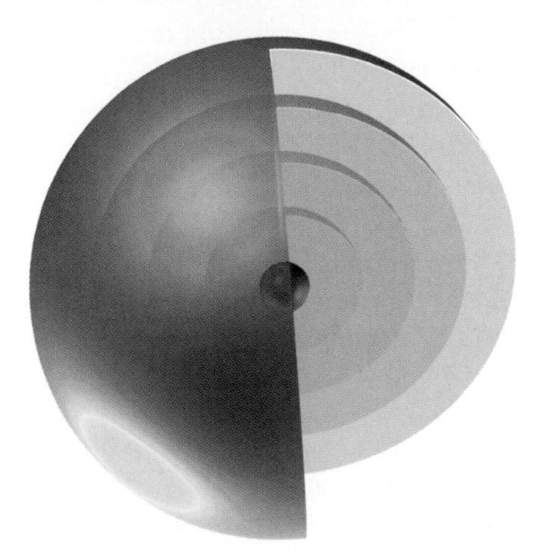

Figure 2
Since orbits are three-dimensional pathways, they are often described as spherical shells arranged concentrically about the nucleus of an atom.

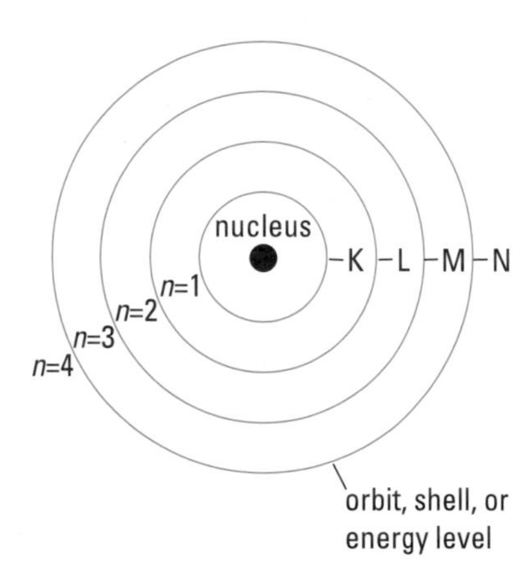

Figure 3
In the Bohr model of an atom, the shell closest to the nucleus is called the K shell. It corresponds to the first energy level or orbit. These orbits or energy levels are indicated by a number: 1, 2, 3, 4, ... up to infinity.

DID YOU KNOW ?

Nobel Prizewinners

Bohr developed his ideas on electrons, energy, and spectra from the work of Max Planck (1858–1947), who first suggested that energy is quantized. This earned Planck the 1918 Nobel Prize in physics. Bohr also made use of Einstein's work on the photoelectric effect (the change in energy of an atom as it absorbs or emits a photon of light). This, not the theory of relativity, was the work for which Einstein received the Nobel Prize in 1921!

Bohr's theory depended on these assumptions:

- that an electron can travel indefinitely within an energy level without losing energy;
- that the greater the distance between the nucleus of the atom and the energy level, the greater the energy required for an electron to travel in that energy level; and
- that an electron cannot exist between orbits, but can move to a higher, unfilled orbit if it absorbs a specific quantity of energy, and to a lower, unfilled orbit if it loses energy.

Bohr theorized that if additional energy is supplied to an electron in a given energy level, then the electron can "jump" to a higher, unfilled energy level, farther away from the nucleus (**Figure 4**). Bohr referred to this kind of jump as a **transition**. The quantity of energy (a quantum) required to cause a transition is equivalent to the difference in energy between the energy levels. When an electron drops back to a lower energy level, it releases an equivalent amount of energy.

When an electron is in the lowest energy level that it can occupy, it is said to be in its **ground state**. Similarly, when all the electrons in an atom are in the lowest possible energy levels, the atom is said to be in its ground state. In addition, Bohr suggested that the properties of the elements could be explained by the arrangement of electrons in orbits around the nucleus. For instance, Bohr

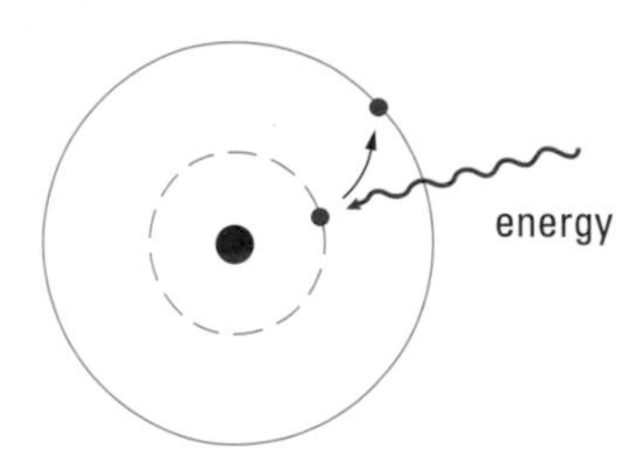

(a) An electron gains a quantum of energy.

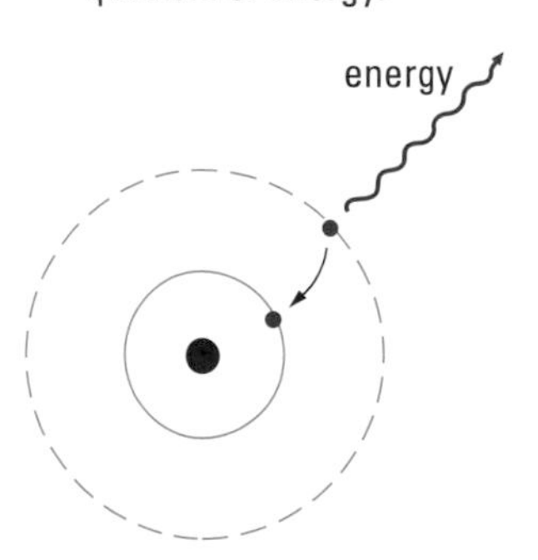

(b) An electron loses a quantum of energy.

Figure 4
(a) Energy is absorbed as electrons rise to a higher energy level.
(b) Energy is released as electrons fall to a lower energy level.

hypothesized that the stable, unreactive nature of certain elements was due to filled outermost energy levels.

The theory developed by Bohr was an attempt to explain empirical evidence obtained from experiments involving hydrogen. When hydrogen gas is heated, it emits a violet light. When this violet light is directed through a prism, it separates into a series of coloured lines called a **line spectrum**. Hydrogen produces a line spectrum consisting of four visible lines: red, blue-green, indigo, and violet (**Figure 5**), although the violet line may not always be visible. The same results can be obtained by passing electricity through a gaseous element at low pressure: The gas will emit light of only certain wavelengths, which we can see as lines if the light is passed through a prism (**Figure 6**, page 40).

These observations contrast with what we see when we shine white light through a glass prism or a spectroscope: In this case, we see a blended pattern of colour or a **continuous spectrum**. The colours of the spectrum (plural: spectra) always appear in the same order: violet, indigo, blue, green, yellow, orange, and finally red. A rainbow is an example of a continuous spectrum.

Bohr proposed the following explanation for the line spectrum of hydrogen: When energy (heat or electricity) is supplied to hydrogen atoms, excited electrons gain a certain quantity of energy. With this extra energy they jump from a lower energy level to a higher energy level. As the electrons drop back to lower energy levels they release energy corresponding to a few precise wavelengths (**Figure 5**). Bohr's theory was important because it explained both the stability of electrons in atoms and atomic emission spectra, and it enabled scientists to cal-

line spectrum: a pattern of distinct lines, each of which corresponds to light of a single wavelength, produced when light consisting of only a few distinct wavelengths passes through a prism or spectroscope

continuous spectrum: the pattern of colours observed when a narrow beam of white light is passed through a prism or spectroscope

Note that these are empirical definitions, based on observations.

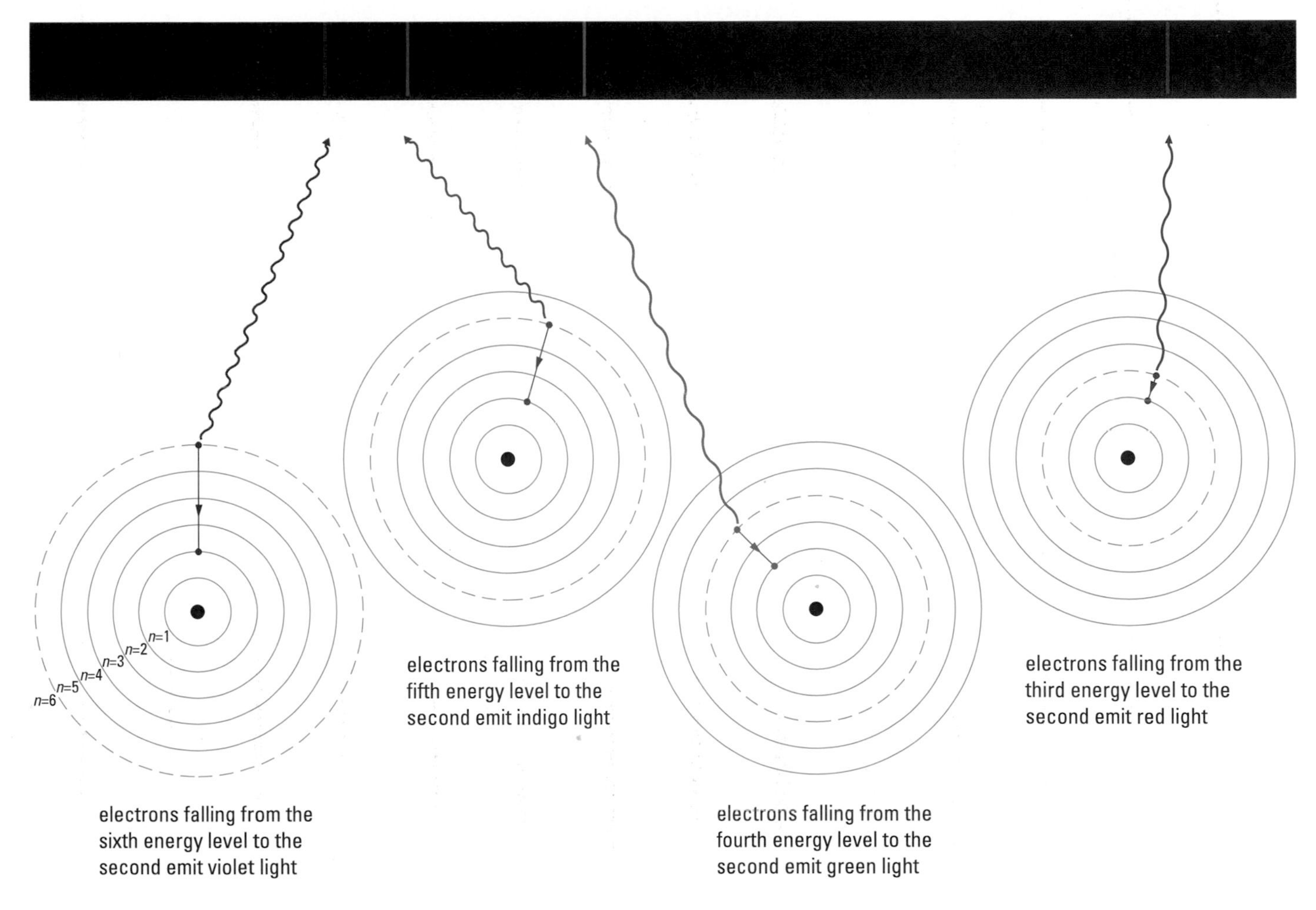

Figure 5
Each line in the spectrum corresponds to a specific quantity of energy released as an electron falls back to the $n = 2$ energy level.

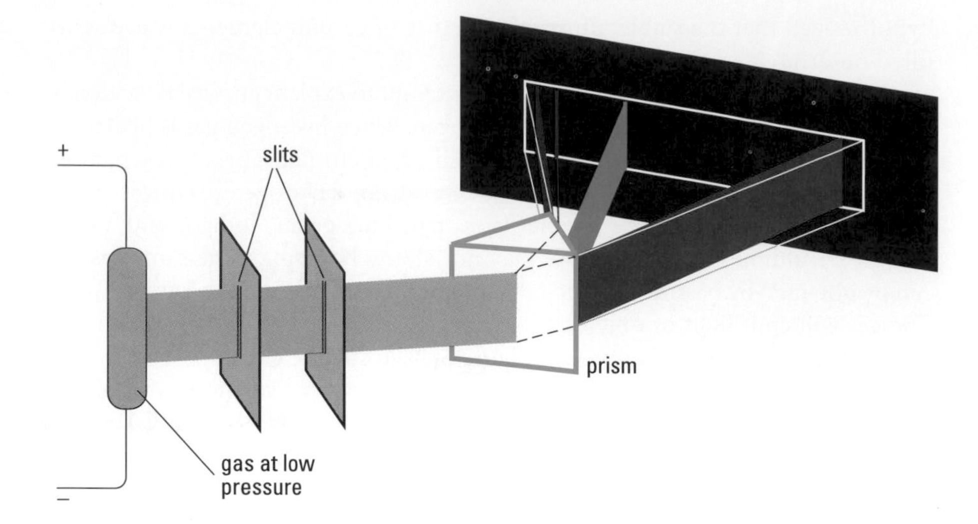

Figure 6
When electricity is passed through a gaseous element, the gas glows as it emits certain wavelengths of light, which can be separated to produce a line spectrum. The apparatus may not show all the lines near the edges of the visible spectrum.

culate the actual energies of the different energy levels within the hydrogen atom. However, Bohr's theory was not able to explain the more complex spectra of other elements.

INQUIRY SKILLS

- ○ Questioning
- ○ Hypothesizing
- ○ Predicting
- ○ Planning
- ○ Conducting
- ● Recording
- ● Analyzing
- ○ Evaluating
- ○ Communicating

Investigation 1.4.1

Atomic Spectra

If the radiation produced by supplying energy to an unknown element is in the visible range, we will see it as light. This light can be separated to produce a line spectrum, which we can use to identify the element.

In this investigation you will observe, record, and compare the continuous spectrum of white light as well as the line spectra of hydrogen and several unknown elements. You will then identify the unknown elements by comparing their observed line spectra with the published line spectra of several elements, found in Appendix C. Complete the **Analysis** and **Synthesis** sections of the lab report.

Question

What unknown substances are held in the gas discharge tubes?

Experimental Design

Different elements will be excited by passing electricity through gaseous samples of each element. In each case, the spectrum produced will be observed with a spectroscope (**Figure 7**) and compared with the continuous spectrum and with line spectra of known elements.

This investigation involves the use of a high-voltage power supply. This power supply should be operated by your teacher only.

Students should stand away from the gas tubes and avoid viewing the tubes directly in case dangerous ultraviolet radiation is produced.

Materials

- eye protection
- incandescent light source (at least 40 W)
- fluorescent light source
- spectroscope
- power supply
- hydrogen gas discharge tube
- other gas discharge tubes

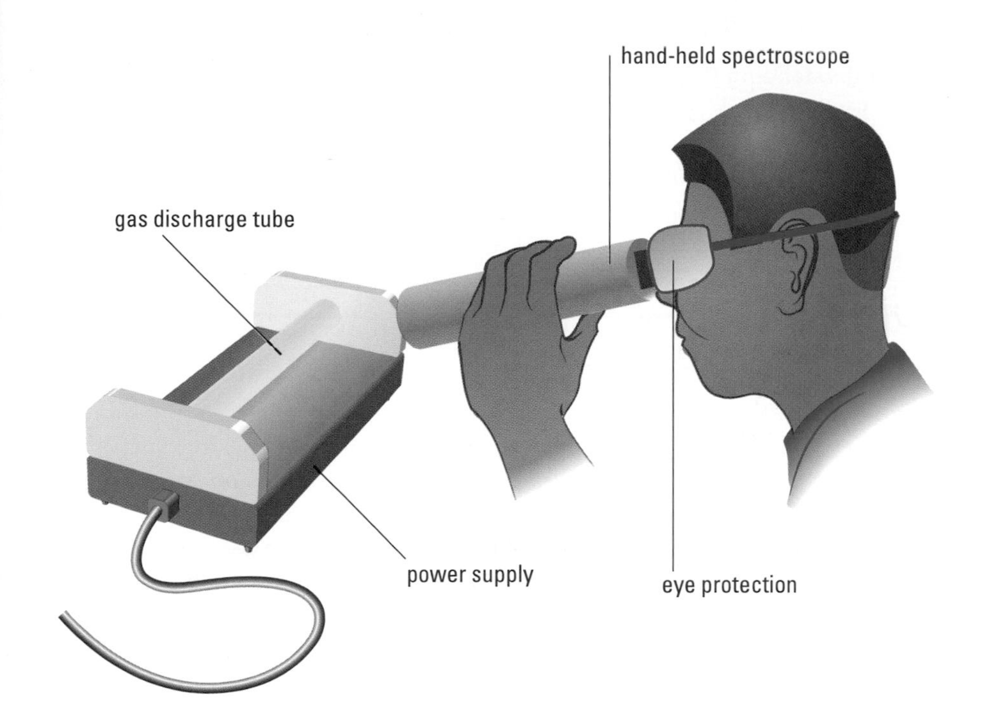

Figure 7
The line spectrum of an element is viewed using a hand-held spectroscope. This kind of spectroscope uses a diffraction grating rather than a prism to separate the lines of the spectrum.

Procedure

1. Use the spectroscope to observe the fluorescent lights in the classroom. Draw and label the spectrum.
2. Turn on the incandescent light and turn off all other lights in the classroom.
3. Use the spectroscope to observe the incandescent light. Draw and label the spectrum that you observe.
4. Connect the hydrogen tube to the power supply and turn off all other lights.
5. Observe the spectrum produced when you pass the light from the hydrogen tube through the spectroscope. Draw the spectrum, indicating the colour of each observed line.
6. Repeat steps 4 and 5 using the gas tubes of unknown elements. If the patterns are complex, note and record the differences in the number, colour, and spacing of the lines.

Analysis

(a) Answer the Question by comparing the line spectra of the unknown substances in the gas discharge tubes with the line spectra in Appendix C. What substances have the line spectra that you observed?

Synthesis

(b) How do the spectra of fluorescent and incandescent light sources compare? What does this suggest about the light produced by each of the two sources?

(c) Which elements produced the largest number of spectral lines? What does this suggest about electron transitions?

Practice

Understanding Concepts

1. When a gas is heated, the gas will emit light. Explain why this phenomenon occurs using the Bohr model of the atom.
2. What do the different colours in a line spectrum represent?
3. Why do different substances show different spectra?
4. Sodium vapour lamps emit a characteristic yellow light. What can we deduce about sodium atoms, based on this observation?

Applications of Emission Spectra: Flame Tests

The coloured lights given off by neon signs, streetlights, signal flares, fireworks, and even distant stars all result from excited electrons moving from higher to lower energy levels in atoms. Each element has a unique emission spectrum, so when elements are heated in a flame, they will give the flame a characteristic colour. We can use this technique to identify elements even when they are parts of compounds. We can also determine the amount of the element present by measuring the intensity of the light emitted. This technique, called flame emission spectroscopy, is used in many situations, including analyzing and establishing the geographic origin of ancient pieces of pottery.

Activity 1.4.1

Creating a Flame Test Key

Some elements show one or more particularly prominent lines in their line spectra. We can make use of this phenomenon to find out which metal is present in an unknown metallic compound. In this test, called a **flame test**, the unknown sample is heated in a flame. The colour of the flame is observed and compared against a chart of known flame colours.

flame test: a diagnostic technique in which a metallic compound is placed in a flame and the colour produced is used to identify the metal in the compound

Your aim in this activity is to observe, and create a key for, the colours produced when solutions of known metal compounds (mostly Group 1 and 2 metals) are heated to high temperature. You will be able to use this key to determine the identity of an unknown metallic compound. To establish this key, you will draw samples of dissolved compounds from solids and solutions using a wire, and then put the end of the wire into a flame.

Materials

Hydrochloric acid is highly corrosive and can burn the skin. Use caution in handling this substance, and use it only in a well-ventilated area.

Wear eye protection and an apron throughout the investigation.

apron
eye protection
nichrome test wire
2 test tubes
test-tube rack
cobalt glass squares
laboratory burner
hydrochloric acid, dilute, 5 mL
sodium chloride solution, 3 mL

samples of the following solids:
- sodium chloride
- sodium nitrate
- calcium chloride
- strontium chloride
- lithium chloride
- potassium chloride
- copper(II) chloride

Procedure

1. Obtain your samples (the solids and 3 mL of the sodium chloride solution in a small test tube) and the 5-mL hydrochloric acid solution (also in a small test tube). Store the test tubes in a rack.
2. Light the laboratory burner and adjust it until it is burning with a blue flame.
3. To clean the nichrome test wire, dip it in the hydrochloric acid, and then hold the wire in the flame of the laboratory burner. Repeat the procedure until the wire adds no colour to the flame.
4. Pick up a small amount of solid sodium nitrate with the wire loop. Hold the end of the wire in the flame. Record your observations.
5. Clean the wire as in step 3.
6. Pick up a small sample of solid sodium chloride on the wire loop. Hold the end of the wire in the flame for several seconds. Record your observations.
7. Clean the wire as in step 3.
8. Dip the nichrome test wire in the sodium chloride solution. Hold the end of the wire in the flame and record your observations.
9. Repeat step 4 with each of the solid samples, remembering to clean the wire after each test. Record your observations.
10. Repeat the flame tests for potassium chloride and sodium chloride, observing the results through cobalt glass. Record your results.

Analysis

(a) Assemble your observations into an identification key that you could use as a quick reference.

Synthesis

(b) Compare the results of your flame tests for solid sodium nitrate, solid sodium chloride, and sodium chloride solution. What do these results indicate?

(c) How did the results compare in testing potassium chloride and sodium chloride with and without the cobalt glass? What purpose does the cobalt glass serve?

(d) **Figure 8** (page 44) shows the results of four flame tests. Using your identification key, identify the metal in each of the compounds.

DID YOU KNOW ?

Nichrome Wire

What properties would you need in a wire that is going to be repeatedly heated and cooled, both wet and dry, and is going to hold materials in a flame so that you can observe their emission spectra? The wire must have a high melting point, be very unreactive, and produce no emission spectrum itself (at least, at the temperatures generally used in the lab). The wire that offers these properties is called nichrome. It is an alloy of almost 80% nickel and 20% chromium, with traces of silicon, manganese, and iron.

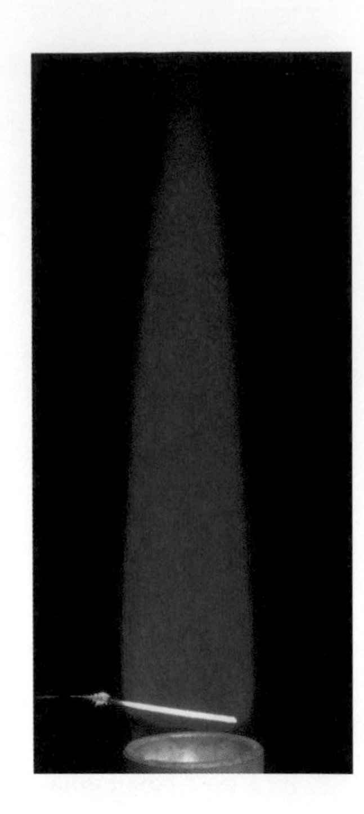

Figure 8
Results of flame tests for four solutions of unknown metallic compounds

INQUIRY SKILLS

- ● Questioning
- ○ Hypothesizing
- ○ Predicting
- ● Planning
- ● Conducting
- ● Recording
- ● Analyzing
- ● Evaluating
- ● Communicating

Investigation 1.4.2

Identifying an Unknown Metal in a Compound with a Flame Test

The purpose of this investigation is to use the flame test key you created in Activity 1.4.1 to identify the unknown metal in a compound provided by your teacher.

You will create your own investigation. Complete a lab report that includes the **Question**, an **Experimental Design, Materials,** a **Procedure, Analysis,** and an **Evaluation** of the design.

Question

(a) Write your own Question, which you will attempt to answer in your investigation.

Experimental Design

(b) Create an outline describing your experiment.
(c) Write a detailed Procedure, including any safety precautions, and have it approved by your teacher.

Materials

unknown chloride

(d) Create a list of additional materials you will need to complete your investigation.

Procedure

1. Carry out your Procedure.

Analysis

(e) Use the Evidence you gathered in your Procedure to answer your Question.

Evaluation

(f) How confident are you that your investigation produced a valid answer to the Question? List some limitations of your Experimental Design, and suggest some improvements.

Practice

Understanding Concepts

5. As far as we know, people from all cultures have been interested in the stars. Modern astronomers use spectroscopy to analyze starlight. What information would these observations provide?
6. According to the Bohr theory, what happens to an electron in an atom as it absorbs energy and as it releases energy?
7. "Spectral lines are the fingerprints of elements." Explain what is meant by this statement.

Making Connections

8. When fireworks explode they produce a variety of colours.
   (a) Provide an explanation for this observation, referring to emission spectra and energy levels.
   (b) Suggest compounds that could be used to produce several specific fireworks colours. Comment on any practical difficulties that might result from your choices.
9. Using the Internet, research the detection and generation of elements in stars and supernovas. Write a brief report, incorporating the results of your research and commenting on the following hypothesis: All elements are conglomerates of hydrogen (stated by William Prout, 1785–1850).

   Follow the links for Nelson Chemistry 11, 1.4.

www.science.nelson.com

The Quantum Mechanical Theory

Bohr made significant contributions to our current understanding of atomic structure, especially in establishing the concept of atomic energy levels. However, his theory of the atom was limited in its ability to predict the line spectra of atoms other than hydrogen, as well as to account for certain details of atomic structure. These shortcomings resulted in intense investigation that led in the 1920s to our modern atomic theory, sometimes known as quantum mechanics theory. A crucial steppingstone along the way was the development of **quantum mechanics**, which builds on the earlier idea that electrons within an atom can possess only discrete quantities of energy. Quantum mechanics theory also

quantum mechanics: a theory of the atom in which electrons are described in terms of their energies and probability patterns

principal quantum number (*n*): a number specifying the theoretical energy level of an electron in an atom

electron cloud: the region of an atom in which electrons are most probably located

retains the concept that electrons fill successive shells, each of which is designated by the **principal quantum number (*n*)**.

One important idea resulting from quantum mechanics, obtained by Werner Heisenberg (1901–1976; famous for his Heisenberg uncertainty principle) in 1927, is that it is impossible to know precisely both the position and motion of an electron at the same time. Unlike Bohr's theory, which describes electrons travelling in spherical orbits at precise distances from the nucleus, quantum mechanics describes electrons within atoms in terms of their energy and probability patterns—that is, the probability of finding an electron of a specified energy within a specific region of space about the nucleus. This region in space is called the **electron cloud** (Figure 9).

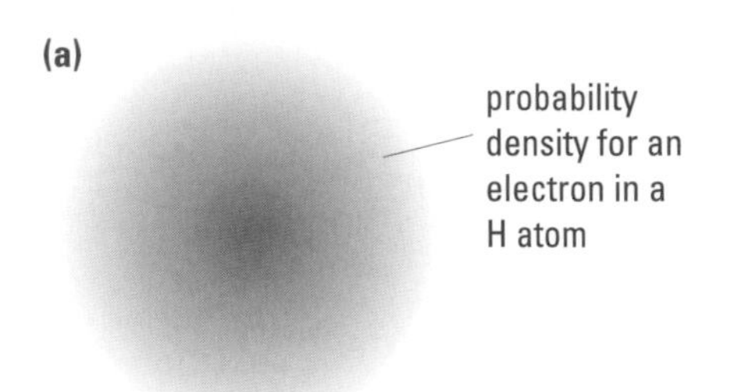

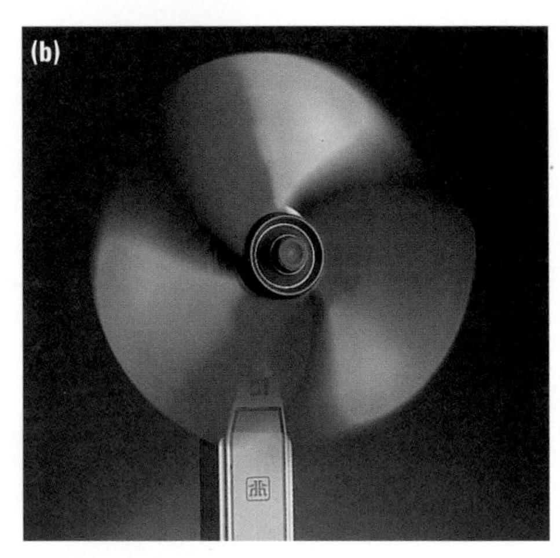

Figure 9
(a) According to quantum mechanics, there is a region around the nucleus of the atom in which there is a high probability of finding the electron (dark). This region is often described as an electron cloud. In every other area, the probability of finding the electron is low (light). The motion of the electron is not known.
(b) Similarly, as the blades of a fan rotate rapidly, the position and motion of an individual blade at any instant are unknown.

It turns out that to fully designate all possible energies and probability patterns, we must modify the atomic model by subdividing each shell or orbit into an increasing number of subshells as the principal quantum number increases (Table 1). Just as there is a maximum number of electrons possible for each shell, there is a maximum number possible for each subshell. In the case of hydrogen, the subshells within any given shell are all equivalent in energy—the same energy as predicted by Bohr's theory. For all other elements, the subshells within a given shell have different energies. This concept becomes important in determining the atomic structure and properties of elements within the periodic table. It also accounts for the spectra of elements other than hydrogen (which Bohr's theory failed to do), since the emission lines result from electrons moving between levels whose energies cannot be calculated solely from the value of *n*.

Table 1: Quantum Mechanical Features of the First Four Shells

n	Symbol	Number of subshells	Maximum number of electrons per shell	Maximum number of electrons per subshell
1	K	1	2	(2)
2	L	2	8	(2 + 6)
3	M	3	18	(2 + 6 + 10)
4	N	4	32	(2 + 6 + 10 + 14)

From his study of the periodic table, Bohr concluded that the maximum number of electrons that can populate a given shell can be calculated from the equation $2n^2$ (where n is the principal quantum number). According to this equation, the first energy level could contain up to $2 \times (1)^2$ or 2 electrons. The second energy level could contain up to $2 \times (2)^2$ or 8 electrons, while the third energy level could contain up to $2 \times (3)^2$ or 18 electrons. According to this restricted theory, the 17 electrons of an atom of chlorine in the ground state would be distributed among three shells. The first shell would be filled with 2 of the 17 electrons. The second, or next highest, shell would be filled with 8 electrons. Finally, with both the first and the second shells filled, the remaining 7 electrons would occupy the third shell. The electrons in the highest occupied energy level are called **valence electrons**.

valence electrons: those electrons that occupy the highest shell of an atom and are used by the atom to form chemical bonds (theoretical definition)

Sample Problem

(a) How many electrons would be in each shell of a fluorine atom?
(b) How many valence electrons does fluorine have?

Solution

(a) Fluorine has atomic number 9, so will have 9 electrons: 2 in the first energy level and 7 in the second.

(b) 7

Notice that the first shell has space for two electrons, and there are two elements in the first period of the periodic table: hydrogen and helium. The second shell has space for eight electrons, and there are eight elements in Period 2 (Li through Ne). The $2n^2$ rule tells us that the $n = 3$ shell can contain a maximum of 18 electrons, so on the basis of Bohr's theory we might expect there to be 18 elements in Period 3. In fact, there are only 8 elements in Period 3 (elements Na through Ar). Why is this? Remember that quantum mechanics theory tells us that the third shell is split into three subshells, each of which has a different energy. The outermost (highest energy) subshell is very close in energy to the first subshell of the next shell ($n = 4$). Experimental measurements show that, once there are 8 electrons in the third shell, the fourth shell is used (in elements K and Ca) before electrons are added to fill up the third shell. Similar effects are also observed in subsequent periods.

According to Bohr's theory, there is a pattern linking electron arrangements to the periodic table for the representative elements:

- for the elements in Groups 1 and 2, the number of valence electrons corresponds to the group number;
- in Groups 13 to 18, the number of valence electrons corresponds to the second digit of the group number. For example, fluorine is found in Group 17 (VIIA) of the periodic table: It contains 7 valence electrons. Both chlorine and fluorine are halogens and have 7 valence electrons. You already know that the halogens share similar physical and chemical properties, which allows them to be placed together in the same chemical group.

In general, representative elements in the same chemical group share not only properties but also the same number of valence electrons—a powerful indicator of the relationship between electron arrangement and periodic trends.

Practice

Understanding Concepts

10. What are valence electrons, and what is their significance?
11. Which two key ideas from previous atomic theories are retained by the theory of quantum mechanics?
12. Use the periodic table and theoretical rules to predict the number of occupied energy levels and the number of valence electrons in each of the following atoms:
    (a) beryllium
    (b) chlorine
    (c) krypton
    (d) iodine
    (e) lead
    (f) arsenic
    (g) cesium

Section 1.4 Questions

Understanding Concepts

1. By custom, hydrogen is put at the top of Group 1 in the periodic table. There are other places it could go. Suggest one other place, using the concepts of periodicity and atomic structure to justify your answer.
2. (a) Copy and complete **Table 2**, using your knowledge of atomic theory and the periodic table.

Table 2: Electron Structure of Selected Elements

Element	Number of electrons	Number of occupied shells	Number of valence electrons
oxygen	?	?	?
sulfur	?	?	?
magnesium	?	?	?
sodium	?	?	?
beryllium	?	?	?
calcium	?	?	?
cesium	?	?	?
nitrogen	?	?	?
chlorine	?	?	?
lithium	?	?	?
helium	?	?	?
bromine	?	?	?
phosphorus	?	?	?
fluorine	?	?	?
potassium	?	?	?

(b) Classify the elements in **Table 2** into groups according to the number of valence electrons.

(c) What common chemical and physical properties would you expect each of the different groups to exhibit?

1.5 Trends in the Periodic Table

Elements within a group have the same number and distribution of valence electrons, so we might expect them to exhibit the same chemical behaviour. Based only on electron distribution, we would expect lithium and sodium, for example, to behave in very much the same way. There are indeed similarities, but chemical and physical properties of elements within groups are not identical. Instead, we can see that there are variations, or trends, within each group. A property will increase or decrease within a group as the atomic number increases. Why is that? And is the change regular, or unpredictable?

While the periodic table helps us to understand the elements, it also raises many questions: Why do the properties of elements vary across the periodic table? Why are groups of elements similar in their physical and chemical properties, but not identical? How do we explain the chemical formulas of compounds formed from elements within a group? Can we use atomic theory to explain trends across the periodic table?

Lab Exercise 1.5.1

Reactivity of Alkali Metals

INQUIRY SKILLS

- ○ Questioning
- ● Hypothesizing
- ○ Predicting
- ○ Planning
- ○ Conducting
- ○ Recording
- ● Analyzing
- ● Evaluating
- ● Communicating

Comparing the chemical reactivity of elements and compounds allows you to establish patterns, such as an order of reactivity, for elements within groups and periods. In this lab exercise, you are provided with Evidence concerning the reactivity of three elements from the same group: lithium, sodium, and potassium. The purpose of this lab exercise is to examine the Evidence and generate a Hypothesis that explains it.

Complete the **Analysis** and **Evaluation** sections of the lab report.

Question

What is the order of reactivity of lithium, sodium, and potassium?

Experimental Design

A small piece of each of lithium, sodium, and potassium is placed in water and the reactivity is observed and recorded.

Materials

safety shield
tongs
tweezers
mineral oil
sharp edge (knife or scoop)
paper towel
petri dish
large beaker
distilled water
wire gauze square
samples of
    lithium
    sodium
    potassium

Procedure

1. Transfer a small piece of lithium into a petri dish containing mineral oil.
2. Using a knife or similar sharp edge, remove the surface layer from one side of the piece of lithium. Observe the fresh surface.
3. Fill the beaker about half full with water.
4. Cut a small piece (about 2 mm cubed) of the element. Remove any oil with paper towel and drop the piece of the element into the beaker. Immediately cover the beaker with the wire gauze. Record observations of any chemical reaction.
5. Repeat steps 1 through 4 using sodium and potassium.

Evidence

Table 1: Observations of Reactions of Alkali Metals

Alkali metal	Observations
lithium	reacts vigorously with water; evidenced by production of light, heat, and sound energy
sodium	very vigorous reaction with water; evidenced by more pronounced production of light, heat, and sound energy
potassium	extremely vigorous reaction with water; evidenced by very pronounced production of light, heat, and sound energy

Analysis

(a) Based on the Evidence, write an answer to the Question as a Hypothesis.

Evaluation

(b) Evaluate the Experimental Design and the Procedure. What flaws do you see?

Synthesis

(c) Based on the observed trends, predict the relative reactivity of each of the following pairs of elements:
  (i) lithium and francium
  (ii) francium and rubidium
  (iii) cesium and sodium
(d) What would you expect to observe if cesium were added to water?

Describing Trends in the Periodic Table

There are many observable patterns in the chemical properties of elements. Within a period, chemical reactivity tends to be high in Group 1 metals, lower in elements toward the middle of the table, and increase to a maximum in the Group 17 nonmetals. (The noble gases, of course, are the least reactive of any group.) Metals react differently from nonmetals. Within a group of representative metals, such as the alkali metals, reactivity increases moving down the group; within a group of nonmetals, reactivity decreases moving down the group (**Figure 1**). Periodic trends such as these demonstrate the periodic nature of

Figure 1
Trends in reactivity within the periodic table

chemical properties. But how can we explain periodicity? *Why* do the properties of elements follow these trends? To answer these questions we must turn to accepted theories of atomic structure and properties of the atom. We will look at five properties—atomic radius, ionic radius, ionization energy, electron affinity, and electronegativity—and try to explain them in terms of number of electrons and energy levels.

atomic radius: a measurement of the size of an atom in picometres (1 pm = 10^{-12} m)

ion: a charged entity formed by the addition or removal of one or more electron(s) from a neutral atom (theoretical definition)

Atomic Radius

Measuring the radius of a CD is easy: With a ruler, measure the distance from its centre to its outside edge. Finding the radius of an atom—its **atomic radius**—is a lot more complicated. For one thing, an atom is so small that all measurements have to be made indirectly. For another, it is very difficult to know exactly where an atom ends. Recall from our discussion of quantum theory that we can only estimate the probability of an electron being in a certain place at a certain time. So where is the outside edge of an atom?

Scientists have proposed several different answers to this question, with different definitions of atomic radius. We will simplify, and assume that the atomic radius is one-half the distance between the nuclei of two atoms of the same element, whether they are bonded (e.g., oxygen and nitrogen, **Figure 2(a)**) or not bonded (e.g., the metallic elements, **Figure 2(b)**).

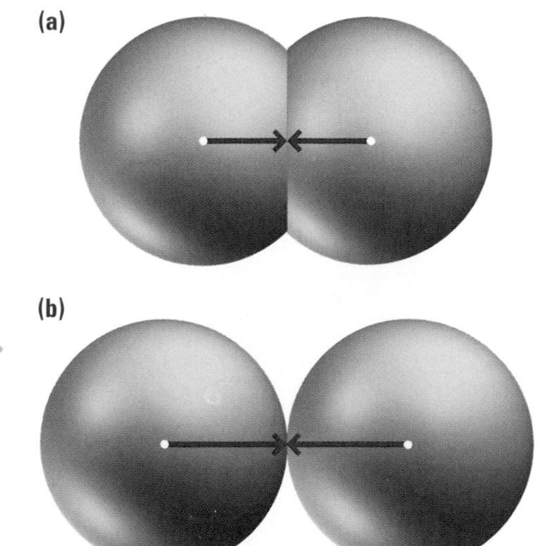

Figure 2
(a) The atomic radius in a molecule
(b) The atomic radius in nonbonded atoms

Figure 4, pages 52–53, shows the atomic radii of elements in the periodic table. Note that, in general, the atomic radii decrease from left to right across each period. We can explain this trend in atomic radius in terms of nuclear charge and energy levels. As we move along a period from Group 1 to Group 17, the number of protons in the nucleus increases in increments of one. Consequently, the nuclear charge increases also. As the nuclear charge increases, the number of filled inner energy levels, electrons shielding the outer electrons from the nucleus, remains the same. Therefore, the strength of the attraction between the nucleus and the valence electrons is stronger as protons are added. This nuclear attraction pulls the electrons closer to the nucleus (decreases the average distance of the valence electrons from the nucleus). Consequently, there is a decrease in the size of the atom.

Chemists hypothesize that the atomic radius increases from top to bottom in a given group because of the increasing number of energy levels. Each additional energy level places electrons at a greater distance from the nucleus. This occurs because the inner energy levels are filled with electrons, which serve to shield the outer electrons from the pull of the nucleus. With each additional energy level that is filled as you move down a group, the attractive force of the nucleus is reduced by this shielding effect and the outer electrons are not as strongly attracted. This decreased nuclear attraction felt by the outer electrons results in a larger radius.

Ionic Radius

The removal of an electron from an atom results in the formation of an **ion**. When a valence electron is removed from an atom of lithium, a lithium ion is formed. This ion has an ionic radius smaller than the atomic radius of the neutral lithium atom (**Figure 3**) because there is one fewer electron shell. The charge on this ion, because it is missing an electron, is positive.

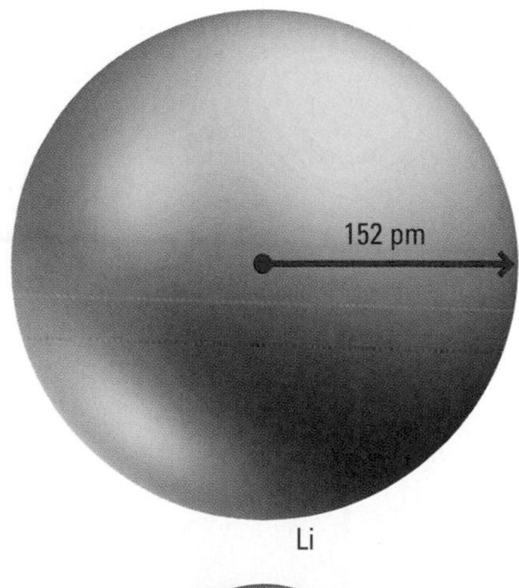

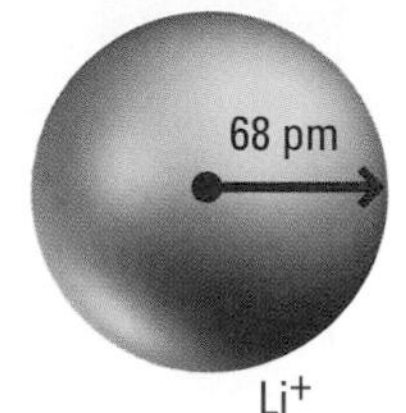

Figure 3
A lithium ion is considerably smaller than a lithium atom.

Most representative elements lose only one or two electrons, to become ions with charges of 1+ and 2+, but a few transition metals can lose even more electrons. The ions in **Figure 4** are the most common ion for each element.

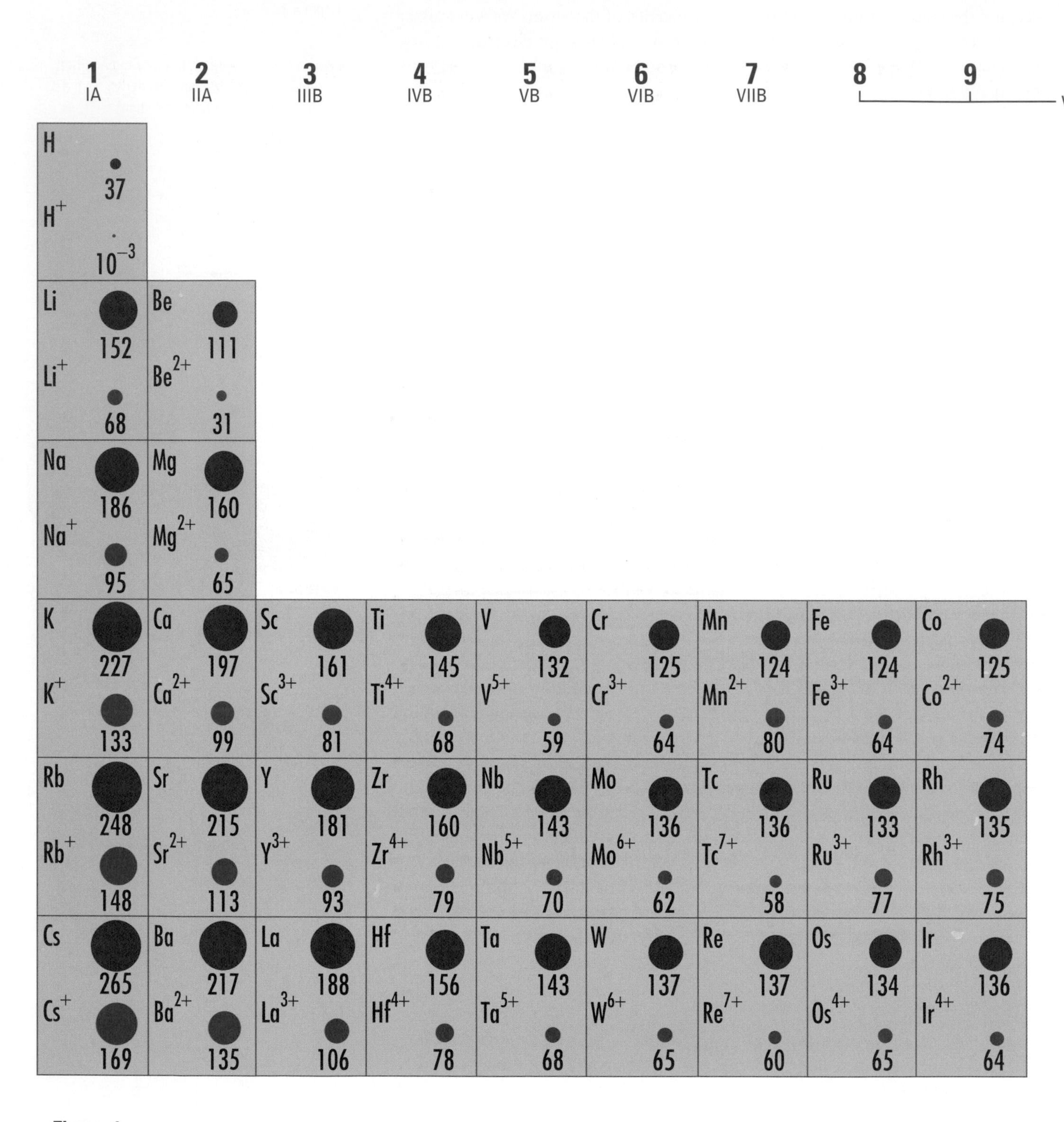

Figure 4
Atomic and ionic radii of selected elements

10	11 IB	12 IIB	13 IIIA	14 IVA	15 VA	16 VIA	17 VIIA	18 VIIIA
							H 37 H^{-} 208	He 50
			B 88	C 77	N 70 N^{3-} 171	O 66 O^{2-} 140	F 64 F^{-} 136	Ne 62
			Al 143 Al^{3+} 50	Si 117	P 110 P^{3-} 212	S 104 S^{2-} 184	Cl 99 Cl^{-} 181	Ar 95
Ni 124 Ni^{2+} 72	Cu 128 Cu^{2+} 72	Zn 133 Zn^{2+} 74	Ga 122 Ga^{3+} 62	Ge 123 Ge^{4+} 53	As 121 As^{3-} 222	Se 117 Se^{2-} 198	Br 114 Br^{-} 196	Kr 112
Pd 138 Pd^{2+} 86	Ag 144 Ag^{+} 126	Cd 149 Cd^{2+} 97	In 163 In^{3+} 81	Sn 140 Sn^{4+} 71	Sb 141 Sb^{3+} 90	Te 137 Te^{2-} 221	I 133 I^{-} 216	Xe 130
Pt 138 Pt^{4+} 70	Au 144 Au^{3+} 91	Hg 160 Hg^{2+} 110	Tl 170 Tl^{+} 144	Pb 175 Pb^{2+} 120	Bi 155 Bi^{3+} 96	Po 167 Po^{4+} 65	At 142 At^{-} 227	Rn 140

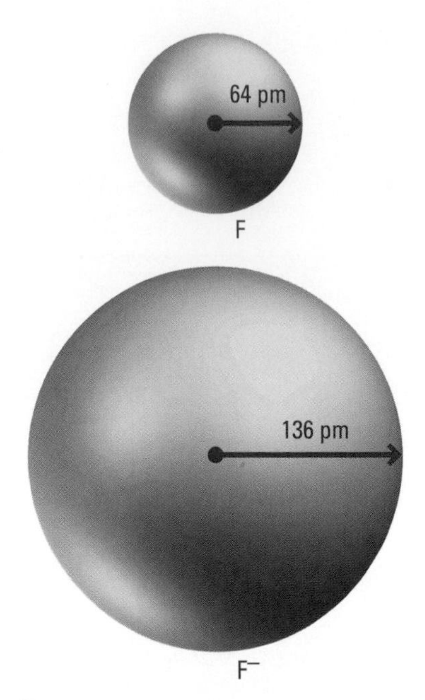

Figure 5
A fluoride ion is considerably larger than a fluorine atom.

How can we explain the observed difference between atomic and ionic radii of positive ions? In theory, each electron repels all of the other electrons. This repulsion forces the electrons in the outer shells farther from the nucleus. Removal of an electron from a neutral atom reduces the strength of these electron–electron repulsions. Positive ions are always smaller than the neutral atom from which they are formed.

Look at the radii of the ions of Group 1 elements (**Figure 4**). See how the size increases as you move down the group. This is because, for each successive ion in the group, the outermost occupied energy level is farther from the nucleus. As we move across Period 3 (from left to right), the values for ionic radii for the positive ions decrease. Na^+, Mg^{2+}, and Al^{3+} ions each contain the same number of electrons. However, for each successive positive ion in the period, the nuclear charge is greater by 1+. Theoretically, each electron in an aluminum ion experiences greater attraction than an electron in a magnesium ion. Similarly, each electron in a magnesium ion experiences greater attraction than each electron in a sodium atom. The attraction of the nucleus for the electrons in positive ions increases across the period, and the ionic radius gets smaller.

When an atom gains an electron to form a negative ion, its radius increases. For instance, the radius of a fluoride ion is over twice that of a fluorine atom (**Figure 5**). In gaining an electron, repulsion among the electrons increases while nuclear charge remains the same. The result is an enlarged electron cloud and a greater ionic radius. Notice that the O^{2-} and N^{3-} ions are larger than the F^- ion. The theoretical explanation is much the same: These ions all have the same number of electrons, but the number of protons is one less for the oxygen ion and two less for the nitrogen ion.

Ionization Energy

ionization energy: the amount of energy required to remove an electron from an atom or ion in the gaseous state

Ionization energy is the amount of energy required to remove an electron from an atom or ion in the gaseous state. This amount of energy is not constant, depending on which electron is being removed. We generally specify which electron we are interested in. The first ionization energy is defined as the amount of energy required to remove the most weakly held electron from a neutral atom. This weakly held electron is always in the valence shell farthest from the nucleus, where it is most shielded from the attractive force of the positive nuclear charge. We can represent the ionization process as follows:

$$X_{(g)} + \text{energy} \rightarrow X^+_{(g)} + e^-$$

The second ionization energy is similarly defined as the energy required to remove a second electron, this one from the gaseous positive ion.

Activity 1.5.1

Graphing First Ionization Energy

It is easier to identify trends if numerical information is presented visually. In this activity, you will create a graph of first ionization energy against atomic number. You may find graphing software useful for this task.

Materials

Table 2, a table of first ionization energies
graph paper (optional)
graphing software (optional)

Table 2: First Ionization Energies of Elements (kJ/mol)

H	1	1312	Fe	26	759	Sb	51	834	Os	76	840
He	2	2372	Co	27	758	Te	52	869	Ir	77	880
Li	3	520	Ni	28	737	I	53	1008	Pt	78	870
Be	4	899	Cu	29	745	Xe	54	1170	Au	79	890
B	5	801	Zn	30	906	Cs	55	376	Hg	80	1007
C	6	1086	Ga	31	579	Ba	56	503	Tl	81	589
N	7	1402	Ge	32	762	La	57	538	Pb	82	716
O	8	1314	As	33	947	Ce	58	528	Bi	83	703
F	9	1681	Se	34	941	Pr	59	523	Po	84	812
Ne	10	2081	Br	35	1140	Nd	60	530	At	85	—
Na	11	496	Kr	36	1351	Pm	61	535	Rn	86	1037
Mg	12	738	Rb	37	403	Sm	62	543	Fr	87	—
Al	13	578	Sr	38	549	Eu	63	547	Ra	88	509
Si	14	786	Y	39	616	Gd	64	592	Ac	89	509
P	15	1012	Zr	40	660	Tb	65	564	Th	90	587
S	16	1000	Nb	41	664	Dy	66	572	Pa	91	568
Cl	17	1251	Mo	42	685	Ho	67	581	U	92	598
Ar	18	1521	Tc	43	702	Er	68	589	Np	93	605
K	19	419	Ru	44	711	Tm	69	597	Pu	94	585
Ca	20	590	Rh	45	720	Yb	70	603	Am	95	578
Sc	21	631	Pd	46	805	Lu	71	524	Cm	96	581
Ti	22	658	Ag	47	731	Hf	72	680	Bk	97	601
V	23	650	Cd	48	868	Ta	73	761	Cf	98	608
Cr	24	653	In	49	558	W	74	770	Es	99	619
Mn	25	717	Sn	50	709	Re	75	760	Fm	100	627

Procedure

1. Plot a point graph of ionization energy against atomic number for the first 38 elements.

Analysis

(a) Describe any trends you detect.

Synthesis

(b) Suggest explanations for the trends.

Table 3: Electron Affinities of Elements 1 to 18

Element	Atomic Number	Electron Affinity (kJ/mol)
hydrogen	1	72.55
helium	2	<0
lithium	3	59.63
beryllium	4	<0
boron	5	26.7
carbon	6	121.85
nitrogen	7	<0
oxygen	8	140.98
fluorine	9	328.16
neon	10	<0
sodium	11	52.87
magnesium	12	<0
aluminum	13	42.5
silicon	14	134
phosphorus	15	72.03
sulfur	16	200.41
chlorine	17	348.57
argon	18	<0

First Ionization Energy and Atomic Radius

Ionization energy and atomic radius are closely related properties. In general, first ionization energies of metals are lower than those of nonmetals. From a graph of first ionization energy versus atomic number, we notice that the ionization energies decrease as you move down a group in the periodic table. This decrease is related to the increase in size of the atoms from top to bottom of a group. As the atomic radius increases, the distance between the valence electrons and the nucleus also increases. As a result, the attraction between the negatively charged electrons and the positive nucleus becomes weaker, so less energy is required to remove the first valence electron.

As we look from left to right across a period, we can see a general increase in ionization energy. Across any given period, the nuclear charge increases while the shielding provided by the non-valence electrons remains the same, decreasing the atomic radius. As a result the valence electrons are more strongly attracted to the nucleus and more energy is required to remove an electron from the atom. This makes sense if we relate it to our empirical knowledge: We know that elements on the left side of the periodic table are much more likely to lose electrons (form positive ions) than are elements on the right side of the periodic table.

There is also a pattern in ionization energies involved in removing second, third, and more electrons from an atom. As we have seen, the removal of an electron will lead to a reduced radius for the atom. The reduced radius strengthens the attraction between the nucleus and its electrons, making the removal of each extra electron progressively more difficult. There is always a large difference between the amount of energy required to remove the last valence electron and the much larger amount required to remove the first electron from the next lower shell: An ion is most chemically stable when it has a full valence shell. This explains why Group 1 elements form 1+ ions readily, but not 2+ ions.

Electron Affinity

As you know, a positively charged ion results from the removal of an electron from a neutral atom. Ions can also be formed by adding an electron to an atom. When an electron is added to a neutral atom in the gaseous state, energy is usually released.

$$X_{(g)} + e^- \rightarrow X^-_{(g)} + \text{energy}$$

electron affinity: the energy change that occurs when an electron is accepted by an atom in the gaseous state

The energy change associated with this ionization process is referred to as **electron affinity**. You can think of electron affinity as the amount of energy an atom (or ion) is willing to pay to buy another electron. To pick an extreme example, when fluorine ionizes to form the fluoride ion (F^-) 328.16 kJ of energy is released for every mole of fluorine atoms ionized. This quantity of energy corresponds to the first electron affinity for fluorine. By convention, a positive electron affinity indicates a release of energy.

The addition of an electron to a neutral atom does not always result in a release of energy (**Table 3**). Among the representative elements, those in Groups 2 and 18 have estimated first electron affinity values that are negative, indicating an absorption of energy. How can these negative first electron affinity values be explained? The electron added in the ionization process is simultaneously attracted to the positive nucleus of the atom and repelled by the atom's negative electrons. When the attractive force outweighs the repulsive force, as is the case for most elements, energy is released as an electron joins the atom to form an ion. When the repulsive force is greater, energy must be injected to form the ion. This is demonstrated well by second electron affinities, which are all negative. When we attempt to add a second electron to an ion with a charge of 1– the repulsive forces experienced by the approaching second electron are always greater than the attractive forces.

We can observe periodic trends in the electron affinities of the elements, even if they are not strictly adhered to. We might expect that when an electron is added to an atom of smaller radius, there is a greater attractive force between the nucleus of atom and the new electron. Electron affinity and atomic radius should, therefore, be inversely related. Since the size of atoms generally decreases across a period, we might predict that electron affinity will increase as we move from left to right within a period. This theoretical statement is supported by our observations of chemical reactions: Group 17 elements have a much greater tendency to form negative ions than do Group 1 elements.

Figure 6
The work of Linus Pauling ranged from physical chemistry to molecular biology. He made significant contributions to the understanding of chemical bonds and molecular structure. In 1954, Pauling received the Nobel Prize in chemistry for his work on molecular structure. He was awarded the Nobel Peace Prize in 1962 for his campaign against the nuclear bomb.

Electronegativity

In 1922, an American chemist, Linus Pauling (1901-1994; **Figure 6**), proposed a way of quantitatively describing the ability of an atom to attract electrons, combining ionization energies, electron affinity, and some other measures of reactivity. This scale is referred to as **electronegativity**. Pauling assigned a value of 4.0 to fluorine, the element considered to have the greatest ability to attract electrons. Electronegativity values for all other elements were assigned relative to the value for fluorine.

The electronegativity of any given element is not constant; it varies depending on the element to which it is bonded. Electronegativity is not a measurement, in the sense that first ionization energy is a measurement. However, electronegativity values do provide a means of predicting the nature of the force that holds a pair of atoms together— a relationship that you will explore further in Chapter 2. Approximate values for the electronegativity of the elements are given in the periodic table at the back of this text.

electronegativity: a number that describes the relative ability of an atom, when bonded, to attract electrons

Activity 1.5.2

Graphing Electronegativity

You will plot a graph of electronegativity against atomic number.

Materials

electronegativity values from the periodic table
graph paper (optional)
graphing software (optional)

Procedure

1. Using the electronegativity values in the periodic table at the back of this book, plot a point graph of electronegativity against atomic number for the first 38 elements.

Analysis

(a) Describe any trends you detect.

Synthesis

(b) Suggest explanations for the trends.

Trends in Electronegativity

If you look at the electronegativity data in the periodic table, you will see that electronegativity generally increases as you go along a period, from left to right. Electronegativity decreases as you look down a given group on the periodic table. The electronegativity values for elements at the bottom of a group are lower than those for elements at the top of a group.

How can these trends be explained? Remember that atomic radii tend to decrease along each period, from left to right, but increase down each group. The attraction between the valence electrons and the nucleus decreases as the distance between them increases. The electronegativity also decreases. We can similarly account for the increasing electronegativity across a period by the decreasing atomic radius. Nonmetallic elements have a strong tendency to gain electrons. Therefore, electronegativity is inversely related to the metallic properties of the elements. While electronegativity increases across a period, metallic properties decrease. Electronegativity decreases as we move down a group. This decrease in electronegativity is accompanied by an increase in metallic properties.

SUMMARY Trends in Periodic Properties

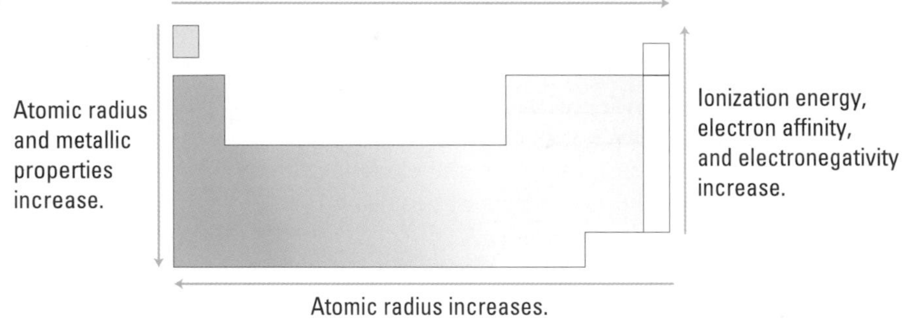

Figure 7

Table 4: Reactivity of Group 1 and Group 17 Elements

Element	Reaction with water
lithium	colourless gas slowly produced
sodium	colourless gas produced rapidly
potassium	colourless gas rapidly produced, explodes in 5 s
rubidium	colourless gas rapidly produced, explodes in 2 s
cesium	instant explosion
	Reaction with sodium
fluorine	extremely rapid reaction, forms white solid
chlorine	very rapid reaction, forms white solid
bromine	rapid reaction, forms white solid
iodine	reacts on heating, forms white solid

Practice

Understanding Concepts

1. Describe what, in theory, happens to the radius of an atom as it
   (a) gains an electron;
   (b) loses an electron.
2. Explain, with diagrams of specific examples, the difference between the ionic radius and the atomic radius of an atom.
3. (a) Describe the periodic trend in ionization energy.
   (b) Provide a theoretical explanation for this trend.
   (c) Give empirical evidence in support of your explanation in (b).
4. Which element in the periodic table would you expect to be
   (a) the most reactive metal? Give your reasons.
   (b) the most reactive nonmetal? Give your reasons.
5. Sketch a graph showing the first, second, and third ionization energies of a Group 2 element.
6. The observations of reactivity shown in **Table 4** were made by placing alkali metals in water and samples of sodium in each of the halogens.
   (a) How does the reactivity of the halogens change as you move up the group?

(b) Do the observations support the prediction that, as you look down a group of representative metals, the reactivity increases?
(c) Explain the observed trends in reactivity for these groups of representative elements in terms of number of valence electrons, nuclear charge, shielding, and distance of valence electrons from the nucleus.
(d) Which pair of elements from the table of observations would you expect to react together most readily?

Applying Inquiry Skills

7. Complete a report for the following investigation, including the Prediction and Experimental Design.

Question

What are the properties of rubidium?

Prediction

(a) Based on your understanding of periodic trends, make Predictions concerning the chemical properties of rubidium.

Experimental Design

(b) Design an experiment to test your Predictions. Include appropriate safety precautions.

8. Design an experiment for the following investigation.

Question

Do the melting points of elements exhibit periodic trends?

Experimental Design

(a) Outline an experiment you could conduct to answer the Question. Describe the methods you would use to gather the Evidence in the investigation.

Reflecting

9. Why would chemical unreactivity be considered a desirable trait in some circumstances? Consider situations in your own life in which you would prefer a material to be unreactive, and situations in which a substance should be reactive.

Section 1.5 Questions

Understanding Concepts

1. Explain how the number of valence electrons is related to the properties of the elements in each group in the periodic table.
2. Describe the relationship between
   (a) ionization energy and electron affinity;
   (b) electron affinity and electronegativity.
3. (a) S, Al, K, Mg, and Sr are all representative elements. Use your understanding of trends in the periodic table to predict their order of increasing ionization energy, atomic radius, electron affinity, and electronegativity.
   (b) Explain each prediction in terms of electronic structure.
4. Would you expect the first ionization energies for two isotopes of the same element to be the same or different? Justify your answer.

(continued)

5. (a) Use the periodic table to predict the most common charges on ions of sodium, magnesium, and aluminum. Provide a theoretical explanation of your answer.
   (b) What would be the trend in ionic radius among these ions? Support your answer with a theoretical argument.
6. The second ionization energy of an unknown element X is about twice as much as its first ionization energy. However, its third ionization energy is many times greater than its second ionization energy.
   (a) How many valence electrons would you expect to be present in an atom of the element?
   (b) What group would you expect it to belong to?
   (c) Based on its location in the periodic table, what other physical and chemical properties would you expect the element to possess?
   (d) Explain the differences in ionization energies.
7. How would you expect a graph of atomic radius against atomic number to compare with one of ionization energy against atomic number? Justify your answer.

Applying Inquiry Skills

8. (a) Using the data available in **Figure 4** (pages 52–53), plot a point graph for the first 38 elements of the periodic table with atomic number on the *x*-axis and atomic radius on the *y*-axis. Label the high and low points on your graph with the symbols for the corresponding elements.
   (b) Name the group that corresponds to the elements indicated by the peaks on your graph. Name the group that corresponds to the elements indicated by the lowest points on your graph.
   (c) What is the trend in the atomic radii as the atomic number increases across a period? Are there any exceptions?
   (d) Use your graph to predict the atomic radius of cesium and barium. Develop a theoretical argument to support your predictions.

Chapter 1 Summary

Key Expectations

Throughout this chapter you have had the opportunity to do the following:

- Demonstrate an understanding of the periodic law, and describe how electron arrangements and forces within atoms can explain periodic trends such as atomic radius, ionic radius, ionization energy, electron affinity, and electronegativity. (1.1, 1.2, 1.4, 1.5)
- Identify chemical substances and reactions in everyday use or of environmental significance. (1.1, 1.3)
- Define and describe the relationship among atomic number, mass number, and atomic mass, isotope, and radioisotope. (1.3)
- Identify and describe careers associated with matter. (1.3)
- Relate the reactivity of a series of elements to their position in the periodic table. (1.5)
- Analyze data involving periodic properties such as ionization energy and atomic radius in order to recognize general trends in the periodic table. (1.5)

Make a Summary

Summarize the concepts that you have learned in this chapter by creating a concept map. Begin with the word "element" and use as many words as possible from the list of key words.

Reflect on your Learning

Revisit your answers to the Reflect on Your Learning questions at the beginning of this chapter.

- How has your thinking changed?
- What new questions do you have?

Key Terms

actinide
alchemy
alkali metal
alkaline earth metal
analogy
atom
atomic mass
atomic number (Z)
atomic radius
carbon-14 dating
compound
continuous spectrum
diagnostic test
electron
electron affinity
electron cloud
electronegativity
element
empirical definition
empirical knowledge
energy level
excited
fission
flame test
ground state
group
half-life
halogen
ion
ionization energy
isotope
IUPAC
lanthanide
law of conservation of mass
law of constant composition
line spectrum
mass number (A)
mass spectrometer
metal
metalloid
model
neutron
neutron number (N)
noble gas
nonmetal
nucleus
orbit
period
periodic law
periodic trend
principal quantum number
proton
quantitative
quantum mechanics
radioactive
radioisotope
representative element
SATP
STP
theoretical knowledge
theory
transition
transition metal
transuranic element
trend
triad
unified atomic mass (u)
valence electrons

Chapter 1 Review

Understanding Concepts

1. What are the major differences between metals and nonmetals?
2. Is the symbol "CA" an acceptable international symbol for calcium? Justify your answer.
3. Why did the periodic law come to be accepted, even though it could not initially be explained?
4. Use the periodic table to answer the following questions.
   (a) At SATP, which elements are liquids and which are gases?
   (b) What is the significance of the "staircase line" that divides the periodic table into two parts?
   (c) Why was the group of elements on the far right of the periodic table discovered relatively late?
   (d) What are the atomic numbers of hydrogen, oxygen, aluminum, silicon, chlorine, and copper?
   (e) Identify an element with six electrons in the outer energy level.
   (f) Identify an element with its second energy level half full.
   (g) Identify an element that would tend to lose two electrons.
   (h) Identify an element that would tend to gain one electron.
5. Mendeleev organized elements according to their atomic mass in his periodic table, but the modern periodic table organizes elements according to their atomic number. Did the order of elements change? Give reasons for your answer.
6. (a) Sketch an outline of the periodic table and label the following: alkali metals, alkaline earth metals, transition metals, staircase line, halogens, noble gases, metals, and nonmetals.
   (b) Determine the number of valence electrons for the alkali metals, alkaline earth metals, and halogens. Explain the trends in number of valence electrons in terms of the Bohr theory of atomic structure.
7. (a) What is the most reactive metal? Nonmetal?
   (b) Explain the trends in reactivity within alkali metals and within halogens.
   (c) What model did you use to explain the trends?
8. What is unusual about the atomic structure of hydrogen compared with other elements?
9. Sketch a model illustrating the atomic theory of each of the following scientists:
   (a) Dalton
   (b) Nagaoka
   (c) Rutherford
   (d) Bohr
10. Why has there been a series of atomic theories?
11. List the three main subatomic particles, including their location in the atom, their relative mass, and their charge.
12. How does the atomic theory proposed by Bohr explain periodic trends?
13. From a representative element's position in the periodic table, how would you determine each of the following?
    (a) number of protons
    (b) number of electrons
    (c) number of valence electrons
    (d) number of occupied energy levels
14. List the number of protons, electrons, and valence electrons in each of the following atoms:
    (a) magnesium
    (b) aluminum
    (c) iodine
15. Write the chemical name and symbol corresponding to each of the following theoretical descriptions:
    (a) 11 protons and 10 electrons
    (b) 18 electrons and a net charge of 3–
    (c) 16 protons and 18 electrons
16. When a radioisotope decays, does it always produce another isotope of the same element? Give reasons for your answer.
17. Determine the number of protons, electrons, and neutrons present in an atom of each of the following isotopes:
    (a) calcium-42
    (b) strontium-90
    (c) cesium-137
    (d) iron-59
    (e) sodium-24
18. Iodine-123, in the compound sodium iodide, is a common radioisotope for medical use.
    (a) How does the mass number of this isotope compare with the atomic mass stated for iodine in the periodic table? How can this difference be explained?
    (b) What does this difference suggest about the abundance of iodine-123 in an average sample of iodine atoms?
19. How does chemical reactivity vary
    (a) among the elements within Groups 1 and 2 of the periodic table?
    (b) among the elements within Groups 16 and 17?
    (c) within period 3?

(d) within Group 18?

20. Describe the trends in the periodic table for each of the following atomic properties, and give a theoretical explanation for each trend:
    (a) atomic radius
    (b) ionization energy
    (c) electronegativity
    (d) electron affinity
21. When a chlorine atom forms an ion its radius increases, but when a sodium atom forms an ion its radius decreases. Explain this apparent contradiction.
22. (a) List the nonmetals of Group 17 from most reactive to least reactive.
    (b) Explain your order in (a) in terms of the tendency of the nonmetals to gain or lose electrons.
23. What characteristics of the noble gas group have made this group especially interesting to chemists? How does the atomic theory put forth by Bohr explain these characteristics?

Applying Inquiry Skills

24. (a) Using **Table 3** on page 56 or Appendix C2, construct a graph of electron affinity against atomic number.
    (b) How does the graph compare with a graph of ionization energy against atomic number? Explain any of the differences or similarities in terms of atomic structure and electron arrangement.
    (c) Use your graph to predict the electron affinity of rubidium.
25. A scientist is working with a sample of an unknown substance to determine its physical and chemical properties. The mass of the sample is 12.4 g. The scientist stops for lunch at noon and returns at 1 p.m. to find that the sample still has a mass of 12.4 g, but it now contains only 6.2 g of the original unknown substance.
    (a) Propose a possible explanation for this change.
    (b) What property can be determined from the information given?
26. A student is investigating the properties of magnesium and barium. After placing magnesium in an acidic solution, the student observes bubbles forming. When barium is placed in the same acidic solution, the reaction is much more vigorous.
    (a) Explain the difference in reactivities.
    (b) Predict the product of the reactions indicated by the bubbles. How could you test the identity of the product?

Making Connections

27. The radioisotope iron-59 is used for the diagnosis of anemia. Iron-59 emits gamma rays. Would it be appropriate to store the radioisotope in a cardboard box? Explain.
28. Calcium is an important element in the human body. For example, it is a major component of bones. Bone is built by drawing from the blood minerals such as calcium, in the form of ions. Calcium is relatively common in our diet. Strontium, also a Group 2 metal, is found only in very small quanitities in our food.
    (a) Based on your understanding of periodic trends, suggest what might happen to any strontium included in the diet of a child. Provide your reasoning.
    (b) Radioactive strontium-90 is one of the products of nuclear explosions. Using the Internet, research and report on how this isotope is produced and the effects it has when ingested.
    (c) Use the information you have gathered in a brief cost/benefit analysis of the testing of nuclear warheads.
    Follow the links for Nelson Chemistry 11, Chapter 1 Review.

GO TO www.science.nelson.com

29. All Group 1 elements are highly reactive, and require special storage and handling. Investigate bottle labels, MSDS (Material Safety Data Sheets) information, and other sources to create a leaflet advising high-school laboratory technicians on safe ways to handle and store each of these elements.

    Follow the links for Nelson Chemistry 11, Chapter 1 Review.

GO TO www.science.nelson.com

Exploring

30. Research and report on the CANDU nuclear power process. Include in your report a simple diagram of a CANDU reactor and describe the function of each part.

    Follow the links for Nelson Chemistry 11, Chapter 1 Review.

GO TO www.science.nelson.com

Chapter

2

Chemical Bonding

In this chapter, you will be able to

- describe the role of electrons in ionic and covalent bonding;
- relate the physical and chemical properties of compounds to the nature of their chemical bonds;
- predict the nature of bonds by comparing electronegativity values;
- use a variety of models to represent the formation and structure of compounds and molecular elements;
- name a variety of compounds using common names, classical names, and IUPAC chemical nomenclature.

In caves we can sometimes see large structures known as stalagmites (**Figure 1**). These formations are made of crystals of calcium carbonate, $CaCO_{3(s)}$, also known as limestone. Calcium carbonate, as its name and formula suggest, is a compound made up of three different elements. In addition to its crystalline structure, calcium carbonate has high melting and boiling points and dissolves to some extent in water. Many other compounds have similar physical and chemical properties, such as ordinary table salt, sodium chloride. Other compounds, such as water, H_2O, and carbon dioxide, CO_2, have significantly different properties. How can we explain these similarities and differences? The answer lies in an understanding of the special forces of attraction, or bonds, that hold atoms together in compounds. These forces form the foundation of chemical properties and reactions.

As you know, we often develop models to help us understand abstract concepts. You are already familiar with several models: Bohr's model of the atom; the water cycle; the collision model. In this chapter we will develop theories and models of chemical bonding (including the character of atomic bonds) to explain the nature and behaviour of matter and to classify compounds.

Reflect on your Learning

1. Why do atoms form compounds? Use examples of compounds you are familiar with in your explanation.
2. Is there more than one type of force present between the atoms in compounds and, if so, how do these forces compare in strength and other properties?
3. How could the forces that hold atoms together in a compound determine the chemical properties of that compound? Again, use examples of familiar compounds and their properties in your speculations.

Throughout this chapter, note any changes in your ideas as you learn new concepts and develop your skills.

Try This Activity

Making Models of Compounds

In Chapter 1 you studied various representations, or models, of the structure of atoms. In this activity, you will build models of a variety of molecules. This exercise will help you to see that different chemicals have different molecular structures.

Materials: molecular models kit

- In your molecular model kit, identify the pieces representing each of the following elements: chlorine, bromine, carbon, nitrogen, oxygen, iodine, and hydrogen.
- Group the different pieces according to the number of holes present.
  (a) To what chemical family or families do these groups of pieces correspond on the periodic table?
  (b) What do the holes in the pieces represent?
  (c) What do the sticks represent in your molecular model kit?
- Construct as many different compounds as possible using the pieces in the kit.
  (d) As best you can, name the compounds you have modelled (later in this chapter you will learn a system for naming compounds).
  (e) How do the connections differ among your models?
  (f) Classify the models by dividing them into groups. Provide a rationale for your classification.
  (g) Organize the models into two more classification schemes. Provide a rationale for each classification.

Figure 1
Stalagmites are shown extending upward from this cavern floor. As you can see, these calcium carbonate formations have the potential to join floor to ceiling in a continuous pillar.

Figure 1
Conductivity is used to distinguish between aqueous solutions of ionic and molecular compounds.

ionic compound: a pure substance formed from a metal and a nonmetal

molecular compound: a pure substance formed from two or more different nonmetals

electrical conductivity: the ability of a material to allow electricity to flow through it

electrolyte: a substance that forms a solution that conducts electricity

2.1 Classifying Compounds

There are many ways to classify substances. You are already familiar with several of them. For example, you know that iron is a solid at SATP (not a liquid or a gas), a metal (not a nonmetal), and an element (not a compound). Similarly, water is a liquid and a compound. In this section, we will look at ways of further classifying compounds.

One category of compounds includes table salt (sodium chloride), $NaCl_{(s)}$, bluestone (copper(II) sulfate), $CuSO_{4(s)}$, and baking soda (sodium bicarbonate), $NaHCO_{3(s)}$. If you think about their chemical formulas, you might notice that each one is made up of a metal joined to a nonmetal. These compounds are called **ionic compounds.**

A second category includes sulfur dioxide, SO_2, carbon dioxide, CO_2, and ammonia, NH_3. Look at their formulas. What do you notice? They are all nonmetals combined with nonmetals, and are called **molecular compounds.**

Many ionic and molecular compounds can be found within your own home. For example, window cleaners, household bleach, antacid tablets, and milk of magnesia contain ionic compounds. Vegetable oil, plastics, and sugar contain molecular compounds.

We can use **electrical conductivity**—the ability to conduct electricity—to distinguish between ionic and molecular compounds: Ionic compounds (many of which dissolve readily in water) form solutions that conduct electricity. However, molecular compounds (some of which dissolve in water) form solutions that generally do not conduct electricity. Of course, there are exceptions to this generalization.

Substances that form solutions that conduct electricity are called **electrolytes.** Their solutions are called electrolytic solutions.

Substances that form solutions that do not conduct electricity are called nonelectrolytes. Their solutions are called nonelectrolytic solutions (**Figure 1**).

INQUIRY SKILLS

- ○ Questioning
- ● Hypothesizing
- ● Predicting
- ○ Planning
- ● Conducting
- ● Recording
- ● Analyzing
- ● Evaluating
- ● Communicating

Investigation 2.1.1

Comparing Ionic and Molecular Compounds

Just as a group of elements can share physical and chemical properties, so too can a class of compounds. For instance, solid ionic compounds generally have very high melting points and are brittle—they can often be crushed fairly easily into a powder. Molecular compounds in solid form tend to have a softer or waxy texture, and many have melting and boiling points so low that they are gases or liquids at room temperature. If we can classify a substance as either ionic or molecular, we should be able to predict some of its properties. Conversely, if we know some of the properties of a substance, we can classify it as ionic or molecular.

There are several properties that we could take into account. Generally, the simplest to investigate are physical properties, such as state at SATP, solubility in water, and ability to conduct electricity in solution.

The purpose of this investigation is to test the generalizations you have encountered about ionic and molecular compounds. Complete the **Hypothesis/Prediction**, **Analysis**, **Evaluation**, and **Synthesis** sections of the lab report.

Question

Which of the following substances are ionic, and which are molecular: sodium nitrate, $NaNO_{3(s)}$; sucrose, $C_{12}H_{22}O_{11(s)}$; sodium chloride, $NaCl_{(s)}$; potassium sulfate, $K_2SO_{4(s)}$; ethanol, $C_2H_5OH_{(l)}$; sodium bicarbonate, $NaHCO_{3(s)}$; and calcium sulfate, $CaSO_{4(s)}$?

Hypothesis/Prediction

(a) Answer the Question, considering the state of matter at room temperature and the chemical formula of each substance. Provide your reasoning for each decision.

(b) Based on your classification, predict whether each substance will dissolve in water and whether any solutions formed will conduct electricity.

Experimental Design

The ionic or molecular nature of several compounds will be determined by applying the tests of solubility and conductivity.

Materials

lab apron
eye protection
8 50-mL beakers
distilled water
wax pencil
scoopula
medicine dropper
stirring rod
sodium nitrate, $NaNO_{3(s)}$
sucrose, $C_{12}H_{22}O_{11(s)}$
sodium chloride, $NaCl_{(s)}$
potassium sulfate, $K_2SO_{4(s)}$
ethanol, $C_2H_5OH_{(l)}$
sodium bicarbonate, $NaHCO_{3(s)}$
calcium sulfate, $CaSO_{4(s)}$
low-voltage conductivity apparatus

Ethanol is flammable; do not use near an open flame. Use only low-voltage conductivity apparatus.

Wear eye protection and an apron, and wash hands thoroughly at the end of the investigation.

Procedure

Part 1: Solubility

1. Obtain a small amount of $NaCl_{(s)}$. Observe and, in a table, record its state at the ambient temperature.
2. Pour about 10 mL of distilled water into a 50-mL beaker. Add a small quantity of the chemical to the water (**Figure 2**).
3. Use a stirring rod to stir the mixture. Note whether the chemical dissolves. Record your observations. Label the beaker with the name of the chemical and put it aside for the conductivity test in Part 2.
4. Repeat steps 1 through 3 for each of the other compounds.

Part 2: Conductivity

5. Obtain a small sample of distilled water in a beaker. Test the electrical conductivity of the sample. A reading of zero should be indicated by the apparatus.
6. Test the conductivity of the mixture of sodium chloride and water that was set aside in Part 1. Record your observations and rinse the probes in distilled water.
7. Repeat step 6 for each of the remaining mixtures.
8. Dispose of the mixtures in the appropriate disposal containers provided by your teacher. Rinse the beakers.

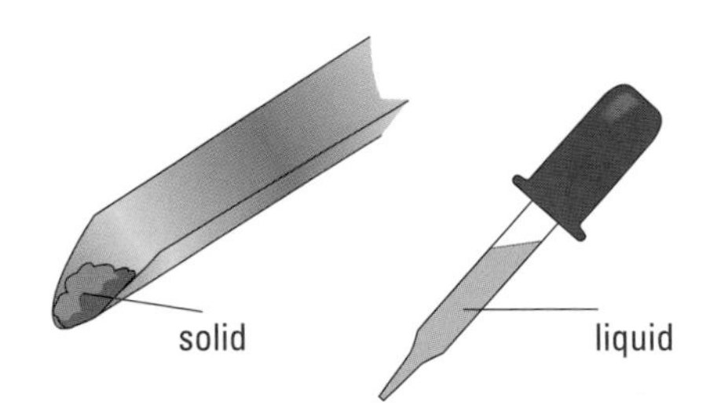

Figure 2
For a solid chemical, use a small quantity in a scoopula as shown. If the chemical is a liquid, use a dropper full of liquid.

Analysis

(c) Use the Evidence you gathered in this experiment to answer the Question (classify each of the substances as ionic or molecular).

Evaluation

(d) Do you have confidence in your observations? Do you feel that they can be used to accurately classify the substances? Explain.

(e) Compare your answer from the Analysis with the answer in your Hypothesis/Prediction (question (a)). How do you account for any differences?

Synthesis

(f) Were you able to accurately predict the properties of the substances based on your initial classification? Why or why not?

(g) What assumptions are being made in this investigation?

Electrolytes and Classification

When table salt (sodium chloride), $NaCl_{(s)}$, dissolves in water it forms a solution that conducts electricity: It is an electrolyte, and dissolving it forms an electrolytic solution. Sodium chloride is a typical ionic compound: Its solution conducts electricity. Sugar will also dissolve in water. However, the sugar solution will not conduct electricity, so sugar is a nonelectrolyte. It is a typical molecular compound: Its solution with water does not conduct electricity. In Unit 3, you will learn more about conductivity and solutions. You will also find out more about molecular acids: important exceptions to the rule that molecular solutions do not conduct electricity.

Practice

Understanding Concepts

1. What types of elements combine to form
   (a) an ionic compound?
   (b) a molecular compound?
2. Briefly describe a diagnostic test for an ionic compound, and give a theoretical explanation for that test.

Applying Inquiry Skills

3. A student hypothesizes that an unknown substance is composed of positive and negative ions held together by the attraction of their opposite charges. Design an experiment that would allow the student to test this hypothesis.
4. Use the evidence in **Table 1** to classify each of the five compounds as ionic or molecular. Provide your reasoning for each classification.

Table 1: Observations of Five Unknown Compounds

Compound	State at SATP	Conductivity of solution
A	solid	yes
B	liquid	no
C	gas	yes
D	solid	yes
E	solid	no

2.2 Ionic Bonding

When sodium (a metal) is put in a vessel containing chlorine (a nonmetal), the two elements combine enthusiastically to form the compound sodium chloride, a substance that you have classified as ionic, as it is an electrolyte. Like some other ionic compounds that you are familiar with, for example, baking soda (sodium hydrogen carbonate) and chalk (calcium carbonate), it is also brittle and has a high melting temperature. How do we explain the formation and properties of this compound?

You will recall, from Chapter 1, that atomic theory describes electrons moving about the nucleus of the atom in energy levels, and that the electrons in the outermost energy level are called the valence electrons. It is the valence electrons of an atom that form **chemical bonds.**

chemical bond: the forces of attraction holding atoms or ions together

According to atomic theory, ionic compounds are formed when one or more valence electrons are transferred from a metal atom to a nonmetal atom. This leaves the metal atom as a positive ion, or cation, and the nonmetal atom as a negative ion, or anion. The two oppositely charged ions are attracted to each other by a force called an **ionic bond** (Figure 1).

ionic bond: the electrostatic attraction between positive and negative ions in a compound; a type of chemical bond

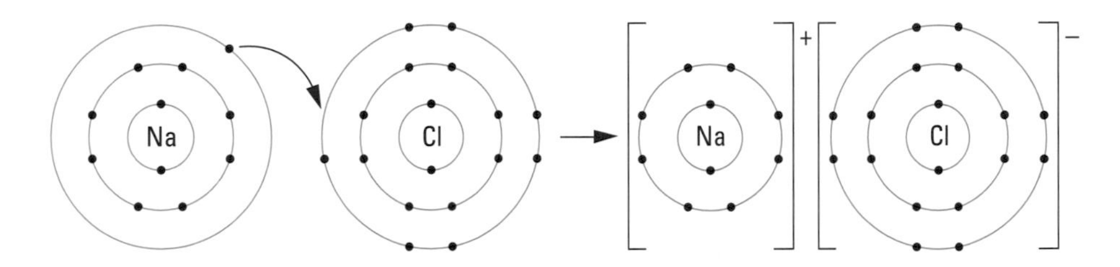

Figure 1
An electron is transferred from sodium to chlorine in the formation of an ionic bond.

Explaining the Properties of Ionic Compounds

Ionic compounds have similar properties: They are solids at SATP with high melting points, and they are electrolytes. We can hypothesize that these properties might be the result of the bonds formed between the ions, holding them firmly in a rigid structure.

Although they are composed of ions, pure ionic compounds are electrically neutral. Therefore, the sum total of the electrical charges on all the ions must be zero. Ionic compounds are made up of a fixed proportion of positive and negative ions. Consequently, ionic compounds can only be identified in terms of the smallest unit of the compound, known as a **formula unit**, that would still have the properties of the compound. In the case of sodium chloride, the sodium and chloride ions are present in a 1:1 ratio, as indicated by its chemical formula, NaCl.

formula unit: the simplest whole-number ratio of atoms or ions of the elements in an ionic compound

The anions and cations in an ionic compound are locked in a regular structure, held by the balance of attractive bonds and electrical repulsion. The most common model of ions shows them as spheres arranged in a regular three-dimensional pattern called a **crystal lattice** (Figure 2, page 70). We can actually see the shape of sodium chloride crystals—an observation that supports our crystal lattice model. We often see similar structures in bridges or scaffolding where struts are joined one to the other in a repeating pattern.

crystal lattice: a regular, ordered arrangement of atoms, ions, or molecules

Not all crystal lattices are square, like that of sodium chloride. Depending on the sizes and charges of the ions that make up the substance, the crystal lattice varies in structure. All lattices are arranged so that each ion has the greatest possible number of oppositely charged ions close, while keeping ions with the same charge as far away as possible. In all cases, each ion is surrounded by ions of opposite charge. In theory, this arrangement of ions creates strong attractions. This theory is supported by empirical evidence such as the hard surfaces and high melting and boiling points of ionic solids.

(a)

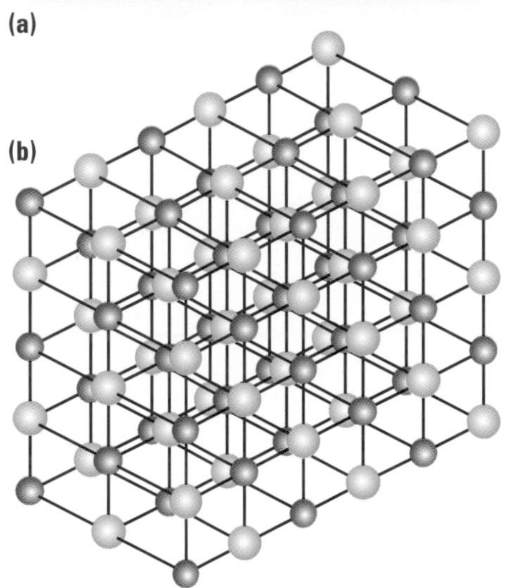

(b)

Figure 2

(a) The cubic structure of table salt crystals provides a clue about the internal structure of sodium chloride.

(b) The arrangement of sodium and chlorine ions in a crystal of sodium chloride. The sodium and chloride ions occupy positions in the crystal lattice known as lattice points.

It requires a great deal of energy to break the strong electrostatic attractions within a crystal lattice. The ions resist any movement, as even a slight shift would cause positive ions to move closer to other positive ions, and negative ions closer to other negative ions, resulting in strong repulsion. We can use this model to explain why ionic substances are hard (the ions resist movement), and also why they are often brittle. Once the lattice is broken, repulsions between ions of the same charge will cause the substance to split into two crystals.

Finally, our model of the structure of ionic compounds can also explain the electrical conductivity of their solutions. When ionic compounds are dissolved in water, the positive and negative ions dissociate:

$$NaCl_{(s)} \rightarrow Na^+_{(aq)} + Cl^-_{(aq)}$$

The ions are responsible for carrying current when charged electrodes are placed in the ionic solution.

The Formation of Ionic Compounds

We classify most simple compounds containing metallic elements as ionic. Elements within a chemical family (group) tend to participate in similar chemical reactions, producing ionic compounds with the same general formula. For example, the metals in Groups 13 to 15, except mercury, will form ionic oxides (compounds composed of a metal and oxygen) when burned in air. In the same way, elements in Groups 1 and 2 form ionic compounds with oxygen: The oxides formed by Group 1 elements have the general formula M_2O while those of Group 2 elements (e.g., magnesium oxide, MgO) have the general formula MO, where M represents a metal ion. Because of its high melting point, magnesium oxide is used to make objects that are exposed to very high temperatures, such as crucibles, furnace linings, and thermal insulation.

From our discussion of trends in electronegativity, it makes sense that Group 1 metals readily react with the elements in Group 17 to form ionic compounds with the general formula MX. These compounds, which are composed of a metal and a halogen, are collectively referred to as ionic halides. Sodium chloride, an example of an ionic halide, is found in large underground deposits in various parts of the world. It is mined from these deposits and used as road salt, table salt, and as a reactant in many industrial processes.

The elements in Group 2 show a similar trend, as they also react with the halogens to produce ionic halides with the general formula MX_2.

In general, the addition of a metal from Group 1 or Group 2 to water will produce hydrogen gas and a basic ionic compound. Calcium hydroxide, $Ca(OH)_2$, is an example of an ionic compound that can be produced in this way. However, the reaction of calcium and water is quite vigorous. A safer means of producing this ionic compound is to react calcium oxide with water. Calcium hydroxide is also referred to as slaked lime and is used to make mortar and plaster for buildings.

Predicting Common Ions of Atoms

stable octet: a full shell of eight electrons in the outer energy level of an atom

Noble gases are stable and virtually inert. Similarly, ions with eight valence electrons appear to have a special stability. This arrangement of electrons is known as a **stable octet**. To reach this stable state, metal atoms of elements in Groups 1, 2, and 3 will lose electrons to form cations, while elements in Groups 15, 16, and 17 will gain electrons to form anions. By looking at an element's position in the periodic table, we can predict the charge on that element's most stable ion. For example, we can predict that a Group 1 element will tend to lose one electron, becoming a cation with a charge of 1+. A Group 16 element, on the other hand, will tend to gain two electrons to complete a stable octet, so it will form an anion with a charge of 2–.

Hydrogen is a special case in that, theoretically, it can either give up an electron to form H^+, which is equivalent to a proton, or gain an electron to form H^-, which has a filled shell like the noble gas helium.

Practice

Understanding Concepts

1. What properties of ionic compounds suggest that ionic bonds are strong?
2. What types of elements form ionic bonds with each other?
3. Which of the representative elements tend to form positive ions? Which tend to form negative ions?
4. What is the minimum number of different ions in the formula of an ionic compound? Explain.
5. Predict the charge on the most stable ion formed by each of the following elements. Indicate the ion by writing the symbol complete with charge.
   (a) sulfur
   (b) barium
   (c) bromine
   (d) chlorine
   (e) calcium
   (f) potassium
   (g) phosphorus
   (h) rubidium
   (i) beryllium

Applying Inquiry Skills

6. You have already discovered that solutions of ionic compounds in water conduct electricity. You might wonder about the conductivity of pure ionic compounds.
   (a) Design an experiment to investigate the conductivity of an ionic solid. With your teacher's approval, conduct your investigation.
   (b) Research the conductivity of molten (liquid) ionic compounds.
   (c) Assemble your findings into a report on the conductivity of ionic compounds in various states.
   (d) Propose a hypothesis for the properties you observe.

Ions and the Human Body

Among the 12 elements that make up more than 99% of the human body (**Table 1**) are five metals: calcium, potassium, sodium, magnesium, and iron. These five metals, which form positively charged ions in solution, are essential for maintaining good health. For example, Mg^{2+}, Na^+, and K^+ are major constituents of blood plasma. Ca^{2+} is vital in the formation of bones and teeth. In addition to positively charged ions, some elements that form negatively charged ions are also essential for life. Chloride ions, Cl^-, are another component of blood, and iodide ions, I^-, are required to prevent a condition called goitre, which results in the enlargement of the thyroid gland.

Table 1: 12 Most Common Elements in the Human Body

Element	Percentage by mass
oxygen	65
carbon	19
hydrogen	9.5
nitrogen	3.2
calcium	1.5
phosphorus	1.0
sulfur	0.3
chlorine	0.2
sodium	0.2
magnesium	0.1
iodine	<0.1
iron	<0.1

Practice

Making Connections

7. Research and report upon the importance of one of the ions that make up the human body. Your report should include: a description of its biological importance; recommended daily minimum requirements; the effects of deficiency or excess on the human body; and some of the naturally occurring sources. Conclude your report with a discussion on whether, and under what circumstances, you would recommend that someone should artificially supplement his/her intake of this ion.

   Follow the links for Nelson Chemistry 11, 2.2.

 www.science.nelson.com

Representing Ionic Bonds

electron dot diagram or **Lewis symbol:** a representation of an atom or ion, made up of the chemical symbol and dots indicating the number of electrons in the valence energy level

We generally find it easier to grasp a new concept if we have a model or mental image. An American chemist, G. N. Lewis, obviously thought the same thing, and developed a model of the valence (bonding) electrons of single atoms or monatomic ions. He represented his models on paper as **electron dot diagrams**, or **Lewis symbols**. These symbols consist of the chemical symbol for the element plus dots representing the number of valence electrons. For example, a sodium atom would be represented as

Na·

The convention for indicating more electrons is as follows:

Li· ·Be· ·Ḃ· ·Ċ· ·N̈· :Ö· :F̤̈· :N̤̈e:

These diagrams can help us to represent the process of ion formation.

Electron dot diagrams also illustrate the theory that ionic bonds tend to produce full outer orbits of electrons: a configuration exactly the same as that of the noble gases. Sodium has one valence electron. By transferring this electron to another entity that has a stronger attraction for the electron, the resulting sodium ion will have the same electron configuration as neon.

$$Na\cdot \rightarrow [Na]^{+} + e^{-}$$

A chlorine atom has seven valence electrons. By attracting an electron from another entity, the resulting chloride ion will have eight electrons in its valence shell and the same number of electrons as its nearest noble gas, argon.

Representative elements around Groups 13 and 14 of the periodic table can also achieve this special stability without losing or gaining electrons, as you will discover later in the chapter.

:Ċ̤l: + e^{-} → [:C̤̈l:]$^{-}$ (compare with :Ä̤r:)

Another way to describe the process of forming ionic compounds is to say that electron transfers result from the large difference in electronegativity between metals and nonmetals. Nonmetals have a strong tendency to gain electrons and metals do not. When an atom of low electronegativity, such as a metal atom, and an atom of high electronegativity, such as a nonmetal atom, are in proximity, one or more electrons are transferred from the atom with low electronegativity, transforming both atoms into ions. The metal ion will be positive and the nonmetal ion will be negative. The resulting oppositely charged ions attract each other and other ions, forming ionic bonds and resulting in a crystal lattice.

Using electron dot diagrams, we can show the formation of an ionic bond between sodium and chlorine.

$$\text{Na}\cdot + \cdot\ddot{\underset{..}{\text{Cl}}}\text{:} \rightarrow [\text{Na}]^+ [\text{:}\ddot{\underset{..}{\text{Cl}}}\text{:}]^-$$

When illustrating the formation of an ionic bond, we place square brackets around the ion to indicate that the charge is not associated with any particular electron and that all the electrons in the valence shell are equivalent.

We can use the periodic table and electron dot diagrams to predict the formulas of other ionic compounds. By finding out how many electrons they tend to lose or gain to reach stable octets, we can figure out what ratio of ions will make an electrically neutral compound.

As an example, suppose we were asked to draw electron dot diagrams to illustrate the formation of calcium fluoride, state the ratio of ions in the compound, and give the formula for the compound. Calcium is in Group 2, so will form an ion with a charge of 2+. Fluorine is in Group 17, so will form an ion with a charge of 1−.

$$\cdot\text{Ca}\cdot + 2\ \text{:}\dot{\underset{..}{\text{F}}}\text{:} \rightarrow [\text{:}\ddot{\underset{..}{\text{F}}}\text{:}]^- [\text{Ca}]^{2+} [\text{:}\ddot{\underset{..}{\text{F}}}\text{:}]^-$$

To form an electrically neutral ionic compound, the ratio of calcium to fluoride ions is 1:2. The formula is CaF_2.

Sample Problem

Draw electron dot diagrams to illustrate the formation of magnesium oxide. Write the ion ratio and the chemical formula.

Solution

$$\cdot\text{Mg}\cdot + \text{:}\ddot{\underset{\cdot}{\text{O}}}\cdot \rightarrow [\text{Mg}]^{2+} [\text{:}\ddot{\underset{..}{\text{O}}}\text{:}]^{2-}$$

The two elements will combine in a ratio of 1:1. The formula is MgO.

Practice

Understanding Concepts

8. (a) How do the electron dot diagrams of metal ions differ from those of nonmetal ions?
   (b) How are the electron dot diagrams of metal ions similar to those of nonmetal ions?
9. Use electron dot diagrams to illustrate the formation of
   (a) lithium iodide
   (b) barium chloride
   (c) potassium oxide
   (d) calcium fluoride
10. Represent each of the following elements using electron dot diagrams:
    (a) nitrogen
    (b) sulfur
    (c) argon
    (d) iodine
    (e) lithium
    (f) cesium
    (g) calcium
    (h) sodium

11. Use electron dot diagrams to determine the ratio in which oxygen will combine with each of the following elements to form an ionic compound. Label each diagram with the chemical formula of the compound.
    (a) calcium
    (b) rubidium
    (c) strontium
    (d) aluminum
12. Represent the five halogens, using electron dot diagrams. How are these diagrams consistent with the concept of a chemical family?
13. Explain, referring to stable octets, how the following ionic compounds are formed from pairs of elements; illustrate the formation of each compound with electron dot diagrams; and predict the formula of each compound.
    (a) magnesium chloride
    (b) sodium sulfide
    (c) aluminum oxide (bauxite)
    (d) barium chloride
    (e) calcium fluoride (fluorite)
    (f) sodium iodide
    (g) potassium chloride (a substitute for table salt)
14. Give the common names for the following chemicals:
    (a) sodium bicarbonate
    (b) NaCl
    (c) calcium carbonate
    (d) $Ca(OH)_2$

Sections 2.1–2.2 Questions

Understanding Concepts

1. Use the concepts of periodic trends and electronegativity to explain why ionic compounds are abundant in nature.
2. Give a theoretical reason why lithium and oxygen combine in the ratio 2:1.
3. How are the topics of ion formation and periodic trends related?
4. Give the correct chemical (IUPAC) names for the following chemicals:
    (a) chalk
    (b) slaked lime
    (c) road salt
    (d) baking soda
5. A Group 1 metal (atomic number 55) is reacted with the most reactive of the halogens. A very vigorous reaction results in the formation of a solid, white compound.
    (a) Represent the formation of the compound with electron dot diagrams.
    (b) Write the formula of the compound formed.
    (c) What type of compound is formed?
    (d) Predict the physical properties of the resulting white compound.
    (e) Explain the properties of the compound in terms of the bonds formed.
    (f) Provide a theoretical explanation for the vigorous reaction.

2.3 Covalent Bonding

Ionic compounds, as you have just learned, contain many ions arranged in a three-dimensional structure. But not all compounds are brittle ionic solids with high melting points. Some, such as paraffin wax, are solids at SATP but have relatively low melting points. Others are liquids or gases, such as water and carbon dioxide. In what way do these substances differ from ionic compounds? Can we explain these differences with atomic theory? The answer is yes, but first we should look at their formulas.

We have written the formulas of ionic compounds as simplest formula ratios. The compound NaCl includes sodium and chloride ions in a 1:1 ratio. We could build a crystal from 8 ions of each element, or 8 million of each and it would form the same structure and would have the same properties. There is only one ionic compound of sodium and chloride ions, and that compound always contains a 1:1 ratio of the two ions.

This is not true of the bonding of nonmetals with each other. For example, consider the simplest ratio formula CH. If carbon and hydrogen formed an ionic compound with this ratio, we would expect that any structure in which these elements were in this ratio would have the same properties. However, this is not the case. Using a mass spectrometer and combustion analysis we can demonstrate that there are several compounds that have this simplest ratio formula. The gas acetylene (ethyne), $C_2H_{2(g)}$, and the liquid benzene, $C_6H_{6(l)}$, both have the simplest ratio formula CH, but they are otherwise much different in their physical and chemical properties. Simplest ratio formulas indicate only the relative numbers of atoms in a molecular compound; they give no information about the actual number of atoms or the arrangement of those atoms in a molecule. To distinguish among molecular compounds, we need to represent them with formulas that describe the molecules that make them up.

Figure 1
The pair of shared electrons between the nuclei of two hydrogen atoms results in a covalent bond.

Molecules can be classified by the number of atoms that they contain. Molecules that contain only two atoms, such as carbon monoxide, CO, are called **diatomic molecules.** If they contain more than two atoms, such as ammonia, NH_3, they are called **polyatomic molecules.**

diatomic molecule: a molecule consisting of two atoms of the same or different elements

polyatomic molecule: a molecule consisting of more than two atoms of the same or different elements

Some elements also exist as molecules. Hydrogen and oxygen are examples of elements composed of diatomic molecules. Sulfur, S_8, and phosphorus, P_4, are polyatomic molecules.

Formation of Covalent Bonds

You already know that hydrogen can form a cation (H^+) by losing a valence electron, or it can form an anion (H^-) by gaining an electron and filling its electron shell. However, two hydrogen atoms can each obtain a stable filled energy level by sharing a pair of electrons. (Remember that the first energy level can only contain two electrons.) The **covalent bond** that results arises from the simultaneous attraction of two nuclei for a shared pair of electrons (**Figure 1**).

covalent bond: the attractive force between two atoms that results when electrons are shared by the atoms; a type of chemical bond

We can use the model of the Lewis symbol, or electron dot diagram, to communicate the theory of covalent bonding. When an electron dot diagram is used to represent covalent bonding, we adapt it slightly and call it a **Lewis structure** (because it illustrates the structure of the molecule). A Lewis structure shows the valence electrons surrounding each of the component atoms as dots, with the exception of the electrons that are shared: These shared electrons are represented by a dash. In effect, this dash represents a covalent bond.

Lewis structure: a representation of covalent bonding based on Lewis symbols; shared electron pairs are shown as lines and lone pairs as dots

$$:\ddot{\underset{\cdot\cdot}{Cl}}\cdot + \cdot\ddot{\underset{\cdot\cdot}{Cl}}: \rightarrow :\ddot{\underset{\cdot\cdot}{Cl}} - \ddot{\underset{\cdot\cdot}{Cl}}:$$

According to the Lewis structure for a chlorine molecule, each Cl atom has three pairs of electrons that are not involved in the formation of a covalent bond. Each pair is referred to as a **lone pair**. There is one shared pair of electrons between the atoms.

lone pair: a pair of valence electrons not involved in bonding

Many elements will form bonds that result in a full valence shell for each atom, so that each atom has the electron structure of an atom of a noble gas when the shared electrons are included. In other words, it has a stable octet. This generalization is referred to as the **octet rule**. There are many exceptions to this rule, but you will not be learning about these in this course.

octet rule: a generalization stating that, when atoms combine, the covalent bonds are formed between them in such a way that each atom achieves eight valence electrons (two in the case of hydrogen)

Sample Problem 1

Draw Lewis symbols for the reaction of two bromine atoms and a Lewis structure for the resulting bromine molecule.

Solution

$$:\underset{\cdot\cdot}{\overset{\cdot\cdot}{Br}}\cdot + \cdot\underset{\cdot\cdot}{\overset{\cdot\cdot}{Br}}: \rightarrow :\underset{\cdot\cdot}{\overset{\cdot\cdot}{Br}} - \underset{\cdot\cdot}{\overset{\cdot\cdot}{Br}}:$$

This method of representing molecules can be further simplified by not indicating lone pairs. This representation is referred to as a **structural formula**. The structural formula for chlorine would be Cl—Cl, and for bromine Br—Br. As you can see, they are easier to write, and are quite similar to the chemical formulas, Cl_2 and Br_2.

structural formula: a representation of the number, types, and arrangement of atoms in a molecule, with dashes representing covalent bonds

Each pair of shared electrons results in a single bond. Elements that need only one more electron to complete their outer shell or energy level tend to form single bonds. Hydrogen and chlorine are typical examples.

Elements in Group 16 are two electrons short of a full outer shell. How, then, can they form covalent bonds with each other and still achieve a stable state? The answer is simple, though not necessarily obvious: They form a double bond, with each pair of atoms sharing two pairs of electrons between them. Oxygen and carbon dioxide are examples of molecules that include double bonds. The double bond is represented, in a Lewis structure and structural formula, as a double dash.

Sample Problem 2

(a) Draw the Lewis structure for a molecule of oxygen.
(b) Give the structural formula for the molecule.

Solution

(a) $:\underset{\cdot}{\overset{\cdot\cdot}{O}}\cdot + \cdot\underset{\cdot}{\overset{\cdot\cdot}{O}}: \rightarrow \underset{\cdot\cdot}{\overset{\cdot\cdot}{O}} = \underset{\cdot\cdot}{\overset{\cdot\cdot}{O}}$
(b) O = O

Can you predict how atoms of nitrogen, which require three electrons each to achieve a stable octet, might form N_2? The two atoms form a triple bond by sharing three electron pairs.

$$:\underset{\cdot}{\overset{\cdot}{N}}\cdot + \cdot\underset{\cdot}{\overset{\cdot}{N}}: \rightarrow :N \equiv N:$$

The structural formula for a molecule of nitrogen is

$$N \equiv N$$

The number of covalent bonds (shared electron pairs) that an atom can form is known as its **bonding capacity** (Table 1). Each atom of nitrogen, we have just

bonding capacity: the number of electrons lost, gained, or shared by an atom when it bonds chemically

learned, shares three electron pairs, so it has a bonding capacity of three. What is the bonding capacity of oxygen? It always shares two electron pairs, so has a bonding capacity of two. It is easy to find the bonding capacity of any element by looking at the Lewis structure of a molecule containing that element: The number of dashes associated with the element is the same as the bonding capacity.

Table 1: Bonding Capacities of Some Common Atoms

Atom	Number of valence electrons	Number of bonding electrons	Bonding capacity
carbon	4	4	4
nitrogen	5	3	3
oxygen	6	2	2
halogens	7	1	1
hydrogen	1	1	1

So far, we have looked at molecules of elements. Because the atoms are the same, each has the same bonding capacity and each contributes the same number of electrons to the covalent bond. The molecules of compounds, however, consist of atoms of two or more different elements, often with different bonding capacities. How do we decide on their arrangement, when we draw structural formulas? The central position in the arrangement is often occupied by the element with the highest bonding capacity. Carbon and nitrogen, for instance, are commonly at the centre of a structural formula. Electronegativity is another means of deciding upon the central atom. When there is a choice of atoms for the central position in the molecule, choose the least electronegative element. Hydrogen is never the central atom since it can only form a single covalent bond. Halides and oxygen are also usually not the central atom.

There are exceptions to these generalizations, but they meet our needs in most cases.

Sample Problem 3

(a) Draw a Lewis structure for a molecule of carbon dioxide.
(b) Give the structural formula for the molecule.

Solution

(a) $:\ddot{\underset{\cdot}{O}}\cdot + \cdot\dot{\underset{\cdot}{C}}\cdot + \cdot\ddot{\underset{\cdot}{O}}: \rightarrow :\ddot{O}=C=\ddot{O}:$
(b) $O=C=O$

Practice

Understanding Concepts

1. Draw a Lewis structure and write the molecular formula for each of the following:
   (a) $F_{2(g)}$
   (b) $H_2O_{(l)}$
   (c) $CH_{4(g)}$
   (d) $PCl_{3(s)}$
   (e) $H_2S_{(g)}$
   (f) $SiO_{2(s)}$

Coordinate Covalent Bonds

Many substances contain a combination of covalent and ionic bonding. Consider the compound ammonium chloride, NH_4Cl. This white, crystalline solid dissolves rapidly in water and is an electrolyte—it dissociates to form a cation and an anion. It has many of the properties of an ionic compound, but it is composed only of nonmetals. We explain the properties of ammonium chloride by describing it as an ionic compound composed of a chloride ion, Cl^-, and a **polyatomic ion**, ammonium, NH_4^+. The bond holding the chloride and ammonium ions together is ionic, but the bonds within the polyatomic ammonium ion are covalent. There are several polyatomic ions, including NH_4^+, SO_4^{2-} (sulphate), and CO_3^{2-}(carbonate), all of which are covalently bonded groups of atoms carrying an overall charge.

polyatomic ion: a covalently bonded group of atoms with an overall charge

How does this arrangement fit with our description of covalent bonds? Molecules that are composed of two or more different elements can sometimes form covalent bonds where both of the electrons making up the bond are provided by the same atom. This type of bond is called a **coordinate covalent bond**. Consider the formation of the ammonium ion from the regular covalent molecule ammonia, NH_3, and a hydrogen ion, H^+. The hydrogen ion does not bring any electrons with it. To achieve a complete outer shell it can borrow two electrons from the atom with which it bonds.

coordinate covalent bond: a covalent bond in which both of the shared electrons come from the same atom

To explain this bond, we can draw Lewis structures. First, we must establish the Lewis structure for ammonia. We can arrange the atoms with nitrogen (which has the highest bonding capacity) at the centre, showing the five valence electrons of nitrogen and the one valence electron of each of the three hydrogen atoms.

$$\begin{array}{c} H\cdot \ \cdot\ddot{N}\cdot \ \cdot H \\ \dot{H} \end{array}$$

We can show the pairs of shared electrons (covalent bonds) between adjacent nitrogen and hydrogen atoms as dashes.

$$\begin{array}{c} H - \ddot{N} - H \\ | \\ H \end{array}$$

Notice the lone pair of electrons in this structure. A hydrogen ion, which has no electrons of its own, can bond to the ammonia molecule by sharing this unbonded pair of electrons. This is the coordinate covalent bond.

$$\begin{array}{c} H^+ \searrow \\ H - \ddot{N} - H \\ | \\ H \end{array} \rightarrow \left[\begin{array}{c} H \\ | \\ H - N - H \\ | \\ H \end{array}\right]^+$$

Once the hydrogen ion is bonded, there is no way to tell which of the hydrogens was the ion. Each of the four hydrogen atoms in the structure are equivalent: The positive charge is not really associated with any particular hydrogen atom. To indicate this, square brackets are placed around the entire ammonium ion and the positive charge is written outside the bracket.

The Strength of Covalent Bonds

Covalent bonds are strong. A large amount of energy is needed to separate the atoms that make up molecules. For this reason, molecules tend to be stable at relatively high temperatures: They do not easily decompose upon heating.

The stronger the bonds within the molecule, the greater the energy required to separate them. The strength of a bond between two atoms increases as the number of electron pairs in the bond increases. Therefore, triple bonds are stronger than double bonds, which are stronger than single bonds between the same two atoms.

SUMMARY Drawing Lewis Structures and Structural Formulas for Molecular Compounds

1. Arrange the symbols of the elements of the compound as you would expect the atoms to be arranged in the compound. The element with the highest bonding capacity is generally written in the central position (**Figure 2(a)**).
2. Add up the number of valence electrons available in each of the atoms (**Figure 2(b)**). If the structure is a polyatomic ion, add one electron for each unit of negative charge, or subtract one for each unit of positive charge.
3. Place one pair of electrons between each adjacent pair of elements (forming single covalent bonds) (**Figure 2(c)**).
4. Place pairs of the remaining valence electrons as lone pairs on the peripheral atoms (not the central atom) (**Figure 2(d)**).
5. If octets are not complete, move lone pairs into bonding position between those atoms and the central atom until all octets are complete (**Figure 2(e)**).
6. If the peripheral atoms all have complete octets and there are pairs of electrons remaining, place these electrons as lone pairs on the central atom.
7. Count the number of bonds between the central atom and the peripheral atoms. If this number exceeds the bonding capacity of the central atom, one or more of the bonds is coordinate covalent. To identify which ones, try removing the peripheral atoms one at a time. If you can do this and leave the central structure with complete octets, you have identified coordinate covalent bonds (**Figure 2(f)**).
8. To give the structural formula, remove the dots representing the lone pairs and replace bond dots with dashes (**Figure 2(g)**).
9. If you are representing a polyatomic ion, place brackets around the entire structure and write the charge outside the brackets.

(a)

O
S
O O

(b)

3(6) + 6 = 24
O S

(c)

O
:
S
O O

(d)

:Ö:
:
S
:O: :O:

(e) Sulfur atom has an incomplete octet.

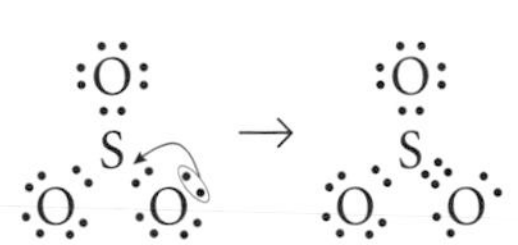

(f) SO_3 includes two coordinate covalent bonds.

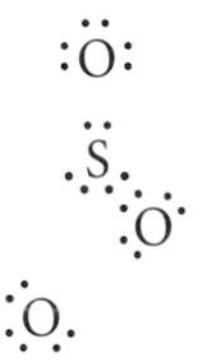

(g)

Figure 2

Practice

Understanding Concepts

2. Draw Lewis structures and structural formulas for each of the following molecules:
   (a) $H_{2(g)}$
   (b) $O_{3(g)}$
   (c) $OF_{2(g)}$
   (d) $NF_{3(g)}$
   (e) $N_2H_{2(g)}$
   (f) $P_2H_{4(g)}$

3. Draw Lewis structures and structural formulas for each of the following polyatomic ions:
   (a) $PO_{4(aq)}^{3-}$
   (b) $OH_{(aq)}^{-}$
   (c) $BrO_{3(aq)}^{-}$
   (d) $ClO_{4(aq)}^{-}$
4. Which of the Lewis structures in questions 2 and 3 include coordinate covalent bonds?

Explaining the Properties of Molecular Compounds

You are familiar with many molecular compounds: propane, $C_3H_{8(g)}$, for the barbecue, water, $H_2O_{(l)}$, in your bathtub, and sugar, $C_{12}H_{22}O_{11(s)}$, for your coffee. As you can see, their physical properties vary greatly. In contrast to ionic compounds, which are all solids at SATP, molecular compounds may be solids, liquids, or gases (**Table 2**).

Table 2: Comparison of Ionic and Molecular Solids

Properties	Ionic	Molecular
melting point	high	low
electrical conductivity		
in the solid state	no conductivity	no conductivity
in the liquid state	conductivity	no conductivity
consistency of solid	hard, brittle	soft, waxy or flexible
Examples	sodium chloride copper(II) sulfate	iodine phosphorus

intramolecular force: the attractive force between atoms and ions within a compound

intermolecular force: the attractive force between molecules

We have discussed the forces that bond atoms and ions together within a compound, the **intramolecular forces** ("intra" means within). These are sufficient to explain the existence of ionic and molecular compounds, and to explain many of the properties of ionic compounds, but they aren't sufficient to explain the physical state of molecular compounds. Why is water a liquid at SATP and a solid at STP? Why isn't it a gas at all temperatures? Something, some force, must hold the molecules together in the solid and liquid states.

Forces between molecules are called **intermolecular forces** ("inter" means between). The evidence indicates that these forces are strong enough to cause molecules to arrange themselves in an orderly fashion to form a lattice structure (similar to that of ionic solids). Both solid water and solid carbon dioxide (dry ice) show such structures when examined by crystallography (**Figure 3**). However, intermolecular forces must be weak compared to covalent bonds. We can deduce this from the observation that it is much easier to melt a molecular solid than it is to cause the same substance to decompose. When water is heated from –4°C to 104°C it changes state from a solid, to a liquid, and then to a gas, but it does not decompose to oxygen and hydrogen. The energy added in the form of heat is sufficient to overcome the intermolecular forces between the molecules, but not the covalent bonds between the atoms. Adding a relatively small amount of heat will cause a solid molecular compound to change state from a solid to a liquid, and then to a gas, but it takes much more energy to break the covalent bonds between the atoms in the compound.

Later in this chapter, you will learn more about intermolecular forces and how they affect the properties of substances.

Practice

Understanding Concepts

5. Distinguish between bonding electrons and lone pairs.
6. Are the following pairs of atoms more likely to form ionic or covalent bonds?
   (a) sulfur and oxygen
   (b) iodine and iodine
   (c) calcium and chlorine
   (d) potassium and bromine
7. (a) List six examples of molecular elements and compounds and six examples of ionic compounds.
   (b) Compare the two lists, referring to bond types to explain the contrasting physical properties.
8. (a) How does the bonding capacity of nitrogen differ from that of chlorine?
   (b) Give a theoretical explanation for your answer to (a).
9. How are coordinate covalent bonds similar to covalent bonds? How are they different?
10. (a) Use an electron dot diagram to explain the formula for nitrogen, N_2.
    (b) Draw the Lewis structure for nitrogen.
    (c) Nitrogen is a fairly inert (unreactive) gas. Explain this, referring to the bonds involved.
11. Illustrate the formation of each of the following molecular compounds, using electron dot diagrams and Lewis structures:
    (a) HCl
    (b) NH_3
    (c) H_2S
    (d) CO_2
12. Use the octet rule to develop a table that lists the bonding capacities for carbon, nitrogen, oxygen, hydrogen, fluorine, chlorine, bromine, and iodine.
13. Use bonding capacities (**Table 1**, page 77) to draw the structural formula of each of the following molecules.
    (a) O_2
    (b) Br_2
    (c) H_2O_2
    (d) C_2H_4
    (e) HCN
    (f) C_2H_5OH
    (g) CH_3OCH_3
    (h) CH_3NH_2
14. (a) Illustrate the structure of a hydronium ion (H_3O^+) by drawing its Lewis structure.
    (b) Name the bonds within the hydronium ion.
    (c) What kinds of bonds is this ion likely to form with other entities?
15. Is it correct for the structural formula of H_2S to be written as H—H—S? Explain, using a diagram.

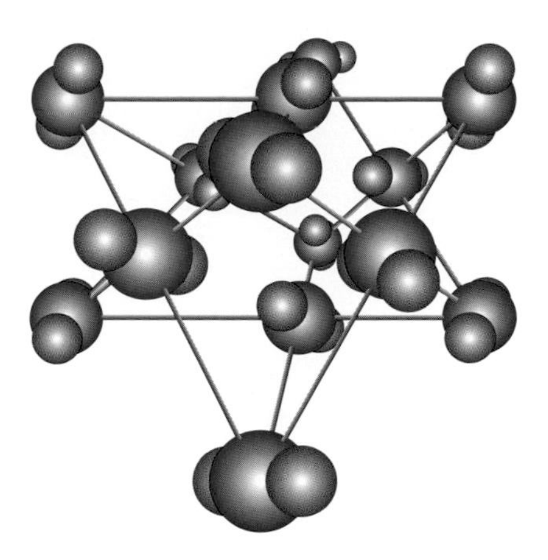

Figure 3
The molecules in solid carbon dioxide, CO_2, form a crystal lattice.

Section 2.3 Questions

Understanding Concepts

1. Compound A is formed when the element with atomic number 3 combines with the element of atomic number 9. Compound B is formed when the element with atomic number 7 combines with the element of atomic number 9.
   (a) Compare the properties of compounds A and B.
   (b) What types of compounds are A and B? Give reasons for your answer.
   (c) Clearly show the structure of each compound formed, using electron dot diagrams and Lewis structures.
2. Oxygen forms ionic bonds with aluminum to form bauxite (or aluminum oxide). However, oxygen forms covalent bonds with carbon to form carbon dioxide. Use the concepts of electronegativity and periodic trends to explain these differences in bonding.
3. An alkali metal M reacts with a halogen X to form a compound with the formula MX.
   (a) Would this compound have ionic or covalent bonds? Explain.
   (b) Predict the physical properties of the compound MX.
4. The compound NaCl and the element Cl_2 are both held together by chemical bonds.
   (a) Classify each substance as either an ionic or a covalent compound.
   (b) Explain, in terms of chemical bonds and attractive forces, why NaCl is a solid at SATP while Cl_2 is a gas at SATP.
5. Using only oxygen and sulfur atoms, create as many compounds as you can and draw Lewis diagrams for each. Which of the compounds contain only multiple bonds?

Applying Inquiry Skills

6. Design an experiment to test your classification of compounds A and B in question 1.
7. Design an experiment to test your predictions of the physical properties of the compound MX in question 3.

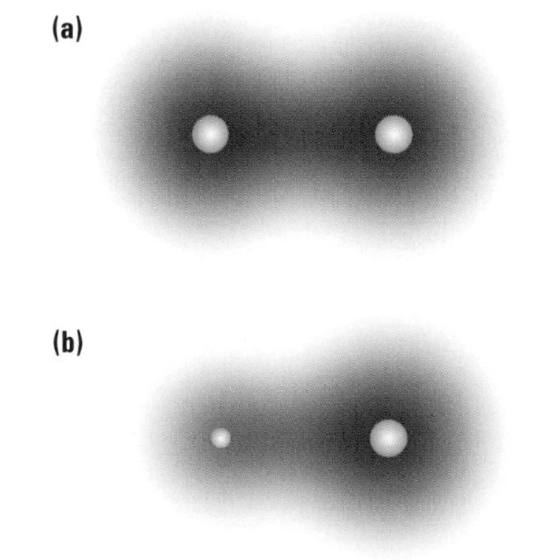

Figure 1
Electron densities of the bonding electrons in two covalent molecules
(a) Cl_2
(b) HCl

2.4 Electronegativity, Polar Bonds, and Polar Molecules

Why are some molecular substances solid, some liquid, and some gaseous at SATP? Why do different liquids have different boiling points? Why can water striders walk across the surface of a pond without falling in? Believe it or not, all these phenomena depend on the bonds and forces between molecules.

Polar Covalent Bonds

So far, we have discussed models for two types of chemical bonding: ionic and covalent. However, when a chemical bond is formed, it is not always exclusively one or the other.

When electrons are shared between two atoms, a covalent bond results. When the atoms are identical, such as in a chlorine molecule, the electrons are shared equally (**Figure 1**(a)). However, this is not the case for a compound like hydrogen chloride, where electrons are shared between two different elements. In this situation, the sharing is unequal, as the bonding electrons spend more time near one atom than near the other (**Figure 1(b)**). The electrons in the H—Cl bond in a hydrogen chloride molecule spend more time near the chlorine atom than near the hydrogen atom. This is because of chlorine's greater attraction for electrons. Due to this unequal sharing of electrons, the hydrogen atom is, on average, slightly positively charged while the chlorine atom is slightly negatively charged (**Figure 2**). We indicate these slight charges by δ^+ and δ^-. (The Greek letter *delta*, δ, indicates "a small difference.") While chlorine shows the greater attraction for the bonding electrons, the attraction is not strong enough to actually bring about an electron transfer, as in an ionic compound. The bond is somewhere between an ionic bond and a covalent bond and is called a **polar covalent bond**.

We can predict which parts of the molecule will have δ^+ and δ^- charges by comparing the electronegativities of the atoms (given in the periodic table at the end of this text). The most electronegative atoms will attract the electron pair strongly, and so will tend to have a δ^- charge in the covalent compounds they form.

Figure 2
Slight positive and negative charges are indicated by the Greek symbol δ, delta.

polar covalent bond: a covalent bond formed between atoms with significantly different electronegativities; a bond with some ionic characteristics

Sample Problem

(a) Draw the structural formula for methane (CH_4, the main constituent of natural gas).
(b) Indicate which atoms are slightly positive and which are slightly negative.

Solution

(a)

$$\begin{array}{ccccc} & & H & & \\ & & | & & \\ H & — & C & — & H \\ & & | & & \\ & & H & & \end{array}$$

(b)

$$\begin{array}{ccccc} & & H^{\delta+} & & \\ & & | & & \\ H^{\delta+} & — & C^{\delta-} & — & H^{\delta+} \\ & & | & & \\ & & H^{\delta+} & & \end{array}$$

Each hydrogen atom carries a partial positive charge while the carbon atom carries a partial negative charge.

To predict whether a chemical bond between two atoms will be ionic, polar covalent, or covalent, we must consider the electronegativities of the elements involved. The absolute value of the difference in electronegativities of two bonded atoms provides a measure of the polarity in the bond: the greater the difference, the more polar the bond (**Table 1**). According to the periodic trends in electronegativity described in Chapter 1, fluorine atoms have the greatest tendency to gain or attract electrons and cesium atoms have the least. The difference between their electronegativities is 3.3 (cesium 0.7 and fluorine 4.0). Since the difference in electronegativity for cesium and fluorine is large, they should (and do) form an ionic bond. In other words, the fluorine atom is strong enough, relative to the cesium atom, to "steal" a cesium electron.

Table 1: Electronegativity Differences of a Selection of Bonds

Bond	Elecronegativity difference
Cl—Cl	0.0
C—Cl	0.3
C—H	0.4
Be—H	0.6
N—H	0.9
C—O	1.0
O—H	1.4
Mg—Cl	1.8
Na—Cl	2.1
Mg—O	2.3
Ba—O	2.6
Ca—F	3.0

The difference in electronegativities of hydrogen (2.1) and chlorine (3.0) is much less. They instead form a polar covalent bond. The greater the difference in electronegativity, the more polar, and then more ionic, the bond becomes. This relationship describes a continuum, a range from covalent to ionic, rather than three different kinds of bonds (**Figure 3**). How do we know when to classify a chemical bond as fully ionic? By convention, a difference in electronegativity greater than 1.7 indicates an ionic bond. However, there are many binary compounds in which the two atoms have an electronegativity difference of less than 1.7, but the compound still has ionic properties. For example, MgI_2 (electronegativity difference 1.3) is an electrolyte, i.e., is an ionic compound.

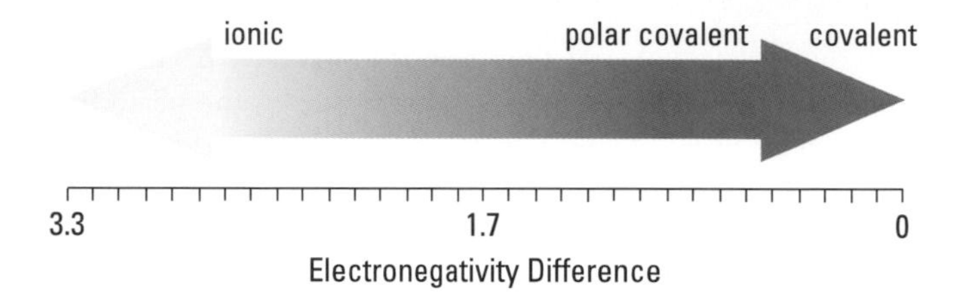

Figure 3
In this model of a bonding continuum, the colour change from right to left indicates the increasing difference in electronegativity and bond polarity.

Practice

Understanding Concepts

1. Describe how we can use electronegativity values to predict the types of bonds that will form within a compound.
2. Which type of bond—ionic or covalent—will form between each of the following pairs of atoms? Of the covalent bonds, indicate which would be the most polar.
   (a) H and Cl
   (b) Si and O
   (c) Mg and Cl
   (d) Li and O
   (e) N and O
   (f) O and O
   (g) I and Cl
   (h) Cr and O
   (i) C and Cl
3. Identify the more polar bond in each of the following pairs.
   (a) H—F; H—Cl
   (b) N—O; C—O
   (c) S—H; O—H
   (d) P—Cl; S—Cl
   (e) C—H; N—H
   (f) S—O; P—O
   (g) C—N; N—N
4. Draw Lewis structures for the following substances. In each case, use appropriate notation to indicate the atoms that are slightly positive and those that are slightly negative.
   (a) H_2O
   (b) Br_2
   (c) HBr
   (d) PCl_3
   (e) OF_2

SUMMARY Bonding Characteristics

Table 2: Summary of Bonding

Intramolecular force	Bonding model
ionic bond	• involves an electron transfer, resulting in the formation of cations and anions • cations and anions attract each other
polar covalent bond	• involves unequal sharing of pairs of electrons by atoms of two different elements • bonds can involve 1, 2, or 3 pairs of electrons, i.e., single (weakest), double, or triple (strongest) bonds
covalent bond	• involves equal sharing of pairs of electrons • bonds can involve 1, 2, or 3 pairs of electrons, i.e., single (weakest), double, or triple (strongest) bonds

Polar Molecules

Molecules of water, ammonia (a reactant in the production of nitrogen fertilizers), and sulfur dioxide (an industrial pollutant contributing to acid rain formation) all have polar covalent bonds holding their atoms together. If a molecule contains polar covalent bonds, the entire molecule may have a positive end and a negative end, in which case it would be classified as a **polar molecule**. Not all molecules containing polar covalent bonds are polar molecules, however. Carbon tetrachloride, $CCl_{4(l)}$, contains four polar covalent bonds and $HCl_{(g)}$ has one. However, $HCl_{(g)}$ is a polar molecule (**Figure 4(a)**), while $CCl_{4(l)}$ is not (**Figure 4(b)**).

polar molecule: a molecule that is slightly positively charged at one end and slightly negatively charged at the other because of electronegativity differences

With practice you will be able to predict which molecules are likely to be polar by looking at their chemical or structural formulas. **Table 3** shows which types of compounds tend to be polar molecules.

Table 3: Guidelines for Predicting Polar and Nonpolar Molecules

	Type	Description	Examples
Polar	AB	diatomic compounds	$CO_{(g)}$
	HA_x	any molecule with a single H	$HCl_{(g)}$
	A_xOH	any molecule with an OH at one end	$C_2H_5OH_{(l)}$
	O_xA_y	any molecule with an O at one end	$H_2O_{(l)}$, $OCl_{2(g)}$
	N_xA_y	any molecule with an N at one end	$NH_{3(g)}$, $NF_{3(g)}$
Nonpolar	A_x	all elements	$Cl_{2(g)}$, $N_{2(g)}$
	C_xA_y	most carbon compounds (including organic solvents, fats, and oils)	$CO_{2(g)}$, $CH_{4(g)}$

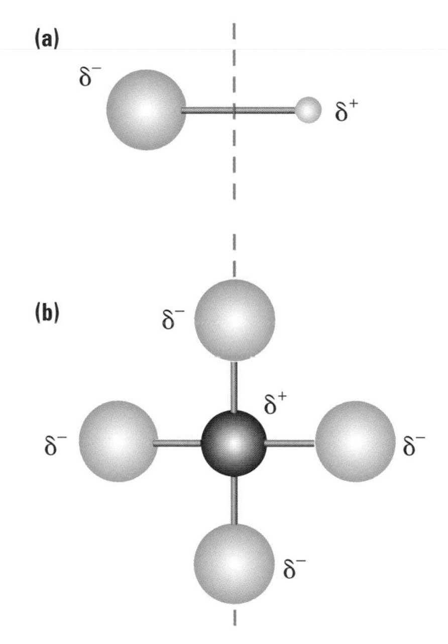

Figure 4
(a) A molecule of HCl acquires oppositely charged ends because of the polar covalent nature of the H-Cl bond.
(b) Since a molecule of CCl_4 is quite symmetrical, it lacks oppositely charged ends and is not polar.

Practice

Understanding Concepts

5. Some molecules contain polar covalent bonds but are not themselves polar. Explain, with diagrams, how this is possible.
6. Use Lewis structures and electronegativity values to explain why methane, $CH_{4(g)}$, is not a polar substance.

Intermolecular Forces

As mentioned earlier, the properties of molecular compounds cannot be explained simply by covalent bonds. If covalent bonds were the only forces at work, most molecular compounds would be gases, as there would be no attraction between the molecules strong enough to order the molecules into solids or liquids. The concepts of the polar molecule and small charges on atoms help explain why these molecular compounds are not all gases at SATP.

The existence of intermolecular forces was first suggested by Johannes van der Waals, toward the end of the 19th century, to explain why gases liquefy when they are cooled. Many different observations, such as surface tension and heat of vaporization, provide evidence that there are three kinds of intermolecular forces, each with different strengths. Two of these are classified as **van der Waals forces** (in honour of Johannes van der Waals): **dipole–dipole forces** and **London dispersion forces**. The third intermolecular force—hydrogen bonding—is generally not grouped with the other two.

van der Waals forces: weak intermolecular attractions, including London dispersion forces and dipole–dipole forces

dipole–dipole force: an attractive force acting between polar molecules

London dispersion force: an attractive force acting between all molecules, including nonpolar molecules

Dipole–dipole forces are the forces of attraction between oppositely charged ends of polar molecules (e.g., HCl). The positive end of each molecule attracts the negative ends of neighbouring molecules—rather like a weak version of ionic bonds. London dispersion forces, by contrast, exist between all molecules—both polar and nonpolar—so are the only intermolecular forces acting between nonpolar molecules. Chemists believe that the weak attractive forces are the result of temporary displacements of the electron "cloud" around the atoms in a molecule, resulting in extremely short-lived dipoles. Because the dipoles last for only tiny fractions of a second, the attraction is continually being lost, so the forces are very weak. Dipole–dipole forces, when they exist, tend to be much stronger than London dispersion forces. You will learn more about these forces later in this course.

Hydrogen Bonds

Water, a polar molecule, consists of one atom of oxygen bound by single covalent bonds to two hydrogen atoms. Its structure is simple, but water exhibits some rather unusual properties: higher than expected melting and boiling points, high vapour pressure, high surface tension, and the ability to dissolve a large number of substances. To explain these properties, we must consider the intermolecular forces that exist between water molecules. As a result of the large difference in electronegativity between hydrogen and oxygen, the O—H bonds in a molecule of water are highly polar covalent. The oxygen atom in a molecule of water carries a slight negative charge while the hydrogen atoms carry a small positive charge (**Figure 5**). As a result, the hydrogen atoms of one water molecule exert a strong force of attraction on the oxygen atom of neighbouring water molecules. This kind of intermolecular force is referred to as a **hydrogen bond.** Hydrogen bonds occur among highly polar molecules containing F—H, O—H, and N—H bonds. Although a hydrogen bond is similar to a dipole–dipole force, it is stronger than any of the van der Waals forces.

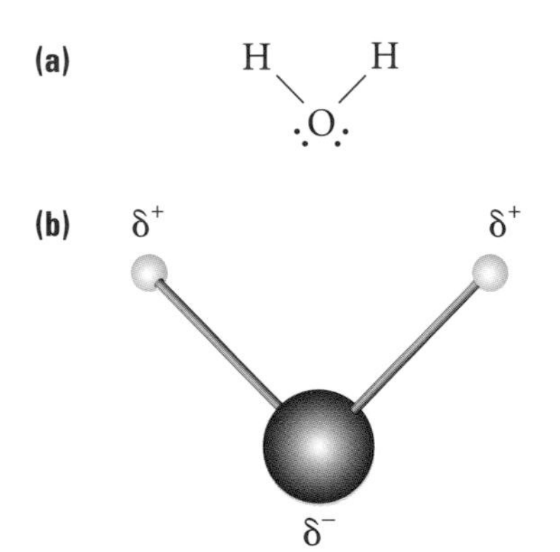

Figure 5
(a) The oxygen atom in a water molecule has two single polar covalent bonds and two lone pairs of electrons. This results in a V-shaped molecule.
(b) A water molecule has a slight negative charge on the oxygen atom and slight positive charges on the hydrogens.

hydrogen bond: a relatively strong dipole–dipole force between a positive hydrogen atom of one molecule and a highly electronegative atom (F, O, or N) in another molecule

Water molecules have a tendency to "stick together" because of hydrogen bonding. This is one of the reasons why water appears to "climb" the sides of a narrow tube (**Figure 6**). We can also explain water's high melting and boiling points using the concept of the hydrogen bond: Large amounts of energy are required to break the hydrogen bonds in the solid and liquid states.

Molecular Models

Molecular models provide a way of representing molecules in three dimensions. In this activity, you will practise various ways of modelling molecules.

Materials: molecular model kits

(a) Draw Lewis structures to represent the following molecules: water, hydrogen sulfide, hydrogen fluoride, methane, carbon dioxide, ammonia, and nitrogen.

(b) Write the structural formula for each compound.

- Assemble models to represent each of the substances.

(c) Which of the molecules required double or triple covalent bonds? Explain why.

(d) Which of the molecules would you expect to be polar?

(e) Give examples of other compounds that you would expect to have similar shapes to water, ammonia, and hydrogen fluoride. Explain your answers using diagrams and models.

Most substances are more dense in the solid state than in the liquid state. However, you have probably seen solid ice cubes floating in a glass of water. If they float, they must be less dense than the liquid. Hydrogen bonds enable us to explain this unusual behaviour: ice is less dense than liquid water because it forms an open lattice structure when it freezes, with a great deal of empty space between the molecules (**Figure** 7). This lower density of water in the solid state enables aquatic life to survive the winter. Instead of freezing from the bottom up, lakes freeze from the top down, creating an insulating layer of ice at the top. This is one of the many reasons why scientists consider water to be indispensable for the existence of life.

Figure 6
The curved surface of water in a container is called the meniscus.

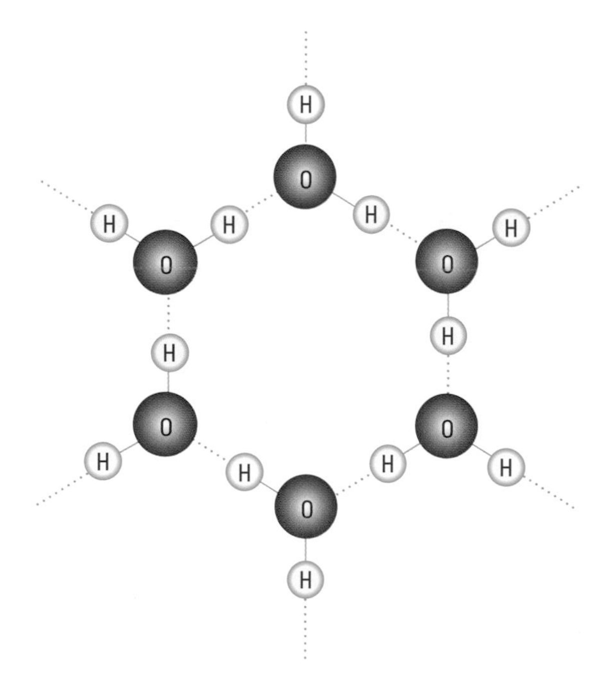

Figure 7
In ice, hydrogen bonds between the molecules result in a regular hexagonal crystal structure. The ··· H– represents a hydrogen (proton) being shared between two pairs of electrons. In liquid water these bonds constantly break and reform as molecules move past each other.

Practice

Understanding Concepts

7. Water is known for its many anomalous properties. Use your knowledge of intermolecular forces and intramolecular bonding to explain theoretically why lakes freeze from top to bottom.
8. Using **Table 3** (page 85), predict whether each of the following molecules would be polar or nonpolar.
   (a) $CH_3OH_{(l)}$
   (b) $I_{2(s)}$
   (c) $HBr_{(g)}$
   (d) $PCl_{3(s)}$
   (e) $HC_2H_3O_{2(aq)}$
   (f) $CCl_{4(l)}$
9. Which compound(s) in question 8 were classified as nonpolar but contain polar covalent bonds? Explain how this is possible.
10. How are hydrogen bonds different from other dipole–dipole forces?
11. Predict the type of intermolecular force that would exist between molecules of the following:
    (a) $I_{2(s)}$
    (b) $H_2O_{(l)}$
    (c) $NH_{3(g)}$
12. Explain each of your predictions in question 11 theoretically.

Applying Inquiry Skills

13. A student was provided with seven sample liquids and asked to investigate whether each liquid is affected by a positively or negatively charged object. Complete the **Prediction**, **Evidence**, **Analysis**, and **Synthesis** sections of the investigation report.

Question

Does a charged object have an effect on a thin stream of the following liquids: NCl_3, H_2O, Br_2, CCl_4, CH_3OH, H_2O_2, and vegetable oil?

Prediction

(a) Using the generalizations of polar and nonpolar molecules, predict an answer to the Question.

Experimental Design

A thin stream of each liquid was tested by holding a positively charged object near the liquid stream. The procedure was repeated with a negatively charged object.

Evidence

Table 4: Effects of Charged Objects on Seven Liquids

Samples	Positive charge	Negative charge
1 - 3	no effect	no effect
4 - 7	stream moved toward charged object	stream moved toward charged object

(b) Which of the seven substances could be samples 1 to 3? Which could be 4 to 7?

Analysis

(c) Use the Evidence to answer the Question.

Synthesis

(d) Provide a theoretical explanation to justify your answer in (b).
(e) Speculate as to why the liquids were affected by both positive and negative charges.

Section 2.4 Questions

Understanding Concepts

1. Explain why the bonding in compounds cannot be described only in terms of covalent bonds and ionic bonds.
2. How does the sharing of the bonding electrons in a molecule of Cl_2 compare to that in a molecule of HCl? Explain your answer, including diagrams.
3. Both boron and nitrogen form compounds with chlorine. In each case, the formula has the general form AX_3.
   (a) Classify each of these compounds as ionic or molecular. Justify your answer.
   (b) How are the B—Cl bonds and N—Cl bonds similar? How are they different?
   (c) What other properties can you predict for the two compounds? Use the concepts of electronegativity and Lewis structures to justify your answers.
4. Carbon tetrachloride, $CCl_{4(l)}$, is a nonpolar substance, although carbon and chlorine have different electronegativities. Explain this lack of polarity.

2.5 The Names and Formulas of Compounds

Prior to the late 1700s, there was no systematic method of naming compounds. Substances were named in a variety of ways. In some cases, the name referred to the use of the compound; in other cases, it incorporated an obvious property, or perhaps referred to the sources of the substance. These common names gave little, if any, information about the composition of the compound.

The classical system of nomenclature based on the Latin names of elements was devised in 1787 by French chemist Guyton de Morveau (1737–1816). As more and more compounds were discovered, however, it became apparent that a more comprehensive and consistent system of nomenclature was needed.

Today, we are aware of a very large number of compounds, and there is every possibility that more will be discovered in the future. We need a system that is easy to use and provides information on the composition of every compound. As a result, scientists now use the system of **chemical nomenclature** chosen by the International Union of Pure and Applied Chemistry (IUPAC) (sometimes just called the IUPAC system) when naming compounds. Just as many immigrants to Canada have learned a new language, you will soon become familiar with the language of chemical nomenclature. Chemical nomenclature provides us with a systematic means of both naming and identifying compounds.

In describing a chemical compound, chemists may use its name or its formula. The formula of an ionic compound tells us the ratio of elements present. The formula of a molecular compound indicates the number of atoms of each element present in a molecule of the compound. Every chemical compound has a unique name. Knowing the names of common chemicals and being able to write their formulas correctly are useful skills in chemistry. The common and IUPAC names for some familiar compounds are shown in **Table 1**.

DID YOU KNOW?

Salt of the Gods

Ammonium chloride was at one time referred to as sal ammoniac. The Greeks built a temple dedicated to the God Ammon after parts of Egypt had been conquered by Alexander the Great (356–323 B.C.). Dried camel manure was used as fuel for fires in this temple. Over many years of manure burning, a white, crystalline, saltlike material was gradually deposited on the walls of the temple. This deposit became known as "salt of Ammon" or sal ammoniac.

chemical nomenclature: a system, such as the one approved by IUPAC, of names used in chemistry

Table 1: Common and IUPAC Names for Some Compounds

Common name	IUPAC name
quicklime	calcium oxide
laughing gas	dinitrogen monoxide
saltpetre	sodium nitrate
potash	potassium carbonate
muriatic acid	hydrochloric acid
baking soda	sodium hydrogen carbonate (sodium bicarbonate)
cream of tartar	potassium hydrogen tartrate
grain alcohol	ethanol or ethyl alcohol
sal ammoniac	ammonium chloride

binary compound: a compound composed of two kinds of atoms or two kinds of monatomic ions

valence: the charge on an ion

Binary Ionic Compounds

The simplest compounds are called **binary compounds.** Binary ionic compounds consist of two types of monatomic ions (ions consisting of one charged atom). In the formula of a binary ionic compound, the metal cation is always written first, followed by the nonmetal anion. (This reflects the periodic table: metals to the left, nonmetals to the right.) The name of the metal is stated in full and the name of the nonmetal ion has an *-ide* suffix; for example, $NaCl_{(s)}$ is sodium chloride and $LiBr_{(s)}$ is lithium bromide. Binary ionic compounds can be made up of more than two ions, providing they are of only two kinds: aluminum oxide is $Al_2O_{3(s)}$.

If we know what ions make up a compound, we can often predict the compound's formula. First, we determine the charges on each type of ion making up the compound. The charge on an ion is sometimes called the **valence.** We then balance the charges to determine the simplest ratio in which they combine. We can predict the charge on the most common ion formed by each representative element by counting the number of electrons that would have to be gained or lost to obtain a stable octet. (Ionic charges for the most common ions of elements are shown in the periodic table at the back of this text.) For example, the compound magnesium chloride, a food additive used to control the colour of canned peas, is composed of the ions Mg^{2+} and Cl^-. For a net charge of zero, the ratio of magnesium to chlorine ions must be 1:2. The formula for magnesium chloride is therefore $MgCl_{2(s)}$.

Sample Problem 1

Predict the formula for magnesium oxide—a source of dietary magnesium when used as a food additive.

Solution

Charge on magnesium ion: 2+
Charge on oxygen ion: 2–
The ratio of magnesium ions to oxide ions that produces a net charge of zero is 1:1.
The formula is therefore MgO.

You may be familiar with a method known as the crisscross rule for predicting the formula of an ionic compound. The crisscross rule works as follows:

1. Write the symbol of each of the elements in the order in which they appear in the name of the compound.
2. Write the valence number (electrons lost or gained in forming that element's most stable ion) above the symbol of each of the elements.
3. Crisscross the numbers written above the symbols such that the valence number of one element becomes a subscript on the other.
4. Divide each subscript by the highest common factor. The resulting subscripts indicate the ratio of ions present in the compound.
5. Omit any subscript equal to 1 from the formula.

For example, if we were to use the crisscross rule to predict the formula of magnesium chloride, we would go through these steps.

1. Write symbols in order:
   Mg Cl

2. Write valences above symbols:

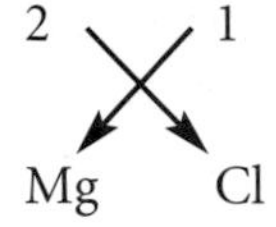

3. Crisscross the valences, making them subscripts:

Mg_1 Cl_2

4. Divide subscripts by highest common factor (= 1):

Mg_1 Cl_2

5. Remove 1 subscripts:

$MgCl_2$

Sample Problem 2

Use the crisscross rule to predict the formula of barium sulfide.

Solution

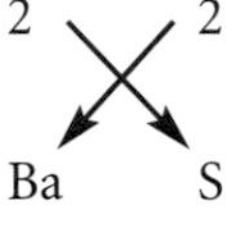

Ba_2 S_2

The formula of barium sulfide is BaS.

Most transition metals and some representative metals can form more than one kind of ion. Metals that can have more than one valence, or charge, are classified as **multivalent**. For example, iron can form an Fe^{2+} ion or an Fe^{3+} ion, although Fe^{3+} is more common. The periodic table at the end of this text shows the most common ion of each element first, with one alternative ion charge below. It does not list all of the possible ions of the element.

multivalent: the property of having more than one possible valence

Copper is a multivalent element. It is capable of bonding with chlorine in two different ratios to form two different chloride compounds: CuCl and $CuCl_2$. How are the names of these compounds different? The IUPAC system of naming compounds containing multivalent ions is very simple. The name of the metal ion includes the charge on the ion, indicated by Roman numerals in brackets. Consequently, $CuCl_{(s)}$ (in which copper has a charge of 1+) is copper(I) chloride, and $CuCl_{2(s)}$ (in which copper has a charge of 2+) is copper(II) chloride. This system of naming is sometimes referred to as the Stock system.

To determine the chemical name of a compound containing a multivalent metal ion, we have to figure out the necessary charge on that ion to yield a net charge of zero. If we are given the formula, we simply have to calculate the equivalent negative charge. The metal's charge is then written, in Roman numerals, after the name of the metal. For example, if you were asked to name the compound MnO_2, using IUPAC nomenclature, you would first look at the charge on the nonmetal ions. In this case, the charge on each O is 2–, so the total negative charge is 4–. The charge on the Mn ion must be 4+. Consequently, the IUPAC name for MnO_2 is manganese(IV) oxide.

If the ion of a multivalent metal is not specified in a name, it is assumed that the charge on the ion is the most common one.

Sample Problem 3

The formula of a compound is found to be $SnCl_2$. What is its IUPAC name?

Solution

Charge on each Cl ion: 1–
Total negative charge is 2–
Charge on Sn ion: 2+
The IUPAC name for $SnCl_2$ is tin(II) chloride.

Sample Problem 4

The formula of a compound is found to be Fe_2O_3. What is its IUPAC name?

Solution

Total negative charge: 6–
Total positive charge: 6+
Charge on each Fe ion: 3+
The IUPAC name of Fe_2O_3 is iron(III) oxide.

The classical nomenclature system has, in the past, been used for naming compounds containing multivalent metals with no more than two possible charges. In this system, the Latin name for the element along with the suffix *-ic* was applied to the larger charge, and the suffix *-ous* was applied to the smaller charge. The compounds formed by copper and chlorine were therefore known as cuprous chloride ($CuCl_{(s)}$) and cupric chloride ($CuCl_{2(s)}$).

In many industries, the classical system is still used extensively. **Table 2** shows a comparison of the classical and IUPAC names of multivalent metal ions.

Table 2: Classical and IUPAC Names of Common Multivalent Metal Ions

Metal	Ion	Classical name	IUPAC name
iron	Fe^{2+} Fe^{3+}	ferrous ferric	iron(II) iron(III)
copper	Cu^{+} Cu^{2+}	cuprous cupric	copper(I) copper(II)
tin	Sn^{2+} Sn^{4+}	stannous stannic	tin(II) tin(IV)
lead	Pb^{2+} Pb^{4+}	plumbous plumbic	lead(II) lead(IV)
antimony	Sb^{3+} Sb^{5+}	stibnous stibnic	antimony(III) antimony(V)
cobalt	Co^{2+} Co^{3+}	cobaltous cobaltic	cobalt(II) cobalt(III)
gold	Au^{+} Au^{2+}	aurous auric	gold(I) gold(II)
mercury	Hg^{+} Hg^{2+}	mercurous mercuric	mercury(I) mercury(II)

Practice

Understanding Concepts

1. How were compounds named before the advent of systematic means of chemical nomenclature?
2. State the common names of each of the following chemicals:
   (a) hydrochloric acid
   (b) sodium hydrogen carbonate
   (c) dinitrogen monoxide
   (d) ethanol
3. How many elements are there in each binary compound?
4. Describe the IUPAC system of naming a binary ionic compound.
5. If an element is described as "multivalent," what characteristic does it have?
6. Write the formula for each of the following compounds:
   (a) calcium fluoride
   (b) sodium sulfide
   (c) aluminum nitride
   (d) aluminum chloride
   (e) potassium oxide
   (f) calcium chloride
   (g) copper(II) sulfide
   (h) lead(II) bromide
   (i) silver iodide
   (j) barium nitride
   (k) iron(II) fluoride
   (l) mercury(I) oxide
   (m) nickel(II) bromide
   (n) zinc oxide
   (o) cobalt(III) chloride
   (p) strontium bromide
   (q) gold(I) fluoride
   (r) lithium chloride
   (s) strontium nitride
   (t) barium bromide
   (u) tin(IV) iodide
7. Write the names of each of the following binary ionic compounds, using the IUPAC system of chemical nomenclature:
   (a) table salt, $NaCl_{(s)}$
   (b) lime, $CaO_{(s)}$
   (c) road salt, $CaCl_{2(s)}$
   (d) magnesia, $MgO_{(s)}$
   (e) bauxite, $Al_2O_{3(s)}$
   (f) zinc ore, $ZnS_{(s)}$
   (g) cassiterite, $SnO_{2(s)}$
   (h) chalcocite, $Cu_2S_{(s)}$
   (i) galena, $PbS_{2(s)}$
   (j) hematite, $Fe_2O_{3(s)}$
   (k) molybdite, $MoO_{3(s)}$
   (l) argentite, $Ag_2S_{(s)}$
   (m) zincite, $ZnO_{(s)}$
8. Write the IUPAC names for the following binary ionic compounds:
   (a) $Na_2O_{(s)}$
   (b) $SnCl_{4(s)}$
   (c) $ZnI_{2(s)}$
   (d) $SrCl_{2(s)}$
   (e) $AlBr_{3(s)}$
   (f) $PbCl_{4(s)}$
   (g) $Ni_2O_{3(s)}$
   (h) $Ag_2S_{(s)}$
   (i) $FeCl_{2(s)}$
   (j) $KBr_{(s)}$
   (k) $CuI_{2(s)}$
   (l) $NiS_{(s)}$
9. Write the chemical formulas and names (using IUPAC chemical nomenclature) for the binary ionic compounds formed by each of the following pairs of elements:
   (a) strontium and oxygen
   (b) sodium and sulfur
   (c) silver and iodine
   (d) barium and fluorine
   (e) calcium and bromine
   (f) lithium and chlorine
10. Write the chemical formulas for the following ionic compounds:
    (a) mercury(II) sulfide, cinnabar ore
    (b) molybdenum(IV) sulfide, molybdenite ore
    (c) manganese(IV) oxide, pyrolusite ore
    (d) nickel(II) bromide

(e) copper(II) chloride
(f) iron(III) iodide

11. Rename each of the following compounds, using the IUPAC system of nomenclature:
(a) ferrous sulfide
(b) plumbic bromide
(c) stannous chloride

Reflecting

12. Why is it sometimes helpful to understand a system of nomenclature that has been replaced by another, more comprehensive, system?

Compounds with Polyatomic Ions

tertiary compound: a compound composed of three different elements

Many familiar compounds (such as sulfuric acid, $H_2SO_{4(aq)}$, used in car batteries, and sodium phosphate, $Na_3PO_{4(aq)}$, a food additive typically found in processed cheese) are composed of three different elements. Compounds of this type are classified as **tertiary compounds**. (Many compounds containing polyatomic ions consist of more than three different elements, but we will not be dealing with these at this stage.) Tertiary ionic compounds are composed of a metal ion and a polyatomic ion (a covalently bonded group of atoms, possessing a net charge). We treat polyatomic ions much like regular monatomic ions when we write them in formulas or chemical equations.

oxyanion: a polyatomic ion containing oxygen

Polyatomic ions that include oxygen are called **oxyanions**. One example is the nitrate ion, NO_3^-. (Compounds involving the nitrate ion are often used in the processing of foods, particularly cured meats, where they are often used to control colour. Potassium nitrate and sodium nitrate are added to foods to control the growth of microorganisms.) In determining the name of a compound containing an oxyanion, the first part of the name is easy: It is the name of the metal cation. The second part requires more thought: We have to consider the three parts of the ion indicated in **Figure 1**.

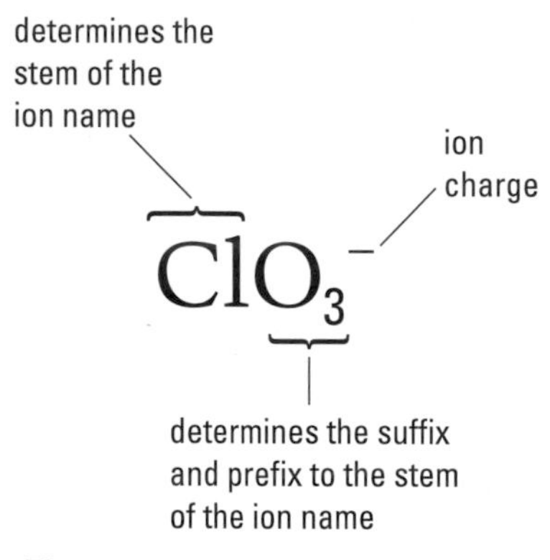

Figure 1
The chlorate anion

There are four polyatomic ions formed from combinations of chlorine and oxygen. Note that all of these oxyanions have the same charge, despite the fact that their formulas are different.

- ClO^- is the hypochlorite ion;
- ClO_2^- is the chlorite ion;
- ClO_3^- is the chlorate ion; and
- ClO_4^- is the perchlorate ion.

Note that in each name the stem is –chlor–. The suffixes and prefixes vary according to the number of oxygen atoms in the ion, as described below.

- The *per–ate* oxyanion has one more oxygen atom than does the *-ate* oxyanion.
- The *-ite* oxyanion has one fewer oxygen atom than does the *-ate* oxyanion.
- The *hypo–ite* oxyanion has one fewer oxygen atom than the *-ite* oxyanion.

Table 3 indicates some polyatomic ions commonly found in compounds. You do not need to memorize each entry in this table. If you become familiar with the *-ate* oxyanions, the most stable and common combination of nonmetal and oxygen, then you need only remember the simple relationship between the names of the ions.

The crisscross method can be applied to predict the formula of ionic compounds involving polyatomic ions. From the ion charges (whether for a single ion or for a polyatomic ion), determine the number of each ion necessary to yield a net charge of zero.

Sample Problem 5

Write the formula of copper(II) nitrate. (Use **Table 3** to find the charge of a nitrate ion.)

Solution

Charge on each Cu ion: 2+
Charge on each NO_3 ion: 1−

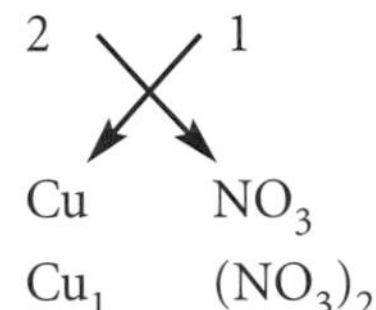

Cu NO_3

Cu_1 $(NO_3)_2$

The formula of copper(II) nitrate is $Cu(NO_3)_2$.

Notice how brackets are used in the formula if there is more than one of the polyatomic ions. Brackets are not required with one polyatomic ion or with simple compounds.

Sample Problem 6

Use IUPAC nomenclature to name $CaCO_3$, which is commonly added to breakfast cereals as a source of calcium.

Solution

$CaCO_3$ is calcium carbonate.

Table 3: IUPAC Names and Formulas of Some Common Polyatomic Ions

Name	Formula
acetate	$C_2H_3O_2^-$
bromate	BrO_3^-
carbonate	CO_3^{2-}
hydrogen carbonate (bicarbonate)	HCO_3^-
hypochlorite	ClO^-
chlorite	ClO_2^-
chlorate	ClO_3^-
perchlorate	ClO_4^-
chromate	CrO_4^{2-}
dichromate	$Cr_2O_7^{2-}$
cyanide	CN^-
hydroxide	OH^-
iodate	IO_3^-
permanganate	MnO_4^-
nitrite	NO_2^-
nitrate	NO_3^-
phosphate	PO_4^{3-}
hydrogen phosphite	HPO_3^{2-}
hydrogen phosphate	HPO_4^{2-}
dihydrogen phosphite	$H_2PO_3^-$
dihydrogen phosphate	$H_2PO_4^-$
sulfite	SO_3^{2-}
sulfate	SO_4^{2-}
hydrogen sulfide (bisulfide)	HS^-
hydrogen sulfite (bisulfite)	HSO_3^-
hydrogen sulfate (bisulfate)	HSO_4^-
thiosulfate	$S_2O_3^-$
ammonium	NH_4^+

Hydrates

Consumers are sometimes surprised to find a tiny white pouch when they open the box containing a newly purchased pair of boots. What is it? What purpose does it serve? The pouch contains a white, crystalline powdered desiccant (a substance that absorbs water) called silica gel ($SiO_{2(s)}$). The pouch keeps the air inside the box dry so mildew and other moulds will not grow. Similarly absorbent compounds are included in such diverse products as powdered foods, talcum powder, and cat litter. Many tertiary ionic compounds form crystals that contain molecules of water within the crystal structure. Such compounds are referred to as **hydrates.** When heat is applied to a hydrate, it will decompose to produce water vapour and an associated ionic compound, indicating that the water is loosely held to the ionic compound. The water molecules are assumed to be electrically neutral in the compound. When this water, called *water of hydration*, is removed, the product is referred to as *anhydrous.* Bluestone, hydrated copper(II) sulfate, is an example of a hydrate. Its formula is written $CuSO_4{\cdot}5\ H_2O_{(s)}$. Notice how the chemical formula includes both the formula of the compound and the formula for water. This is true of the chemical formulas for all hydrated compounds. The formula for bluestone indicates the association of five water molecules with each unit of copper(II) sulfate (**Figure 2**, page 96). The IUPAC names for ionic hydrates indicate the number of water molecules by a Greek prefix (**Table 4**, page 96), so bluestone, or $CuSO_4{\cdot}5H_2O_{(s)}$, is called copper(II) sulfate pentahydrate.

hydrate: a compound that contains water as part of its ionic crystal structure (theoretical definition); a compound that decomposes to an ionic compound and water vapour when heated (empirical definition)

DID YOU KNOW ?

Naming Hydrates

When researching scientific references, you may encounter an alternative naming system for hydrates, in which $CuSO_4{\cdot}5H_2O_{(s)}$ is called copper(II) sulfate-5-water.

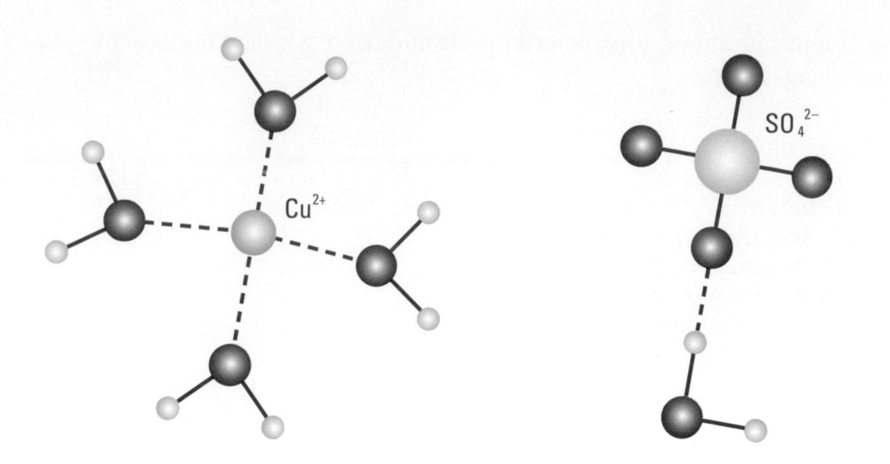

Figure 2
In a model of the compound copper(II) sulfate pentahydrate, the copper(II) ions are surrounded by four water molecules. The fifth water molecule is hydrogen-bonded to the sulfate ion.

Table 4: Prefixes Used When Naming Hydrated Compounds

Number of water molecules in chemical formula	Prefix in chemical nomenclature
1	mono
2	di
3	tri
4	tetra
5	penta
6	hexa
7	hepta
8	octa
9	nona
10	deca

Practice

Understanding Concepts

13. How many elements are there in a tertiary compound?
14. Use each of the following terms correctly in a sentence about the formation of compounds:
    (a) polyatomic ion
    (b) oxyanion
    (c) hydrate
15. Write the IUPAC name for each of the following ionic compounds:
    (a) $NaNO_{3(s)}$ (found in tobacco)
    (b) $NaNO_{2(s)}$ (a meat preservative)
    (c) $Cu(NO_3)_{2(s)}$ (forms a blue solution in water)
    (d) $CuNO_{3(s)}$ (forms a green solution in water)
    (e) $Al_2(SO_3)_{3(s)}$ (a food additive in pickles)
    (f) $Ca(OH)_{2(s)}$ (firming agent in fruit products)
    (g) $PbCO_{3(s)}$ (cerussite, a mineral popular with collectors)
    (h) $Sn_3(PO_4)_{2(s)}$ (use to fix paints to silk)
    (i) $Fe_2(SO_4)_{3(s)}$ (a mineral found on Mars)
16. Write the chemical formula for each of the following ionic compounds:
    (a) calcium carbonate (active ingedient in antacids)
    (b) sodium bicarbonate (a foaming agent added to foods)
    (c) sodium hypochlorite (a component of bleach)
    (d) calcium sulfate (plaster of Paris)
    (e) ammonium nitrate (used in fertilizers)
    (f) ammonium phosphate (a leavening agent added to foods)
    (g) copper(II) sulfate (used as a fungicide)
    (h) sodium hydroxide (a strong base used as a washing agent)
    (i) potassium permanganate (a traditional antiseptic)
17. Use IUPAC chemical nomenclature to name each of the following ionic compounds containing polyatomic ions:
    (a) $LiClO_{3(s)}$
    (b) $BaSO_{4(s)}$
    (c) $Hg_2CO_{3(s)}$
    (d) $Mg(NO_3)_{2(s)}$
    (e) $Fe(BrO_3)_{3(s)}$
    (f) $Na_3PO_{4(s)}$
    (g) $NH_4IO_{3(s)}$
    (h) $AuC_2H_3O_{2(s)}$
    (i) $Zn_3(PO_4)_{2(s)}$
    (n) $Ag_2SO_{4(s)}$
    (o) $Hg(BrO_3)_{2(s)}$
    (p) $Fe_2(CO_3)_{3(s)}$
    (q) $NH_4ClO_{(s)}$
    (r) $Au(NO_3)_{3(s)}$
    (s) $Mg(BrO_3)_{2(s)}$
    (t) $NaIO_{(s)}$
    (u) $Zn(ClO_2)_{2(s)}$
    (v) $SnCO_{3(s)}$

(j) $Sb(ClO_3)_{5(s)}$
(k) $MnSO_{3(s)}$
(l) $KBrO_{(s)}$
(m) $AlPO_{5(s)}$
(w) $SrSO_{3(s)}$
(x) $NiPO_{4(s)}$
(y) $Cu(C_2H_3O_2)_{2(s)}$
(z) $Ba_3(PO_5)_{2(s)}$

18. Write the IUPAC name and chemical formula for each of the following ionic compounds containing polyatomic ions:
    (a) cuprous hypophosphite
    (b) stannic chlorite
    (c) ferrous bromate
    (d) ferric chlorite
    (e) plumbic sulfate
19. Name each of the following hydrated ionic compounds:
    (a) bluestone, $CuSO_4{\cdot}5\ H_2O_{(s)}$
    (b) $Na_2SO_4{\cdot}10\ H_2O_{(s)}$
    (c) $MgSO_4{\cdot}7\ H_2O_{(s)}$
20. Write the chemical formulas for the following ionic hydrates:
    (a) iron(III) oxide trihydrate (rust)
    (b) aluminum chloride hexahydrate (component of antiperspirant)
    (c) sodium thiosulfate pentahydrate (photographic "hypo")
    (d) cadmium(II) nitrate tetrahydrate (photographic emulsion)
    (e) lithium chloride tetrahydrate (in fireworks)
    (f) calcium chloride dihydrate (deicer)
21. How would you convert a hydrate to an anhydrous compound?

Naming Molecular Compounds

So far, in this section, we have been looking at the names and formulas of ionic compounds. We now turn to molecular compounds, for which a different system is used.

If a binary compound is formed from two nonmetals, it is classified as a molecular compound. Even though there are fewer nonmetals than metals, there is a wide variety of compounds formed from the combination of two nonmetals, because two nonmetals may combine to form more than one compound. For example, N_2O, NO, and NO_2 are three of the several binary compounds that can be formed from nitrogen and oxygen. Each compound has different properties, and so different uses.

In naming compounds formed from two nonmetals, a Greek prefix is attached to the name of each element in the binary compound indicating the number of atoms of that element in the molecule. The common prefixes and their numerical equivalences are shown in **Table 5**. If there is only one of the first type of atom, we leave out the prefix "mono."

Suppose you are asked to write the IUPAC name for the chemical compound represented by the formula N_2O. Looking at the first element, you can see that the subscript after the nitrogen is two, so the prefix for nitrogen is "di." Looking at the second element, you can see that there is only one oxygen atom, so the prefix for oxygen will be "mono." Therefore, the formula's IUPAC name is dinitrogen monoxide.

Table 5: Prefixes Used When Naming Binary Covalent Compounds

Subscript in chemical formula	Prefix in chemical nomenclature
1	mono
2	di
3	tri
4	tetra
5	penta
6	hexa
7	hepta
8	octa
9	nona
10	deca

Sample Problem 7

What is the IUPAC name for the chemical compound CF_4?

Solution

C: carbon (not monocarbon)
F: tetrafluoride
The IUPAC name for CF_4 is carbon tetrafluoride.

Once again, hydrogen is an exception to this rule: The common practice is not to use the prefix system for hydrogen. For example, we do not call H_2S dihydrogen sulfide, but simply hydrogen sulfide.

Practice

Understanding Concepts

22. Write the chemical formula for each of the following molecules:
    (a) nitrogen
    (b) carbon dioxide
    (c) carbon monoxide
    (d) nitrogen dioxide
    (e) nitrogen monoxide
    (f) dinitrogen oxide
    (g) dinitrogen tetroxide
    (h) sulfur dioxide
    (i) diiodine pentoxide
    (j) silicon tetrafluoride
    (k) boron trifluoride
    (l) phosphorus triiodide
    (m) diphosphorus pentoxide
    (n) sulfur tetrafluoride
    (o) phosphorus pentachloride
    (p) disulfur dichloride
    (q) carbon tetrachloride
    (r) sulfur trioxide
    (s) sulfur hexafluoride
    (t) chlorine dioxide
    (u) dinitrogen pentoxide
    (v) phosphorus trichloride
    (w) silicon tetrachloride
    (x) carbon disulfide
    (y) phosphorus pentabromide
    (z) carbon tetrafluoride
23. Name the compound indicated by each of the following formulas:
    (a) $SF_{6(g)}$
    (b) $N_2O_{3(g)}$
    (c) $NO_{2(g)}$
    (d) $PCl_{3(l)}$
    (e) $PCl_{5(s)}$
    (f) $IF_{7(g)}$
    (g) $BF_{3(g)}$
    (h) $P_2S_{5(s)}$
    (i) $P_2O_{5(s)}$

Naming Acids

You are already familiar with many acids and bases. You may have taken acetylsalicylic acid (also known as ASA, or Aspirin) to treat a headache, poured a little acetic acid solution (vinegar) on your fish and chips, or used an ammonia solution to clean the smears from a mirror. You also know one of the tests for acids and bases: acids turn blue litmus red, and bases turn red litmus blue. But you might have been somewhat mystified about the naming of these substances.

Acids are well-known, long-established chemicals. They were originally named decades or even centuries ago, and the use of traditional names persists. The International Union of Pure and Applied Chemistry suggests that names of acids should be derived from the IUPAC name for the compound. According to this rule, sulfuric acid would be called aqueous hydrogen sulfate. However, it is difficult to get people to change to new names when the old, familiar names are so widely used.

Let us take, as a simple example, $HCl_{(g)}$. It is a binary compound formed from a combination of hydrogen and a halogen. When a gas, it is named hydrogen chloride. When it is dissolved in water, the resulting aqueous solution displays a set of specific properties called acidic, and the name of the substance changes. Binary acids are classically named by using the prefix *hydro-* with the stem of the name of the most electronegative element and the ending *-ic*. The name "hydrogen" does not appear. Instead, the word "acid" is added after the *hydro*-stem-*ic* combination, as indicated in **Table 6**. Consequently, the classical name for $HCl_{(aq)}$ is hydrochloric acid. Its IUPAC name is aqueous hydrogen chloride. Similarly, $HBr_{(aq)}$ is hydrobromic acid or aqueous hydrogen bromide.

Note that the difference between the solution and the pure binary compound is indicated by the presence or absence of the subscript (aq) in the formula.

Table 6: Naming Systems for Binary Acids

Formula	Classical name	IUPAC name
$HF_{(aq)}$	hydrofluoric acid	aqueous hydrogen fluoride
$HCl_{(aq)}$	hydrochloric acid	aqueous hydrogen chloride
$HBr_{(aq)}$	hydrobromic acid	aqueous hydrogen bromide
$HI_{(aq)}$	hydroiodic acid	aqueous hydrogen iodide
$H_2S_{(aq)}$	hydrosulfuric acid	aqueous hydrogen sulfide

A second group of acids is named (by the classical system) in the same way as binary acids. In this group, the IUPAC names for the polyatomic ions end in *-ide* (e.g., the cyanide ion, CN^-). Looking at **Table 6**, you can see that the classical name for the acidic solution $HCN_{(aq)}$ will be hydrocyanic acid.

A third group of acids is formed from various combinations of oxyanions (negative polyatomic ions consisting of a nonmetal plus oxygen) with hydrogen. Perhaps the best-known example is $H_2SO_{4(aq)}$, or sulfuric acid, which is one of the most widely produced industrial chemicals in the world. It is used to make pharmaceuticals, detergents, and dyes, and is a component of car batteries. Phosphoric acid, $H_3PO_{4(aq)}$, is another example of an acid formed from an oxyanion and hydrogen. Phosphoric acid is an ingredient in soft drinks, is a reagent in the manufacture of fertilizers, and can be used as a rust remover. These acids are classified as **oxyacids** because they incorporate oxyanions. **Table 7** compares the classical name to the IUPAC name for a series of acids derived from a chlorine-based oxyanion.

oxyacid: an acid containing oxygen, hydrogen, and a third element

Table 7: Classical and IUPAC Nomenclature System for Chlorine-Based Oxyacids

Classical name	IUPAC name	Formula
perchloric acid	aqueous hydrogen perchlorate	$HClO_{4(aq)}$
chloric acid	aqueous hydrogen chlorate	$HClO_{3(aq)}$
chlorous acid	aqueous hydrogen chlorite	$HClO_{2(aq)}$
hypochlorous acid	aqueous hydrogen hypochlorite	$HClO_{(aq)}$
hydrochloric acid	aqueous hydrogen chloride	$HCl_{(aq)}$

The classical names for oxyacids can be derived according to the simple rules in **Table 8**.

Table 8: Rules for Naming Acids and Oxyanions

Name of oxyanion	Example	Formula	Classical name of acid	Example
per–ate	persulfate	SO_5^{2-}	*per–ic* acid	persulfuric acid
-ate	sulfate	SO_4^{2-}	*-ic* acid	sulfuric acid
-ite	sulfite	SO_3^{2-}	*-ous* acid	sulfurous acid
hypo–ite	hyposulfite	SO_2^{2-}	*hypo–ous* acid	hyposulfurous acid

When naming oxyacids, we omit the word hydrogen and add the word "acid." For example, to name the acidic solution with the formula $HNO_{2(aq)}$, we would first consider the IUPAC name: hydrogen nitrite (from **Table 3**, or from the Table of Polyatomic Ions in Appendix C). "Nitrite" changes to "nitrous," we drop the "hydrogen" from the front of the name, and add "acid" to the end. Thus, $HNO_{2(aq)}$ is called nitrous acid.

If you were asked to write the formula for an acid, you would first have to figure out the names of the ions involved, then their symbols or formulas, then their ratio. For example, what is the formula for phosphoric acid? The *-ic* ending indicates the presence of the *-ate* oxyanion of phosphorus: phosphate. The phosphate oxyanion is PO_4^{3-} with a charge of 3–. The cation in oxyacids is always hydrogen, which has a charge of 1+. To find the ratio of the ions, use the crisscross method:

1 3

H (PO_4)

This gives the subscripts for each ion.

H_3 $(PO_4)_1$

Divide each subscript by the highest common factor—in this case, 1:

H_3 $(PO_4)_1$

The hydrogen ions and the phosphate oxyanions will combine in a ratio of 3:1. Therefore, the correct formula is $H_3PO_{4(aq)}$.

Sample Problem 8

What is the formula for the oxyacid sulfurous acid?

Solution

sulfurous indicates sulfite ion: SO_3^{2-}

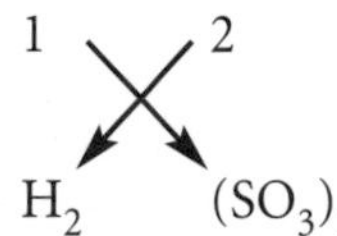

The formula for sulfurous acid is $H_2SO_{3(aq)}$.

Sample Problem 9

What is the formula for the oxyacid hypochlorous acid?

Solution

hypochlorous indicates one fewer oxygen atom than a chlorite ion: ClO^-

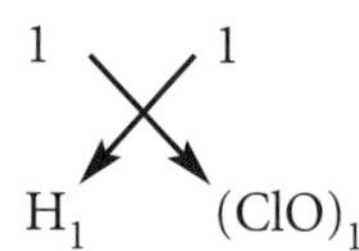

The formula for hypochlorous acid is $HClO_{(aq)}$.

Naming Bases

Chemists have discovered that all aqueous solutions of ionic hydroxides are bases. (You will learn more about bases in Chapter 8.) Other solutions have also been classified as bases, but for the time being we will restrict our exploration of bases to aqueous ionic hydroxides such as $NaOH_{(aq)}$ and $Ba(OH)_{2(aq)}$. Notice that these bases are formed of a combination of a metal cation with one or more hydroxide anions. The name of the base is the name of the ionic hydroxide: in this case, aqueous sodium hydroxide and aqueous barium hydroxide.

Practice

Understanding Concepts

24. Write the chemical formulas for the following acids:
    (a) aqueous hydrogen chloride
    (b) hydrochloric acid
    (c) aqueous hydrogen sulfate
    (d) sulfuric acid
    (e) aqueous hydrogen acetate
    (f) acetic acid
    (g) aqueous hydrogen nitrite
    (h) nitric acid
    (i) hydrobromic acid
    (j) hyposulfurous acid
    (k) hydroiodic acid
    (l) aqueous hydrogen perchlorate
25. Name each of the following compounds, using both the classical and the IUPAC nomenclature systems:
    (a) $H_2SO_{3(aq)}$
    (b) $H_3PO_{4(aq)}$
    (c) $HCN_{(aq)}$
    (d) $H_2CO_{3(aq)}$
    (e) $H_2S_{(aq)}$
    (f) $HCl_{(aq)}$
    (g) $HCN_{(aq)}$
    (h) $H_2SO_{4(aq)}$
    (i) $H_3PO_{4(aq)}$
26. Write the names of the following bases:
    (a) $KOH_{(aq)}$
    (b) $CaOH_{2(aq)}$
27. Write the formulas of the following bases:
    (a) aqueous magnesium hydroxide
    (b) aqueous sodium hydroxide
    (c) aqueous aluminum hydroxide

Section 2.5 Questions

Understanding Concepts

1. The atmosphere of Saturn contains traces of ammonia, while hydrogen cyanide is a component of the atmosphere of one of Saturn's many moons.
    (a) Write formulas for each of these compounds.
    (b) Classify each substance as ionic or covalent.
    (c) Which compound contains both ionic and covalent bonds? How could you verify this prediction?
    (d) If hydrogen cyanide is added to water, what substance is formed? If the substance were tested with litmus paper, what would you expect the results to be?
2. Caustic potash (KOH) is a component of oven cleaners. Muriatic acid (HCl) is used to clean mortar from bricks.
    (a) Classify each of these compounds. Provide your reasoning.
    (b) Create electron dot diagrams and structural formulas for each compound.
    (c) Use the IUPAC system to name each compound.
    (d) What can you predict about the reactivity of these substances? How would this affect the way you handle the substances? Explain your answer.
3. (a) Write the IUPAC names and formulas for as many compounds as you can, using only the following elements: K, C, H, F, Mg, O, Cl, and Na.
    (b) State the common name, if there is one, for each compound.
    (c) Classify each compound as polar or nonpolar. In each case, use Lewis structures and periodic trends to justify your classification.
    (d) Classify the compounds as binary, tertiary, acidic, or basic.
    (e) Indicate which compounds, if any, contain both ionic and covalent bonding.

Chapter 2 Summary

Key Expectations

Throughout this chapter, you have had the opportunity to do the following:

- Identify chemical substances and reactions in everyday use or of environmental significance. (all sections)
- Explain how different elements combine to form covalent and ionic bonds, using the octet rule. (2.2)
- Demonstrate an understanding of the formation of ionic and covalent bonds and explain the properties of the products. (2.2, 2.3)
- Predict the ionic character or polarity of a given bond using electronegativity values, and represent the formation of ionic and covalent bonds using diagrams. (2.2, 2.3, 2.4)
- Draw Lewis structures, construct molecular models, and give the structural formulas for compounds containing single and multiple bonds. (2.3)
- Relate common names of substances to their systematic (IUPAC) names. (all sections)
- Write, using IUPAC or traditional (classical) systems, the formulas of binary and tertiary compounds, including those containing elements with multiple valences, and recognize the formulas in various contexts. (2.5)
- Use appropriate scientific vocabulary to communicate ideas related to chemical reactions. (all sections)

Key Terms

binary compound
bonding capacity
chemical bond
chemical nomenclature
coordinate covalent bond
covalent bond
crystal lattice
diatomic molecule
dipole–dipole force
electrical conductivity
electrolyte
electron dot diagram
formula unit
hydrate
hydrogen bond
intermolecular force
intramolecular force
ionic bond
ionic compound
Lewis structure
Lewis symbol
London dispersion force
lone pair
molecular compound
multivalent
octet rule
oxyacid
oxyanion
polar covalent bond
polar molecule
polyatomic ion
polyatomic molecule
stable octet
structural formula
tertiary compound
valence
van der Waals forces

Make a Summary

Summarize the concepts presented in this chapter by constructing and completing a table like **Table 1**. Try to include as many examples of compounds as possible for each type of intramolecular bond.

Table 1: Summarizing Bonds and Forces

Compound	Properties	Lewis structure	Intramolecular bond type	Polarity	Intermolecular forces

Reflect on your Learning

Revisit your answers to the Reflect on Your Learning questions at the beginning of this chapter.

- How has your thinking changed?
- What new questions do you have?

Chapter 2 Review

Understanding Concepts

1. Describe the relationship between the positions of two reacting elements in the periodic table and the type of compound and bond they form.
2. How does an ionic bond differ from a covalent bond?
3. (a) Briefly summarize and explain the properties of ionic and molecular compounds.
   (b) Explain why electrical conductivity is generally a suitable test for ionic compounds.
4. (a) What is intramolecular bonding? Describe the main types with examples.
   (b) What is the relationship between the type of bond and the properties of a substance?
5. For each chemical bond or force listed below, indicate which types of entities are involved.
   (a) covalent bond
   (b) dipole–dipole force
   (c) hydrogen bonds
   (d) ionic bond
6. How does the information conveyed by the chemical formulas for ionic compounds differ from the information conveyed by the chemical formulas for molecular compounds? Include examples with your explanation.
7. Explain why halogens tend to form diatomic molecules.
8. Sketch the bonding continuum. Indicate where Na—Cl and Cl—Cl bonds would fit along the continuum. Provide a theoretical explanation of your answer.
9. Draw Lewis symbols (electron dot diagrams) to represent a single atom of each of the following elements:
   (a) Ca
   (b) Al
   (c) K
   (d) N
   (e) S
   (f) Br
   (g) Ne
10. For each of the following substances, predict whether the bonds present are ionic, covalent, or polar covalent:
    (a) $I_{2\,(s)}$
    (b) $SO_{2(g)}$
    (c) $OCl_{2(g)}$
    (d) $Fe_2O_{3(s)}$
    (e) $KBr_{(s)}$
    (f) $SrO_{(s)}$
11. The most common oxides of Period 3 elements are as follows: $Na_2O_{(s)}$, $MgO_{(s)}$, $Al_2O_{3(s)}$, $SiO_{2(s)}$, $P_2O_{5(s)}$, $SO_{2(g)}$, and $Cl_2O_{(g)}$.
    (a) Classify the oxides as either ionic or molecular.
    (b) Use electron dot diagrams or Lewis structures to show the formation of each compound.
    (c) What differences would you expect to observe in the properties of each of the compounds?
    (d) How is the difference in electronegativity of the constituent elements related to the properties of the compound?
12. Create a table like **Table 1** and complete it for the following compounds: $HF_{(g)}$, $BCl_{3(g)}$, $SiH_{4(g)}$, $CCl_{4(l)}$, $NCl_{3(g)}$, $H_2O_{2(l)}$, $CO_{2(g)}$, $HCN_{(g)}$

Table 1: Structures of Covalent Compounds

Compound	Lewis structure	Structural formula	Types of bonds

13. Use electronegativity values to predict the polarity of the bond formed by each of the following pairs of elements. In each case, represent the formation of the bonds using diagrams.
    (a) carbon and chlorine
    (b) calcium and fluorine
    (c) aluminum and chlorine
    (d) silicon and oxygen
    (e) carbon and oxygen
14. Use Lewis structures and structural formulas to represent the following compounds. Indicate which compounds, if any, involve coordinate covalent bonds.
    (a) $NH_4Cl_{(s)}$
    (b) $BF_{3(g)}$
    (c) $NH_{3(g)}$
    (d) $H_2O_{(l)}$
    (e) $NH_3BF_{3(g)}$
15. Write the chemical formula (plus state) for each of the following substances:
    (a) sodium hydrogen sulfate (toilet bowl cleaner)
    (b) sodium hydroxide (lye, drain cleaner)
    (c) carbon dioxide (dry ice, soda pop)
    (d) acetic acid (vinegar)
    (e) sodium thiosulfate pentahydrate (photographic "hypo")
    (f) sodium hypochlorite (laundry bleach)
    (g) octasulfur (vulcanizing rubber)
    (h) potassium nitrate (meat preservative)
    (i) phosphoric acid (rust remover)
    (j) iodine (disinfectant)
    (k) aluminum oxide (alumina, aluminum ore)
    (l) potassium hydroxide (caustic potash)
    (m) aqueous hydrogen carbonate (carbonated beverages)
16. Write the chemical formula for each of the following substances:
    (a) magnesium bromide
    (b) carbon disulfide
    (c) mercury(II) nitrite

(d) hydrochloric acid
(e) lithium hydroxide
(f) silver carbonate
(g) aluminum perchlorate
(h) copper(II) sulfate
(i) sulfur trioxide
(j) nickel(III) phosphate
(k) magnesium oxide
(l) dinitrogen monoxide
(m) iron(II) persulfate
(n) carbonic acid
(o) calcium hydroxide
(p) zinc hypochlorite
(q) lead(IV) perchlorate
(r) phosphorous pentabromide
(s) arsenic(V) chloride
(t) bismuth(III) nitrate
(u) sodium hypochlorite
(v) oxygen dichloride
(w) tin(II) bromide
(x) sulfuric acid
(y) potassium hydroxide
(z) barium carbonate

17. Write the chemical formula for each of the following substances:
(a) ammonium dihydrogen phosphite
(b) lithium hydrogen sulfite
(c) potassium hydrogen sulfate
(d) barium chloride trihydrate
(e) sodium dihydrogen phosphate
(f) sodium hydrogen carbonate

18. Give the names of the following substances, using IUPAC chemical nomenclature:
(a) $CaCO_{3(s)}$ (marble, limestone, chalk)
(b) $P_2O_{5(s)}$ (fertilizer)
(c) $MgSO_4{\cdot}7\ H_2O_{(s)}$ (Epsom salts)
(d) $N_2O_{(g)}$ (laughing gas, an anesthetic)
(e) $Na_2SiO_{3(s)}$ (water glass)
(f) $Ca(HCO_3)_{2(s)}$ (hard-water chemical)
(g) $HCl_{(aq)}$ (muriatic acid, gastric fluid)
(h) $CuSO_4{\cdot}5\ H_2O_{(s)}$ (copperplating, bluestone)
(i) $H_2SO_{4(aq)}$ (acid in car battery)
(j) $Ca(OH)_{2(s)}$ (slaked lime)
(k) $SO_{3(g)}$ (a cause of acid rain)
(l) $NaF_{(s)}$ (toothpaste additive)

19. Give the IUPAC names of the following substances:
(a) $NaCl_{(s)}$
(b) $P_2O_{3(s)}$
(c) $HNO_{3(aq)}$
(d) $Pb(C_2H_3O_2)_{2(s)}$
(e) $NH_4OCl_{(s)}$
(f) $Sn(BrO_3)_{4(s)}$
(g) $Sb_2O_{3(s)}$
(h) $Zn(IO_3)_{2(s)}$
(i) $Fe(NO_4)_{2(s)}$
(j) $Ca(OH)_{2(s)}$
(k) $KI_{(s)}$
(l) $SF_{2(g)}$
(m) $HBr_{(aq)}$
(n) $CuCO_{3(s)}$
(o) $Al_2(SO_3)_{3(s)}$
(p) $NH_4OH_{(l)}$
(q) $Ba(C_2H_3O_2)_{2(s)}$
(r) $ICl_{(s)}$
(s) $AuCl_{3(s)}$
(t) $MgS_{(s)}$
(u) $N_2F_{2(g)}$
(v) $NiSO_{4(s)}$
(w) $H_2S_{(aq)}$
(x) $AgBrO_{3(s)}$
(y) $LiClO_{4(s)}$

20. Give the names of the following substances, using IUPAC chemical nomenclature:
(a) $CaHPO_{4(s)}$
(b) $CuSO_4{\cdot}7\ H_2O_{(s)}$
(c) $Na_2HPO_{4(s)}$
(d) $LiHCO_{3(s)}$
(e) $KHSO_{4(s)}$

21. Write the name and formula (with state at SATP) for the compound formed by each of the following pairs of elements. Where a molecular compound is formed, give the structural formula. For ionic compounds, assume the most common ion charges for the ions.
(a) potassium and bromine
(b) silver and iodine
(c) lead and oxygen
(d) zinc and sulfur
(e) copper and oxygen
(f) lithium and nitrogen

Applying Inquiry Skills

22. A forensic chemist was given samples of four unknown solutions, the identity of which could affect the outcome of a court case involving an electrocution. The chemist had reason to believe that the four substances were $KCl_{(aq)}$, $C_2H_5OH_{(aq)}$, $HCl_{(aq)}$, and $Ba(OH)_{2(aq)}$. The investigation was designed to identify the chemicals. Complete the Analysis and Synthesis sections of the report.

Question
What is the identity of each of the four substances, labelled 1 to 4?

Experimental Design
Each of the samples was dissolved in water and tested with a conductivity apparatus and litmus paper. Taste tests were not carried out.

Evidence

Table 2: Evidence for Identifying Unknown Solutions

Solution	Conductivity	Litmus
water	none	no change
1	high	no change
2	high	blue to red
3	none	no change
4	high	red to blue

Analysis

(a) Answer the Question.

Synthesis

(b) Why was the water used to prepare the solutions also tested?

(c) Which of the solutions, 1, 2, 3, or 4, could have been involved in somebody getting electrocuted? Use electron dot diagrams and the concept of electronegativity to explain your conclusion.

Making Connections

23. In historical dramas you may have seen someone brought back to consciousness by having a bottle of "smelling salts" waved under their noses. Smelling salts were made by mixing perfume and the active ingredient, which was once called sal volatile.
    (a) What is the IUPAC name and chemical formula for sal volatile?
    (b) Based on the nature of the bonds in this substance, what properties would you predict that it would demonstrate?
    (c) Does your prediction fit with the evidence that smelling salts have a strong, sharp, ammonia smell? Explain.
    (d) Using the Internet, find out how smelling salts used to be administered, how they worked, and comment on the safety concerns that this use would raise today.

    Follow the links for Nelson Chemistry 11, Chapter 2 Review.

    GO TO www.science.nelson.com

24. Boron nitride (BN) is a compound that involves a large number of covalent bonds. Use the Internet to research the structure of boron nitride. Use your findings to make predictions about the physical properties of boron nitride. How do the properties of boron nitride and diamond compare? Could boron nitride serve as a substitute for diamond in some practical way? Provide a theoretical argument that supports your answer.

    Follow the links for Nelson Chemistry 11, Chapter 2 Review.

    GO TO www.science.nelson.com

25. Glycerol is a sweet-tasting, colourless, viscous liquid that is considered (by many national food safety organizations) to be quite safe to eat. The food industry adds glycerol (sometimes called glycerine) to a wide range of processed cheese and meat, baked goods, and candies.
    (a) Look for glycerol on ingredients labels at home or on the shelves of stores. Make a list of foods in which glycerol is used.
    (b) Research the physical and chemical properties of glycerol.
    (c) Research the chemical formula and molecular shape of glycerol. Use this information, as well as your knowledge of intra- and intermolecular bonding, to explain at least three of the properties of glycerol.
    (d) Write a short report, outlining why glycerol is so useful to the food industry.

    Follow the links for Nelson Chemistry 11, Chapter 2 Review.

    GO TO www.science.nelson.com

Exploring

26. London dispersion forces are named for Fritz London, who explained the origin of intermolecular forces between nonpolar molecules in 1928. Research London's work and write a brief report on his major findings.

    Follow the links for Nelson Chemistry 11, Chapter 2 Review.

    GO TO www.science.nelson.com

Chapter

3

Chemical Reactions

In this chapter, you will be able to

- show that you understand the relationship between various types of chemical reactions and the nature of the reactants;
- use both the periodic table and experimentation to predict the reactivity of a series of elements and compare the reactivity of metals and alloys;
- predict the products of various types of reactions, write chemical equations to represent the reactions, and test your predictions experimentally, demonstrating safe handling of chemicals;
- identify chemical compounds and reactions in everyday use or of environmental significance.

Plastics, synthetic fabrics, pharmaceuticals, chemical fertilizers, and pesticides, all of which are very much a part of modern society, are each produced as a result of our knowledge and understanding of chemical reactions. These products illustrate the value of chemicals and the reactions that produce them. However, chemical reactions also have the potential to produce hazardous substances. On July 9, 1997, 400 t (4.00×10^5 kg) of plastic caught fire at a plastics manufacturing site in Hamilton, Ontario (**Figure 1**). The fire raged for more than three days, engulfing parts of Hamilton in thick black smoke. As a result of the fire, many people were exposed to hydrochloric acid, benzene, and other such potentially hazardous compounds. The Hamilton fire was an example of chemical reactions allowed to proceed in an uncontrolled fashion.

The fire was also an example of the release of energy that accompanies many chemical reactions. Energy, whether produced or absorbed, is an important aspect of chemical reactions. When fossil fuels such as oil, natural gas, gasoline, and diesel fuel are burned, energy is released in the form of heat. Combustion converts chemical energy to thermal energy, which allows us to drive cars, fly airplanes, and heat our homes. Not all chemical reactions produce energy. Some, including photosynthesis, require energy. Photosynthesis is perhaps the most important chemical reaction on Earth. In green plants, energy from the Sun is used to convert carbon dioxide and water into starch and oxygen. This process converts light energy to chemical energy and is largely responsible for the maintenance of life on Earth.

The reactions in the Hamilton fire were obviously bad for people and the environment. We work to avoid such catastrophes. The consequences of other reactions require a more balanced approach. Environmental problems such as acid rain and the greenhouse effect have been linked to the products of everyday combustion reactions, the ones that keep us warm and mobile. Although resolving such issues involves more than an understanding of science concepts, we will be in a better position to decide how to act if we understand chemical reactions.

Reflect on your Learning

1. How do you know whether or not a chemical reaction has taken place? What clues do you use to help you decide?
2. What types of chemical reactions are there? Describe as many categories as you can.
3. If you put solid aluminum in a solution that contains dissolved iron(III) nitrate, the following reaction will take place:

   $Al_{(s)} + Fe(NO_3)_{3(aq)} \rightarrow Fe_{(s)} + Al(NO_3)_{3(aq)}$

   However, if you put solid lead in an iron(III) nitrate solution, the following reaction does *not* occur:

   $3\ Pb_{(s)} + 2\ Fe(NO_3)_{3(aq)} \rightarrow 2\ Fe_{(s)} + 3\ Pb(NO_3)_{2(aq)}$

   Why do some reactions occur, but others do not?

Observing Chemical Change

All around us, and even inside us, substances are combining to produce new substances. Sometimes the products are useful, and other times not so useful. Sometimes they may even be very hazardous. For instance, common household bleach can react with acidic toilet bowl cleaners to produce poisonous chlorine gas.

In some reactions, the change is almost undetectable. Other reactions produce a visible product, a colour change, or a temperature change. In this activity, you will observe what happens when an ordinary iron nail is placed in a solution of copper(II) sulfate.

Materials: eye protection, apron, iron nail, copper(II) sulfate solution, test tube, test-tube rack

- Place an iron nail in the test tube.
- Add 5 to 10 mL of copper(II) sulfate solution to the test tube.
- Place the test tube in a test-tube rack. Examine the test tube periodically over the next 20 min. Note any changes.

(a) Which of your observations provide evidence of a chemical change?
(b) What products may have formed?
(c) Suggest additional steps that you could take to identify the products.

Copper(II) sulfate is harmful if swallowed.

Wash hands thoroughly after this activity.

Figure 1
In 1997, a chemical fire burned out of control for several days, exposing Hamilton residents to noxious fumes.

3.1 Recognizing and Understanding Chemical Changes

chemical change: any change in which a new substance is formed

With the ever-growing number of motor vehicles on our roads, gasoline is in greater demand today than ever before. When gasoline is burned at high temperature in the engine of a motor vehicle, a **chemical change** occurs that releases large volumes of gases and a considerable amount of heat, which together drive the movement of the engine's pistons. The exhaust gases also cause considerable environmental problems (**Figure 1**).

Figure 1
Photochemical smog—a form of air pollution caused by the action of sunlight on other pollutants—is a major problem in large cities where there are many vehicles.

How do you know whether the change you are observing is a chemical change or a physical change? The heat and gases produced by burning gasoline are only two of the indicators of a chemical change. In general, there are several clues that can help us classify change as either chemical or physical (**Table 1**). However, we cannot always rely upon simple observation to distinguish between a physical and a chemical change, as both physical and chemical changes often involve changes in state and energy. For example, when gasoline evaporates (a physical change) it absorbs heat from its surroundings. As you can see, both the physical change and the chemical change involve energy flows.

Table 1: Evidence of Chemical Reactions

Evidence	Description
change in colour	Final product(s) may have a different colour than the colours of the starting material(s).
change in odour	Final material(s) may have a different odour than the odours of the starting material(s).
formation of gas/solid	Final material(s) may include a substance in a state that differs from the starting material(s); commonly, a gas or a solid (precipitate) is produced.
release/absorption of heat	Energy (light, electricity, sound, or, most commonly, heat) is absorbed in an endothermic reaction and released in an exothermic reaction.

Figure 2
On a winter morning, we can see that some kind of gas is produced by car engines. Is this sufficient evidence of a chemical change?

On a cold winter day, the hot gases produced by combustion engines become easily visible as the water vapour in the exhaust condenses as it cools (**Figure 2**). However, the products of a chemical reaction are not always easily detected. In some cases, we may have to perform a chemical analysis (perform diagnostic tests) to confirm that a new substance has been produced (**Figure 3**).

So far, we have discussed only qualitative observations of chemical change. Can we use quantitative measurements to obtain more evidence? Quantitative evidence could include measurements of mass, temperature, and volume. Unfortunately, such measurements are not always diagnostic. Consider mass and the combustion of gasoline. At first glance it may appear that mass has disappeared when gasoline burns. However, if we are careful to collect all the products and measure their mass we would find that the total mass of the products is the same as the total mass of the reactants. In chemical changes, the total mass of matter present before the change is always the same as the total mass present after the change, no matter how different the new substances appear. This evidence, compiled from many careful measurements, supports the atomic theory of matter. If we think of a chemical change as a rearrangement of particles at the molecular level, then it is simple to argue that the mass must be constant. The individual particles do not change, except in the ways they are associated with each other. Any model that we use to represent chemical reactions must follow the law of conservation of mass.

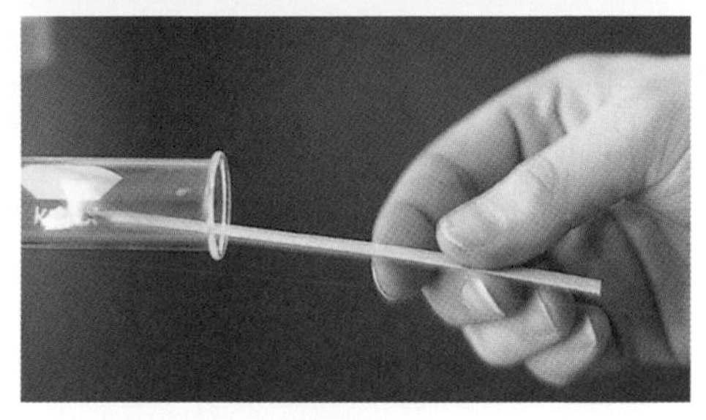

Figure 3
In the presence of oxygen, a glowing splint will ignite and burn brightly. This is a qualitative diagnostic test for oxygen.

A Mechanism for Chemical Change

You might recall the **kinetic molecular theory** from previous grades, when you used it to explain changes of state and physical change. The central idea of this theory is that the smallest particles of a substance are in continuous motion. As they move about, they collide with each other and with objects in their path. Very tiny objects, such as specks of carbon in car exhaust, are buffeted by these particles, and move about erratically (**Figure 4**). A temperature increase indicates that these particles are moving more quickly. We can make use of the kinetic molecular theory as we search for a better understanding of chemical change.

You have applied the kinetic molecular theory to examples of physical change. For example, it can explain why the smell of spilled gasoline is so strong near a filling station. When a substance absorbs heat from its surroundings, the particles in the substance (in this example, the molecules in the spilled gasoline) gain energy and move more rapidly. Some of the faster-moving molecules in the spill will leave the surface of the spilled liquid—evaporate—so they can be detected by our noses.

Octane, $C_8H_{18(l)}$, one of the components of gasoline, is a highly volatile liquid at room temperature. Liquid octane does not react with oxygen. However, at higher temperatures, gaseous octane will react with oxygen to form gaseous carbon dioxide and water vapour. Can we use kinetic molecular theory to explain this chemical change? We can, in a combination of kinetic molecular theory and atomic theory known as the **collision–reaction theory**.

kinetic molecular theory: a theory stating that all matter is made up of particles in continuous random motion; temperature is a measure of the average speed of the particles.

collision–reaction theory: a theory stating that chemical reactions involve collisions and rearrangements of atoms or groups of atoms, and that the outcome of collisions depends on the energy and orientation of the collisions

The KMT in Action

Particles in liquids and gases are always in motion. Usually we cannot see this motion, because the particles are very small. Here is a way to see the kinetic molecular theory in action.

Materials: 250-mL beaker, tap water, food colouring

- Pour about 200 mL of tap water into the beaker and let it settle for a few seconds.
- Add a drop of food colouring. Observe.

(a) Describe what you see, or draw a series of sketches to illustrate your observations.

(b) Explain your observations in terms of the kinetic molecular theory.

Figure 4
The erratic motion of tiny particles was first observed and described by Robert Brown in 1827. You can view each motion with a flashlight and a heterogeneous mixture.

According to kinetic molecular theory, the motion of all particles results in random collisions. At room temperature, the colliding molecules of octane and oxygen simply bounce off one another unchanged. However, at higher temperatures, the molecules collide at greater speeds, and therefore with greater energy. Atomic theory suggests that in such a collision, the repulsive forces of the electrons in the two molecules could be overcome. In some collisions, the valence shells of the **reactants** will overlap. When this occurs, their electrons can be rearranged to form new bonds.

reactants: the substances that combine in a chemical reaction

Collisions do not always result in a chemical change. The orientation (relative positions) of the reactants is important. In the case of burning octane, the carbon and hydrogen atoms in the octane must be oriented in such a way that the atoms in the colliding oxygen molecule can form C–O and H–O bonds simultaneously.

The burning of another common fuel, methane, $CH_{4(g)}$, can be explained in the same way. When methane is burned in a laboratory burner, the reactants (methane and oxygen molecules) collide with enough energy and at the proper orientation for successful rearrangements to occur. As a result, new substances are formed: gaseous carbon dioxide and water vapour. The rearrangement of atoms that has occurred (**Figure 5**) happens in only a tiny percentage of all the collisions; however, there is an enormous number of collisions and the reaction takes place rapidly.

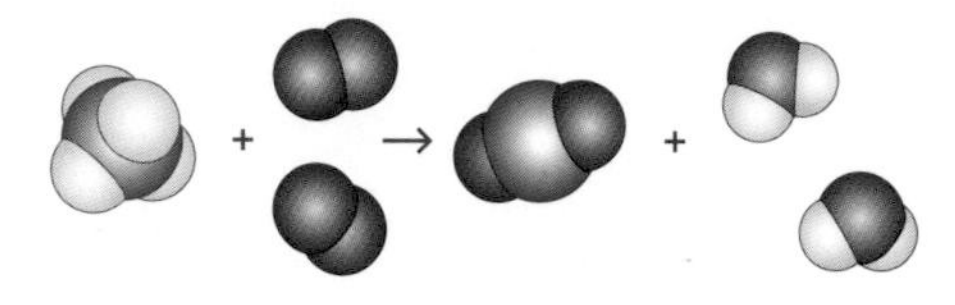

Figure 5
Collision–reaction theory suggests that particles of methane and oxygen must collide with the correct orientation and at sufficient speed to react. However, chemists do not yet fully understand the mechanism of even this relatively simple reaction. The process may go through as many as 100 intermediate steps!

Try This **Activity**

A Model for the Collision-Reaction Theory

In this activity, you will simulate the motion and behaviour of reactants in a chemical reaction. The reactants are represented by different coloured Ping-Pong balls with Velcro affixed on one side (**Figure 6**).

Materials: shallow box with lid, about 20 coloured balls with "hook" part of Velcro, about 20 white balls with "loop" part of Velcro

- Place an equal number of coloured and white balls in the box. Place the lid on the box and shake the box gently for 1 min.

  (a) What type of motion do the "particles" exhibit when the box is gently shaken?
  (b) Open the box and count the number of free and combined particles. What proportion of the particles have formed a "compound"?

- Separate any joined reactants, put the lid back on the box, and shake it vigorously for 1 min.

  (c) What does this vigorous shaking represent?
  (d) Open the box and count the number of free and combined particles. What proportion of the particles have formed a compound?
  (e) Did all of the particles react?
  (f) Explain your observations in terms of the collision-reaction theory.

- Separate any joined reactants and remove the same number of both types of reactants. Put the lid on the box and shake it vigorously for 1 min.

  (g) Open the box and count the number of free and combined particles. Has the proportion of particles that formed a compound changed? Explain why this might be so.

Figure 6
Only brown and white balls have the possibility of joining together upon collision, since they are each equipped with the opposite parts of the Velcro.

Practice

Understanding Concepts

1. When an Alka-Seltzer tablet is added to a glass of water, bubbles are produced and the tablet eventually disappears. Is this a chemical or physical change? Give reasons for your answer.
2. Give an example of a physical change that may look like a chemical change. Why is it important that conclusions regarding chemical change are not based solely on visual observation?
3. As gasoline will burn only when it is a gas, the internal-combustion engine is designed to convert liquid gasoline to the gas phase. Explain, using the kinetic molecular theory, how gaseous gasoline is available for combustion in an engine.
4. According to the collision-reaction theory, what are the requirements for a chemical reaction to take place?

Representing Chemical Change

We have looked at the evidence for a chemical reaction taking place, and the mechanism by which it has taken place. The next piece of the puzzle is developing a way to represent exactly what happens during the chemical change: what the reactants are and what they change into.

Over the years, chemists have developed two formats that effectively communicate information about chemical reactions: the **word equation**, in which only the chemical names of the reactants and products are shown, and the **chemical equation**, in which the proportions of reactants and products are shown by writing **coefficients** in front of the chemical symbols for each substance.

word equation: a representation of a chemical reaction using only the names of the chemicals involved

chemical equation: a representation of a chemical reaction that indicates the chemical formulas, relative number of entities, and states of matter of the reactants and products

coefficient: a whole number indicating the ratio of molecules or formula units of each substance involved in a chemical reaction

The word equation for the combustion of octane reads

octane + oxygen → carbon dioxide + water

The reactants are written on the left of the arrow; the products are written on the right. In every equation, an arrow is used to take the place of the word "yields" or "produces."

The chemical equation for this reaction conveys much more information:

$$2\ C_8H_{18(g)} + 25\ O_{2(g)} \rightarrow 16\ CO_{2(g)} + 18\ H_2O_{(g)}$$

The symbols tell us the precise composition of each substance. Placing coefficients in front of each symbol tells us the amount of each substance, or how many molecules (in ionic substances, formula units) of each substance are involved. In our combustion example, 2 molecules of octane and 25 molecules of oxygen react to produce 16 molecules of carbon dioxide and 18 molecules of water. Finally, the subscript letters tell us the state of each reactant and product (all gases in this reaction).

A method for writing balanced chemical equations is developed in Unit 2.

Activity 3.1.1

Understanding Chemical Reactions

In this activity, you will build structural models to represent three chemical reactions: methane, $CH_{4(g)}$, combines with oxygen, $O_{2(g)}$, to produce carbon dioxide, $CO_{2(g)}$, and water, $H_2O_{(g)}$; water is dissociated to produce hydrogen, $H_{2(g)}$, and oxygen, $O_{2(g)}$; carbon, $C_{(s)}$, combines with oxygen, $O_{2(g)}$, to produce carbon dioxide, $CO_{2(g)}$.

Materials

molecular model kit

Procedure

1. Construct molecular models of the reactants in the methane reaction. Record the structure of each molecule.
2. Rearrange the same pieces to make models of the products. (If necessary, make more of the reactant molecules first.) All the pieces you use should be part of both the reactants and the products.
3. Repeat steps 1 and 2 for the second and the third reactions.

Analysis

(a) Write a balanced chemical equation for each reaction.

Synthesis

(b) Which law are you following by balancing the equations?

DID YOU KNOW ?

Lead Poisoning

The platinum and palladium in catalytic converters are "poisoned" and become ineffective if they are exposed to lead. This is one factor that encouraged the Canadian government to ban the use of leaded gasoline.

Catalysts and Collisions

In an internal-combustion engine more reactions take place than just the oxygen–octane reaction described earlier. The reason is that there are many more than just two substances in the engine. Some of these substances are present in gasoline (such as benzene and various other carbon compounds, and some sulfur and other impurities) and some are present in air (such as nitrogen). As a result of these additional reactions and incomplete combustion of the fuel, exhaust gases contain several products other than carbon dioxide and water, including carbon monoxide, nitrogen oxides, and unburned hydrocarbons. Many cars today are equipped with **catalytic converters** to decrease the quantity of pollutants released when gasoline is burned. The metals platinum and palladium and nickel oxide can be used as catalysts in a converter. A **catalyst** is a substance that speeds up a chemical reaction, but is still present at the end of the reaction. Catalysts do participate in reactions; they are consumed and regenerated as the reaction proceeds. For this reason catalysts appear as neither product nor reactant in chemical equations. Instead, the catalyst involved (in this case nodules of platinum and palladium) is written above the arrow. Using this convention, we can represent the conversion of carbon monoxide to less harmful carbon dioxide in a catalytic converter:

catalytic converter: a device that uses catalysts to convert pollutant molecules in vehicle exhaust to less harmful molecules

catalyst: a substance that speeds up the rate of a reaction without undergoing permanent change itself

$$2\ CO_{(g)} + O_{2(g)} \xrightarrow{Pt/Pd} 2\ CO_{2(g)}$$

Explore an Issue

Take a Stand: Catalytic Converters

Catalytic converters are built into the exhaust systems of many cars, and can be retrofitted to older models. The purpose of the catalytic converter is to catalyze reactions between pollutant gases and oxygen. The products of the catalyzed reactions are considerably less harmful to the environment when they are released. However, this extra technology does not come without a financial cost. Is it worth the price?

(a) Research the design and working of catalytic converters, their benefits, and the price of making and installing them.

(b) Assemble the arguments for and against making catalytic converters compulsory in all vehicles. Decide on your position.

(c) Put together a presentation, perhaps a pamphlet or a video, to be distributed to people who are about to purchase a used vehicle. In your presentation, outline your position and support it with reasoned arguments and evidence.

Follow the links to Nelson Chemistry 11, 3.1.

 www.science.nelson.com

DECISION-MAKING SKILLS

- ● Define the Issue
- ○ Identify Alternatives
- ● Research
- ● Analyze the Issue
- ● Defend a Decision
- ○ Evaluate

Practice

Understanding Concepts

5. (a) List four changes that can be used as evidence for chemical reactions.
   (b) For each change, provide two examples from everyday life.
6. Define each of the following terms with respect to chemical equations:
   (a) reactant
   (b) product
   (c) coefficient
   (d) balanced
7. Which of the following chemical equations adhere to the law of conservation of mass? Copy and balance those that do not.
   (a) $H_{2(g)} + O_{2(g)} \rightarrow H_2O_{(g)}$
   (b) $2\ NaOH_{(aq)} + Cu(ClO_3)_{2(aq)} \rightarrow Cu(OH)_{2(s)} + 2\ NaClO_{3(aq)}$
   (c) $Pb_{(s)} + AgNO_{3(aq)} \rightarrow Ag_{(s)} + Pb(NO_3)_{2(aq)}$
   (d) $2\ NaHCO_{3(s)} \rightarrow Na_2CO_{3(s)} + CO_{2(g)} + H_2O_{(l)}$

Section 3.1 Questions

Understanding Concepts

1. Chlorine is produced industrially in huge amounts because it is used in a great variety of products, many of which are used for cleaning and disinfecting. Compounds of chlorine are used to sanitize and disinfect municipal water supplies and swimming pools. Household bleaches and many cleaning products also contain chlorine compounds. About 95% of the chlorine produced comes from the electrolysis of sodium chloride or brine solutions. The

(continued)

net reaction shows that aqueous sodium chloride and water are converted into chlorine gas, hydrogen gas, and aqueous sodium hydroxide.

(a) Write a word equation to represent the reaction described above.
(b) Write a balanced chemical equation for the reaction.
(c) This reaction takes place inside sealed containers. Suggest why.
(d) If you could watch this reaction taking place, what would you expect to observe as evidence of chemical change?
(e) Classify each of the reactants and products as ionic compounds, molecular compounds, or molecular elements.
(f) Which of the reactants and products could be classified as polar molecules?
(g) Chlorine is a poisonous yellow-green gas at SATP. Explain why chlorine compounds can be safely used to clean municipal water supplies.

3.2 Combustion, Synthesis, and Decomposition Reactions

In Chapters 1 and 2, you found out how useful it is to be able to classify substances. Chemists also find it convenient to be able to classify chemical reactions. There are thousands of potential reactions, each one of them unique (**Figure 1**).

Figure 1
(a) Potassium reacts vigorously with water.
(b) The burning of steel wool in pure oxygen produces light and heat.
(c) When silver nitrate is added to sodium chloride, an insoluble substance is formed.
(d) When an Alka-Seltzer tablet is added to water, a gas is produced.

Some have elements as reactants and compounds as products; some the reverse. Some produce huge amounts of heat; some absorb heat. Some take place only in aqueous solution; some will take place whenever the reactants come in contact with each other. How can we make sense of such a variety?

Once again, we analyze evidence obtained from many chemical reactions and look for patterns. For generations, chemists have been doing this, and have provided a basis upon which we can classify chemical reactions. While there are exceptions, most reactions can be classified as one of the types listed in **Table 1**.

Table 1: Types of Chemical Reactions

Reaction type	Generalization
combustion	AB + oxygen → commmon oxides of A and B
synthesis	A + B → AB
decomposition	AB → A + B
single displacement	A + BC → AC + B
double displacement	AB + CD → AD + CB

Combustion Reactions

Every time we flick a lighter, strike a match, fire up the barbecue, or stoke a campfire we are starting a **combustion reaction**. It is, of course, more commonly known as burning, and is easily identified because of its most useful feature: the production of heat. For combustion to proceed, three things must be present: fuel, oxygen, and heat.

combustion reaction: the reaction of a substance with oxygen, producing oxides and energy

In a complete combustion reaction, a reactant burns with oxygen to produce the most common oxides of the elements making up the substance that is burned. For example, gasoline often contains trace amounts of elemental sulfur. When the gas is burned, the sulfur combines with oxygen to produce sulfur dioxide.

$$S_{(s)} + O_{2(g)} \rightarrow SO_{2(g)}$$

To successfully predict the products of a complete combustion reaction, we must know the most common oxides (**Table 2**). For metals, the most common oxide is the one formed with the metal's most common ion.

Table 2: Common Oxides Resulting from Combustion Reactions

Combustion situation	Element in reactant	Common oxide
coal in a coal-fired electricity generator	carbon	$CO_{2(g)}$
burning of rocket fuel	hydrogen	$H_2O_{(g)}$
commercial production of sulfuric acid	sulfur	$SO_{2(g)}$
lightning strikes and volcanoes	nitrogen	$NO_{2(g)}$

DID YOU KNOW ?

Cellular Respiration

In cellular respiration, which is a combustion reaction, glucose is "burned" to supply living cells with a usable form of energy. The products of the reaction are carbon dioxide and water—the same ones you get when you burn methane in a laboratory burner.

Rain is normally slightly acidic due to the presence of carbon dioxide, which dissolves in atmospheric moisture and reacts to form very dilute carbonic acid, $H_2CO_{3(aq)}$. Nitrogen oxides from lightning strikes and plant decay, and sulfur oxides from volcanic eruptions also contribute to the natural acidity of rain, forming nitric and sulfuric acid, respectively, in reactions with atmospheric water. However, human intervention has considerably increased the quantities of oxides being released into the atmosphere. Our everyday activities involve a great many combustion reactions, from riding buses to heating our homes. Gaseous oxides of nitrogen (e.g., nitrogen monoxide, nitrogen dioxide) and sulfur (e.g., sulfur dioxide, sulfur trioxide) are released from sources such as automobiles and coal-burning power plants. These oxides join the naturally produced oxides in the atmosphere, react with water vapour to form acids, and are responsible for the increased acidity of precipitation known as **acid rain**.

acid rain: any form of natural precipitation that has an abnormally high acidity

INQUIRY SKILLS

- ○ Questioning
- ○ Hypothesizing
- ● Predicting
- ○ Planning
- ● Conducting
- ● Recording
- ● Analyzing
- ● Evaluating
- ● Communicating

Investigation 3.2.1

The Combustion of Butane

When a substance burns, we call the process a combustion reaction. The generalization for a complete combustion reaction is that the substance reacts with air to produce light and heat, and the most common oxides of the elements present. The purpose of this investigation is to test the ability of this generalization to predict the products of a combustion reaction.

This investigation involves the combustion of a sample of butane under an Erlenmeyer flask. Complete a lab report, including the **Prediction**, **Analysis**, and **Evaluation**.

Question

What are the products of the combustion of butane?

Prediction

(a) Predict the products of the combustion reaction of butane, $C_4H_{10(g)}$, with oxygen, $O_{2(g)}$. Use the combustion generalization to justify your Prediction.

Experimental Design

A sample of butane in a disposable plastic lighter is burned under an Erlenmeyer flask filled with air. In addition to direct observations, diagnostic tests for water (using cobalt chloride paper) and carbon dioxide gas (using limewater) are conducted before and after the reaction.

Materials

Butane is highly flammable. Use butane lighter with care. Wear an apron and eye protection.

The neck of the Erlenmeyer flask becomes hot. When handling the flask after combustion, use insulated mitts or tongs.

lab apron
eye protection
tweezers
cobalt(II) chloride paper
2 dry 250-mL Erlenmeyer flasks
stoppers for Erlenmeyer flasks
utility stand and clamp
insulated mitts or tongs
disposable plastic lighter containing butane
limewater solution
10-mL graduated cylinder

Procedure

1. Using tweezers, place a dry strip of cobalt(II) chloride paper in a dry flask. Stopper the flask, shake it, and record your observations of the colour of the strip. Remove the paper from the flask.
2. Using a graduated cylinder, add a few millilitres of limewater solution to the dry flask. Stopper, shake, and observe.
3. Clamp the second dry Erlenmeyer flask upside down, as shown in **Figure 2**.
4. Light the lighter and hold it under the flask, with most of the flame inside the opening. Let the butane burn for about 20 s, then remove the lighter from the flask. Immediately insert the rubber stopper in the mouth of the flask.

5. Turn the flask the right way up. Use tweezers to obtain a dry strip of $CoCl_2$ paper. Quickly remove the stopper, drop the strip into the flask, and replace the stopper.
6. Shake the flask and record your observations of the colour of the strip.
7. Obtain a few millilitres of limewater solution in a graduated cylinder. Again remove the stopper, add the limewater solution, and replace the stopper. Shake the flask and record your observations.

Figure 2
The reactants and products of this reaction can be determined by diagnostic tests.

Analysis

(b) What products were formed as a result of the combustion of butane? What observations provide Evidence to support your answer?
(c) Express the reaction observed in this investigation in the form of a word equation.
(d) Write a balanced chemical equation for the reaction observed.

Evaluation

(e) How confident are you in your experimental Evidence? Give your reasons.
(f) If you have confidence in the accuracy of your Evidence, compare it to your Prediction. Was your Prediction verified? Comment on the validity of the generalization.

Combustion and the Atmosphere

Carbon dioxide and water are products of the combustion of carbon compounds, including fossil fuels and wood in forest fires. These gases are also released in volcanic eruptions and by living organisms during cellular respiration. All of these sources have been present for billions of years, to greater or lesser degree, and have had a significant effect on the atmosphere, including making it warmer than it would otherwise be. Scientists have proposed a theory to explain the effect of carbon dioxide and water vapour on the average temperature on Earth, a phenomenon known as the **greenhouse effect**. Visible sunlight is able to pass through the atmosphere and reach the Earth's surface, where much of it is absorbed. The heated surface emits infrared radiation (heat) into the atmosphere. However, atmospheric carbon dioxide and water vapour absorb some of this outgoing infrared radiation and then re-emit it randomly in all directions, including back toward Earth. This has the effect of preventing some of the heat from escaping, thereby further heating the air and Earth's surface.

greenhouse effect: a theory stating that heat is trapped near Earth's surface by carbon dioxide gas, atmospheric water vapour, and some other gases

Our increased burning of fossil fuels in vehicles, in furnaces, and in industrial processes over the past century is widely blamed for the increasing concentration of atmospheric carbon dioxide. Average global temperatures have also been on the rise. Many (but not all) scientists are convinced that these two observations are linked: that the artificial increase in the amount of carbon dioxide in the atmosphere is like an extra blanket on the bed, causing the temperature to rise. Continued emission of carbon dioxide at the same rate may lead to continued global warming and dramatic climate changes. You will find out more about the greenhouse effect in Chapter 11.

Practice

Understanding Concepts

1. What are the reactants, necessary conditions, and products of a combustion reaction?

2. List the most common oxides of carbon, hydrogen, sulfur, nitrogen, and iron.
3. Describe at least two environmental issues related to combustion reactions.

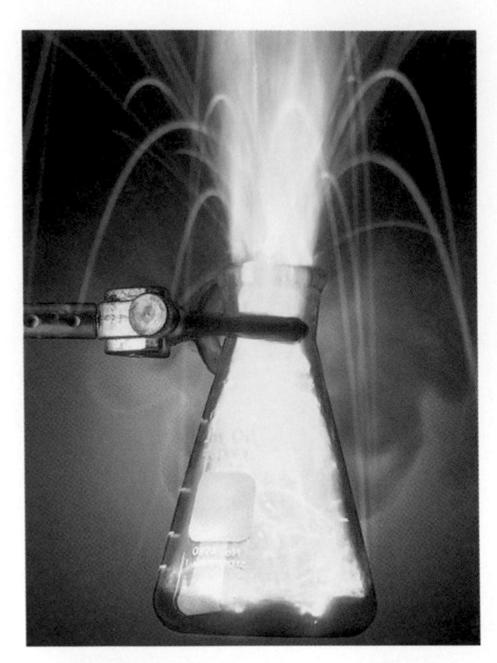

Figure 3
What evidence is there that a chemical reaction is taking place?

synthesis reaction: a chemical reaction in which two or more substances combine to form a more complex substance

Synthesis Reactions

The ionic compound sodium bromide is, as you know, made up of the elements sodium and bromine. These two elements are sometimes combined in the lab to produce sodium bromide. When sodium metal is added to a container of bromine liquid, the reaction is quite spectacular (**Figure 3**).

Ammonia is an important industrial product. It is prepared by combining nitrogen gas and hydrogen gas in what is referred to as the Haber process. This process was initially developed in Germany as a step in the manufacture of explosives, but is now used around the world to produce agricultural fertilizer.

Each of these reactions is referrred to as a **synthesis reaction** (sometimes also called a combination reaction). In a synthesis reaction, two or more substances combine to form a single product. The compound formed by a synthesis reaction can be either ionic or molecular. Synthesis reactions can be represented by the general equation

$$A + B \rightarrow AB$$

Several synthesis reactions are associated with the combustion of gasoline, and the vehicle exhaust. For example, the burning of gasoline at high temperature causes nitrogen in the air to react with oxygen to produce nitrogen monoxide.

nitrogen + oxygen → nitrogen monoxide

$$N_{2(g)} + O_{2(g)} \rightarrow 2\ NO_{(g)}$$

In cases where the reactants are elements, it is easy to recognize the reaction as a synthesis reaction. After all, two elements reacting together can only form a compound. However, a reaction in which one or both of the reactants is a compound can still be a synthesis reaction. For example, when gasoline is burned in air, the nitrogen monoxide produced can further react with oxygen to produce nitrogen dioxide.

nitrogen monoxide + oxygen → nitrogen dioxide

$$2\ NO_{(g)} + O_{2(g)} \xrightarrow{Pt} 2\ NO_{2(g)}$$

Sample Problem 1

Sulfur trioxide is a byproduct of the combustion of gasoline in car engines. In the atmosphere it reacts with condensed water on dust particles, producing sulfuric acid. Write a word equation and a balanced chemical equation for the reaction.

Solution

sulfur trioxide + water → sulfuric acid

$$SO_{3(g)} + H_2O_{(l)} \rightarrow H_2SO_{4(aq)}$$

As you know, photosynthesis is an important chemical reaction for life on Earth. In photosynthesis, energy from the Sun is used to convert carbon dioxide and water to glucose and oxygen.

$$6\ CO_{2(g)} + 6\ H_2O_{(l)} \xrightarrow{\text{sunlight/chlorophyll}} C_6H_{12}O_{6(aq)} + 6\ O_{2(g)}$$

From the chemical equation for photosynthesis, it is apparent that a complex compound (glucose) is synthesized from simpler compounds (carbon dioxide and water). In this regard, one might be tempted to classify the reaction as a synthesis reaction. However, oxygen is also formed. Therefore, despite its name, we cannot classify photosynthesis as a true synthesis reaction. Photosynthesis is an important exception to the classification of chemical reactions that we are discussing in this chapter.

Decomposition Reactions

In some urban areas, during periods of warm, dry weather, nitrogen dioxide (from industrial processes and vehicle exhaust) can accumulate in the atmosphere, forming a brownish haze.

Ultraviolet radiation from the Sun can cause NO_2 to decompose to NO and O, in what is referred to as a **decomposition reaction.** The oxygen atom, O, produced by this reaction is very reactive. It can combine with other chemicals to cause serious health effects, including breathing disorders.

decomposition reaction: a chemical reaction in which a compound is broken down into two or more simpler substances

Decomposition reactions can be represented by this general equation:

$$AB \rightarrow A + B$$

In this type of reaction, a compound decomposes to form two or more substances. The products could be elements or compounds, depending on the nature of the reactants. Decomposition reactions of simple binary compounds generally yield, as products, the two elements that make up the compound. In earlier times, decomposition reactions were used by chemists to discover the chemical formulas of substances. Today, decomposition reactions are the basis for industrial production of many elements from naturally available compounds. For example, the electrolysis of molten sodium chloride will yield liquid sodium and chlorine gas. This is an important method for the production of the element sodium.

sodium chloride → sodium + chlorine

$$2\ NaCl_{(l)} \xrightarrow{\text{electricity}} 2\ Na_{(l)} + Cl_{2(g)}$$

The decomposition of water produces oxygen and hydrogen (**Figure 4**).

water → hydrogen + oxygen

$$2\ H_2O_{(l)} \xrightarrow{\text{electricity}} 2\ H_{2(g)} + O_{2(g)}$$

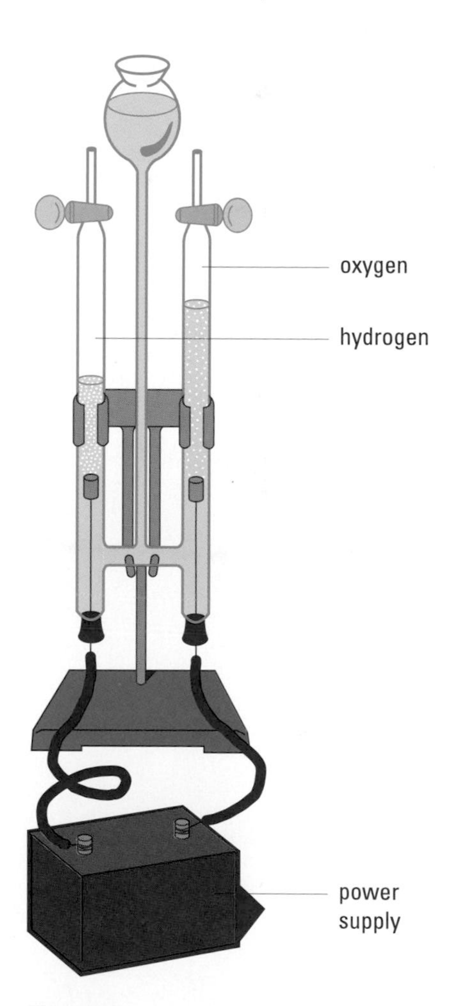

Figure 4
The Hoffman apparatus is used for the decomposition of water.

Compounds consisting of more than two elements often decompose to form simpler compounds. For example, when a hydrated salt is heated, the products are the anhydrous salt and water.

copper(II) sulfate pentahydrate → copper(II) sulfate + water

$$CuSO_4{\cdot}5\ H_2O_{(s)} \rightarrow CuSO_{4(s)} + 5\ H_2O_{(g)}$$

Sample Problem 2

The high-temperature combustion of gasoline in an engine produces a serious air pollutant, nitrogen monoxide. The platinum/palladium catalytic converter built into today's automobiles catalyzes the decomposition of nitrogen monoxide into harmless nitrogen and oxygen. Write a word equation and a balanced chemical equation to represent this reaction.

Solution

nitrogen monoxide → nitrogen + oxygen

$$2\ NO_{(g)} \xrightarrow{Pt/Pd} N_{2(g)} + O_{2(g)}$$

thermal decomposition: a decomposition reaction that occurs when the reactant is heated

When a compound decomposes as a result of heating, we call it a **thermal decomposition** reaction. Many carbonates and hydrogen carbonates undergo thermal decomposition. For example, sodium hydrogen carbonate (baking soda or sodium bicarbonate) will decompose when heated to produce sodium carbonate, water, and carbon dioxide in the following reaction:

$$2\ NaHCO_3 \rightarrow Na_2CO_3 + H_2O + CO_2$$

Clearly, this is not a case of a binary compound splitting into two elements. Rather, the sodium hydrogen carbonate has split into three simpler compounds upon decomposing. The more complex the reactant, the more difficult it becomes to predict the products of the reaction, but the production of an oxide and water is a common result of decomposition. The thermal decompositions of carbonic acid and calcium hydroxide illustrate this generalization:

$$H_2CO_{3(aq)} \rightarrow CO_{2(g)} + H_2O_{(g)}$$
$$Ca(OH)_{2(s)} \rightarrow CaO_{(s)} + H_2O_{(l)}$$

Sample Problem 3

Zinc carbonate undergoes thermal decomposition to produce zinc oxide and carbon dioxide. Represent this reaction in a word equation and a balanced chemical equation.

Solution

zinc carbonate → zinc oxide + carbon dioxide

$$ZnCO_{3(s)} \rightarrow ZnO_{(s)} + CO_{2(g)}$$

SUMMARY Combustion, Synthesis, Decomposition Generalizations

Table 3: Reaction Type Generalizations

Reaction type	Reactants	Products
combustion	metal + oxygen nonmetal + oxygen fossil fuel + oxygen	metal oxide nonmetal oxide carbon dioxide + water
synthesis	element + element element + compound compound + compound	compound more complex compound more complex compound
decomposition	binary compound complex compound	element + element simpler compound + simpler compound or simpler compound + element(s)

Practice

Understanding Concepts

4. Classify each of the following reactions as synthesis or decomposition. Give reasons for your answers.
   (a) $2\ KBr_{(s)} \rightarrow 2\ K_{(s)} + Br_{2(l)}$
   (b) $6\ K_{(s)} + N_{2(g)} \rightarrow 2\ K_3N_{(s)}$
   (c) $16\ Al_{(s)} + 3\ S_{8(s)} \rightarrow 8\ Al_2S_{3(s)}$
5. Classify the following reactions as synthesis, decomposition, or combustion. Predict the products of the reactions, using balanced chemical equations.
   (a) $Al_{(s)} + F_{2(g)} \rightarrow$
   (b) $KCl_{(s)} \rightarrow$
   (c) $S_{8(g)} + O_{2(g)} \rightarrow$
   (d) methane + oxygen $\rightarrow$
   (e) aluminum oxide $\rightarrow$
   (f) $Hg_{(l)} + O_{2(g)} \rightarrow$
   (g) iron(III) bromide $\rightarrow$
6. Compare the reactants that would be involved in each of the following reaction types.
   (a) synthesis reaction
   (b) decomposition reaction
   (c) combustion reaction
7. For each of the following reactions, predict the products by writing a balanced chemical equation to represent the reaction. Classify the reaction type.
   (a) Very reactive sodium metal reacts with the poisonous gas chlorine to produce an inert, edible chemical.
   (b) For many thousands of years, molten copper has been produced by heating the ore $CuO_{(s)}$.
   (c) A frequent technological problem associated with the operation of swimming pools is that copper pipes react with aqueous chlorine, added to kill bacteria.

Applying Inquiry Skills

8. An investigation is planned and carried out to test the generalization for decomposition reactions: $AB \rightarrow A + B$. Complete the **Prediction**.

   Question
   What are the products of the thermal decomposition of lithium oxide, magnesium oxide, zinc chloride, and magnesium hydroxide?

   Prediction
   (a) Predict the answer to the Question.

9. A forensic chemist takes a sample of a white powder from a crime scene. She needs to know the identity of the elements that make up the powder. Complete the **Evaluation** and **Synthesis** sections of her laboratory report.

   Question
   What is the composition of the white powder?

   Experimental Design
   A sample of the white powder is heated in a test tube. Any changes in appearance are noted. Heating is continued until no further changes are observed.

   Evidence
   A white solid and a clear, colourless gas are produced.

   Evaluation
   (a) Comment on the Experimental Design. Is it sufficient to answer the Question? If not, suggest an improved experimental design that would allow the forensic chemist to answer the Question.

   Synthesis
   (b) Classify the reaction and write a general equation for it.
   (c) If the reactant was calcium carbonate, write word and balanced chemical equations for the reaction.

Making Connections

10. Vehicle exhausts produce a rich mix of polluting chemicals that are linked to a whole range of health risks and environmental effects. Use the Internet to investigate one of these exhaust pollutants. Describe the chemical reactions that produce it and its harmful effects on living organisms. Develop a list of suggestions for reducing the production of this pollutant. Discuss the pros and cons of implementing these suggestions from the perspective of a car manufacturer and from the perspective of an environmentalist.

    Follow the links for Nelson Chemistry 11, 3.2.

    GO TO www.science.nelson.com

11. Some vehicles use diesel fuel instead of gasoline. Use the Internet to investigate the properties of each of these fuels and the products of their combustion. Which fuel poses a greater threat to the environment?

    Follow the links for Nelson Chemistry 11, 3.2.

    GO TO www.science.nelson.com

12. Natural gas and propane are alternative fuels for vehicles. Research the chemical composition of these two fuels. How would you expect their products of combustion to compare with those of gasoline?

    Follow the links for Nelson Chemistry 11, 3.2.

    GO TO www.science.nelson.com

Section 3.2 Questions

Understanding Concepts

1. (a) Why are there signs at gas stations warning motorists to turn off their vehicles' engines, refrain from smoking, and avoid causing sparks?
   (b) What type of reaction is the concern?
   (c) What are the necessary conditions for this reaction to proceed?
2. (a) Classify the types of chemical reactions in which the following reactants are likely to participate, giving reasons:
   (i) butane, C_4H_{10}
   (ii) calcium carbonate
   (iii) Li and Br
   (iv) bluestone, $CuSO_4{\cdot}5\ H_2O_{(s)}$
   (b) Write word and balanced chemical equations for each reaction.

Making Connections

3. Sulfur oxides are produced at the high temperatures found in internal-combustion engines. Sulfur dioxide reacts with oxygen to form sulfur trioxide, which in turn reacts with liquid water to form sulfuric acid.
   (a) Represent the formation of the oxide in an engine and its subsequent reaction in the atmosphere, using balanced chemical equations. Categorize each reaction as synthesis, decomposition, or combustion.
   (b) How is combustion related to acid rain?
4. The catalytic converter, which is part of the muffler system of automobiles, catalyzes the decomposition of nitrogen monoxide.
   (a) Predict the products of this decomposition reaction and represent it with a balanced chemical equation.
   (b) Explain the value of the catalytic converter in reducing air pollution.

3.3 Single Displacement Reactions

Most metals occur naturally as ores—metal-containing minerals. To isolate silver from its ore, it is sometimes first reacted with a solution of cyanide, to form aqueous silver cyanide, which is then reacted with another metal, such as zinc.

$$Zn_{(s)} + 2\ AgCN_{(aq)} \rightarrow 2\ Ag_{(s)} + Zn(CN)_{2(aq)}$$

This type of reaction is referred to as a **single displacement reaction.** Single displacement reactions are typical in metallurgical processes. In a single displacement reaction one element replaces another similar element in a compound (**Figure 1**, page 124). We can write a generalization for such reactions:

single displacement reaction: the reaction of an element with a compound to produce a new element and a new compound

$$A + BC \rightarrow AC + B$$

Figure 1
In single displacement reactions, like displaces like—a metallic element takes the place of a metal in a compound; a nonmetallic element takes the place of a nonmetal in a compound.

INQUIRY SKILLS

- ○ Questioning
- ○ Hypothesizing
- ○ Predicting
- ○ Planning
- ● Conducting
- ● Recording
- ● Analyzing
- ○ Evaluating
- ○ Communicating

Investigation 3.3.1

A Single Displacement Reaction

Early photographic developing techniques made use of copper plates coated with silver nitrate, a light-sensitive compound. In this investigation, you will carry out and observe the reaction that follows when copper is placed in a solution of silver nitrate, $Ag(NO_3)_2$ (**Figure 2**). Complete the **Analysis** section of the report.

Experimental Design

Copper metal is placed in a silver nitrate solution.

Materials

eye protection
protective gloves
lab apron
test tube
silver nitrate solution
copper wire
test-tube rack

Silver nitrate is toxic and can cause stains on skin and clothing. Wear eye protection, protective gloves, and a laboratory apron.

Dispose of chemicals according to your teacher's instructions.

Procedure

1. Pour about 5 to 10 mL of silver nitrate solution into the test tube.
2. Take a 30-cm length of copper wire, and twist 20 cm at one end of it around a pencil to form a horizontal coil. Make a vertical hook from the last 10 cm of wire.
3. Place the copper wire in the silver nitrate solution. Stand the entire assembly in a test-tube rack. Record your observations as frequently as possible over the next two days.

Analysis

(a) Write a chemical equation for the reaction. Include your reasoning in support of your equation.

Figure 2
What does the blue colour of the solution suggest?

Activity Series

Silver can be produced from copper and a solution of silver nitrate.

copper + silver nitrate → silver + copper(II) nitrate
(metal) (compound) (metal) (compound)

$$Cu_{(s)} + 2\,AgNO_{3(aq)} \rightarrow 2\,Ag_{(s)} + Cu(NO_3)_{2(aq)}$$

Similarly, iodine can be produced from chlorine and aqueous sodium iodide.

chlorine	+ sodium iodide	→	iodine	+ sodium chloride
(nonmetal)	(compound)		(nonmetal)	(compound)
$Cl_{2(aq)}$	+ 2 $NaI_{(aq)}$	→	$I_{2(s)}$	+ 2 $NaCl_{(aq)}$

The examples above and many others show us that, in single displacement reactions, a metal will sometimes displace another metal from an ionic compound, while a nonmetal will sometimes displace another nonmetal. But which elements displace which other elements? Is there a pattern? Using empirical evidence gathered in many experiments, scientists have been able to list the elements in order of their reactivity. Separate ordered lists, **activity series**, have been produced for metals and nonmetals.

activity series: a list of elements arranged in order of their reactivity, based on empirical evidence gathered from single displacement reactions

Investigation 3.3.2

Developing an Activity Series

INQUIRY SKILLS

- ● Questioning
- ● Hypothesizing
- ○ Predicting
- ● Planning
- ● Conducting
- ● Recording
- ● Analyzing
- ● Evaluating
- ● Communicating

Whether or not one element displaces another element in a compound depends on the relative reactivities of the two elements. The more reactive element will replace the less reactive element. The purpose of this investigation is to develop your own activity series of metals, much as was done many years ago. You will test two elements at a time (one as an element and one as a compound in solution), to find out which displaces which from a compound, and, therefore, which is more reactive. Because elements generally have low solubility, you will be able to tell when one element is displacing another by the formation of a precipitate. A colour change in the solution may also provide clues that a reaction has taken place.

You will have available six metals and six solutions of aqueous ionic compounds to build your activity series using the techniques learned in Investigation 3.3.1.

Complete an investigation report, including the **Question**, **Experimental Design**, **Procedure**, **Analysis**, **Evaluation**, and **Synthesis** sections.

Question

(a) Write a Question that you will attempt to answer during the investigation.

Experimental Design

Many pairs of metals will be tested (one as an element, one as a compound in solution), to see which is displaced (i.e., is less reactive).

(b) From the list of Materials provided, plan an investigation using pairs of elements and solutions that will answer your Question.

(c) Using the Procedure below as a guide, write a full Procedure for your investigation and have it approved by your teacher. Include safety precautions.

Materials

apron
eye protection
microtray
test-tube rack
steel wool or sandpaper

Tin(IV) chloride can irritate skin and eyes. Wear eye protection.

Copper(II) sulfate is toxic. Avoid contact with skin and wear eye protection.

Dispose of all chemical waste according to your teacher's instructions.

pieces of: copper, iron, zinc, aluminum, magnesium, and tin solutions of
- copper(II) sulfate
- iron(II) sulfate
- zinc sulfate
- aluminum nitrate
- magnesium nitrate
- tin(IV) chloride

(d) List any other materials you will require.

Procedure

1. Place your microtray on a sheet of paper, and label the rows and columns according to which substances you will place in each.
2. Place two or three drops of your chosen solutions in the appropriate microtray wells, as required by your Experimental Design.
3. Obtain samples of the metals you have chosen to test. Clean the surface of each sample with steel wool or sandpaper. Place each metal sample in the appropriate solution.
4. Make and record your observations before, during, and after the reactions.

Analysis

(e) In which wells did a reaction take place? In each case, what is your Evidence? What does a reaction suggest?

(f) In which wells did a chemical reaction not occur? What does this suggest?

(g) Based on your observations, make a statement comparing the reactivity of the pairs of metals present in each well of the microtray.

(h) Compile your results into an activity series by ranking the metals you tested in order of reactivity.

Evaluation

(i) Evaluate your Experimental Design. Were you able to answer your Question? How could you have improved your design?

Synthesis

(j) Write word and chemical equations to represent each of the reactions you observed. Be sure to indicate the states of all reactants and products.

(k) Referring to the periodic table, create a hypothesis to explain your activity series.

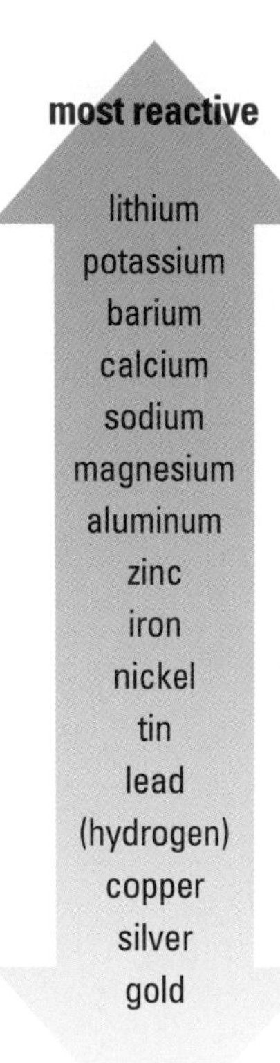

Figure 3
In the activity series of metals, each metal will displace any metal listed below it. This series is based on experiments like Investigation 3.3.2. Other series, using different criteria, may be slightly different. Hydrogen is usually included in activity series, even though it is not a metal, because hydrogen can form positive ions, just like the metals.

Single Displacement Reactions and the Activity Series

An activity series of metals appears in **Figure 3**. It consists of a list of metallic elements arranged in order of decreasing reactivity. Whether or not a single displacement reaction will proceed can usually (but not always) be predicted by referring to the activity series. Each metal element will displace a metal ion that appears below it in the series.

We can predict, by looking at **Figure 3**, that there are several metals that will displace hydrogen from its compounds with nonmetals, including water and acids. Experimental evidence supports this prediction. Magnesium, aluminum, zinc, iron, nickel, tin, and lead will all react with aqueous solutions of acids, displacing hydrogen. The very reactive elements (potassium, barium, calcium, sodium, and lithium) will react vigorously with water and acids at room temperature to produce hydrogen gas. The other product of these reactions with water is an aqueous solution that contains hydroxides (strong bases). Since these reactions are violent and dangerous, this is not the procedure used to manufacture the metal hydroxides of these elements. Rather, they are prepared by reacting the metal oxide with water.

Copper, silver, and gold do not react with acids or water. This evidence supports their placement below hydrogen in the activity series.

The concept of electronegativity was created to help predict properties of elements like the activity series of the metals. For the metals, the lower the electronegativity, the more reactive the metal should be. As you can see (**Table 1**) electronegativity works reasonably well in predicting the activity series, with some exceptions.

Table 1: Electronegativity of Selected Elements

Element	Electronegativity
lithium	1.0
potassium	0.8
barium	0.9
calcium	1.0
sodium	0.9
magnesium	1.2
aluminum	1.5
zinc	1.6
iron	1.8
nickel	1.8
tin	1.8
lead	1.8
hydrogen	2.1
copper	1.9
silver	1.9
gold	2.4

Sample Problem 1

In welding operations, aluminum is reacted with iron oxide in what is called the Thermite process. Use the activity series to predict the products of this reaction and represent the reaction in a balanced chemical equation.

Solution

According to the activity series, aluminum is more reactive than iron and will displace it from iron oxide. The reaction will produce iron and aluminum oxide:

$$2\ Al_{(s)} + Fe_2O_{3(s)} \rightarrow 2\ Fe_{(s)} + Al_2O_{3(s)}$$

Sample Problem 2

Use the activity series to predict whether zinc reacts with magnesium nitrate and, if so, the products of this reaction. Represent the reaction in a balanced chemical equation.

Solution

Zinc is listed below magnesium in the activity series so it will not displace magnesium from the compound magnesium nitrate. Therefore the reaction will not take place.

$$Zn_{(s)} + Mg(NO_3)_{2(aq)} \rightarrow NR$$

Note that NR is written in place of the products in the reaction in Sample Problem 2, indicating No Reaction—zinc and magnesium nitrate do not react.

The halogens also have their own activity series (**Figure 4**). A halogen will be displaced from a compound by a halogen listed above it in this activity series. We might predict this sequence based on the concept of electronegativity. As you move down the group of halogens, electronegativity decreases. Iodine is the least reactive halogen and fluorine is the most reactive in this group.

Trends in reactivity within the same period are also evident among the nonmetals. Electronegativity increases as you move from left to right within the same period, and the nonmetals show a general tendency to increase in reactivity as you move in the same direction across a period.

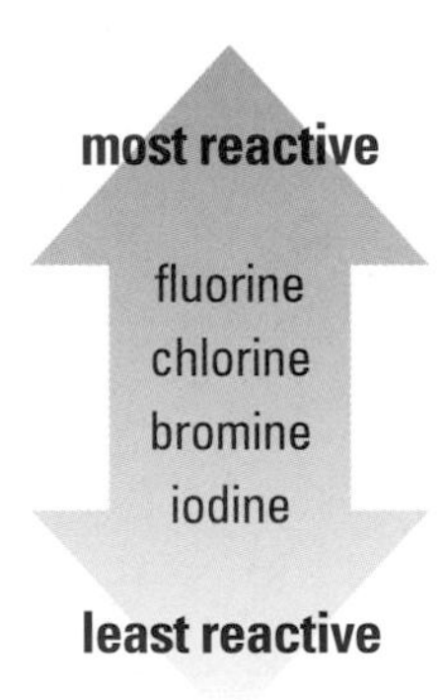

Figure 4
An activity series for the halogens

Sample Problem 3

Predict the products of the reaction (if any) between chlorine gas and a solution of sodium bromide obtained from seawater (brine). Represent the reaction in a balanced equation.

Solution

The products are sodium chloride and bromine.

$$Cl_{2(g)} + 2\ NaBr_{(aq)} \rightarrow 2\ NaCl_{(aq)} + Br_{2(g)}$$

Sample Problem 4

Predict the products of the reaction (if any) between iodine and a solution of sodium chloride. Represent the reaction in a balanced equation.

Solution

The reaction will not proceed.

$$I_{2(s)} + NaCl_{(aq)} \rightarrow NR$$

Practice

Understanding Concepts

1. What types of reactants are likely to be involved in a single displacement reaction?
2. What is meant by the term "activity series"?
3. For each of the following, use an activity series to determine which single displacement reactions will proceed. For the reactions that do occur, predict the products and complete and balance the equation. Note reactions that do not occur with NR.
   (a) $Zn_{(s)} + CuCl_{2(aq)} \rightarrow$
   (b) $Br_{2(aq)} + CaCl_{2(aq)} \rightarrow$
   (c) $Pb_{(s)} + HCl_{(aq)} \rightarrow$
   (d) $Cl_{2(aq)} + NaI_{(aq)} \rightarrow$
   (e) $Ca_{(s)} + H_2O_{(l)} \rightarrow$
   (f) $Au_{(s)} + ZnSO_{4(aq)} \rightarrow$
   (g) $Sn_{(s)} + AgNO_{3(aq)} \rightarrow$
   (h) $Al_{(s)} + H_2O_{(l)} \rightarrow$
   (i) $Br_{2(aq)} + MgI_{2(aq)} \rightarrow$
   (j) $Al_{(s)} + ZnSO_{4(aq)} \rightarrow$
4. Write a generalization about the order of reactivity of a group of elements in the periodic table. Support your generalization with examples.
5. Moving from left to right within the same period, you will observe a general decrease in the reactivity of the metallic elements. However, a general increase in reactivity is observed among the nonmetals. Explain this observation.
6. Chromium, manganese, and barium are all examples of metals that can be prepared in a pure form by combining oxides of these elements with reactive metals. When aluminum is chosen as the reactive metal, the process is referred to as the Goldschmidt process or the Thermite process.
   (a) What type of reaction is the Goldschmidt process?
   (b) Write the formulas for the common oxides formed by chromium, manganese, and barium.
   (c) Predict the products formed by each of the three oxides in a Goldschmidt process, using a balanced chemical equation.
   (d) Explain briefly why pure metals are produced in these reactions.

7. Metallurgy is the study of metals and those processes involved in their extraction from ores. In a metallurgical process, titanium tetrachloride, $TiCl_{4(s)}$, is prepared by the reaction of titanium dioxide, carbon, and chlorine. The product is purified in a distillation process and further processed by reacting it with magnesium at a high temperature.
   (a) Write word equations for the reactions described above.
   (b) Represent the reaction of titanium tetrachloride with magnesium as a chemical equation. What type of chemical reaction is this?
   (c) Explain why the reaction in (b) takes place.

DID YOU KNOW ?

Hi-Tech Role for an Ancient Metal

Because of its lack of reactivity and high conductivity, gold is used for coating the electrical contacts on computer memory chips. There is an industry growing up around the recycling of computer parts, to reclaim the gold and other valuable components.

Extracting Metals

As a result of their greater reactivity, many of the metals listed above hydrogen in the activity series are most commonly found in compounds, ores, rather than as elements. Many can be mined from the Earth's crust. The most common types of metal ores are oxides, sulfides, halides, carbonates, sulfates, and silicates.

Copper is sometimes found relatively pure as an element, but usually it is mined as carbonate or oxide ores. These ores are treated with dilute sulfuric acid to leach the copper out as a copper(II) sulfate solution. This solution is then reacted with iron. Since iron is higher in the activity series than copper, the iron will displace the copper from the solution to yield solid copper and a solution of iron(II) sulfate.

$$Fe_{(s)} + CuSO_{4(aq)} \rightarrow FeSO_{4(aq)} + Cu_{(s)}$$

According to the activity series, silver and gold are among the least reactive elements, so they would tend to be displaced from any compound by almost any other metal, leaving the precious metals as elements in the Earth's crust. This is largely the case—gold is often found as small pieces of elemental metal (nuggets) inside rock. As gold-containing ore is eroded by wind and rain, small pieces of gold break free and are washed into nearby streams and rivers. During the Yukon gold rush of 1897–1898, thousands of people, hoping to make their fortunes, panned for gold in the riverbeds close to where the metal had previously been discovered (**Figure 5**).

Figure 5
The gold-panners' technique was simple, requiring only a shallow pan. The gold panners placed sand from the riverbed in the pan with some water, then swirled the pan so that the water gradually washed away the lighter sand. Heavier flecks and nuggets of gold were left behind.

Rock that contains eight grams of gold per tonne of rock is considered good ore, so larger-scale methods of extracting gold from ore involve processing huge masses of rock. The older method involved mixing crushed ore with mercury, which formed an **alloy** (mercury alloys are often called amalgams) with the gold that was separated from the remaining rock. This amalgam was then heated to remove the mercury (which would be reused), leaving "gold sponge." A more modern method of gold extraction involves grinding the ore, mixing it with lime, reacting the mixture with cyanide (to encourage the gold to ionize), adsorbing the gold onto carbon, and washing off the gold with superheated water.

alloy: a homogeneous mixture (a solution) of two or more metals

Similar processes can be used to recover silver from its native ore.

Not surprisingly, there are environmental concerns with both of these methods of extracting precious metals: mercury and cyanide are both extremely toxic; the volumes of ore that are subjected to these procedures result in huge open-pit mines and large heaps of leftover tailings; and nearby rivers and streams can be damaged by the chemical and particulate runoff from the mines.

Practice

Understanding Concepts

8. The activity series plays an important role in many mining processes, especially when metals are extracted from their ores. Elements that are high in the activity series, such as calcium and sodium, are never extracted from compounds in an aqueous environment. Provide a possible explanation for this.

Making Connections

9. (a) Choose a metallic element that is mined in Canada. Use the Internet to research how and where it is mined, extracted, and purified, and the uses of the metal.
   (b) Compile your findings about the metal's production and uses in a table, listing positive and negative aspects. Use your table to help you decide on changes that could be made in the way we use the metal or the way we obtain it. Report on your findings.

   Follow the links for Nelson Chemistry 11, 3.3.

www.science.nelson.com

Figure 6
A copper-plated welding rod and an acetylene torch can be used to join two metal objects.

Corrosion and Protection

corrosion: the unwanted reaction of metals with chemicals in the environment

The protection of metals against **corrosion** also has its basis in the activity series. Corrosion, which can result in the destruction of metal products, is a costly problem. The most common form of corrosion occurs when metals react with atmospheric oxygen in the presence of moisture. The rusting of iron, forming hydrated iron(III) oxide, is a familiar example of this process. The rust is slightly soluble, and fairly soft, so tends to flake away, leaving more iron exposed to the air and ready to rust further. To reduce this problem, a thin layer of a less reactive metal is sometimes plated onto iron to form a barrier between the iron and the air. For example, a welder might use a welding rod consisting of low-carbon steel (an alloy of iron with a small amount of carbon) plated with a thin layer of copper (**Figure 6**). Copper is listed below iron in the activity series, so is less likely than iron to corrode.

galvanizing: the coating of iron or steel with zinc to prevent rusting

Another means of preventing corrosion is **galvanizing**, in which iron is coated with another element. This time, however, a *more* reactive metal—zinc—is used to coat the iron. The zinc reacts quickly with oxygen to form hard, insoluble zinc oxide. This oxide protects the iron beneath by forming a barrier between the iron and the air. Galvanized iron nails are often used in the construction industry. Galvanized iron is also found in water tanks, roofs, and chain-link fences (**Figure 7**). Eventually, rusting may still occur as erosion of the zinc oxide slowly leads to the exposure of the iron.

Figure 7
The metal in this fence is galvanized iron—iron coated with zinc oxide.

Finally, iron can be made resistant to corrosion by combining it with small amounts of other metals to form stainless steel, an alloy. In addition to iron, stainless steel contains carbon and metals such as chromium. The carbon increases the hardness of the iron (allowing it, for example, to take and hold a sharp edge). Chromium is added to increase iron's resistance to corrosion. Alloying iron with chromium decreases the reactivity of iron.

Investigation 3.3.3

Testing the Activity Series

INQUIRY SKILLS

- ○ Questioning
- ○ Hypothesizing
- ● Predicting
- ● Planning
- ● Conducting
- ● Recording
- ● Analyzing
- ● Evaluating
- ● Communicating

The activity series of metals is a useful tool for predicting whether a particular single displacement reaction will occur. Its other uses include helping us to predict how quickly a reaction will occur.

In this investigation, you will put the activity series to the test. You will use the activity series to predict the reactivity of various metals in dilute hydrochloric acid, and then evaluate your predictions. Complete the lab report, including the **Prediction, Experimental Design, Analysis,** and **Evaluation** sections.

Question

What is the order of reactivity of copper, iron, magnesium, and zinc with dilute hydrochloric acid?

Prediction

(a) Use the activity series to order the provided metals from most reactive to least reactive.

Experimental Design

(b) Design an investigation to test your Prediction, including a description of variables. How will you ensure that the reactivity of each metal is tested fairly?

(c) Complete a Procedure for your design, including the numbered steps that you will follow and any necessary safety procedures. Have your Procedure approved by your teacher before carrying it out.

Materials

lab apron
eye protection
equal-sized pieces of the following metals:
  copper
  iron
  magnesium
  zinc
dilute hydrochloric acid
steel wool or sandpaper
6 large test tubes
6 one-hole rubber stoppers
test-tube rack
medicine dropper
250-mL beaker

(d) Add Materials you think will be necessary.

Hydrochloric acid is corrosive. Eye protection and lab aprons should be worn at all times.

Dispose of the acid in the sink with lots of running water.

Procedure

1. Carry out your Procedure. Be sure to record all of your observations.

Analysis

(e) Comment on the vigour of the reactions, and what this tells you about the reactivity of the reactants.

(f) Use your observations to organize the metals in order of reactivity from most to least reactive.

Evaluation

(g) Evaluate your evidence. Was there anything in the design or execution of your investigation that could have resulted in poor or unreliable evidence?
(h) Compare your empirical activity series, the one arrived at in your Analysis, with the theoretical series in your Prediction. How useful is the accepted activity series for predicting the relative reactivities of a series of metals?

Synthesis

(i) Suggest some explanations for any discrepancies between your empirical activity series and the series shown in **Figure 3**, page 126.
(j) In the reactions that did proceed, what type(s) of reaction took place? Write balanced chemical equations to represent these reactions.

Alloys

Alloys consist of metals (and sometimes other substances, such as carbon) that have been mixed while the principal metals are in liquid phase and then cooled to the solid state. They are, therefore, solid solutions. Alloys often have characteristics different from the "parent" metals. They are almost exclusively made to address a certain technological demand—one that could not be met by the properties of a single pure metal. For example, pure aluminum is very light, making it a popular material for the engines and bodies of vehicles and aircraft. However, in its pure form it is also rather weak. When small amounts of other metals are added, it becomes much stronger while remaining relatively light. Consequently, parts used in the construction of vehicles and aircraft are often made of aluminum alloys. The most common aluminum alloys contain metals such as copper, magnesium, and manganese. These alloys are also used to make the cans for carbonated beverages, bicycle frames, and the siding on homes. Because aluminum is a very useful material, and expensive to produce from ore, there is a strong market for recycled aluminum. It is one of the materials that is collected by many of Ontario's blue-box household-recycling programs.

INQUIRY SKILLS

- ○ Questioning
- ○ Hypothesizing
- ○ Predicting
- ○ Planning
- ○ Conducting
- ○ Recording
- ● Analyzing
- ● Evaluating
- ○ Communicating

Lab Exercise 3.3.1

Corrosion in Seawater

When designing a plant that extracts ionic compounds from seawater, the engineers must select materials that will not be damaged by the seawater, which is quite corrosive. To help in the choice of an appropriate material for pipes, an engineer investigated the corrosion resistance of various metal alloys in seawater.

There are, of course, different costs associated with different types of metal: Mild steel, cast iron, and steel coated with cement are relatively inexpensive options. Higher-cost choices are copper-based alloys and stainless steel. These costs would have to be considered when recommending an appropriate material. Higher-diameter pipes (which could be used to reduce the velocity of the seawater flowing through them while delivering the same amount of material) cost more than low-diameter pipes because more material is needed to make them.

Complete the **Experimental Design**, **Analysis**, **Evaluation**, and **Synthesis** sections of the report.

Question

Which alloy is the best choice for use in pipes conducting seawater to a processing plant, and should the pipes be of low diameter or high diameter?

Experimental Design

Surfaces of each of the different alloys (**Table 2**) are exposed for approximately one year to seawater moving at two different velocities—8.2 m/s to simulate high-diameter pipe and 35-42 m/s to simulate low-diameter pipe. The amount of corrosion is determined by measuring the thickness of the alloys before the experiment and at its end. Measurements made in this experiment are compared to literature values for corrosion in still seawater.

(a) Identify independent and dependent variables in this experiment.

Evidence

Table 3: Effect of Velocity on the Corrosion of Metals in Seawater

Alloy	Quiet seawater (literature)	Moving seawater 8.2 m/s	35-42 m/s
	corrosion rate mm/a	corrosion rate mm/a	corrosion rate mm/a
carbon steel	0.075	-	4.5
grey cast iron	0.55	4.4	13.2
admiralty gunmetal	0.027	0.9	1.07
copper alloy	0.017	1.8	1.32
nickel resist cast iron	0.02	0.2	0.97
nickel-aluminum bronze	0.055	0.22	0.97
copper-nickel alloy	<0.02	0.12	1.47
type 316 stainless steel	0.02	<0.02	<0.01
6% Mo stainless steel	0.01	<0.02	<0.01
nickel-copper alloy 400	0.02	<0.01	0.01

Source: http://marine.copper.org/2-corros.html#up

Table 2: Composition of Test Metals

Low-Cost Alloys	Composition
carbon steel	composition not provided (iron with some carbon)
grey cast iron	composition not provided (mostly iron)
admiralty gunmetal	88% copper 10% tin 2% zinc
Higher-Cost Alloys	**Composition**
copper alloy	85% copper 10% zinc 5% lead
nickel resist cast iron (type 1B)	2.0–3.0% carbon 1.0–2.8 % silicon 1.0–1.5 % manganese 3.5– 7.5 % nickel 5.5–7.5 % copper 2.75–3.5% chromium remainder iron
nickel-aluminum bronze	80% copper 10% aluminum 5% iron 5% nickel
copper-nickel alloy	67% copper 30% nickel 1% iron 2% chromium
type 316 stainless steel	composition not provided (iron with some carbon and other metals)
6% Mo stainless steel	6% molybdenum (remainder not provided)
nickel-copper alloy 400	66% nickel 31.5% copper 1.35% iron 0.9% manganese

Analysis

(b) In general, is there a relationship between the speed of seawater and the amount of corrosion? Support your answer using the Evidence in **Table 3**.

(c) Describe a solution to the engineer's problem that you would expect to result in the low corrosion of pipes.

(d) Describe the best cheap solution.

Evaluation

(e) Is the Evidence sufficient to decide on a best material for the pipes? Provide your reasoning. If the Evidence is insufficient, suggest a further experiment that would clear up any remaining uncertainties.

Synthesis

(f) Under what circumstances might the engineer recommend the use of a cheap option? What might be the benefits and risks of this choice?

(g) Are there any circumstances under which the engineer should recommend the use of a more expensive option? Explain.

Figure 8
By law, all gold jewellery sold in Canada must be stamped with its degree of purity.

Familiar Alloys

You are probably most familiar with stainless steel in the kitchen. Chopping knives, cutlery, pans, cutting blades in the food processor—all are made of stainless steel. Steel is ideal for these uses: It is rigid and strong, keeps a sharp edge, does not react with the acids in food, and does not react with oxygen in the air (it won't rust if you leave it wet in a drying rack).

Gold is a soft metal. In pure form, a fine gold chain would stretch and break under stress. In the manufacture of fine gold jewellery, metals such as silver, copper, and zinc can be added to gold to form alloys harder than pure gold, but with a similar appearance. In North America, the karat system is used to indicate the purity of gold. When jewellery is manufactured it is stamped with a number followed by the letter K (meaning "parts per 24") (**Figure 8**). An object stamped 24K is 100% gold, but such objects are extremely rare—they would scratch and deform easily. Fine jewellery is often manufactured in 18K (75% pure) gold, which is harder and so more practical. 14K (58.3% pure) and 10K (41.7% pure) gold are even harder.

The compositions of some common alloys are shown in **Table 4**.

Table 4: Compostions of Common Alloys

Alloy	Constituent metals by mass	Property
yellow brass	Cu (70%) Zn (30%)	harder and more resistant to corrosion than copper
stainless steel (Grade 430)	Fe (81.06-79.06%) Cr (16-18%) Ni (0.75%) C (0.12%) P (0.04%) S (0.03%) Si (1.00%)	more resistant to corrosion than iron
18K yellow gold	Au (75%) Ag (13-15%) Cu (10-12%)	harder and less malleable than 24K gold
14K yellow gold	Au (58.33%) Ag (13-15%) Cu (26.75%) Zn (0.14%)	harder than 18K gold

The reasons for the special properties of alloys are very complicated, but have to do with the bonding among the elements within the solution. Pure metals are malleable because the atoms are, in effect, floating in a sea of free electrons. When other elements are added, the bonds change, electrons become locked in place, and the entire structure becomes much more rigid. This accounts for the hardness of stainless steel, for example, and its ability to keep a sharp edge (**Figure 9**).

Figure 9
The strong bonds in stainless steel keep the metal rigid. Its cutting edge stays sharp for years of use.

Practice

Understanding Concepts

10. Briefly describe two alloys, including their properties. Compare the properties of the alloys with those of their components.

Making Connections

11. Women sometimes have more acidic skin than men.
    (a) Develop an explanation for the observation that most men wear 14K gold rings, but most women wear 18K gold rings.
    (b) Suggest what reaction might be occurring when a person's skin develops a greenish stain beneath a 14K gold bracelet.
12. The word "steel" covers a huge range of alloys of iron. Use the Internet to research the composition of various steels. Choose one alloy and list its properties and its applications. Write a short "infomercial," advertising the benefits of this material to potential users. Include any precautions necessary for its safe use.

    Follow the links for Nelson Chemistry 11, 3.3.

    GO TO **www.science.nelson.com**
13. In Kitimat, British Columbia, hydroelectric power is used to supply the energy needed in the production of aluminum metal from $Al_2O_{3(s)}$. The demand for electricity has resulted in the damming of many rivers in British Columbia for hydroelectric power. Use the Internet to research the applications of aluminum and its alloys and the environmental issues surrounding aluminum production. Use your findings to comment on the following statement: Risks to the environment posed by mining and refining aluminum are outweighed by the technological benefits of aluminum alloys.

    Follow the links for Nelson Chemistry 11, 3.3.

    GO TO **www.science.nelson.com**

Careers in Metallurgy

Careers in metallurgy involve the study and use of properties and characteristics of metals. There are many different employment opportunities associated with metallurgy, involving planning, design, and development of machinery and methods to extract, refine, and process metals and alloys. Many metallurgical processes involve chemical reactions such as single displacement reactions.

Hydrometallurgist

The work of a hydrometallurgist can involve a variety of reactions in solution. Most often, however, a hydrometallurgist is specifically required to develop and oversee the process of metal recovery from solutions. By manipulating temperature, pressure, concentration, and reactants, the aim is to achieve the maximum yield of the desired products as efficiently as possible. With our society's increasing focus on recycling metals, this career path is gaining in importance.

Metallurgical Engineer

This is a specialized branch of materials engineering. Metallurgical engineers work in a variety of areas, from processes and methods for extracting metals from ore and mineral resources, to the study of metals and alloys in useful products.

Corrosion Engineer

A corrosion engineer might focus on ships, bridges, car bodies, or household appliances. The main task involves designing ways to reduce corrosion with paints, other coatings, or by restricting electrical contact with other materials. Research, communication, and teamwork skills are also important.

Practice

Making Connections

14. Research one of the careers listed or a related career and write a report that
    (a) provides a general description of the nature of the work and how chemical reactions are involved;
    (b) describes the educational background and the length of study required to obtain employment in this field;
    (c) gives examples of programs offered by educational institutions leading to this career;
    (d) forecasts employment trends for this field; and
    (e) describes working conditions and salary.

3.4 Double Displacement Reactions

Magnesium chloride, a soluble ionic compound, occurs naturally in seawater. In the large-scale commercial preparation of magnesium, calcium hydroxide is added to large quantities of seawater, resulting in the production of a white powder, magnesium hydroxide, $Mg(OH)_2$:

$$MgCl_{2(aq)} + Ca(OH)_{2(aq)} \rightarrow Mg(OH)_{2(s)} + CaCl_{2(aq)}$$

double displacement reaction: a reaction in which aqueous ionic compounds rearrange cations and anions, resulting in the formation of new compounds

The reaction used to extract magnesium from seawater is known as a **double displacement reaction.** This type of reaction involves two reactants recombining to form two new compounds (**Figure 1**). We can write a generalization for such reactions:

$$AB + CD \rightarrow AD + CB$$

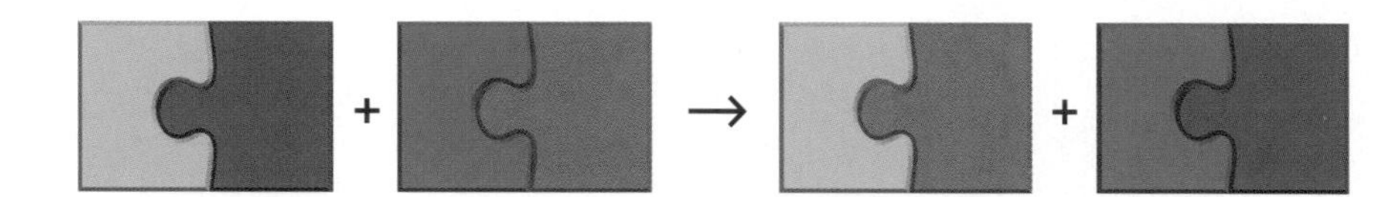

Figure 1
In a double displacement reaction, the cations from the two compounds switch over, making two new compounds.

The reaction involves two ionic compounds as reactants. As the equation above shows, the ions "change partners" to form the products. This type of reaction commonly occurs in aqueous solutions. In order to predict products of single and double displacement reactions, you need to understand the nature of solutions and learn a method of determining whether a substance dissolves in water to an appreciable extent.

solute: a substance that dissolves in a solvent to form a solution

solvent: the medium in which a solute dissolves

solubility: a property of a substance; a measure of the maximum amount of a solute that will dissolve in a known amount of solvent at a given temperature and pressure

precipitate: a solid formed from a reaction that takes place in solution

Solubility

A solution is a homogeneous mixture of a **solute** and a **solvent.** Every solute has its own **solubility** in a given solvent. You will learn more about solubility in Unit 3, but it can be simply described as the maximum quantity of a solute that will dissolve in a solvent at a given temperature and pressure. For example, if you add salt (sodium chloride) repeatedly to water, stirring after each addition, the salt will dissolve until the salt solution reaches its maximum concentration. After this point, any salt you add will simply remain in the solid state and settle to the bottom of the container.

We can take advantage of the individual solubilities of different substances in water. If the product of a reaction is very soluble, it will stay in solution, so we assign it the subscript (aq). However, if the product is only slightly soluble, effectively insoluble, it is likely that more will be produced than can dissolve. If the product is a solid, it will be visible as a **precipitate** and will be assigned the subscript (s) (**Figure 2**). If it is a gas, it will appear as bubbles and be given the subscript (g).

Figure 2
A precipitate is formed when aqueous solutions of calcium chloride and sodium carbonate are mixed.

Sometimes an experimenter wants to know whether the ions of a certain element are present in a solution. A quick way of testing is to try to make the ions react with another substance to form a product that is only slightly soluble. The solid will precipitate out and become visible. But how do we know which substances are likely to have high solubilities, and which low solubilities? A solubility table outlines the solubilities of a large number of ionic compounds (**Table 1**). You can use this table, or the table in Appendix C, to predict whether an ionic compound, formed as the product of a chemical reaction in solution, is likely to form a solid precipitate (i.e., have low solubility) or remain in solution (i.e., have high solubility). You find the solubility of a compound by looking up the anion first, then finding the cation in the body of the table, and following across to the left, to read the solubility.

Table 1: Solubility of Ionic Compounds at SATP

		Anions						
		Cl^-, Br^-, I^-	S^{2-}	OH^-	SO_4^{2-}	CO_3^{2-}, PO_4^{3-}, SO_3^{2-}	$C_2H_3O_2^-$	NO_3^-
Cations	High solubility (aq) ≥0.1 mol/L (at SATP)	most	Group 1, NH_4^+ Group 2	Group 1, NH_4^+ Sr^{2+}, Ba^{2+}, Tl^+	most	Group 1, NH_4^+	most	all
		All Group 1 compounds, including acids, and all ammonium compounds are assumed to have high solubility in water.						
	Low Solubility (s) <0.1 mol/L (at SATP)	Ag^+, Pb^{2+}, Tl^+, Hg_2^{2+}, (Hg^+), Cu^+	most	most	Ag^+, Pb^{2+}, Ca^{2+}, Ba^{2+}, Sr^{2+}, Ra^{2+}	most	Ag^+	none

Suppose a technician has to test the water in a stream for a variety of substances, including sulfates. The planned test would produce calcium sulfate, $CaSO_4$. What is the solubility of $CaSO_4$? Would calcium sulfate, if it forms, precipitate out? To solve this problem, we can refer to the solubility table (**Table 1**, and in Appendix C). Looking at the top rows of the table, we can locate the column containing the anion SO_4^{2-}. We then look in the two boxes below this anion to find out whether the sulfate containing the cation Ca^{2+} has high solubility or low solubility. The table tells us its solubility is low, and so it will tend to precipitate out of any solution in which it is present.

If we had checked for the solubility of copper(II) sulfate, $CuSO_4$, we would discover it belongs in "most" category for the sulfate anion: high solubility. Copper(II) sulfate would tend to stay dissolved in any aqueous solution in which it forms.

Sample Problem 1

A pyrotechnist is setting up a fireworks display on a barge floating in Lake Ontario. A firework containing a carefully measured quantity of potassium nitrate accidentally falls into the lake. Will the potassium nitrate dissolve to any significant degree?

Solution

All nitrates have high solubility. A significant amount of the potassium nitrate is likely to dissolve.

Practice

Understanding Concepts

1. Which group of monatomic ions forms compounds that all have high solubility in water?
2. Which positive polyatomic ion forms compounds that all have high solubility in water?
3. Use **Table 1** (page 137) to predict the solubility of the following ionic compounds. Write the chemical formula and indicate its expected state in water in a subscript, either (aq) or (s).
   (a) KCl (fertilizer)
   (b) $Ca(NO_3)_2$ (fireworks)
   (c) Na_2SO_4 (Glauber's salt)
   (d) $AgC_2H_3O_2$ (an oxidizing agent)
   (e) ammonium bromide (fireproofing)
   (f) barium sulfide (vulcanizing)
   (g) lead(II) iodide (photography)
   (h) calcium hydroxide (slaked lime)
   (i) iron(III) hydroxide (rust)
   (j) lead(II) sulfate (car battery)
   (k) calcium phosphate (rock phosphorus)
   (l) potassium permanganate (fungicide)
   (m) ammonium nitrate (fertilizer)
   (n) cobalt(II) chloride (humidistat)
   (o) calcium carbonate (limestone)

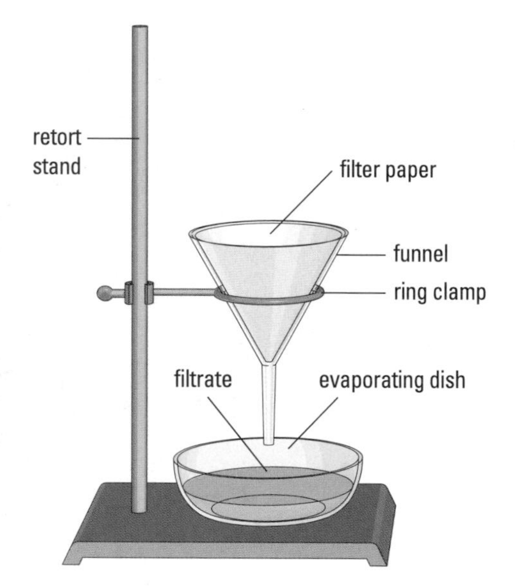

Figure 3
The filtration apparatus is designed to separate mixtures of solids and liquids, and is suitable for isolating a precipitate from a liquid. The filtrate can be allowed to evaporate, revealing any more soluble compounds that it contained.

Types of Double Displacement Reactions

There are three clues that tell us that a reaction is proceeding: if a precipitate is formed, if a gas is produced, or if the solution becomes more neutral.

Precipitation Reactions

When separate aqueous solutions of calcium chloride and sodium carbonate are mixed together, calcium carbonate and sodium chloride are produced.

$$\underset{\text{(dissolved)}}{CaCl_{2(aq)}} + \underset{\text{(dissolved)}}{Na_2CO_{3(aq)}} \rightarrow \underset{\text{(solid precipitate)}}{CaCO_{3(s)}} + \underset{\text{(dissolved)}}{2\ NaCl_{(aq)}}$$

As a result of its low solubility, the calcium carbonate formed by the reaction appears as a solid. Sodium chloride, a highly soluble compound, will remain in solution. Many double displacement reactions will form products that have low solubility. If one of the products has low solubility, it may form a precipitate. Precipitates can be removed from a solution by filtration (**Figure 3**).

To predict whether a precipitate might form from a double displacement reaction, we need to identify the ionic compounds that will be formed by the reaction and determine whether they will be soluble or insoluble. This information can be obtained from the solubility table (**Table 1** and in Appendix C). Recall that insoluble substances produced in the reaction will be visible as a precipitate, while soluble products will remain in solution.

Sample Problem 2

Write a balanced equation to represent the reaction of an aqueous solution of barium chloride with an aqueous solution of potassium sulfate. Indicate the physical state of the reactants and products involved.

Solution

$$BaCl_{2(aq)} + K_2SO_{4(aq)} \rightarrow 2\ KCl_{(aq)} + BaSO_{4(s)}$$

Reactions Producing a Gas

Some double displacement reactions result in a gas product. For example, sodium sulfide solution and hydrochloric acid react to produce hydrogen sulfide gas:

$$Na_2S_{(aq)} + 2\ HCl_{(aq)} \rightarrow 2\ NaCl_{(aq)} + H_2S_{(g)}$$

Neutralization Reactions

Finally, an acid can react with a base in a third type of double displacement reaction, sometimes referred to as a **neutralization reaction.** (You will learn more about acids and bases in Chapter 8.) In a neutralization reaction, the products formed are neither acidic nor basic. They are typically water and an ionic compound (referred to as a "salt"). The reaction between magnesium hydroxide and hydrochloric acid is an example of this type of double displacement reaction:

neutralization reaction: a double displacement reaction between an acid and a base to produce an ionic compound (a salt) and usually water

$$Mg(OH)_{2(s)} + 2\ HCl_{(aq)} \rightarrow MgCl_{2(aq)} + 2\ H_2O_{(l)}$$

You may have an aqueous suspension of magnesium hydroxide in your medicine cabinet at home. It is more commonly known as the antacid "milk of magnesia." The above reaction explains how milk of magnesia relieves acid indigestion by neutralizing excess stomach acid.

SUMMARY Predicting Double Displacement Reactions

Double displacement reactions are likely to proceed if

- one of the products is a precipitate;
- one of the products is a gas; or
- the reactants are an acid and a base, and the result is a neutralization reaction producing a salt and water.

Investigation 3.4.1

A Double Displacement Reaction

INQUIRY SKILLS

- ○ Questioning
- ○ Hypothesizing
- ● Predicting
- ● Planning
- ● Conducting
- ● Recording
- ● Analyzing
- ● Evaluating
- ● Communicating

When double displacement reactions occur among ions in solution, the products often include one of the following: water, a precipitate, or a gas. The formation of a precipitate, or the evolution of a gas, along with the nature of the reactants, provide evidence that a double displacement reaction has occurred.

The purpose of this investigation is to use the generalizations you have learned about double displacement reactions and solubilities to predict the products of a reaction. Complete the **Prediction**, **Experimental Design**, **Analysis**, **Evaluation**, and **Synthesis** sections of the investigation report.

Question

Will aqueous ammonium phosphate and iron(III) nitrate react and, if so, what will be the products of the reaction?

Prediction

(a) Predict whether aqueous ammonium phosphate and iron(III) nitrate will react, using a chemical equation.

Experimental Design

The reactants will be mixed and observed. Evidence will be obtained from these observations to support or falsify the Prediction.

(b) State what Evidence you will look for to confirm that a reaction is proceeding.

(c) Write a Procedure for your investigation, including how you plan to separate any products that might form and how you intend to identify those products. Do not carry out your Procedure until you have your teacher's approval.

Materials

Wear eye protection and a lab apron at all times. Avoid spills.

Dispose of all waste materials as directed by your teacher.

eye protection
laboratory apron
funnel
filter paper
retort stand
ring clamp
evaporating dish
distilled water (in wash bottle)
solutions of:
- ammonium phosphate
- iron(III) nitrate

test-tube rack
2 small test tubes

Procedure

1. With your teacher's approval, carry out your Procedure.

Analysis

(d) Analyze your observations and use them to answer the Question.

Evaluation

(e) Compare your answer in (d) to your Prediction. Does your Evidence support your Prediction? If your Evidence failed to support your Prediction, try to account for the discrepancy.

Synthesis

(f) Write a word equation and a balanced chemical equation outlining the reaction that, according to your Evidence, took place.

Investigation 3.4.2

Identifying an Unknown

INQUIRY SKILLS

- Questioning (filled)
- Hypothesizing (open)
- Predicting (filled)
- Planning (filled)
- Conducting (filled)
- Recording (filled)
- Analyzing (filled)
- Evaluating (filled)
- Communicating (filled)

Analytical chemists often make use of the potential for precipitate formation in double displacement reactions to determine the presence of a particular chemical in a sample. This type of analysis is referred to as a qualitative analysis, as only the identity of the chemical is determined, not the amount of chemical present.

In this investigation, you will use the solubility table (**Table 1**, page 137, or in Appendix C), along with experimental techniques learned in Investigation 3.4.1, to identify an unknown solution as containing either silver nitrate or calcium nitrate. Your task is to design the investigation and carry out the analysis. Your report should include completed **Question**, **Prediction**, **Experimental Design**, **Materials**, **Analysis**, and **Evaluation** sections.

Question

(a) Write the Question that you will attempt to answer.

Prediction

(b) Write a prediction in the form of one or more diagnostic tests; for example, If [procedure] and [observation], then [analysis].

Experimental Design

(c) Write a Procedure to determine the identity of the unknown compound. Include any safety precautions that should be taken, and instructions for disposal of waste materials. Submit your Procedure to your teacher for approval before beginning your experiment.

Materials

solution of an unknown ionic compound

(d) List the Materials you will need to conduct your tests.

Procedure

1. Carry out your Procedure.

Analysis

(e) Interpret your observations to answer your Question.

(f) Write word equations and chemical equations describing any reactions that took place during your tests.

Evaluation

(g) Did your Experimental Design and Procedure allow you to discover the identity of the compound efficiently and with reasonable certainty? How could you improve your design to increase the certainty of the identification?

Practice

Understanding Concepts

4. What types of reactants are involved in double displacement reactions?

5. Predict the products formed by double displacement reactions in aqueous solutions of each of the following pairs of compounds. In each case, write a balanced chemical equation indicating the physical state of the products formed:
   (a) copper(II) nitrate and magnesium chloride
   (b) barium hydroxide and iron(III) sulfate
   (c) magnesium hydroxide and sulfuric acid
   (d) ammonium sulfide and iron(II) sulfate
6. For double displacement reactions, write the generalizations that can be used to classify the reaction and predict the products.
7. Predict the solubility in water of each of the following compounds by writing the chemical formula along with a state of matter subscript:
   (a) potassium nitrate (saltpetre)
   (b) calcium chloride (desiccant)
   (c) magnesium hydroxide (antacid)
   (d) aluminum sulfate (used in water treatment)
   (e) lead(II) iodide (yellow pigment)
   (f) calcium phosphate (rock phosphorus)
   (g) ammonium carbonate (smelling salts)
8. For each of the products in question 7, suggest a pair of ionic compounds that would produce the product in a double displacement reaction. Write a balanced chemical equation for each reaction.
9. Classify each reaction as single or double displacement:
   (a) $Cu_{(s)} + 2\ AgNO_{3(aq)} \rightarrow 2\ Ag_{(s)} + Cu(NO_3)_{2(aq)}$
   (b) $KI_{(aq)} + AgNO_{3(aq)} \rightarrow AgI_{(s)} + KNO_{3(aq)}$
10. Rewrite each of the following reactions as a chemical equation (including subscripts indicating state), and classify it as single or double displacement.
   (a) chlorine + aqueous sodium bromide $\rightarrow$
   bromine + aqueous sodium chloride
   (b) sulfuric acid + aqueous sodium hydroxide $\rightarrow$
   water + aqueous sodium sulfate
   (c) aqueous calcium nitrate + aqueous sodium phosphate $\rightarrow$
   solid calcium phosphate + aqueous sodium nitrate
11. For each of the following reactions, classify the reaction type as single or double displacement, predict the products of the reaction, and write a complete chemical equation. (Assume the most common ion charge and SATP state of matter, unless indicated otherwise.)
   (a) Silver metal is recovered in a laboratory by placing aluminum foil in aqueous silver nitrate.
   $Al_{(s)} + AgNO_{3(aq)} \rightarrow$
   (b) Bromine is mined from the ocean by bubbling chlorine gas through ocean water containing sodium bromide.
   $Cl_{2(g)} + NaBr_{(aq)} \rightarrow$
   (c) A traditional laboratory method of producing hydrogen gas is to react zinc metal with sulfuric acid.
   zinc + sulfuric acid $\rightarrow$
   (d) The presence of the chloride ion in a water sample is indicated by the formation of a white precipitate when aqueous silver nitrate is added to the sample.
   aqueous silver nitrate + aqueous magnesium chloride $\rightarrow$
   (e) An analytical chemist uses sodium oxalate to precipitate calcium ions in a water sample from an acid lake.
   $Na_2C_2O_{4(aq)} + CaCl_{2(aq)} \rightarrow$
   (f) Sodium metal reacts vigorously with water to produce a flammable gas and a basic (hydroxide) solution.
   sodium + water $\rightarrow$
   (g) When aqueous potassium hydroxide is added to a sample of well

water, the formation of a rusty-brown precipitate indicates the presence of an iron(III) compound in the water.
$KOH_{(aq)} + FeCl_{3(aq)} \rightarrow$

12. One of the major commercial uses of sulfuric acid is in the production of phosphoric acid and calcium sulfate in a double displacement reaction. The phosphoric acid is then isolated from the products and used for fertilizer.
    (a) Identify the other reactant that would be combined with sulfuric acid in this reaction.
    (b) Write a chemical equation describing the double displacement reaction for the formation of phosphoric acid. Indicate the state of all reactants and products at SATP.
    (c) Describe a simple procedure that can be employed to isolate phosphoric acid from the other product of the double displacement reaction.
    (d) Why would the fertilizer manufacturers prefer to use the compound identified in part (a), rather than a compound containing sodium as the cation?

Sections 3.3–3.4 Questions

Understanding Concepts

1. (a) Create a table in your notebook composed of 9 columns and 9 rows. Cross out the cell in the uppermost left-hand corner of the table; you will not be completing this cell. In the remaining cells of the first horizontal row of the table, list the following 8 anions: Cl^-, Br^-, NO_3^-, $C_2H_3O_2^-$, PO_4^{3-}, SO_4^{2-}, OH^-, and SO_3^{2-}. In the cells of the first vertical column on the left, list the following 8 cations: K^+, Ca^{2+}, Al^{3+}, Cu^+, Pb^{2+}, Fe^{3+}, Mg^{2+}, and NH_4^+. Complete the remaining empty cells of the table by writing formulas for each of the 64 compounds the pairs of ions will form.
    (b) Name each of the compounds using the IUPAC system of nomenclature.
    (c) Note which compounds have high solubility in water and which have low solubility by identifying the state of the compound as either solid (s) or aqueous (aq) in water solution.
    (d) Which of the compounds are binary ionic and which are tertiary ionic?
    (e) Predict whether a single displacement reaction would occur if solid zinc were combined with aqueous solutions of each of the soluble compounds. Write word and chemical equations for each of these reactions, indicating the products formed and their states.
2. List at least five of the chemical reactions that can result from the burning of gasoline in a combustion engine. Classify them as combustion, synthesis, decomposition, single displacement, or double displacement reactions. Write chemical equations for each reaction. Note any safety, environmental, or health concerns for each reaction.

Applying Inquiry Skills

3. Elements X and Y are located in the same group in the periodic table. Elements Y and Z are located in the same period. An experiment was carried out to place the three unknown elements in order of reactivity.

(continued)

Experimental Design

Each of the elements, X, Y, and Z, were reacted separately with water and acid.

Evidence

Table 2: Observations of Elements X, Y, and Z in Water and Acid

Element	Reaction with water	Reaction with acid
X	no evidence of chemical change	no evidence of chemical change
Y	bubbles observed	many bubbles observed
Z	no evidence of chemical change	bubbles observed

Analysis

(a) In what order would these elements appear in the activity series?

(b) Relate the position of each of the elements to that of hydrogen in the activity series. Which of the three elements is most likely to be hydrogen? Use the empirical evidence to justify your choice. Provide a theoretical rationale for your choice.

Synthesis

(c) In what order would X and Y appear as you descend the group? In what order would Y and Z appear as you move from left to right within the same period? Justify your answers, using empirical and theoretical knowledge.

(d) You are told that X and Y are in Group 1 and that Y and Z are consecutive elements in Period 3. Write balanced chemical equations representing each reaction that took place.

(e) Classify each of the reactions in part (d). Suggest a diagnostic test that you could use to support your classification.

Making Connections

4. Beryllium was first isolated as a relatively pure metal in 1823, when investigators heated solid beryllium chloride with solid potassium in a platinum crucible.
   (a) Predict the products of this reaction in a balanced equation, and classify the reaction.
   (b) Beryllium chloride, which is still used in the industrial production of beryllium, is a sweet-tasting substance. It can cause lung disease if small particles of the substance are inhaled and can damage skin with repeated contact. It also reacts with water vapour in the air to form $HCl_{(g)}$, a highly corrosive substance. Suggest some safety precautions that would be needed in the industrial storage and use of this substance.
5. Silicon can be obtained from SiO_2, found in ordinary sand. Liquid silicon dioxide is reacted with solid carbon (coke) in a furnace, producing liquid silicon as one of the products. When cooled, the solid silicon is purified by reacting it with chlorine gas, producing liquid silicon tetrachloride. This compound is then reacted with solid magnesium to produce a purer form of solid silicon.
   (a) Predict all of the products of each reaction described above by writing chemical equations for each reaction. Classify each reaction.
   (b) Silicon is now produced in large quantities. Which of the reactions would you consider most hazardous to the environment? Provide your reasoning.

Chapter 3 Summary

Key Expectations

Throughout this chapter, you have had the opportunity to do the following:

- Predict the products of, and use balanced chemical equations to represent, simple chemical reactions, and test the prediction through experimentation. (3.1, 3.2, 3.3)
- Classify chemical reactions as synthesis, decomposition, combustion, single displacement or double displacement, and relate the type of reaction to the nature of the reactants. (3.2, 3.3, 3.4)
- Demonstrate an understanding of the need for the safe use of chemicals in everyday life. (3.2, 3.3, 3.4)
- Relate the reactivity of a series of elements to their positions in the periodic table. (3.3)
- Evaluate and compare the reactivity of metals and alloys, and explain why most metals are found in nature as compounds. (3.3)
- Investigate through experimentation the reaction of elements to produce an activity series. (3.3)
- Use appropriate terms to communicate ideas and information about chemical reactions. (3.3, 3.4)
- Explore careers related to chemical reactions. (3.3)
- Identify chemical substances and reactions in everyday use or of environmental significance. (3.1, 3.2, 3.3, 3.4)

Reflect on your Learning

Revisit your answers on the Reflect on Your Learning questions, at the beginning of this chapter.

- How has your thinking changed?
- What new questions do you have?

Key Terms

acid rain
activity series
alloy
catalyst
catalytic converter
chemical change
chemical equation
coefficient
collision-reaction theory
combination reaction
combustion reaction
corrosion
decomposition reaction
double displacement reaction
galvanizing
greenhouse effect
incomplete combustion
kinetic molecular theory
neutralization reaction
precipitate
reactant
single displacement reaction
solubility
solute
solvent
synthesis reaction
thermal decomposition
word equation

Make a Summary

Starting with the term "chemical reactions," create a concept map to summarize what you have learned in this chapter. Use the key terms and key expectations to help you create the map. Include in your map as many definitions, generalizations, examples, and applications as possible.

Chapter 3 Review

Understanding Concepts

1. Use the kinetic molecular theory to describe the motion of particles that make up solids, liquids, and gases.
2. According to the collision-reaction theory, chemical reactions occur when particles of reactants collide. Under what circumstances would collisions *not* result in a chemical reaction?
3. Classify each of the following equations as synthesis, decomposition, combustion, single displacement, or double displacement. In each case, balance the equation.
   (a) $S_8 + O_2 \rightarrow SO_2$
   (b) $HBr + NaOH \rightarrow NaBr + H_2O$
   (c) $N_2 + H_2 \rightarrow NH_3$
   (d) $PtCl_4 \rightarrow Pt + Cl_2$
   (e) $MgO + Si \rightarrow Mg + SiO_2$
   (f) $Na_2S + HCl \rightarrow NaCl + H_2S$
   (g) $P_4 + O_2 \rightarrow P_4O_{10}$
   (h) $Zn + HCl \rightarrow ZnCl_2 + H_2$
4. Write word and balanced chemical equations (including states) to represent each of the following chemical reactions. In each case, classify the reaction as one of the five types studied in this chapter.
   (a) When heated, potassium chlorate forms potassium chloride and oxygen.
   (b) Hydrogen gas and sodium hydroxide are formed when sodium is added to water.
   (c) Carbon dioxide gas is produced when solid carbon burns in air.
   (d) When zinc is added to aqueous sulfuric acid, zinc sulfate and hydrogen gas are produced.
   (e) When a solution of silver nitrate is added to a solution of potassium iodide, silver iodide and potassium nitrate are formed.
   (f) Aqueous solutions of sodium sulfate and barium chloride react to form barium sulfate and sodium chloride.
   (g) Iron(III) oxide is produced by reacting solid iron with oxygen gas.
   (h) Sulfur trioxide, a gas produced by motor vehicles, reacts with atmospheric water vapour to produce sulfuric acid, contributing to acid rain.
5. Which of the following products of double displacement reactions would precipitate out of aqueous solution?
   (a) barium hydroxide
   (b) copper(II) chloride
   (c) calcium sulfate
   (d) potassium nitrate
   (e) potassium sulfate
   (f) ammonium sulfide
   (g) sodium chloride
   (h) silver iodide
   (i) copper(I) chloride
   (j) lead(II) sulfate
6. For each of the products in question 5 that formed a precipitate, suggest a pair of ionic compounds that would produce the product in a double displacement reaction.
7. On what basis do you predict whether the product of a single displacement reaction will be a metal or a nonmetal?
8. Indicate which of the following reactions will proceed. For each reaction that proceeds, predict the products and write a balanced equation. Write NR for reactions that do not proceed.
   (a) $Li_{(s)} + H_2O_{(l)} \rightarrow$
   (b) $K_{(s)} + H_2O_{(l)} \rightarrow$
   (c) $Cu_{(s)} + AgNO_{3(aq)} \rightarrow$
   (d) $Fe_{(s)} + NaCl_{(aq)} \rightarrow$
   (e) $Mg_{(s)} + Ca(NO_3)_{2(aq)} \rightarrow$
   (f) $Al_{(s)} + HCl_{(aq)} \rightarrow$
   (g) $Pb_{(s)} + Cu(NO_3)_{2(aq)} \rightarrow$
   (h) $F_{2(g)} + HCl_{(aq)} \rightarrow$
   (i) $I_{2(s)} + NaBr_{(aq)} \rightarrow$
9. As part of an experimental procedure, a lab technician bubbles chlorine gas into a solution of potassium iodide. A colour change is observed.
   (a) What type of chemical reaction has occurred?
   (b) Predict the products of the reaction.
   (c) Write a chemical equation to represent the reaction.
   (d) Give a theoretical explanation of why the reaction proceeded.
10. Place each of the following groups of elements in order from least reactive to most reactive. In each case, justify your order.
   (a) bromine, nitrogen, and fluorine
   (b) potassium, magnesium, and rubidium

Applying Inquiry Skills

11. As part of an experiment to try to improve the effectiveness of galvanization, a corrosion engineer placed a sample of a soft, silvery solid into a test tube containing a colourless, odourless liquid. Complete the **Analysis** and **Evaluation** sections of the report.

Evidence

The reaction produced gas bubbles and a colourless solution.

The gas produced a "pop" sound when ignited.
Red litmus turned blue in the final solution.
The final solution produced a bright red flame in a flame test.

Analysis

(a) Identify the reactants and products, with reasons, and write an appropriate chemical equation.

Evaluation

(b) How sure are you of your identification? Suggest further tests to confirm the identity of each of the reactants. How would the reactions in these tests be classified?

12. A metallurgical engineer conducting a study on the properties and characteristics of metals carried out a series of tests involving metals and metal compounds. Complete the **Prediction** and **Analysis** sections of the report.

Prediction

(a) Read the Procedure. Predict which substances should be formed in each of the tests.

Procedure

Part 1: Nickel(II) Sulfate and Sodium Hydroxide

1. About 5 mL of nickel(II) sulfate solution is poured into a test tube. Several drops of sodium hydroxide solution are added to the solution.
2. The mixture is filtered into a separate test tube. Several drops of barium chloride are added to the filtrate.

Part 2: Heating Magnesium Ribbon

3. A strip of magnesium metal is loosely coiled and placed into a crucible equipped with a lid. The mass of the crucible containing the magnesium, with lid in place, is measured.
4. The open crucible is placed on a clay triangle and the bottom is heated to red heat. When no further changes are observed, heating is terminated.
5. The crucible is allowed to cool. With lid in place, the mass is again measured.

Part 3: Zinc and Sulfuric Acid

6. An inverted test tube filled with water is placed in a 250-mL beaker containing 30 mL of water. Zinc metal is added to the beaker and the mouth of the test tube is placed over the zinc. 10 mL of concentrated sulfuric acid is carefully added to the beaker.
7. When all the water is displaced from the test tube, it is carefully removed from the beaker at a 45° angle. The gas inside is tested with a burning splint held to the mouth of the test tube.

Evidence

Table 1: Observations of Tests on Metals and Metal Compounds

Reaction	Step number and observation
Nickel(II) sulfate and sodium hydroxide	1. A very pale yellow precipitate forms. 2. Filtrate becomes cloudy.
Heating magnesium ribbon	3. 38.7 g 4. Fine, white powder forms on the magnesium. 5. 39.1 g
Zinc and sulfuric acid	6. Bubbles form in the test tube, displacing the water. 7. A "pop" sound is heard.

Analysis

(b) Write equations representing each reaction observed. Classify each reaction.

Making Connections

13. Barium ions in solution are very poisonous. However, for diagnosis of ulcers and tumours, patients are often given a suspension of barium sulfate in order to X-ray digestive organs. Explain why this medical application of barium is safe.

14. Roasting and smelting are two common metallurgical processes. These processes are often used to obtain elements such as copper, nickel, zinc, and iron from their ores. Research the roasting and smelting processes for one of these metals. Include a discussion of the chemical reactions involved in each step of the process.

    Follow the links for Nelson Chemistry 11, Chapter 3 Review.

    GO TO www.science.nelson.com

15. Safety matches have been designed and manufactured to make a potentially dangerous combustion reaction occur whenever and wherever we want, without placing the user in constant danger. Research safety matches, their components, and their chemical reactions. Create a poster display showing the various layers of chemicals, and how they react together to produce the desired effect: heat.

    Follow the links for Nelson Chemistry 11, Chapter 3 Review.

    GO TO www.science.nelson.com

New Elements

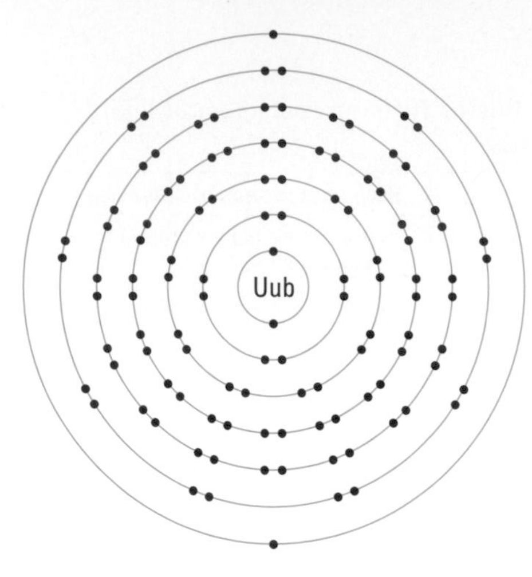

Figure 1
Ununbium was synthesized in an experiment involving the fusion of a zinc atom with a lead atom. Fusion of the atoms was accomplished by repeatedly accelerating a zinc atom toward a lead target, finally generating the specific velocity required for the reaction. At the time of its discovery in 1996, it was considered to be the heaviest atom produced. An atom of ununbium consists of 112 protons, 112 electrons, and 165 neutrons. The electrons are arranged in seven energy levels. The seventh energy level is occupied by two valence electrons.

Elements such as carbon, sulfur, copper, iron, and gold were known to ancient civilizations. However, most of the remaining elements have been discovered since the early part of the 18th century. The end of the 20th century witnessed the discovery of transition metals with atomic numbers 110, 111, 112, 114, 116, and 118. The International Union for Pure and Applied Chemistry has assigned temporary names and symbols to newly discovered elements (**Table 1**).

Table 1: Classical Origins of Temporary Names of Newly Discovered Elements

Atomic number	Digits	Latin or Greek words	Interim name
110	one, one, zero	unus, unus, nil	ununnilium
111	one, one, one	unus, unus, unus	unununium
112	one, one, two	unus, unus, bi	ununbium
114	one, one, four	unus, unus, quattuor	ununquadium
116	one, one, six	unus, unus, hex	ununhexium
118	one, one, eight	unus, unus, oct	ununoctium

Elements 110, 111, and 112 (**Figure 1**) were identified between 1994 and 1996 by an international research team working in Germany. Ununquadium (114) was synthesized in Dubna, Russia, in 1998, and both ununhexium and ununoctium (116 and 118) in California, U.S.A., in 1999. The elements are all extremely unstable and short-lived, rapidly decomposing to form other, lighter elements.

Imagine that you are a member of an international research group attempting to produce or identify the undiscovered elements with atomic numbers 117 and 119. Your task is to write a two-part research paper. In the first part of your paper, you will propose names and symbols for each of the elements, and predict

(a) their atomic structures,

(b) their physical and chemical properties, and

(c) the compounds that they would form.

This part of your paper should include a rationale for your predictions complete with chemical nomenclature, structural formulae, and chemical equations as required to support the rationale.

Investigation

The Properties of Elements 117 and 119

In the second part of your paper, you will outline an investigation by which to test your Predictions of the physical and chemical properties of the two new elements. (You will have to assume that they are relatively stable, and will not decompose within milliseconds of their formation.)

Question

(d) Write the specific Questions that you would like to answer.

Hypothesis/Prediction

(e) Predict answers to your Questions. As part of the explanation of your Predictions, include balanced chemical equations and expected observations for any tests you will use.

Experimental Design

(f) Design an experiment that would answer one or more of the Questions and test your Predictions.
(g) Write the Procedure, including any safety and disposal precautions.
(h) Prepare a complete list of Materials, including any safety equipment.

Assessment

Your completed task will be assessed according to the following criteria:

Process

- Use atomic theory to predict atomic structure.
- Use the periodic table to predict physical and chemical properties.
- Use atomic theory, the periodic table, and relevant concepts to predict the compounds formed and their structural formulas.
- Develop Predictions and Hypotheses to support your Predictions.
- Develop an Experimental Design.
- Choose appropriate tools, equipment, and Materials.

Product

- Prepare a suitable paper.
- Demonstrate an understanding of the relevant concepts, principles, laws, and theories.
- Use terms, symbols, equations, chemical nomenclature, and structural formulas to communicate effectively.

Unit 1 Review

Understanding Concepts

1. Calculate the number of protons, neutrons, and electrons present in one atom of each of the following:
   (a) iodine-127
   (b) phosphorus-32
   (c) Cu-64
   (d) Hg-203
2. Compare alpha particles, beta particles, and gamma rays in terms of their mass, speed, and charge.
3. A radioactive substance has a half-life of 12 days. Construct a graph to illustrate the decay of 200 g of the substance over a period of 60 days. Use your graph to determine the mass of the original substance remaining in the sample after 21 days.
4. Justify each of the following statements in terms of electronic arrangement:
   (a) Bromine and fluorine are members of the same chemical family.
   (b) The first ionization energy of magnesium is lower than its second ionization energy.
   (c) Oxygen forms anions with a charge of 2–.
   (d) Argon is an unreactive element.
   (e) Fluorine is a very reactive element.
5. Describe the role of radioactivity in the development of a model of atomic structure.
6. Classify each of the following elements as alkali metal, alkaline earth metal, halogen, transition metal, or noble gas:
   (a) a metal that is a liquid at SATP
   (b) a highly reactive gas that forms diatomic molecules
   (c) an element that reacts vigorously with water to produce hydrogen
   (d) elements that form oxides with general formula XO
   (e) an element with seven valence electrons
   (f) elements with very high first ionization energies
7. How would you experimentally distinguish between
   (a) a metal and a nonmetal?
   (b) an acid and a base?
   (c) an element and a compound?
   (d) an ionic compound and a molecular compound?
8. Use the periodic table and references in this unit to determine which entity in each of the following pairs would have the greater radius:
   (a) ions of sodium and rubidium
   (b) ions of sodium and sulfur
   (c) atoms of magnesium and calcium
   (d) atoms of magnesium and phosphorus
9. Use the periodic table to identify which entity, in each of the following pairs, would have the greater electronegativity:
   (a) atoms of rubidium and magnesium
   (b) atoms of boron and fluorine
10. Justify each of the choices made in questions 8 and 9, with reference to factors such as shielding, nuclear charge, number of valence electrons, and the distance of valence electrons from the nucleus.
11. Explain the relationship between ionization energy and
    (a) reactivity in metals and nonmetals;
    (b) the type of chemical bond formed between entities of two elements.
12. Oxygen will react with lithium to form lithium oxide and will also react with nitrogen to form nitrogen dioxide. Lithium oxide is a solid at SATP, while nitrogen dioxide is a gas at SATP. Explain this difference in physical properties of the two oxides in terms of chemical bonding. In what other ways would you expect their physical properties to differ?
13. Classify each of the following as ionic or molecular compounds:
    (a) $CS_{2(l)}$
    (b) $SiBr_{4(l)}$
    (c) $MgCl_{2(s)}$
    (d) $SO_{3(g)}$
    (e) $Li_2O_{(s)}$
14. For each of the compounds classified as molecular in question 13, draw electron dot diagrams and Lewis structures of the compound.
15. Carbon tetrachloride ($CCl_{4(l)}$) is a nonpolar solvent that was at one time used in the dry-cleaning industry. Explain why carbon tetrachloride is classified as nonpolar.
16. Name each of the following compounds, using IUPAC chemical nomenclature:
    (a) $FeO_{(s)}$
    (b) $HgI_{2(s)}$
    (c) $SnS_{2(s)}$
    (d) $SbCl_{5(l)}$
    (e) $Fe_2(SO_4)_{3(s)}$
    (f) $CaCO_{3(s)}$
    (g) $CuSO_4{\cdot}5H_2O_{(s)}$
    (h) $KMnO_{4(s)}$
    (i) $N_2O_{(g)}$
    (j) $BF_{3(g)}$

17. Write formulas for each of the following compounds:
   (a) copper(II) nitrate
   (b) magnesium carbonate
   (c) boron trioxide
   (d) zinc chloride
   (e) sodium hydroxide
   (f) zinc hypochlorite
   (g) ammonium chloride
   (h) hydrogen sulfate
   (i) hydrobromic acid
   (j) sulfur dioxide
18. Predict the products of each of the following reactions. In each case, write a balanced chemical equation to represent the reaction.
   (a) $ZnCO_{3(s)}$ decomposes when heated
   (b) $Li_{(s)} + H_2O_{(l)} \rightarrow$
   (c) $Ca_{(s)} + O_{2(g)} \rightarrow$
   (d) $Pb(NO_3)_{2(aq)} + NaCl_{(aq)} \rightarrow$
   (e) $Sn_{(s)} + Cl_{2(g)} \rightarrow$
   (f) $Zn_{(s)} + HCl_{(aq)} \rightarrow$
   (g) $NaOH_{(aq)} + H_2SO_{4(aq)} \rightarrow$
   (h) $BaCl_{2(aq)} + K_2SO_{4(aq)} \rightarrow$
   (i) $CH_{4(g)} + O_{2(g)} \rightarrow$
   (j) $C_6H_{12}O_{6(l)} + O_{2(g)} \rightarrow$
19. Predict whether the following reactions will proceed. Write a balanced chemical equation (including states) to represent each reaction that proceeds.
   (a) $Al_{(s)} + H_2O_{(aq)} \rightarrow$
   (b) $Ni_{(s)} + HCl_{(aq)} \rightarrow$
   (c) $(NH_4)_2CO_{3(s)} + CaCl_{2(aq)} \rightarrow$
   (d) $Ca(NO_3)_{2(aq)} + Na_2SO_{4(aq)} \rightarrow$
   (e) $I_{2(s)} + HCl_{(aq)} \rightarrow$
   (f) $Sn_{(s)} + H_2O_{(l)} \rightarrow$
   (g) $Fe_{(s)} + H_2O_{(l)} \rightarrow$
   (h) $Ba(OH)_{2(aq)} + NaNO_{3(aq)} \rightarrow$
   (i) $CuSO_{4(aq)} + NaNO_{3(aq)} \rightarrow$
20. Classify each of the reactions from questions 18 and 19 as combustion, synthesis, decomposition, single displacement, or double displacement.
21. Which of the reactions from question 18 and 19 involved the formation of a precipitate? In each case, identify the precipitate formed.

Applying Inquiry Skills

22. Two elements, X and Y, form oxides with formulas X_2O and Y_2O, respectively. Design an experimental Procedure to determine whether the two elements belong to the same group.
23. A student carried out an experiment to investigate the effect of heating a compound known as Epsom salts. Complete the **Analysis** section of the report.

Question
What type of compound is Epsom salts?

Experimental Design
Epsom salts are placed in a test tube and heated over a flame.

Evidence
Mass of Epsom salts before heating: 5.00 g
Mass of solid remaining after heating: 2.45 g
During heating, clear, colourless droplets formed at the top of the test tube. When tested with cobalt chloride paper, the drops changed the colour of the paper from blue to pink.

Analysis
   (a) Write a word equation to describe the reaction.
   (b) Classify the reaction. Justify your answer.
   (c) What type of compound is Epsom salts? Justify your answer.
24. Manganese dioxide, MnO_2, is a black, powdery solid that is a common component of the type of dry-cell batteries used in flashlights and radios. It is a toxic chemical, and is known to react violently with powdered aluminum. A chemist conducted the following experiment as part of a study of the chemical properties of manganese dioxide. Complete the **Analysis** section of the report.

Experimental Design
A sample of manganese dioxide is placed in an unknown liquid bleaching agent in a test tube. Any solid material apparent after the reaction is removed by filtration, dried, and its mass measured. The identity of the filtrate is established by a diagnostic test.

Evidence
Initial volume of unknown liquid: 10 mL
Initial mass of manganese dioxide: 0.30 g
Initial appearance of unknown liquid: clear, colourless liquid
When manganese dioxide is added to the liquid, bubbles form.
A glowing splint subsequently held to the mouth of the test tube burns brightly.
Filtrate was clear and colourless, and changed the colour of cobalt chloride paper from blue to pink.
Mass of the remaining black, powdered residue: 0.29 g

Analysis

(a) What is the evidence of chemical change?
(b) What might be the identity of the bleaching agent? Justify your answer, using a chemical equation to communicate the reaction.
(c) What is the role of manganese dioxide in the reaction?

25. Analytical chemists conduct qualitative analyses to determine the identity or presence of a substance. Qualitative analysis may involve a number of tests to reveal physical and chemical properties of the unknown substance. An analytical chemist carried out this investigation to determine whether or not an unknown compound is calcium carbonate. Complete the **Question** and **Analysis** sections of the report.

Question

(a) Write a Question for this investigation.

Procedure

Small samples of the unknown compound were subjected to the following tests:

1. A sample was added to water and observed for solubility.
2. A second sample, about 2 g, of the compound was added to 5 mL of dilute hydrochloric acid in a test tube.
3. The gas produced in step 2 was bubbled through limewater.
4. A nichrome wire was dipped in the solution produced in step 2, and heated in a flame.
5. A third sample of the compound was added to a test tube to a depth of about 5 mm. The test tube was heated directly in a burner flame until no further changes occurred (**Figure 1**).
6. The gas produced in step 5 was bubbled through limewater.

Evidence

1. The sample did not appear to dissolve readily in water.
2. Bubbles of gas formed around the compound.
3. The limewater changed from clear to cloudy.
4. The solution imparted a red colour to the flame.
5. A white, lumpy powder remained in the test tube.
6. The limewater changed from clear to cloudy.

Analysis

(b) What does the Evidence from step 1 indicate about the classification of the unknown compound?
(c) What does the Evidence from step 4 indicate about the composition of the unknown substance?

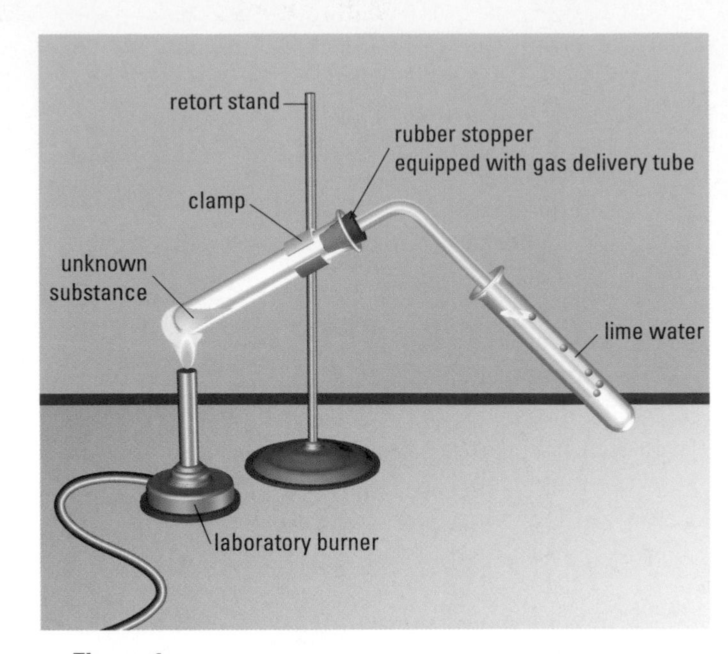

Figure 1
The compound was heated gently and the resulting gas collected.

(d) What does the Evidence from steps 3, 5, and 6 indicate about the composition of the unknown substance? Justify your answer using chemical equations.
(e) Answer your Question.

Making Connections

26. Calcium oxide is a compound used in the process of extracting magnesium from seawater. Calcium oxide is obtained by heating oyster shells, which are mostly composed of $CaCO_{3(s)}$. Are there any environmental hazards associated with the industrial process of heating oyster shells? Use chemical equations to justify your answer.

27. Lead is considered to be a serious environmental pollutant. The health effects of exposure to lead include brain damage, convulsions, behavioural disorders, and death. During the height of the Roman Empire, wealthy Romans used lead and lead compounds in many ways, including as a wine sweetener, as a colourful glaze on pottery, for water pipes, and as a cosmetic. Could the erratic behaviour of the Roman Emperor Nero be attributed to lead poisoning? Research the evidence and present your answer as if you were a physician explaining Nero's behaviour to a secret Senate inquiry.

    Follow the links for Nelson Chemistry 11, Unit 1 Review.

GO TO www.science.nelson.com

28. Research the compounds used to produce various colours in fireworks. Find at least five different compounds, and write chemical equations for their combustion.

Follow the links for Nelson Chemistry 11, Unit 1 Review.

GO TO www.science.nelson.com

29. Archaeologists think that humans started using metals around 9000 B.C., starting with copper. Since that time, the evolution of metallurgical processes has been quite remarkable. Research various methods of extracting and preparing metals that have been used over the past thousand years. From your research, prepare a brief report comparing the various developments in metallurgy. Can any one development be considered responsible for the dawn of the age of steel?

Follow the links for Nelson Chemistry 11, Unit 1 Review.

GO TO www.science.nelson.com

30. Steel can be manufactured to meet a variety of technological demands: resistance to corrosion, hardness, malleability, and tensile strength. Research the various alloying elements responsible for these properties and the means by which each property is achieved. Present your findings in the form of an information pamphlet to be distributed to potential customers by a hardware manufacturer.

Follow the links for Nelson Chemistry 11, Unit 1 Review.

GO TO www.science.nelson.com

31. Antimony is a grey metalloid. It is usually found in nature as the mineral stibnite, Sb_2S_3. Pure antimony can be obtained in either of two ways. In the first method, stibnite is reacted with iron. The second method involves two reactions. First, stibnite reacts with oxygen in a process called roasting, which forms antimony(IV) oxide, Sb_2O_4, and sulfur dioxide. The antimony(IV) oxide is then reacted with carbon to displace solid antimony from the compound.
    - (a) Predict the products of the first method of extraction. Write a balanced chemical equation to represent the process. Classify the reaction as one of the five types of reaction studied in this unit.
    - (b) Write a chemical equation to represent the process of roasting.
    - (c) Predict the products of the reaction that follows roasting, and represent the reaction as a balanced chemical equation.
    - (d) What environmental concerns would be associated with each method of extraction? Which do you think would be the most environmentally friendly method? Give reasons for your position, including equations to illustrate any environmental links.

Exploring

32. Research and report upon the links among lead pollution, its health effects, and the political decision to outlaw the use of leaded gasoline in Canada. What steps must be taken for a law such as this to be passed?

Follow the links for Nelson Chemistry 11, Unit 1 Review.

GO TO www.science.nelson.com

33. Radioisotopes have many practical uses, including medical technology. Research the involvement of radioisotopes in current medical diagnostic tools, and present your findings in a brief report written for a popular science magazine.

Follow the links for Nelson Chemistry 11, Unit 1 Review.

GO TO www.science.nelson.com

34. In the study of atomic structure, the mass spectrometer provides a means of determining atomic mass. This technology has a wide range of applications. How is the mass spectrometer useful in forensic science? Prepare a presentation of your findings.

Follow the links for Nelson Chemistry 11, Unit 1 Review.

GO TO www.science.nelson.com

35. In various chemical nomenclatures, different names can be given to the same compound. Research the various naming systems that are used in the pharmaceutical industry. Briefly explain the purpose of each type of naming system. Is there likely to be any confusion with these nomenclatures? Are there any other disciplines in which multiple naming systems are commonplace? Present your findings as one or more warning posters to be distributed to schools where students are trained in the various disciplines.

Follow the links for Nelson Chemistry 11, Unit 1 Review.

GO TO www.science.nelson.com

150
130
110
90
70
50

Unit 2

Quantities in Chemical Reactions

"As a biochemist and molecular biologist, my work involves human identification using recombinant DNA (deoxyribonucleic acid). We use robotic workstations to do most of the manual processing steps, such as dilution and working with chemical reagents. The DNA itself is carefully extracted from biological samples in a very clean area to eliminate problems from contamination. In the future, as the technology moves forward, we expect to see more of the chemistry and biology of the process performed on biochips or silicon wafers.

DNA has been called the silent witness, since it is often left indiscriminately at the crime scene. The ability to find and extract minute quantities of DNA from biological samples has helped solve numerous crimes and exonerated the innocent. The quest is to gain more information from smaller samples in the most efficient manner possible. I like challenges. In forensic science we have the best of both worlds—that constant challenge, combined with the satisfaction of seeing the direct benefit of our research efforts."

Ron Fourney, Officer in Charge of the National DNA Data Bank of Canada, RCMP, Ottawa

Overall Expectations

In this unit, you will be able to

- use the mole concept in the analysis of chemical systems;
- carry out experiments and complete calculations based on quantitative relationships in balanced chemical reactions;
- develop an awareness of the importance of quantitative chemical relationships in the home and in industry.

Unit

2

Quantities in Chemical Reactions

Are You Ready?

Knowledge and Understanding

1. **Figure 1** shows the model of an atom. Identify the element represented by the model.

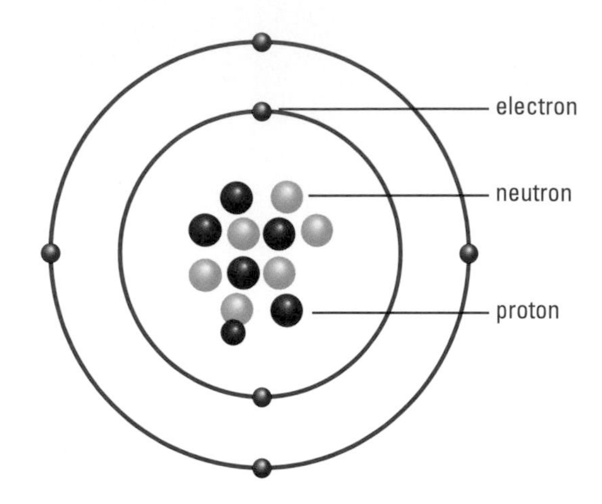

Figure 1
The Bohr-Rutherford model of an atom (not to scale)

2. Complete **Table 1** with the correct number of protons, neutrons, or the isotope indicated.

Table 1

Isotope	Number of protons	Number of neutrons
^{13}C	?	?
?	92	146
^{3}H	?	2

3. Give an example of each of the following entities:
   - an element
   - an ion
   - a molecule
   - an ionic compound
4. In **Table 2** fill in the correct International Union of Pure and Applied Chemistry (IUPAC) name or formula, as required.

Table 2

IUPAC name	Chemical formula
barium chloride	?
sodium hydroxide	?
diphosphorus pentoxide	?
copper(II) sulfate	?
?	$SiCl_4$
?	Fe_2O_3
?	$PbBr_2$
?	CO

5. For each of the following, write the correct skeleton equation and then balance it to form a chemical equation.
   (a) copper(II) oxide + hydrogen → copper + water

(b) lead(II) nitrate + potassium iodide $\rightarrow$ lead(II) iodide + potassium nitrate
(c) calcium + water $\rightarrow$ calcium hydroxide + hydrogen gas
(d) hydrogen sulfide $\rightarrow$ hydrogen + sulfur

Inquiry and Communication

6. Complete the information in **Figure 2**.

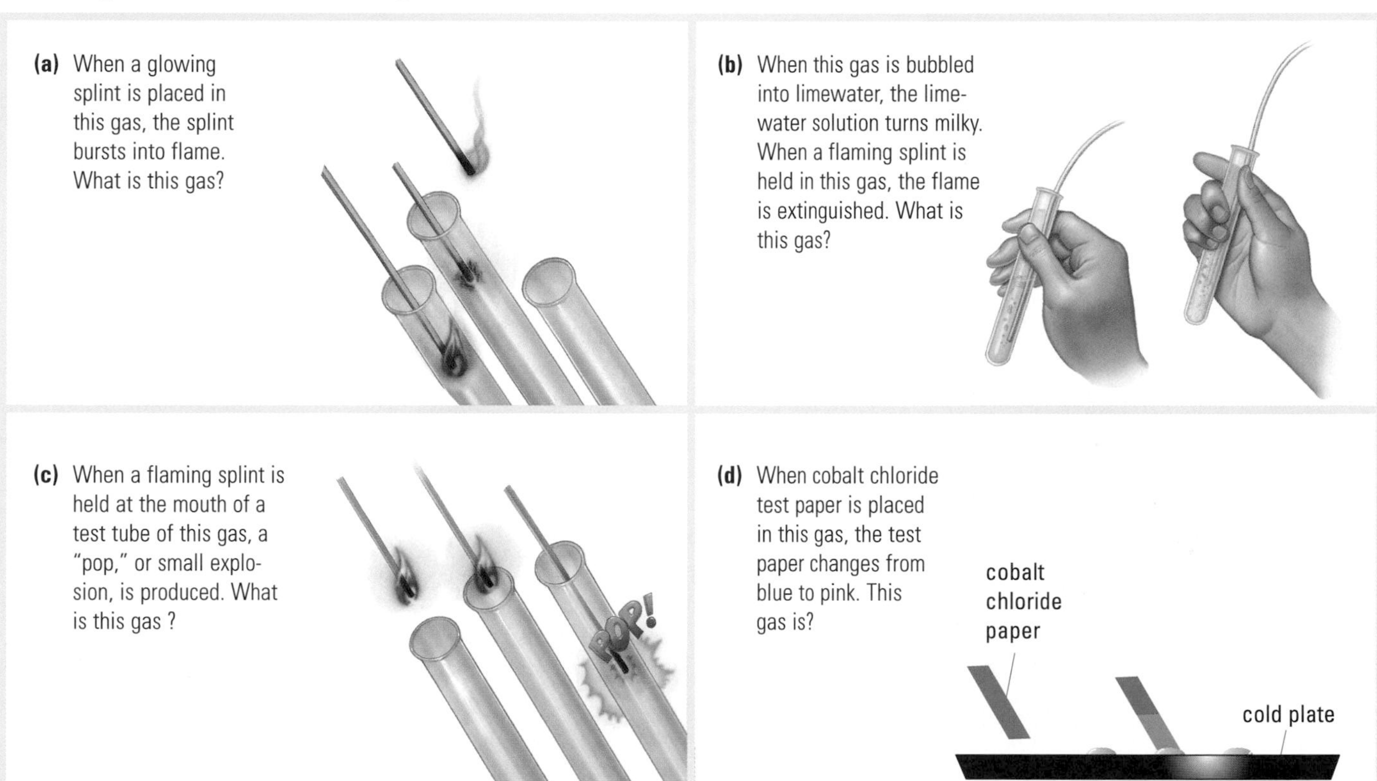

Figure 2

Mathematical Skills

7. Complete the following calculations using the correct number of significant digits and units:
   (a) Given $x/y = 3/2$, find y when $x = 5.12$.
   (b) There are 460 girls and 523 boys at a school; calculate the percentage of girls at the school.
   (c) $\dfrac{5.12 \times 10^{15}\ \text{g}}{8.336\ \text{mL}} = ?$
   (d) $\dfrac{1.000\ \text{m}}{2.43 \times 10^{17}\ \text{s}} = ?$
8. (a) Graph the data in **Table 3**, plotting the volume values on the x-axis and the mass values on the y-axis, and draw a line of best fit. Include a title for the graph and labelled axes.
   (b) The measurements were made to determine the identity of an unknown element. Based on this evidence, which element could it be?

Table 3

mass (g)	volume (mL)
1.0	2.9
2.3	6.9
3.5	10.6
5.2	15.4
7.4	22.4

Chapter

4

Quantities in Chemical Formulas

In this chapter, you will be able to

- explain the law of definite proportions and the significance of different proportions of elements in compounds;
- explain the relationship between isotopic abundance and relative atomic mass;
- describe and explain Avogadro's constant, the mole concept, and the relationship between the mole and molar mass;
- determine empirical and molecular formulas using percentage composition obtained from given data and through experimentation;
- develop technological skills for quantitative analysis;
- solve problems involving quantities in moles, numbers of particles, and mass numbers.

When you read the list of ingredients on a package of cereal, do you also notice how much of each ingredient is contained in a serving? We can compare the quantities of sugar or fat or the percentage of daily requirements of vitamins and minerals in different brands (**Figure 1**). This quantitative information helps us decide which product to select to suit our needs.

Quantities in chemical formulas offer similar important information about the composition and properties of compounds. For example, water ($H_2O_{(l)}$) and hydrogen peroxide ($H_2O_{2(l)}$) both contain the same types of atoms. The only difference is in the number of oxygen atoms. This difference, which appears small, actually results in significant differences in the properties of the two compounds.

Water is very stable, can be stored safely for indefinite periods, and can be used for drinking and washing. It is an important ingredient of living cells and is essential for all life on Earth. Hydrogen peroxide, on the other hand, is so unstable that it must be stored in darkened containers to slow its decomposition. Concentrated solutions of hydrogen peroxide must be used with caution, because the chemical reacts readily with other substances and will cause blistering of the skin on contact. Because hydrogen peroxide is toxic it is used to kill bacteria—in low concentrations (0.03%) it is used as an antiseptic to treat minor cuts and abrasions. At higher concentrations (6%) it is used to bleach hair, pulp and paper, and synthetic and natural fibres. At even higher concentrations, its bactericidal properties can be applied as part of the treatment of waste water. At sufficiently high concentrations it is explosive.

How can we determine the chemical composition of a substance? Identifying the substance's properties is one method. In this chapter, you will learn other methods to do this. You will also learn how to measure and communicate quantities when dealing with entities as small as atoms, ions, and molecules.

Reflect on your Learning

1. The label on a bag of jelly beans states that the bag contains 40% large jelly beans and 60% small jelly beans, by mass. Do you think this information is sufficient if we want to know how many jelly beans of each size there are in the bag? Explain your answer.
2. Given your knowledge of chemical reactions, list reasons why you think it is important to be able to communicate information about the number of atoms, ions, or molecules that are reacting or that are produced.
3. Many fossil fuels that are burned in factories contain sulfur; when sulfur reacts with oxygen in the air, sulfur oxides—known air pollutants—are produced. Technicians use different methods to predict the masses of these and other chemicals released into the air. Suggest reasons why it is important to be able to make these predictions. Speculate on how technicians can make this kind of analysis.

Throughout this chapter, note any changes in your ideas as you learn new concepts and develop your skills.

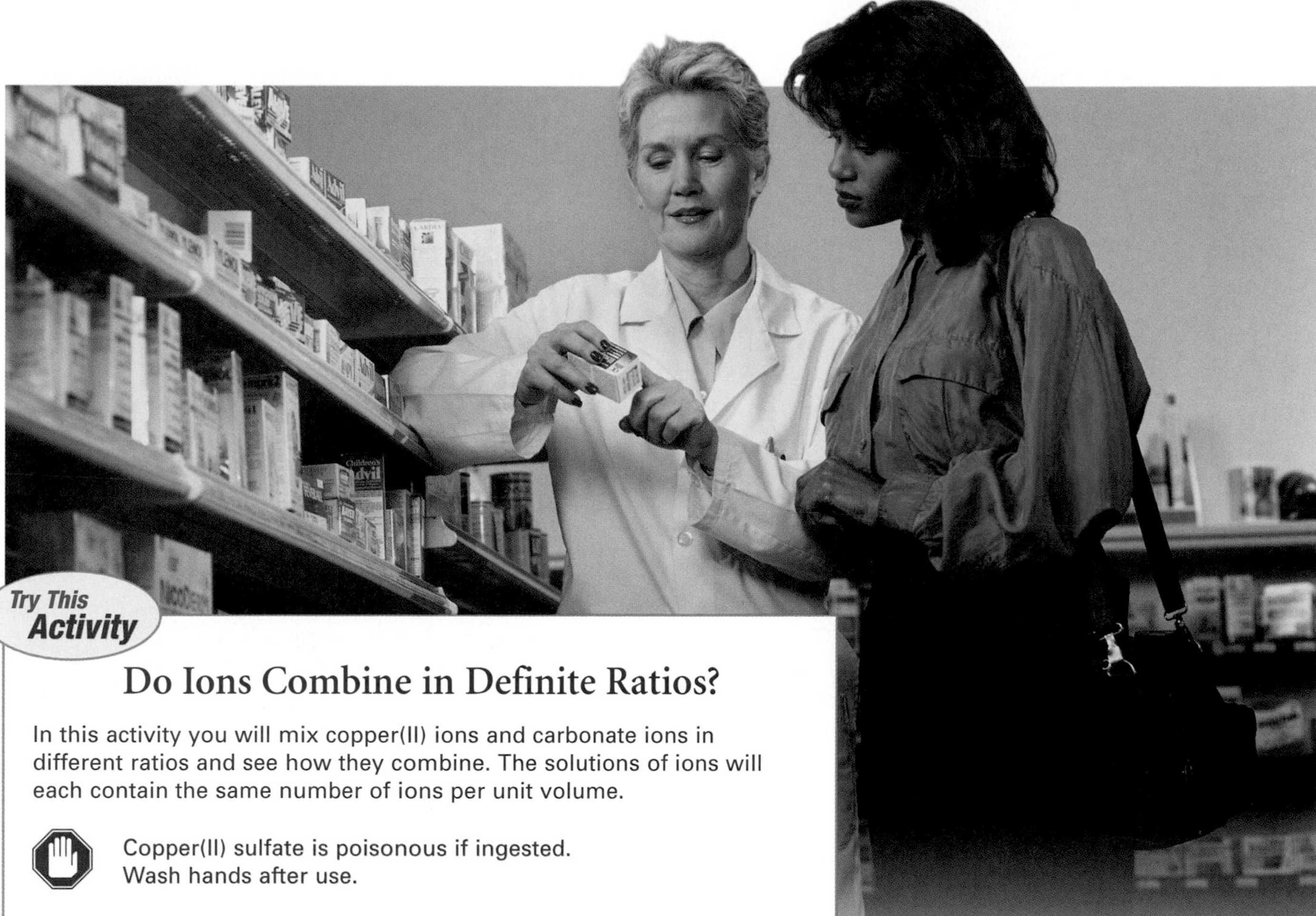

Try This Activity

Do Ions Combine in Definite Ratios?

In this activity you will mix copper(II) ions and carbonate ions in different ratios and see how they combine. The solutions of ions will each contain the same number of ions per unit volume.

Copper(II) sulfate is poisonous if ingested. Wash hands after use.

Materials: lab apron, eye protection, 5 small test tubes of equal size, test-tube rack, eyedropper, 10 mL copper(II) sulfate solution (0.10 mol/L), 10 mL sodium carbonate solution (0.10 mol/L), distilled water

- Number the test tubes from 1 to 5, and place in the test-tube rack.
- Using the dropper, add drops of copper(II) sulfate solution to each test tube, according to **Table 1**.
- Wash the dropper thoroughly with distilled water and use the same dropper to add drops of sodium carbonate solution to each test tube, according to **Table 1**. After you have finished putting drops in each tube, the test tubes should be filled to equal depth since they contain the same number of drops (10 drops total).
- Swirl each test tube gently to mix the contents. Allow the precipitates to settle for about 5 min.
- Wash your hands thoroughly.
  (a) What is the ratio of Cu^{2+} ions to CO_3^{2-} ions in each test tube?
  (b) Which test tube contains the most precipitate ($CuCO_{3(s)}$)? What is the ratio of Cu^{2+} ions to CO_3^{2-} ions in this test tube?
  (c) From what you learned in Unit 1 about ionic bonds, does this ratio agree with a prediction of how copper(II) ions and carbonate ions would combine to form a compound?
  (d) Which test tubes contain the smallest amount of precipitate? Suggest reasons why the ratios of ions in these test tubes produced the least amounts.
  (e) What evidence is there that some copper ions remain unused in solution in some tubes?
  (f) Explain how the evidence suggests that ions combine in definite ratios.
- Dispose of the materials according to your teacher's instructions.

Figure 1
Quantities in a list of ingredients help us compare and select products to suit our needs.

Table 1: Mixing Ions

	Test tube				
	1	2	3	4	5
drops $Cu^{2+}_{(aq)}$	1	3	5	7	9
drops $CO^{2-}_{3(aq)}$	9	7	5	3	1

4.1 Proportions in Compounds

Figure 1
Ice Storm '98 in Kingston, Ontario

In January 1998, nearly double the average annual total of freezing rain fell in one week in eastern Ontario and Quebec, creating the worst ice storm in decades: Ice Storm '98 (**Figure 1**). A study undertaken by the Ontario Ministry of Health to determine effects associated with the ice storm revealed a striking rise in carbon monoxide poisoning. The most common sources of poisoning were gas-powered generators and barbecues inappropriately used indoors, in locations such as basements and garages.

The chemical formula for carbon monoxide, CO, is similar to that for carbon dioxide, CO_2 (**Figure 2**). The difference in the number of oxygen atoms per molecule, however, is responsible for the very different properties of the two gases (**Table 1**).

CO molecule

CO_2 molecule

Figure 2
Molecular models of CO and CO_2

Table 1: Comparison of the properties of $CO_{(g)}$ and $CO_{2(g)}$

	CO	**CO_2**
physical properties	colourless, odourless gas, poisonous	colourless, odourless gas, slightly acid-tasting
melting point (°C)	–199.0	–78.5
boiling point (°C)	–191.5	–78.5*
density (g/L)	1.250	1.977
solubility in water (g/100 mL)	0.0044	0.339

*sublimes

Carbon Monoxide

We cannot sense carbon monoxide gas, just as we cannot sense carbon dioxide when we exhale. Both are colourless and odourless gases at room temperature and pressure. However, carbon monoxide is much deadlier than carbon dioxide. Carbon oxides are produced whenever fossil fuels are burned. Carbon atoms combine with oxygen atoms to form compounds of carbon and oxygen. When the supply of oxygen is plentiful, each carbon atom bonds with two atoms of oxygen to form carbon dioxide. But when the supply of oxygen is low, each carbon atom bonds with only one atom of oxygen, forming carbon monoxide.

The deadly effects of carbon monoxide are due to its effect on the blood's ability to carry oxygen to body tissues and vital organs such as the heart and brain (**Table 2**). Carbon monoxide competes with oxygen when binding to

Table 2: Symptoms of Carbon Monoxide Poisoning

Levels of $CO_{(g)}$		
Low	**Moderate**	**High**
shortness of breath	severe headaches	loss of consciousness
mild nausea	dizziness	brain damage
mild headaches	confusion	death
	nausea	
	faintness	
	death (if these levels persist for a long time)	

hemoglobin. It binds to hemoglobin so readily that the number of oxygen molecules that are able to bind is drastically reduced. When oxygen molecules cannot be transported by the hemoglobin to all the cells of the body, the cells are deprived of the oxygen needed for cellular respiration, resulting in cell death, tissue damage, and organ failure (**Figure 3**).

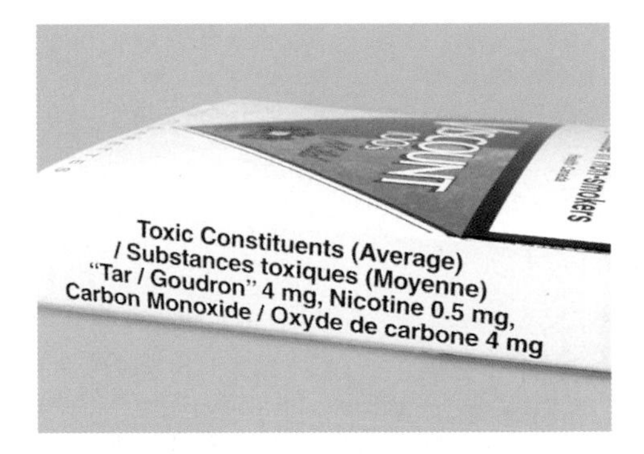

Figure 3
Inhaled cigarette smoke contains about 400 parts per million (ppm) of CO. Because of the effect of CO on the blood's ability to transport oxygen, the heart is forced to work harder to supply enough oxygen to body tissues. This may result in increased incidence of heart disease and heart attacks.

Carbon Dioxide

Carbon dioxide is normally present in our atmosphere and is harmless at low concentrations. It is essential to the growth of green plants and so is essential to all life on Earth. When food molecules are broken down in a cell in a human body, carbon dioxide is produced and is carried away from the body tissues by the bloodstream, to be exhaled (**Figure 4**). Unlike carbon monoxide, carbon dioxide does not bind to hemoglobin, but it takes on an interesting and important role: its chemical properties enable it to control the acidity of the blood, preventing sudden life-threatening fluctuations in the pH level.

Many everyday applications make use of some of the physical and chemical properties of CO_2. Solid CO_2 is used in refrigeration as dry ice. Gaseous CO_2 can be dissolved in water under pressure to form carbonated beverages, which release bubbles of the gas when the pressure drops. Carbon dioxide is denser than air and does not support combustion. Because of this property, the foam in some fire extinguishers contains CO_2; this blankets the fire, thereby starving the fire of oxygen. This property can also be extremely hazardous: at high concentrations, there is a danger of suffocation if the thickness of the layer of $CO_{2(g)}$ is too great and one cannot escape for air.

Figure 4
If an unknown gas is bubbled through a limewater solution and the mixture becomes cloudy, then the gas most likely contains CO_2. The limewater diagnostic test provides evidence for CO_2 in the breath you exhale.

Practice

Understanding Concepts

1. In a table, list the similarities and the differences between the compounds carbon monoxide and carbon dioxide. Include physical and chemical properties and the information derived from their chemical formulas.
2. Explain why carbon monoxide exerts a toxic effect in the body.
3. Carbon dioxide has many useful applications in everyday products. Choose two products, and explain the physical and chemical properties of carbon dioxide that are applicable in each.

The Law of Definite Proportions

The carbon monoxide and carbon dioxide comparison illustrates that not only do the properties of substances depend on the elements they are composed of but also on the quantities of each element as well. If we decomposed the same mass of each gas, we would get carbon and oxygen from each, but the masses of each element collected would be different. Conversely, if we were to synthesize each of these gases, they would both require carbon and oxygen but would require different amounts of each element. The composition of a compound is constant, regardless of how it was synthesized. Indeed, carbon would always combine in the same proportions with oxygen to produce carbon dioxide, whether the carbon burned is charcoal or diamond. Recall from the activity at the beginning of this chapter that $Cu^{2+}_{(aq)}$ ions and $CO^{2-}_{3(aq)}$ ions always combined in a 1:1 ratio, even when there was an excess of one of the ions.

law of definite proportions: a specific compound always contains the same elements in definite proportions by mass

These observations are summarized in the **law of definite proportions**, first stated by Joseph Proust (1734–1794):

A specific compound always contains the same elements in definite proportions by mass.

stoichiometry: the study of the relationships between the quantities of reactants and products involved in chemical reactions

Proust was one of the first scientists to apply quantitative analysis to chemical reactions. He and others started an area of study known as stoichiometry. This word is derived from the Greek words *stoikheion,* meaning "element," and *metron,* meaning "measure." **Stoichiometry** is the study of the relationships between the quantities of reactants and products involved in chemical reactions, important knowledge in the application of chemistry in many fields, such as industry, medicine, and ecology.

Practice

Understanding Concepts

4. Using your own words and some examples, describe the law of definite proportions.
5. What is the meaning of the term "stoichiometry"?
6. The oxides of hydrogen and carbon all contain only nonmetal atoms.
   (a) Suggest a reason why so many compounds with different combining proportions are formed when nonmetals are bonded to nonmetals.
   (b) Predict which of the following elements would combine with oxygen to form compounds with different combining proportions: sodium, aluminum, sulfur, and phosphorus.
7. From your knowledge of multivalent ions, write formulas to show how each of the following pairs of elements combine to form two compounds of different combining proportions.
   (a) copper and chlorine
   (b) iron and oxygen
   (c) lead and sulfur

Making Connections

8. The Try This Activity on p. 159 is an introduction to the type of quantitative analysis used by chemists in many fields. List as many careers as you can in which quantitative analysis of unknown substances is important. One example is given in **Figure 5**.

Figure 5
A forensic chemist may be asked to determine whether a document has been tampered with. The dye components in an ink can be determined by testing minute samples, and, from the ink formulation, the manufacturer and the year in which the ink was first made available can be traced. In cases where several entries in a document were written with the same ink, the forensic chemist can determine the age of each entry by performing quantitative analysis to determine the quantity of volatile components remaining on the paper.

Section 4.1 Questions

Understanding Concepts

1. Explain which of the following sentences illustrates the law of definite proportions:
   (a) When 2.0 g of a reactant combines with 3.0 g of another reactant, 5.0 g of a single product is formed.
   (b) A compound is formed by the combination of 3.0 g of magnesium and 2.0 g of oxygen. The same compound is formed by the combination of 6.0 g of magnesium and 4.0 g of oxygen.
   (c) When a compound is decomposed, it always forms products in the same ratio by mass.
   (d) Two elements may combine in different ratios, but they will form the same product.

Applying Inquiry Skills

2. Design an experiment to determine the proportions in which silver (Ag^+) ions will combine with chromate (CrO_4^{2-}) ions, given that silver chromate is insoluble in water.
   (a) Describe the procedure and method you would use to analyze the evidence.
   (b) Explain how this experiment may be used to test the law of definite proportions.

Making Connections

3. Nitrogen and oxygen combine to form several different oxides of nitrogen. Research the oxides of nitrogen on the Internet and prepare a table that shows the different properties and uses of $NO_{(g)}$, $NO_{2(g)}$, and $N_2O_{(g)}$.
   Follow the links for Nelson Chemistry 11, 4.1.

GO TO www.science.nelson.com

DID YOU KNOW ?

CO_2 Disaster

On August 21, 1986, an 80-m fountain of CO_2 gas erupted in Lake Nyos in northern Cameroon, Central Africa, killing 1700 people. Carbon dioxide generated in volcanic processes rose from below and dissolved in ground water that flowed into Lake Nyos. Over time, the mineral-rich and therefore dense ground water formed a CO_2-saturated layer at the bottom of the lake, where the pressure is high. When a break in this layer was triggered, the pressure was released, and so was the CO_2, much like a carbonated beverage bubbles when its top is popped. The gas flowed out at over 70 km/h and formed a layer several metres thick at ground level. This low-oxygen layer displaced the relatively oxygen-rich layer of air, essentially suffocating people and animals in its path.

4.2 Relative Atomic Mass and Isotopic Abundance

When you look at the periodic table, you can find the value of atomic mass for each element. How is this value useful? Beginning with the mass of atoms, we can calculate the mass of larger numbers of atoms and molecules and predict the mass of reactants and products for any chemical reaction. The values for atomic mass are important information if we are to use our knowledge of chemistry in practical applications.

How were these atomic masses determined, and why are most of the atomic mass values in the periodic table not whole numbers? The concept of determining atomic masses was, in fact, initially based on the law of definite proportions. In the early 19th century, John Dalton proposed that atoms of different elements had different masses and that all atoms of the same element had the same mass. At the time, there was no method of determining the actual mass of an atom. Based on what they did know, however (that elements always combine in the same proportions when they form compounds), chemists were able to determine the masses of elements that reacted with each other. By comparing these masses, Dalton devised a system of assigning relative atomic masses to the known elements. To do this, Dalton assigned an atomic mass to atoms of a particular element, and then determined the atomic masses of atoms of other elements relative to the first element.

Figure 1
Canada's standard kilogram measure is kept in a vault at the National Research Council in Ottawa.

The concept of relative mass is not uncommon. When we refer to the mass of any object on Earth, we are comparing the mass of that object to the mass of a reference, or standard, mass. In fact, there exists a whole bureau, the International Bureau of Weights and Measures, in France, whose purpose is to ensure worldwide uniformity of measurements. The bureau houses, among other things, the international standard mass for one kilogram. The National Research Council of Canada houses our standard mass for the kilogram (see **Figure 1**).

Relative atomic mass is the mass of each element needed to react with a fixed mass of the element chosen to be the standard, currently carbon-12. When Dalton prepared his first table of relative atomic masses, he assigned an atomic mass of 1 to hydrogen, the lightest element known. He then determined the mass

relative atomic mass: the mass of an element that would react with a fixed mass of a standard element, currently carbon-12

of other elements that reacted with a fixed mass of hydrogen. Since formulas of compounds were not known, errors were made in assigning relative mass. For example, when water was analyzed, it was found that 1 g of hydrogen combines with 8 g of oxygen, leading to oxygen being incorrectly assigned a relative atomic mass of 8. At the time, Dalton did not know that, in fact, two atoms of hydrogen had combined with each atom of oxygen, and that the relative atomic mass of oxygen should be 16.

Later, oxygen was chosen as the standard and was assigned an atomic mass of 16. Other elements were then assigned relative atomic masses according to the mass needed to react with a fixed mass of oxygen. For example, experimental evidence shows that 24 g of magnesium atoms are needed to react with 16 g of oxygen atoms. Similarly, 40 g of calcium atoms are needed to react with 16 g of oxygen atoms. Therefore, if we assign a relative atomic mass of 16 to oxygen, the relative atomic mass of magnesium would be 24, and the relative atomic mass of calcium would be 40. All other elements can be assigned relative atomic masses in a similar way. Sixteen was chosen as the relative atomic mass for oxygen because that gives the lightest element, hydrogen, a relative atomic mass of 1.

In 1961, the scientific community accepted the modern scale of relative atomic mass based on carbon, more specifically, the carbon-12 (C-12) isotope. There are two reasons for choosing C-12 as the standard. First, carbon is a very common element that is available to scientists everywhere. Second, and more important, by assigning a value of 12 to C-12, the atomic masses of nearly all the other elements are very close to whole numbers, with the lightest atom having a mass of very close to 1.

atomic mass unit (u): a unit of mass defined as 1/12 the mass of a carbon-12 atom

One **atomic mass unit** (SI symbol, u) is defined as exactly 1/12 the mass of a C-12 atom. In other words, one atom of C-12 is defined as having a mass of 12.0 u. Although the mass of 1 u was not known when the atomic mass scale was first developed, it has since been determined experimentally:

$$1\ \text{u} = 1.660\ 540\ 2 \times 10^{-27}\ \text{kg}$$

This IUPAC and SI convention is now used by scientists worldwide and is an example of the international cooperation between scientists that makes it possible for them to communicate effectively with one another.

Practice

Understanding Concepts

1. Explain why it was necessary to assign atomic masses relative to a standard atomic mass.
2. State the definition and SI symbol of the atomic mass unit.
3. What three elements have been used, at one time or another, as the reference element for communicating relative atomic mass?
4. The mass of a sodium atom is 1.92 times that of a C-12 atom. Calculate the atomic mass of sodium in u.

Making Connections

5. When one communication system is used throughout the international scientific community, it needs to serve many purposes and be accepted by all. Suggest some criteria that you think this communication system must meet, and assess whether the SI and IUPAC systems meet these criteria.

Answers

4. 23.04 u

Isotopic Abundance

We have noted that the atomic mass unit is defined as 1/12 the mass of a C-12 atom. The reason for this detailed description is that there are several isotopes of carbon, each with a different mass. The C-12 atom has 6 protons and 6 neutrons, and the C-13 atom has 6 protons and 7 neutrons; as a result, their mass is different. Any sample of carbon will contain atoms of both isotopes. The existence of isotopes explains why most elements have atomic masses that are not whole numbers.

For most elements, the isotopic composition of any given sample is constant. Any sample of naturally occurring carbon has the same percentage of the two non-radioactive isotopes of carbon, C-12 and C-13. Because of this constant isotopic composition, we can use an *average* value for the atomic mass of carbon, taking into account the percentage of each isotope in a typical sample. The percentage of an isotope in a sample of an element is called its **isotopic abundance**. Naturally occurring samples of carbon consist of 98.89% C-12 and 1.11% C-13. Measurements show the atomic mass of carbon atoms to be 12.01 u, rather than the 12.00 u that would be expected if all carbon atoms were of the C-12 isotope. When we use the isotopic abundance of each isotope in the carbon sample to calculate the atomic mass, we obtain a value very close to the measured value of 12.01 u. Let's see how the calculation is done.

isotopic abundance: the percentage of an isotope in a sample of an element

Our carbon sample has a mixture of atoms of different mass, C-12 and C-13. Since atoms cannot be divided into fractional parts, we consider a large number of them in our calculation to avoid calculating fractions of atoms. Let's assume we have 10 000 atoms. To obtain an average mass (**Figure 2**) for the mixture of atoms (m_{av}), we need to find the total mass (m_{tot}),and then divide by the number of atoms.

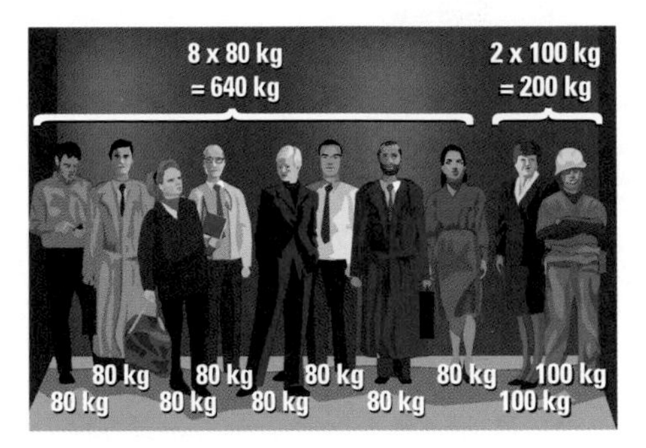

Figure 2
What is the average mass of the people in the elevator?

$$m_{tot} = 640\text{ kg} + 200\text{ kg}$$
$$= 840\text{ kg}$$
$$m_{av} = 840\text{ kg/10 persons}$$
$$= 84\text{ kg/person}$$

Note that 84 kg is an average value and that no person in the elevator actually has a mass of 84 kg.

98.89%, or 9889 atoms, are C-12

1.11%, or 111 atoms, are C-13

$$m_{C\text{-}12} = 9889 \text{ atoms} \times 12\text{ u}$$
$$= 118\ 668\text{ u}$$
$$m_{C\text{-}13} = 111 \text{ atoms} \times 13\text{ u}$$
$$= 1443\text{ u}$$
$$m_{tot} = (118\ 668 + 1443)\text{ u}$$
$$= 120\ 111\text{ u}$$
$$m_{av} = \frac{120\ 111\text{ u}}{10\ 000}$$
$$m_{av} = 12.01\text{ u}$$

The average atomic mass of carbon is 12.01 u.

Note that 12.01 u is an average value, and that no atom in the sample has a mass of 12.01 u.

Sample Problem

Natural chlorine contains two isotopes, Cl-35 and Cl-37. In a sample, 75.53% of the atoms are Cl-35, and 24.47% of the atoms are Cl-37. Calculate the average atomic mass of chlorine. Consider a sample of 10 000 atoms of chlorine.

Solution

75.53%, or 7553 atoms, are Cl-35

24.47%, or 2447 atoms, are Cl-37

$$m_{tot} = (7553 \times 35\ u) + (2447 \times 37\ u)$$
$$= 354\ 894\ u$$
$$m_{av} = \frac{354\ 894\ u}{10\ 000}$$
$$= 35.49\ u$$

The average atomic mass of chlorine is 35.49 u.

Practice

Understanding Concepts

6. Explain the following concepts:
   (a) average atomic mass
   (b) isotope
   (c) isotopic abundance
7. What is the composition of the isotope C-12?
8. Explain why the relative atomic mass of carbon, as referenced on the periodic table, is not exactly 12.
9. Why would you want to avoid calculating the mass of fractions of an atom?
10. Natural potassium consists of 93.10% K-39 and 6.90% K-41. Do these relative values confirm the accepted average atomic mass for natural potassium?
11. Natural argon contains 99.60% Ar-40, 0.34% Ar-36, and 0.06% Ar-38. Do these relative values confirm the accepted average atomic mass for argon?

Answers

10. 39.14 u
11. 39.99 u

INQUIRY SKILLS

- ○ Questioning
- ○ Hypothesizing
- ○ Predicting
- ○ Planning
- ○ Conducting
- ○ Recording
- ● Analyzing
- ● Evaluating
- ○ Communicating

Lab Exercise 4.2.1

Determination of Relative Atomic Mass

Oxides of nitrogen are formed in any high-temperature combustion that involves air. Thus, car engines produce nitrogen oxides that add to air pollution. In this exercise, mass determinations of each element reacted in the synthesis of an oxide of nitrogen are used to determine the atomic mass of nitrogen relative to oxygen. This is similar to the method of determining relative atomic mass. Use the evidence given to complete the **Analysis** and **Evaluation** sections of a lab report.

Experimental Design

A known mass of nitrogen is allowed to completely react with oxygen, and the oxide of nitrogen produced is collected and its mass determined. The relative atomic mass of nitrogen is determined from analysis of the evidence and by making assumptions of combining ratios of the two elements.

Evidence

mass of nitrogen reacted = 3.75 g
mass of oxide of nitrogen produced = 12.34 g

Analysis

(a) What is the mass of oxygen reacted?
(b) What is the ratio of the mass of nitrogen reacted to the mass of oxygen reacted?
(c) Using the mass ratio in (b), calculate the mass of nitrogen that would react with 16.00 g of oxygen.
(d) Using a system of relative atomic mass, if we assign an atomic mass to oxygen of 16, what would be the relative atomic mass of nitrogen? Assume 1 atom of nitrogen reacts with 1 atom of oxygen.
(e) Repeat (d) using the evidence that the formula for this oxide of nitrogen is NO_2.

Evaluation

(f) Was the information given in the evidence sufficient to determine the atomic mass of nitrogen relative to oxygen?
(g) What additional evidence was needed?

Section 4.2 Questions

Understanding Concepts

1. Through experiment, it is determined that an element has a relative atomic mass that is nine times that of a C-12 atom. What element is that atom likely to be?
2. Distinguish between the terms *atomic mass unit* and *atomic mass.*
3. Why is knowledge of the combining proportions of elements in compounds essential to the assigning of relative atomic masses?
4. Explain, using an example, how the isotopic abundance of the isotopes of an element is related to the relative atomic mass of the element.
5. A naturally occurring sample of boron consists of 19.8% B-10 and 80.2% B-11. Calculate the average atomic mass of this sample of boron.

4.3 The Mole and Molar Mass

Since atoms, ions, and molecules are much too small to see, observable changes in chemical reactions must involve extremely large numbers of these entities. If we want to predict the amount of product formed from a chemical reaction, we must know the mass of enormous numbers of the atoms and molecules involved. In other words, we need a precise and convenient way of counting large numbers of entities and determining their mass. In this section, we will learn how to do this, using atomic masses and a unit of measurement called the mole.

Table 1: Convenient Numbers

Quantity	Number	Example
pair	2	shoes
dozen	12	eggs
gross	144	pencils
ream	500	sheets of paper
mole	6.02×10^{23}	molecules

Avogadro's constant: the number of entities in one mole: 6.02×10^{23}; SI symbol N_A

mole: the amount of substance containing 6.02×10^{23} entities

Avogadro's Constant and the Mole

You are already familiar with some terms used to define convenient numbers (**Table 1**). For example, a dozen is a convenient number referring to items such as eggs or doughnuts. The number used by chemists to define numbers of entities as small as atoms is the *mole* (SI symbol, mol). Twelve eggs or doughnuts make a dozen, but how many atoms, molecules, or ions make a mole? Modern methods of estimating this number of entities have led to the value 6.02×10^{23} entities/mol. This value is called **Avogadro's constant** (SI symbol N_A), named after the Italian physicist Amedeo Avogadro (1776–1856), who researched the idea (although an Austrian scientist, J. J. Loschmidt, is credited with determining the first reasonable estimate of the number). A **mole** is the amount of substance containing 6.02×10^{23} of anything, just as a dozen is the amount of substance containing 12 of anything:

- one mole of sodium atoms is 6.02×10^{23} sodium atoms;
- one mole of chlorine molecules is 6.02×10^{23} chlorine molecules;
- one mole of sodium chloride is 6.02×10^{23} formula units of NaCl;
- one mole of elephants is 6.02×10^{23} elephants.

Although the mole represents an extraordinarily large number, a mole of atoms, ions, or molecules is an amount that is observable and convenient to measure and handle. **Figure 1** shows a mole of each of three common substances: an element, an ionic compound, and a molecular compound. In each case, a mole of entities is a sample size that is convenient for laboratory work.

Practice

Understanding Concepts

1. Explain in your own words what is meant by a mole of a substance.
2. What is the numerical value of Avogadro's constant?
3. Avogadro's constant is usually written in scientific notation. Express the number in extended form.
4. How many molecules are there in 3.00 mol of carbon dioxide?
5. How many atoms are there in 0.500 mol of $Ar_{(g)}$?
6. (a) If one dozen oranges has a mass of 1.43 kg, calculate the mass of a single orange.
   (b) If one mole of hydrogen atoms has a mass of 1.01 g, calculate the mass of a single hydrogen atom.

Answers

4. 1.81×10^{24} molecules
5. 3.01×10^{23} atoms
6. (a) 11.9 g
   (b) 1.67×10^{-24} g

Figure 1
These amounts of carbon, table salt, and sugar each contain about a mole of entities (atoms, formula units, molecules, respectively) of the substance. The mole represents a convenient and specific quantity of a chemical.

Molar Mass

Why is this particular number, 6.02×10^{23}, so useful for counting atoms, molecules, and other entities? It is useful because scientists have determined experimentally that there are 6.02×10^{23} carbon atoms in 12 g of C-12. In other words, each *atom* of C-12 has a mass of 12 u, and each *mole* of C-12 atoms (6.02×10^{23} atoms) has a mass of 12 g. Isn't that convenient? Notice how the unit of mass changes from atomic unit (u) to grams (g) as we change from talking about a single atom to talking about a mole of atoms. The mole allows us to easily convert atomic mass into mass in grams. Let's see how this conversion works.

Consider a single Mg-24 atom. It has a mass of 24 u, twice the mass of a single C-12 atom. Since a mole of C-12 atoms has a mass of 12 g, it follows that a mole of Mg-24 atoms must also have twice the mass, 24 g. We can recognize a relationship between the atomic mass and the mass of a mole of atoms of an element. If the atomic mass is 12 u, the mass of a mole of atoms is 12 g. If the atomic mass is 24 u, the mass of a mole of atoms is 24 g. This relationship applies to all elements:

- mass of one C-12 atom is 12 u; mass of one mole of C-12 atoms is 12 g;
- mass of one Al-27 atom is 27 u; mass of one mole of Al-27 atoms is 27 g;
- mass of one N-14 atom is 14 u; mass of one mole of N-14 atoms is 14 g;
- mass of one O-16 atom is 16 u; mass of one mole of O-16 atoms is 16 g.

The relationship between the mass of atoms and the mass of moles also applies to molecules and ions. Take, for example, a molecule of water, H_2O. It consists of two atoms of hydrogen and one atom of oxygen. The average mass of a single water molecule is the sum of the average masses of its components—2.02 u (2 H) and 16.00 u (O)—totalling 18.02 u. Since the average molecule of water has a mass of 18.02 u, one mole of water molecules has a mass of 18.02 g. When we measure 18.02 g of water in a container, using a balance, we have in effect "counted" 6.02×10^{23} molecules of H_2O in the container (**Figure 2**).

Figure 2
The mass of one mole of water is 18.02 g; 18.02 g of water contains 6.02×10^{23} molecules.

Since electrons have negligible mass compared to protons and neutrons, ions have essentially the same mass as their corresponding atoms. Thus a chloride ion (Cl^-) has the same mass as a chlorine atom, 35.49 u. A mole of chloride ions has the same mass as a mole of chlorine atoms, 35.49 g. When ions combine to form ionic compounds, the mass of one formula unit is the sum of the masses of the ions in the unit. Thus, the mass of one formula unit of sodium chloride is the sum of the mass of one sodium ion and one chloride ion.

- The mass of an average water molecule (H_2O) is $(2 \times 1.01 + 16.00)$ u = 18.02 u; the mass of one mole of water molecules is 18.02 g.
- The mass of an average ammonia molecule (NH_3) is $(14.01 + 3 \times 1.01)$ u = 17.04 u; the mass of one mole of ammonia molecules is 17.04 g.
- The mass of an average hydroxide ion (OH^-) is $(16.00 + 1.01)$ u = 17.01 u; the mass of one mole of OH^- ions is 17.01 g.
- The mass of an average formula unit of NaCl is $(22.99 + 35.45)$ u = 58.44 u; the mass of one mole of NaCl is 58.44 g.

DID YOU KNOW ?

How Many People in One Mole?

world population (6 billion)	6 000 000 000
1 mol of people	602 000 000 000 000 000 000 000

The mass in grams of one mole of a substance is called its **molar mass** (SI unit, g/mol; SI symbol *M*). One mole of carbon atoms has a molar mass of 12.01 g/mol. Similarly, one mole of water molecules has a molar mass of 18.02 g/mol.

molar mass: the mass, in grams per mole, of one mole of a substance. The SI unit for molar mass is g/mol.

When we communicate amounts in moles, it is important to specify the entity being measured. Note the two values for the molar mass of oxygen:

molar mass of oxygen *atoms* = 16.00 g/mol

molar mass of oxygen *molecules* = 32.00 g/mol

In the first value of molar mass, oxygen atoms are being counted; in the second, oxygen molecules are being counted. The molar mass of oxygen *atoms* (O) is 16.00 g/mol, and the molar mass of oxygen *molecules* (O_2) is 32.00 g/mol. The difference in molar mass reflects the difference between the mass of an oxygen atom and that of an oxygen molecule. As we normally encounter it,

diatomic: composed of two atoms

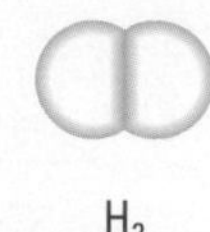

H_2

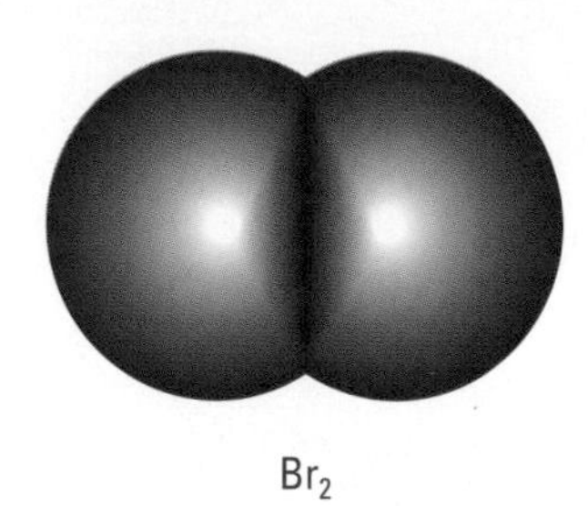

Br_2

Figure 3
Models representing diatomic elements H_2 and Br_2

oxygen is **diatomic**, composed of two atoms (**Figure 3**). The usual form of the element is O_2, so a mole of oxygen usually means a mole of oxygen molecules rather than oxygen atoms. The same interpretation applies to the other elements that form diatomic molecules: H_2, N_2, F_2, Cl_2, Br_2, and I_2. Unless otherwise stated, a mole of each of these elements is taken to mean a mole of molecules. To avoid ambiguity, however, it is always best to specify whether the entity discussed is an atom or a molecule.

Sample Problem

What is the molar mass of ammonium phosphate, $(NH_4)_3PO_4$?

Solution

$$M_{(NH_4)_3PO_4} = 3(N + 4H) + P + 4O$$
$$= 3N + 12H + P + 4O$$
$$= (3 \times 14.01) + (12 \times 1.01) + (1 \times 30.97) + (4 \times 16.00)$$
$$M_{(NH_4)_3PO_4} = 149.12 \text{ g/mol}$$

The molar mass of ammonium phosphate is 149.12 g/mol.

SUMMARY Calculating Molar Mass

1. Write the chemical formula for the substance.
2. Determine the number of atoms or ions of each element in one formula unit of the substance.
3. Use the atomic molar masses from the periodic table and the amounts in moles to determine the molar mass of the chemical.
4. Communicate the molar mass in units of grams per mole (g/mol). In this text, we will calculate molar mass to two decimal places where possible.

Practice

Understanding Concepts

7. Define the term *molar mass* and state its SI unit.
8. What is the molar mass of calcium hydroxide, $Ca(OH)_{2(s)}$, a key ingredient in some antacid tablets?
9. What is the molar mass of chlorine, $Cl_{2(g)}$, a poisonous and reactive nonmetal that is used in solution as an industrial and domestic bleach?
10. What is the molar mass of hydroxide ions, OH^-, present in many bases?
11. If the molar mass of atoms of an element was found to be 197.0 g/mol, what element could it be?
12. If the molar mass of a substance is 67.2 g/mol, what is the mass of 8.0 mol of the substance?
13. (a) What does the term *diatomic* mean?
    (b) Name seven elements that form diatomic molecules.
    (c) How many molecules are present in one mole of any diatomic element?

Answers

8. 74.10 g
9. 70.90 g
10. 17.01 g
12. 0.54 kg

Section 4.3 Questions

Understanding Concepts

1. Define Avogadro's constant, and explain its significance in quantitative analysis.
2. Distinguish between the terms *atomic mass* and *molar mass*.
3. Calculate the mass of a molecule of sucrose (sugar, $C_{12}H_{22}O_{11(s)}$) and the mass of a mole of sucrose.
4. What is the molar mass of octane, $C_8H_{18(l)}$, a component of gasoline?

Making Connections

5. One mole of any gas at 0°C and 101 kPa occupies the volume 22.4 L. Use this information and your knowledge of molar mass to determine the density (in grams per litre) of gaseous hydrogen, helium, nitrogen, oxygen, and carbon dioxide in those conditions.

Reflecting

6. You are learning how to calculate molar mass. How do you think this skill will be useful?

DID YOU KNOW ?

Mole Day

Mole Day, which is held in recognition of the mole concept, is celebrated yearly on October 23 between 6:02 a.m. and 6:02 p.m.

4.4 Calculations Involving the Mole Concept

In this section, we will apply the mole concept in a number of situations that we encounter frequently in the study of chemical reactions. Chemical formulas and equations are expressed using amount in moles. In a laboratory, we measure mass. So we constantly need to convert amount in moles into mass and vice versa.

Think about the term *molar mass*. It contains the concepts of both moles and mass. We always use molar mass in connecting the two measurements of amount in moles and mass.

Converting Mass to Amount in Moles

To calculate an amount in moles, we take the given mass in grams and divide by the molar mass. Let's take an example. Each mole of carbon atoms has a mass of 12 g. If we have 24 g of carbon atoms, what is the amount in moles? Since 24 g is twice 12 g, we have 2.0 moles of carbon atoms.

Mathematically, when we divide the mass we are given, 24 g, by the molar mass, 12 g/mol, we get 2.0 mol.

$$\text{Amount in moles} = \frac{\text{mass (g)}}{\text{molar mass (g/mol)}}$$

$$n_c = \frac{24\text{ g}}{12\text{ g/mol}}$$

$$= 24\text{ }\cancel{\text{g}} \times \frac{1\text{ mol}}{12\text{ }\cancel{\text{g}}}$$

$$n_c = 2.0\text{ mol}$$

In SI symbols, the relationship of amount (n), mass (m), and molar mass (M) is expressed as

$$n = \frac{m}{M}$$

Sample Problem 1

Convert a mass of 1.5 kg of calcium carbonate to an amount in moles.

Solution

$m_{CaCO_{3(s)}} = 1.5 \text{ kg}$

$M_{CaCO_{3(s)}} = (1 \times 40.08) + (1 \times 12.01) + (3 \times 16.00) = 100.09 \text{ g/mol}$

$$n_{CaCO_{3(s)}} = 1.5 \text{ kg} \times \frac{1 \text{ mol}}{100.09 \text{ g}}$$

$$= 1500 \text{ g} \times \frac{1 \text{ mol}}{100.09 \text{ g}}$$

$$n_{CaCO_3} = 15 \text{ mol}$$

A mass of 1.5 kg of calcium carbonate is equal to 15 mol of calcium carbonate.

Practice

Understanding Concepts

1. Convert a mass of 2.5 g of table salt (sodium chloride) to an amount in moles.
2. Convert a mass of 1.0 kg of glucose, $C_6H_{12}O_{6(s)}$, to an amount in moles.
3. What is the amount in moles of 25.0 g of oxygen gas?
4. A clear, colourless liquid when decomposed produced 24.0 g of carbon atoms, 6.0 g of hydrogen atoms, and 16.0 g of oxygen atoms.
   (a) Calculate the amount in moles of each element in the compound.
   (b) Determine the ratio of the number of atoms of carbon to hydrogen to oxygen, that is, the mole ratio of C:H:O.
   (c) From your answer in (b), suggest a formula for the clear, colourless liquid.

Answers

1. 0.043 mol
2. 5.6 mol
3. 0.781 mol
4. (a) 2.00 mol C, 5.94 mol H, 1.00 mol O
   (b) 2:6:1

Converting Amount in Moles to Mass

If we know the amount in moles of a reactant or a product, we can calculate the mass by using the molar mass. Consider the amount of 2.0 mol of sodium hydroxide, $NaOH_{(s)}$. The molar mass of $NaOH_{(s)}$ is (22.99 + 16.00 + 1.01) g/mol, or 40.00 g/mol. The mass of 2.0 mol of $NaOH_{(s)}$ would be exactly double the mass of 1.0 mol of $NaOH_{(s)}$. Therefore, the mass of 2.0 mol of $NaOH_{(s)}$ is double 40 g, which is 80 g.

Mathematically, in order to calculate the mass of a substance, we multiply the amount in moles by its molar mass.

$$\text{mass} = \text{amount in moles (mol)} \times \text{molar mass (g/mol)}$$

$$m_{NaOH} = 2.0 \text{ mol} \times \frac{40.00 \text{ g}}{1 \text{ mol}}$$

$$= 80 \text{ g}$$

This relationship can be represented by

$$m = nM$$

which is simply a rearrangement of the relationship we used earlier:

$$n = \frac{m}{M}$$

Sample Problem 2

Convert a reacting amount of 0.346 mol of sodium sulfate into mass in grams.

Solution

$n_{Na_2SO_4} = 0.346\ \text{mol}$

$M_{Na_2SO_4} = 142.04\ \text{g/mol}$

$$m_{Na_2SO_4} = 0.346\ \cancel{\text{mol}} \times \frac{142.04\ \text{g}}{1\ \cancel{\text{mol}}}$$

$$m_{Na_2SO_4} = 49.1\ \text{g}$$

The mass of 0.346 mol of $Na_2SO_{4(s)}$ is 49.1 g.

(a)

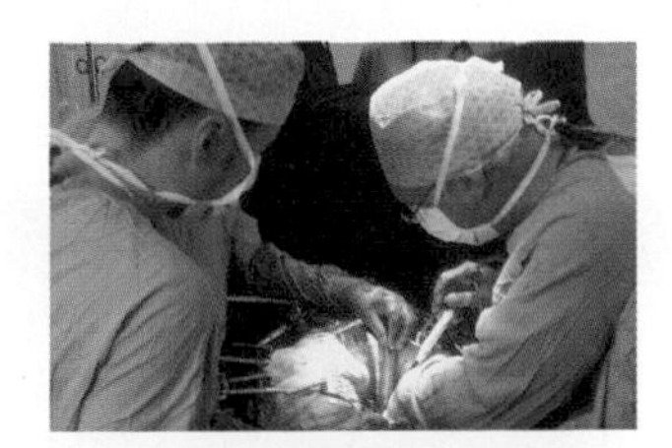

(b)

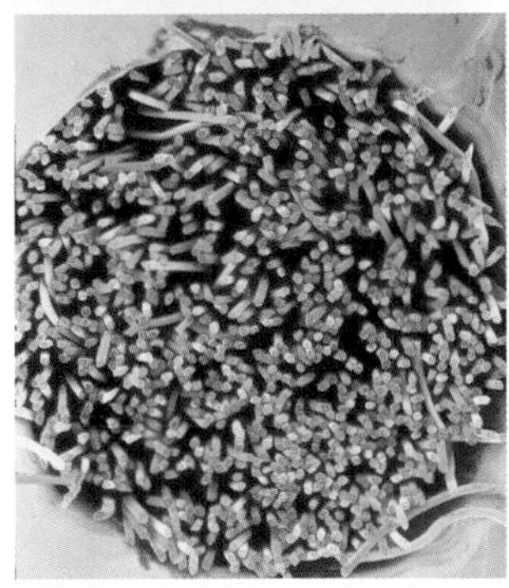

Figure 1
Dacron is a type of polyester made from 1,4-benzendioic acid.
(a) Dacron tubes reinforce a damaged artery.
(b) Dacron fibres are used as insulation in sleeping bags.

Practice

Understanding Concepts

5. Magnesium hydroxide is a base that is used in some antacids. What is the mass in grams of 0.45 mol of magnesium hydroxide?
6. Ammonia is a gas that, dissolved in water, is used as a cleaning agent. Convert 87 mmol of ammonia, $NH_{3(g)}$, into mass in grams.
7. 1,4-benzenedioic acid, $C_8H_6O_{4(s)}$, is a raw material used in the manufacture of Dacron, a synthetic fibre (**Figure 1**). What is the mass in grams of 63.28 mol of 1,4-benzenedioic acid?
8. If a patient is prescribed 1.0×10^{-3} mol of acetylsalicylic acid (Aspirin, $C_9H_8O_{4(s)}$), what mass of Aspirin should she take?

Answers

5. 26 g
6. 1.5 g
7. 10.51 kg
8. 0.18 g

Calculations Involving Number of Entities

We often need to know the number of entities in a sample of a substance. When dealing with gases, for example, the number of atoms or molecules in a container is related to other factors, such as the temperature and the pressure of the gas.

We will first look at how we can convert an amount in moles into the number of entities. We know there are always 6.02×10^{23} entities in a mole of substance. If we have 2.0 mol of copper, then there will be $2.0 \times 6.02 \times 10^{23}$ copper atoms. The number of entities, N, in an amount in moles is calculated by multiplying the amount in moles by 6.02×10^{23}, Avogadro's constant:

$$N = nN_A.$$

In the case of 2.0 mol of copper atoms,

$$N_{Cu} = 2.0\ \cancel{\text{mol}} \times \frac{6.02 \times 10^{23}\ \text{atoms}}{1\ \cancel{\text{mol}}}$$

$$= 1.2 \times 10^{24}$$

There are 1.2×10^{24} atoms in 2.0 mol of copper metal.

Now let's look at calculating the number of entities in a given mass. How many copper atoms do you think there are in a penny? (We'll assume our pennies are dated before 1997, since after this time they were no longer made of copper, but copper-plated zinc.) The calculations are similar to the preceding example; however, since we are not told the amount in moles, we need to first convert the mass into amount in moles.

Our strategy is first to measure the mass of the penny (**Figure 2**). Then we convert the mass into amount in moles. Finally, we can convert the amount in moles to number of atoms. Mass and number of entities are both easily converted to amount in moles, so the middle step in these conversions is always to determine the amount in moles.

Figure 2
To reduce error in determining the mass of a single penny, we can find the mass of 150 pennies and divide the total mass by 150.

Suppose we measured the mass of a penny and found it to be 2.63 g. We will convert this mass to amount in moles of copper atoms.

$$M_{Cu(s)} = 63.55 \text{ g/mol}$$

Therefore,

$$n_{Cu(s)} = 2.63 \cancel{\text{g}} \times \frac{1 \text{ mol}}{63.55 \cancel{\text{g}}}$$

$$n_{Cu(s)} = 0.0414 \text{ mol}$$

We now simply multiply the amount in moles by N_A:

$$N_{Cu(s)} = 0.0414 \cancel{\text{mol}} \times \frac{6.02 \times 10^{23}}{1 \cancel{\text{mol}}}$$

$$N_{Cu(s)} = 2.49 \times 10^{22}$$

There are 2.49×10^{22} atoms in a copper penny.

Sample Problem 3

Determine the number of chloride ions in 0.563 mol of calcium chloride, $CaCl_{2(s)}$.

Solution

$$n_{CaCl_2} = 0.563 \text{ mol}$$

$$N_{CaCl_2} = 0.563 \cancel{\text{mol}} \times \frac{6.02 \times 10^{23}}{1 \cancel{\text{mol}}}$$

$$N_{CaCl_2} = 3.39 \times 10^{23}$$

There are 3.39×10^{23} formula units in 0.563 mol of $CaCl_2$.

Since each formula unit ($CaCl_2$) contains two chloride ions, there are $2 \times (3.39 \times 10^{23})$, 6.78×10^{23}, chloride ions in 0.563 mol of calcium chloride.

Sample Problem 4

How many sugar (sucrose, $C_{12}H_{22}O_{11(s)}$) molecules are there in a 1.000 kg bag of sugar?

Solution

$M_{C_{12}H_{22}O_{11(s)}} = [(12 \times 12.01) + (22 \times 1.01) + (11 \times 16.00)] = 342.34 \text{ g/mol}$

$m = 1.000 \text{ kg}$

$$n_{C_{12}H_{22}O_{11}} = 1000.0 \text{ g} \times \frac{1 \text{ mol}}{342.34 \text{ g}}$$

$$= 2.9211 \text{ mol}$$

$$N_{C_{12}H_{22}O_{11}} = 2.9211 \text{ mol} \times \frac{6.02 \times 10^{23}}{1 \text{ mol}}$$

$$N_{C_{12}H_{22}O_{11}} = 1.76 \times 10^{24}$$

There are 1.76×10^{24} sugar molecules in a 1.000 kg bag of sucrose.

Sample Problem 5

What is the mass of one water molecule (**Figure 3**)?

Solution

$M_{H_2O} = (2 \times 1.01) + (1 \times 16.00) = 18.02 \text{ g/mol}$

Since there are 6.02×10^{23} molecules/mol,

$$m_{H_2O} = \frac{18.02 \text{ g}}{1 \text{ mol}} \times \frac{1 \text{ mol}}{6.02 \times 10^{23}}$$

$$m_{H_2O} = 2.99 \times 10^{-23} \text{ g}$$

The mass of one water molecule is 2.99×10^{-23} g.

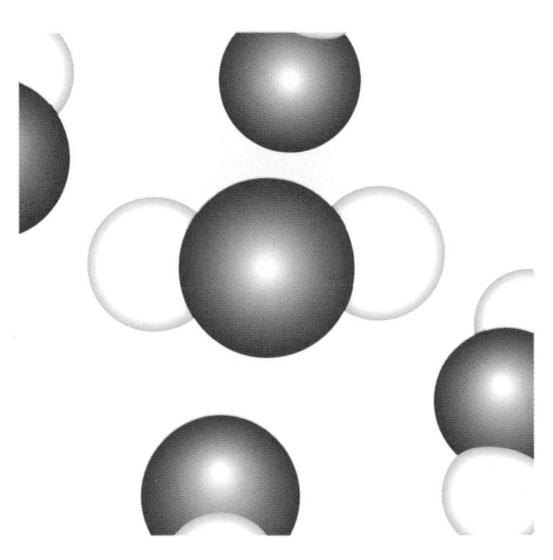

Figure 3
The mass of one water molecule is an unimaginable 2.99×10^{-23} g.

SUMMARY Calculating Mass, Amounts in Moles, and Number of Entities

1. n represents the amount in moles, m the mass measured, M the molar mass, N the number of entities, and N_A Avogadro's constant (**Table 1**).
2. $n = \frac{m}{M}$
3. $m = nM$
4. $N = nN_A$

Table 1: Stoichiometry, Symbols and Units

Symbol	Quantity	Unit
n	amount in moles	mol
m	mass	mg, g, kg
M	molar mass	g/mol
N	number of entities	atoms, ions, formula units, molecules
N_A	Avogadro's constant, 6.02×10^{23}	—

Counting Atoms, Molecules, and Other Entities

Using the materials and equipment supplied, calculate and measure the quantities described in each of the steps below. Write an explanation of your calculations.

Materials: balance, graduated cylinder, beaker, disposable cups, copper pennies, iron nails, granulated sugar, table salt, chalk, water

- Determine the mass of a drop of water by measuring the mass of 50 drops of water. Place a single drop of water on the lab bench and record the time it takes for it to completely evaporate. Calculate how many molecules of water evaporate per second.
- Calculate the number of copper atoms in a penny and use that number to calculate the value of each atom of copper in the penny.
- Measure into a graduated cylinder half a mole of sucrose molecules ($C_{12}H_{22}O_{11(s)}$). Record the reading on the graduated cylinder.
- Measure into a graduated cylinder the quantity of sugar that contains two moles of carbon atoms. Record the reading on the graduated cylinder.
- Measure the mass of a piece of chalk. Use the piece of chalk to write your full name on the chalkboard. Measure the mass of the chalk again. Calculate the number of atoms that were needed to write your name (assume chalk is made entirely of calcium carbonate).
- Dissolve 3.00 g of table salt (assume $NaCl_{(s)}$) in 200 mL of water. Calculate the number of sodium ions in the salt solution.
- Calculate the number of iron atoms in the iron nail.
- Calculate the number of years to span a mole of seconds.

Practice

Understanding Concepts

9. Calculate the molar mass of each of the following substances:
   (a) $H_2O_{(l)}$ (water)
   (b) $CO_{2(g)}$ (respiration product)
   (c) $NaCl_{(s)}$ (pickling salt, sodium chloride)
   (d) $C_{12}H_{22}O_{11(s)}$ (table sugar, sucrose)
   (e) $(NH_4)_2Cr_2O_{7(s)}$ (ammonium dichromate)
10. Calculate the amount of pure substance present (in moles) in each of the following samples of pure substances:
   (a) a 10.00-kg bag of table sugar
   (b) a 500-g box of pickling salt
   (c) 40.0 g of propane, $C_3H_{8(g)}$, in a camp-stove cylinder
   (d) 325 mg of acetylsalicylic acid (Aspirin), $HC_9H_7O_{4(s)}$, in a headache-relief tablet
   (e) 150 g of 2-propanol (rubbing alcohol), $CH_3CH_2OHCH_{3(l)}$, from a pharmacy
11. Calculate the mass of each of the following substances:
   (a) 4.22 mol of ammonia in a window-cleaning solution
   (b) 0.224 mol of sodium hydroxide in a drain-cleaning solution
   (c) 57.3 mmol of water vapour produced by a laboratory burner
   (d) 9.44 kmol of potassium permanganate fungicide
   (e) 0.77 mol of ammonium sulfate fertilizer
12. Calculate the number of entities in each of the following samples:
   (a) 15 mol of solid carbon dioxide, in dry ice
   (b) 15 g of ammonia gas, in household cleaners

Answers

9. (a) 18.02 g/mol
   (b) 44.01 g/mol
   (c) 58.44 g/mol
   (d) 342.34 g/mol
   (e) 252.10 g/mol
10. (a) 29.21 mol
   (b) 8.56 mol
   (c) 0.907 mol
   (d) 1.80×10^{-3} mol
   (e) 2.45 mol
11. (a) 71.9 g
   (b) 8.96 g
   (c) 1.03 g
   (d) 1.49×10^3 kg
   (e) 1.0×10^2 g
12. (a) 9.0×10^{24} molecules
   (b) 5.3×10^{23} molecules

(c) 15 g of hydrogen chloride gas, in hydrochloric acid
(d) 15 g of sodium chloride, in table salt

13. Calculate the mass, in grams, of the characteristic entity in each of the following samples:
(a) carbon dioxide from respiration
(b) glucose from photosynthesis
(c) oxygen from photosynthesis

14. How many water molecules are in a 1.000 L bottle of water? (Recall the density of water is 1.00 g/mL.)

Answers

12. (c) 2.5×10^{23} molecules
(d) 1.5×10^{23} formula units

13. (a) 7.31×10^{-23} g
(b) 2.99×10^{-22} g
(c) 5.32×10^{-23} g

14. 3.34×10^{25} molecules

Section 4.4 Questions

Understanding Concepts

1. (a) Calculate the number of oxygen molecules in 1.5 mol of oxygen gas, $O_{2(g)}$.
(b) Calculate the number of atoms in 1.5 mol of $O_{2(g)}$.

2. A daily vitamin tablet contains 90 mg of vitamin C. The chemical name for vitamin C is ascorbic acid, $H_2C_6H_6O_6$. How many molecules of vitamin C are you taking each day if you take a daily vitamin tablet?

3. A thyroid condition called goitre can be treated by increasing iodine in the diet. Iodized salt contains calcium iodate, $Ca(IO_3)_{2(s)}$, which is added to table salt.
(a) How many atoms of iodine are in 1.00×10^{-2} mol of calcium iodate?
(b) What is the mass of calcium iodate that contains that many atoms of iodine?

4. A recipe for a sweet-and-sour sauce calls for
500 g water
200 g sugar ($C_{12}H_{22}O_{11(s)}$)
25 g vinegar (assume acetic acid), $HC_2H_3O_{2(l)}$
15 g citric acid ($C_6H_8O_{7(s)}$)
5 g salt ($NaCl_{(s)}$)
Convert the recipe into amounts in moles.

5. If necessary, use density as a conversion factor to answer the following questions. (The density of elements can be referenced on the periodic table.)
(a) If the density of pure ethanol is 0.789 g/mL, how many ethanol molecules are there in a 17-mL sample of ethanol, the approximate quantity of ethanol in a bottle of beer?
(b) How many nickel atoms are there in 0.72 cm^3 of a nickel sample, the approximate volume of a Canadian quarter?
(c) How many water molecules are there in a 100-mL sample of pure water?

Applying Inquiry Skills

6. Silver ions in waste solutions can be recovered by immersing copper metal in the solution (**Figure 4**). The solid copper loses mass as copper goes into solution as ions. Crystals of silver are deposited on the copper metal. Design an experiment to determine the ratio of the amount in moles of copper ions dissolved to

(continued)

Figure 4
When copper metal is placed in a solution of silver ions, a single displacement reaction occurs. Copper ions go into solution and silver crystals are formed.

the amount in moles of silver atoms formed. Describe the procedure, materials, and equipment used, safety procedures, and explain the calculations needed.

Making Connections

7. Suppose that there is a prestigious award given by the Academy of Science each year to the most significant scientific concept. Write a paper nominating the mole concept for this award, citing the mole's role and importance in the application of chemical reactions in society, industry, and the environment.

4.5 Percentage Composition

Using molar mass values from a periodic table, we can calculate the mass of reactants and products in chemical reactions—if we already know the chemical formulas. However, when a new substance is produced, we first need to determine its chemical formula. To do this, we need to determine the chemical composition of the compound, that is, what elements it is made of and the quantities of each element.

In this section, we will experimentally determine the composition by mass of a substance and then convert the mass amounts to percentages, to give us the percentage composition. We can then use atomic mass and molar mass to determine the correct chemical formula.

Figure 1
This experimental car burns hydrogen as a fuel, producing water vapour as an exhaust. The dish collects solar energy, which is used to dissociate water into hydrogen and oxygen.

Before we begin, we will practise the mathematical steps in the calculation of percentage composition from mass measurements of reactants and products. Let's consider 20 g of red jelly beans mixed with 30 g of green jelly beans. What is the percentage composition of the mixture, by mass? The percentage of red jelly beans by mass would be 20 g of the total 50 g, which is 20 g/50 g × 100%, or 40%. Similarly, the percentage of green jelly beans would be 30 g of the total 50 g, which is 60%.

Now let's look at the percentage composition of water. Water is formed when hydrogen is allowed to react with oxygen, a reaction that gives off large amounts of energy (**Figure 1**). The results of an experiment revealed that 2.5 g of hydrogen, when completely reacted, produced 22.5 g of water. What is the percentage composition of water by mass?

Since 2.5 g of hydrogen combined with an amount of oxygen to produce 22.5 g of water, we can calculate the mass of oxygen by subtraction:

$$m_{H} = 2.5 \text{ g}$$

$$m_{H_2O} = 22.5 \text{ g}$$

$$m_{O} = (22.5 - 2.5) \text{ g} = 20.0 \text{ g}$$

$$\% \text{ H} = \frac{m_{H}}{m_{H_2O}} \times 100\%$$

$$\% \text{ H} = \frac{2.5 \not{\text{g}}}{22.5 \not{\text{g}}} \times 100\% = 11.1\%$$

Similarly,

$$\% \text{ O} = \frac{20.0 \not{\text{g}}}{22.5 \not{\text{g}}} \times 100\%$$

$$\% \text{ O} = 88.9\%$$

We will see later how this percentage composition, together with the atomic mass of hydrogen and oxygen, will allow us to determine a possible formula for water.

Sample Problem 1

Sodium is a very reactive alkali metal, and chlorine is a poisonous green gas. When they are allowed to react, they combine to form sodium chloride, the stable and usually harmless table salt, an ionic compound. In an experiment, 3.45 g of sodium metal reacted with 5.33 g of chlorine gas to give 8.78 g of sodium chloride. Calculate the percentage composition by mass of sodium chloride.

Solution

$m_{Na^+} = 3.45 \text{ g}$

$m_{Cl^-} = 5.33 \text{ g}$

$m_{NaCl} = 8.78 \text{ g}$

$$\% \text{ Na}^+ = \frac{3.45 \text{ g}}{8.78 \text{ g}} \times 100\%$$

$$\% \text{ Na}^+ = 39.3\%$$

$$\% \text{ Cl}^- = \frac{5.33 \text{ g}}{8.78 \text{ g}} \times 100\%$$

$$\% \text{ Cl}^- = 60.7\%$$

Therefore, the percentage composition by mass of sodium chloride is 39.3% sodium and 60.7% chlorine. (Note that the two percentages total 100%.)

Practice

Understanding Concepts

1. A 27.0-g sample of a compound contains 7.20 g of carbon, 2.20 g of hydrogen, and 17.6 g of oxygen. Calculate the percentage composition of the compound.
2. Carbon will burn in sufficient oxygen to produce carbon dioxide. In an experiment, 8.40 g of carbon reacts with oxygen and 30.80 g of carbon dioxide is produced.
   (a) What mass of oxygen reacted with 8.40 g of carbon?
   (b) Calculate the percentage composition by mass of carbon dioxide.
3. In one sample of a compound of copper and oxygen, 3.12 g of the compound contains 2.50 g of copper and the remainder is oxygen. In another sample of a compound of copper and oxygen, 1.62 g of the compound contains 1.44 g of copper and the remainder is oxygen.
   (a) Calculate the percentage composition of each compound.
   (b) Are the two samples of the same compound? Give reasons for your answer.

Answers

1. 26.7%, 8.1%, 65.2%
2. (a) 22.40 g
   (b) 27.3% C, 72.7% O
3. (a) 80.1% Cu, 19.9% O; 88.9% Cu, 11.1% O

INQUIRY SKILLS

- ○ Questioning
- ○ Hypothesizing
- ● Predicting
- ○ Planning
- ● Conducting
- ● Recording
- ● Analyzing
- ● Evaluating
- ● Communicating

Wear eye protection.

Use crucible tongs to transfer hot crucible.

Care is required in using a laboratory burner and handling hot apparatus.

Magnesium ribbon burns with a hot flame and an extremely bright light. If the magnesium ignites, do not look directly at the flame because it may damage your eyes.

DID YOU KNOW?

Magnesium Sparkles

Finely ground magnesium powder readily ignites upon heating in air, giving a dazzling white light. This property of magnesium is used in devices such as flares, incendiary bombs, and fireworks. Magnesium dust is a common component of fireworks, including sparklers, a nonexplosive type of fireworks constructed by coating a slurry of chemicals on a wire. The slurry consists of a fuel containing magnesium powder, an oxidizer, iron or steel powder, and a binder, all of which are proportioned to burn slowly, giving off a shower of bright shimmering sparks.

Investigation 4.5.1

Percentage Composition by Mass of Magnesium Oxide

In this investigation, you will test the law of definite proportions. To do this, you will determine the composition by mass of magnesium oxide, calculate the percentage composition, and compare your results and those of other students to the predicted values.

Magnesium is a silvery metal that burns with such a bright flame that it was once used in flashbulbs for photography. Magnesium oxide, a white powder, is produced as a result of the synthesis reaction of combusting magnesium. Here, you will conduct a slow combustion of magnesium and then complete the **Evidence**, **Analysis**, and **Evaluation** sections of a lab report.

Questions

What is the percentage composition by mass of magnesium oxide?
Is this percentage constant?

Prediction

(a) Make a prediction based on the law of definite proportions. What should the percentage composition by mass of magnesium oxide be?

Experimental Design

A known mass of magnesium metal is heated in a crucible over a laboratory burner. The mass of the magnesium oxide formed is used to determine the mass of the oxygen that reacted and the percentage composition by mass of the two components. Some of the magnesium also reacts with nitrogen in air to form a nitride. This compound is converted to magnesium oxide by adding water and reheating the solid.

Materials

lab apron
eye protection
centigram or analytical balance
7–8 cm magnesium ribbon
steel wool
porcelain crucible and lid
laboratory burner
retort stand
ring stand and clamp
clay triangle
crucible tongs
glass stirring rod
distilled water

Procedure

1. Using a balance, determine the mass of a clean, dry porcelain crucible and lid.
2. Polish the magnesium ribbon with the steel wool and fold the ribbon to fit into the bottom of the crucible.
3. Determine the mass of the crucible, lid, and magnesium ribbon.
4. Place the crucible securely on the clay triangle. Set the lid slightly off-centre on the crucible to allow air to enter but to prevent the magnesium oxide from escaping.
5. Place the laboratory burner under the crucible, light it, and begin heating with a gentle flame.
6. Gradually increase the flame intensity until all the magnesium turns into a white powder.
7. Cut the flow of gas to the burner and allow the crucible, lid, and contents to cool.
8. Using the stirring rod, crush the contents of the crucible into a fine powder. Carefully add about 10 mL of distilled water to the powder. Use some of the water to rinse any powder on the stirring rod into the crucible.
9. Heat the crucible and contents, with the lid slightly ajar, gently for 3 min and strongly for another 7 min.
10. Allow the crucible and contents to cool.
11. Using the balance, determine the mass of the cooled crucible, lid, and contents.

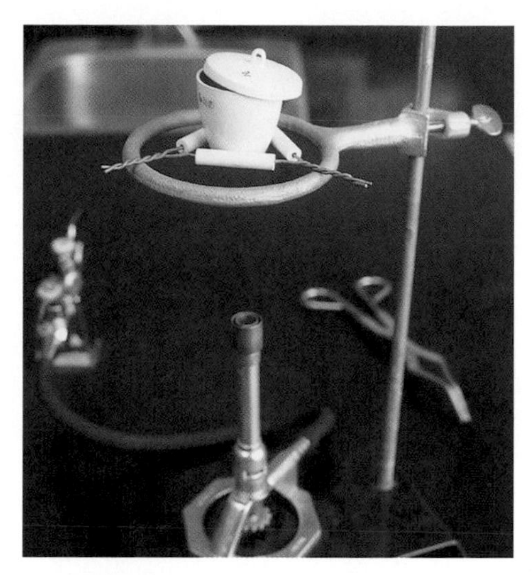

Figure 2
Apparatus for heating in a crucible

Analysis

(b) What evidence do you have that a chemical reaction took place?
(c) Calculate the mass of oxygen that reacted with the magnesium.
(d) Use your evidence to calculate the percentage composition by mass of magnesium oxide.
(e) Based on the evidence (your results and those of your classmates), what are the answers to the Questions?

Evaluation

(f) If some of the magnesium oxide had escaped from the crucible, would your percentage composition calculation of magnesium be too high or too low? Explain.
(g) If the magnesium had reacted with some other component in the air, would your percentage composition calculation of magnesium be too high or too low? Explain.
(h) The magnesium ribbon was polished to remove any white film on its surface before beginning the experiment. Explain why this is necessary.
(i) Suggest a modification in the procedure to ensure that all of the magnesium completely reacts with oxygen.
(j) Evaluate your prediction. Based on the evidence obtained from several groups, is the law of definite proportions valid?

Practice

Making Connections

4. Use the Internet to research information about tires. When you look at the sidewall of a tire, you will see a lot of information, such as the name of the tire, its size, whether it is tubeless or a tube type, the grade, and the speed rating. It also gives important safety information such as the maximum load and maximum inflation for the tire. In addition, the composition of the tire can be obtained from the manufacturer. For example, a Goodyear all-season passenger tire contains approximately

   1.8 kg of 8 types of natural rubber
   2.3 kg of 8 types of carbon black
   0.5 kg of steel cord for belts
   0.5 kg of polyester and nylon
   0.5 kg of steel bead wire
   1.4 kg of 40 kinds of chemicals, waxes, oils, pigments
   2.7 kg of 5 different types of synthetic rubber

   (a) Calculate the percentage composition of this tire.
   (b) Research and compare the percentages of synthetic and natural rubber used in various types of tires, for example, tires for light trucks, racecars, and off-highway vehicles.
   (c) Relate the characteristics of synthetic and natural rubber mixes to their application in the different types of tires.
   (d) Research the contributions of Goodyear in the composition of rubber, and assess the impact of the development of rubber tires on society and the environment.
   Follow the links for Nelson Chemistry 11, 4.5.

Answers

4. (a) 18.6%, 23.7%, 5.2%, 5.2%, 5.2%, 14.4%, 27.8%

What Makes Popcorn Pop?

In each kernel of popping corn, there is a small drop of water in a circle of soft starch. When heated, the water expands and builds up pressure against the hard outer surface, eventually exploding and turning the kernel inside out.

Materials: popping corn, hot-air popcorn popper, balance

- Measure the mass of some unpopped popping corn.
- Pop the popping corn.
- Allow the popcorn to cool and measure the mass again.
- Assume that any difference in masses is caused by loss of water from the kernels. Calculate the percentage of water in the sample of popcorn.
- Repeat the activity with kernels of popping corn that have been cut in half either lengthwise or crosswise.
- Record the percentage of popped kernels from each cutting method.

(a) Do the results confirm the given reason why popcorn pops? Explain.

Percentage Composition Calculations from a Formula

We have seen how to determine the percentage composition of a compound through experimentation. Empirical information is often used to determine the formula of a compound.

Sometimes we also need to calculate the percentage composition of a compound whose formula we already know. For example, we may wish to verify the purity of a compound by comparing its percentage composition obtained experimentally to the theoretical value, calculated from the compound's formula. Percentage composition also has commercial uses, for example, in fertilizers. The chemical formula can be used to determine the percentage by mass that is contributed by each element in the compounds that make up the fertilizer. Nitrogen is one of the key elements delivered to plants by fertilizers; it is important to calculate the percentage of nitrogen in fertilizer compounds to determine the correct quantities of fertilizer to apply.

If we know the chemical formula of the compound, calculating percentage composition is straightforward. Essentially, we want to calculate the "contribution" of each element to the total mass of the compound. Thus, we first calculate the mass of all the atoms of each element. Then, we calculate the total mass of all the elements in the compound. To obtain the percentage contribution of each element, we divide the mass of each element by the total mass.

Sample Problem 2

Determine the percentage composition of sodium carbonate, Na_2CO_3, also known as soda ash.

Solution

$$m_{Na} = 22.99 \text{ u} \times 2 = 45.98 \text{ u}$$

$$m_{C} = 12.01 \text{ u} \times 1 = 12.01 \text{ u}$$

$$m_{O} = 16.00 \text{ u} \times 3 = 48.00 \text{ u}$$

$$m_{Na_2CO_{3(s)}} = 105.99 \text{ u}$$

$$\% \text{ Na} = \frac{45.98 \not{\text{u}}}{105.99 \not{\text{u}}} \times 100\%$$

$$\% \text{ Na} = 43.38\%$$

$$\% \text{ C} = \frac{12.01 \not{\text{u}}}{105.99 \not{\text{u}}} \times 100\%$$

$$\% \text{ C} = 11.33\%$$

$$\% \text{ O} = \frac{48.00 \not{\text{u}}}{105.99 \not{\text{u}}} \times 100\%$$

$$\% \text{ O} = 45.29\%$$

The percentage composition of $Na_2CO_{3(s)}$ is 43.38% sodium, 11.33% carbon, and 45.29% oxygen.

Practice

Answers

5. 2.1% H, 32.7% S, 65.2% O
6. 41.6% Mg
7. 77.7% Fe, 22.3% O; 69.9% Fe, 30.1% O
8. 28.2% N

Understanding Concepts

5. Calculate the percentage composition by mass of sulfuric acid, $H_2SO_{4(aq)}$, used in car batteries.
6. Calculate the percentage by mass of magnesium in magnesium hydroxide, $Mg(OH)_{2(s)}$, used in some antacids.
7. Iron and oxygen combine to form two different compounds. The formulas of the compounds are $FeO_{(s)}$ and $Fe_2O_{3(s)}$. Calculate the percentage composition of each compound.
8. Calculate the percentage of nitrogen in ammonium phosphate, $(NH_4)_3PO_4$, a compound used in fertilizers.

Section 4.5 Questions

Understanding Concepts

1. Explain why it is necessary to determine the percentage composition of a new compound by experiment.
2. In a compound consisting of potassium and chlorine, 33.5 g of potassium combined with 30.4 g of chlorine. Calculate the percentage composition of the compound.
3. The following evidence was obtained in an experiment to determine the percentage composition of a compound containing sodium, sulfur, and oxygen:
   mass of Na atoms = 23.0 g
   mass of S atoms = 16.0 g
   mass of O atoms = 32.0 g

   Calculate the percentage composition of this compound.
4. Ammonium nitrate, $NH_4NO_{3(s)}$, and ammonium sulfate, $(NH_4)_2SO_{4(s)}$, are both compounds used as fertilizers. Determine which compound contains the greater percentage by mass of nitrogen.
5. Calcium sulfate dihydrate, $CaSO_4{\cdot}2H_2O_{(s)}$, is commonly called gypsum and is used in building materials such as drywall. It contains water of crystallization, some of which is lost on heating, leaving $(CaSO_4)_2{\cdot}H_2O_{(s)}$. Compare the percentage by mass of water in each compound.

Applying Inquiry Skills

6. In this lab exercise, a synthesis reaction is used to determine the percentage composition of a compound of copper and sulfur.

   Experimental Design

   When heated strongly in a crucible, copper wire or turnings react with an excess of sulfur to produce a solid, a sulfide of copper.

   Procedure

   (a) Design a Procedure, based on the Experimental Design, to obtain the evidence needed.

   Analysis

   (b) Explain how the evidence gathered in your procedure would be used to calculate the percentage composition of the product.

Evaluation

(c) Evaluate the experimental design.

Making Connections

7. Name two products that you might find around the house where percentage compositions are given
   (a) in mass;
   (b) in measurements other than mass.

4.6 Empirical and Molecular Formulas

If we are given an unknown substance and we want to find its chemical formula, how do we begin? We need to identify the elements that are in the compound as well as the number of atoms of each element. To begin with, we can determine its percentage composition by mass. Then, we simply convert the mass values into amounts in moles, which gives us the subscripts in the chemical formula of the substance.

A formula derived in this way is called an **empirical formula**, which means that it is derived from observations in an experiment rather than from theory. An empirical formula tells us the simplest ratio of the combining elements.

empirical formula: simplest whole-number ratio of atoms or ions in a compound

Currently, many new substances that are synthesized are organic compounds, that is, contain mainly carbon and hydrogen atoms. One of the technologies available to measure the percentage composition of these compounds uses combustion analyzers (**Figure 1**). In this process, several milligrams of a compound are burned inside a combustion chamber. When the compound is burned, oxygen combines with the carbon atoms to form carbon dioxide and with the hydrogen atoms to form water vapour. Any other elements present are similarly converted to their oxides. The quantities of these combustion products are precisely measured, and computer analysis discloses the percentage by mass of each element detected in the compound. This percentage composition is then used to calculate the empirical formula of the compound.

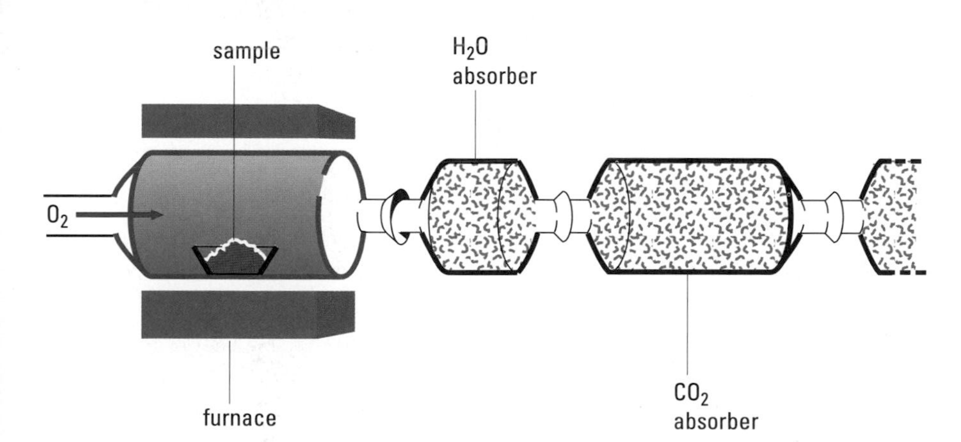

Figure 1
A substance burned in a combustion analyzer produces oxides that are captured by absorbers in chemical traps. The initial and final masses of each trap indicate the masses of the oxides produced. These masses are then used in the calculation of the percentage composition of the substance burned.

An empirical formula does not necessarily provide the correct information about the number of atoms in a molecule. That is, an empirical formula may not be the correct formula. Different molecules may have the same percentage composition but contain different numbers of atoms in a molecule. For example, ethyne (also known as acetylene), used in welders' torches, and benzene, used as a solvent, both have the same percentage composition by mass, so they also have the same empirical formula, CH. A **molecular formula** is needed to tell us the actual number and kind of atoms in a molecule of the substance. The molecular formula for ethyne is C_2H_2, and the molecular formula for benzene is C_6H_6. **Table 1** compares the properties of ethyne and benzene.

molecular formula: a group of chemical symbols representing the number and kind of atoms covalently bonded to form a single molecule

Table 1: Comparison of Ethyne and Benzene

	Ethyne	Benzene
empirical formula	CH	CH
molecular formula	$C_2H_{2(g)}$	$C_6H_{6(l)}$
physical properties	colourless, odourless, flammable gas	colourless, flammable liquid
melting point	–81°C	5.5°C
boiling point	–57°C	80.1°C
density	0.621 g/L	0.877 g/mL
solubility	acetone, benzene, chloroform	alcohol, ether, acetone, acetic acid

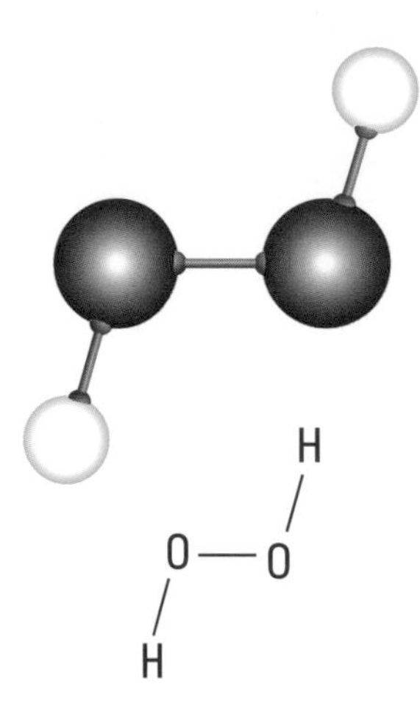

(a) The empirical formula of hydrogen peroxide is HO; its molecular formula is H_2O_2.

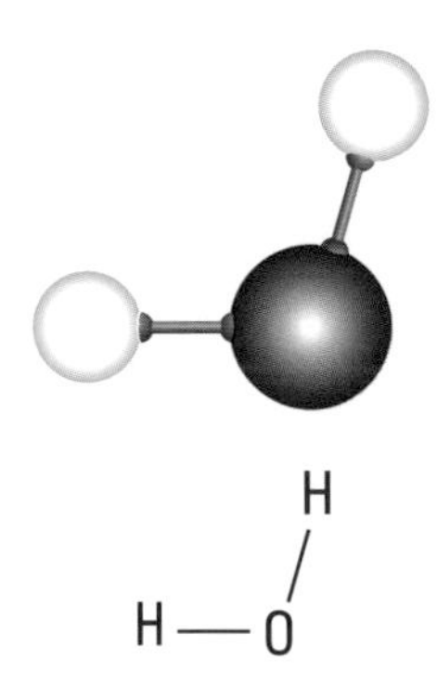

(b) The empirical formula of water is H_2O; its molecular formula is also H_2O.

Figure 2

To further illustrate the difference between empirical and molecular formulas, let's revisit the compound hydrogen peroxide. Observations from decomposition or synthesis experiments reveal that hydrogen peroxide contains hydrogen atoms and oxygen atoms in a combining ratio of 1:1. The empirical formula for hydrogen peroxide is, therefore, HO. When other information such as the mass of a molecule is taken into account, we find that each molecule of hydrogen peroxide contains two hydrogen atoms and two oxygen atoms (**Figure 2**). Thus, the molecular formula for hydrogen peroxide is H_2O_2.

As you can see, H_2O_2 is a multiple of the empirical formula HO. The empirical formula HO gives the combining ratio in its simplest form, and the molecular formula H_2O_2 gives the same ratio in the actual number of combining atoms. In this case, the subscripts in the molecular formula H_2O_2 are double the subscripts in the empirical formula HO. In other cases, the empirical and molecular formulas may be the same. Water is an example in which the empirical formula, H_2O, is the same as the molecular formula, H_2O.

Practice

Understanding Concepts

1. Define the term "empirical."
2. What information does a molecular formula provide that is not provided by an empirical formula?
3. Can two different compounds have the same percentage composition? Explain your answer.
4. If CH_3 is the empirical formula of a compound, what is a possible molecular formula of this compound?
5. Write the empirical formula of each of the following:
   (a) dinitrogen tetroxide, $N_2O_{4(g)}$
   (b) carbon dioxide, $CO_{2(g)}$

(c) acetic acid (vinegar), $HC_2H_3O_2$

(d) polychlorinated biphenyl (PCBs) used in transformers, $(C_6H_4Cl_2)_{2(l)}$

6. Does an ionic compound such as sodium chloride have an empirical formula and a molecular formula? Explain.

Distinguish Between Empirical and Molecular Formulas

In this activity you will examine the structure of some compounds and distinguish between their empirical and molecular formulas.

Materials: molecular model kits

- Assemble a model of each of the following organic compounds based on the structural formulas shown in **Figure 3**: ethane, butane, hexane, ethene, butene, hexene.
- In a table record the name, empirical formula, and molecular formula of each compound.

```
    H   H
    |   |
H — C — C — H
    |   |
    H   H
  ethane

    H   H   H   H
    |   |   |   |
H — C — C — C — C — H
    |   |   |   |
    H   H   H   H
       butane

    H   H   H   H   H   H
    |   |   |   |   |   |
H — C — C — C — C — C — C — H
    |   |   |   |   |   |
    H   H   H   H   H   H
             hexane

    H   H
    |   |
H — C = C — H
  ethene

    H   H   H   H
    |   |   |   |
H — C — C = C — C — H
    |           |
    H           H
       butene

    H   H   H   H   H   H
    |   |   |   |   |   |
H — C — C — C = C — C — C — H
    |   |           |   |
    H   H           H   H
             hexene
```

Figure 3
Structural formulas for some hydrocarbons

4.7 Calculating Chemical Formulas

In section 4.5, you synthesized magnesium oxide by heating magnesium ribbon in a crucible in the presence of oxygen. By analyzing the masses of each element reacted and the mass of the compound formed, you calculated the percentage composition. We will now use that information to illustrate how an empirical formula can be determined from percentage composition.

Calculating an Empirical Formula

Suppose that the results of Investigation 4.5.1 gave us the following percentage composition of magnesium oxide: 60% magnesium and 40% oxygen. This means, in effect, that in 100.00 g of the compound, there are 60.00 g of magnesium atoms and 40.00 g of oxygen atoms. Now we can determine how many atoms there are in 60.00 g of magnesium and how many atoms there are in 40.00 g of oxygen. We will then know the ratio of the number of magnesium atoms to oxygen atoms.

First we convert 60.00 g of magnesium to an amount in moles:

$M_{Mg} = 24.31 \text{ g/mol}$

$$n_{Mg} = 60.00 \not{g} \times \frac{1 \text{ mol}}{24.31 \not{g}}$$

$$n_{Mg} = 2.47 \text{ mol}$$

Similarly, we can determine the amount in moles of oxygen atoms:

$M_O = 16.00 \text{ g/mol}$

$$n_O = 40.00 \not{g} \times \frac{1 \text{ mol}}{16.00 \not{g}}$$

$$n_O = 2.50 \text{ mol}$$

The mole ratio of Mg:O is 2.47:2.50, or 1:1.

Since the mole ratio of the two elements is 1:1, the empirical formula of magnesium oxide is MgO. Since magnesium oxide is an ionic compound, the empirical formula indicates that the ratio of magnesium ions to oxide ions is 1:1.

From the percentage composition, we can merely tell the *ratio* of atoms, or ions, not the actual *number*. Based only on the empirical evidence from the investigation, the formula for this compound could be MgO, Mg_2O_2, Mg_3O_3, or any other multiple. We would need to know the molar mass of the compound to be able to determine the molecular formula.

Sample Problem 1

What is the empirical formula for a compound whose percentage composition is 21.6% sodium, 33.3% chlorine, and 45.1% oxygen?

Solution

$m_{Na} = 21.6\% \times 100.0\text{ g Na} = 21.6\text{ g}$ $\quad M_{Na} = 22.99\text{ g/mol}$

$m_{Cl} = 33.3\% \times 100.0\text{ g Cl} = 33.3\text{ g}$ $\quad M_{Cl} = 35.45\text{ g/mol}$

$m_O = 45.1\% \times 100.0\text{ g O} = 45.1\text{ g}$ $\quad M_O = 16.00\text{ g/mol}$

$$n_{Na} = 21.6\text{ g} \times \frac{1\text{ mol}}{22.99\text{ g}}$$

$$n_{Na} = 0.940\text{ mol}$$

$$n_{Cl} = 33.3\text{ g} \times \frac{1\text{ mol}}{35.45\text{ g}}$$

$$n_{Cl} = 0.939\text{ mol}$$

$$n_O = 45.1\text{ g} \times \frac{1\text{ mol}}{16.00\text{ g}}$$

$$n_O = 2.82\text{ mol}$$

The mole ratio (Na:Cl:O) is 0.940:0.939:2.82. Dividing by 0.939 to obtain the lowest ratio, we get the mole ratio of Na:Cl:O = 1:1:3.

The empirical formula of the compound is $NaClO_3$.

SUMMARY Calculations to Determine Empirical Formula

1. Find the mass of each element in 100 g of the compound, using percentage composition.
2. Find the amount in moles of each element by converting the mass in 100 g to moles, using the molar mass of the element.
3. Find the whole-number ratio of atoms in 100 g to determine the empirical formula. Reduce the ratio to lowest terms.

Practice

Understanding Concepts

1. What information do you need to determine the empirical formula of a compound?

2. What information, in addition to the empirical formula, is needed in order to determine the molecular formula of a compound?
3. Potassium persulfate is a chemical used in the process of photodeveloping to remove hypo (hyposulfite or sodium thiosulfate, a fixer) from photographic paper. By mass, potassium persulfate contains 28.9% potassium, 23.7% sulfur, and the remainder is oxygen. What is the empirical formula of this compound?
4. The percentage compositions of two antibiotics are given below. Find the empirical formula of each:
   (a) chloromycetin: 40.87% carbon, 3.72% hydrogen, 8.67% nitrogen, 24.77% oxygen, and 21.98% chlorine.
   (b) sulfanilamide: 41.86% carbon, 4.65% hydrogen, 16.28% nitrogen, 18.60% oxygen, and 18.60% sulfur.
5. Phosphorus combines with oxygen to form two oxides. Find the empirical formula for each oxide of phosphorus if the percentage composition for each is
   (a) 43.6% oxygen
   (b) 56.6% oxygen
6. Propane is a hydrocarbon that is used as fuel in barbecues and some cars. In a 26.80-g sample of propane, 4.90 g is hydrogen and the remainder is carbon. What is the empirical formula of propane?

Answers

3. KSO_4
4. (a) $C_{11}H_{12}N_2O_5Cl_2$
   (b) $C_6H_8N_2O_2S$
5. (a) P_2O_3
   (b) P_2O_5
6. C_3H_8

Applying Inquiry Skills

7. An oxide of copper reacts with carbon to produce carbon dioxide and copper metal. High temperatures are required for the reaction to occur, and sufficient carbon must be supplied to allow all the oxide of copper to react. One of the products formed in this reaction is a gas, so only the masses of the starting oxide of copper and the product, copper metal, can be easily determined.
   (a) Design an experiment to determine the empirical formula of the copper oxide. Complete a report outlining your experimental design and the procedure you would follow, create a blank table in which you could record the evidence you would need to obtain, and describe how you would analyze the evidence.
   (b) Predict the evidence you would obtain if the molecular formula of the compound is Cu_2O.

Calculating a Molecular Formula

The empirical formula tells us the *ratio* of atoms in a molecule, but does not tell us the *number* of atoms in a molecule of the compound. The molar mass takes into account all of the atoms in a molecule. Therefore, if we know the molar mass of a compound and its empirical formula, we can determine its molecular formula.

Let's return to our example of water and hydrogen peroxide. The percentage composition of water, determined by experiment, is 11.1% hydrogen and 88.9% oxygen. This gives us a mole ratio of H:O of 2:1 and an empirical formula of H_2O. If this were the correct formula for water, the molar mass of water would be 18.02 g/mol. We find that when the molar mass is determined, it is indeed 18.02 g/mol, which suggests that the empirical formula we obtained is also the molecular formula.

In the case of hydrogen peroxide, the percentage composition is 5.9% hydrogen and 94.1% oxygen, giving us a mole ratio of H:O of 1:1 and an empirical formula of HO. If HO were the molecular formula, the molar mass would be 17.01 g/mol. The molar mass, determined by measurement, is actually 34.02 g/mol, double the empirical molar mass. Therefore, the molecular formula must be H_2O_2.

Sample Problem 2

A compound with a molar mass of 30.00 g/mol has an empirical formula of CH_3. Determine its molecular formula.

Solution

$$M_{CH_3} = (1 \times 12.01) + (3 \times 1.01)$$

$$M_{CH_3} = 15.04 \text{ g/mol}$$

The actual molar mass is twice the molar mass predicted by the empirical formula. Therefore, the molecular formula is C_2H_6.

Determining Molar Mass of New Compounds

For new substances the molecular formula is unknown, so we cannot calculate molar masses from the atomic masses in the periodic table. The molar mass of a new substance has to be determined experimentally. Of the number of laboratory methods available to determine molar mass, chemists most often rely on a mass spectrometer.

In a mass spectrometer, a small gaseous sample is bombarded by a beam of electrons, which causes the molecules to break up into charged fragments. For example, the two main fragments for a particular compound are shown in **Figure 1**. The fragments are accelerated by an electric field and then deflected by a magnetic field. The amount of deflection depends on the mass and the charge of the fragment. From the amount of deflection, the molar mass of the original sample can be determined. The mass spectrum (the evidence from a mass spectrometer test) of the compound shown in **Figure 2** shows several fragments.

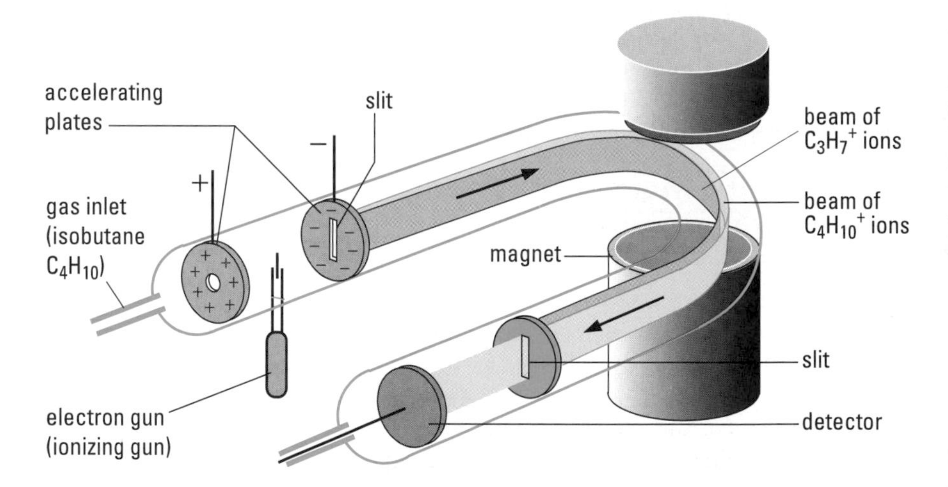

Figure 1
A mass spectrometer is used to determine the masses of ionized entities by measuring the amount of deflection in their path as they pass through a magnetic field.

Let's consider an empirical analysis of ascorbic acid (vitamin C), designed to determine its molecular formula. Experiment reveals that it is a compound of C, H, and O. A combustion analyzer reveals that the combustion of 1.000 g of ascorbic acid produces 1.500 g of CO_2 and 0.405 g of H_2O.

First, we will find the mass of the C atoms (12.01 g/mol) in 1.500 g of the carbon oxide that was formed, CO_2 (44.01 g/mol):

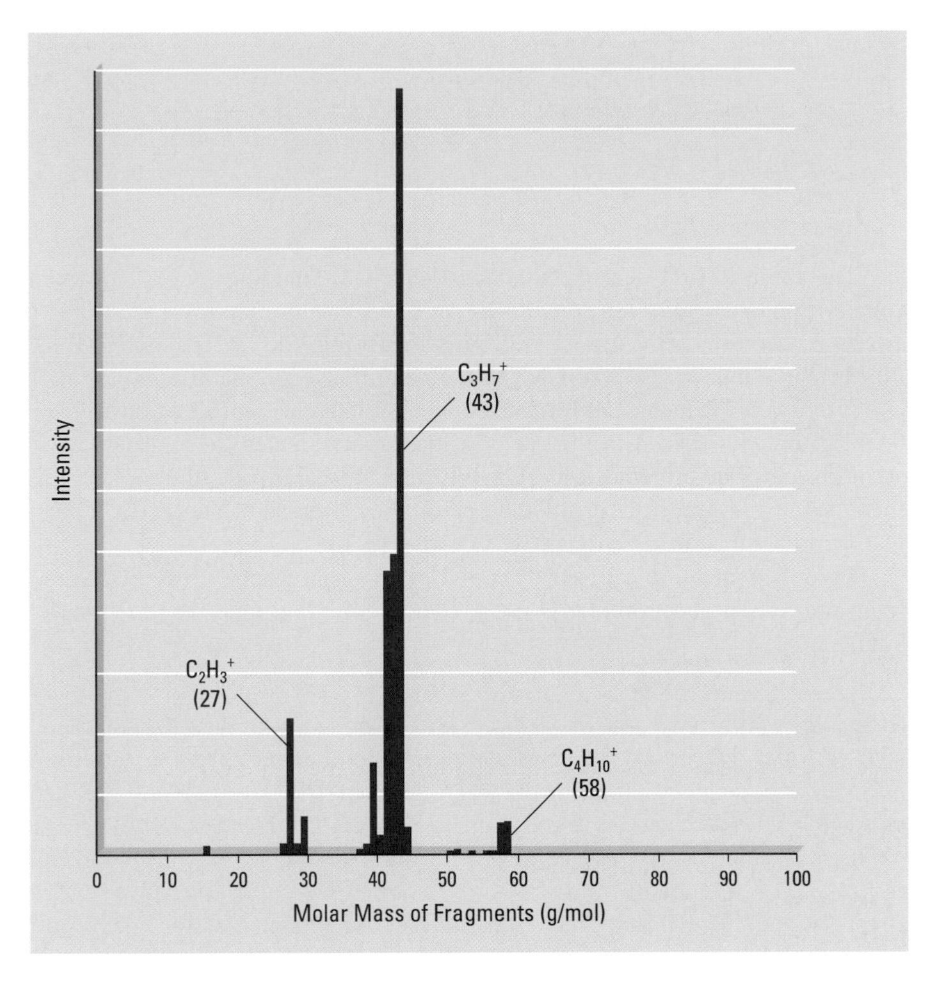

Figure 2
A mass spectrum of lighter fluid (butane) is used to determine the molar mass of the chemical in the fluid. Together with the known empirical formula C_2H_5, the molar mass provides a molecular formula of C_4H_{10} for butane.

$$m_C = 1.500 \text{ g} \times \frac{12.01 \text{ g/mol}}{44.01 \text{ g/mol}}$$

$$m_C = 0.409 \text{ g}$$

Similarly, we will find the mass of the 2 H atoms (2.02 g/mol) in the hydrogen oxide that was formed in the combustion, H_2O (18.02 g/mol):

$$m_H = 0.405 \text{ g} \times \frac{2.02 \text{ g/mol}}{18.02 \text{ g/mol}}$$

$$m_H = 0.045 \text{ g}$$

Some of the oxygen in the products came from ascorbic acid, and some came from the oxygen present in the combustion chamber. However, there is an easy way to find the mass of oxygen in the ascorbic acid. We simply subtract the calculated masses of carbon and hydrogen from the original mass of ascorbic acid, and what remains must be the mass of the oxygen:

$$m_O = 1.00 \text{ g} - (0.409 \text{ g} + 0.045 \text{ g})$$

$$m_O = 0.546 \text{ g}$$

We can now convert the mass of each element into the amount in moles:

$$n_C = 0.409 \text{g} \times \frac{1 \text{ mol}}{12.01 \text{g}}$$

$$n_C = 0.034 \text{ mol}$$

$$n_H = 0.045 \cancel{g} \times \frac{1 \text{ mol}}{1.01 \cancel{g}}$$

$$n_H = 0.045 \text{ mol}$$

$$n_O = 0.546 \cancel{g} \times \frac{1 \text{ mol}}{16.00 \cancel{g}}$$

$$n_O = 0.034 \text{ mol}$$

The ratio of C:H:O is thus 0.034:0.045:0.034. Dividing by the common value, we obtain 1:1.32:1. To convert the ratio to whole numbers, we multiply by 3 (chosen because it's the lowest multiplier that brings the 1.32 close to a whole number) to obtain 3:3.96:3, or 3:4:3, and an empirical formula of $C_3H_4O_3$.

If this were the molecular formula, the molar mass of ascorbic acid would be 88.07 g/mol. However, with the use of a mass spectrometer, the molar mass of ascorbic acid is determined to be 176.14 g/mol, which is twice the molar mass predicted by the empirical formula. Based on the empirical evidence, the molecular formula for ascorbic acid is $C_6H_8O_6$.

The following sample problem shows how evidence from a combustion analyzer and a mass spectrometer is used to determine the molecular formula of butane.

Sample Problem 3

What is the molecular formula of the fluid in a butane lighter?

Solution

Evidence from combustion analysis:

% C = 82.5%

% H = 17.5%

Evidence from mass spectrometry:

$$M_{butane} = 58.0 \text{ g/mol}$$

To make the calculation easier, work with one mole (58.0 g) of the compound:

$$m_C = \frac{82.5}{100} \times 58.0 \text{ g}$$

$$m_C = 47.8 \text{ g}$$

$$n_C = 47.8 \cancel{g} \times \frac{1 \text{ mol}}{12.01 \cancel{g}}$$

$$n_C = 3.98 \text{ mol}$$

$$m_H = \frac{17.5}{100} \times 58.0 \text{ g}$$

$$m_H = 10.2 \text{ g}$$

$$n_H = 10.2 \cancel{g} \times \frac{1 \text{ mol}}{1.01 \cancel{g}}$$

$$n_H = 10.0 \text{ mol}$$

or, in one step

$$n_C = \frac{82.5}{100} \times 58.0 \cancel{g} \times \frac{1 \text{ mol}}{12.01 \cancel{g}} = 3.98 \text{ mol}$$

$$n_H = \frac{17.5}{100} \times 58.0 \cancel{g} \times \frac{1 \text{ mol}}{1.01 \cancel{g}} = 10.0 \text{ mol}$$

The mole ratio of carbon atoms to hydrogen atoms in the compound analyzed is

4:10. According to the evidence, the molecular formula of the fluid in a butane lighter is C_4H_{10}.

SUMMARY Calculations to Determine a Molecular Formula

From the empirical formula and measured molar mass:

1. Calculate the molar mass of the empirical formula.
2. Compare the measured molar mass of the substance with the molar mass derived from the empirical formula and increase subscripts in the empirical formula by the multiple needed to make the two molar masses equal.

From percentage composition and measured molar mass:

1. Find the mass of each element in one mole of the compound by multiplying the percentage by the molar mass of the compound.
2. Use the molar mass of the element to convert the mass of the element to amount in moles.
3. The mole ratio of the elements in the compound provides the subscripts in the molecular formula.

Practice

Understanding Concepts

8. A key ingredient of nail polish remover is found to have the following percentage composition by mass: 62.0% carbon, 10.4% hydrogen, and 27.5% oxygen. If its molar mass is 58.1 g/mol, what is its molecular formula?
9. Analysis of an air pollutant reveals that the compound is 30.4% nitrogen and 69.6% oxygen, by mass. The mass spectrograph for the pollutant shows that its molar mass is 92.0 g/mol. Determine the molecular formula and the chemical name of the polluting compound.

Applying Inquiry Skills

10. Carbohydrates, an important source of food energy, are compounds that contain carbon, hydrogen, and oxygen. A food chemist extracts a carbohydrate from honey and submits a sample to a spectroscopy lab for analysis. Complete the **Analysis** section of the lab report.

Experimental Design

A sample of the carbohydrate is burned in a combustion analyzer to determine the percent by mass of each element in the compound. Another sample is analyzed by a mass spectrometer to determine the molar mass of the carbohydrate.

Evidence

percent by mass of carbon = 40.0%

percent by mass of hydrogen = 6.8%

percent by mass of oxygen = 53.2%

molar mass = 180.2 g/mol

Analysis

(a) Calculate the molecular formula of the carbohydrate.

Answers

8. C_3H_6O
9. N_2O_4
10. $C_6H_{12}O_6$

Figure 3
Rosehips are the fruit formed from rose flowers. Rosehip shells are a source of vitamin C, often made into a drink. The oil extracted from rose seeds is used in skin care products that claim to reduce wrinkles and scars.

Natural Vitamins

Earlier we analyzed ascorbic acid (vitamin C) empirically, using evidence from a combustion analyzer and a mass spectrometer. The percentage composition and molecular mass confirmed the molecular formula $C_6H_8O_6$. Does ascorbic acid from natural sources, such as from rosehips (**Figure 3**) and some fruits, have the same molecular formula and percentage composition as chemically synthesized ascorbic acid, that is, from a laboratory where atoms and molecules have been poured and mixed? The law of definite proportions tells us that a compound always has the same composition, no matter how it is made.

Store shelves are full of vitamin and mineral supplements; usually both natural and chemically synthesized products are available. But the cost of supplements from natural sources is generally several times more than the synthetic substitutes. Judging by the abundance of brands and stores specializing in natural health products, people are happy to pay the extra cost for the natural product. Are natural vitamins worth the money?

Depending on who is asked, the answer to that question varies. The manufacturer of the natural products might tell you that its products are indeed worth more because they cost more to make (natural materials must be harvested and processed, instead of being synthesized by mass production). They also offer greater health benefits. Natural products contain more than just the vitamin, for example, that a pharmaceutical company sells. It is claimed that these ingredients work together naturally to enhance the positive effects of the product, including by speeding or increasing absorption and by providing therapeutic effects not offered by the pure chemical. Another benefit claimed for natural products is that many of them have been used for centuries in traditional medicine (ginseng, echinacea, and St. Johns wort are all examples), and that over that time they have proved their worth.

A manufacturer of chemical compounds would offer a different answer: vitamin C is vitamin C, $C_6H_8O_6$, whether it is made by a rose or by a factory. It might claim that the extra ingredients in the natural products are a cause of concern, in that their effects are sometimes not well known and could be harmful. For example, many plants produce toxins to discourage animals from eating their seeds and leaves. These toxins could find their way into a natural product based on plant material. Another criticism of the natural products is that, unlike the pharmaceutical product in which each pill is identical, natural products sometimes vary in their composition, so the dosage is less well known.

Practice

Understanding Concepts

11. What difference may exist between a product from a natural source and a chemically synthesized product?

Making Connections

12. Using the Internet, research vitamin D deficiency. Find out the symptoms, historical treatment of the symptoms, use of cod-liver oil, the availability of vitamin D tablets, as well as the addition of vitamin D in milk. Use this information to help you form an opinion on whether synthetic vitamins should be added to food products.

    Follow the links for Nelson Chemistry 11, 4.7.

GO TO **www.science.nelson.com**

Explore an Issue

Debate: Are Natural Vitamins Better for Your Health?

- Visit a health food store and a drugstore and note the sources and costs of different brands of vitamin and mineral supplements. If possible, interview the owner of the store, the pharmacist, and several consumers to find out their opinions.
- In small groups, discuss the issue from several perspectives. Keep notes and organize your ideas into supportive arguments for each side.
- Your teacher will divide the class into two teams for the debate on the resolution "Natural vitamins are better for your health."
- Return to your first group; discuss the issue again and arrive at a position that is agreed upon by every member of the group.
  (a) Prepare a one-page summary of your group's position on the issue.
  Follow the links for Nelson Chemistry 11, 4.7.

GO TO www.science.nelson.com

DECISION-MAKING SKILLS

- Define the Issue
- Identify Alternatives
- Research
- Analyze the Issue
- Defend a Decision
- Evaluate

Investigation 4.7.1

Determining the Formula of an Unknown Hydrate

INQUIRY SKILLS

- Questioning
- Hypothesizing
- Predicting
- Planning
- Conducting
- Recording
- Analyzing
- Evaluating
- Communicating

When crystals form from the evaporation of an aqueous solution of a salt, water molecules often become incorporated into the crystal structure. When these crystals are heated, these loosely held water molecules are lost. This type of crystal is called a *hydrate*, and the water is called *water of hydration*.

Gypsum (**Figure 4**), the common name for the chemical used in plaster casts and sculptures, is a hydrate of calcium sulfate. Its formula is $CaSO_4{\cdot}2H_2O$, a dihydrate (or 2-water), which means that it has two molecules of water for each $CaSO_4$ formula unit. When the dihydrate is heated, it loses some of its water of hydration and forms crystals of $(CaSO_4)_2{\cdot}H_2O$. When these crystals are ground and moistened, they absorb water and are converted back to the dihydrate, solidifying within minutes.

In this investigation, you will determine the formula of an unknown hydrate by determining the amount in moles of water of hydration that is lost from a known amount of a hydrate. Conduct the **Procedure**, and complete the **Analysis** and **Evaluation** sections of a lab report.

Figure 4
Alabaster, the sculptor's name for gypsum, was the stone used for this sculpture.

Question

What is the chemical formula of copper(II) sulfate hydrate?

Experimental Design

A known mass of a hydrate of copper(II) sulfate is heated until all water of hydration has been removed. The mass of water of hydration is calculated and evidence is analyzed to determine the amount in moles of hydrate and water of hydration.

Materials

apron
eye protection

$CuSO_4$ is poisonous if swallowed. Aprons and eye protection must be worn.

Care is required in handling hot materials and apparatus.

4 g of $CuSO_4{\cdot}xH_2O_{(s)}$
crucible and lid
crucible tongs
clay triangle
iron ring
ring stand
laboratory burner
wire gauze
centigram balance

Figure 5
Apparatus for heating a substance in a porcelain crucible

Procedure

1. Place a clean, dry crucible and lid on a clay triangle on a ring stand, with the lid slightly ajar (**Figure 5**). Using the laboratory burner, heat strongly for 3 min. Using the tongs, place the crucible and lid on the wire gauze to cool. Determine the mass of the crucible and lid using the balance.
2. Place the hydrate in the crucible and determine its mass to the nearest centigram.
3. Heat the crucible and contents gently, with the lid slightly ajar to allow water to escape. Heat for about 10–15 min, until contents turn to a white powder. Allow to cool on wire gauze.
4. Determine the mass of crucible, lid, and contents.
5. Repeat steps 3 and 4 until the mass remains constant within 0.03 g.
6. Follow your teacher's instructions for disposal of the copper salt. Wash your hands.

Analysis

(a) Use the evidence obtained to calculate the mass of hydrate that you used.
(b) Calculate the mass of water of hydration lost and the mass of dehydrated salt remaining ($CuSO_{4(s)}$).
(c) Determine the amount in moles of the water of hydration and the amount in moles of the dehydrated salt.
(d) Determine the formula of the hydrate.

Evaluation

(e) What was the purpose of heating the empty crucible and lid in step 1?
(f) Why were the crucible and contents heated until the mass remained constant?

Practice

Understanding Concepts

13. Explain briefly what information can be obtained from a mass spectrometer.
14. The empirical formula for ethyne (acetylene), a hydrocarbon used in welders' torches (**Figure 6**), is CH. If its molar mass is 26 g/mol, what is its molecular formula?
15. An important raw material for the petrochemical industry is ethane, which is extracted from natural gas. Determine the molecular formula for ethane using the following evidence:

    percent by mass of carbon = 79.8%
    percent by mass of hydrogen = 20.2%
    molar mass = 30.1 g/mol

Answers

14. C_2H_2
15. C_2H_6

Figure 6
Ethyne (acetylene) releases a great deal of heat when it combusts and is used as a fuel in welding.

Applying Inquiry Skills

16. Many ionic solids composed of polyatomic ions are thermally unstable, that is, they sometimes decompose when heated to give simpler compounds. When one such compound is heated, it decomposes into two products, CaO and CO_2. Design an experiment to determine the ratio of the moles of CaO to CO_2. List the materials and equipment used, the procedure, and an explanation of the calculations needed.

Reflecting

17. Suppose that you are given a new compound for which the molecular formula is unknown. What analytical methods have you learned that you can use toward determining the molecular formula of this compound?

Sections 4.6–4.7 Questions

Understanding Concepts

1. Explain the difference between the empirical formula of a compound and its molecular formula.
2. What information is needed to calculate a molecular formula from an empirical formula?
3. Explain how percentage composition can be used to determine the empirical formula of a compound.
4. The molecular formulas of some substances are given below. Write their empirical formulas.

   naphthalene, $C_{10}H_{8(l)}$
   ethanol, $C_2H_6O_{(l)}$
   hydrazine, $N_2H_{4(s)}$
   ozone, $O_{3(g)}$
   carotene, $C_{40}H_{56(s)}$
   sulfanilamide, $C_6H_8N_2O_2S_{(s)}$
5. Does the concept of empirical formulas and molecular formulas apply to ionic compounds? Explain.
6. Lactic acid is the acid responsible for the sour taste in sour milk. On the basis of combustion analysis, its percentage composition by mass is 40.00% carbon, 6.71% hydrogen, and *x*% oxygen.
   (a) Calculate the empirical formula of lactic acid.
   (b) The molar mass of lactic acid is measured at 90 g/mol in a mass spectrometer. Determine the molecular formula of the compound.

Applying Inquiry Skills

7. As an analogy to determining empirical formulas, we can determine the simplest ratio of two different coins.
   (a) Suppose your school collected a large number of pennies and quarters. Design a procedure that you could use to determine the simplest ratio of the number of pennies to the number of quarters that have been collected. Write the steps of the procedure, including the equipment needed and the calculations involved.
   (b) Draw a flowchart comparing your procedure to the procedure a scientist might follow in determining the empirical formula of a new compound.

Chapter 4 Summary

Key Expectations

Throughout this chapter, you have had the opportunity to do the following:

- Explain how different stoichiometric combinations of elements in compounds can produce substances with different properties. (4.1)
- Explain the law of definite proportions. (4.1)
- Explain the relationship between isotopic abundance and relative atomic mass. (4.2)
- Demonstrate an understanding of Avogadro's constant, the mole concept, and the relationship between the mole and molar mass. (4.3)
- Use appropriate scientific vocabulary to communicate ideas related to chemical calculations. (4.3, 4.4, 4.5, 4.6, 4.7)
- Solve problems involving quantity in moles, number of particles, and mass. (4.4, 4.5, 4.6, 4.7)
- Determine percentage composition of a compound through experimentation, as well as through analysis of the formula and a table of relative atomic masses. (4.5)
- Give examples of the application of chemical quantities and calculations. (4.5, 4.6, 4.7)
- Distinguish between the empirical formula and the molecular formula of a compound. (4.6, 4.7)
- Determine empirical formulas and molecular formulas, given molar masses and percentage composition or mass data. (4.6, 4.7)

Key Terms

atomic mass unit
Avogadro's constant
diatomic
empirical formula
isotopic abundance
law of definite proportions
molar mass
mole
molecular formula
relative atomic mass
stoichiometry

Make a Summary

In this chapter, you learned the mole concept and many terms related to it. To summarize your learning, try this activity using index cards.

- Prepare an index card for each of these terms: atomic mass unit, Avogadro's constant, relative atomic mass, molar mass, amount in moles, mass in grams, number of entities, percentage composition, empirical formula, and molecular formula. On the front of each card, write the term and its common unit. On the reverse, write a definition of the term.
- Place the index card for "amount in moles" in the centre of your desk and arrange the other cards to show how the terms are related to the mole and to each other. Use the units on each card to check how each term can be calculated from the other terms.
- Shuffle the cards, randomly select another term to place in the centre, and repeat the process.

Reflect on your Learning

Revisit your answers to the Reflect on Your Learning questions at the beginning of this chapter.

- How has your thinking changed?
- What new questions do you have?

Chapter 4 Review

Understanding Concepts

1. In your own words, explain the law of definite proportions, and relate how this law was applied in the determination of relative atomic mass of the elements.
2. If an atom of an element has a mass double that of a C-12 atom, what would be its relative atomic mass?
3. If the mass of a C-12 atom were reassigned a value of 18 u, what would be the relative mass of a hydrogen atom?
4. In what way was the knowledge of molecular formulas important in the assignment of relative atomic mass?
5. Explain the term "isotopic abundance," and explain why the relative atomic masses of the elements are not whole numbers.
6. The isotopic abundance of a naturally occurring sample of silicon is given in **Table 1**. Calculate the relative atomic mass of silicon.

Table 1

Isotope	Percentage abundance
Si-28	92.21%
Si-29	4.70%
Si-30	3.09%

7. (a) What is the mass, in atomic mass units, of a single C-12 atom?
   (b) What is the mass, in grams, of a mole of C-12 atoms?
   (c) What is Avogadro's constant and how is it related to the term mole?
   (d) Identify the term and the SI symbol used to represent the mass of a single mole of a substance.
8. What is the molar mass of each of the following chemicals?
   (a) calcium carbonate (limestone)
   (b) dinitrogen tetroxide (pollutant)
   (c) sodium carbonate decahydrate (washing soda)
9. Convert the following masses into amounts (in moles):
   (a) 1.000 kg of table salt
   (b) 1.000 kg of dry ice
   (c) 1.000 kg of water
10. Convert the following amounts into masses:
    (a) 1.50 mol of liquid oxygen
    (b) 1.50 mmol of liquid mercury
    (c) 1.50 kmol of liquid bromine
11. How many molecules are there in each of the following masses or amounts?
    (a) 0.42 mol of hydrogen acetate (acetic acid, vinegar)
    (b) 7.6×10^{-4} mol of carbon monoxide (poisonous gas)
    (c) 100 g of carbon tetrachloride (poisonous fluid)
    (d) 100 g of dihydrogen sulfide (rotten egg gas)
12. Aspartame is an artificial sweetener marketed under the brand name NutraSweet. It has the molecular formula $C_{14}H_{18}N_2O_{5(s)}$.
    (a) What is the molar mass of aspartame?
    (b) What is the amount of aspartame in moles contained in one package of NutraSweet (35 mg of aspartame)?
    (c) How many hydrogen atoms are in 35 mg of aspartame?
13. Calculate the percentage by mass of the indicated element in each of the following compounds:
    (a) sodium in sodium azide, $NaN_{3(s)}$, a compound that rapidly decomposes, forming nitrogen gas. This reaction is used in automobile inflatable airbags;
    (b) aluminum in aluminum oxide, $Al_2O_{3(s)}$, the naturally occurring mineral corundum;
    (c) nitrogen in dopamine, $C_8H_{11}O_2N_{(aq)}$, a neurotransmitter in the brain.
14. (a) What is the difference between an empirical formula and a molecular formula?
    (b) Explain why an empirical formula can be determined from percentage composition of a compound, but a molecular formula requires further evidence.
15. Caffeine, a stimulant found in coffee and some soft drinks, consists of 49.5% C, 5.15% H, and 28.9% N by mass; the rest is oxygen.
    (a) Determine the empirical formula of caffeine.
    (b) Given the molar mass of caffeine is 195 g/mol, determine its molecular formula.
16. Many compounds in a family called esters have pleasant odours and are used in the perfume industry. When one of these esters was analyzed, the information in **Table 2** was obtained.

Table 2

mass of sample of ester analyzed	4.479 g
mass of carbon in sample	3.161 g
mass of hydrogen in sample	0.266 g
mass of oxygen in sample	1.052 g

    (a) Calculate the empirical formula of the ester.

(b) Mass spectrometer analysis indicates the ester has a molar mass of 136 g/mol. Determine the molecular formula of this ester.

17. Vanillin, the compound responsible for the flavour vanilla, has a percentage composition of 63.2% C, 5.26% H, and 31.6% O by mass. Determine the empirical formula of vanillin.

18. What is the percentage composition of sodium phosphate, commonly known as TSP, used for cleaning hands of grease and oil?

19. A forensic scientist analyzes a sample for sodium arsenate, $Na_3AsO_{4(s)}$, a source of arsenic. Calculate the percentage composition of sodium arsenate.

Applying Inquiry Skills

20. A chloride of zinc is synthesized in the reaction of zinc metal and hydrochloric acid. Complete the **Analysis** and **Evaluation** sections of a lab report, using appropriate scientific vocabulary and calculations.

Question

What is the empirical formula of zinc chloride?

Experimental Design

Concentrated hydrochloric acid is added to zinc metal in a crucible and lid. When all of the zinc has reacted, the crucible and contents are heated in a fume hood to evaporate any excess acid.

Evidence

mass of crucible and lid = 35.603 g
mass of crucible, lid, and zinc = 36.244 g
mass of crucible, lid, and chloride of zinc = 36.933 g

Analysis

(a) Based on the Evidence, answer the Question.

Evaluation

(b) Evaluate the experimental design. Is the Evidence likely to be valid?

21. When an iron nail is immersed in a copper(II) sulfate solution, copper atoms coat the surface of the nail (see **Figure 1**).
    (a) Plan an experiment to determine the number of copper atoms that are coating the nail after 10 min of immersion.
    (b) Explain what type of experimental evidence you need to obtain and what other information is needed.
    (c) Can this experimental design be used or modified to determine the percentage of copper in copper(II) sulfate? Explain.

Figure 1

22. In experiments in which a crucible is used to strongly heat a substance, the empty crucible and lid are first heated in a hot flame.
    (a) What is the reason for heating the empty crucible and lid?
    (b) List all safety precautions needed in handling hot equipment and materials.

23. In an experiment to determine the percentage composition of an oxide, the element is allowed to burn in air and the oxide formed is collected. Describe how we can determine the mass of the oxygen in the air that was reacted.

Making Connections

24. Propane is a hydrocarbon that burns to produce carbon dioxide or carbon monoxide, depending on the availability of oxygen. Using a propane-fuelled gas barbecue as an example, illustrate in a paragraph the importance of understanding that different stoichiometric combinations of elements can produce compounds with different properties.

25. Suggest reasons why the mole is the unit used to represent quantities in many medical products, such as intravenous fluids, while mass units are generally used when representing common food products.

26. Look in the business section of telephone directories and identify five or more businesses where
    (a) the skills of quantitative laboratory techniques are used (e.g., in photodeveloping)
    (b) analysis of unknown substances is performed (e.g., in medical laboratories).

Exploring

27. Many new substances are discovered, sometimes accidentally, and are found to have useful applications in society or in industry. Examples are Teflon, used to coat pots and pans, Kevlar, used in bullet-proof vests (**Figure 2**), and Aspirin, which is derived from a compound, salcylic acid, present in willow bark. Research one of these or another discovery, and present a report on how the molecular formula of the compound was determined.

    Follow the links for Nelson Chemistry 11, Chapter 4 Review.

    GO TO www.science.nelson.com

Figure 2
Kevlar is a silky, soft, and light synthetic fibre that is stronger than steel—an extraordinary combination of properties. Chemical engineers have used their imagination and expertise to apply these properties in designing products where lightweight strength is essential, for example, in bullet-proof vests, protective gloves, tires, racing shoes, fibre optics, and aircraft.

28. Using the Internet, research and prepare a one-page report on how the present value for Avogadro's constant was determined.

    Follow the links for Nelson Chemistry 11, Chapter 4 Review.

    GO TO www.science.nelson.com

29. The process of cheesemaking began thousands of years ago but is today an industrial process using technology that has been investigated by chemists in many countries (**Figure 3**). Enzymes act as chemical catalysts that change the milk fats and proteins into cheese; moisture in the final product is controlled by various methods of salting, moulding, and pressing. Using the Internet, research the "ripening" process in cheesemaking, and report in one page your findings about how the percentage of moisture and milk fats in different cheeses is related to their characteristic shape, texture, and flavour.

    Follow the links for Nelson Chemistry 11, Chapter 4 Review.

    GO TO www.science.nelson.com

Figure 3
There are more than 400 different types of cheeses and several different ways to classify them.

Chapter

5

Quantities in Chemical Equations

In this chapter, you will be able to

- balance chemical and nuclear equations;
- solve problems involving quantity in moles, mass, or molecules of any reactant or product in a chemical equation;
- determine the limiting reagent in a chemical reaction and solve problems involving percentage yield;
- analyze experimental results to compare actual yield and percentage yield and evaluate sources of experimental error;
- give examples of the application of chemical quantities and calculations;
- identify careers related to the study of chemistry.

One of the most useful applications of chemistry is the synthesis of new substances that benefit society. New medicines are developed every year, and new materials for clothing, sports equipment, and cleaning products are advertised daily.

That is why the discovery of new and better ways of synthesizing existing chemicals is so exciting. New procedures often follow the development of new technologies and always require an understanding of both qualitative and quantitative aspects of chemical reactions. Industries are particularly interested in the costs of their chemical processes, and quantitative analysis is key to a company's profitability.

Let's take the example of the production of soda ash (sodium carbonate, $Na_2CO_{3(s)}$). You may have used sodium carbonate as a cleaner, commonly called washing soda. It was first extracted from plant ashes and used to make soap. In 1775, as the demand for soap exceeded the supply of sodium carbonate, the French Academy of Science offered a prize for the development of an industrial method of making sodium carbonate from common substances. As a result of this offer, the LeBlanc process was developed. To make sodium carbonate using the LeBlanc process, sodium chloride and sulfuric acid are heated; then limestone and coal are added and heated again, to a high temperature. Quantitatively, the LeBlanc process was inefficient. It required burning a lot of coal, which was expensive and generated hydrogen chloride, a severe air pollutant. One of the byproducts was an insoluble residue that had no commercial value. In other words, a great deal of resources and energy was put into the reaction, and only a small proportion of the materials was recovered as a useful product.

It was not until 1867—after approximately a hundred years of using the LeBlanc process—that a new process was developed for the production of sodium carbonate. You will learn more about this Solvay process later in the chapter (**Figure 1**); you will also learn how to calculate amounts of substances taking part in chemical reactions and how science and technology are linked in the search for better products and processes.

Reflect on your Learning

1. (a) Based on what you learned in previous chapters, summarize all of the qualitative and quantitative information that is communicated in a balanced chemical equation.
   (b) Of all the different types of information in your summary, can you suggest one that is less important than the others? Justify your answer.
2. In what way does an understanding of the quantities of reactants and products in chemical reactions increase profitability in an industrial process?
3. What different strategies and equipment have you used in quantitative applications, at home or at school (e.g., measuring spoon, centigram balance, thermometer)? Using a scale of 1 to 10, rank each strategy or piece of equipment for accuracy of measurement.

Throughout this chapter, note any changes in your ideas as you learn new concepts and develop your skills.

How Much Gas?

When sodium hydrogen carbonate, $NaHCO_{3(s)}$, commonly called baking soda, reacts with an acid such as acetic acid (in vinegar), carbon dioxide is produced. In this activity, you will compare the amounts of gas produced when a fixed amount of sodium hydrogen carbonate is mixed with increasing amounts of acetic acid.

Materials: eye protection, sodium hydrogen carbonate (100 mL, approximately 5 tbsp.), 300 mL vinegar (5% acetic acid), 3 small, sealable plastic bags, graduated cylinder, measuring spoon

Acetic acid in vinegar may irritate the eyes. If acetic acid comes in contact with the eye, immediately wash the eye under cold running water.

- Using the measuring spoon, carefully place 15 mL (approximately 3 tsp.) of sodium hydrogen carbonate into a bottom corner of one bag. Fold the corner upward and tape it in place so that the solid is kept out of contact with other ingredients to be added to the bag.
- Carefully pour 30 mL of vinegar into the other bottom corner of the bag, keeping it out of contact with the solid.
- Flatten and press the bag gently to remove as much air as possible, and seal the bag.
- Remove the tape and shake the bag to allow the sodium hydrogen carbonate to mix well with the vinegar.
- Allow the reaction to continue until no more bubbles are produced. Keep the bag sealed for later observations.
- Repeat the procedure with another bag, using the same amount of sodium hydrogen carbonate and increasing the amount of vinegar to 60 mL.
- Repeat, increasing the amount of vinegar to 120 mL.

  (a) Record the *relative* volumes of gas in the three sealed bags. Is there evidence that the amount of product formed is related to the amount of reactants used?
  (b) Is there evidence that some of the sodium hydrogen carbonate had not completely reacted in one or more of the bags?

- Open each bag and add 15 mL of vinegar to each.

  (c) Is there evidence that a shortage of vinegar was limiting the production of carbon dioxide gas in some bags?

- Add 15 mL of sodium hydrogen carbonate to each bag.

  (d) Is there evidence that a lack of sodium hydrogen carbonate was limiting the production of carbon dioxide gas in some bags?
  (e) When sodium hydrogen carbonate, a bitter-tasting base, is used to make lemon cake, it reacts with the citric acid in the lemon juice to produce bubbles of carbon dioxide. Should the amount of sodium hydrogen carbonate be chosen to limit the production of bubbles, or should it be used in excess amounts? Explain your answer.
  (f) Some recipes give measurements in capacity units (e.g., millilitres or teaspoons of baking soda) while others use mass units (e.g., grams of flour). Explain which of these measurements you think is more accurate.

- Clean up according to instructions from your teacher. Wash your hands.

Figure 1
Soap making was a yearly event undertaken by early Canadian settlers. Waste cooking grease and wood ashes were collected year-round and boiled together in quantities determined by trial and error, with uncertain results. Advances in the industrial production of soaps led to great improvements in personal hygiene and public health.

5.1 Quantitative Analysis

Figure 1
A breathalyzer provides on-the-spot quantitative analysis of the alcohol content in a sample of exhaled air.

When a driver is asked to breathe into a breathalyzer, the instrument analyzes the breath sample and measures the quantity of alcohol in the exhaled air (**Figure 1**). At a swimming pool, a lifeguard may take a sample of water to analyze the quantity of acid or amount of chemicals needed to disinfect the water. Soil and well water samples may be sent to a laboratory to be analyzed for the quantity of toxic ions or bacterial content. In all of these cases, the quantity of a substance in a sample is measured. This type of analysis is called **quantitative analysis**; it requires a combination of scientific concepts, precision equipment, and technical expertise (**Figure 2**).

Many different strategies are used in quantitative analysis. For instance, if the substance we wish to analyze is in solution, we can allow the substance to completely react to form a precipitate. We can then collect and measure the quantity of the precipitate and use our knowledge of the chemical reaction to calculate the quantity of the substance that was in the original solution.

quantitative analysis: measuring the quantity of a substance

limiting reagent: the reactant that is completely consumed in a chemical reaction

excess reagent: the reactant that is present in more than the required amount for complete reaction

In this strategy, we need to ensure that the substance we wish to quantify is completely reacted and that none of it remains in solution, unaccounted for. In order to do this, we need to supply enough of any other reactant that is taking part in the reaction. In fact, we usually supply more than enough, or an excess, of the other reactants. In this way, the reaction will continue until all of the substance we wish to quantify has been used up. The quantity of this substance is the limiting factor in the reaction, and this substance is called the **limiting reagent**. The other reactants that are supplied in excess are called **excess reagents.**

Recall the activity at the beginning of this chapter, reacting sodium hydrogen carbonate with different amounts of acetic acid, producing carbon dioxide gas. After the reaction stopped in the first bag, we could still see solid sodium hydrogen carbonate remaining in the bag, unreacted. The reaction must have stopped because it ran out of the other reactant, the acetic acid. When more acetic acid was added to the bag, the bubbling began again as carbon dioxide was again produced. This shows that acetic acid had indeed been completely used up and was the limiting reagent in this case.

Figure 2
Quantitative analysis often requires careful and precise work. It is also important to have attitudes that include perseverance and openness to unexpected results.

In the third bag, when the reaction stopped, all of the sodium hydrogen carbonate had disappeared. When more sodium hydrogen carbonate was added to the bag, the reaction resumed. This confirmed that, in this bag, the absent reactant causing the reaction to stop was sodium hydrogen carbonate, which was thus the limiting reagent. Acetic acid had been present in excess, as evidenced by the resumption of the reaction without any more acetic acid being added to the bag.

We often use household substances in excess or as limiting reagents, as needed. When we wash greasy dishes, for example, we usually add an excess of detergent to the dishwater to react with and remove the grease. If droplets of grease remain visible on the dishes, the detergent has become the limiting reagent and it is time to add more detergent if we wish to remove all the grease. In another example, calcium and magnesium ions present in hard water react with soap molecules to form an insoluble scum. To soften the water for laundry, the calcium and magnesium ions can be precipitated out by adding sodium carbonate (washing soda). The sodium carbonate is used in excess to ensure that all of the ions causing hard water are removed from the solution before soap is added.

We sometimes select the limiting reagent in a reaction in order to control the amount of a product formed. In winemaking, for example, the alcohol content of a wine can be controlled by making the sugar the limiting reagent. The sugar is a reactant in the fermentation process, and when the sugar is used up, the reaction stops. In baking, the amount of baking soda used is often the limiting reagent in the production of carbon dioxide bubbles, thus controlling the texture of a chewy cookie or a spongy cake. Quantities of limiting reagents must be carefully calculated and measured because they determine the quantity of all the products formed.

Practice

Understanding Concepts

1. If there are only two reactants in a quantitative chemical reaction, is there always a limiting reagent?
2. If you are trying to determine the amount of substance A by reacting it with a solution of substance B, which substance must be the limiting reagent and which must be in excess? Explain your reasoning.

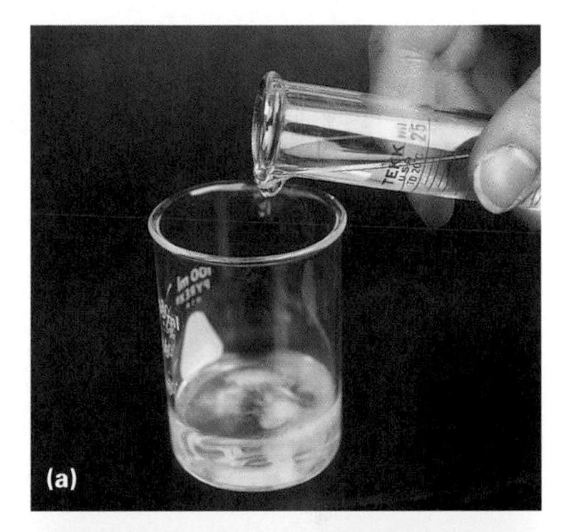

Figure 3
(a) Once the precipitate settles and the top layer appears clear, you can test for the completeness of the reaction.
(b) Carefully run a drop or two of the excess reagent down the side of the beaker and watch for additional precipitation, indicating that some of the limiting reagent remains in the solution.When adding several drops of the excess reagent solution no longer creates cloudiness, the reagent is indeed in excess.

Excess Reagents in Quantitative Analysis

We can usually decide on the amount of limiting reagent to use in a chemical analysis, but how do we predict the quantity of excess reagent to use with it? A procedure involving trial and error may be required. For a precipitation reaction, increasing amounts of the excess reagent may be added to a known trial quantity of the limiting reagent, until no more precipitate forms. The quantity of excess reagent added gives an estimate of the quantity required for the experiment. This procedure is detailed below.

1. Precisely measure a convenient volume (e.g., 80 mL) of the limiting reagent solution.
2. Slowly add a small volume (e.g., 20 mL) of the excess reagent solution.
3. Allow the precipitate to settle enough so that you can see a clear solution at the top of the mixture.
4. Add a few more drops of the excess reagent (**Figure 3**).
5. If more cloudiness appears, which indicates that the reaction has not yet been completed, add another small volume (e.g., 20 mL) and repeat steps 3 and 4.
6. Continue until no more cloudiness appears on adding a few drops of the excess reagent. Record the volumes of both reactants and use this information to calculate the volumes needed in the experiment.

Alternatively, the mixture may be filtered and a diagnostic test performed on the filtrate to determine if any unreacted chemical remains. For example, a flame test might be used to test for any unreacted limiting reagent that remains.

Another strategy used in quantitative analysis is to compare, using graphs, empirical observations against known values. For example, in a medical laboratory, laboratory technologists routinely analyze blood and urine samples for specific chemicals, such as cholesterol and sugar. To help with their interpretation, they compare their results against graphs that have been prepared using known values. This type of graph is called a **standard curve**; it lets us match values from unknown samples to values from known samples previously determined under the same conditions.

standard curve: a graph used for reference, plotted with empirical data obtained using known standards

To illustrate the use of a standard curve, **Table 1** shows reference data obtained in the double displacement reaction of aqueous lead(II) nitrate and potassium iodide. The relationship between the mass of the reactant, $Pb(NO_3)_{2(aq)}$, and the mass of the product, $PbI_{2(s)}$, is graphed as a standard curve (**Figure 4**). An unknown solution of $Pb(NO_3)_{2(aq)}$ can then be analyzed by using the graph to match the mass of $PbI_{2(s)}$ that is precipitated under the same conditions.

Table 1: Reference Data Reaction of Lead(II) Nitrate and Potassium Iodide

Mass of $PbI_{2(s)}$ produced (g)	Mass of $Pb(NO_3)_{2(aq)}$ reacting (g)
1.39	1.00
2.78	2.00
4.18	3.00
5.57	4.00
6.96	5.00

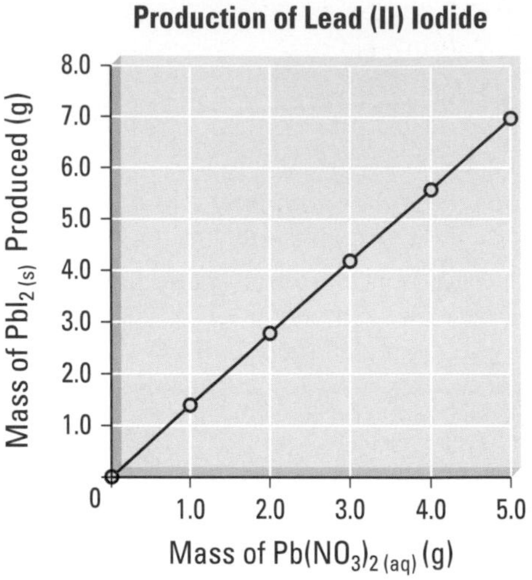

Production of Lead (II) Iodide

Figure 4
Standard curve showing data from **Table 1**

INQUIRY SKILLS

- ○ Questioning
- ○ Hypothesizing
- ○ Predicting
- ○ Planning
- ● Conducting
- ● Recording
- ● Analyzing
- ● Evaluating
- ● Communicating

Investigation 5.1.1

Quantitative Analysis of Sodium Carbonate Solution

In this investigation, the skills used by chemical technicians in the quantitative analysis of a substance by precipitation are practised.

You will determine the mass of sodium carbonate in a solution by using experimental data and a standard curve, the data for which are supplied. **Conduct** the investigation, **Analyze** the evidence, and **Evaluate** the experimental design.

Question

What is the mass of sodium carbonate present in a 50.0-mL sample of a solution?

Experimental Design

The mass of sodium carbonate present in the sample solution is determined by having it react with an excess quantity of calcium chloride in solution. A double displacement reaction takes place; one of the products formed, calcium carbonate, is insoluble in water. The only other product formed is sodium chloride, which is soluble in water. The mass of calcium carbonate precipitate produced is used to determine the mass of sodium carbonate that reacted. The value is read from a standard curve plotted using reference data in **Table 2**.

Table 2: Reference Data Reaction of Sodium Carbonate and Calcium Chloride

Mass of $Na_2CO_{3(aq)}$ reacting (g)	Mass of $CaCO_{3(s)}$ produced (g)
0.50	0.47
1.00	0.94
1.50	1.42
2.00	1.89
2.50	2.36

Materials

lab apron
eye protection
50 mL of $Na_2CO_{3(aq)}$ (approximately 0.50 mol/L)
100 mL of $CaCl_{2(aq)}$ (approximately 0.30 mol/L)
wash bottle of pure water
100-mL graduated cylinder
two 250-mL beakers
stirring rod
medicine dropper
filter paper
filter funnel, rack, and stand
centigram balance
paper towel

Procedure

1. Measure 40 mL of $Na_2CO_{3(aq)}$ into the graduated cylinder and pour into a clean 250-mL beaker.
2. Rinse the graduated cylinder with the pure water and measure 80 mL of $CaCl_{2(aq)}$.
3. To the $Na_2CO_{3(aq)}$ slowly add, while stirring, 10 mL of the $CaCl_{2(aq)}$.
4. Allow the mixture to settle. When the top layer of the mixture becomes clear, add a few extra drops of $CaCl_{2(aq)}$ using the dropper.
5. If cloudiness is visible, repeat steps 3 and 4 until the solution remains clear.
6. Measure and record the mass of a piece of filter paper.
7. Filter the mixture into the other beaker, using the pure water to rinse the beaker and stirring rod into the filter paper. Discard the filtrate into the sink.
8. Dry the precipitate and filter paper overnight on a folded paper towel.
9. Measure and record the mass of the dried filter paper plus precipitate.

Analysis

(a) What is the mass of $CaCO_{3(s)}$ produced in your sample?
(b) What is the shape of the standard curve plotted with the reference data? What type of relationship exists between the mass of reactant and mass of product in this reaction?
(c) From the standard curve, what was the mass of $Na_2CO_{3(aq)}$ in your sample?

Evaluation

(d) Compare your results with those of other groups in the class. Suggest reasons why there may be some variation in the results obtained.
(e) Refer to any variations in results within your class where all groups used the same solutions and equipment, and evaluate the reliability of using the reference data collected in a different laboratory by other groups.

Synthesis

(f) Suppose the precipitate formed in a similar reaction is slightly soluble in water. Can a standard curve still be used to determine the mass of reactant?

CAREER

Quantitative Careers

There are many interesting and rewarding careers that combine analysis of unknown substances with other areas of study. Find out more about one of the examples outlined on this page or another career in quantitative analysis that matches your interests.

Art Conservationist

An art conservationist combines the technology of chemical analysis with a knowledge of art history to determine the extent of damage to artwork and methods of restoration. For example, one technique may involve the application of minimal amounts of solvent to a historical painting, followed by quantitative analysis of the swab by infrared spectroscopy. Other procedures may include the use of microscopes and X-ray machines to identify the chemical composition of materials used in the painting. Working closely with art historians, an art conservationist may use this information to assess the authenticity of a piece of artwork.

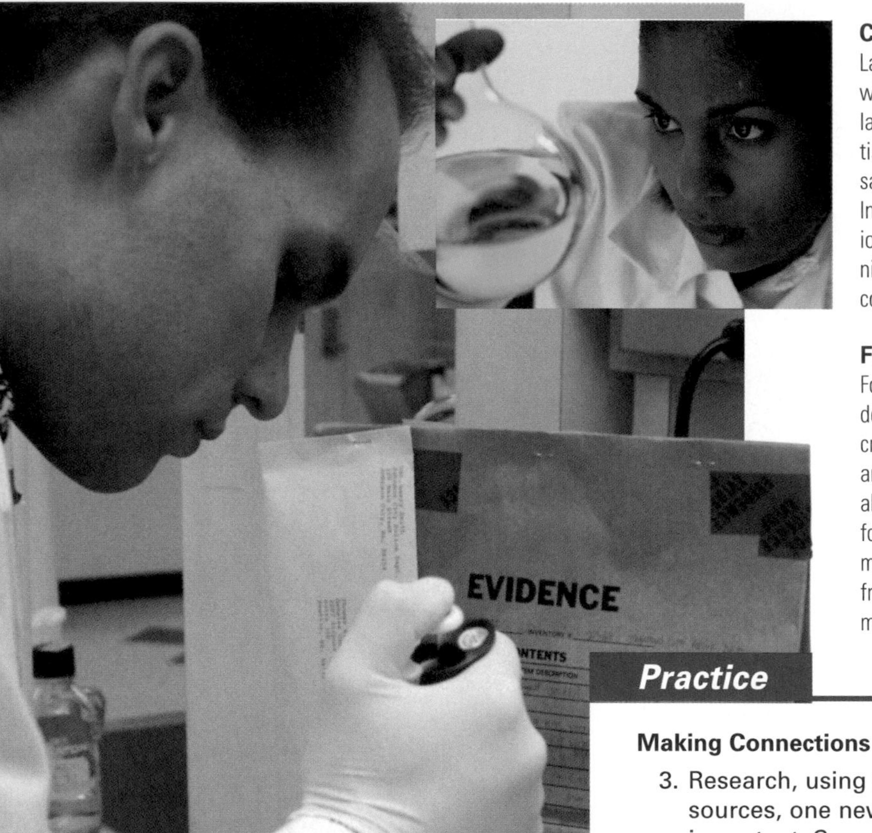

Chemical Laboratory Technician

Laboratory technicians are an important part of any analytical team, whether it is in medical diagnosis or industrial research. In a medical laboratory, technicians run analytical procedures on blood, urine, or tissue samples. In pharmaceutical companies, they analyze product samples to ensure that the quantities of active ingredients are exact. In industry, technicians maintain or improve product quality, using various separation and purification techniques. In addition to good technical skills and precision instrumentation, technicians need excellent communication skills to prepare reports on the results of the analyses.

Forensic Chemist

Forensic chemists work with police forces, customs officers, and fire department investigators to analyze materials that are related to a criminal offence. They prepare reports that may become part of a trial and often appear in court as expert witnesses, answering questions about the analysis of the materials on exhibit. To identify materials, forensic chemists rely on a variety of technologies, including microscopy, gas chromatography, infrared spectroscopy, and X-ray diffraction. An understanding of statistics is also important in determining the significance of an analysis.

Practice

Making Connections

3. Research, using newspaper or magazine articles or electronic sources, one news item in which analysis of unknown substances is important. Some suggestions are forensics, water or air purity, blood sugar, and gold mining. Write a report relating the story, discussing the technology involved, and describing a related science- and technology-based career.
4. Identify and describe a career that requires knowledge of chemical analysis from those above or from searching the Internet. What secondary school and post-secondary courses and training would you need to enter this career?

   Follow the links for Nelson Chemistry 11, 5.1.

GO TO www.science.nelson.com

Practice

Applying Inquiry Skills

5. What is a standard curve, and how is it used in quantitative chemical analysis?
6. When lead(II) nitrate solution reacts with sodium sulfide solution, a black precipitate of lead(II) sulfide is produced (**Figure 5**).
   (a) Describe a procedure to determine a reasonable volume of sodium sulfide solution to use in excess with 10.0 mL of lead(II) nitrate solution.
   (b) What reference data can be graphed on the axes of a standard curve to determine the mass of lead(II) nitrate in a sample solution?

Making Connections

7. At the post office, a reference list matches the mass of a parcel with the postage required. Explain whether the information on this list can be considered reference data for a standard curve for determining postage.

Figure 5
When lead(II) nitrate and sodium sulfide solutions are mixed, a precipitate of lead(II) sulfide is formed. Sodium nitrate remains dissolved in water.

Section 5.1 Questions

Understanding Concepts

1. Suppose 1.0 mol of substance A reacts with 1.0 mol of substance B to produce 2.0 mol of substance C. In a reaction, 2.4 mol of substance A is allowed to react with 2.1 mol of substance B.
   (a) Which substance is the limiting reagent?
   (b) What amount in moles of substance C is produced?
2. Each mole of methane, CH_4, reacts with 2 mol of oxygen gas to completely combust. If 32 g of methane is the limiting reagent in the reaction, what mass of oxygen gas would provide approximately a 10% excess?
3. Explain why a standard curve can be plotted even if a balanced equation may not be known for a chemical reaction.

Applying Inquiry Skills

4. In a quantitative analysis, which substance requires accurate measurement, the limiting reagent or the excess reagent? Explain your reasoning.
5. Describe how you can estimate the quantity of excess reagent to use in an experiment in which one of the products formed is insoluble in water.
6. Aluminum metal reacts with copper(II) sulfate solution to produce copper metal and an aqueous solution of aluminum sulfate. Describe the steps you would use to gather evidence and prepare a standard curve for the mass of copper produced from a mass of aluminum reacted.

Making Connections

7. There are many situations in which quantitative analysis relies on comparing measurements to reference data in a table or on a standard curve. Identify two everyday situations and two work-related contexts in which this is performed.

5.2 Balancing Chemical Equations

Before we can analyze a chemical reaction quantitatively, we have to know the quantitative relationship between reactants and products. Is there a simple "recipe" for a chemical reaction that tells us how much material we need to start with and how much product we will end up with? Yes, indeed there is. All the information we need is incorporated in a balanced chemical equation. Writing a balanced equation is almost always the first step in a quantitative analysis, and we will briefly review the skills you learned and practised earlier. Let's use the method of balancing equations "by inspection," which means that we will simply count the number of atoms of each element before and after the reaction and adjust coefficients to make sure the numbers are the same for each element.

Let's examine an equation for the reaction between copper metal and silver nitrate solution. First, we observe metallic copper wire, $Cu_{(s)}$, being immersed in a colourless solution of silver nitrate, $AgNO_{3(aq)}$. The solution gradually changes colour to blue, and sparkling crystals form on the copper wire. Diagnostic tests confirm that the blue solution is due to formation of copper(II) nitrate, $Cu(NO_3)_{2(aq)}$, which goes into solution, and the crystals are metallic silver, $Ag_{(s)}$. This reaction is used industrially to recover and recycle silver by trickling waste solutions containing silver ions over scrap copper.

Write a reaction equation, including states of matter:

$$? \, Cu_{(s)} + ? \, AgNO_{3(aq)} \rightarrow ? \, Ag_{(s)} + ? \, Cu(NO_3)_{2(aq)}$$

First, balance atoms that are not in polyatomic ions, and are not O or H. (Oxygen and hydrogen atoms may be constituents of many compounds in a reaction. Balancing O and H early could be counterproductive, as they might need to be rebalanced again at a later stage.) In this case, we can start with Cu and Ag, but as stated they are already balanced (one atom of each on each side of the equation). Our equation remains unchanged.

$$? \, Cu_{(s)} + ? \, AgNO_{3(aq)} \rightarrow ? \, Ag_{(s)} + ? \, Cu(NO_3)_{2(aq)}$$

Next, balance any polyatomic ions. In this case, there are two $(NO_3)^-$ ions in the product, but only one in the reactant, so the quantity of the reactant compound needs to be doubled:

$$? \, Cu_{(s)} + 2 \, AgNO_{3(aq)} \rightarrow ? \, Ag_{(s)} + ? \, Cu(NO_3)_{2(aq)}$$

Doubling $AgNO_3$ requires doubling of the Ag after reaction:

$$? \, Cu_{(s)} + 2 \, AgNO_{3(aq)} \rightarrow 2 \, Ag_{(s)} + ? \, Cu(NO_3)_{2(aq)}$$

Next, balance O atoms and H atoms. We have none in this case.

Balance any remaining elements, such as H_2 or O_2. Again, we have none, so we can state our final equation:

$$Cu_{(s)} + 2 \, AgNO_{3(aq)} \rightarrow 2 \, Ag_{(s)} + Cu(NO_3)_{2(aq)}$$

Final check:
Before reaction: 1 Cu, 2 Ag, 2 (NO_3)
After reaction: 1 Cu, 2 Ag, 2 (NO_3)
The equation is balanced as stated.

This systematic approach generally reduces the time spent in a random trial-and-error method. For reactions that involve many reactants and products, a more detailed system of accounting for atoms may be used. You will learn that method in later courses.

There is one more tip on balancing some reaction equations that appear to require fractional coefficients. For example, the following balanced equation requires 7 atoms of oxygen contained in $\frac{7}{2}$ molecules of O_2.

$$2\ NH_{3(g)} + \frac{7}{2}\ O_{2(g)} \rightarrow 3\ H_2O + 2\ NO_{2(g)}$$

To avoid fractional coefficients, we double all coefficients, thereby obtaining a balanced equation of whole number coefficients while retaining the same mole ratio for the reaction.

$$4\ NH_{3(g)} + 7\ O_{2(g)} \rightarrow 6\ H_2O + 4\ NO_{2(g)}$$

Sample Problem 1

Write a balanced equation for the reaction of aqueous sodium phosphate and aqueous calcium chloride to produce sodium chloride in solution and a precipitate of calcium phosphate.

Solution

$$Na_3PO_{4(aq)} + \ CaCl_{2(aq)} \rightarrow \ Ca_3(PO_4)_{2(s)} + \ NaCl_{(aq)}$$

$$2\ Na_3PO_{4(aq)} + 3\ CaCl_{2(aq)} \rightarrow Ca_3(PO_4)_{2(s)} + 6\ NaCl_{(aq)}$$

SUMMARY Balancing Chemical Equations by Inspection

1. Write the chemical formula for each reactant and product, including the state of matter for each one.
2. Try balancing any atom that is not in a polyatomic ion and is not oxygen or hydrogen.
3. If possible, balance polyatomic ions as a group.
4. Balance the remaining atoms and molecules, taking into account the oxygen and hydrogen atoms and water; leave until last any elements such as H_2 or O_2.
5. Check the final reaction equation to ensure that there is the same number of each type of atom before and after the reaction.

Practice

Understanding Concepts

1. Explain why a balanced chemical equation is required by the law of conservation of mass.
2. Balance the following equations by inspection:
   (a) $Mg_{(s)} + HCl_{(aq)} \rightarrow MgCl_{2(aq)} + H_{2(g)}$

Answers

2. (a) 1, 2, 1, 1

Answers

2. (b) 2, 2, 2, 1
   (c) 1, 2, 1, 1, 1
   (d) 1, 4, 1, 2, 2
   (e) 2, 9, 6, 6
3. (a) 2, 1, 2
   (c) 1, 2, 2, 1

(b) $Na_{(s)} + H_2O_{(l)} \rightarrow NaOH_{(aq)} + H_{2(g)}$
(c) $CaCO_{3(s)} + HCl_{(aq)} \rightarrow CaCl_{2(s)} + H_2O_{(l)} + CO_{2(g)}$
(d) $Cu_{(s)} + HNO_{3(aq)} \rightarrow Cu(NO_3)_{2(aq)} + NO_{(g)} + H_2O_{(l)}$
(e) $C_3H_{6(g)} + O_{2(g)} \rightarrow CO_{2(g)} + H_2O_{(g)}$

3. Which of the following chemical equations is balanced correctly? Write the correct equation for any incorrect ones.
   (a) $H_{2(g)} + O_{2(g)} \rightarrow H_2O_{(g)}$
   (b) $2NaOH_{(aq)} + Cu(ClO_3)_{2(aq)} \rightarrow Cu(OH)_{2(s)} + 2NaClO_{3(aq)}$
   (c) $Pb_{(s)} + AgNO_{3(aq)} \rightarrow Ag_{(s)} + Pb(NO_3)_{2(aq)}$
   (d) $2NaHCO_{3(s)} \rightarrow Na_2CO_{3\,(s)} + CO_{2\,(g)} + H_2O_{(l)}$
4. Write a balanced equation for the reaction between iron(III) nitrate and lithium hydroxide solutions to produce lithium nitrate solution and iron(III) hydroxide precipitate.

Mole Ratios in Balanced Chemical Equations

A balanced equation tells us the ratio of the amounts of reactants and products taking part in a chemical reaction. This ratio applies to the number of entities or, more usefully, to the amount in moles.

Let's consider a simple reaction to produce ammonia, $NH_{3(g)}$. You may recognize the characteristic pungent smell of ammonia since it is used in some household cleansing agents, such as window cleaner. It is also used as fertilizer, commonly applied directly to the soil from tanks containing the liquefied gas. As you can see from its formula, ammonia consists of nitrogen and hydrogen atoms; in fact, it can be synthesized directly from nitrogen gas and hydrogen gas. The balanced equation is given below, together with various interpretations of the coefficients for each reactant and product:

$N_{2(g)}$	+	$3\ H_{2(g)}$	$\rightarrow$	$2\ NH_{3(g)}$
1 molecule		3 molecules		2 molecules
1 dozen molecules		3 dozen molecules		2 dozen molecules
1 mol nitrogen		3 mol hydrogen		2 mol ammonia
$1\ (6.02 \times 10^{23})$ molecules		$3(6.02 \times 10^{23})$ molecules		$2(6.02 \times 10^{23})$ molecules
6.02×10^{23} molecules		1.81×10^{24} molecules		1.20×10^{24} molecules

mole ratio: the ratio of the amount in moles of reactants and/or products in a chemical reaction

Note that the numbers in each row are in the same ratio (1:3:2) whether individual molecules, moles, or large numbers of molecules are considered. When moles are used to express the coefficients in the balanced equation, the ratio of reacting amounts is called the **mole ratio.**

The process to produce ammonia described by the equation above is called the Haber process, invented by the German chemist Fritz Haber in 1913. Later, during World War I, the Germans used this process to create explosives. Before this process was developed, the source of the nitrates needed to make explosives (and fertilizers) was a huge deposit of guano (sea bird droppings) along the coast of Chile. But the Germans anticipated a British blockade that would cut off this supply, hence the Haber process. The existence of the Haber process in effect prolonged the war for several years, indirectly causing the loss of many more lives. However, the large-scale availability of ammonia as a chemical fertilizer has indirectly saved many people from food shortages and perhaps even starvation.

Haber understood and applied many chemical principles to make the commercial production of ammonia possible. Today, the Haber process and modifications of it are still used (**Figure 1**).

Figure 1
The Haber process is employed to produce large quantities of ammonia for use as a chemical fertilizer.

Practice

Understanding Concepts

5. Explain how the mole ratio applies in a chemical equation.
6. The total number of atoms in reactants equals the total number of atoms in products in a balanced equation. Does this also mean that the total amount in moles before and after a chemical reaction is the same? Give an example to illustrate your answer.
7. Write a balanced equation for the reaction between nitrogen dioxide gas and water to produce nitric acid and nitrogen monoxide gas. State the mole ratios of reactants and products.
8. Explain why the mole ratio for a chemical reaction is important in industrial chemical processes.

How Much Is Too Much?

Earlier we discussed the need to supply excess amounts of reactants to ensure that the desired chemical reaction can go to completion. We often use excess reagents in everyday situations where a little may be sufficient, but more may be better. When we wash our laundry, for example, we may use a little extra detergent to make sure our clothes are completely rid of dirt. When we nourish our crops, we may use a little extra fertilizer to make sure that we reap the largest harvest possible. Is there any harm in using a little excess? In the case of fertilizers, it seems that a little excess can accumulate to a large excess, with accompanying consequences.

As we learned in our discussion of the Haber process, fertilizers supply nitrogen to the soil. Nitrogen is needed for plant growth, and nitrates are the main source of this element for higher plants. Bacteria in the soil convert atmospheric nitrogen into nitrites (NO_2^-), which in turn are converted into nitrates (NO_3^-). The nitrates are released into the soil, where they are absorbed by the roots of the plants and incorporated into organic nitrogen compounds for use in other parts of the plant.

However, nitrates are in short supply in most soils, so farmers and gardeners add synthetic chemical fertilizers to their soil. The manufacture of these fertilizers has become a multimillion-dollar industry that employs many chemists and technologists, producing millions of tonnes of nitrogen-rich chemicals. Homeowners and farmers use huge quantities of these fertilizers, and the unused excess amounts are readily washed from the soil and from roadways by rain and irrigation into ground water, rivers, and lakes. In areas of intensive agriculture, or where a waterway collects effluent from several communities, the fertilizer not taken up by the crop plants is proving harmful to local ecosystems.

This problem was brought to light by the plight of the Mississippi River in the United States. A study showed that 25% of all fertilizer applied in the states of Minnesota, Wisconsin, Iowa, Illinois, and Missouri is washed off farm fields into the upper Mississippi River. In addition to the substantial damage to the environment, the financial costs of dredging the river, habitat preservation, and the loss of excess applied fertilizer are enormous.

You might think that a little excess fertilizer would do no harm to the rivers and lakes, and would even nourish the plant life in those environments. In fact, the opposite takes place. Once in a lake, the nitrates do what they were intended to do: stimulate plant growth. When the concentration of nitrates is high, however, it encourages rapid growth of plants and algae on the surface of the water.

When the algae die and decay, the decomposing bacteria consume large quantities of dissolved oxygen from the water, depriving fish and other oxygen-dependent organisms; and the overnourished growth on the water surface becomes susceptible to insect and disease invasion. Entire habitats covering huge areas have been destroyed by the nutrient runoff that forms a sludge at the bottom of waterways.

To address the problem of what to do about nutrient runoff, groups of researchers have proposed some form of government initiative to regulate the amount of fertilizers used by farmers and homeowners. Many groups have special interests in this proposal:

- City officials are concerned about the cost and safety of the water supply.
- Environmentalists are concerned that the contamination of the water by nutrient runoff is upsetting the balance of natural resources, which will affect the long-term survival of these ecosystems.
- Representatives from the tourism industry, whose livelihood depends on fishing and boating, are concerned about lost revenue resulting from contamination of natural resources.
- Others are concerned about the amount of money poured into the manufacture of fertilizers, money that could be better spent in other ways, such as on land management and wildlife preservation.
- Farmers cannot calculate exact amounts of fertilizers needed, nor can they predict weather conditions that cause runoff from the soil. Excess fertilizers are necessary for maximizing yield and ensuring a successful harvest, particularly in Canada's short growing season.
- Farmers also feel that regulating the amount of chemical fertilizer applied only partially addresses the problem. Excess nutrients also come from livestock manure and from industrial waste. Plant growers should not be targeted for a much wider problem.
- Some government officials believe that any regulation would be difficult to enforce. The government does not regulate the amount of gasoline we use in our cars or the number of cigarettes we smoke, both of which produce pollutants. Rather, governments should encourage voluntary limits of fertilizer use.
- Others insist that setting up regulatory systems to monitor and enforce limits of use would be a very costly endeavour. The money can be better spent in other ways, such as on land management and wildlife preservation.

As you can see, the issue is not cut-and-dried.

Practice

Understanding Concepts

9. How have fertilizers helped society?
10. How does the overuse of fertilizers cause damage to wildlife habitats?

Making Connections

11. Find out more on the work of environmental groups and their efforts to protect and preserve our natural resources. Choose a group and prepare a short presentation on the group's effect.

    Follow the links for Nelson Chemistry 11, 5.2.

GO TO www.science.nelson.com

Explore an Issue

Role Play: Controlling the Use of Fertilizers

In your area, a group of concerned citizens wants the government to regulate the amount of fertilizers used. The mayor of your town has organized a panel to discuss this and has asked for representation from everyone with an opinion on the issue so that she can present a balanced report and recommendation to the Minister of Agriculture and Agri-Food Canada.

You are a member of this panel. You might be a farmer, a homeowner, a member of a conservation group, a representative of a fertilizer manufacturer, a fisher, a member of the town council, a high-school student, a member of a government agricultural agency, a restaurant owner in a tourist location, a parent of young children, or anyone else you think would have an opinion on the subject.

(a) Research the issue, then select a role and a position. Defend that position in the panel discussion.
Follow the links for Nelson Chemistry 11, 5.2

www.science.nelson.com

DECISION-MAKING SKILLS

- ● Define the Issue
- ● Identify Alternatives
- ● Research
- ● Analyze the Issue
- ● Defend a Decision
- ○ Evaluate

Selection 5.2 Questions

Understanding Concepts

1. Explain what is meant by the phrase "balancing an equation by inspection."
2. Translate the following sentences into internationally understood balanced chemical equations:
   (a) Two moles of solid aluminum and three moles of aqueous copper(II) chloride react to form three moles of solid copper and two moles of aqueous aluminum chloride. (This reaction does not always produce the expected products.)
   (b) One mole of solid copper reacts with two moles of hydrochloric acid to produce one mole of hydrogen gas and one mole of copper(II) chloride. (When tested in the laboratory, this prediction is found to be false.)
   (c) Two moles of solid mercury(II) oxide decompose to produce two moles of liquid mercury and one mole of oxygen gas. (This decomposition reaction is a historical but dangerous method of producing oxygen.)
   (d) Methanol (antifreeze and fuel) is produced from natural gas in large quantities by the following reaction series:
      (i) One mole of methane gas reacts with one mole of steam to produce one mole of carbon monoxide gas and three moles of hydrogen gas.
      (ii) One mole of carbon monoxide gas reacts with two moles of hydrogen gas to produce one mole of liquid methanol.
3. The following equations communicate reactions that occur before, during, and after the formation of acid rain; balance them by inspection:
   (a) $C_{(s)} + O_{2(g)} \rightarrow CO_{2(g)}$
   (b) $S_{8(s)} + O_{2(g)} \rightarrow SO_{2(g)}$
   (c) $Cu(OH)_{2(s)} + H_2SO_{4(aq)} \rightarrow H_2O_{(l)} + CuSO_{4(aq)}$
   (d) $CaSiO_{3(s)} + H_2SO_{3(aq)} \rightarrow H_2SiO_{3(aq)} + CaSO_{3(s)}$
   (e) $CaCO_{3(s)} + HNO_3 \rightarrow H_2CO_{3(aq)} + Ca(NO_3)_{2(aq)}$
   (f) $Al_{(s)} + H_2SO_{4(aq)} \rightarrow H_{2(g)} + Al_2(SO_4)_{3(aq)}$

(continued)

(g) $SO_{2(g)} + H_2O_{(l)} \rightarrow H_2SO_{3(aq)}$
(h) $Fe_{(s)} + H_2SO_{3(aq)} \rightarrow H_{2(g)} + Fe_2(SO_3)_{3(s)}$
(i) $N_{2(g)} + O_{2(g)} \rightarrow NO_{(g)}$
(j) $CO_{2(g)} + H_2O_{(l)} \rightarrow H_2CO_{3(aq)}$
(k) $CH_{4(g)} + O_{2(g)} \rightarrow CO_{2(g)} + H_2O_{(g)}$
(l) $C_4H_{10(g)} + O_{2(g)} \rightarrow CO_{2(g)} + H_2O_{(g)}$
(m) $FeS_{(s)} + O_{2(g)} \rightarrow FeO_{(s)} + SO_{2(g)}$
(n) $H_2S_{(g)} + O_{2(g)} \rightarrow H_2O_{(g)} + SO_{2(g)}$
(o) $CaCO_{3(s)} + SO_{2(g)} + O_{2(g)} \rightarrow CaSO_{4(s)} + CO_{2(g)}$

4. Write balanced equations for each of the following reactions, and state the mole ratio for the reactants and products:
   (a) Nitromethane, $CH_3NO_{2\,(l)}$, is a fuel commonly burned in drag-racing vehicles. It combusts in oxygen to produce carbon dioxide gas, water vapour, and nitrogen dioxide gas.
   (b) Aluminum metal reacts in copper(II) chloride solution to produce aluminum chloride solution and copper metal.

Making Connections

5. The Haber process facilitated the production of chemical fertilizers such as ammonium nitrate and has had a dramatic impact on crop yields. Investigate and describe the type of calculations a farmer must make to predict the quantity and cost of fertilizer to purchase and apply, in order to attain a certain crop yield.

Figure 1
Marie Curie coined the term *radioactivity*. Searching for a topic for her doctoral thesis, Curie became intrigued by a discovery by Henri Becquerel. In 1896, Becquerel had discovered that compounds of uranium spontaneously emitted rays, or radiation, that exposed photographic plates. She decided to investigate the emissions from uranium and to find out if this property was exhibited by other elements. She named this property "radioactivity."

5.3 Balancing Nuclear Equations

So far in this unit, we have been discussing chemical reactions in which the bonds that are broken and formed are between atoms or ions. In this section, we will discuss nuclear reactions, in which the changes involved are within the nucleus of an atom. The particles in the nucleus—protons and neutrons—are called **nucleons.** Of all energy changes, nuclear reactions involve the greatest quantities of energy. Energy changes in nuclear reactions are a result of potential energy changes as bonds between the nucleons are broken or formed.

Based on empirical evidence, scientists classify nuclear reactions into four types:

- radioactive decay
- artificial transmutation
- fission
- fusion

For each of these nuclear reactions, balanced nuclear equations can be written. But since the changes occur within the nucleus, when we balance a nuclear equation only the nucleons are represented.

nucleon: any particle in the nucleus of an atom

radioactive decay: the spontaneous decomposition of a nucleus

Radioactive Decay

As you learned in Chapter 1, certain atoms are radioactive: the unstable nucleus spontaneously decomposes. When this happens it is said to have decayed, or undergone **radioactive decay** (Figure 1). Depending on the type of particle emitted, radioactive decay is *alpha decay* or *beta decay.* In each case, a more stable nucleus results, and extra energy is given off as *gamma rays.*

When a nucleus undergoes a change that changes the number of protons, its identity is necessarily changed. It is no longer the same element. This change in the identity of a nucleus is called a **transmutation**. For example, the radioisotope uranium-238 contains 92 protons and 146 neutrons. It spontaneously decomposes into two fragments.

transmutation: the changing of one element into another as a result of radioactive decay

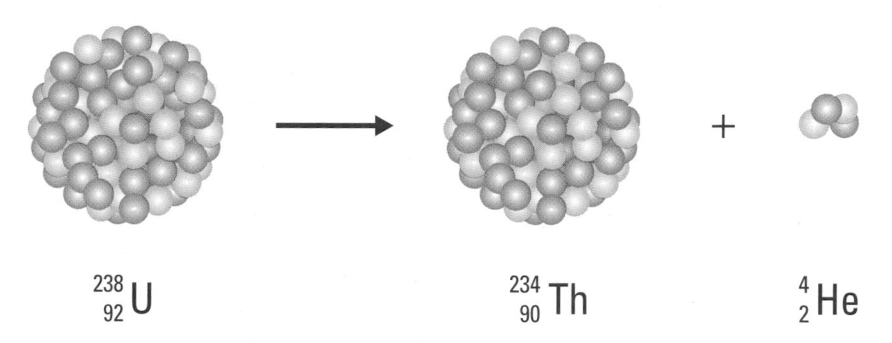

Figure 2
Uranium-238 transmutes to a thorium-234 atom plus an alpha particle.

The uranium-238 atom loses a helium-4 nucleus, leaving a fragment with 90 protons and 144 neutrons, a total of 234 nucleons. Atoms with 90 protons are thorium atoms; therefore, a thorium-234 atom is formed. In other words, uranium has been transmuted to thorium. The other particle emitted, the helium-4 nucleus, is called an **alpha particle** (Figure 3); a stream of alpha particles is called an **alpha ray**, and the process in which alpha particles are emitted is called **alpha decay**. Alpha particles carry a 2+ charge, which is usually not shown in nuclear equations.

alpha particle: the nucleus of a helium atom

alpha ray: a stream of alpha particles

alpha decay: the process in which alpha particles are emitted

Notice in the nuclear equation above that the sums of the atomic numbers on both sides of the equation are the same (92 = 90 + 2). The sums of the mass numbers on both sides of the equation are also the same (238 = 234 + 4). Since no protons or neutrons are created or destroyed in a nuclear reaction, the atomic numbers and mass numbers in nuclear equations are always balanced.

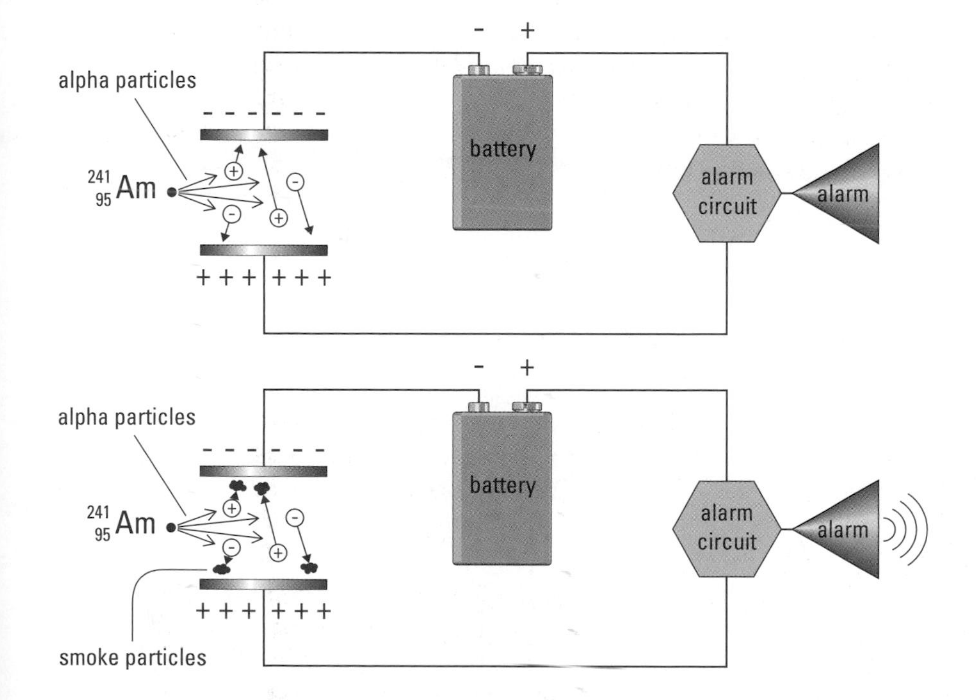

Figure 3
Some smoke detectors contain a small amount of the radioisotope americium-241, which emits alpha particles. An electric alarm is connected to a pair of electrodes, and the air in the space between the electrodes is ionized by the alpha particles, allowing a small current to flow. In the event of a fire, smoke particles enter the space between the electrodes and absorb the radiation-induced ions, thus interrupting the current flow. An electronic sensor in the alarm circuitry responds to the stoppage of current by triggering the alarm.

Sample Problem 1

Write an equation for the emission of an alpha particle from $^{226}_{88}Ra$.

Solution

$$^{226}_{88}Ra \rightarrow \quad + \, ^{4}_{2}He$$

$$^{226}_{88}Ra \rightarrow \, ^{222}_{86} \quad + \, ^{4}_{2}He$$

$$^{226}_{88}Ra \rightarrow \, ^{222}_{86}Rn + \, ^{4}_{2}He$$

beta particle: an electron emitted by certain radioactive atoms

beta ray: a stream of beta particles

beta decay: the process in which beta particles are emitted

Some unstable nuclei may spontaneously emit a negatively charged electron, called a **beta particle**. A stream of beta particles is called a **beta ray**, and this type of radioactive decay is called **beta decay**. An electron is represented by $^{0}_{-1}e$: the mass of an electron is shown as zero because it is exceedingly small and therefore negligible compared to the mass of nucleons, and the electron has a charge of 1–. When we follow the uranium-238 alpha decay further, we see that the resultant thorium-234 nucleus is itself unstable. It undergoes beta decay and is transmuted to a protactinium-234 nucleus (**Figure 4**):

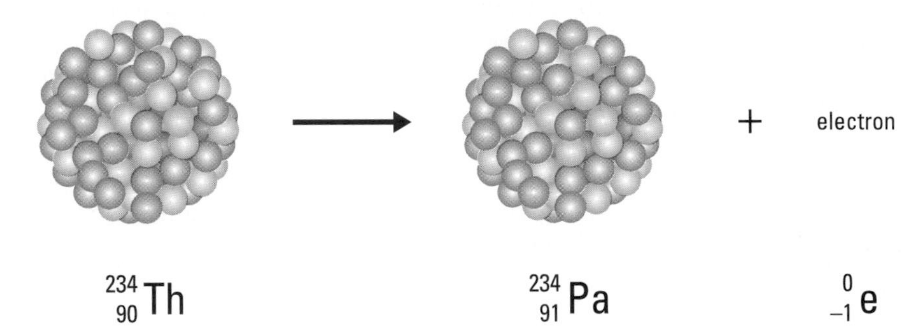

$$^{234}_{90}Th \qquad ^{234}_{91}Pa \qquad ^{0}_{-1}e$$

Figure 4
Uranium-238 transmutes to a thorium-234 atom, which is also unstable; the thorium-234 atom transmutes to a protactinium-234 atom plus a beta particle.

Notice in the nuclear equation above that the sums of the atomic numbers (234 = 234 + 0) and the mass numbers (90 = 91 – 1) remain the same on both sides of the equation. In this case, the thorium nucleus seems to have gained a proton, although the total number of protons and neutrons remains constant at 234. In effect, some scientists believe, one of the neutrons in the thorium-234 nucleus has transformed into a proton ($^{1}_{1}p$) and an electron at the moment of decay. Thus, the electron emitted did not originally reside in the nucleus because electrons are not found in nuclei. Rather, it was created by the nuclear reaction:

$$^{1}_{0}n \rightarrow \, ^{1}_{1}p + \, ^{0}_{-1}e$$

gamma rays: high-energy (short wavelength) electromagnetic radiation emitted during radioactive decay

In alpha decay, a more stable nucleus is produced by decreasing the mass of a heavy nucleus, and $^{4}_{2}He$ particles are emitted. In beta decay, a more stable nucleus is produced by conversion of a neutron into a proton, and $^{0}_{-1}e$ particles are emitted. In both cases, high-energy electromagnetic radiation is given off as **gamma rays**, the energy emitted when the nucleons reorganize into more stable arrangements. Gamma rays are represented as $^{0}_{0}\gamma$—zero mass and zero charge—because this radiation changes neither the atomic number nor the mass number of a nucleus. Generally, gamma rays are not shown when writing nuclear equations just as released heat energy is often absent from chemical equations. A summary of particles common in nuclear equations is given in **Table 1**, and a summary of the properties of alpha, beta, and gamma rays is given in **Table 2**.

Table 1: Summary of Particles Common in Nuclear Equations

Particle	Symbol
neutron	$^{1}_{0}n$
proton	$^{1}_{1}p$ or $^{1}_{1}H$
electron	$^{0}_{-1}e$
alpha particle	$^{4}_{2}He$
beta particle	$^{0}_{-1}e$

Table 2: Summary of the Properties of Alpha, Beta, and Gamma Rays

Property	Alpha (α)	Beta (β)	Gamma (γ)
nature of radiation	$^{4}_{2}He$ nuclei	$^{0}_{-1}e$ (electrons)	high-energy radiation
charge	2+	1–	0
mass	4 u	0	0
penetrating ability	stopped by 4 cm of air or a sheet of paper	stopped by about 12 cm of air or several millimetres thickness of paper	intensity is decreased by about 10% by 3 cm of lead

Sample Problem 2

Write an equation for the emission of a beta particle from $^{214}_{82}Pb$.

Solution

$$^{214}_{82}Pb \rightarrow {}^{214}_{83} \quad + {}^{0}_{-1}e$$

$$^{214}_{82}Pb \rightarrow {}^{214}_{83}Bi + {}^{0}_{-1}e$$

Figure 5
A badge dosimeter measures the total quantity of radiation received by the wearer over a period of exposure. The central component of the dosimeter contains different filters made of plastic, aluminum, and lead–tin alloy, thus registering the exposure to radiations of various energies and penetrating abilities: X rays, beta rays, and gamma rays.

Practice

Understanding Concepts

1. Describe the difference in meaning between each of the following pairs of terms:
   (a) chemical reaction and nuclear reaction
   (b) radioactive decay and nuclear transmutation
   (c) alpha decay and beta decay
2. What is a gamma ray and why is it generally not shown in nuclear equations?
3. Write a balanced nuclear equation for the radioactive decay of thorium-230 with the emission of an alpha particle.
4. Write an equation for the beta decay of thorium-231.
5. Write an equation for the beta decay of $^{93}_{38}Sr$.
6. A $^{214}_{84}Po$ nucleus emits an alpha particle, followed by the emission of a beta particle. Write two nuclear equations to illustrate the two consecutive reactions.

Making Connections

7. Research, using print or electronic resources, some applications of natural radioactivity; an example is the determination of the age of an old object. Share your research in a short presentation.

Artificial Transmutation

artificial transmutation: the artificial bombardment of a nucleus by a small entity such as a helium nucleus, a proton, or a neutron

A nucleus can also change its identity in another way, called **artificial transmutation**. In an artificial transmutation, a nucleus is struck by a neutron or by another nucleus. The symbol for a neutron is $^{1}_{0}n$, indicating that it has no charge and a mass number of 1. To show an example of an artificial transmutation, say a chlorine-35 nucleus being struck by a neutron, the nuclear equation would be

$$^{35}_{17}Cl + ^{1}_{0}n \rightarrow ^{35}_{16}S + ^{1}_{1}p$$

In this collision, a sulfur-35 nucleus and a proton, a hydrogen-1 nucleus, are produced. Notice again that the sums of the atomic numbers ($17 + 0 = 16 + 1$) and the mass numbers ($35 + 1 = 35 + 1$) are balanced in the equation.

The process of artificial transmutation can be used to produce atoms with a specified number of protons and neutrons. For example, researchers may require an atom with 15 protons and 15 neutrons for a medical application. They choose what particles to use for both the nucleus to be bombarded and the particles to do the bombarding to make the desired atom. The particles used to bombard a nucleus must be travelling at very high speeds in order to overcome the electrostatic repulsion between them and the target nucleus. This is accomplished by machines called particle accelerators, which use magnetic and electric fields, for example, the TRIUMF cyclotron at the University of British Columbia (**Figure 6**).

Figure 6
Canada's TRIUMF cyclotron at the University of British Columbia. Charged particles are accelerated around a spiral ring by strong magnetic and electrostatic fields.

Sample Problem 3

Write a nuclear equation for the bombardment of nitrogen-14 by an alpha particle, giving off a proton and the desired isotope.

Solution

$$^{14}_{7}N + ^{4}_{2}He \rightarrow \quad + ^{1}_{1}p$$

$$^{14}_{7}N + ^{4}_{2}He \rightarrow ^{17}_{8} \quad + ^{1}_{1}p$$

$$^{14}_{7}N + ^{4}_{2}He \rightarrow ^{17}_{8}O + ^{1}_{1}p$$

The isotope produced is $^{17}_{8}O$.

Practice

Understanding Concepts

8. Distinguish between the terms *radioactive decay* and *artificial transmutation.*
9. Write the nuclear equation for the bombardment of beryllium-9 by an alpha particle, producing carbon-12 and a particle.
10. Uranium-239 can be produced by the bombardment of uranium-238. Write a nuclear equation to show the type of particle used in this artificial transmutation.

Making Connections

11. Research the type of research and applications supported at Canada's TRIUMF cyclotron. Choose one application and write a one-page report.

    Follow the links for Nelson Chemistry 11, 5.3.

    GO TO www.science.nelson.com

Figure 7
The Bruce Nuclear Generating Station generates energy by nuclear fission of uranium. The energy released by the fission of 1 kg of uranium-235 is equivalent to the energy released from burning 2.6 million litres of fuel oil. This photo was taken before the reactor went into operation.

Fission

Another important nuclear reaction is fission. When a uranium-235 nucleus undergoes fission, it absorbs a neutron and splits into two smaller nuclei. In the process, two or three neutrons are also given off. Notice in the nuclear equation below that the atomic numbers and mass numbers are again balanced on both sides of the equation:

$$^{235}_{92}U + ^{1}_{0}n \rightarrow ^{141}_{56}Ba + ^{92}_{36}Kr + 3\,^{1}_{0}n$$

Nuclear fission reactions release tremendous amounts of energy. The reaction requires the absorption of one neutron, but three neutrons are produced. These neutrons will encounter other uranium-235 nuclei and cause more fission. A chain reaction results, and huge quantities of energy are released instantaneously. Nuclear fission reactions provide the energy for nuclear power generating stations, such as the Bruce Nuclear Generating Station (**Figure 7**).

Sample Problem 4

Write a balanced nuclear equation for the fission of $^{235}_{92}U$ when struck by a neutron, to produce $^{97}_{40}Zr$, two neutrons, and another isotope.

Solution

$$^{1}_{0}n + ^{235}_{92}U \rightarrow \quad + ^{97}_{40}Zr + 2\,^{1}_{0}n$$

$$^{1}_{0}n + ^{235}_{92}U \rightarrow ^{137}_{52} \quad + ^{97}_{40}Zr + 2\,^{1}_{0}n$$

$$^{1}_{0}n + ^{235}_{92}U \rightarrow ^{137}_{52}Te + ^{97}_{40}Zr + 2\,^{1}_{0}n$$

fusion: a nuclear change in which small nuclei combine to form a larger nucleus accompanied by the release of very large quantities of energy

Fusion

The nuclear reactions that occur in the Sun are important to us because they supply the energy that sustains life on Earth (**Figure 8**). There are many different nuclear reactions taking place in the Sun, as in other stars. In one of the main reactions, four hydrogen atoms fuse, producing one helium atom. This type of reaction—light atoms combining to form heavier atoms—is called **fusion**. A fusion reaction releases enormous amounts of energy, more energy than is released in a fission reaction:

$$4\,^{1}_{1}H + 2\,^{0}_{-1}e \rightarrow ^{4}_{2}He$$

Notice that the atomic numbers and mass numbers are balanced on both sides of the equation.

Figure 8
Direct solar radiation provides the energy required for green plants to produce food and oxygen daily. Indirectly, solar energy is also the source of energy from winds, water, and fossil fuels. According to current theory, fossil fuels are the remains of plants and animals that originally depended on sunlight for energy. Fossil fuels are therefore considered a stored form of solar energy.

Fusion reactions require positively charged nuclei to combine; to do this they must collide at sufficiently high speeds to overcome their mutual repulsion. Thus, fusion can only take place at the extremely high temperatures needed to achieve these speeds, such as in the Sun and other stars, approximately 20 million Celsius degrees. However, scientists and engineers are researching the possibility of fusion of two isotopes of hydrogen for use in nuclear fusion reactors on Earth.

Sample Problem 5

Write a nuclear reaction for the fusion of tritium (hydrogen-3) with deuterium (hydrogen-2) to produce helium and a neutron.

Solution

$${}^{3}_{1}H + {}^{2}_{1}H \rightarrow {}^{4}_{2}He + {}^{1}_{0}n$$

Practice

Understanding Concepts

12. Explain why a chain reaction is released in the nuclear fission of uranium-235.
13. Why are enormously high temperatures necessary for fusion reactions to take place?
14. One of the fusion reactions believed to occur in the Sun involves the fusion of two helium-3 nuclei to form two protons and one other particle. Write a nuclear equation to represent this reaction, and identify the particle.
15. Research the principles of the nuclear reactions used in the CANDU reactors, and write nuclear equations for each step.

Making Connections

16. The heat generated by nuclear power stations is often a source of environmental concern. Write a one-page report on the causes and effects of this form of thermal pollution.
17. Canada is regarded internationally as a leader in the peaceful uses of nuclear energy. These uses include providing energy through nuclear generators and developing useful radioactive isotopes for medical use. Research the development of the Atomic Energy of Canada Limited (AECL) and list its major contributions in nuclear research. Follow the links for Nelson Chemistry 11, 5.3.

GO TO www.science.nelson.com

Section 5.3 Questions

Understanding Concepts

1. Why does a nuclear reaction often result in the conversion of one element into another?
2. What is a fission reaction?
3. Write nuclear equations
   (a) for the alpha decay of ${}^{226}_{88}Ra$;
   (b) for the beta decay of bromine-84;
   (c) for the emission of an alpha particle from plutonium-242;
   (d) to illustrate the collision of a proton with a nitrogen-14 nucleus, forming an alpha particle and an isotope of carbon;

(e) to illustrate the absorption of a neutron by an oxygen-18 nucleus, emitting a beta particle and an isotope.

4. (a) Complete the following nuclear reaction:
   $^{235}_{92}U + ^{1}_{0}n \rightarrow ^{147}_{59}Pr + ? + 3\ ^{1}_{0}n$
   (b) What type of nuclear reaction can this be classified as?
5. Complete the following nuclear equations and state which type of nuclear reaction is illustrated by each equation:
   (a) $^{31}_{15}P + ^{1}_{1}p \rightarrow ^{28}_{14}Si + ?$
   (b) $^{27}_{13}Al + ^{4}_{2}He \rightarrow ? + ^{1}_{1}P$
   (c) $^{235}_{92}U + ^{1}_{0}n \rightarrow ^{90}_{37}Rb + ? + 2\ ^{1}_{0}n$
   (d) $^{12}_{5}B \rightarrow ? + ^{0}_{-1}e$
   (e) $^{228}_{90}Th \rightarrow ? + ^{4}_{2}He$
   (f) $2\ ^{3}_{1}H \rightarrow ^{4}_{2}He + ?$
6. Complete and balance the following nuclear equations by supplying the missing particle:
   (a) $^{32}_{16}S + ^{1}_{0}n \rightarrow ? + ^{1}_{1}p$
   (b) $^{235}_{92}U + ^{1}_{0}n \rightarrow ^{135}_{54}Xe + ? + 2^{1}_{0}n$
   (c) $? \rightarrow ^{187}_{76}Os + ^{0}_{-1}e$
   (d) $^{11}_{5}B + ^{1}_{1}p \rightarrow 3\ ?$
   (e) $^{98}_{42}Mo + ^{2}_{1}H \rightarrow ? + ^{1}_{0}n$
7. The $^{235}_{92}U$ nucleus naturally decays through a series of steps and ends with the formation of the stable $^{207}_{82}Pb$. The decays proceed through a series of alpha particle and beta particle emissions. How many of each type of particle is emitted in this series of radioactive decay?

Making Connections

8. Nuclear radiation is known to be a cause of some cancers, yet radiation is also used in certain treatments for cancer. Research this topic and prepare a report on the science and technology of radiation in medicine.

   Follow the links for Nelson Chemistry 11, 5.3.

GO TO www.science.nelson.com

5.4 Calculating Masses of Reactants and Products

Whether we are baking cookies in the kitchen or manufacturing fertilizers in a factory, we need to be able to calculate the quantity of reactants needed to produce a certain quantity of product. In industry, a manufacturer needs to know how much raw material to buy to make the products ordered by its customers. An extraction metallurgist needs to know the amount of ore needed to produce a certain amount of pure metal. A car manufacturer needs to know the amount of pollutant that will be produced in the exhaust gases in order to design an efficient system to deal with the problem. In this section, we will apply the mole concept from Chapter 4 to balanced chemical equations to make quantitative predictions.

To predict masses of reactants and products, we always begin with a balanced equation for the reaction. Then we use the mole ratios given by the coefficients in the equation and convert the relationships to masses as needed. If a reactant or product is a gas, the amount in moles can also be converted to volume if desired, using mole–volume conversion techniques that you will learn in Chapter 9.

gravimetric stoichiometry: the procedure for calculating the masses of reactants or products in a chemical reaction

The procedure for calculating the *masses* of reactants or products in a chemical reaction is called **gravimetric stoichiometry**. This term is quite a mouthful but easily understood simply by taking one word at a time: *gravimetric* refers to mass measurement, and *stoichiometry* refers to the relationships between the quantities of reactants and products involved in chemical reactions. In general, in using gravimetric stoichiometry, we look at the mole ratios in balanced equations and use them to predict masses required or produced.

Let's consider the synthesis reaction that occurs when zinc metal is heated with sulfur, $S_{8(s)}$ (**Figure 1**).

Figure 1
The reaction of powdered zinc and sulfur is rapid and gives off high amounts of heat. Because of the numerous safety precautions necessary to make zinc sulfide, this reaction is not usually demonstrated in school laboratories.

$$8\ Zn_{(s)} + S_{8(s)} \rightarrow 8\ ZnS_{(s)}$$

As the balanced equation indicates, 8 mol of zinc reacts with 1 mol of sulfur to produce 8 mol of zinc sulfide. Suppose that we wish to predict the mass of ZnS that would be produced from 1.00 g of $Zn_{(s)}$. First, we convert the mass (1.00 g) of $Zn_{(s)}$ into the amount in moles. Then we use the mole ratio given in the equation and predict the amount in moles of $ZnS_{(s)}$ that is prescribed by the mole ratio. Finally, we convert the predicted amount in moles of ZnS into mass.

We begin by writing the balanced equation and writing the mass of $Zn_{(s)}$ and the molar masses of $Zn_{(s)}$ and $ZnS_{(s)}$ beneath their formulas:

$$8\ Zn_{(s)} + S_{8(s)} \rightarrow 8\ ZnS_{(s)}$$

1.00 g
65.38 g/mol   97.44 g/mol

Next we convert the mass (1.00 g) into amount in moles (65.38 g/mol):

$$n_{Zn} = 1.00\ \cancel{g} \times \frac{1\ mol}{65.38\ \cancel{g}} = 0.0153\ mol$$

Enter that information beneath the equation:

$$8\ Zn_{(s)} + S_{8(s)} \rightarrow 8\ ZnS_{(s)}$$

1.00 g
65.38 g/mol   97.44 g/mol
0.0153 mol

Now we use the mole ratio given in the equation. The mole ratio prescribes 8 mol of $ZnS_{(s)}$ for each 8 mol of $Zn_{(s)}$, a 1:1 ratio. Therefore,

$$n_{ZnS} = 0.0153\ \cancel{mol\ Zn} \times \frac{8\ mol\ ZnS}{8\ \cancel{mol\ Zn}} = 0.0153\ mol$$

Enter this information beneath ZnS in the equation.

$$8\ Zn_{(s)} + S_{8(s)} \rightarrow 8\ ZnS_{(s)}$$

1.00 g
65.38 g/mol   97.44 g/mol
0.0153 mol   0.0153 mol

Since we want to know the mass of the $ZnS_{(s)}$ produced, and we know that there are 0.0153 mol produced, we convert to mass:

$$m_{ZnS} = 0.0153\ \cancel{mol} \times \frac{97.44\ g}{1\ \cancel{mol}} = 1.49\ g$$

The mass of zinc sulfide produced by the reaction of 1.00 g of zinc with sulfur is 1.49 g (**Figure 2**). The certainty here, three significant digits, is determined by the least certain value, 1.00 g. As you can see, it is possible to predict the mass of product without actually conducting the experiment.

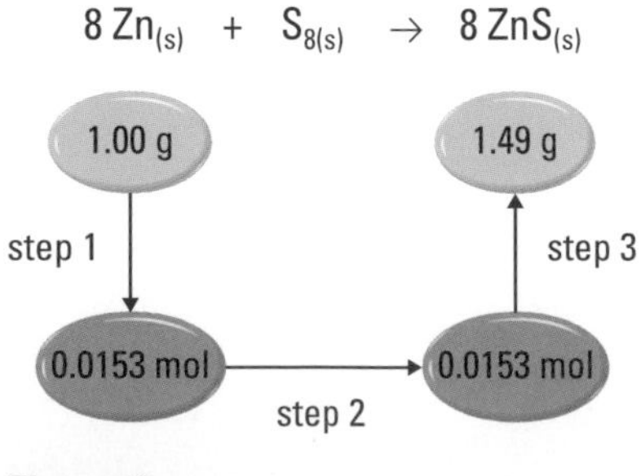

$8\ Zn_{(s)} + S_{8(s)} \rightarrow 8\ ZnS_{(s)}$

1.00 g   1.49 g
step 1   step 3
0.0153 mol   0.0153 mol
step 2

Figure 2
Steps showing calculations

Sample Problem 1

Iron is the most widely used metal in North America (**Figure 3**). It may be produced by the reaction of iron(III) oxide, from iron ore, with carbon monoxide to produce iron metal and carbon dioxide. What mass of iron(III) oxide is required to produce 100.0 g of iron?

Figure 3
Wrought iron is a pure form of iron. The ornate gates on Parliament Hill in Ottawa are made of wrought iron.

Solution

$$Fe_2O_{3(s)} + 3\ CO_{(g)} \rightarrow 2\ Fe_{(s)} + 3\ CO_{2(g)}$$

m	100.0 g
159.70 g/mol	55.85 g/mol

mole ratio Fe_2O_3:Fe = 1:2

$$n_{Fe} = 100.0\ \cancel{g} \times \frac{1\ mol}{55.85\ \cancel{g}}$$

$$= 1.790\ mol$$

$$n_{Fe_2O_3} = 1.790\ \cancel{mol}\ Fe \times \frac{1\ mol\ Fe_2O_3}{2\ \cancel{mol}\ Fe}$$

$$= 0.8952\ mol$$

$$m_{Fe_2O_3} = 0.8952\ \cancel{mol} \times \frac{159.70\ g}{1\ \cancel{mol}}$$

$$m_{Fe_2O_3} = 143.0\ g$$

or, in one step

$$m_{Fe_2O_3} = 1.790\ \cancel{mol\ Fe_{(s)}} \times \frac{1\ \cancel{mol\ Fe_2O_{3(s)}}}{2\ \cancel{mol\ Fe_{(s)}}} \times \frac{159.70\ g}{1\ \cancel{mol\ Fe_2O_{3(s)}}}$$

$$m_{Fe_2O_3} = 143.0\ g$$

The mass of iron(III) oxide required to produce 100.0 g of iron is 143.0 g (**Figure 4**).

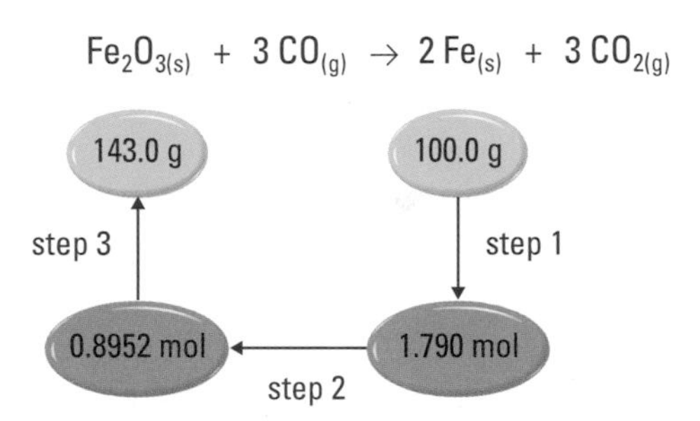

Figure 4
Steps showing calculations

SUMMARY Calculating Mass of Reactants and Products

Begin with a balanced chemical equation, with the measured mass of reactant or product written beneath the corresponding formula.

1. Convert the measured mass into an amount in moles.
2. Use the mole ratio in the balanced equation to predict the amount in moles of desired substance.
3. Convert the predicted amount in moles into mass (**Figure 5**).

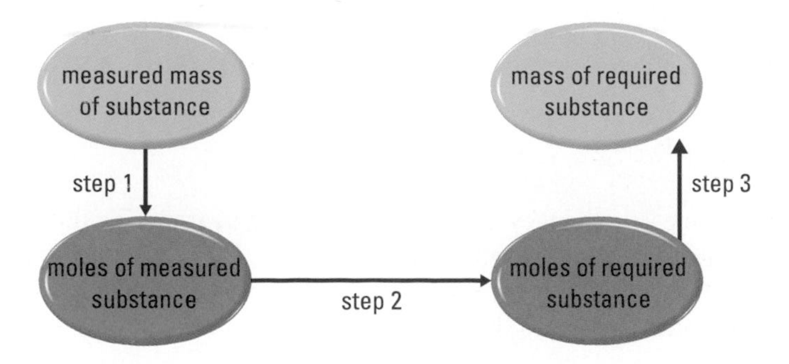

Figure 5
Summary of steps in gravimetric stoichiometry

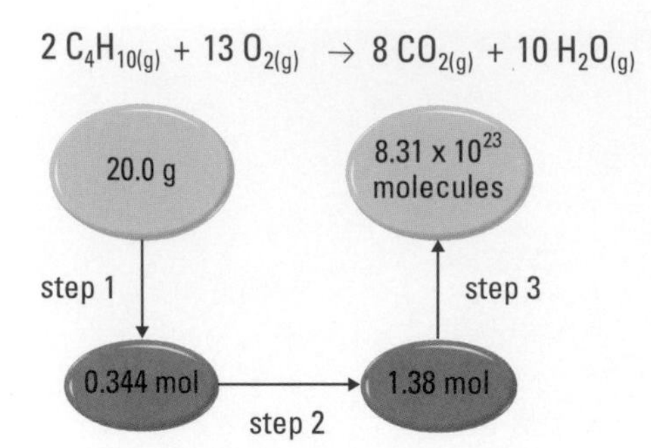

Figure 6
Steps showing calculations

This stoichiometric procedure can also be extended to determine the number of molecules (N) that have reacted or are produced. For example, if 20.0 g of butane, $C_4H_{10(g)}$, is completely burned in a lighter, how many molecules of carbon dioxide are produced? We begin again with writing the chemical equation for this combustion, with the mass of butane beneath its formula:

$$2\ C_4H_{10(g)} + 13\ O_{2(g)} \rightarrow 8\ CO_{2(g)} + 10\ H_2O_{(g)}$$

20.0 g

58.14 g/mol

Next we convert 20.0 g of butane into amount in moles:

$$n_{C_4H_{10}} = 20.0\ \text{g} \times \frac{1\ \text{mol}}{58.14\ \text{g}}$$

$$n_{C_4H_{10}} = 0.344\ \text{mol}$$

Now we use the mole ratio in the equation to predict the amount in moles of $CO_{2(g)}$ produced:

$$n_{CO_2} = 0.344\ \text{mol}\ C_4H_{10} \times \frac{8\ \text{mol}\ CO_2}{2\ \text{mol}\ C_4H_{10}}$$

$$n_{CO_2} = 1.38\ \text{mol}$$

Now we can convert 1.38 mol $CO_{2(g)}$ into the desired quantity, in this case, the number of molecules:

$$N_{CO_2} = 1.38\ \text{mol} \times \frac{6.02 \times 10^{23}}{1\ \text{mol}}$$

$$N_{CO_2} = 8.31 \times 10^{23}$$

If 20.0 g of butane is completely burned, 8.31×10^{23} molecules of carbon dioxide are produced (**Figure 6**).

Sample Problem 2

Magnesium metal reacts with hydrochloric acid to produce magnesium chloride in solution and hydrogen gas. How many hydrogen molecules are produced when 60.0 g of magnesium is reacted in excess hydrochloric acid?

Solution

$$Mg_{(s)} + 2\ HCl_{(aq)} \rightarrow MgCl_{2(aq)} + H_{2(g)}$$

60.0 g

24.31 g/mol

$$n_{Mg} = 60.0\ \text{g} \times \frac{1\ \text{mol}}{24.31\ \text{g}}$$

$$n_{Mg} = 2.47\ \text{mol}$$

$$n_{H_2} = n_{Mg} = 2.47\ \text{mol}$$

$$N_{H_2} = 2.47 \text{ mol} \times \frac{6.02 \times 10^{23}}{1 \text{ mol}}$$

$$N_{H_2} = 1.49 \times 10^{24}$$

When 60.0 g of magnesium are reacted, 1.49×10^{24} molecules of hydrogen are produced.

Practice

1. Bauxite ore contains aluminum oxide, which is decomposed using electricity to produce aluminum metal and oxygen. What mass of aluminum metal can be produced from 125 g of aluminum oxide?
2. Determine the mass of oxygen required to completely burn 10.0 g of propane, $C_3H_{8(g)}$.
3. Calculate the mass of lead(II) chloride precipitate produced when 2.57 g of sodium chloride in solution reacts in a double displacement reaction with excess aqueous lead(II) nitrate.
4. Predict the mass of hydrogen gas produced when 2.73 g of aluminum reacts in a single displacement reaction with excess sulfuric acid.
5. What mass of copper(II) hydroxide precipitate is produced by the double displacement reaction in solution of 2.67 g of potassium hydroxide with excess aqueous copper(II) nitrate?
6. How many molecules of oxygen are produced from the simple decomposition of 25.0 g of water?

Answers

1. 66.2 g
2. 36.3 g
3. 6.12 g
4. 0.306 g
5. 2.32 g
6. 4.17×10^{23}

Lab Exercise 5.4.1

Testing Gravimetric Stoichiometry

INQUIRY SKILLS

- ○ Questioning
- ○ Hypothesizing
- ● Predicting
- ○ Planning
- ○ Conducting
- ○ Recording
- ● Analyzing
- ● Evaluating
- ● Communicating

The most rigorous test of any scientific concept is whether or not it can be used to make accurate predictions. If the prediction is shown to be valid, then the concept is judged to be acceptable. If the percentage difference between the actual and the predicted values is considered to be too great (e.g., more than 10%), then the concept may be judged unacceptable.

The purpose of this lab exercise is to test the validity of the stoichiometric method. To do so, we will use the percentage difference between an experimental value and the value predicted by gravimetric stoichiometry.

Prepare a lab report to present your findings. Provide a **Prediction** of the mass of lead produced in the single displacement reaction between zinc metal and aqueous lead(II) nitrate. In the **Analysis** of the evidence presented, answer the question. In your **Evaluation**, evaluate the experimental design, the prediction, and the method of gravimetric stoichiometry.

Question

What is the mass of lead produced by the reaction of 2.13 g of zinc with an excess of lead(II) nitrate in solution (**Figure 7**)?

Prediction

(a) Write a balanced equation for the reaction.
(b) Use the gravimetric stoichiometric method to predict the mass of lead produced.

Figure 7
Zinc reacts with a solution of lead(II) nitrate.

Experimental Design

A known mass of zinc is placed in a beaker with an excess of lead(II) nitrate solution. The lead produced in the reaction is separated by filtration and dried. The mass of the lead is determined. Assume that the reagents used are pure and that the technical skills used in carrying out the experiment are adequate for the experiment. An excess of one reactant is used to ensure the complete reaction of the limiting reagent.

Evidence

In the beaker, crystals of a shiny black solid were produced and all of the zinc reacted.
mass of filter paper = 0.92 g
mass of dried filter paper plus lead = 7.60 g

Analysis

(c) Use the evidence to calculate the actual mass of lead produced.
(d) Calculate a percentage difference between the experimental value and the predicted (theoretical) value of the mass of lead. The percentage difference is calculated by dividing the difference in mass by the predicted mass.

Evaluation

(e) Why was an excess of lead(II) nitrate in solution used in this experiment?
(f) How can we tell when the lead(II) nitrate is in excess?
(g) Based on the evidence and your evaluation of the experimental design, is the gravimetric stoichiometric method valid?

Practice

Understanding Concepts

7. How are scientific concepts tested?
8. Explain why it is necessary to convert mass of reactant or product to amount in moles before applying the coefficients in balanced chemical equations.
9. When octane, C_8H_{18}, in gasoline burns, it combines with oxygen to produce carbon dioxide and water according to the following equation:

$$2\ C_8H_{18(g)} + 25\ O_{2(g)} \rightarrow 16\ CO_{2(g)} + 18\ H_2O_{(g)}$$

   (a) What is the mole ratio of octane:carbon dioxide?
   (b) What mass of carbon dioxide is produced if 22.8 g of octane is completely combusted in oxygen?
   (c) What is the mole ratio of octane:oxygen?
   (d) What mass of oxygen is required to completely combust 22.8 g of octane?
   (e) From your knowledge of the properties of different stoichiometric combinations of carbon and oxygen to produce oxides of carbon, explain why oxygen should not be the limiting reagent when burning octane.
10. Titanium is a metal that is used in a variety of products, including jewellery, cookware, and golf clubs. Titanium is purified from the mineral titanium dioxide, $TiO_{2(s)}$, in a two-step process:

$$TiO_{2(s)} + 2\ Cl_{2(g)} + 2\ C_{(s)} \rightarrow TiCl_{4(g)} + 2\ CO_{(g)}$$

$$TiCl_{4(g)} + 2\ Mg_{(s)} \rightarrow Ti_{(s)} + 2\ MgCl_{2(s)}$$

   How many grams of titanium can be obtained from 1.00 kg of titanium dioxide?

Answers

9. (a) 2:16
   (b) 70.2 g
   (c) 2:25
   (d) 79.8 g
10. 599 g

Section 5.4 Questions

Understanding Concepts

1. When using gravimetric stoichiometry, should you combine the two reactants in the same ratio by mass as they appear in the balanced chemical equation? Explain your reasoning.
2. Metallic iron can be obtained by heating iron ore, $Fe_2O_{3(s)}$, with hydrogen. The balanced equation for the reaction is given below:

   $$Fe_2O_{3(s)} + 3\ H_{2(g)} \rightarrow 2\ Fe_{(s)} + 3\ H_2O_{(g)}$$

   (a) What mass of iron is produced from 500 kg of iron ore?
   (b) What mass of hydrogen gas is required to convert 1.0 t of iron ore into iron?
   (c) If 220 kg of water is formed, what mass of iron ore was used up?
3. In the industrial synthesis of nylon, one of the starting materials, adipic acid, $C_6H_{10}O_{4(s)}$, is made by the oxidation of cyclohexane, $C_6H_{12(l)}$:

   $$2\ C_6H_{12(l)} + 5\ O_{2(g)} \rightarrow 2\ C_6H_{10}O_{4(s)} + 2\ H_2O_{(l)}$$

   (a) What mass of cyclohexane is required to produce 280 kg of adipic acid?
   (b) What mass of oxygen is required to react with 125 kg of cyclohexane?
   (c) If 200 g of oxygen is available, what is the maximum mass of cyclohexane that can be reacted?
4. Write a balanced equation for the reaction in which solid iron reacts with oxygen gas to produce iron(II) oxide, a solid. Calculate the mass of iron(II) oxide produced from 56.8 g of iron in excess oxygen.

Applying Inquiry Skills

5. When sodium oxalate is added to hard water, the calcium ions in the water react to form insoluble calcium oxalate, which is precipitated. The balanced equation is:

   $$Ca^{2+}{}_{(aq)} + Na_2C_2O_{4(aq)} \rightarrow CaC_2O_{4(s)} + 2\ Na^{+}{}_{(aq)}$$

   Complete the following sections of a lab report for the following investigation, using appropriate scientific vocabulary and SI units: Prediction, Experimental Design, Materials, and Procedure.

   Question

   What amount, in moles, of calcium ions are present in a sample of hard water?

   Prediction

   (a) Use gravimetric stoichiometry to make a Prediction.

   Experimental Design

   (b) Write the Experimental Design, including a description of variables. Select appropriate techniques for your investigation.
   (c) Include an explanation of how you will calculate the amount in moles of Ca^{2+} ions present in a sample of hard water, including correct SI units.

(continued)

Materials

(d) Select appropriate Materials.

Procedure

(e) Write a full Procedure. In your Procedure, include steps to experimentally determine the amount of excess reagent to use, and steps to determine the mass of precipitate formed.

Making Connections

6. Give examples of the applications of gravimetric stoichiometry that an industry (e.g., a manufacturer of aluminum cans) may use in
   (a) purchasing raw materials;
   (b) pricing its products;
   (c) predicting the amount of air pollutants that may be produced as a result of the manufacturing process;
   (d) predicting packaging, transportation, and waste removal needs.

5.5 Calculating Limiting and Excess Reagents

When reactants in a chemical reaction are mixed together, the quantities that react are determined by the mole ratios in a balanced equation for the reaction. It is important to know the minimum quantities of each reactant needed for a desired quantity of product. Generally, industrial processes and laboratory procedures are designed so that one of the reactants is used as the limiting reagent and the other reactants are used in slight excess.

Using gravimetric stoichiometry, we can now calculate the exact quantities of reactants needed for a reaction. It is also possible, therefore, to calculate the quantities needed for a reactant to be present in excess. As a general guideline, an excess of 10% should be sufficient to ensure that the limiting reagent is completely consumed.

Let's look at the reaction of methane, $CH_{4(g)}$, burning in sufficient oxygen to produce carbon dioxide and water according to the following equation:

$$CH_{4(g)} + 2\ O_{2(g)} \rightarrow CO_{2(g)} + 2\ H_2O_{(g)}$$

If we are burning 2.0 mol of $CH_{4(g)}$, how much $O_{2(g)}$ would be needed? The mole ratio in the balanced equation tells us that the ratio of CH_4:O_2 = 1:2. Therefore, 2.0 mol of $CH_{4(g)}$ would require 2.0 mol × 2 = 4.0 mol of $O_{2(g)}$.

To provide excess oxygen, we would add 10% excess (0.4 mol) and so provide 4.4 mol of oxygen.

Since we cannot directly measure moles, we can convert the amount in moles into mass. The molar mass of O_2 is 32.00 g/mol. Therefore, the mass of 4.4 mol of $O_{2(g)}$ is equal to 4.4 mol × 32.00 g/mol, which is 140.00 g of $O_{2(g)}$.

Sample Problem 1

In an experiment to test the stoichiometric method, 2.00 g of copper(II) sulfate solution is reacted with an excess of sodium hydroxide in aqueous solution. What mass of solid sodium hydroxide is reasonable to use?

Solution

$CuSO_{4(aq)}$	+	$2\ NaOH_{(aq)} \rightarrow Cu(OH)_{2(s)} + Na_2SO_{4(aq)}$
2.00 g		
159.61 g/mol		40.00 g/mol

$$n_{CuSO_4} = 2.00\ \not{g} \times \frac{1\ \text{mol}}{159.61\ \not{g}}$$
$$= 0.0125\ \text{mol}$$

$$n_{NaOH} = 0.0125\ \text{mol}\ \not{CuSO_4} \times \frac{2\ \text{mol NaOH}}{1\ \text{mol}\ \not{CuSO_4}}$$
$$= 0.0250\ \text{mol}$$

$$m_{NaOH} = 0.0250\ \not{\text{mol}} \times \frac{40.00\ \text{g}}{1\ \not{\text{mol}}}$$
$$= 1.00\ \text{g}$$

$$\text{excess NaOH}_{(aq)} = 1.00\ \text{g} + (10\% \times 1.00\ \text{g})$$
$$= 1.10\ \text{g}$$

In this experiment, 1.10 g of $NaOH_{(s)}$ would be a reasonable amount to use.

SUMMARY Calculating Excess Reagent

1. Determine the amount in moles of the limiting reagent.
2. Apply the mole ratios in the balanced equation to calculate the minimum amount in moles of other reagent(s).
3. Add 10% to the minimum amount in moles of excess reagent(s).
4. Convert the excess amount in moles to mass.

Practice

Understanding Concepts

1. Recall that in Chapter 4 we discussed the dangers of burning fossil fuels in insufficient oxygen. In addition to carbon dioxide, a small quantity of the much more toxic carbon monoxide is produced. The balanced equation for the combustion of methane to form carbon monoxide is given below:

   $$2\ CH_{4(g)} + 3\ O_{2(g)} \rightarrow 2\ CO_{(g)} + 4\ H_2O_{(g)}$$

   (a) Calculate the mass of methane, supplied in excess, that will react with 3.0 mol of oxygen gas to produce carbon monoxide.
   (b) Compare your answer in (a) to the mass of methane that will react with 3.0 mol of oxygen to produce carbon dioxide and water.
   (c) When a propane barbecue is used in a closed garage, which reactant in the combustion of propane is the limiting reagent?
2. Zinc and sulfur react to form zinc sulfide according to the following balanced equation:

   $$8\ Zn_{(s)} + S_{8(s)} \rightarrow 8\ ZnS_{(s)}$$

   If 5.00 g of zinc is to be completely reacted, what is a reasonable mass of sulfur to use in excess in the reaction?

Answers

1. (a) 32 g
   (b) 24 g
2. 2.70 g (2.45 g + 10% excess)

Answers

3. 1.3 g (1.2 g + 10% excess)
4. 2.43 g (2.21 g + 10% excess)

3. Sodium and chlorine gas react according to the following balanced equation:

$$2\ Na_{(s)} + Cl_{2(g)} \rightarrow 2\ NaCl_{(s)}$$

If 0.75 g of sodium is to be completely reacted, what is a reasonable mass of chlorine gas to use in the reaction?

4. Calcium hydroxide is used in some antacid tablets to react with stomach acid, which is mainly hydrochloric acid. Calcium chloride and water are produced. The balanced equation for the reaction is given below:

$$Ca(OH)_{2(s)} + 2\ HCl_{(aq)} \rightarrow CaCl_{2(aq)} + 2\ H_2O_{(l)}$$

If 2.17 g of hydrochloric acid need to be completely reacted, what is a reasonable mass of calcium hydroxide needed?

INQUIRY SKILLS

- ○ Questioning
- ○ Hypothesizing
- ● Predicting
- ● Planning
- ○ Conducting
- ○ Recording
- ● Analyzing
- ● Evaluating
- ● Communicating

Lab Exercise 5.5.1

Designing and Testing Gravimetric Stoichiometry: Calculating an Excess Reagent

In this lab exercise, you are given a reaction to use to test the method of gravimetric stoichiometry. Write a **Prediction** and the **Procedure**. Prepared evidence is provided for use in the **Analysis** and **Evaluation**.

Question

What mass of precipitate is formed when 2.98 g of sodium phosphate in solution reacts with an excess of aqueous calcium nitrate?

Prediction

(a) Calculate the mass of excess reagent that should be used in this investigation and predict the mass of product that will be formed.

Experimental Design

The given mass of sodium phosphate is dissolved in water and reacted with a calculated excess mass of calcium nitrate in solution. One of the products formed is calcium phosphate, which is insoluble in water. The other product formed is sodium nitrate, which remains in solution. The stoichiometric method is tested by comparing the predicted mass of precipitate formed with the actual mass obtained.

Evidence

mass of filter paper = 0.93 g
mass of dried filter paper plus precipitate = 3.82 g

Analysis

(b) Answer the Question: What mass of precipitate is formed when 2.98 g of sodium phosphate in solution reacts with an excess of aqueous calcium nitrate?

Evaluation

(c) What is the difference between the predicted and actual mass of precipitate formed?
(d) What are some possible causes for any difference in the predicted and actual mass of precipitate formed?
(e) Is there any observable test to determine if the amount of calcium nitrate used was in excess?
(f) Based on the evidence and your evaluation of the design, is the stoichiometric method an acceptable method for predicting masses of reactants and products?

Which Reagent Is Limiting?

Suppose we have, in a container, 2.5 mol of methane and 6.0 mol of oxygen gas. Is one of the reactants limiting? If so, which one? This information is important if we wish to know the amount of product that will be formed, and we can use gravimetric stoichiometry to find the answer.

The balanced equation for the combustion of methane to form carbon dioxide and water is given below:

$$CH_{4(g)} + 2\ O_{2(g)} \rightarrow CO_{2(g)} + 2\ H_2O_{(g)}$$

To determine which reactant is the limiting one, we can restate the question as: How many moles of oxygen does 2.5 mol of methane need? We then compare that amount to the amount we have and see if there is enough.

According to the mole ratio in the balanced equation, 2.5 mol of methane would need 5.0 mol of oxygen because the mole ratio of methane:oxygen is 1:2. Do we have at least 5.0 mol of oxygen? Yes, we do. In fact, we have 6.0 mol of oxygen, more than enough. Therefore, we can conclude that oxygen is not the limiting reagent; methane is (**Figure 1**).

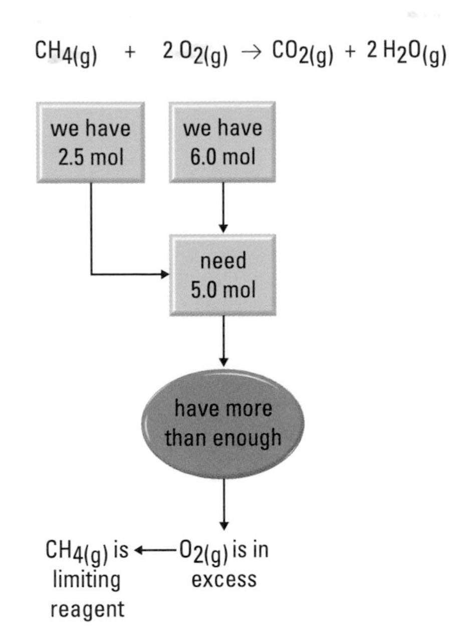

Figure 1
Steps showing limiting reagent

We could also have taken a similar approach in the other direction: we could have restated the question as: How many moles of methane does 6.0 mol of oxygen need? We would have come out with the same answer. Let's follow the calculations to confirm this.

According to the mole ratio in the balanced equation, 6.0 mol of oxygen would need 3.0 mol of methane because the mole ratio of methane:oxygen is 1:2. Do we have at least 3.0 mol of methane? No, we don't. We have only 2.5 mol of methane. This means that when all of the 2.5 mol of methane is used up, no more product can be formed. Only 5.0 mol of oxygen will have been used, and the remaining 1.0 mol of unused oxygen will not have any methane to react with. Therefore, methane is limiting the amount of product formed. We conclude, as before, that methane is the limiting reagent in this system (**Figure 2**).

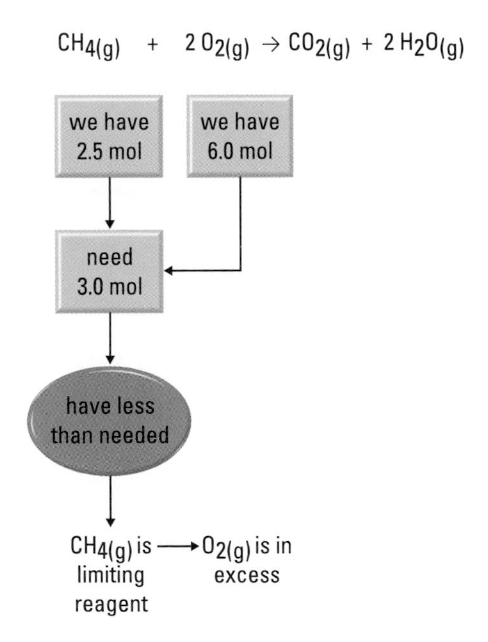

Figure 2
Steps showing limiting reagent

Once we have determined which reagent is limiting, we can use gravimetric stoichiometry to predict the amount of any product formed in the reaction. The amount of limiting reagent is *always* the deciding factor in the amount of products formed. The amount of the excess reagent includes some quantities that will be left over, unreacted, and therefore cannot be used to predict other quantities in a reaction.

For example, in the combustion of magnesium metal in oxygen gas, 6.73 g of magnesium and 8.15 g of oxygen gas are available for reaction. Identify the limiting reagent in this reaction.

$$2\ Mg_{(s)} + O_{2(g)} \rightarrow 2\ MgO_{(s)}$$

First, determine the mass of oxygen gas needed to react with 6.73 g of magnesium:

$$n_{Mg} = 6.73\ g \times \frac{1\ mol}{24.3\ g}$$

$$n_{Mg} = 0.277\ mol$$

mole ratio $Mg{:}O_2 = 2{:}1$

$$n_{O_2}\ \text{needed} = 0.277\ \cancel{mol\ Mg} \times \frac{1\ mol\ 0_2}{2\ \cancel{mol\ Mg}}$$

$$n_{O_2}\ \text{needed} = 0.138\ mol$$

$$n_{O_2}\ \text{available} = 8.15\ \cancel{g} \times \frac{1\ mol}{32.0\ \cancel{g}} = 0.255\ mol$$

More O_2 is available than is needed. Oxygen is in excess. The limiting reagent is magnesium.

Note that the same conclusion is reached if we had started with the amount in moles of oxygen instead of magnesium. Try it out for yourself.

Next, determine the mass of magnesium oxide formed.

The limiting reagent is magnesium.

$$n_{Mg} = 0.277\ mol$$

mole ratio of $Mg{:}MgO = 2{:}2$

$$n_{MgO} = n_{Mg(s)} = 0.277\ mol$$

$$m_{MgO} = 0.277\ \cancel{mol} \times \frac{40.3\ g}{1\ \cancel{mol}}$$

$$m_{MgO} = 11.2\ g$$

The mass of magnesium oxide formed is 11.2 g.

SUMMARY Determining the Limiting Reactant

1. Write a balanced equation for the reaction.
2. Select one of the reactants and calculate the amount in moles *available*.
3. Use mole ratios in the balanced equation to calculate the amount in moles *needed* of the *other* reactants.
4. Calculate the *available* amount in moles of the *other* reactants. If the available amount of a reactant is more than sufficient, it is in excess. If the available amount is insufficient, it is limiting.

Practice

Understanding Concepts

5. Why do we not use the amount of excess reagent to predict the amount of product formed in a reaction?
6. Complete balanced equations for the following reactions and identify the limiting and excess reagents for each of the pairs of reactants. How much excess is present in each case?
   (a) $Zn_{(s)}$ + $CuSO_{4(aq)}$ →
   0.42 mol 0.22 mol
   (b) $Cl_{2(aq)}$ + $NaI_{(aq)}$ →
   10 mmol 10 mmol
   (c) $AlCl_{3(aq)}$ + $NaOH_{(aq)}$ →
   20 g 19 g
7. In a reaction to produce zinc sulfide, 5.00 g of zinc and 3.00 g of sulfur, S_8, are available. The balanced equation is given below:

   $$8\,Zn_{(s)} + S_{8(s)} \rightarrow 8\,ZnS_{(s)}$$

   (a) Identify the limiting reagent.
   (b) Calculate the mass of zinc sulfide formed.
8. Propane, $C_3H_{8(g)}$, used in gas barbecues, burns in oxygen to produce carbon dioxide and water according to the following equation:

   $$C_3H_{8(g)} + 5\,O_{2(g)} \rightarrow 3\,CO_{2(g)} + 4\,H_2O_{(g)}$$

   In a reaction, 100.7 g of propane and 367.4 g of oxygen gas are available.
   (a) Identify the limiting reagent in this reaction.
   (b) Calculate the mass of water formed.
9. One of the reactions in the formation of acid rain is the synthesis reaction of sulfur dioxide, $SO_{2(g)}$, with $O_{2(g)}$ to form sulfur trioxide, $SO_{3(g)}$, as the only product.
   (a) Write a balanced equation for this reaction.
   (b) If 4.55 kg of sulfur dioxide and 2.88 kg of oxygen gas are available, what mass of sulfur trioxide will be produced?
10. When solid iron(III) oxide is heated in the presence of carbon monoxide gas, iron metal and carbon dioxide gas are produced.
    (a) Write a balanced equation for this reaction.
    (b) If 74.2 kg of iron(III) oxide and 40.3 kg of carbon monoxide are available, what is the mass of iron metal produced?

Applying Inquiry Skills

11. Design an experimental procedure to determine whether the excess reagent in a reaction has indeed been provided in excess in a reaction in which one of the products formed is a solid.

Answers

6. (a) 0.20 mol excess $Zn_{(s)}$
   (b) 5 mmol excess $Cl_{2(aq)}$
   (c) 0.025 mol excess $NaOH_{(aq)}$
7. (a) $Zn_{(s)}$
   (b) 7.45 g
8. (a) $C_3H_{8(g)}$
   (b) 164.6 g
9. (b) 5.69 kg
10. (b) 51.9 kg

INQUIRY SKILLS

- ○ Questioning
- ○ Hypothesizing
- ● Predicting
- ● Planning
- ● Conducting
- ● Recording
- ● Analyzing
- ● Evaluating
- ● Communicating

Investigation 5.5.1

Which Reagent Is Limiting and How Much Precipitate Is Formed?

The purpose of this investigation is to test gravimetric stoichiometry by predicting and determining the mass of precipitate produced by the reaction of aqueous strontium nitrate and aqueous copper(II) sulfate pentahydrate. This investigation is similar in design to Lab Exercise 5.5.1. However, you will design the experiment, including determining which reagent you use is in excess, and conduct it as well, obtaining your own evidence. Complete a report that includes a **Prediction**, the **Materials** list, **Procedure, Evidence, Analysis,** and **Evaluation** of the experiment.

Strontium chloride is moderately toxic. Copper(II) sulfate is a strong irritant and is toxic if ingested. Lab aprons and eye protection must be worn.

Question

What is the mass of precipitate produced by the reaction of 2.00 g of strontium chloride with 2.00 g of copper(II) sulfate pentahydrate in 75 mL of water?

Prediction

(a) Determine which reactant is the limiting reagent. Predict the mass of precipitate formed.

Experimental Design

Strontium chloride reacts with copper(II) sulfate pentahydrate to produce strontium sulfate as a precipitate. The other product formed is copper(II) chloride, which remains in solution.

(b) Write a procedure. Describe safety precautions and include disposal instructions, which you can obtain from your teacher. Have your procedure approved by your teacher before beginning.

Procedure

1. Carry out your procedure.

Analysis

(c) Analyze the evidence you obtained. What is the answer to the Question?

Evaluation

(d) Evaluate your prediction based on the evidence you obtained and on an evaluation of the experimental design and your procedure.

(e) Evaluate the stoichiometric method as used in the prediction of masses of reactants and products.

Practice

Applying Inquiry Skills

12. Aluminum metal reacts with copper(II) sulfate solution to produce copper metal and aluminum sulfate solution. The copper(II) sulfate solution is blue and the aluminum sulfate solution is colourless.

Describe any evidence that would indicate that
(a) aluminum metal was the limiting reagent;
(b) copper(II) sulfate solution was the limiting reagent.

13. When magnesium metal reacts with hydrochloric acid, aqueous magnesium chloride and hydrogen gas are produced. Describe any evidence that would indicate that
(a) magnesium was the limiting reagent;
(b) hydrochloric acid was the limiting reagent.

Section 5.5 Questions

Understanding Concepts

1. Why is the limiting reagent not necessarily the reagent available in the least amount by mass?
2. Why is the amount of limiting reagent—not the excess reagent—always used to predict the mass of products formed?
3. When testing stoichiometric relationships experimentally, why is it necessary to provide one reactant in excess?
4. A quick, inexpensive source of hydrogen gas is the reaction of zinc with hydrochloric acid. If 0.35 mol of zinc is placed in 0.60 mol of hydrochloric acid,
(a) which reactant will be completely consumed?
(b) what mass of the other reactant will remain after the reaction is complete?
5. A chemical technician is planning to react 3.50 g of lead(II) nitrate with excess potassium bromide in solution.
(a) What would be a reasonable mass of potassium bromide to use in this reaction?
(b) Predict the mass of precipitate expected.
6. In a reaction of copper metal with silver nitrate in solution, crystals of silver and copper(II) nitrate are formed according to the following equation:

$$Cu_{(s)} + 2\ AgNO_{3(aq)} \rightarrow 2\ Ag_{(s)} + Cu(NO_3)_{2(aq)}$$

In a reaction, 10.0 g of copper is placed in a solution containing 20.0 g of silver nitrate.
(a) Identify the limiting reagent.
(b) Predict the mass of silver crystals formed.
7. In an experiment, 26.8 g of iron(III) chloride in solution is combined with 21.5 g of sodium hydroxide in solution:

$$FeCl_{3(aq)} + 3\ NaOH_{(aq)} \rightarrow Fe(OH)_{3(s)} + 3\ NaCl_{(aq)}$$

(a) Which reactant is in excess and by how much?
(b) What mass of each product will be obtained?

Applying Inquiry Skills

8. If none of the products formed is a solid, is it still possible to determine experimentally whether sufficient excess reagent has been made available for the reaction to go to completion? Give examples to illustrate your answer.

5.6 The Yield of a Chemical Reaction

In all of the predictions of masses of reactants and products so far, we have based our calculations on the balanced equations for the chemical reactions. We must remember that the quantities that we calculate in our predictions are *theoretical* quantities. They are the quantities of products that *should* be produced, according to the numbers in the balanced equation. They are not necessarily the quantities that we *actually* get when we carry out the reaction, mixing the reactants and collecting the products.

yield: the amount of product that is obtained in a chemical reaction

theoretical yield: the amount of product that we predict will be obtained, calculated from the equation

actual yield: the amount of product that is actually obtained at the end of a procedure

The amount of product that is obtained in a chemical reaction is called the **yield**. It may be measured in mass or in moles. The amount of product that we predict we should get by stoichiometric calculations using the balanced equation is called the **theoretical yield**. When we carry out the reaction, whether it is in a school or in an industrial laboratory, the amount of product that is obtained and measured at the end of the procedure is called the **actual yield**.

Often the actual yield in a chemical reaction turns out to be less than the theoretical yield. Theoretically, each and every atom, ion, and molecule proceeds through the reaction according to the balanced equation and is accounted for and measured in one of the products. In actual practice, there are several reasons why all of the materials do not end up in the collected product. The most common loss of product is as a result of experimental procedures, such as in transferring solutions, filtering precipitates, and splattering during heating. These losses can be reduced by improving technical skills or the equipment used, or by reducing the number of steps in the experimental design.

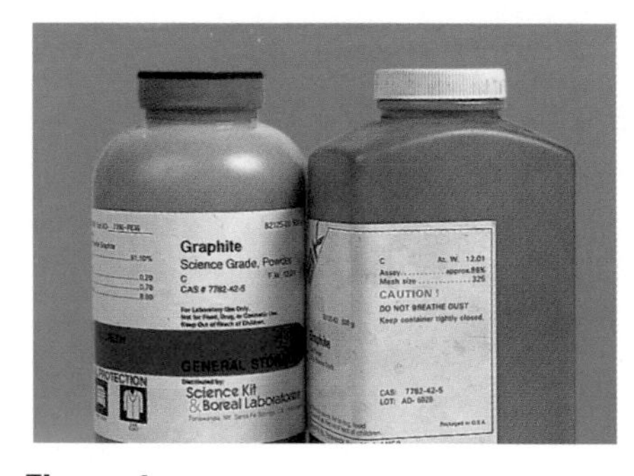

Figure 1
Chemicals come in a wide variety of grades (purities). The purity of a chemical can significantly affect experimental results in a chemical reaction.

The poor yield could also be due to impurities in the reagents used. Chemicals come in a wide variety of grades, or purities (**Figure 1**). Some low-purity or technical grades may be only 80% to 90% pure; if this factor is not accounted for in the amount of reactant used, the actual yield will again be less than the theoretical yield. Impurities may also be a result of other processes. For example, metals such as magnesium readily react with air to form a layer of metal oxide on the surface. These impurities are included in the mass of reactants but do not proceed to form the products collected, thus causing the actual yield to differ from the theoretical yield.

Another cause of low yield is a side reaction, forming other than the desired product. When magnesium ribbon is heated in a crucible, it reacts with oxygen in the air to form magnesium oxide. However, air contains a mixture of gases, including nitrogen. A side reaction may occur in which some of the magnesium reacts with nitrogen to form magnesium nitride. If the product collected is presumed to be magnesium oxide, but in fact also contains the nitride, the actual yield will be different from the theoretical yield. To correct this discrepancy, additional steps can be introduced to convert all products to the desired product.

In yet other cases, the conditions may not be ideal for the reaction to go to completion. This happens when, as more and more products are formed, the reverse reaction occurs, in which the products of the reaction become the reactants of the reverse reaction. A certain amount of the products are thus being used up at the same time that they are being produced. In such cases, the actual yield is the amount of product present and is always less than the theoretical yield. To minimize these losses, the conditions for the reaction may need to be changed to allow the reaction to go to completion.

percentage yield: the ratio, expressed as a percentage, of the actual or experimental quantity of product obtained (actual yield) to the maximum possible quantity of product (theoretical yield) derived from a gravimetric stoichiometry calculation

A comparison of the actual yield and the theoretical yield gives an indication of the *efficiency* of a chemical reaction. We can calculate this efficiency as a **percentage yield**, by dividing the actual yield by the theoretical yield:

$$\text{percentage yield} = \frac{\text{actual yield}}{\text{theoretical yield}} \times 100\%$$

For example, for a certain reaction, if the theoretical yield is 10.0 kg and the actual yield is 9.0 kg, the percentage yield would be 90%:

$$\text{percentage yield} = \frac{9.0 \text{ kg}}{10.0 \text{ kg}} \times 100\% = 90\%$$

Sample Problem

Arsenates, used in some pesticides, are compounds of arsenic. The most common ore of arsenic, $FeSAs_{(s)}$, can be heated to produce arsenic according to the following equation:

$$FeSAs_{(s)} \rightarrow FeS_{(s)} + As_{(s)}$$

When 250 kg of the ore was processed industrially, 95.3 kg of arsenic was obtained. Calculate the percentage yield of arsenic in the process.

Solution

mole ratio FeSAs:As = 1:1

actual yield of $As_{(s)}$ = 95.3 kg

$$n_{FeSAs} = 250\ 000 \text{ g} \times \frac{1 \text{ mol}}{162.83 \text{ g}}$$
$$= 1535 \text{ mol}$$
$$n_{As} = 1535 \text{ mol}$$
$$m_{As} = 1535 \text{ mol} \times \frac{74.92 \text{ g}}{1 \text{ mol}}$$
$$m_{As} = 115.0 \text{ kg}$$

$$\text{percentage yield} = \frac{95.3 \text{ kg}}{115.0 \text{ kg}} \times 100\% = 82.8\%$$

The percentage yield of arsenic in the process was 82.8%.

Practice

Understanding Concepts

1. Describe the distinction between the terms *actual yield* and *theoretical yield.*
2. Can the actual yield ever be greater than the theoretical yield? Explain.
3. In an experiment, 5.00 g of silver nitrate is added to a solution containing an excess of sodium bromide. It was found that 5.03 g of silver bromide is obtained.
   (a) Write a balanced equation for the reaction.
   (b) What is the theoretical yield of silver bromide?
   (c) What is the actual yield of silver bromide in the experiment?
   (d) What is the percentage yield of the experiment?
4. In an experiment, when 16.1 g of FeS reacted with 10.8 g of O_2, 14.1 g of $Fe_2O_{3(g)}$ was produced. The balanced equation for the reaction is given below:
   $$4\ FeS_{(s)} + 7\ O_{2(g)} \rightarrow 2\ Fe_2O_{3(s)} + 4\ SO_{2(g)}$$
   (a) Identify the limiting reagent in the experiment.

Answers

3. (b) 5.53 g
   (c) 5.03 g
   (d) 91.0%
4. (a) $FeS_{(s)}$

Answers

4. (b) 14.6 g
   (c) 96.6%
5. 90.8%
6. 74.9%

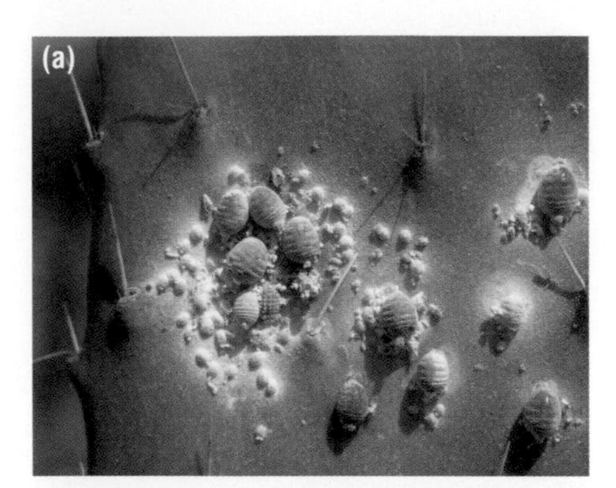

(b) Calculate the theoretical yield.
(c) Calculate the percentage yield of $Fe_2O_{3(s)}$ in the experiment.

5. Iron is produced from its ore, hematite, $Fe_2O_{3(s)}$, by heating with carbon monoxide in a blast furnace. If the industrial process produced 635 kg of iron from 1000 kg of hematite, what is the percentage yield of iron in the process? The equation for the reaction is given below:

$$Fe_2O_{3(s)} + 3\ CO_{(g)} \rightarrow 2\ Fe_{(s)} + 3\ CO_{2(g)}$$

6. Methyl salicylate, $C_8H_8O_{3(l)}$, is the chemical responsible for the wintergreen flavouring. It can be prepared by heating salicylic acid, $C_7H_6O_{3(s)}$, with methanol, $CH_3OH_{(l)}$, according to the equation below:

$$C_7H_6O_{3(s)} + CH_3OH_{(l)} \rightarrow C_8H_8O_{3(l)} + H_2O_{(l)}$$

If 2.00 g of salicylic acid is reacted with excess methanol, and the yield of oil of wintergreen is 1.65 g, what is the percentage yield?

Applying Inquiry Skills

7. In an experiment to recover a precipitate that was formed in a chemical reaction, a chemistry student followed this procedure:
   - The mass of a reactant is determined using weighing paper on a centigram balance.
   - The reactant is transferred to a large beaker.
   - A second aqueous reactant, used in excess, is measured using a graduated cylinder and added to the beaker.
   - The mixture is stirred and heated to dryness in an evaporating dish over a laboratory burner.
   - The precipitate is transferred from the evaporating dish to weighing paper and the mass is determined.

   Suggest ways in which the procedure can be modified to improve the yield.

Making Connections

8. When you drink a beverage or eat a candy artificially coloured with red food dye, you may be ingesting a chemical that was produced by a tiny red insect and extracted by a process developed by two Canadian chemists. Several synthetic red dyes have been found to be carcinogenic (cancer-causing), but a vivid red dye called carmine, which is made from the cochineal insect (**Figure 2**), has been approved for use in foods, drugs, and cosmetics. The method of production has been improved to increase the purity and the yield of the product. Use the Internet to research the following and summarize your findings in a one-page report:
   (a) the modifications in the procedure for producing the dye carmine from the insects, thus increasing yield;
   (b) the effect that the industrial production of carmine has had on the people of Peru, where the insects are found.

   Follow the links for Nelson Chemistry 11, 5.6.

GO TO www.science.nelson.com

Figure 2
(a) The cochineal lives in prickly pear cacti in the high desert plains of the Peruvian Andes.
(b) (c) A vivid red dye is made of the dried and crushed bodies of female cochineal insects.

Investigation 5.6.1

Determining Percentage Yield in a Chemical Reaction

INQUIRY SKILLS

- ○ Questioning
- ○ Hypothesizing
- ● Predicting
- ○ Planning
- ● Conducting
- ● Recording
- ● Analyzing
- ● Evaluating
- ● Communicating

In this investigation, mass relationships in a chemical reaction will be studied by performing an experiment, collecting and recording evidence, and analyzing the evidence. The actual yield, theoretical yield, and percentage yield of copper in a single displacement reaction are determined and the experimental procedure evaluated.

Write a report to present the **Prediction, Evidence, Analysis,** and **Evaluation**.

Copper chloride dihydrate is toxic and must not be ingested.

Care must be taken when handling hot equipment.

Eye protection and lab aprons must be worn.

Question

What mass of copper is formed when excess aluminum is reacted with a given mass of copper(II) chloride dihydrate?

Prediction

(a) Calculate the theoretical yield for this experiment.

Experimental Design

A known mass of a copper salt is dissolved in water and is reacted with an excess of aluminum. The mass of copper formed in the reaction is determined. Percentage yield is calculated.

The reaction is represented by this balanced equation:

$$3\ CuCl_2{\cdot}2H_2O_{(aq)} + 2\ Al_{(s)} \rightarrow 3\ Cu_{(s)} + 2\ AlCl_{3(aq)} + 6\ H_2O_{(l)}$$

Materials

eye protection
aluminum foil, 8 cm × 8 cm
copper(II) chloride dihydrate, 2.00 g
two 150-mL beakers
50-mL graduated cylinder
stirring rod
ruler
forceps
hot plate
ring stand
iron ring
wire gauze
watch glass
crucible tongs
centigram balance

Procedure

1. Measure a mass of 2.00 g of the copper salt, to 0.01 g, and dissolve in 50 mL of water in a beaker.
2. Fold the aluminum foil lengthwise to make a strip 2 cm × 8 cm. Coil the strip loosely to fit into the copper chloride solution in the beaker, making sure that the strip is entirely immersed.
3. Heat the beaker *gently* on the hot plate for 5 min or longer, until all blue colour in the solution has disappeared. Continue heating gently for another 5 min. Allow to cool.
4. Use the forceps to shake loose all copper formed on the aluminum foil. Carefully transfer the copper to a weighed beaker and rinse the copper with water.
5. Pour off as much of the rinse water as possible. Spread the copper on the bottom of the beaker.
6. Cover the beaker containing the wet copper with a watch glass and gently heat the beaker to drive off the water. Reduce heat if the copper begins to turn black.
7. When the copper is dry, determine the mass of copper.
8. Wash hands thoroughly after the experiment, and follow your teacher's instructions for disposal of the waste materials.

Analysis

(b) Answer the Question.
(c) Identify the limiting reagent and the excess reagent in this reaction. What visible evidence is there to confirm your identification?
(d) Determine the actual yield of copper.
(e) Determine the percentage yield of copper in this experiment.

Evaluation

(f) If your percentage yield is less than 100%, suggest specific techniques or equipment that may account for the loss of product.

Synthesis

(g) Suppose that the percentage yield was greater than 100%. Suggest one or more specific factors in this experiment that may account for this.
(h) What steps did you take to ensure that the reaction went to completion?
(i) If you wanted to use aluminum as the limiting reagent, what changes in the procedure would be needed? What visible evidence would you look for to ensure that the reaction had gone to completion?

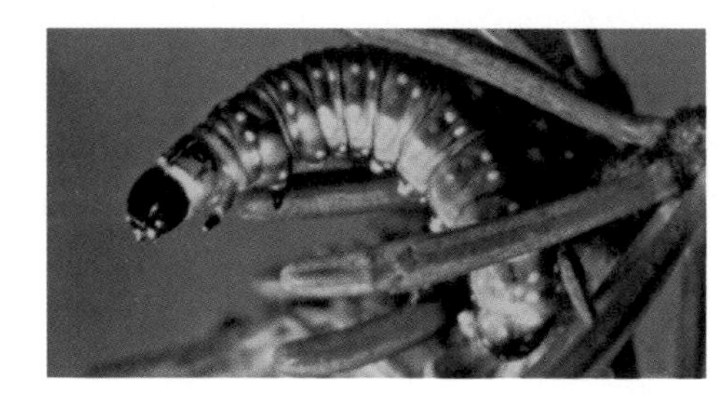

Figure 3
The spruce budworm survives harsh Canadian winters by producing an antifreeze protein that lowers the freezing point of its body fluids. Scientists are interested in the possible use of these antifreeze proteins to help preserve organs at low temperatures prior to transplant operations. To increase the yield of these antifreeze proteins, biochemists have isolated the spruce budworm gene responsible for making the proteins and incorporated it into bacteria. Since bacteria grow rapidly, they are ideal "factories" for mass production of the selected proteins.

Yield in Industrial Chemical Reactions

In industrial applications, it is important for manufacturers of a product to achieve as close to 100% yield as possible (**Figure 3**). The chemists or engineers must first determine the percentage yield of the operation and analyze all aspects of the process to look for ways of improving efficiency. This may require changing the conditions of the reaction, such as the temperature or pressure, which would in turn change the final equilibrium conditions of the reaction. It may require changing some steps in the procedure to reduce loss due to inadequate equipment or poor technique. The final decision will rest on an analysis of the costs of the changes required to improve the yield as well as the increased profit from an improved yield. The decision may also depend on whether byproducts or unused reactants can be recycled or used in other profitable processes.

A model case of striving for maximum yield is that of the synthesis of ibuprofen, an analgesic (painkiller) sold under several brand names, such as Motrin and Advil (**Figure 4**). The company that manufactures ibuprofen, BHC Company, researched and refined its chemical process to produce a more efficient synthesis, creating less waste and fewer byproducts.

The traditional industrial synthesis of ibuprofen was developed in the 1960s and involved a six-step process that resulted in large quantities of unwanted chemicals that needed to be disposed of. Even if the percentage yield was at an acceptable level, the yield of products from raw materials was low due to the reactions used in the experimental design. In this process, dubbed the "brown" process, 40% of the total atoms present in the reactants were recovered in the desired product.

In 1991, a new three-step process was implemented that dramatically reduced the quantities of waste chemicals produced and increased to 77% the recovery of atoms from reactants to desired product. The shorter "green" process also offered the advantage of producing larger quantities of ibuprofen in less time and with less capital expenditure. These improvements not only increased profitability for the company, but benefited the environment also by reducing the need to dispose of millions of kilograms of waste materials.

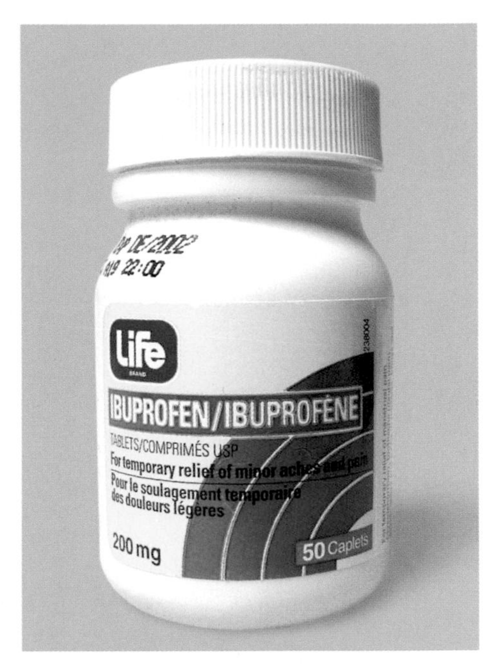

Figure 4
The analgesic ibuprofen is part of a booming pharmaceutical business with estimated sales of U.S. $124.6 billion in 1998.

Practice

Understanding Concepts

9. What are some factors that may contribute to less than 100% percentage yield in a chemical reaction?
10. One of the reactions used in the smelting of copper ores to produce copper involves reacting copper(I) oxide with copper(I) sulfide. The balanced equation for the reaction is given below:

$$2\ Cu_2O_{(s)} + Cu_2S_{(s)} \rightarrow 6\ Cu_{(s)} + SO_{2(g)}$$

When 250 kg of copper(I) oxide is heated with 129 kg of copper(I) sulfide, 285 kg of copper is recovered.
(a) Determine the limiting reagent.
(b) Calculate the theoretical yield of copper.
(c) Determine the percentage yield of copper.

11. The carbon in coal can be converted into methane, $CH_{4(g)}$, by first heating the coal powder with steam and oxygen, followed by heating with carbon monoxide and hydrogen. The overall process is summarized below:

$$C_{(s)} + 2\ H_{2(g)} \rightarrow CH_{4(g)}$$

(a) When 10.0 kg of coal is used in the process, 4.20 kg of methane is produced. What is the percentage yield of methane, assuming the coal is pure carbon?
(b) Further analysis shows that the coal contains only 40.0% carbon by mass. Recalculate the percentage yield of methane, taking into account the purity of the coal.

Making Connections

12. Maximizing percentage yield is not the only factor to consider in designing a chemical process. A reaction must be assessed for its efficiency, potential effect on the environment, and many other factors. For each of the following factors, select the "greener" option and give reasons for your selection.

Answers

10. (a) Cu_2S
(b) 309 kg
(c) 92.2%

11. (a) 31.4%
(b) 78.6%

(a) a reaction that requires the use of an organic solvent versus one that uses water as a solvent
(b) a reaction that takes place at high temperature versus one that takes place at room temperature
(c) a reaction that requires the product to be dried versus one that does not require a drying agent
(d) a reaction that requires the product to be purified versus one that requires no purification of the product
(e) a reaction that uses starting material derived from crude oil versus one that uses material derived from plant or animal matter

Section 5.6 Questions

Understanding Concepts

1. What are some ways of improving percentage yield in
   (a) a school chemistry experiment?
   (b) an industrial chemical process?
2. Acetylsalicylic acid (ASA), $C_9H_8O_{4(s)}$, is the chemical name for an analgesic whose common name is Aspirin. It is manufactured by heating salicylic acid, $C_7H_6O_{3(s)}$, with acetic anhydride, $C_4H_6O_{3(s)}$, according to the equation below:

$$C_7H_6O_{3(s)} + C_4H_6O_{3(s)} \rightarrow C_9H_8O_{4(s)} + C_2H_4O_{2(s)}$$

   (a) If 2.00 g of salicylic acid is heated with 4.00 g of acetic anhydride, what is the theoretical yield?
   (b) If the actual yield is 2.09 g, what is the percentage yield?

Applying Inquiry Skills

3. In this lab exercise, iron(III) silicate, $Fe_2(SiO_3)_{3(s)}$, is to be synthesized in a chemical reaction, in a simulation of an industrial process. Iron(III) silicate is produced as a yellow-orange precipitate in the reaction of sodium silicate and excess iron(III) nitrate.
   (a) Design an experiment to determine the actual yield of iron(III) silicate using this reaction. Write a report that includes the procedure, safety procedures, and an evaluation of the chemical process used in the synthesis.
   (b) Assume that the percentage yield in the experiment is 80.0%. Evaluate the experimental process with regard to maximizing yield.

Making Connections

4. The efforts of the BHC Company in improving its process in the manufacture of ibuprofen were recognized and the company received several awards. Industrial designs of products and processes that are efficient and benign to the environment and to human health are referred to as "green" chemistry. Using the Internet, research other "green" chemistry projects and briefly summarize the major features of one such project.
   Follow the links for Nelson Chemistry 11, 5.6.

GO TO www.science.nelson.com

5. Chemical engineering is a discipline that was developed when the discovery of oil led to the need for engineers who understood its chemistry and who could convert the chemicals from oil into

useful products. Chemical engineers now work in many areas of science, especially in industry. They design and develop industrial processes to make different consumer goods; they scale up laboratory experiments into industrial-size operations; they analyze data using electronics and computers. They are also responsible for the design of an efficient and safe chemical plant that protects the natural environment. Overall, a chemical engineer is a versatile problem-solver.

Research some specific fields in chemical engineering and the university courses that lead to a degree in chemical engineering. If possible, interview a chemical engineer to learn more about a career in this field. Prepare a report on your findings.

Follow the links for Nelson Chemistry 11, 5.6

GO TO www.science.nelson.com

5.7 Chemistry in Technology

What is the difference between science and technology? Are they the same thing? How are they related? Science is the study of the natural world to describe, predict, and explain changes and substances; technology encompasses the skills, processes, and equipment required to make useful products or to perform useful tasks. Sometimes technology is a practical application of science. However, often a certain technology existed long before the scientific principles behind it were understood. In this section we will look at the technology of glassmaking and how a scientific understanding of chemical reactions led to the industrial production of one of its key ingredients.

Glass is formed by heating a mixture of sand, sodium carbonate (soda), and calcium oxide (lime) to a high temperature—1425°C to 1600°C, depending upon the exact composition—at which point the mixture takes on a molten (liquid) state. The sand provides silicon dioxide (silica), which makes up the largest percentage of the mixture. Sodium carbonate lowers the temperature at which the sand will melt, and calcium oxide makes the glass strong and water-resistant. The percentage composition of this glass mixture determines the properties of the resultant glass. When cooled, the molten glass mixture becomes a "supercooled" liquid that retains its liquid shape as a solid. Glass making is known to have begun in Egypt about 5000 years ago, long before chemical formulas and reactions were understood, a case of technology preceding scientific knowledge. In Canada, glass making dates back to the 1800s, with the production of mostly industrial glassware, such as lantern and streetlight globes, lenses for railway and ship lanterns, and telephone line insulators.

The advances in atomic theory and the development of systematic quantitative analysis in the early 1800s brought with them an understanding of chemical formulas and equations. The application of chemical analysis and calculations led to improved properties of glass as well as the establishment of industrial chemical processes (**Figure 1**).

The LeBlanc Process

At the beginning of this chapter we talked about the LeBlanc process used during the 1700s and 1800s to produce sodium carbonate. Not only was the process inefficient but it was harmful also to the environment. The LeBlanc process begins with the reaction of salt and sulfuric acid, releasing hydrogen chloride:

DID YOU KNOW ?

Science in Glass Making

The development of systematic quantitative chemical analysis in the early 19th century, followed by chemical formulas and equations, contributed a great deal to the large-scale industrial supply of raw materials such as soda ash, used in glass making. In 1830, the French chemist Jean-Baptiste-André Dumas (1800–1884) showed that soda-lime-silica glass was most durable when the mass ratio of the three was 1:1:6, a ratio that is still used in modern soda-lime-silica glass.

Figure 1
Glass-making technology has changed greatly in the last two centuries.

$$NaCl_{(s)} + H_2SO_{4(l)} \xrightarrow{\text{heat}} NaHSO_{4(s)} + HCl_{(g)}$$

The residue that remains is mostly a mixture of the sodium hydrogen sulfate and any unreacted sodium chloride. Further heating produces sodium sulfate and more hydrogen chloride gas:

$$NaHSO_{4(s)} + NaCl_{(s)} \xrightarrow{\text{heat}} Na_2SO_{4(s)} + HCl_{(g)}$$

The sodium sulfate is then mixed with limestone ($CaCO_3$) and coal and heated to a high temperature. Complex changes occur, and the net result is summarized in the equation below:

$$Na_2SO_{4(s)} + CaCO_{3(s)} + 2C_{(s)} \xrightarrow{\text{heat}} Na_2CO_{3(s)} + CaS_{(s)} + 2\ CO_{2(g)}$$

The desired sodium carbonate is extracted from this mixture by leaching with warm water, followed by filtration and evaporation.

Let's take a look at the chemical reactions and conditions needed in the LeBlanc process and identify some of its problems. Hydrogen chloride, an air pollutant that has a devastating effect on the environment, is generated as a product in two of the reactions. Each step in the process requires heat; the cost of fuel used in the heating is considerable. In addition to fuel, a reactant in the final step is coal, making the process unhealthy and dirty. The residue that is left at the end of the process is an insoluble mixture that has no commercial use or value. Overall, the LeBlanc process was an expensive and environmentally harmful procedure, although at the time, the pollution was not considered a problem because this "dirty" process provided a plentiful source of alkali for soap-making. Personal hygiene became more affordable, resulting in a considerable positive impact on public health in Europe.

It took further pure scientific research to provide the knowledge necessary to develop a new technology capable of improving the yield and addressing the environmental concerns in this industrial process.

Practice

Understanding Concepts

1. Using the equations of the chemical reactions in the LeBlanc process, calculate (in moles) the amount of the air pollutant hydrogen chloride released for each mole of sodium carbonate produced.
2. Explain why the LeBlanc process was expensive.
3. Suggest reasons why air pollutants were not a cause for concern in the 17th and 18th centuries, but are a major concern now.

Answer

1. 2 mol

The Solvay Process

In 1865 Ernest Solvay (**Figure 2**), a Belgian chemist, began to perfect a new process, the Solvay process, for the production of sodium carbonate, and in 1867 Solvay's process was installed for the first time in his small factory in Belgium.

Since the LeBlanc process was so firmly established, the new Solvay process did not gain immediate acceptance. However, the cost of the new process was one-third the cost of the old LeBlanc process, so Solvay processing plants were eventually built in every major industrialized country. The wide use of his invention brought Solvay a great deal of money, much of which he channelled into charitable work in Brussels.

The overall, or net, reaction in the Solvay process, involving calcium carbonate and sodium chloride, is one that does not occur spontaneously at room temperature:

$$CaCO_{3(s)} + 2\ NaCl_{(aq)} \rightarrow Na_2CO_{3(aq)} + CaCl_{2(aq)}$$

Imagine adding chalk to a salt solution—no reaction occurs. How then can this reaction be implemented industrially to produce large quantities of sodium carbonate? Solvay's design involved an indirect route with a series of intermediate reactions. His major breakthrough includes a reaction between ammonium hydrogen carbonate and sodium chloride that at first glance seems improbable:

$$NH_4HCO_{3(aq)} + NaCl_{(aq)} \rightarrow NH_4Cl_{(aq)} + NaHCO_{3(s)}$$

What Solvay discovered by experimentation is that in cold water ammonium chloride has a higher solubility than sodium hydrogen carbonate. As a result, sodium hydrogen carbonate can be separated out of the solution by crystallization. This separation by solubility allows the $NaHCO_{3(s)}$ to be separated and sold as baking soda or to be decomposed into $Na_2CO_{3(s)}$ as washing soda.

The ingenuity of Solvay's design becomes apparent when you write out the reactions and see that all of the intermediate products are recycled as reactants in other reactions. Nothing is left as a byproduct except calcium chloride, which today is sold as road salt and as a drying agent (desiccant). See for yourself in the following questions.

Figure 2
Ernest Solvay (1838–1922) solved the practical problems of large-scale commercial production of sodium carbonate with his invention of the Solvay carbonating tower. In it an ammonia–salt solution could be mixed with carbon dioxide. He also founded several scientific institutes and conferences which, in turn, contributed to the development of theories on quantum mechanics and atomic structure.

Practice

Understanding Concepts

4. Write and balance the reaction equations for the Solvay process from the word equations below:
   (a) Limestone, $CaCO_{3(s)}$, is decomposed by heat to form calcium oxide (lime) and carbon dioxide.
   (b) Carbon dioxide reacts with aqueous ammonia and water to form aqueous ammonium hydrogen carbonate.
   (c) In the same vessel, the aqueous ammonium hydrogen carbonate reacts with brine, $NaCl_{(aq)}$, to produce aqueous ammonium chloride and solid baking soda, $NaHCO_{3(s)}$.
   (d) Heating the separated baking soda decomposes it into solid washing soda, water vapour, and carbon dioxide.
   (e) The first of two recycling reactions involves the reaction of lime, $CaO_{(s)}$, with water to produce slaked lime, $Ca(OH)_{2(s)}$.
   (f) Next, the slaked lime is added to the aqueous ammonium chloride (an intermediate product) to produce ammonia, aqueous calcium chloride, and water.
   (g) Write the net (overall) reaction for the Solvay process.
5. *Intermediate products* are produced partway through a process and become reactants in a later reaction. Cross out all intermediate products in the reaction equations you have written for the Solvay process. Don't be concerned about quantities of reactants and products. What you have left should combine to give you the net unbalanced reaction for the Solvay process.
6. *Raw materials* are the materials that are consumed in the net reaction. What are the raw materials for the Solvay process? Where are

these raw materials obtained? What makes these materials suitable for a large-scale chemical process?

7. The *primary products* and the *byproducts* of a chemical process depend on how marketable the products are. What are the primary products and byproducts of the Solvay process?
8. What intermediate product in the Solvay process is highly marketable? What are some consequences of removing this intermediate from the system of reactions?
9. Resources other than chemical and technological resources are required for most chemical processes. What additional natural resources are needed for the Solvay process?
10. Large quantities are used in newer processing plants that are built on what is called a "world scale" to produce quantities for international distribution. Why do you think chemical plants are being built on an increasingly large scale?

The Cost of Technology

As we said at the beginning of this section, technology encompasses the skills, processes, and equipment required to make useful products or to perform useful tasks. When technology is applied in industry, these skills, processes, and equipment pose considerable cost. Since the underlying objective of any industry is to make a profit, the most inexpensive process is naturally preferred.

However, the most inexpensive process is not necessarily the most desirable environmentally. We learned that the LeBlanc process was costly not only in terms of the cost of raw materials and fuel, but also in terms of the damage to the environment and that, even though the newer Solvay process was more efficient economically and more environmentally friendly, it took many years before it was adopted. What if the Solvay process had been more expensive? Would it still have replaced the LeBlanc process? If becoming more environmentally friendly means making less profit, the loss of profit will likely be passed on to the consumer in the form of higher prices. Are we willing to pay the cost?

We do indeed find these choices presented to us on store shelves. Paper products, for example, are made from wood pulp that must be bleached with chlorine or compounds containing chlorine, all of which are harmful to the environment. We can buy unbleached paper products, such as brown coffee filters, but we may be surprised to find that they cost more than the bleached variety. For other items, such as paper towels or envelopes, we may pay more for the products made from recycled materials than for those made from new materials. At the service station, we may choose to fill up our cars with gasohol, a blend of gasoline and alcohol. The alcohol is made from organic materials such as corn, so the need for nonrenewable fossil fuels is reduced. The combustion of alcohol also produces smaller amounts of pollutants such as carbon monoxide and sulfur oxides. You may find, however, that the cost of gasohol is slightly higher than that of gasoline in the price per litre; gasohol also has a lower fuel efficiency.

There are several reasons why these choices cost more. A major obstacle in the way of changing technologies is the large start-up costs of building new plants, buying new equipment, and restructuring existing processes. Other reasons may be the low consumer demand for the "green" product, limiting production to quantities that are less than optimal in efficiency. Industries may be willing or may be required to develop and implement environmentally friendlier technology. Are the higher prices charged to consumers justified? Should we as consumers share the costs of these new technologies?

Consider the following points:

- Without consumers buying their products in sufficient quantities, industries would not be able to stay in business. The consumer ultimately determines what products are marketed and should be willing to share the cost of better technology.
- If industries are regulated to adhere to environmental standards, the costs of which are not recovered in increased price, they may move to other countries where restrictions are less stringent. This results in no benefit to the global environment and a loss of jobs to other countries.
- It is the role of governments and industries to ensure that the best technology is used to protect the environment. For the consumer, paying for essentials such as food and shelter is more crucial than supporting industries that are making large profits.
- Rather than developing expensive new technologies, we should change our consumer attitudes and reduce the quantities of harmful substances used. For example, instead of driving, we should walk to school or shopping malls wherever possible; instead of disposable paper towels, we should use cloth towels.

Explore an Issue

Take a Stand: Should We Be Willing to Pay for Better Technology?

Who should pay the costs of improved technology?

Working in small groups, choose one industry (e.g., pulp and paper, nickel smelting, gasohol) and research different companies in that industry to see what efforts have been made to develop environmentally friendly solutions. Your research should include the following components:

- the reason for changing technologies
- the benefits of the new technology
- the costs involved in changing to the new technology
- the consumer demand or support for the new technology

(a) Choose one product made by the industry you researched and take a stand on whether we should be willing to pay the cost for improved technology. Explain your stand on the issue in a pamphlet that could be distributed to other consumers.

DECISION-MAKING SKILLS

- ● Define the Issue
- ● Identify Alternatives
- ● Research
- ● Analyze the Issue
- ● Defend a Decision
- ○ Evaluate

Practice

Making Connections

11. List three reasons why industries are reluctant to change technologies even if new processes offer the possibility of economic or other benefits.
12. In what way do consumers have input in determining the chemical processes used by industry?

13. Using the Internet, identify three specific examples of the application of chemical quantities and calculations in an industrial process, and discuss their importance (e.g., percentage composition in manufacture of gasohol or glass).

    Follow the links for Nelson Chemistry 11, 5.7.

GO TO www.science.nelson.com

Section 5.7 Questions

Understanding Concepts

1. What is the difference between science and technology?
2. Classify each of the following as a scientific or technological question:
   (a) What coating on a nail can prevent corrosion?
   (b) Which chemical reactions are involved in the corrosion of iron?
   (c) What is the accepted explanation for the chemical formula of water?
   (d) What process produces nylon thread continuously?
   (e) Why is a copper(II) sulfate solution blue?
   (f) How can automobiles be redesigned to achieve safer operation?
3. What factors spurred the development of the LeBlanc process for producing sodium carbonate?
4. Why was the LeBlanc process costly, both economically and environmentally?
5. The Solvay process was ingenious because all of the primary products and byproducts are marketable. Discuss three of these products or byproducts and describe their uses.

Chapter 5 Summary

Key Expectations

Throughout this chapter, you have had the opportunity to do the following:

- Identify everyday situations and work-related contexts in which analysis of unknown substances is important. (5.1)
- Identify and describe science- and technology-based careers related to the subject area under study. (5.1)
- Use appropriate scientific vocabulary to communicate ideas related to chemical calculations. (5.1, 5.4, 5.5, 5.6)
- Balance chemical equations by inspection. (5.2)
- Balance simple nuclear equations. (5.3)
- State the quantitative relationships expressed in a chemical equation. (5.4)
- Calculate, for any given reactant or product in a chemical equation, the corresponding mass or quantity in moles or molecules of any other reactant or product. (5.4)
- Solve problems involving percentage yield and limiting reagents. (5.5, 5.6)
- Demonstrate the skills required to plan and carry out investigations using laboratory equipment safely, effectively, and accurately. (5.5)
- Compare, using laboratory results, the theoretical yield of a reaction to the actual yield, calculate the percentage yield, and suggest sources of experimental error. (5.6)
- Give examples of the application of chemical quantities and calculations. (5.7)

Key Terms

actual yield
alpha decay
alpha particle
alpha ray
artificial transmutation
beta decay
beta particle
beta ray
excess reagent
fission
fusion
gamma rays
gravimetric stoichiometry
limiting reagent
mole ratio
nucleons
percentage yield
quantitative analysis
radioactive decay
standard curve
theoretical yield
transmutation
yield

Make a Summary

In this chapter, you learned to calculate quantities using balanced chemical equations. To summarize your learning, design a simple board game in which a player is asked to move from a square such as "mass of reactant A" by selecting the simplest path to reach another square, such as "mass of product C." The intermediate steps should offer a choice of squares such as "amount in moles of reactant A," "determine limiting reagent," "find mole ratio from balanced equation," etc. Some possible formats are the games Monopoly and Snakes and Ladders. Play the game with a partner, selecting a chemical reaction for each player and performing sample numerical calculations for each step.

Reflect on your Learning

Revisit your answers to the Reflect on Your Learning questions at the beginning of this chapter.

- How has your thinking changed?
- What new questions do you have?

Chapter 5 Review

Understanding Concepts

1. Explain the relationships between each of the following pairs of terms:
   (a) limiting reagent and excess reagent
   (b) chemical reaction and nuclear reaction
   (c) alpha decay and beta decay
   (d) actual yield and theoretical yield
   (e) empirical formula and molecular formula
2. Write a balanced chemical equation for each of the following reactions. Assume that substances are pure.
   (a) Research indicates that sulfur dioxide gas reacts with oxygen in the atmosphere to produce sulfur trioxide gas.
   (b) Sulfur trioxide gas travelling across international boundaries causes disagreements between governments.

   sulfur trioxide + water → sulfuric acid

   (c) The means exist for industry to reduce sulfur dioxide emissions; for example, by treatment with lime.

   calcium oxide + sulfur dioxide + oxygen → calcium sulfate

   (d) Restoring acid lakes to normal is expensive; for example, by adding lime to lakes from the air.

   calcium oxide + sulfurous acid → water + calcium sulfite

   (e) Fish in acidic lakes may die from mineral poisoning due to the leaching of minerals from lake bottoms.

   solid aluminum silicate + sulfuric acid → aqueous hydrogen silicate + aqueous aluminum sulfate

3. Write a nuclear equation for each of the following steps of reactions in a nuclear reactor:
   (a) Thorium-233 undergoes beta decay to form protactinium-233.
   (b) Protactinium-233 undergoes beta decay to form uranium-233.
4. Write a balanced nuclear equation for the beta decay of iodine-131.
5. Complete and balance the following nuclear equations by supplying the missing particle:
   (a) ${}^{122}_{53}I \rightarrow {}^{122}_{54}Xe + ?$
   (b) ${}^{59}_{26}Fe \rightarrow ? + {}^{0}_{-1}e$
   (c) ${}^{222}_{86}Rn \rightarrow {}^{218}_{84}Po + ?$
   (d) ${}^{252}_{98}Cf + {}^{10}_{5}B \rightarrow ? + 3\ {}^{1}_{0}n$
   (e) ${}^{239}_{94}Pu + ? \rightarrow {}^{242}_{96}Cm + {}^{1}_{0}n$
6. The ${}^{237}_{93}Np$ nucleus naturally decays through a series of steps and ends with the formation of the stable ${}^{209}_{83}Bi$. The decays proceed through a series of alpha and beta particle emissions. How many of each type of particle is emitted in this series of radioactive decay?
7. Complete the following nuclear equations:

   ${}^{190}_{75}Re \rightarrow {}^{190}_{76}Os + ?$

   ${}^{9}_{3}Li \rightarrow {}^{8}_{3}Li + ?$

   ${}^{214}_{83}Bi \rightarrow ? + {}^{4}_{2}He$

   ${}^{162}_{69}Tm \rightarrow ? + {}_{-1}^{0}e$

   ${}^{120}_{49}In \rightarrow ? + {}_{-1}^{0}e$
8. Isooctane, $C_8H_{18(l)}$, is one of the main constituents of gasoline. Calculate the mass of carbon dioxide gas produced by the complete combustion of 692 g of isooctane.
9. The metal tungsten is used to make the filament in incandescent light bulbs. Tungsten can be obtained by heating tungsten(VI) oxide with hydrogen:

   $$WO_{3(s)} + 3\ H_{2(g)} \rightarrow W_{(s)} + 3\ H_2O_{(g)}$$

   (a) What mass of tungsten(VI) oxide is needed to produce 5.00 g of tungsten?
   (b) What mass of water vapour is produced?
10. The compound used as artificial pineapple flavour in foods is an ester called ethylbutanoate, $C_4H_7O_2C_2H_{5(l)}$. The reaction for the synthesis is shown below:

    $$HC_4H_7O_{2(l)} + C_2H_5OH_{(l)} \rightarrow C_4H_7O_2C_2H_{5(l)} + H_2O_{(l)}$$

    butanoic acid   ethanol   ethylbutanoate

    If 30.0 g of butanoic acid is mixed with 18.0 g of ethanol, find
    (a) the limiting reagent for the reaction;
    (b) the mass of ethylbutanoate formed.
11. Nitric acid can be manufactured from ammonia in a series of reactions called the Ostwald process, developed in 1902 by German chemist Wilhelm Ostwald. The reactions are shown below:

    $$4\ NH_{3(g)} + 5\ O_{2(g)} \rightarrow 4\ NO_{(g)} + 6\ H_2O_{(l)}$$
    $$2\ NO_{(g)} + O_{2(g)} \rightarrow 2\ NO_{2(g)}$$
    $$3\ NO_{2(g)} + H_2O_{(l)} \rightarrow 2\ HNO_{3(aq)} + NO_{(g)}$$

Calculate the mass of nitric acid that is produced if 4.00 mol of ammonia is completely reacted through the Ostwald process.

12. Nitroglycerin, $C_3H_5(NO_3)_{3(s)}$, an explosive that is used in dynamite, can be made from the reaction of glycerol, $C_3H_5(OH)_{3(l)}$, and nitric acid. Water is the only other product formed. In one experiment, a chemist reacted 10.4 g of glycerol and 19.2 g of nitric acid, to produce 22.6 g of nitroglycerin.
    (a) Write a balanced equation for the above reaction.
    (b) What is the limiting reagent in the reaction?
    (c) What is the theoretical yield of nitroglycerin?
    (d) What is the percentage yield of nitroglycerin?
13. List three reasons why the actual yield in a reaction may be less than the theoretical yield.
14. Silica is a name for silicon dioxide, $SiO_{2(s)}$, a chief component of glass. Silica can react with hydrofluoric acid, $HF_{(aq)}$, to produce silicon tetrafluoride, $SiF_{4(g)}$, which is a gas at room temperature:

    $SiO_{2(s)} + 4\ HF_{(aq)} \rightarrow SiF_{4(g)} + 2\ H_2O_{(l)}$

    (a) What mass of silicon tetrafluoride can be produced from 6.80 g of silica?
    (b) What mass of hydrofluoric acid is required to completely react with 53.2 g of silica?
    (c) What mass of water is produced when 10.6 g of silica completely reacts in this reaction?
15. Gasohol is a fuel available at gas stations for use in cars in place of gasoline. The alcohol in gasohol is ethanol, $C_2H_5OH_{(l)}$.
    (a) Write a balanced equation to represent the complete combustion of ethanol in oxygen gas to produce gaseous carbon dioxide and water vapour as the only products.
    (b) What is the mass of carbon dioxide produced when 450.0 g of ethanol is completely combusted?
    (c) What is the mass of water produced when 450.0 g of ethanol is completely combusted?
    (d) What is the mass of oxygen required to completely combust 450.0 g of ethanol?
    (e) Do your answers in (b), (c), and (d) agree with the law of conservation of mass?
16. After copper(II) hydroxide carbonate (malachite) is decomposed, the next step in the production of copper metal is the reaction of copper(II) oxide and carbon to produce copper metal and carbon dioxide. Determine the mass of carbon required to react with 50.0 kg of copper(II) oxide.

Applying Inquiry Skills

17. Describe a sequence of steps that you would carry out to determine whether a reagent in a precipitation reaction is present in excess.
18. The purpose of this lab exercise is to design an experiment that will determine the purity of a sample of sodium sulfate by precipitation with barium chloride. Sodium sulfate is dissolved in water and reacted with aqueous barium chloride to form sodium chloride and insoluble barium sulfate. Present your work in a report, using appropriate scientific vocabulary and SI units.

Prediction

(a) Use gravimetric stoichiometry to predict the yield.

Experimental Design

(b) Describe an experimental design, selecting appropriate materials, techniques, and steps to determine the amount of excess reagent to use.
(c) Write a procedure, including the steps to determine the mass of precipitate formed. Identify any safety concerns and include control measures necessary to reduce the risk.

Analysis

(d) Include an explanation of how you would calculate the purity of the sample of sodium sulfate.

Evaluation

(e) Evaluate the experimental design. Where might errors arise that would change the actual yield?

Making Connections

19. The Haber process facilitated the production of chemical fertilizers such as ammonium nitrate and has had a dramatic impact on crop yields. Since the 19th century, average crop yields per hectare have increased almost five-fold for corn and eight-fold for wheat. However, runoff from fertilized fields is a source of water pollution. Also, the high cost of chemical fertilizers has driven some farmers into debt. In a brief report, evaluate the costs and benefits of technology in this situation from several perspectives.

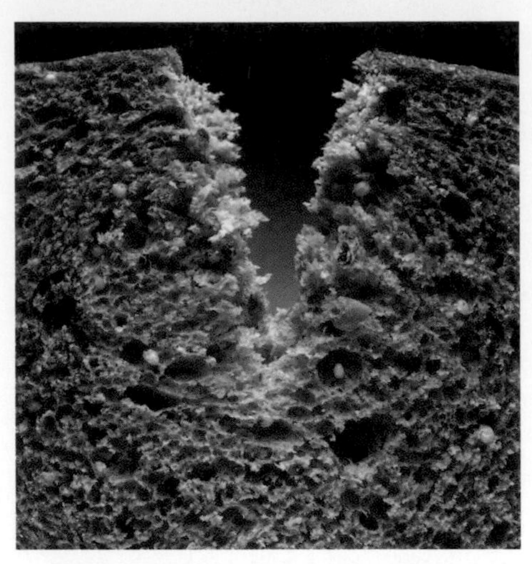

Figure 1
Tiny bubbles are formed in the process of baking, making the dough rise.

Quantitative Analysis of a Reaction

The active ingredient in baking soda is sodium hydrogen carbonate, $NaHCO_{3(s)}$. Upon heating, this ingredient releases bubbles of a gas that give baked goods a light, spongy consistency (**Figure 1**). Your task is to use your knowledge and skills of quantitative analysis to identify the decomposition reaction of sodium hydrogen carbonate from five possibilities. In each of the possible reactions listed below, one or more gases are produced and a solid remains:

1. sodium hydrogen carbonate → sodium (solid), water vapour, carbon monoxide gas, oxygen gas
2. sodium hydrogen carbonate → sodium carbonate (solid), water vapour, carbon dioxide gas
3. sodium hydrogen carbonate → sodium hydroxide (solid), carbon dioxide gas
4. sodium hydrogen carbonate → sodium oxalate, $Na_2C_2O_{4(s)}$, water vapour, oxygen gas
5. sodium hydrogen carbonate → sodium oxide (solid), water vapour, carbon dioxide gas

Investigation

Quantitative Analysis of a Chemical Reaction

You will quantitatively analyze the decomposition of sodium hydrogen carbonate and then identify the chemical reaction that has taken place from the five reactions given above. In order to do so, you will design and perform an experiment to determine the mass of solid product formed when a known mass of sodium hydrogen carbonate is completely decomposed by heating.

Write a dzetailed report to communicate the **Prediction, Experimental Design, Materials, Procedure, Evidence, Analysis,** and **Evaluation** of your investigation.

Question

Which of the five reactions is correct?

Prediction

(a) Balance each of the five reaction equations.
(b) Use stoichiometric calculations to predict the mass of solid product formed in each of the five reactions from the known mass of sodium hydrogen carbonate used.

Experimental Design

(c) Design an experiment that will allow you to answer the Question. Outline the variables you will measure and any controls you will need. Include sample observation tables.

(d) When you are satisfied with your design, write a complete Procedure. Describe any safety precautions needed and include a step to appropriately dispose of materials. Your design and procedure must be approved by your teacher.

Materials

sodium hydrogen carbonate, 3.0 g

(e) Complete the Materials list. The equipment you select should be commonly available. You are required to measure mass to 0.01 g.

Procedure

1. Carry out your Procedure, taking care to record your observations in correct SI units.

Analysis

(f) Analyze the evidence you obtained and compare it with your mass predictions. Identify the correct decomposition reaction. Include your reasoning.

Evaluation

(g) Using the reaction you identified, calculate the percentage yield of the solid product, and suggest reasons why the actual yield may be different from the theoretical yield in this experiment.

Synthesis

(h) From your knowledge of the properties of the two oxides of carbon, which of the five possible reactions is not likely a chemical reaction designed for use in the home? Explain.

(i) Describe other evidence or diagnostic tests that may be performed on the products of the reaction to rule out one or more of the reactions under consideration.

(j) It is often recommended that we keep a box of baking soda on hand in the kitchen for extinguishing small grease fires. Relate this application to the properties of the products formed in the reaction you identified.

(k) Sodium hydrogen carbonate is a weak base that reacts with acids to produce carbon dioxide gas, water, and a salt. It is used in cooking recipes where an acidic ingredient is present, and in common household products such as toothpaste, antacid tablets, and cleaners. Evaluate the importance of accurate chemical quantities and calculations in each of these applications.

Assessment

Your completed task will be assessed according to the following criteria:

Process

- Develop predictions.
- Develop an appropriate experimental design.
- Choose and safely use appropriate equipment and materials.
- Carry out the approved investigation.
- Record observations with appropriate precision.
- Analyze the results.
- Evaluate the experimental design and experimental error.

Product

- Prepare a suitable lab report, including appropriate tables for evidence.
- Demonstrate an understanding of the relevant concepts, principles, laws, and theories.
- Use terms, symbols, equations, and SI metric units correctly.

Understanding Concepts

1. Calculate the mass of each of the following:
   (a) 34.2 mol of baking soda, $NaHCO_{3(s)}$
   (b) 2.17 mol of hydrogen peroxide, used in solution as a bleach
   (c) 6.13×10^{-2} mol of magnesium hydroxide in an antacid
   (d) 4.10×10^{-3} mol of carbon monoxide produced in the combustion of methane
   (e) 1.19×10^{3} mol of iron
   (f) 5.99 mol of vitamin D, $C_{56}H_{88}O_{2(s)}$
2. Calculate the amount in moles of each of the following:
   (a) 10.5 g of silver in a bracelet
   (b) 8.55 g of ethanol, $C_2H_5OH_{(l)}$
   (c) 6.74×10^{3} g of ammonium nitrate, used in fertilizers
   (d) 50.0 g of hydrogen acetate (acetic acid), $HC_2H_3O_{2(aq)}$
   (e) 38.9 g of cholesterol, $C_{27}H_{46}O_{(s)}$
   (f) 1.25×10^{-2} g of sodium fluoride, $NaF_{(s)}$, an ingredient in toothpaste to reduce tooth decay
3. Which of the following samples contains the greatest number of atoms? Show your calculations.
   (a) 5.00 g of gold
   (b) 5.0×10^{-2} mol of gold
   (c) 5.00 g of oxygen gas
   (d) 5.0×10^{-2} mol of oxygen gas
4. Rolaids tablets contain a base, calcium carbonate, $CaCO_{3(s)}$, which is used to neutralize excess acid in the stomach. Calculate the amount of this ingredient in moles in a tablet of Rolaids containing 335 mg of this compound.
5. A car battery contains lead electrodes in sulfuric acid. A compound formed on the lead plates was analyzed and was found to consist of 68.3% lead, 10.6% sulfur, and 21.1% oxygen by mass. What is the empirical formula of this compound?
6. Find the percentage composition of each of the following chemicals that are used to produce colours in ceramic glazes (**Figure 1**):
   (a) barium chromate, $BaCrO_{4(s)}$, which produces yellow to green colours
   (b) cobalt(II) carbonate, $CoCO_{3(s)}$, which produces blue colours
   (c) iron(III) chloride hexahydrate, $FeCl_3{\cdot}6H_2O_{(g)}$, which produces gold colours
7. Write a sentence to describe each of the following balanced chemical equations, including coefficients (in moles) and states of matter. State the mole ratio for the complete reaction equation.

Figure 1

   (a) $2\ NiS_{(s)} + 3\ O_{2(g)} \rightarrow 2\ NiO_{(s)} + 2\ SO_{2(g)}$
   (b) $2\ Al_{(s)} + 3\ CuCl_{2(aq)} \rightarrow 2\ AlCl_{3(aq)} + 3\ Cu_{(s)}$
   (c) $2\ H_2O_{2(l)} \rightarrow 2\ H_2O_{(l)} + O_{2(g)}$
8. For each of the following reactions, classify the reaction and balance the equation:
   (a) $NaCl_{(s)} \rightarrow Na_{(s)} + Cl_{2(g)}$
   (b) $Na_{(s)} + O_{2(g)} \rightarrow Na_2O_{(s)}$
   (c) $Na_{(s)} + H_2O_{(l)} \rightarrow H_{2(g)} + NaOH_{(aq)}$
   (d) $AlCl_{3(aq)} + NaOH_{(aq)} \rightarrow Al(OH)_{3(s)} + NaCl_{(aq)}$
   (e) $Al_{(s)} + H_2SO_{4(aq)} \rightarrow H_{2(g)} + Al_2(SO_4)_{3(aq)}$
   (f) $C_8H_{18(l)} + O_{2(g)} \rightarrow CO_{2(g)} + H_2O_{(g)}$
9. For each pair of reactants, classify the reaction type, complete the chemical equation, and balance the equation. Also, state the mole ratio for each equation.
   (a) $Ni_{(s)} + S_{8(s)} \rightarrow$
   (b) $C_6H_{6(l)} + O_{2(g)} \rightarrow$
   (c) $K_{(s)} + H_2O_{(l)} \rightarrow$
10. Chlorine gas is bubbled into a potassium iodide solution and a colour change is observed. Write the balanced chemical equation for this reaction and describe a diagnostic test for one of the products.
11. For each of the following reactions, translate the information into a balanced chemical equation. Then classify the main perspective—scientific, technological, ecological, economic, or political—suggested by the introductory statement.
    (a) Oxyacetylene torches are used to produce high temperatures for cutting and welding metals such as steel. This involves burning acetylene, $C_2H_{2(g)}$, in pure oxygen.
    (b) In chemical research conducted in 1808, Sir Humphry Davy produced magnesium metal by

decomposing molten magnesium chloride using electricity.

(c) An inexpensive application of single replacement reactions uses scrap iron to produce copper metal from waste copper(II) sulfate solutions.

(d) The emission of sulfur dioxide into the atmosphere creates problems across international borders. Sulfur dioxide is produced when zinc sulfide is roasted in a combustion-like reaction in a zinc smelter.

(e) Burning leaded gasoline added toxic lead compounds to the environment, which damaged both plants and animals. Leaded gasoline contained tetraethyl lead, $Pb(C_2H_5)_{4(l)}$, which undergoes a complete combustion reaction in a car engine.

12. Balance each of the following nuclear equations:
    (a) the radioactive decay of thorium-230 with the emission of an alpha particle
    (b) lead-214 undergoes decay to produce bismuth-214 and a beta particle
    (c) nitrogen-14 is bombarded by an alpha particle to give off a proton and a desired isotope

13. Determine the molecular formula for nicotine from the following evidence:
    molar mass = 162.24 g/mol
    percent by mass of carbon = 74.0%
    percent by mass of hydrogen = 8.7%
    percent by mass of nitrogen = 17.3%

14. Consider a quantitative analysis in which a sample reacts with another chemical to produce a precipitate.
    (a) Which substance is the limiting reagent?
    (b) What is the purpose of using an excess quantity of the second reactant?

15. (a) How is a graph used in a quantitative analysis?
    (b) What are some advantages of using a graph?

16. In all stoichiometry, why is it necessary to always convert to or convert from amounts in moles?

17. (a) What is a percentage yield?
    (b) List some reasons why the percentage yield of product in a chemical reaction is generally less than 100%.

18. How is chemical science different from chemical technology in terms of emphasis and scope?

19. In a chemical analysis, 3.00 g of silver nitrate in solution was reacted with excess sodium chromate to produce 2.81 g of precipitate. What is the percentage yield?

20. A solution containing 9.80 g of barium chloride is mixed with a solution containing 5.10 g of sodium sulfate.
    (a) Which reactant is in excess?
    (b) Determine the excess mass.
    (c) Predict the mass of the precipitate.

21. A solution containing 18.6 g of chromium(III) chloride is reacted with a 15.0-g piece of zinc to produce chromium metal.
    (a) Which reactant is in excess?
    (b) Determine the excess mass.
    (c) If 5.10 g of chromium metal is formed, what is the percentage yield?

22. Iron(III) phosphate is a hydrated salt. When a sample of this salt was heated to drive off the water of crystallization, the following evidence was obtained:
    mass of crucible and lid = 24.80 g
    mass of crucible, lid, and hydrated salt = 29.93 g
    mass of crucible, lid, and anhydrous salt = 28.27 g
    What is the formula of the hydrate of iron(III) phosphate?

23. When silver jewellery or cutlery is exposed to the air, the small amount of hydrogen sulfide, $H_2S_{(g)}$, in the air reacts with the silver to produce a layer of silver sulfide, resulting in a tarnish on the silver. The equation for the reaction is

$$4\ Ag_{(s)} + 2\ H_2S_{(g)} + O_{2(g)} \rightarrow 2\ Ag_2S_{(s)} + 2\ H_2O_{(g)}$$

    (a) What mass of silver sulfide is formed from the reaction of 0.120 g of silver?
    (b) What mass of hydrogen sulfide is needed in the same process?

24. When concentrated sulfuric acid is added to sucrose, $C_{12}H_{22}O_{11(s)}$, a dehydration reaction occurs where water and carbon are formed (**Figure 2**). The equation for the reaction is

$$C_{12}H_{22}O_{11(s)} + H_2SO_{4(l)} \rightarrow 12\ C_{(s)} + 11\ H_2O_{(l)} + H_2SO_{4(aq)}$$

Figure 2

(a) What mass of carbon is formed from 20.0 g of sucrose, according to this equation?
(b) What mass of water is formed in the same process?

25. Through the process of photosynthesis, plants produce glucose, $C_6H_{12}O_{6(aq)}$, from carbon dioxide and water, according to the following equation:

$$6\ CO_{2(g)} + 6\ H_2O_{(g)} \rightarrow C_6H_{12}O_{6(aq)} + 6\ O_{2(g)}$$

(a) What mass of carbon dioxide is needed for the plant to produce 100.0 g of glucose?
(b) What mass of oxygen is produced in the same process?

26. If a beaker of ammonium hydroxide and a beaker of hydrochloric acid are placed side by side, a white solid forms on the beakers (**Figure 3**). The white solid is ammonium chloride, $NH_4Cl_{(s)}$, which is produced when ammonia gas readily combines with hydrogen chloride gas. Write a balanced equation for this reaction, and calculate the mass of ammonium chloride that will be formed when 2.00 g of ammonia gas reacts with 2.00 g of hydrogen chloride gas.

Figure 3

Applying Inquiry Skills

27. The purpose of this exercise is to test the method of stoichiometry. Complete the **Prediction**, **Experimental Design**, **Analysis**, and **Evaluation** sections of an investigation report.

Question
What is the mass of precipitate formed when 3.43 g of barium hydroxide in solution reacts with an excess of sulfuric acid?

Prediction
(a) Calculate the theoretical yield of barium sulfate.

Experimental Design
(b) Describe the experiment, including the mass of excess reagent to be used.
(c) Write a procedure and choose materials. Include any safety precautions that are necessary.

Evidence
A white precipitate formed when the barium hydroxide solution was mixed with the sulfuric acid.

mass of filter paper = 0.96 g
mass of dried filter paper plus precipitate = 5.25 g

Analysis
(d) Answer the Question.

Evaluation
(e) Based on the evidence and your evaluation of the experimental design, evaluate the stoichiometric method.

28. For the experiment in question 27, describe a test you could use to determine whether the reagent used in excess was indeed in excess.

29. Discuss some reasons that would explain evidence showing that the actual yield in an experiment is greater than the theoretical yield.

Making Connections

30. Research the training and requirements involved in becoming an analytical chemist. Include a list of workplaces where analytical chemists are employed. If possible, interview someone to learn more about this occupation.

31. Smog is a major problem in many large cities. Find out what chemicals contribute to this problem and write chemical reaction equations for the production and further reactions of two of the chemicals. What are the ecological implications of the presence of smog? What might be done to help solve this problem? In your answer, consider a variety of perspectives.

32. Chemical industries provide many useful and essential products and processes. The manufacture of sulfuric acid in the contact process yields an annual worldwide production of about one trillion tonnes, making it the most commonly used acid in the world. Research the main chemical reactions involved in the contact process and list some byproducts of processes involving sulfuric acid. What precautions are necessary when handling concentrated sulfuric acid?

33. Aluminum is the most abundant metallic element in Earth's crust, found at the surface mostly as aluminum compounds in clay. The most extractable source of aluminum is the ore bauxite, which consists mainly of aluminum oxide, $Al_2O_{3(s)}$. Canada has several aluminum smelters, located in Quebec (**Figure 4**) and in British Columbia, processing bauxite shipped from South America, Jamaica, and West Africa. Using the Internet, research the chemical reactions used in the extraction of aluminum from its ore, the reasons for the locations of the smelters, the waste products of the industrial process, and environmental issues arising from disposal of these waste products. Present your findings in the form of a report.

    Follow the links for Nelson Chemistry 11, Unit 2 Review.

    GO TO www.science.nelson.com

Figure 4

Exploring

34. Analytical chemists work in many varied fields where analysis of unknown substances is important. For example, chemical "detectives" in food science have developed sensitive tests to identify beet or corn sugars that juice manufacturers have fraudulently used in fruit juices. Toxicologists have traced an outbreak of food poisoning to a toxin found in mussels from Prince Edward Island. At pharmaceutical companies, analytical chemists study the purity of new medications and how well they are absorbed by the body.

    Using the Internet, identify and research one of these or other work contexts that interest you, and prepare a report to describe

    (a) a company or an industry that does this type of work;
    (b) the type of analytical techniques used;
    (c) the level of education and qualifications needed;
    (d) why this is a career that you may be interested in.

    Follow the links for Nelson Chemistry 11, Unit 2 Review.

    GO TO www.science.nelson.com

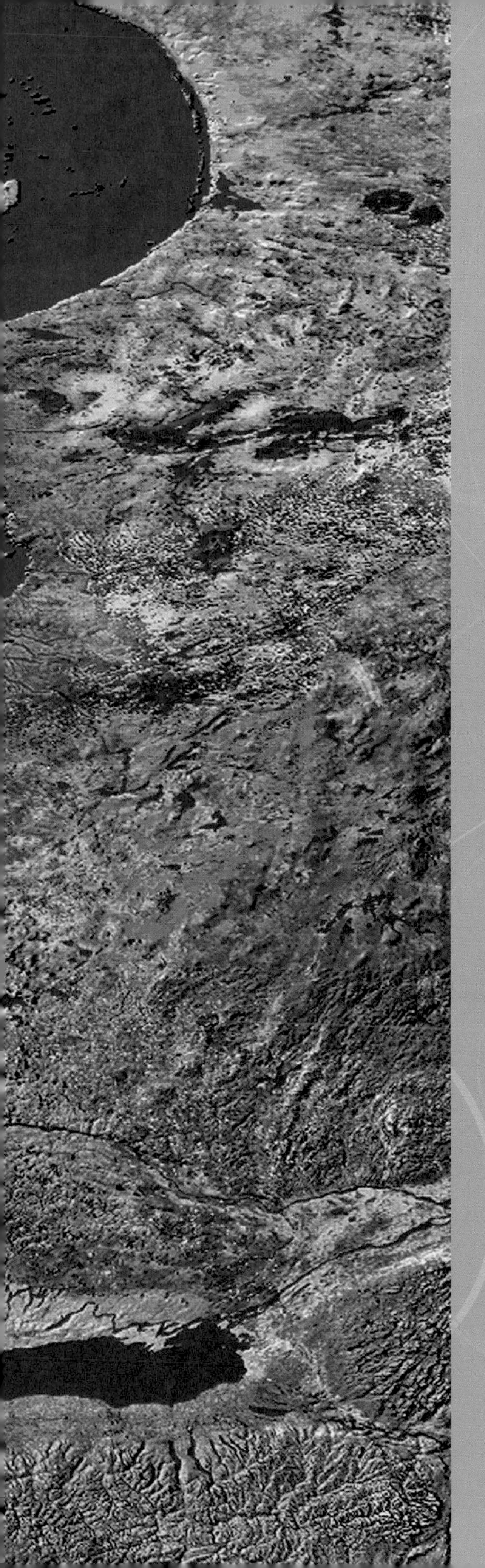

Unit 3

Solutions and Solubility

"I take great joy in being a scientist, not least because I was diagnosed as an epileptic when I was just 15, in Grade 10. I had been told that because of the epilepsy I could never go to university, much less study science. However, I did succeed in science. Being an epileptic is one of the major reasons that I directed my research and studies in the direction that I did. In studying aqueous solutions, I was not dealing with chemicals that were explosive and dangerous to handle. My studies also fit my inclinations. I love to think in three dimensions and contemplate how crystals form from solutions, and also how they dissolve. As a university researcher, I have the freedom to study what I am interested in, whereas in an industrial position I would be told what to research. My current focus is on solutions formed from crystalline aluminum hydroxide, which is amphoteric (it can act as an acid or a base). In acid solutions, its chemistry is important for environmental reasons; in basic solutions it is important in the extraction of aluminum from ore."

Susan Bradley, Assistant Professor, Department of Chemistry, University of British Columbia

Overall Expectations

In this unit, you will be able to

- understand the properties of solutions, the concept of concentration, and the importance of water as a solvent;
- prepare, analyze, and react solutions using qualitative and quantitative methods;
- relate the scientific knowledge of solutions and solubility to a variety of technological, societal, and environmental examples, including water quality

Unit

3

Solutions and Solubility

Are You Ready?

Technical Skills and Safety

1. In this unit you will work with many different solutions.
   (a) What should you do immediately if some solution is spilled on your hand?
   (b) Draw or describe the WHMIS symbol for a corrosive substance.

Knowledge and Understanding

2. Copy and complete the classification scheme in **Figure 1**.

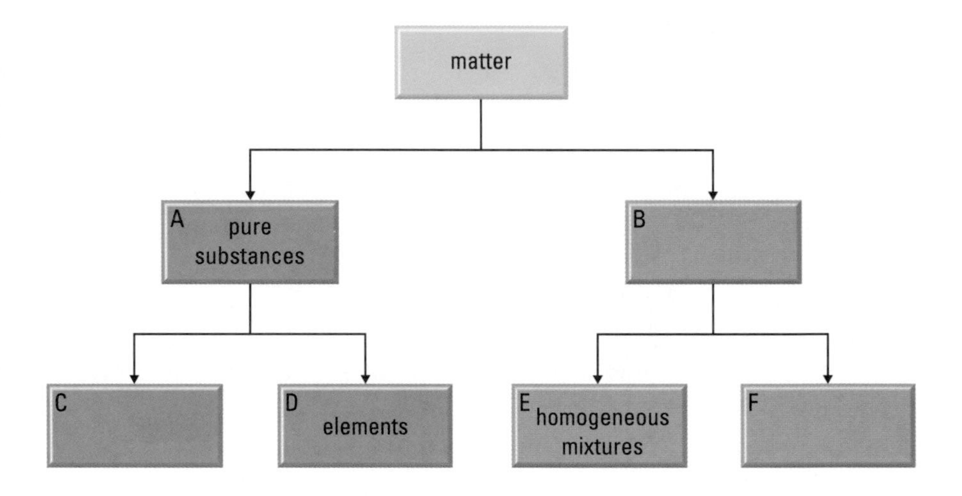

Figure 1
A classification of matter

3. Match each of the substances in **Table 1** to the classification categories illustrated in **Figure 1**.

Table 1

Substance	A or B	C or D or E or F
(a) vinegar		
(b) pure water		
(c) sulfur		
(d) air		
(e) milk		

4. Distinguish between ionic and molecular compounds based on their
   (a) chemical name or formula
   (b) empirical (observable) properties

5. In **Table 2**, match each term in Column A with its corresponding description in Column B.

Table 2

A	B
(a) compound	A. Cannot be broken down into simpler substances
(b) solution	B. Contains two or more visible components
(c) element	C. Can be identified with a single chemical formula
(d) heterogenous mixture	D. A mixture of two or more pure substances with a single visible component

6. Write the missing words from the following statement in your notebook.

   According to modern atomic theory, an atom contains a number of positively charged __________, determined by the ______ _______ of the element, and an equal number of negatively charged __________.

7. Atoms of the representative elements form ions. Using calcium and fluorine as examples, draw electron dot diagrams showing atoms and ions.
8. Draw a diagram to illustrate the model of a small sample (a few particles) of
   (a) sodium chloride
   (b) water
9. What is the type of bond between the atoms of a water molecule? What are the types of bonds between the molecules of water in a sample?
10. Refer to the list of substances in **Table 3** to answer the following questions.
    (a) Which substances have London (dispersion) forces present between the molecules?
    (b) Classify each substance as polar or nonpolar.
    (c) Which substance would be expected to have hydrogen bonding as part of the intermolecular forces between the molecules?
    (d) Distinguish between intermolecular and intramolecular forces.

Table 3

Substance	Chemical formula	Use
propane	$C_3H_{8(g)}$	propane barbeques
ethanol	$C_2H_5OH_{(l)}$	in gasohol (gasoline-alcohol fuel)
dichloromethane	$CH_2Cl_{2(l)}$	paint stripper

11. According to the kinetic molecular theory, how does increasing temperature affect the speed of the particles that make up a substance?
12. Write balanced chemical equations, including states of matter at SATP, for reactions involving the following pairs of reactants.
    (a) aqueous iron(III) chloride and aqueous sodium hydroxide
    (b) aqueous silver nitrate and copper
    (c) sulfuric acid and aqueous potassium hydroxide
    (d) aqueous chlorine and aqueous sodium bromide
13. Classify each of the reactions in question 12 as a single or a double displacement reaction.
14. What does the coefficient represent in a chemical equation?
15. To dispose of a sodium sulfide solution, an excess of aqueous aluminum sulfate is added. What mass of precipitate can be obtained from a solution containing 12.5 g of sodium sulfide?

Chapter

6

The Nature and Properties of Solutions

In this chapter, you will be able to

- describe and explain the properties of water, and demonstrate an understanding of its importance as a universal solvent;
- explain the formation of solutions involving various solutes in water and nonpolar solutes in non-aqueous solvents;
- use the terms solute, solvent, solution, electrolyte, concentration, standard solution, stock solution, and dilution;
- solve solution concentration problems using a variety of units;
- develop the technological skills for the preparation of solutions;
- determine qualitative properties of solutions;
- provide examples of solutions involving all three states;
- provide consumer and commercial examples of solutions, including those in which the concentration must be precisely known;
- explain the origins of pollutants in natural water.

Is there such a thing as pure, natural water? Certainly it can't be found in the oceans. In Samuel Taylor Coleridge's classic poem, *The Rime of the Ancient Mariner*, written in 1798, an old seafarer describes the desperation of becalmed sailors, drifting without fresh water under the fierce sun, driven mad with thirst:

> *Water, water, everywhere,*
> *Nor any drop to drink.*

Drinking the water of the sea, which is rich in dissolved solutes, can be fatal. Today, seagoing ships carry distillation equipment to convert salt water into drinking water by removing most of those solutes.

Fresh water from lakes and rivers, which we depend on for drinking, cooking, irrigation, electric power generation, and recreation, is also impure. Even direct from a spring, fresh water is a solution that contains dissolved minerals and gases. So many substances dissolve in water that it has been called "the universal solvent." Many household products, including soft drinks, fruit juices, vinegar, cleaners, and medicines, are aqueous (water) solutions. ("Aqueous" comes from the Latin *aqua* for "water," as in aqueduct and aquatics.) Our blood plasma is mostly water, and many substances essential to life are dissolved in it, including glucose and carbon dioxide.

The ability of so many materials to dissolve in water also has some negative implications. Human activities have introduced thousands of unwanted substances into water supplies. These substances include paints, cleaners, industrial waste, insecticides, fertilizers, salt from highways, and other contaminants. Even the atmosphere is contaminated with gases produced when fossil fuels are burned. Rain, falling through these contaminants, may become acidic. Learning about aqueous solutions and the limits to purity will help you understand science-related social issues forming around the quality of our water.

Reflect on your Learning

1. (a) List some substances that can dissolve in water.
   (b) Classify the substances into two or more categories.
2. Are there any types of substances that generally do not dissolve in water? Why not?
3. Both table salt and table sugar dissolve in water to produce clear, colourless solutions. Using your present knowledge, what is similar in the formation of these two solutions? What is different?

Try This **Activity**

Substances in Water

Even the "universal solvent" doesn't dissolve all solutes equally well. See what happens when you add different substances to water.

Materials: 5 petri dishes; water; 5 substances, such as potassium permanganate crystals, a marble chip (calcium carbonate), a sugar cube, a few drops of alcohol (ethanol), a few drops of vegetable oil

- Pour a few millilitres of water into each of five petri dishes, then add one of the substances to each dish. Record your observations.
  - (a) Which substances dissolved in water?
  - (b) How certain are you about each substance in (a)? Give your reasons.
  - (c) Which substances do not appear to dissolve?
  - (d) How certain are you about your answer in (c)?
  - (e) Do the mixtures all have the same properties? Other than visible differences, hypothesize how they might differ.
  - (f) Design some tests for your hypotheses.

Figure 1
The waterfall is clear and clean, but is it pure?

6.1 Defining a Solution

DID YOU KNOW?

Misleading Labelling

Milk is sometimes labelled as "homogenized," meaning that the cream is equally distributed throughout the milk. This use of the word does not match the chemistry definition of homogeneous. Using the strict chemistry definition, milk is not a homogeneous mixture, but a heterogeneous mixture. Milk is not a solution.

Many of the substances that we use every day come packaged with water. We buy other substances with little or no water, but then mix water with them before use. For example, we may purchase syrup, household ammonia, and pop with water already added, but we mix baking soda, salt, sugar, and powdered drinks with water. Most of the chemical reactions that you see in high school occur in a water environment. Indeed, most of the chemical reactions necessary for life on our planet occur in water.

Because so many substances dissolve in it, water is often referred to as the universal solvent. Of course, this is an exaggeration. Not all things dissolve in water. Imagine if they did; we would not be able to find a container for water.

Before restricting our study to mixtures involving water, we will review the more general definition and types of a solution.

Solutions

solution: a homogeneous mixture of substances composed of at least one solute and one solvent

homogeneous mixture: a uniform mixture of only one phase

solute: a substance that is dissolved in a solvent (e.g., salt, NaCl)

solvent: the medium in which a solute is dissolved; often the liquid component of a solution (e.g., water)

Solutions are **homogeneous mixtures** of substances composed of at least one **solute** and one **solvent**. Liquid-state and gas-state solutions are clear (transparent)—you can see through them; they are not cloudy or murky in appearance. Solutions may be coloured or colourless. Opaque or translucent (cloudy) mixtures, such as milk, contain undissolved particles large enough to block or scatter light waves. These mixtures are considered to be heterogeneous.

It is not immediately obvious whether a clear substance is pure or is a mixture, but it is certainly homogeneous. Homogeneous mixtures in the liquid state and the gas state are always clear with only one phase present. If you were to do a chemical analysis of a sample of a homogeneous mixture (i.e., a solution), you would find that the proportion of each chemical in the sample remains the same, regardless of how small the sample is. This is explained by the idea that there is a uniform mixture of particles (atoms, ions, and/or molecules) in a solution. Empirically, a solution is homogeneous; theoretically, it is uniform at the atomic and molecular level.

Both solutes and solvents may be gases, liquids, or solids, producing a number of different combinations (**Table 1**). In metal alloys, such as bronze or the mercury amalgam used in tooth fillings, the dissolving has taken place in liquid form before the solution is used in solid form. Common liquid solutions that have a solvent other than water include varnish, spray furniture polish, and gasoline. Gasoline, for example, is a mixture of as many as 400 different hydrocarbons and other compounds (**Figure 1**). These substances form a solution—a homogeneous mixture at the molecular level. There are many such hydrocarbon solutions, including kerosene (a Canadian-invented fuel for lamps and stoves), and turpentine (used for cleaning paintbrushes). Most greases and oils will dissolve in hydrocarbon solvents.

Table 1: Classification of Solutions

Solute in solvent	Example of solution
gas in gas	oxygen in nitrogen (in air)
gas in liquid	oxygen in water (in most water)
gas in solid	oxygen in solid water (in ice)
liquid in gas	water in air (humidity)
liquid in liquid	methanol in water (in antifreeze)
liquid in solid	mercury in silver (in tooth fillings)
solid in liquid	sugar in water (in syrup)
solid in solid	tin in copper (in bronze)

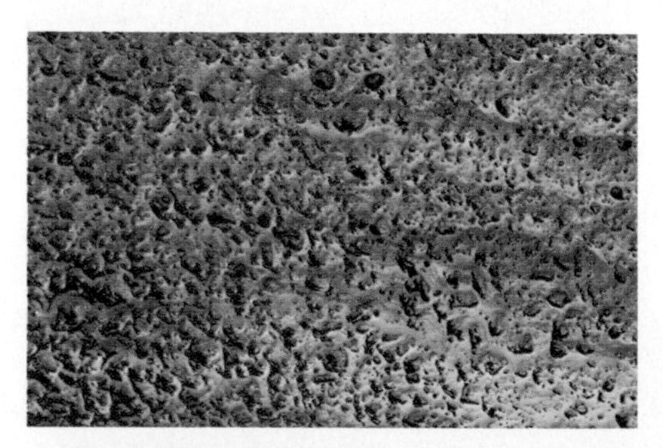

Figure 1
Gasoline, shown here in a spill on asphalt, is a nonaqueous solution containing many different solutes (mostly hydrocarbons such as benzene and paraffin) in an octane solvent. The composition of gasoline is not fixed: It varies with the source of the raw material, the manufacturer, and the season.

Other examples of liquids and solids dissolving in solvents other than water include the many chemicals that dissolve in alcohols. For example, solid iodine dissolved in ethanol (an alcohol) is used as an antiseptic (**Figure 2**). Aspirin (acetylsalicylic acid, ASA) dissolves better in methanol (a poisonous alcohol) than it does in water, for example, when doing chemical analyses. Of course, it should never be mixed with alcohol when ingested. Some glues and sealants make use of other solvents: acetic acid is used as a solvent of the components of silicone sealants. You can smell the vinegar odour of acetic acid when sealing around tubs and fish tanks.

Figure 2
Tincture of iodine is a solution of the element iodine and the compound potassium or sodium iodide dissolved in ethanol. It is used to prevent the infection of minor cuts and scrapes.

The chemical formula representing a solution specifies the solute by using its chemical formula and shows the solvent by using a subscript. For example,

$NH_{3(aq)}$	ammonia gas (solute) dissolved in water (solvent)
$NaCl_{(aq)}$	solid sodium chloride (solute) dissolved in water (solvent)
$I_{2(al)}$	solid iodine (solute) dissolved in alcohol (solvent)
$C_2H_5OH_{(aq)}$	liquid ethanol (solute) dissolved in water (solvent)

By far the most numerous and versatile solutions are those in which water is the solvent (**Figure 3**). Water can dissolve many substances, forming many unique solutions. All **aqueous solutions** have water as the solvent and are clear (transparent). They may be either coloured or colourless. Although water solutions are all different, they have some similarities and can be classified or described in a number of ways. This chapter deals primarily with the characteristics of aqueous solutions.

aqueous solution: a solute dissolved in water

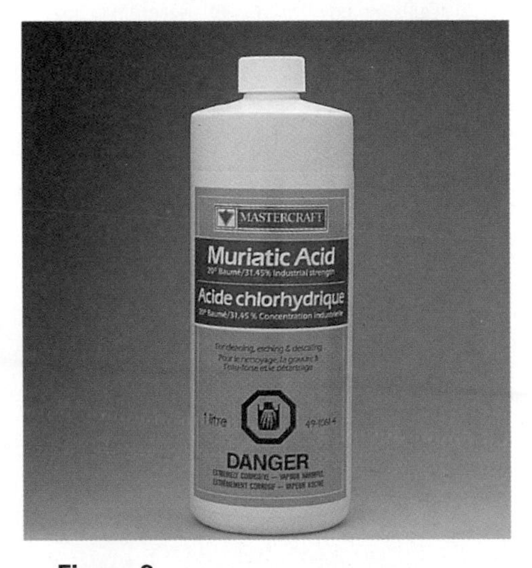

Figure 3
Concentrated hydrochloric acid (often sold under its common name, muriatic acid) contains hydrogen chloride gas dissolved in water. It is used to etch concrete before painting it, clean rusted metal, and adjust acidity in swimming pools.

Properties of Aqueous Solutions

Compounds can be classified as either electrolytes or nonelectrolytes. Electrolytes are solutes that form solutions that conduct electricity. At this point we will restrict ourselves to compounds in aqueous solutions. Compounds are **electrolytes** if their aqueous solutions conduct electricity. Compounds are **nonelectrolytes** if their aqueous solutions do not conduct electricity. Most household aqueous solutions, such as fruit juices and cleaning solutions, contain electrolytes. The conductivity of a solution is easily tested with a simple conductivity apparatus (**Figure 4**) or an ohmmeter. This evidence also provides a diagnostic test to determine the class of a solute—electrolyte or nonelectrolyte. This very broad classification of compounds into electrolyte and nonelectrolyte categories can be related to the main types of compounds classified in Chapter 2. Electrolytes are mostly highly soluble ionic compounds (e.g., $KBr_{(aq)}$), including bases such as ionic hydroxides (e.g., sodium hydroxide, $NaOH_{(aq)}$). Most molecular

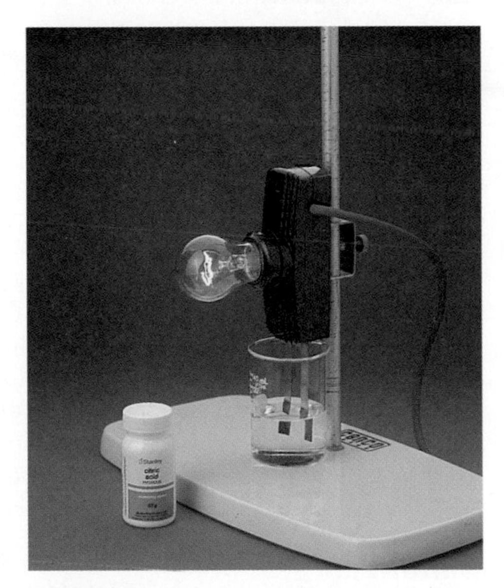

Figure 4
The bulb in this conductivity apparatus lights up if the solute is an electrolyte.

electrolyte: a compound that, in an aqueous solution, conducts electricity

nonelectrolyte: a compound that, in an aqueous solution, does not conduct electricity

compounds (e.g., ethanol, $C_2H_5OH_{(aq)}$) are nonelectrolytes, with the exception of acids. Acids (e.g., nitric acid, $HNO_{3(aq)}$) are molecular compounds that, in aqueous solution, conduct electricity.

Another empirical method of classifying solutions uses litmus paper as a test to classify solutes as **acids**, **bases**, or **neutral** substances. Acids form acidic solutions, bases form basic solutions, and most other ionic and molecular compounds form neutral solutions (**Table 2**).

acid: a substance that, in aqueous solution, turns blue litmus paper red

base: a substance that, in aqueous solution, turns red litmus paper blue

neutral: a substance that, in aqueous solution, has no effect on either red or blue litmus paper; neither acidic nor basic

Note: these are empirical definitions, based on the results of the litmus test. Later in this unit, you will encounter theoretical definitions.

Table 2: Properties of Solutes and Their Solutions

Type of solute	Conductivity test
electrolyte	light on conductivity apparatus glows; needle on ohmmeter moves compared to the control
nonelectrolyte	light on conductivity apparatus does not glow; needle on ohmmeter does not move compared with position for the control
Type of solution	**Litmus test**
acidic	blue litmus turns red
basic	red litmus turns blue
neutral	no change in colour of litmus paper

Investigation 6.1.1

Qualitative Chemical Analysis

INQUIRY SKILLS

- ○ Questioning
- ○ Hypothesizing
- ○ Predicting
- ● Planning
- ● Conducting
- ● Recording
- ● Analyzing
- ● Evaluating
- ● Communicating

In this investigation you will design and carry out a chemical analysis of four compounds (calcium chloride, citric acid, glucose, and calcium hydroxide) to find out which is which. Complete the **Experimental Design** and **Materials** section of the report. You will use the diagnostic tests discussed so far: the conductivity test and the litmus test. After conducting your tests, complete the **Analysis** and **Evaluation** sections.

Question

Which of the white solids labelled 1, 2, 3, and 4 is calcium chloride, citric acid, glucose, and calcium hydroxide?

Experimental Design

You may use litmus paper and the conductivity apparatus. Like all such tests, the conductivity and litmus tests require control of other variables. For example, the temperature of the solution and the quantity of dissolved solute must be kept the same for all substances tested, to allow valid analysis of any evidence collected.

(a) In a short paragraph, plan your tests for this experiment, including independent, dependent, and important controlled variables.

(b) Write a Procedure in which you will use your tests, including safety precautions. (If necessary, refer to MSDS information for the four solids.) Have your Procedure approved by your teacher.

Calcium hydroxide is corrosive. Do not touch any of the solids. Wear eye protection, gloves, and an apron.

Materials

(c) Prepare a list of all materials, including chemicals and equipment. Note that this experiment can easily be done on a small scale using a spot or well plate. Be sure to include safety and disposal equipment.

Procedure

1. Carry out your Procedure, recording your observations in a suitable format.

Analysis

(d) Using the evidence you have collected in your experiment, answer the Question: Which of the white solids labelled 1, 2, 3, and 4 is calcium chloride, citric acid, glucose, and calcium hydroxide?

Evaluation

(e) Evaluate the evidence by critiquing the Experimental Design, Materials, and Procedure. Look for any flaws, sources of error, and possible improvements. Overall, how certain are you about your answer to the Question?

Practice

Understanding Concepts

1. Classify the following mixtures as heterogeneous or homogeneous. Justify your answers.
   (a) fresh-squeezed orange juice
   (b) white vinegar
   (c) red wine
   (d) an antique bronze dagger
   (e) a stainless steel knife
   (f) an old lead water pipe
   (g) humid air
   (h) a cloud
   (i) a dirty puddle
2. Which of the following are solutions and which are *not* solutions?
   (a) milk
   (b) apple juice
   (c) the gas in a helium-filled balloon
   (d) pop
   (e) pure water
   (f) smoke-filled air
   (g) silt-filled water
   (h) rainwater
   (i) 14K gold in jewellery
3. State at least three ways of classifying solutions.
4. (a) What is an aqueous solution?
   (b) Give at least five examples of aqueous solutions that you can find at home.
5. Using the information in **Table 3**, classify each of the compounds as either an electrolyte or a nonelectrolyte. Provide your reasoning.
6. (a) What types of solutes are electrolytes?
   (b) Write a definition of an electrolyte.
7. Describe the solutes in the following types of solutions:
   (a) acidic
   (b) basic
   (c) neutral
8. Electrolytes are lost during physical activity and in hot weather through sweating. The body sweats in order to keep cool—cooling by

Table 3

Compound	Class
methanol	molecular
sodium chloride	highly soluble ionic
hydrochloric acid	acid (molecular)
potassium hydroxide	base (ionic hydroxide)

evaporation of water. Sweating removes water and the substances dissolved in the water, such as salts and other electrolytes. We replace lost electrolytes by eating and drinking. By law the ingredients of a food item are required to be placed on the label in decreasing order of quantity, as they are in Gatorade (**Figure 5**), a noncarbonated drink.

(a) Classify the ingredients of Gatorade as electrolytes or nonelectrolytes. How does the number and quantity of electrolytes and nonelectrolytes compare?
(b) Which ingredients contain sodium ions? Which contain potassium ions? Are there more sodium or potassium ions in the drink?
(c) Does the most energy in the drink come from proteins, carbohydrates, or fats (oils)?
(d) What three chemical needs does the drink attempt to satisfy?

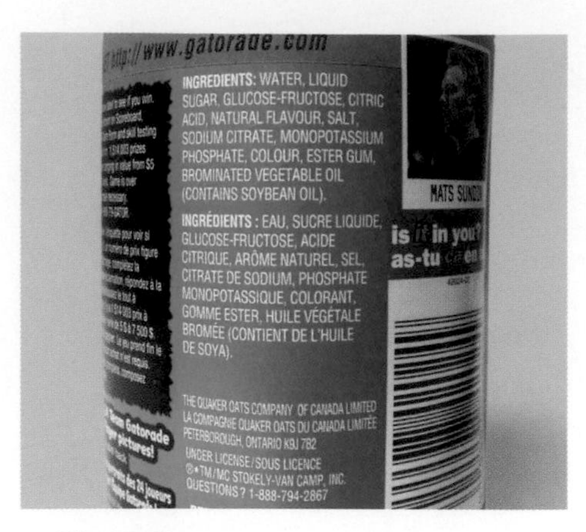

Figure 5
Gatorade is a drink that its manufacturer recommends to athletes, to restore electrolytes to the body.

INQUIRY SKILLS

- ○ Questioning
- ○ Hypothesizing
- ○ Predicting
- ○ Planning
- ○ Conducting
- ○ Recording
- ● Analyzing
- ○ Evaluating
- ○ Communicating

Lab Exercise 6.1.1

Identification of Solutions

For this investigation assume that the labels on the four containers have been removed (perhaps washed off in a flood). Your task as a laboratory technician is to match the labels to the containers, using a litmus indicator and conductivity apparatus to identify the solutions.

You are provided with the Evidence gathered. Complete the **Analysis** section of this report.

Question

Which of the solutions labelled 1, 2, 3, and 4 is hydrobromic acid, which is ammonium sulfate, which is lithium hydroxide, and which is methanol?

Experimental Design

Each solution is tested with both red and blue litmus paper and with conductivity apparatus. The temperatures of the solutions and the procedures are controlled.

Evidence

Table 4: Properties of the Unknown Solutions

Solution	Red litmus	Blue litmus	Conductivity
1	red	blue	none
2	red	red	high
3	red	blue	high
4	blue	blue	high

Analysis

(a) Analyze the Evidence and use it to answer the Question: Which of the solutions labelled 1, 2, 3, and 4 is hydrobromic acid, which is ammonium sulfate, which is lithium hydroxide, and which is methanol? Justify your answer.

Section 6.1 Questions

Understanding Concepts

1. (a) Construct a table that has a column heading Solute (with sub-headings of Solid, Liquid, and Gas), and a row heading Solvent (with sub-headings of Solid, Liquid, and Gas). Complete the table, including examples for as many categories as possible.
   (b) What three kinds of solutions, in your experience, are most common?
2. Include the following terms in a concept map built around the subject "Solutions": solute, solvent, gas, liquid, solid, water, homogeneous mixture, aqueous solution, electrolyte, nonelectrolyte, acid, base, and neutral.
3. Kerosene is a hydrocarbon solution. Name two more examples.
4. Paints are generally sold as alkyd (oil-based) or latex (water-based). To dilute these paints or to clean the paintbrushes, what solvents must be used for each type? Be specific, and explain why each solvent is appropriate.
5. Classify each compound as an electrolyte or a nonelectrolyte:
   (a) sodium fluoride (in toothpaste)
   (b) sucrose (table sugar)
   (c) calcium chloride (a road salt)
   (d) ethanol (in wine)
6. Based upon your current knowledge, classify each of the following compounds (**Figure 6**) as forming an acidic, basic, or neutral aqueous solution, and predict the colour of litmus in each solution.
   (a) $HCl_{(aq)}$ (muriatic acid for concrete etching)
   (b) $NaOH_{(aq)}$ (oven and drain cleaner)
   (c) methanol (windshield washer antifreeze)
   (d) sodium hydrogen carbonate (baking soda)

Applying Inquiry Skills

7. Imagine that you are to plan an investigation to discover which of the compounds listed in **Table 5** are acids, bases, or neutral, and which are electrolytes or nonelectrolytes. You will first use the chemical formulas to make a **Prediction** for each compound, and then plan an **Experimental Design** to test your predictions.

   Question
   Which of the chemicals listed in **Table 5** are acids, bases, neutral, electrolytes, nonelectrolytes?

   Prediction
   (a) Predict the answer to the Question by copying and completing **Table 5**.

Figure 6
Everyday chemicals form acidic, basic, or neutral solutions.

Table 5: Predicting Properties of Compounds

Substance	Acidic/Basic/Neutral	Electrolyte/Nonelectrolyte
$C_3H_7OH_{(l)}$ (a rubbing alcohol)	?	?
calcium hydroxide (slaked lime)	?	?
$H_3PO_{4(aq)}$ (for manufacturing fertilizer)	?	?
glucose (a product of photosynthesis)	?	?
sodium fluoride (in toothpaste)	?	?

(continued)

Experimental Design

(b) Write an experimental design for this investigation.

Making Connections

8. Since grease dissolves in gasoline, some amateur mechanics use gasoline to clean car, bicycle, or motorcycle parts in their basements. Why is this an unsafe practice? What precautions would make the use of gasoline for this purpose safer?

Reflecting

9. Consider what our lives would be like if water really was a universal solvent. What would our planet look like? What would we look like?

6.2 Explaining Solutions

Water is common and familiar, but it is also a unique chemical and solvent. Aqueous solutions can be found everywhere—in nature, in homes, in stores, in laboratories, and in industries. There are many other liquids; some, like hydrocarbons, are found naturally and many others are synthetic. Can anything else act as a solvent? What determines which solute dissolves in which solvent?

INQUIRY SKILLS

- ○ Questioning
- ● Hypothesizing
- ● Predicting
- ○ Planning
- ○ Conducting
- ○ Recording
- ● Analyzing
- ● Evaluating
- ○ Communicating

Lab Exercise 6.2.1

Testing a Hypothesis on Dissolving

Your task is to create and test a hypothesis concerning what kinds of substances dissolve in each other. To be scientific, a hypothesis must be able to be tested empirically against the real world. For ease of testing, choose the liquids as solvents.

You are to complete the **Hypothesis**, **Prediction, Analysis,** and **Evaluation** sections of the report.

Question

Do water, table salt, table sugar, motor oil, gasoline, and ammonia dissolve in each other?

Hypothesis

(a) Create a testable hypothesis about what classes of substances dissolve in each other.

Prediction

(b) Write predictions based upon your Hypothesis—which of the provided chemicals will dissolve in each other? For each Prediction, use your Hypothesis to provide the reasoning behind the Prediction.

Experimental Design

Each of the solutes is mixed with the specified liquid solvent to determine if it dissolves (**Figure 1**). The quantities of solute and solvent, the temperature, and the stirring are controlled variables.

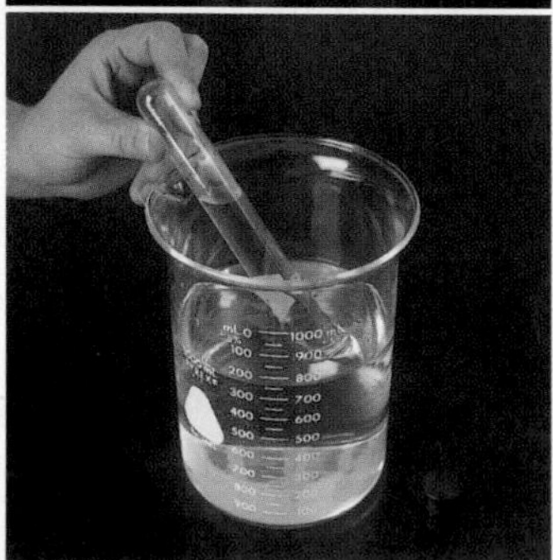

Figure 1
If a gas dissolves in water, then the water rises in the test tube.

Evidence

Table 1

	Solvent	
Solute	**Water**	**Gasoline**
salt	soluble	not soluble
sugar	soluble	not soluble
motor oil	not soluble	soluble
ammonia	soluble	not soluble

Analysis

(c) Answer the Question using a summary of the evidence collected.

Evaluation

(d) What are some sources of error or uncertainty in this experiment?
(e) Suggest some improvements you can make to the Materials and Procedure if this experiment were to be repeated.
(f) Are your predictions verified, falsified, or inconclusive? Justify your answer by comparing the experimental and predicted results.
(g) Based on your answer to (f), is your Hypothesis acceptable?

Practice

Understanding Concepts

1. How can you tell from a molecular formula whether a substance is polar? List four categories, with examples of each.
2. How can you tell from a molecular formula whether a substance is nonpolar? List two categories, with examples of each.

Explaining Water Mixtures

How can we explain the properties of solutes and solvents? We could use models, such as the concepts of molecules, atoms, electrons, and bonds or forces. In Chapter 2 we used the concept of **intramolecular forces** (i.e., covalent bonds) to describe, explain, and predict the chemical formulas of molecular substances. We use another concept to explain the physical properties of substances: the concept that there are forces holding molecules close to each other. These forces act between molecules, and so are called **intermolecular forces.**

intramolecular force: a specific attraction within a molecule

intermolecular force: an attraction between molecules

Now we have the challenge of explaining why some substances dissolve in a given solvent, but others do not. Let us start with aqueous solutions. Why do only some chemicals dissolve in water? Why are some chemicals mutually attracted to one another?

Molecular Substances in Water

We have gathered evidence and made some generalizations about which substances, including water, contain polar molecules (Chapter 2). Let us now find out if polarity has an effect on solubility. For example, we could test the hypothesis that polar substances dissolve in polar substances and nonpolar solutes in nonpolar solvents ("like dissolves like"). In other words, do like-polarity substances dissolve in each other?

DID YOU KNOW ?

Intra/Inter

The prefix "intra" is used in "intramural" as well as in "intramolecular." It means "within."

The prefix "inter" means "between," as in "interschool competitions," "the Internet," and "intermolecular forces."

INQUIRY SKILLS

- ○ Questioning
- ○ Hypothesizing
- ○ Predicting
- ● Planning
- ● Conducting
- ● Recording
- ● Analyzing
- ● Evaluating
- ● Communicating

Investigation 6.2.1

Polar and Nonpolar Solutes

The purpose of this investigation is to test the hypothesis that molecular substances with similar polarity dissolve in each other—"like dissolves like."

Complete the **Materials** and **Procedure** sections. Be sure to include all necessary safety precautions and equipment. When these have your teacher's approval, carry out the investigation, record your **Evidence**, and complete the **Analysis** and **Evaluation** sections of the report.

Question

Do water ($H_2O_{(l)}$), ethylene glycol ($C_2H_4(OH)_{2(l)}$), toluene ($C_7H_{8(l)}$), and mineral oil (a mixture of hydrocarbons, $C_xH_{y(l)}$) dissolve in each other (**Figure 2**)?

Hypothesis/Prediction

According to the hypothesis that like-polarity substances dissolve in each other, water and ethylene glycol will dissolve in each other, but not in toluene or mineral oil, and toluene and mineral oil will dissolve in each other. The reasoning behind this prediction is that water and propanol contain polar molecules while toluene and mineral oil contain nonpolar molecules.

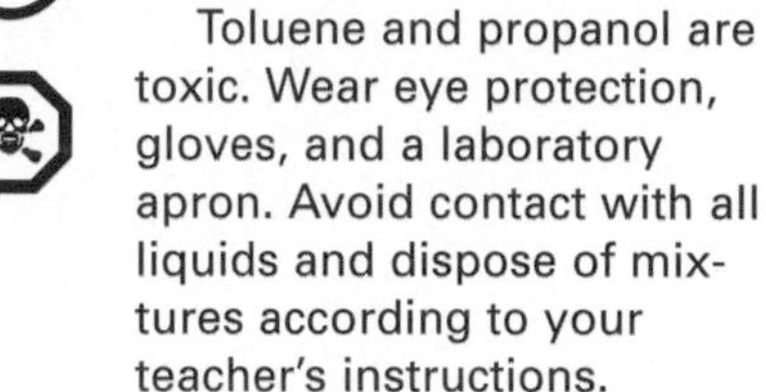

Toluene, ethylene glycol, and mineral oil are flammable.

Toluene and propanol are toxic. Wear eye protection, gloves, and a laboratory apron. Avoid contact with all liquids and dispose of mixtures according to your teacher's instructions.

Experimental Design

Water, ethylene glycol, toluene, and mineral oil are mixed in pairs (**Figure 3**). The independent variable is the kind of chemicals used (polar or nonpolar); the dependent variable is whether the substances dissolve; and the controlled variables are volume, temperature, and mixing.

(a) Plan and write a Procedure as a numbered list of steps. Be sure to include safety considerations and a disposal step.

Figure 3
If two liquid layers are apparent, then the liquids have low solubility.

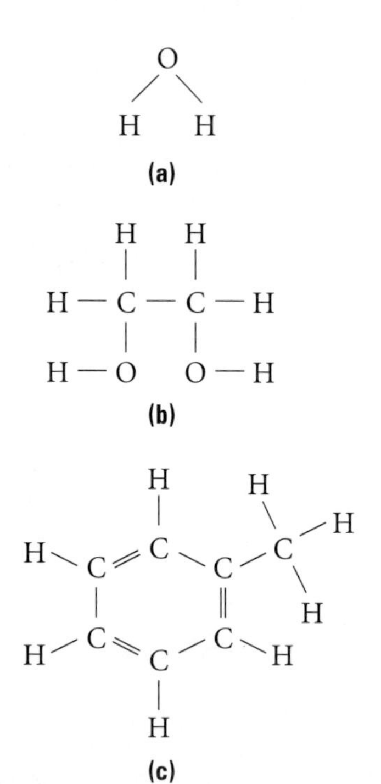

Figure 2
(a) Water
(b) Ethylene glycol
(c) Toluene

Materials

(b) List all materials, including sizes and approximate quantities.

Procedure

1. With your teacher's approval, carry out your procedure. Record your observations in a suitable format.

Analysis

(c) Do water, propanol, toluene, and mineral oil dissolve in each other? Indicate which substances dissolve in each other and which do not.

Evaluation

(d) Evaluate the evidence by critiquing the Experimental Design, Materials, and Procedure. Look for any flaws, sources of error, and opportunities for improvement. Overall, how would you judge the quality of the evidence?
(e) Assuming that the evidence was of suitable quality, judge the prediction and hypothesis. Are they valid?

The Effects of Polarity and Hydrogen Bonds

Ethylene glycol, a chemical with polar molecules, has a high solubility in water. Chemists have a theoretical explanation for this: Polar solute molecules are surrounded and suspended in solution by polar solvent molecules (**Figure 4**).

How do chemists explain the extraordinary solubility in water of solutes with hydrogen bonding capability, such as ethylene glycol? In Chapter 2, you saw that the higher than expected boiling points of some pure substances could be explained by hydrogen bonding. Recall that any substance with a hydrogen (H) atom covalently bonded to a N, O, or F atom can "hydrogen bond" to its own molecules to increase the intermolecular forces beyond the strength of London and dipole–dipole forces. Also recall that water has two hydrogen atoms bonded to an oxygen atom, which, in turn, has two lone-pairs of electrons to take part in hydrogen bonding with other water molecules.

Now consider that, when water is the solvent, there is a potential for solutes with N, O, or F lone-pairs or with a H—N, H—O, or H—F bond to form hydrogen bonds with water. For example, ammonia, $NH_{3(g)}$, has a very high solubility in water (as in household ammonia). Ammonia fulfills both criteria: It has a lone pair of electrons that could accept a positively charged hydrogen for sharing; and it has three hydrogens to share with the two lone-pairs on a nearby water molecule (**Figure 5**). When multiple hydrogen bonding is possible, we would predict an especially high solubility. More importantly, if both the solute and the solvent have H—N, H—O, or H—F bonds, then evidence suggests that the attractions and the solubility are high.

The explanation for the high solubility of ammonia is both logical and consistent, and can be tentatively accepted on that basis. The next test for any concept is its ability to predict an outcome. Can our concept for explaining solubility successfully predict solubilities in an experiment?

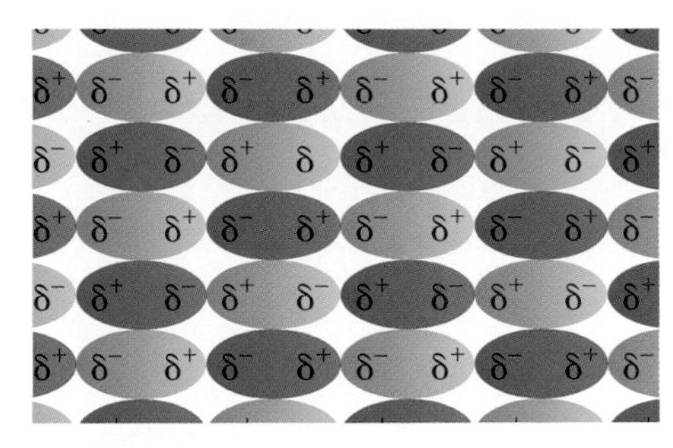

Figure 4
Polar solute molecules (in red) are surrounded by polar solvent molecules (in green). The dipole–dipole attractions explain the increase in solubility.

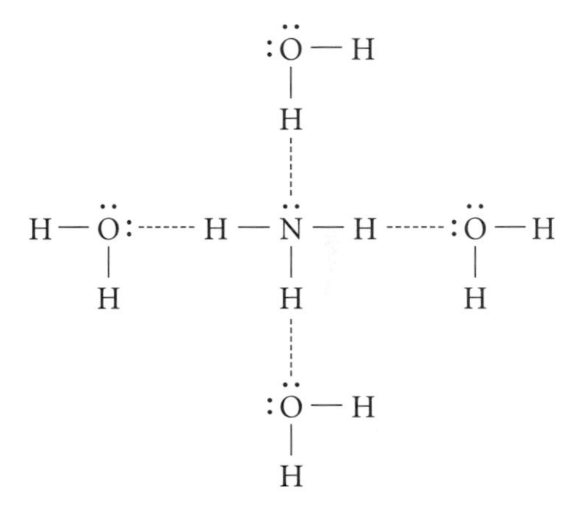

Figure 5
Multiple hydrogen bonds between ammonia and water result in a very high solubility for ammonia, much higher than expected from London and dipole–dipole intermolecular forces. Hydrogen bonding is the main reason for this high solubility, but there are other factors, which you will discover in Chapter 8.

Lab Exercise 6.2.2

Predicting High and Low Solubilities

We have created a concept concerning the solubilities of molecular substances: polar solutes dissolve in polar solvents, and nonpolar solutes dissolve in nonpolar solvents. In this exercise, we will use this concept and the information in **Table 2** and **Figure 6** (page 276) to predict the solubilities of a variety of substances in water. The procedure has been carried out, and the observations of solubilities are presented as a table of evidence. Complete the **Prediction**, **Analysis**, and **Evaluation** sections of the report.

INQUIRY SKILLS

- ○ Questioning
- ○ Hypothesizing
- ● Predicting
- ○ Planning
- ○ Conducting
- ○ Recording
- ● Analyzing
- ● Evaluating
- ○ Communicating

(a) acetic acid

```
        H
        |         O
        |       //
  H  —  C  —  C        H
        |       \     /
        |         O
        H
```

(b) carbon dioxide

```
  O = C = O
```

(c) dimethyl ether

```
        H
        |
  H  —  C  —  O        H
        |       \     /
        |         C
        H       /   \
              H       H
```

(d) methanol

```
        H
        |        H
        |       /
  H  —  C  —  O
        |
        H
```

(e) oxygen

```
  O = O
```

(f) propane

```
        H     H     H
        |     |     |
  H  —  C  —  C  —  C  —  H
        |     |     |
        H     H     H
```

(g) tetrachloromethane

```
         Cl
         |
  Cl  —  C  —  Cl
         |
         Cl
```

Figure 6
Structural formulas of some molecular substances

Table 2: Intermolecular Forces in Water

Chemical	Formula	Intermolecular force
acetic acid	$HC_2H_3O_{2(l)}$	LF, D-D, many H-B
carbon dioxide	$CO_{2(g)}$	LF, some H-B
dimethyl ether	$CH_3OCH_{3(g)}$	LF, D-D, some H-B
methanol	$CH_3OH_{(l)}$	LF, D-D, many H-B
oxygen	$O_{2(g)}$	LF, some H-B
propane	$C_3H_{8(g)}$	LF
tetrachloromethane	$CCl_{4(l)}$	LF

Key:
LF London forces D-D dipole–dipole forces H-B hydrogen

Question

Which of the substances listed in **Table 2** have high solubility in water? Which have low solubility in water?

Prediction

(a) Use the concepts of polarity and hydrogen bonding to predict the relative solubility in water for each substance listed under Materials. You may use general terms such as low and high, or simply rank the solubilities. Provide your reasoning for your prediction.

Materials

acetic acid, $HC_2H_3O_{2(l)}$
carbon dioxide, $CO_{2(g)}$
dimethyl ether, $CH_3OCH_{3(g)}$
methanol, $CH_3OH_{(l)}$
oxygen, $O_{2(g)}$
propane, $C_3H_{8(g)}$
tetrachloromethane, $CCl_{4(l)}$
water
test tubes
stirring rods

Evidence

Table 3: Solubility of Molecular Compounds in Water

Chemical	Solubility
acetic acid	very high
carbon dioxide	low
dimethyl ether	high
methanol	very high
oxygen	low
propane	very low
tetrachloromethane	very low

Analysis

(b) Answer the Question: Which of the substances have high solubility in water? Which have low solubility in water?

Evaluation

(c) What additional observations would be useful to improve the quality of the Evidence?
(d) Compare the experimental and predicted results and evaluate the Prediction.

Practice

Understanding Concepts

3. Distinguish between intramolecular and intermolecular forces.
4. Suppose someone spilled some gasoline while filling a gas tank on a rainy day.
   (a) If some gasoline ran into a puddle of water, would it dissolve in the water? What evidence would support your prediction?
   (b) What rule did you use to predict whether dissolving will occur?
   (c) How does this rule apply to the gasoline-water mixture?
5. Windshield washer fluid contains methanol dissolved in water.
   (a) Why does methanol dissolve well in water? Explain in terms of intermolecular forces.
   (b) Draw a Lewis structure of a methanol molecule and several water molecules to show possible hydrogen bonds. (Use dashed lines to represent H-bonds.)
   (c) What would you expect to be the relationship between the number of possible hydrogen bonds and the solubility? Why?

Ionic Compounds in Water

Water is the most important solvent on Earth. The oceans, lakes, rivers, and rain are aqueous solutions containing many different ionic compounds and a few molecular solutes. As you know, there are some ionic compounds that dissolve only very slightly in water, such as limestone (calcium carbonate) buildings and various other rocks and minerals. Nevertheless, many more ionic compounds dissolve in water than in any other known solvent.

Why are ionic compounds so soluble in water? The key to the explanation came from the study of electrolytes. Electrolytes were first explained by Svante Arrhenius who was born in Wijk, Sweden, in 1859. While attending the University of Uppsala, he became intrigued by the problem of how and why some aqueous solutions conduct electricity, but others do not. This problem had puzzled chemists ever since Sir Humphry Davy and Michael Faraday experimented over half a century earlier by passing electric currents through chemical substances.

Faraday believed that an electric current produces new charged particles in a solution. He called these electric particles ions (a form of the Greek word for "to go"). He could not explain what ions were, or why they did not form in solutions of substances such as sugar or alcohol dissolved in water.

In 1887 Arrhenius proposed a new hypothesis: that particles of a substance, when dissolving, separate from each other and disperse into the solution. Nonelectrolytes disperse electrically neutral particles throughout the solution. As **Figure 7** shows, molecules of sucrose (a nonelectrolyte) separate from each other

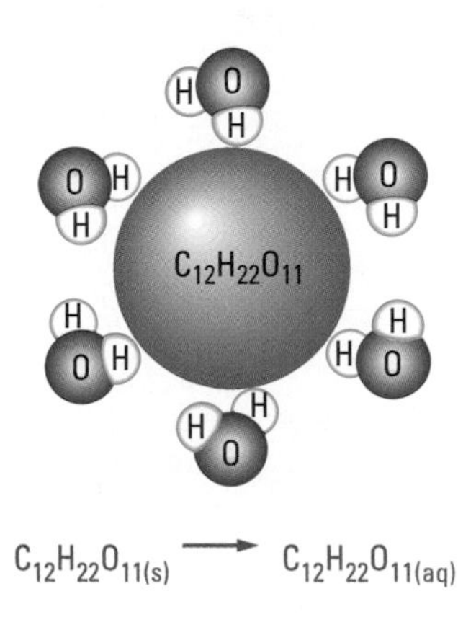

$$C_{12}H_{22}O_{11(s)} \rightarrow C_{12}H_{22}O_{11(aq)}$$

Figure 7
This model illustrates sucrose dissolved in water. The model, showing electrically neutral particles in solution, agrees with the evidence that a sucrose solution does not conduct electricity.

and disperse in an aqueous solution as individual molecules of sucrose surrounded by water molecules.

But what about the conductivity of solutions of electrolytes? Arrhenius' explanation for this observation was quite radical. He agreed with the accepted theory that electric current involves the movement of electric charge. Ionic compounds form conducting solutions. Therefore, according to Arrhenius, electrically charged particles must be present in their solutions. For example, when a compound such as table salt dissolves, it **dissociates** into individual aqueous ions (**Figure 8**). The positive ions are surrounded by the negative ends of the polar water molecules, while the negative ions are surrounded by the positive ends of the polar water molecules. Dissociation equations, such as the following examples, show this separation of ions.

dissociation: the separation of ions that occurs when an ionic compound dissolves in water

$$NaCl_{(s)} \rightarrow Na^{+}_{(aq)} + Cl^{-}_{(aq)}$$

$$(NH_4)_2SO_{4(s)} \rightarrow 2\ NH^{+}_{4(aq)} + SO^{2-}_{4(aq)}$$

Notice that the formula for the solvent, $H_2O_{(l)}$, does not appear as a reactant in the equation. Although water is necessary for the process of dissociation, it is not consumed and hence is not a reactant. The presence of water molecules surrounding the ions is indicated by the subscript $_{(aq)}$.

DID YOU KNOW?

Getting Your Vitamins

Some vitamins, whose molecules are non-polar, are soluble in fats. Fat-soluble vitamins include Vitamins A, D, and E. Other vitamins, whose molecules are polar, are soluble in water. Water-soluble vitamins include Vitamin C, the B vitamins, and pantothemic and folic acids. The fat soluble vitamins must be ingested with fatty foods. Like dissolves in like.

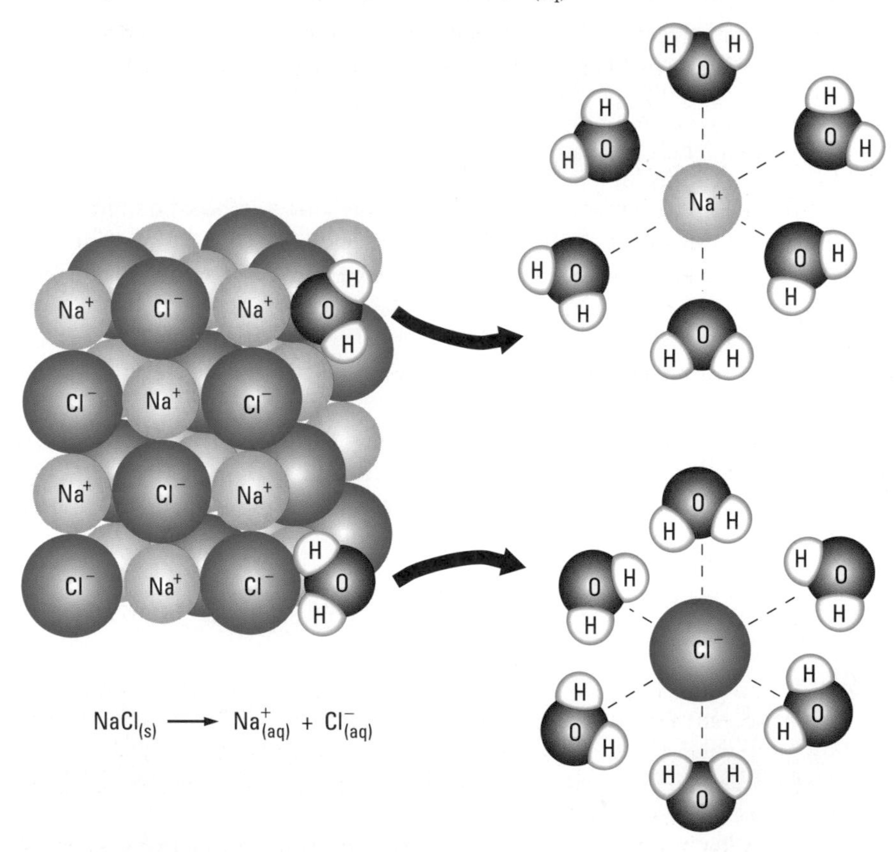

Figure 8
This model represents the dissociation of sodium chloride into positive and negative ions.

Practice

Understanding Concepts

6. A sugar cube and a lump of table salt are put into separate glasses of water.
   (a) Note as many observations as you can about each mixture.

(b) According to theory, how is the dissolving process similar for both solutes?
(c) According to theory, how are the final solutions different?
(d) What theoretical properties of a water molecule help to explain the dissolving of both solutes?

7. Write equations to represent the dissociation of the following ionic compounds when they are placed in water:
   (a) sodium fluoride
   (b) sodium phosphate
   (c) potassium nitrate
   (d) aluminum sulfate
   (e) ammonium hydrogen phosphate
   (f) cobalt(II) chloride hexahydrate

Explaining Nonaqueous Mixtures

We now have an explanation for the solubility of ionic compounds (electrolytes) and molecular substances (nonelectrolytes) in water. The next question is, why do nonpolar solutes (e.g., grease) dissolve in nonpolar solvents (e.g., kerosene)? These chemicals do not show evidence of either polar molecules (dipoles) or hydrogen bonding. The third intermolecular force that we can look to for an explanation is the London (dispersion) force. London forces are weak intermolecular forces, and can explain the relatively low boiling points of nonpolar and non-hydrogen bonding compounds. Similarly, they can explain why these same compounds dissolve in each other to form solutions (**Figure 9**).

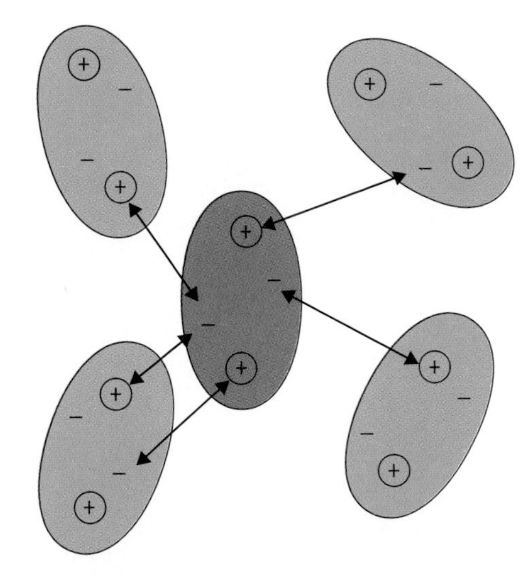

Figure 9
London (dispersion) forces are believed to be responsible for nonpolar solutes (red) dissolving in nonpolar solvents (blue).

Water—"The Universal Solvent"

Seventy-five percent of Earth is covered with water—both liquid and solid. All forms of life, including animals and plants, depend upon the life-supporting fluid called water. Water dissolves so many more substances than any other liquid that it is not surprising that water is often referred to as the "universal solvent." This success as a solvent is due to three features of water molecules: their small size; their highly polar nature; and their considerable capacity for hydrogen bonding.

Practice

Understanding Concepts

8. Describe as many properties of water as you can, using diagrams where necessary. Classify the properties as either empirical or theoretical.
9. If you get some grease from a bicycle chain on your pants, what kind of solvent would be best to dissolve the grease? Explain your choice.
10. (a) Which of $C_6H_{14(l)}$ and $C_6H_{12}O_{6(s)}$ is more soluble in water, and why?
    (b) Which of $C_2H_5OH_{(l)}$ and $CH_3OCH_{3(g)}$ (**Figure 10**) is more soluble in water, and why?
    (c) Which of $Na_2CO_{3(s)}$ and $CO_{2(g)}$ is more soluble in water, and why?
11. There is a series of gaseous compounds in which one carbon molecule is bonded to four other atoms: CH_4, CH_3Cl, CH_2Cl_2, $CHCl_3$, and CCl_4.
    (a) Which of these five compounds are most likely to dissolve in water?
    (b) Which are most likely to dissolve in a nonpolar solvent, such as hexane? Briefly justify your answers.

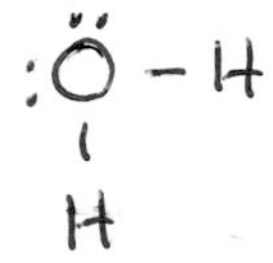

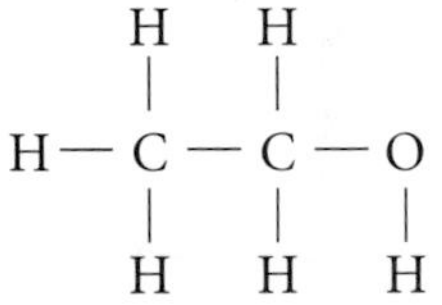

ethanol, $C_2H_5OH_{(l)}$

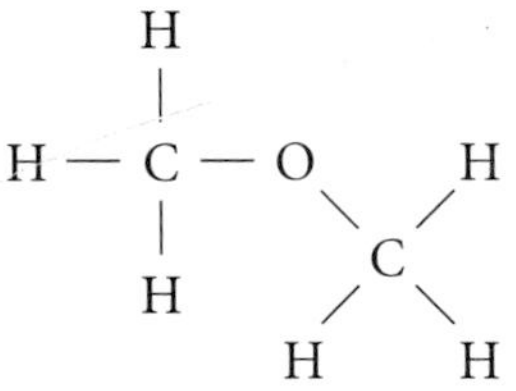

dimethyl ether, $CH_3OCH_{3(g)}$

Figure 10
The shapes of molecules can often help us predict their solubility.

12. Laboratory evidence indicates that butanol, $C_4H_9OH_{(l)}$, has a higher solubility in gasoline than methanol, $CH_3OH_{(l)}$. Create a theoretical hypothesis to explain this evidence.

Applying Inquiry Skills

13. For an explanation to be accepted by the scientific community, what criteria are used to judge the explanation?

Making Connections

14. Water and soap are not always sufficient to clean clothes, especially when they are stained with grease or oil. Dry cleaners clean clothes without water.
    (a) What solvents are used and what are their properties?
    (b) What health regulations relate to these solvents and their use in the dry cleaning industry?

    Follow the links for Nelson Chemistry 11, 6.2.

GO TO www.science.nelson.com

Section 6.2 Questions

Understanding Concepts

1. Why is water such a good solvent for so many substances? In your answer refer to the properties of water, and describe them at the molecular level.
2. (a) From your own observations, which of the following substances dissolve in each other and which do not?
    (i) water and gasoline
    (ii) oil and vinegar
    (iii) gasoline and grease
    (iv) sugar and water
    (v) water and alcohol

    (b) Were your answers based on empirical or theoretical knowledge? Provide your reasoning.
3. Which of the following substances dissolve in each other and which do not? Give reasons for your predictions.
    (a) $H_2O_{(l)}$ and $C_8H_{18(l)}$
    (b) $HC_2H_3O_{2(l)}$ and $H_2O_{(l)}$
    (c) $CCl_{4(l)}$ and $HCl_{(aq)}$
    (d) $H_2O_{(l)}$ and $C_3H_7OH_{(l)}$
    (e) $N_{2(g)}$ and $H_2O_{(l)}$
4. Order the chemicals—ammonia, methane, and methanol—in terms of increasing solubility in water. Provide your reasoning.

Applying Inquiry Skills

5. Aqueous solutions of salt and sugar are mixed to determine whether salt dissolves in sugar. Critique this experimental design.
6. Plan an investigation to discover whether $C_6H_{6(l)}$ or $C_6H_5OH_{(l)}$ has a higher solubility in water. Provide the Question, Prediction, Experimental Design, and Materials.

Making Connections

7. A scientist is developing a glue suitable for children to use. What properties should the glue have? What should be the chemical properties of the compounds in the glue?

6.3 Solution Concentration

Most aqueous solutions are colourless, so there is no way of knowing, by looking at them, how much of the solute is present in the solution. As we often need to know the amount of solute in the solution, it is important that solutions be labelled with this information. We use a ratio that compares the quantity of solute to the quantity of the solution. This ratio is called the solution's **concentration**. Chemists describe a solution of a given substance as **dilute** if it has a relatively small quantity of solute per unit volume of solution (**Figure 1**). A **concentrated** solution, on the other hand, has a relatively large quantity of solute per unit volume of solution.

concentration: the quantity of a given solute in a solution

dilute: having a relatively small quantity of solute per unit volume of solution

concentrated: having a relatively large quantity of solute per unit volume of solution

In general, the concentration, c, of any solution is expressed by the ratio

$$\text{concentration} = \frac{\text{quantity of solute}}{\text{quantity of solution}}$$

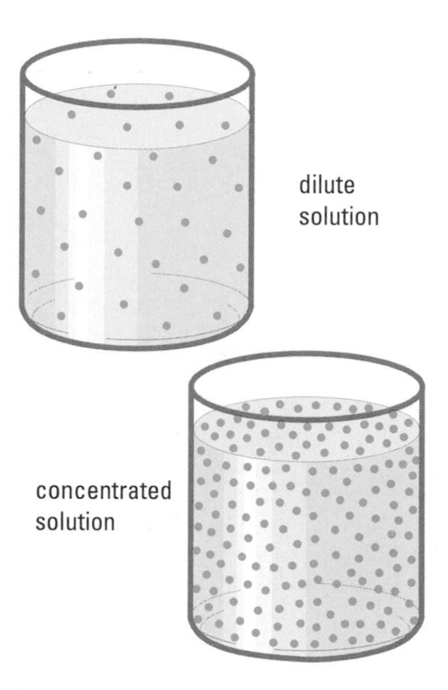

Figure 1
The theoretical model of the dilute solution shows fewer solute entities (particles) per unit volume compared with the model of the concentrated solution.

Percentage Concentration

Many consumer products, such as vinegar (acetic acid), are conveniently labelled with their concentration ratios expressed as percentages (**Figure 2**). A vinegar label listing "5% acetic acid (by volume)" means that there are 5 mL of pure acetic acid dissolved in every 100 mL of the vinegar solution. This type of concentration is often designated as % V/V, percentage volume by volume, or percentage by volume.

$$c_{HC_2H_3O_2} = \frac{5\text{ mL}}{100\text{ mL}} = 5\%\text{ V/V}$$

In general, a percentage by volume concentration may be defined as

$$c = \frac{v_{\text{solute}}}{v_{\text{solution}}} \times 100\%$$

Sample Problem 1

A photographic "stop bath" contains 140 mL of pure acetic acid in a 500-mL bottle of solution. What is the percentage by volume concentration of acetic acid?

Solution

$$v_{HC_2H_3O_{2(l)}} = 140\text{ mL}$$
$$v_{HC_2H_3O_{2(aq)}} = 500\text{ mL}$$
$$c_{HC_2H_3O_2} = \frac{140\text{ mL}}{500\text{ mL}} \times 100$$
$$c_{HC_2H_3O_2} = 28.0\%\text{ V/V}$$

The percentage by volume concentration of acetic acid is 28.0%.

Another common concentration ratio used for consumer products is "percentage weight by volume" or % W/V. (In consumer and commercial applications, "weight" is used instead of "mass," which explains the W in the W/V label.) For example, a hydrogen peroxide topical solution used as an antiseptic is 3% W/V (**Figure 2**). This means that 3 g of hydrogen peroxide is in every 100 mL of solution.

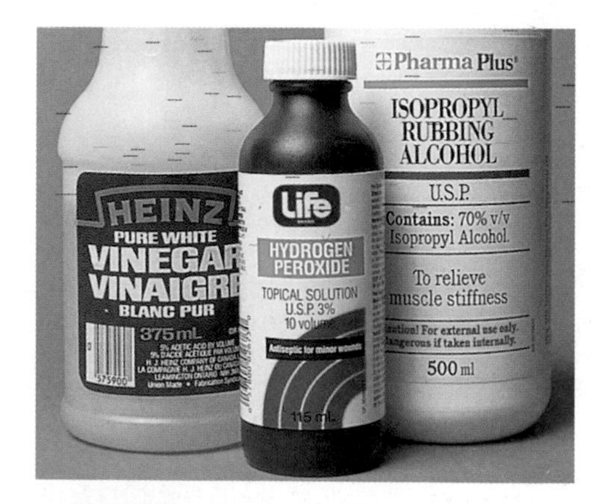

Figure 2
The concentrations of different consumer products depend on the product and sometimes the brand name. Concentrations are usually expressed as a percentage.

$$c_{H_2O_2} = \frac{3\text{ g}}{100\text{ mL}}$$

$$c_{H_2O_2} = 3\%\text{ W/V}$$

In general, we write a percentage weight by volume concentration as

$$c = \frac{m_{solute}}{v_{solution}} \times 100\%$$

A third concentration ratio is the "percentage weight by weight," or % W/W:

$$c = \frac{m_{solute}}{m_{solution}} \times 100\%$$

Sample Problem 2

A sterling silver ring has a mass of 12.0 g and contains 11.1 g of pure silver. What is the percentage weight by weight concentration of silver in the metal?

Solution

$m_{Ag} = 11.1\text{ g}$

$m_{alloy} = 12.0\text{ g}$

$$c_{Ag} = \frac{11.1\text{ g}}{12.0\text{ g}} \times 100\%$$

$$c_{Ag} = 92.5\%\text{ W/W}$$

The ring is 92.5 % W/W silver.

Very Low Concentrations

parts per million: unit used for very low concentrations

In studies of solutions in the environment, we often encounter very low concentrations. For very dilute solutions we choose a concentration unit to give reasonable numbers for very small quantities of solute. For example, the concentration of toxic substances in the environment or of chlorine in a swimming pool is usually expressed as **parts per million** (ppm, $1{:}10^6$) or even smaller ratios, such as parts per billion (ppb, $1{:}10^9$) or parts per trillion (ppt, $1{:}10^{12}$). These ratios are used for liquid and solid mixtures and are a special case of the weight by weight (W/W) ratio. One part per million of chlorine in a swimming pool corresponds to 1 g of chlorine in 10^6 g of pool water. Because very dilute aqueous solutions are very similar to pure water, their densities are considered to be the same: 1 g/mL. Therefore, 1 ppm of chlorine is 1 g in 10^6 g or 10^6 mL (1000 L) of pool water, which is equivalent to 1 mg of chlorine per litre of water. Small concentrations such as ppm, ppb, and ppt are difficult to imagine, but are very important in environmental studies and in the reporting of toxic effects of substances (**Table 1**).

Table 1: Parts per Million, Billion, and Trillion

1 ppm	1 drop in a full bathtub
1 ppb	1 drop in a full swimming pool
1 ppt	1 drop in 1000 swimming pools

We can express the parts per million (ppm) concentration using a variety of units. Choose the one that matches the information given in the example you are calculating. For aqueous solutions,

$$\begin{aligned} 1\text{ ppm} &= 1\text{ g}/10^6\text{ mL} \\ &= 1\text{ g}/1000\text{ L} \\ &= 1\text{ mg/L} \\ &= 1\text{ mg/kg} \\ &= 1\ \mu\text{g/g} \end{aligned}$$

Sample Problem 3

Dissolved oxygen in natural waters is an important measure of the health of the ecosystem. In a chemical analysis of 250 mL of water at SATP, 2.2 mg of oxygen was measured. What is the concentration of oxygen in parts per million?

Solution

$m_{O_2} = 2.2 \text{ mg}$

$v_{O_2} = 250 \text{ mL or } 0.250 \text{ L}$

$c_{O_2} = \frac{2.2 \text{ mg}}{0.250 \text{ L}}$

$= 8.8 \text{ mg/L}$

$c_{O_2} = 8.8 \text{ ppm}$

The oxygen concentration is 8.8 ppm.

mole: the amount of a substance; the number of particles equivalent to Avogadro's number (6.02×10^{23}); the number of carbon atoms in exactly 12 g of a carbon-12 sample; the unit of stoichiometry

molar concentration: the amount of solute, in moles, dissolved in one litre of solution

Molar Concentration

Chemistry is primarily the study of chemical reactions, which we communicate using balanced chemical equations. The coefficients in these equations represent amounts of chemicals in units of **moles**. Concentration is therefore communicated using **molar concentration**. Molar concentration, *C*, is the amount of solute in moles dissolved in one litre of solution.

$$\text{molar concentration} = \frac{\text{amount of solute (in moles)}}{}$$

$$C = \frac{n}{}$$

The units of molar concentration (mol/L) come directly from this ratio. The symbol *C* denotes a molar quantity, just as *M* is molar mass, but *m* is mass.

Molar concentration is sometimes indicated by the use of square brackets. For example, the molar concentration of sodium hydroxide in water could be represented by $[NaOH_{(aq)}]$.

DID YOU KNOW?

What's M?

Although it is not an SI unit, some chemists use the unit "molar," M, to express molarity or molar concentration (**Figure 3**). 1 M = 1 mol/L.

Sample Problem 4

In a quantitative analysis, a stoichiometry calculation produced 0.186 mol of sodium hydroxide in 0.250 L of solution. Calculate the molar concentration of sodium hydroxide.

Solution

$n_{NaOH} = 0.186 \text{ mol}$

$v_{NaOH} = 0.250 \text{ L}$

$C_{NaOH} = \frac{0.186 \text{ mol}}{0.250 \text{ L}}$

$C_{NaOH} = 0.744 \text{ mol/L}$

The sodium hydroxide molar concentration is 0.744 mol/L.

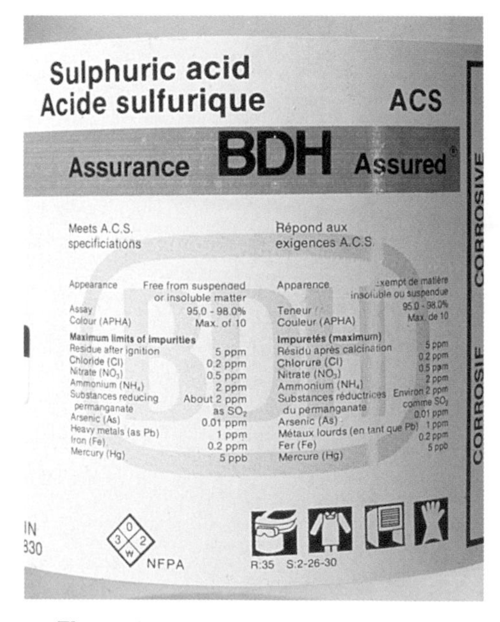

Figure 3
Sulfuric acid in the laboratory may have a molar concentration of 17.8 mol/L.

Practice

Answers

2. 7.5% V/V
3. 32% W/V
4. 4.8% W/W
5. 8 mg
6. 5.4 ppm
7. (a) 1/1000
   (c) 30μg
8. 1.8 mol/L

Understanding Concepts

1. What are three different ways of expressing the concentration of a solution?
2. Gasohol, which is a solution of ethanol and gasoline, is considered to be a cleaner fuel than just gasoline alone. A typical gasohol mixture available across Canada contains 4.1 L of ethanol in a 55-L tank of fuel. Calculate the percentage by volume concentration of ethanol.
3. Solder flux, available at hardware and craft stores, contains 16 g of zinc chloride in 50 mL of solution. The solvent is aqueous hydrochloric acid. What is the percentage weight by volume of zinc chloride in the solution?
4. Brass is a copper-zinc alloy. If the concentration of zinc is relatively low, the brass has a golden colour and is often used for inexpensive jewellery. If a 35.0 g pendant contains 1.7 g of zinc, what is the percentage weight by weight of zinc in this brass?
5. If the concentration of oxygen in water is 8 ppm, what mass of oxygen is present in 1 L of water?
6. Formaldehyde, $CH_2O_{(g)}$, is an indoor air pollutant that comes from synthetic materials and cigarette smoke. Formaldehyde is controversial because it is a probable carcinogen. If a 500-L indoor air sample with a mass of 0.59 kg contained 3.2 mg of formaldehyde, this would be considered a dangerous level. What would be the concentration of formaldehyde in parts per million?
7. Very low concentrations of toxic substances sometimes require the use of the parts per billion (ppb) concentration.
   (a) How much smaller is 1 ppb than 1 ppm?
   (b) Use the list of equivalent units for parts per million to make a new list for parts per billion.
   (c) Copper is an essential trace element for animal life. An average adult requires the equivalent of a litre of water containing 30 ppb of copper a day. What mass of copper is this per kilogram of solution?
8. A plastic dropper bottle for a chemical analysis contains 0.11 mol of calcium chloride in 60 mL of solution. Calculate the molar concentration of calcium chloride.

Making Connections

9. Toxicity of substances for animals is usually expressed by a quantity designated as "LD_{50}." Use the Internet to research the use of this quantity. What does LD_{50} mean? What is the concentration in ppm for a substance considered "extremely toxic" and one considered "slightly toxic"?
   Follow the links for Nelson Chemistry 11, 6.3.

GO TO www.science.nelson.com

Reflecting

10. How is your report card mark in a subject like a concentration? What other ratios have you used that are similar to concentration ratios?

Calculations Involving Concentrations

Solutions are so commonly used in chemistry that calculating concentrations might be the primary reason why chemists pull out their calculators. In associated calculations, chemists and chemical technicians also frequently need to calculate a quantity of solute or solution. Any of these calculations may involve percentage concentrations, very low concentrations, or molar concentrations. When we know two of these values—quantity of solute, quantity of solution, and concentration of solution—we can calculate the third. Because concentration is a ratio, a simple procedure is to use the concentration ratio (quantity of solute/quantity of solution) as a conversion factor. This approach parallels the one you followed when using molar mass as a conversion factor.

Suppose you are a nurse who needs to calculate the mass of dextrose, $C_6H_{12}O_{6(s)}$, present in a 1000-mL intravenous feeding of D5W, which is a solution of 5.0 % W/V dextrose in water. The conversion factor you need to use is the mass/volume ratio.

$$m_{C_6H_{12}O_6} = 1000 \cancel{\text{mL}} \times \frac{5.0 \text{ g}}{100 \cancel{\text{mL}}}$$

$$m_{C_6H_{12}O_6} = 50 \text{ g}$$

In some calculations you may want to find the quantity of solution, in which case you will have to flip the ratio to quantity of solution/quantity of solute. This is then the appropriate conversion factor.

For example, what volume of 30.0% W/V hydrogen peroxide solution can be made from 125 g of pure hydrogen peroxide? You know that the answer must be greater than 100 mL because 125 g is greater than 30.0 g (the quantity in 100 mL). Notice how the units cancel to produce the expected volume unit, millilitres, when we use the volume/mass ratio.

$$v_{H_2O_2} = 125 \cancel{\text{g}} \times \frac{100 \text{ mL}}{30.0 \cancel{\text{g}}}$$

$$v_{H_2O_2} = 417 \text{ mL}$$

Thinking about the quantity given and the concentration ratio helps to ensure you are calculating correctly. This method also works for other concentration ratios.

The quantities of solute and solution may be expressed as mass and volume respectively, so the appropriate conversion ratio would be a mass/volume ratio.

Sample Problem 5

A box of apple juice has a fructose (sugar) concentration of 12 g/100 mL (12% W/V) (**Figure 4**). What mass of fructose is present in a 175-mL glass of juice? (The chemical formula for fructose is $C_6H_{12}O_6$.)

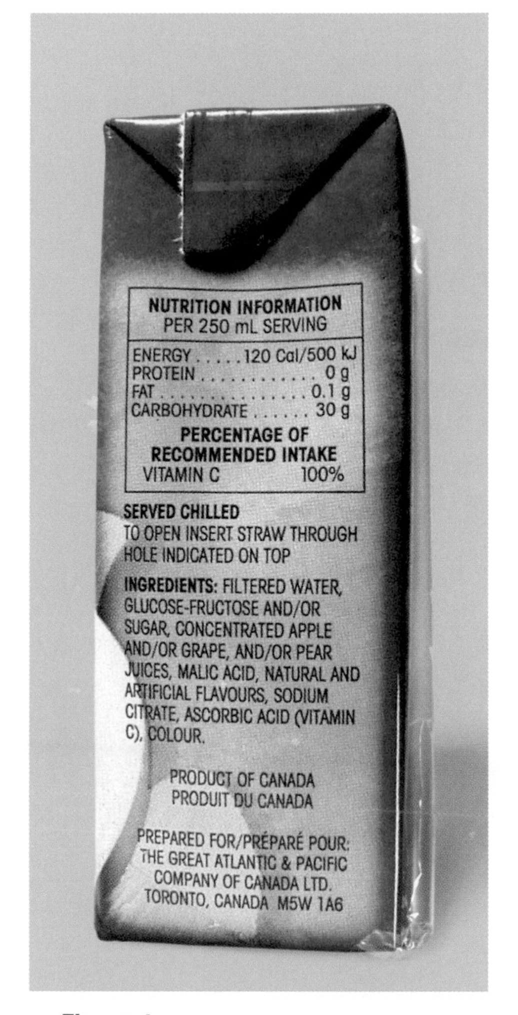

Figure 4
The label on a box of apple juice gives the ingredients and some nutritional information, but not the concentration of the various solutes.

Solution

$$C_{C_6H_{12}O_6} = \frac{12 \text{ g}}{100 \text{ mL}}$$

$$v_{\text{apple juice}} = 175 \text{ mL}$$

$$m_{C_6H_{12}O_6} = 175 \cancel{\text{mL}} \times \frac{12 \text{ g}}{100 \cancel{\text{mL}}}$$

$$m_{C_6H_{12}O_6} = 21 \text{ g}$$

There are 21 g of fructose in each glass of apple juice.

Sample Problem 6

People with diabetes have to monitor and restrict their sugar intake. What volume of apple juice could a diabetic person drink, if his sugar allowance for that beverage was 9.0 g? Assume that the apple juice has a sugar concentration of 12 g/100 mL (12% W/V), and that the sugar in apple juice is fructose.

Solution

$$c_{C_6H_{12}O_6} = \frac{12\ g}{100\ mL}$$

$$m_{C_6H_{12}O_6} = 9.0\ g$$

$$v_{C_6H_{12}O_6} = 9.0\ \cancel{g} \times \frac{100\ mL}{12\ \cancel{g}}$$

$$v_{C_6H_{12}O_6} = 75\ mL$$

The person could drink 75 mL of apple juice.

When you are given a concentration in parts per million (ppm) you may choose from among a variety of conversion factors. If, for example, you are given a value of 99 ppm for biomagnification of DDT in a 2-kg gull, what mass of DDT is present? Since the mass, rather than the volume, of the gull is given, use 99 ppm as 99 mg/kg rather than 99 mg/L. The concentration ratio is 99 mg/kg. Note the cancellation of kilograms.

$$m_{DDT} = 2\ \cancel{kg} \times \frac{99\ mg}{1\ \cancel{kg}}$$

$$m_{DDT} = 0.2\ g \text{ (rounded from 198 mg)}$$

Sample Problem 7

A sample of well water contains 0.24 ppm of dissolved iron(III) sulfate from the surrounding rocks. What mass of iron(III) sulfate is present in 1.2 L of water in a kettle?

Solution

$$c_{Fe_2(SO_4)_3} = 0.24\ ppm$$

$$= \frac{0.24\ mg}{1\ L}$$

$$v = 1.2\ L$$

$$m_{Fe_2(SO_4)_3} = 1.2\ \cancel{L} \times \frac{0.24\ mg}{1\ \cancel{L}}$$

$$m_{Fe_2(SO_4)_3} = 0.29\ mg$$

The mass of iron dissolved in the water in the kettle is 0.29 mg.

Sample Problem 8

A sample of laboratory ammonia solution has a molar concentration of 14.8 mol/L (**Figure 5**). What amount of ammonia is present in a 2.5 L bottle?

Solution

$C_{NH_3} = \frac{14.8\ \text{mol}}{1\ \text{L}}$

$v_{NH_3} = 2.5\ \text{L}$

$$n_{NH_3} = 2.5\ \cancel{\text{L}} \times \frac{14.8\ \text{mol}}{1\ \cancel{\text{L}}}$$

$$n_{NH_3} = 37\ \text{mol}$$

The amount of ammonia present in the bottle is 37 mol.

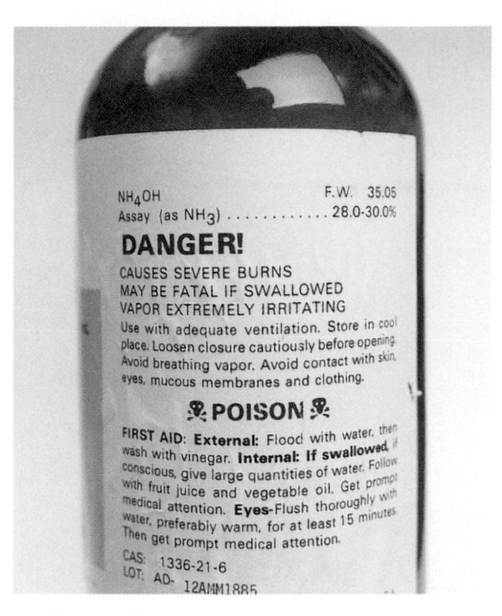

Figure 5
Aqueous ammonia is purchased for science laboratories as a concentrated solution. What is the concentration of the solute?

You should always check that your answer makes sense. For example, in Sample Problem 8, 14.8 mol/L means that there is 14.8 mol of ammonia in 1 L of solution. Therefore, 2.5 L, which is greater than 1 L, must produce an amount greater than 14.8 mol.

In some situations you may know the molar concentration and need to find either the volume of solution or amount (in moles) of solute. In these situations use either the volume/amount or amount/volume ratio. Notice that the units of the quantity you want to find should be the units in the numerator of the conversion factor ratio.

Sample Problem 9

What volume of a 0.25 mol/L salt solution in a laboratory contains 0.10 mol of sodium chloride?

Solution

$C_{NaCl} = \frac{0.25\ \text{mol}}{1\text{L}}$

$n_{NaCl} = 0.10\ \text{mol}$

$$v_{NaCl} = 0.10\ \cancel{\text{mol}} \times \frac{1\ \text{L}}{0.25\ \cancel{\text{mol}}}$$

$$v_{NaCl} = 0.40\ \text{L}$$

You need 0.40 L of salt solution to provide 0.10 mol of sodium chloride.

Practice

Understanding Concepts

11. Rubbing alcohol, $C_3H_7OH_{(l)}$, is sold as a 70.0% V/V solution for external use only. What volume of pure $C_3H_7OH_{(l)}$ is present in a 500-mL (assume three significant digits) bottle?
12. Suppose your company makes hydrogen peroxide solution with a generic label for drugstores in your area. What mass of pure hydrogen peroxide is needed to make 1000 bottles each containing 250 mL of 3.0% W/V $H_2O_{2(aq)}$?
13. The maximum acceptable concentration of fluoride ions in municipal water supplies is 1.5 ppm. What is the maximum mass of fluoride ions you would get from a 0.250-L glass of water?

Answers

11. 0.350 L
12. 7.5 kg
13. 0.38 mg

Answers

14. 4.1 mol
15. 0.25 mol
16. 0.403 L
17. 54 mL

14. Seawater contains approximately 0.055 mol/L of magnesium chloride. What amount, in moles, of magnesium chloride is present in 75 L of seawater?
15. A bottle of 5.0 mol/L hydrochloric acid is opened in the laboratory, and 50 mL of it is poured into a beaker. What amount of acid is in the beaker?
16. A household ammonia solution (e.g., a window-cleaning solution) has a concentration of 1.24 mol/L. What volume of this solution would contain 0.500 mol of $NH_{3(aq)}$?
17. A student needs 0.14 mol of $Na_2SO_{4(aq)}$ to do a quantitative analysis. The concentration of her solution is 2.6 mol/L $Na_2SO_{4(aq)}$. What volume of solution does she need to measure?

Making Connections

18. When shopping for a floor cleaner, you have a choice between a dilute solution that is used directly and a concentrated solution of the same chemical that is diluted with water at home. How would you decide which one to buy? Describe several criteria you would use.

Mass, Volume, and Concentration Calculations

While the mole is a very important unit, measurements in a chemistry laboratory are usually of mass (in grams) and of volume (in millilitres). A common chemistry calculation involves the mass of a substance, the volume of a solution, and the molar concentration of that solution. This type of calculation requires the use of two conversion factors—one for molar mass and one for molar concentration. Calculations using molar mass are just like the ones you did in Unit 2. For example, a chemical analysis requires 2.00 L of 0.150 mol/L $AgNO_{3(aq)}$. What mass of silver nitrate solid is required to prepare this solution? First, you must determine the amount of silver nitrate needed, in moles.

$$n_{AgNO_3} = 2.00\ \cancel{L} \times \frac{0.150\ \text{mol}}{1\ \cancel{L}}$$

$$n_{AgNO_3} = 0.300\ \text{mol}$$

You can then convert this amount into a mass of silver nitrate by using its molar mass, *M*. The molar mass of silver nitrate is 169.88 g/mol.

$$m_{AgNO_3} = 0.300\ \cancel{\text{mol}} \times \frac{169.88\ \text{g}}{1\ \cancel{\text{mol}}}$$

$$m_{AgNO_3} = 51.0\ \text{g}$$

The solution requires 51.0 g of solid silver nitrate.

If you clearly understand these two steps, you could combine them into one calculation.

$$m_{AgNO_3} = 2.00\ \cancel{L} \times \frac{0.150\ \cancel{\text{mol}}}{1\ \cancel{L}} \times \frac{169.88\ \text{g}}{1\ \cancel{\text{mol}}}$$

$$m_{AgNO_3} = 51.0\ \text{g}$$

In order to successfully combine the steps into one operation, as shown above, you need to pay particular attention to the units in the calculation. Cancelling the units will help you to check your procedure.

Sample Problem 10

To study part of the water treatment process in a laboratory, a student requires 1.50 L of 0.12 mol/L aluminum sulfate solution. What mass of aluminum sulfate must she measure for this solution?

Solution

$v_{Al_2(SO_4)_{3(aq)}} = 1.50\ L$

$C_{Al_2(SO_4)_3} = 0.12\ mol/L$

$M_{Al_2(SO_4)_3} = 342.14\ g/mol$

$$n_{Al_2(SO_4)_3} = 1.50\ \cancel{L} \times \frac{0.12\ mol}{1\ \cancel{L}}$$

$$= 0.180\ mol$$

$$m_{Al_2(SO_4)_3} = 0.180\ \cancel{mol} \times \frac{342.14\ g}{1\ \cancel{mol}}$$

$$m_{Al_2(SO_4)_3} = 61.6\ g$$

or

$$m_{Al_2(SO_4)_3} = 1.50\ \cancel{L} \times \frac{0.12\ \cancel{mol}}{1\ \cancel{L}} \times \frac{342.14\ g}{1\ \cancel{mol}}$$

$$m_{Al_2(SO_4)_3} = 61.6\ g$$

The student will need 61.6 g of aluminum sulfate.

Another similar calculation involves the use of a known mass and volume to calculate the molar concentration of a solution. This is similar to the examples given above, in reverse order.

Sample Problem 11

Sodium carbonate is a water softener that is a significant part of the detergent used in a washing machine. A student dissolves 5.00 g of solid sodium carbonate to make 250 mL of a solution to study the properties of this component of detergent. What is the molar concentration of the solution?

Solution

$m_{Na_2CO_3} = 5.00\ g$

$v_{Na_2CO_3} = 250\ mL$

$M_{Na_2CO_3} = 105.99\ g/mol$

$$n_{Na_2CO_3} = 5.00\ \cancel{g} \times \frac{1\ mol}{105.99\ \cancel{g}}$$

$$= 0.0472\ mol$$

$$C_{Na_2CO_3} = \frac{0.0472\ mol}{0.250\ L}$$

$$C_{Na_2CO_3} = 0.189\ mol/L$$

The molar concentration of the solution is 0.189 mol/L.

Practice

Understanding Concepts

Answers

19. 15.0 g
20. 0.16 kg
21. (a) 355 mg
    (b) 8.07 mmol/L
22. (a) 7.83% W/V
    (b) 1.34 mol/L

19. A chemical technician needs 3.00 L of 0.125 mol/L sodium hydroxide solution. What mass of solid sodium hydroxide must be measured?
20. Seawater is mostly a solution of sodium chloride in water. The concentration varies, but marine biologists took a sample with a molar concentration of 0.56 mol/L. What mass of sodium chloride was there in the biologists' 5.0-L sample?
21. Acid rain may have 355 ppm of dissolved carbon dioxide, which contributes to its acidity.
    (a) What mass of carbon dioxide is present in 1.00 L of acid rain?
    (b) Calculate the molar concentration of carbon dioxide in the acid rain sample.
22. A brine (sodium chloride) solution used in pickling contains 235 g of pure sodium chloride dissolved in 3.00 L of solution.
    (a) What is the percent concentration (%W/V) of sodium chloride?
    (b) What is the molar concentration of sodium chloride?

SUMMARY Concentration of a Solution

Type	Definition	Units
percentage		
%V/V	$c = \frac{v_{solute}}{v_{solution}} \times 100\%$	mL/100 mL
%W/V	$c = \frac{m_{solute}}{v_{solution}} \times 100\%$	g/100 mL
%W/W	$c = \frac{m_{solute}}{m_{solution}} \times 100\%$	g/100 g
very low (number)	$c = \frac{m_{solute}}{m_{solution}}$	ppm, ppb, ppt
molar	$C = \frac{n_{solute}}{v_{solution}}$	mol/L

Section 6.3 Questions

Understanding Concepts

1. What concentration ratio is often found on the labels of consumer products? Why do you think this unit is used instead of moles per litre?
2. Bags of a D5W intravenous sugar solution used in hospitals contain a 5.0% W/V dextrose-in-water solution.
   (a) What mass of dextrose is present in a 500.0-mL bag?
   (b) What is the concentration of D5W expressed in parts per million?
3. The maximum concentration of salt in water at 0°C is 31.6 g/100 mL. What mass of salt can be dissolved in 250 mL of solution?
4. An Olympic-bound athlete tested positive for the anabolic steroid nandrolone. The athlete's urine test results showed one thousand times the maximum acceptable level of 2 mg/L. What was the test result concentration in parts per million?
5. Bald eagle chicks living around Lake Superior were found to contain PCBs (polychlorinated biphenyls) at an average

concentration of 18.9 mg/kg. If a chick had a mass of 0.6 kg, what mass of PCBs would it contain?

6. If the average concentration of PCBs in the body tissue of a human is 4.00 ppm, what mass of PCBs is present in a 64.0-kg person?
7. Each 5-mL dose of a cough remedy contains 153 mg ammonium carbonate, 267 mg potassium bicarbonate, 22 mg menthol, and 2.2 mg camphor. What is the concentration of each of these ingredients in grams per litre?
8. To prepare for an experiment using flame tests, a student requires 100 mL of 0.10 mol/L solutions of each of the following substances. Calculate the required mass of each solid.
   (a) $NaCl_{(s)}$
   (b) $KCl_{(s)}$
   (c) $CaCl_{2(s)}$
9. An experiment is planned to study the chemistry of a home water-softening process. The brine (sodium chloride solution) used in this process has a concentration of 25 g/100 mL. What is the molar concentration of this solution?
10. What volume of 0.055 mol/L glucose solution found in a plant contains 2.0 g of glucose, $C_6H_{12}O_{6(aq)}$?

Making Connections

11. What would be the implications of selling medicines in much more concentrated solutions? Present points both in favour and against.
12. Why is it important for pharmacists, nurses, and doctors to establish a common system for communicating the concentration of solutions? Use the Internet to find out what system(s) they use and create a brief pamphlet containing advice and precautions concerning the concentrations of medicines. Your target audience will be medical professionals in training.

    Follow the links for Nelson Chemistry 11, 6.3.

GO TO www.science.nelson.com

6.4 Drinking Water

We land mammals have a very biased view of our planet: we think almost exclusively about its solid surface. The fact is, over 70% of Earth is covered with a dilute aqueous solution averaging about 4 km deep. That represents an absolutely staggering amount of water, and it is useful to us for a great number of things—*except* drinking. Only about 0.02% of the water on Earth is fresh water in lakes and rivers, and about 0.6% more is ground water soaked into the soil and porous rock of the planet's crust.

In Canada we are extremely fortunate to have access to the most abundant supply of fresh water in the world. Our natural water supply can be divided into two types: surface water and ground water. Our surface water is in the Great Lakes, the thousands of smaller lakes, and the rivers, streams, and springs that make Canada such a wonderful place to live. Our ground water is a huge, hidden underground resource made up of thousands of **aquifers**. All of these bodies of water are potential sources of drinking water.

aquifer: an underground formation of permeable rock or loose material that can produce useful quantities of water when tapped by a well

Water is continually on the move: flowing downhill, moving between surface water and ground water, and following the hydrologic cycle (**Figure 1**). Given this abundant natural resource of water, it seems ironic that we have to be concerned about the state of our drinking water, but this is indeed the case. Canada's population is very concentrated, with more and more people living in urban or suburban areas in the southern portion of the country than elsewhere. Where people live in high concentrations, human activity almost invariably has an effect on the water supply.

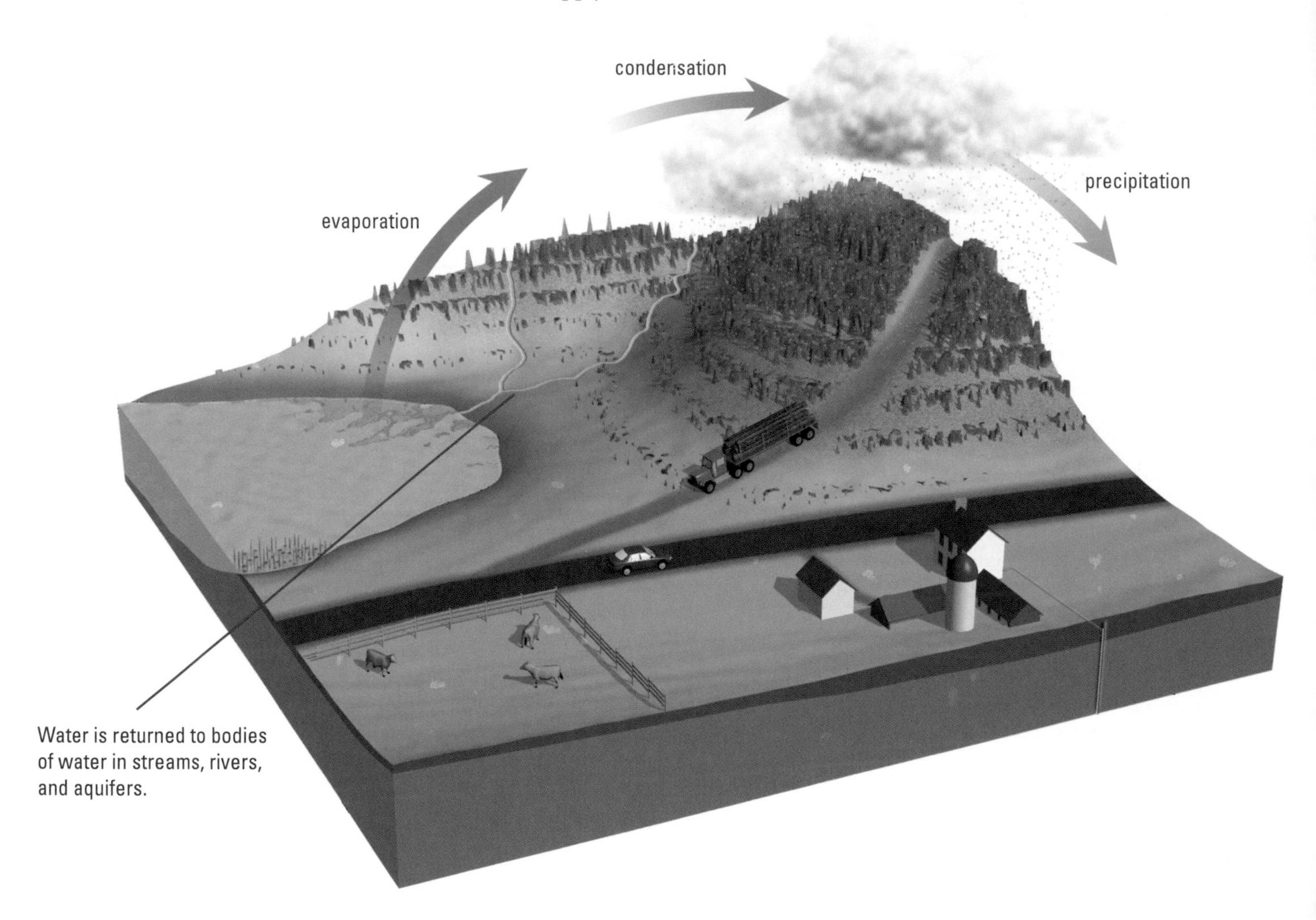

Figure 1
The hydrologic cycle provides the fresh water used by agriculture, mining, industry, and municipalities.

Of course, even completely natural, uncontaminated water contains many substances besides water molecules: living organisms, suspended particles, and a wide range of naturally occurring dissolved chemicals from the surrounding rocks and minerals. Calcium, magnesium, and sodium are the most common cations, while hydrogen carbonate, sulfate, and chloride are the most common anions. The concentration of naturally occurring ions varies widely from one body of water to another. For example, the ion concentration is 48 mg/L in Lake Superior, and 329 mg/L in the Grand River in southern Ontario.

Water Contamination

The advent of people, particularly large numbers of people living in fairly small areas, has changed the quantities and types of substances that find their way into the water. We call these substances, collectively, contaminants, and classify them into three categories: biological (including viruses, bacteria, and algae), physical (suspended particles that make the water appear "dirty"), and chemical (all dissolved substances, including metals, manufactured chemicals, and even artificially high levels of natural minerals).

Water has always contained contaminants, and always threatened the health of those who depend on it. So why do we need to be more concerned about contamination, and particularly chemical contamination, now than in the past? Quite simply, we are producing, transporting, and dumping larger quantities of a greater variety of chemicals now than at any time in the past. While biological contamination was most likely a leading cause of death and illness in earlier centuries, the 20th century's Industrial Revolution has placed a completely different set of stresses on our water supply: chemical contaminants. We are now using chemicals in every aspect of our lives, from growing our food to running our hand-held computers. And not all of these chemicals are staying neatly where we put them (**Figure 2**). They leak, they run off, they leach out, and they are dumped, and they end up in our water.

Figure 2
In this stream, pollution is visible, but this is not always the case.

The environment does have a certain tolerance for pollution. Many substances, if present only in low concentrations, can be broken down naturally. But in the quantities released and discarded by today's society, the overload of nitrates, phosphates, and industrial wastes eventually puts the aquatic ecosystem out of balance, leading to low oxygen levels. Fortunately, however, given enough time, this damage is reversible.

Sometimes the damage caused by pollutants in our waterways, including PCBs, dioxins, heavy metals, and some pesticides, cannot be reversed. These persistent chemicals are not broken down by natural processes. Even in tiny amounts some of these substances can cause serious harm. PCBs and dioxins are carcinogenic (cancer-causing), and heavy metals have been linked to impaired brain development in vertebrates. Such toxic chemicals seriously contaminate the Great Lakes, the Fraser River, and the St. Lawrence River. We can be certain that they are also in our ground water.

The most common sources of water pollution are listed in **Table 1**. Others include industrial leaks or spills; highway or railway accidents resulting in spills; byproducts from coal-fired power plants, old coal gasification sites, and petroleum refineries; underground waste disposal; and contaminants in rain, snow,

landfill leachate: water that has filtered through or under garbage in a landfill site, picking up pollutants during its passage

Table 1: Common Sources and Examples of Water Contamination

Source	Typical contamination
inadequate septic systems or leaky sewer lines	untreated sewage containing bacteria, nitrates, and phosphates
landfill leachate	heavy metals, e.g., mercury, lead, cadmium; bacteria; acids; organic compounds, e.g., benzene ($C_6H_{6(l)}$) and tetrachloroethylene ($C_2Cl_{4(l)}$)
road salt storage areas or road salt runoff	sodium, potassium, and calcium chlorides
livestock wastes	nitrates, heavy metals, bacteria
agricultural and residential use of fertilizers, including treated sewage sludge	nitrates and phosphates
crop and forest spraying	pesticides
leaky tanks or pipelines containing petroleum products	gasoline; other organic compounds
mining and mine tailings	sulfides; cyanides; sulfuric acid; toxic heavy metals, e.g., lead, mercury, cadmium, arsenic

and dry atmospheric fallout. The possible dangers of the various pollutants are listed in **Table 2**.

Table 2: Environmental and Health Effects of Some Water Contaminants

Contaminant	Environmental or health effect
acid	kills soil bacteria; reduces plant growth
bacteria	cause infection, possibly resulting in illness or death
heavy metals	interfere with brain and nerve development in vertebrates
mineral solids	make water cloudy, inhibiting aquatic plant growth
nitrates and phosphates	encourage plant growth, sometimes resulting in algal blooms causing deoxygenation of water
organic compounds	poisonous or carcinogenic; sometimes interfere with oxygen diffusion into surface water
pesticides	toxic to many invertebrates; may bioaccumulate to levels toxic to vertebrates
salt	kills freshwater organisms; makes water unsuitable for drinking

We use both surface water and ground water as sources for drinking water. Large towns and cities generally use surface water for the huge volumes of water they supply to people and industries. Smaller towns and more isolated rural settlements are more often supplied by municipal wells drilled down into a suitable aquifer. Farms, cottages, and rural homes may be supplied by their own small private wells.

As a source of drinking water, ground water has traditionally been considered safer than surface water because of its filtration through soil (removing the suspended solids and many of the dissolved chemicals) and its long residence underground (killing the majority of microorganisms that cannot survive for more than a few days outside a host organism).

Water obtained privately, whether from wells or from nearby surface water, may be completely untreated. The quality of the water from these wells is the responsibility of the well's owner. However, all municipal supplies of drinking water have to be monitored and tested regularly, both before and after treatment, to ensure their safety.

All levels of government in Canada have some degree of responsibility in ensuring the safety of Canada's drinking water, resulting in a patchwork of policies. The provinces and territories are responsible for setting and enforcing standards to ensure adequate drinking water treatment, while municipal governments have the responsibility for supplying safe water to their residents as an essential public service. Federal–provincial water quality guidelines list the maximum acceptable concentration (MAC) of a large number of chemicals in our treated drinking water (**Table 3**). Water treatment procedures set up by municipal public utilities commissions must conform to these water quality guidelines. Chemical technicians at water-testing labs assess water quality by measuring the amounts of various substances in the water.

Practice

Understanding Concepts

1. List the three categories of contaminants, and give at least two examples of each.

2. Briefly describe some of the ways by which contaminants enter ground water or surface water.
3. What are the potential environmental and health hazards of leaky sewer pipes?
4. Describe how pollutants might enter our drinking water from a landfill site. Include a brief discussion of what the pollutants might be, and their possible effects.
5. Create a flow chart illustrating how three different contaminants enter the ground water.
6. If a water source contained the maximum acceptable concentration of tetrachloroethylene, given in **Table 3**, what mass of the chemical would be present in 250 L of bath water?
7. The values for maximum acceptable concentration in **Table 3** are based on an average daily intake of 1.5 L of drinking water. If a community's drinking water contained the maximum acceptable concentrations of cadmium, lead, and mercury, how much of each metal would an average person consume in a year?
8. A 10.00-mL sample of drinking water was tested and found to contain 5.4 mg of nitrate ion. How does the nitrate ion concentration compare with the maximum acceptable concentration given in **Table 3**?

Answers

6. 8 mg
7. 3 mg Cd, 5.5 mg Pb, 0.5 mg Hg
8. 12:1

Making Connections

9. Using the Internet, research at least three sources for information on the problem of ground water contamination. Write a brief article outlining the sources, results, and prevention of such contamination.

   Follow the links to Nelson Chemistry 11, 6.4.

   GO TO **www.science.nelson.com**

10. Propose a solution to reduce the pollution of ground water. Pick one source of contamination and suggest how its effect could be minimized.

Table 3: MAC of Selected Chemicals in Canadian Drinking Water

Substance	Typical source	MAC (ppm)
arsenic	mining waste, industrial effluent	0.025
benzene	industrial effluent, spilled gasoline	0.005
cadmium	leached from landfill	0.005
cyanide	mining waste, industrial effluent	0.2
lead	leached from landfill, old plumbing	0.010
mercury	industrial effluent, **agricultural runoff**	0.001
nitrate	agricultural runoff	45.0
tetrachloroethylene	dry cleaners	0.03
trichloromethane	water chlorination	0.1

agricultural runoff: the surface water, with its load of pollutants in solution and in suspension, that drains off farmland

Water Treatment

What water treatment procedures are in place to achieve the increasingly high quality that we demand of our drinking water? Water from a natural source is often treated in a series of steps (**Figure 3**, page 296) to make it potable, or safe to drink. The most serious concern is safeguarding health, so disinfecting to kill harmful microorganisms is the most important part of the treatment process. Besides safety, the consumer is generally most concerned about appearance, taste, and odour.

1. Collection: The raw water is pumped in from intake pipes deep in the surface water, or from a well. Large particles and debris are removed by travelling screens as the water enters the treatment plant.

2. Coagulation, Flocculation, and Sedimentation: Coagulation is the process of rapidly mixing chemicals known as coagulants with the water to make the small particles in the water clump together. Flocculation is the gentle mixing to form a light, fluffy, flocculent (wool-like, or jelly-like) precipitate, called a floc. During sedimentation the floc settles very slowly, sinking and carrying suspended particles with it and thereby clearing the water.

3. Filtration: In this stage, the remaining floc, other chemical and physical impurities, and most of the biological impurities (bacteria, etc.) are removed. The water flows by gravity through efficient filters made up of layers of sand and anthracite (carbon), and is then collected via an under-drain system.

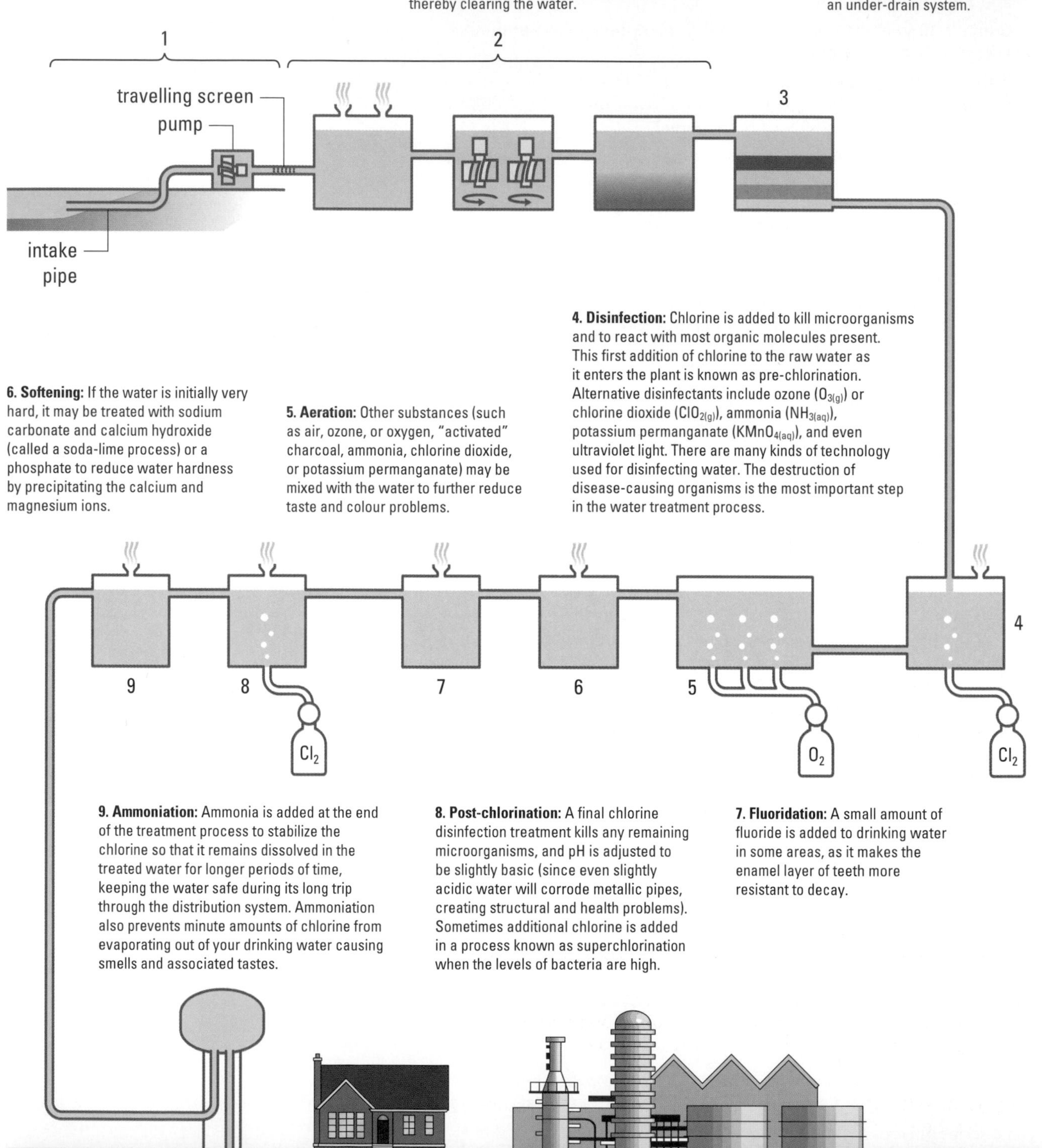

Figure 3
Although not all water treatment plants use all of these steps, all municipal water goes through some form of treatment.

When the water has passed through all these processes, it is then pumped out through the distribution system to industries, homes, and businesses.

This is, of course, an expensive process (**Figure 4**). Municipalities tend to try to find the least expensive way of achieving safe water, so occasionally water is distributed that is insufficiently treated.

Figure 4
Safe water treatment involves dealing with huge volumes daily with no margin of error allowable.

Simulated Water Treatment

This simple activity lets you see what happens during the settling part of the water treatment process. You will use a widely available chemical known as alum, often sold at pharmacies. The term alum used commercially refers to one of the hydrates of: aluminum sulfate, $Al_2(SO_4)_{3(s)}$; sodium aluminum sulfate, $NaAl(SO_4)_{2(s)}$; potassium aluminum sulfate, $KAl(SO_4)_{2(s)}$; or ammonium aluminum sulfate, $NH_4Al(SO_4)_{2(s)}$. The container may or may not specify which compound—and it doesn't really matter for purposes of this activity. The substance is relatively harmless, and may be safely disposed of down the sink with lots of water.

Materials: alum, soil, household ammonia, 3 drinking glasses, 3 teaspoons

- Place 5 mL (about a level teaspoonful) of alum into a glass nearly full of local tap water at room temperature. Stir gently until solid crystals are no longer visible. Add a teaspoon of household ammonia and stir. What do you observe?
- Repeat with a second glass, but this time also add about 5 mL of soil to the glass before adding the alum.
- Stir only a teaspoon of soil into the water in the third glass.
- Place the glasses where they will not be disturbed and record your observations over the next several hours (preferably overnight, or a full 24 h).

(a) What differences can you see among the three tests, after several hours? Explain these differences.

(b) Why is the gelatinous precipitate normally separated in water treatment simply by allowing it lots of time to settle, rather than by filtration?

Figure 5
The contamination of Walkerton's water supply raised concerns about the processes in place to ensure a high quality of drinking water in Canada.

In Canada we have historically had small populations sharing huge water resources, so traditionally we have treated all our municipal water supplies, even though less than 2% of the daily water use of a typical Canadian is for drinking. This may change in the future. In fact, sales of bottled water for home consumption are increasing dramatically, indicating that Canadians are starting to think seriously about different water quality requirements. Perhaps it is not necessary that the water we use to wash our cars or water our gardens be as clean as the water we want to drink, cook in, or use to brush our teeth.

Ontarians received a wake-up call about our drinking water in the spring of 2000. There were unusually heavy rains in the southern part of the province, causing minor flooding in many areas. The rain saturated the ground, mixing surface contaminants with the water in the aquifers. In the town of Walkerton (**Figure 5**), this situation resulted in contamination of the drinking water, which turned out to be fatal for seven people and caused serious illness in hundreds more who suffered with nausea, severe cramps, and bloody diarrhea. The symptoms usually cleared in a few days, but about 5% of those affected developed complications, including kidney failure. The very young and the elderly were particularly susceptible. For months afterward, residents of Walkerton were required to boil their drinking water and to add household bleach to water used for washing hands and dishes.

The events in Walkerton sparked extensive investigations to determine what happened: what went wrong, who was to blame, and how such a tragedy could be avoided in the future. The town's water supply comes from three wells that draw from the ground water stored in the aquifer lying under the town. Engineers investigated the flow of ground water in the Walkerton area and established that, while it appears that the main problem was with one specific well, there were pathways to all three wells that would allow contamination to enter. Indications are that the Walkerton tragedy was due to agricultural runoff from cattle manure, which often carries bacteria. Usually, impurities are filtered out as the water percolates down, but it seems likely that this time water carrying a particularly virulent strain of *E. coli* bacteria found its way into the aquifer from which the town's water supply was drawn. This would not have been tragic had the town's water treatment system been working properly. Unfortunately, the chlorination system was not functioning, and contaminated water was delivered to all the homes depending on the town water supply. Although routine laboratory tests had shown that there were bacteria in the water, this information was not relayed to the people of Walkerton in time to prevent widespread infection. What is perhaps even more alarming is that similar contamination and chlorination failure had occurred several times during the previous five years. It seems that the incident in Walkerton was a disaster waiting to happen.

This unfortunate series of failures, in a system that most of us take for granted, has raised concerns about the safety of our drinking water. We are starting to pay more attention to our water, and asking questions about who is responsible for the quality of water in Canada.

It is clear that we can no longer take safe drinking water for granted. Our water supply is being polluted and people are becoming sick and dying from unsafe drinking water. While this has been happening for thousands of years, and is certainly still the case in many parts of the world, it is a situation that Canadians are unwilling to accept. Between 1970 and 1990 there was a five-fold increase in water consumption in Canada, so it is all the more important to consider improvements to municipal water treatment practices across the country. A major challenge in designing a better water policy for Canada is balancing the need to regulate and enforce drinking-water standards with the need to contain treatment costs.

Practice

Understanding Concepts

11. What is the most important step in freshwater treatment?
12. Prepare a two-column list of the steps in the purification of drinking water. In the first column, list the usual steps, and in the second column, list optional steps that would depend on local conditions.

Making Connections

13. Drinkable water is an important concern when hiking and camping in the backcountry; the natural water available may look clear and clean, but can be dangerous to your health. Use the Internet to research some portable technologies that are used to purify water. Include examples of both physical and chemical treatments.

    Follow the links for Nelson Chemistry 11, 6.4.

14. Some bottles of "spring" water and "mineral" water are (perfectly legally) filled directly from a tap somewhere, and have not been further purified in any way. Look at several different brands of bottled water to find out what kinds of information are provided on the labels. Using this evidence plus any other evidence you want to introduce, decide whether it is worthwhile to buy bottled water. Outline and defend your position in a note for your school cafeteria.
15. Identify and describe a career that is related to water testing and treatment. Include a brief description of the job and the training required.

    Follow the links for Nelson Chemistry 11, 6.4.

GO TO **www.science.nelson.com**

Take a Stand: Safe to Drink?

The Walkerton incident prompted a province-wide inspection of water treatment plants in Ontario. The inspectors found numerous problems, including substandard wells, insufficient sampling and analysis, problems with water treatment equipment, underqualified operators, and poor commmunication between the testing labs and the authorities.

How can you reduce the chances of your community being supplied with contaminated drinking water? Working in small groups, research the current state of the drinking water in your community. Your research should include the following components:

- the source and quality of the water before it is treated;
- possible sources of contaminants;
- testing standards for safe drinking water;
- treatment standards for municipal water.

Follow the links for Nelson Science 11, 6.4.

(a) Choose one aspect of the delivery of safe water to your tap. How might it be improved upon? Draft a letter to your local municipality, commending it on the strong points of its current water delivery practice while recommending improvements where you think they are needed.

DECISION-MAKING SKILLS

- ● Define the Issue
- ● Identify Alternatives
- ● Research
- ● Analyze the Issue
- ● Defend a Decision
- ○ Evaluate

Section 6.4 Questions

Understanding Concepts

1. List at least ten pollutants of water, and classify them as chemical, physical, or biological contaminants.
2. What immediate steps could a municipal water treatment facility take if it found that one of its water sources had become contaminated with a potentially harmful bacterium?
3. Why is it important that the maximum acceptable concentration of contaminants not be exceeded in drinking water? Give some specific examples in your answer.
4. Choose one specific contaminant of drinking water. Write a presentation on this contaminant. Include a diagram showing its possible sources, and information about its maximum acceptable concentration and its potentially damaging effects.
5. A technician in a water-testing laboratory finds that a sample of well water contains lead in a concentration of 0.01 g/L. Does the source of this water exceed the maximum acceptable concentration for drinking water?

Making Connections

6. On a small scale, such as in a laboratory, the main steps following the collection in the water purification sequence can be done in one or two beakers. On a large scale in a water treatment plant, a different design is used. Describe the general technological design for the sequence of water treatment and state why this design is used.
7. "Contamination of ground water is a human problem, rather than a wider environmental problem." Take a position on this statement and support it with research and reasoned arguments. Either write a letter to the editor of a local newspaper, or create a presentation that could be used at a town hall meeting of citizens in an area where the safety of the ground water is under threat.
8. Very precise technical equipment is used to test water samples. Why is this necessary?

6.5 Solution Preparation

When you mix up a jug of iced tea using a package of crystals and water, you are preparing a solution from a solid solute (actually, from several solid solutes). But when you mix the tea from a container of frozen concentrate, you are preparing a solution by dilution. Scientists use both of these methods to prepare solutions. In this course you will be preparing only aqueous solutions. The knowledge and skills for preparing solutions are necessary to complete some of the more complex laboratory investigations that come later in this course.

Preparation of Standard Solutions from a Solid

standard solution: a solution for which the precise concentration is known

Solutions with precisely known concentrations, called **standard solutions**, are routinely prepared in both scientific research laboratories and industrial processes. They are used in chemical analysis as well as for the precise control of chemical reactions. To prepare a standard solution, precision equipment is required to measure the mass of solute and volume of solution. Electronic balances are used for precise and efficient measurement of mass (**Figure 1**). For measuring a precise volume of the final solution, a container called a volumetric flask is used (**Figure 2**).

Figure 1
In many school laboratories, electronic balances that measure masses to within 0.01 g or 0.001 g have replaced mechanical balances.

Figure 2
Volumetric glassware comes in a variety of shapes and sizes. The Erlenmeyer flask on the far left has only approximate volume markings, as does the beaker. The graduated cylinders have much better precision, but for high precision a volumetric flask (on the right) is used. The volumetric flask shown here, when filled to the line, contains 100.0 mL ± 0.16 mL at 20 °C. This means that a volume measured in this flask is uncertain by less than 0.2 mL at the specified temperature.

Activity 6.5.1

A Standard Solution from a Solid

In this activity you will practise the skills required to prepare a standard solution from a pure solid. You will need these skills in many investigations in this book.

Materials

lab apron
eye protection
copper(II) sulfate pentahydrate, $CuSO_4{\cdot}5H_2O_{(s)}$
150-mL beaker
centigram balance
laboratory scoop
stirring rod
wash bottle of **pure water**
100-mL volumetric flask with stopper
small funnel
medicine dropper
meniscus finder

Wear eye protection and a laboratory apron. Copper(II) sulfate is harmful if swallowed.

pure water: deionized or distilled water

Procedure

1. Calculate the mass of solid copper(II) sulfate pentahydrate needed to prepare 100.0 mL of a 0.5000 mol/L solution.
2. Obtain the calculated mass of copper(II) sulfate pentahydrate in a clean, dry 150-mL beaker.
3. Dissolve the solid in 40 to 50 mL of pure water.
4. Transfer the solution into a 100-mL volumetric flask. Rinse the beaker two or three times with small quantities of pure water, transferring the rinsings into the volumetric flask.
5. Add pure water to the volumetric flask until the volume is 100.0 mL.
6. Stopper the flask and mix the contents thoroughly by repeatedly inverting the flask.

Note: Store your solution for the next activity.

Answers

1. 3.55 g
2. 200 g
4. (a) 33.2 g
5. (a) 5.93 g

Figure 3
Hard-water deposits such as calcium carbonate can seriously affect water flow in a pipe.

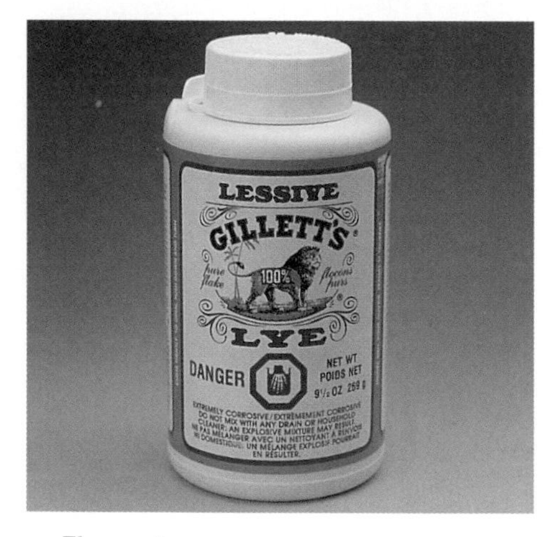

Figure 4
Solutions of sodium hydroxide in very high concentration are sold as cleaners for clogged drains. The same solution can be made less expensively by dissolving solid lye (a commercial name for sodium hydroxide) in water. The pure chemical is very caustic and the label on the lye container recommends rubber gloves and eye protection.

dilution: the process of decreasing the concentration of a solution, usually by adding more solvent

stock solution: a solution that is in stock or on the shelf (i.e., available); usually a concentrated (possibly even saturated) solution

Practice

Understanding Concepts

1. To test the hardness of water (**Figure 3**), an industrial chemist performs an analysis using 100.0 mL of a 0.250 mol/L standard solution of ammonium oxalate. What mass of monohydrate ammonium oxalate, $(NH_4)_2C_2O_4 \cdot H_2O_{(s)}$, is needed to make the standard solution?
2. Calculate the mass of solid lye (sodium hydroxide) needed to make 500 mL of a 10.0 mol/L strong cleaning solution (**Figure 4**).
3. List several examples of solutions that you prepared from solids in the last week.

Applying Inquiry Skills

4. You have been asked to prepare 2.00 L of a 0.100 mol/L aqueous solution of cobalt(II) chloride dihydrate for an experiment.
   (a) Show your work for the pre-lab calculation.
   (b) Write a complete specific procedure for preparing this solution, as in Activity 6.5.1. Be sure to include all necessary precautions.
5. (a) A technician prepares 500.0 mL of a 0.0750 mol/L solution of potassium permanganate as part of a quality-control analysis in the manufacture of hydrogen peroxide. Calculate the mass of potassium permanganate required to prepare the solution.
   (b) Write a laboratory procedure for preparing the potassium permanganate solution. Follow the conventions of communication for a procedure in a laboratory report.

Preparation of Standard Solutions by Dilution

A second method of preparing solutions is by **dilution** of an existing solution. You apply this process when you add water to concentrated fruit juice, fabric softener, or a cleaning product. Because dilution is a simple, quick procedure, it is common scientific practice to begin with a **stock solution** or a standard solution, and to add solvent (usually water) to decrease the concentration to the desired level.

While there are no hard and fast rules, we often describe solutions with a molar concentration of less than 0.1 mol/L as dilute, while solutions with a concentration of greater than 1 mol/L may be referred to as concentrated.

Calculating the new concentration after a dilution is straightforward because the quantity (mass or amount) of solute is not changed by adding more solvent. Therefore, the mass (or amount) of solute before dilution is the same as the mass (or amount) of solute after dilution.

$m_i = m_f$ or $n_i = n_f$

m_i = initial mass of solute

m_f = final mass of solute

n_i = initial amount of solute (in moles)

n_f = final amount of solute (in moles)

Using the definitions of solution concentration ($m = vc$ or $n = vC$), we can express the constant quantity of solute in terms of the volume and concentration of solution.

$v_iC_i = v_fC_f$

C_i = initial concentration

C_f = final concentration
v_i = initial volume before dilution
v_f = final volume after dilution

This means that the concentration is inversely related to the solution's volume. For example, if water is added to 6% hydrogen peroxide disinfectant until the total volume is doubled, the concentration becomes one-half the original value, or 3%.

Any one of the values expressed may be calculated for the dilution of a solution, provided the other three values are known. (Note that the dilution calculation for percentage weight by weight (%W/W) will be slightly different because the mass of solution is used; i.e., $m_{solute} = m_{solution}c$.)

Many consumer and commercial products are purchased in concentrated form and then diluted before use. This saves on shipping charges and reduces the size of the container—making the product less expensive and more environmentally friendly. Citizens who are comfortable with dilution techniques can live more lightly on Earth (**Figure 5**).

Figure 5
You can save money and help save the environment by diluting concentrated products.

Sample Problem 1

Water is added to 0.200 L of 2.40 mol/L $NH_{3(aq)}$ cleaning solution, until the final volume is 1.000 L. Find the molar concentration of the final, diluted solution.

Solution

$v_i = 0.200$ L
$v_f = 1.000$ L
$C_i = 2.40$ mol/L

$$v_iC_i = v_fC_f$$

$$C_{f\,NH_3} = \frac{v_iC_i}{v_f}$$

$$C_{f\,NH_3} = \frac{0.200\ \cancel{L} \times 2.40\ \text{mol}/\cancel{L}}{1.000\ \text{L}}$$

$$= 0.480\ \frac{\text{mol}}{\text{L}}$$

The molar concentration of the final, diluted solution is 0.480 mol/L.

Alternatively, this problem can be solved another way:

$$n_{NH_3} = 0.200\ \cancel{L} \times \frac{2.40\ \text{mol}}{1.00\ \cancel{L}}$$

$$= 0.480\ \text{mol}$$

$$C_{NH_3} = \frac{0.480\ \text{mol}}{1.000\ \text{L}}$$

$$C_{NH_3} = 0.480\ \text{mol/L}$$

Again, the molar concentration of the final, diluted solution is 0.480 mol/L.

Sample Problem 2

A student is instructed to dilute some concentrated $HCl_{(aq)}$ (36%) to make 4.00 L of 10% solution (**Figure 6**). What volume of hydrochloric acid solution should the student initially measure to do this?

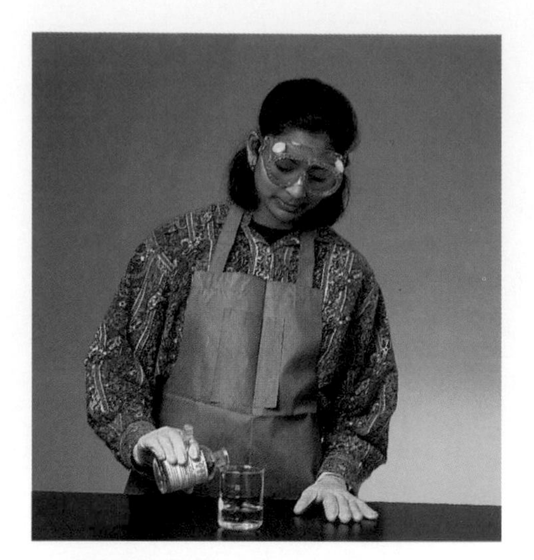

Figure 6
When diluting all concentrated reagents, especially acids, always add the concentrated reagent to water. When acids are mixed with water, heat is often produced. If a large amount of acid is added to a small amount of water, the water might boil and spatter the reagent out of the container. Remember "AAA": Always Add Acid.

Solution

$c_i = 36\%$
$v_f = 4.00\ \text{L}$
$c_f = 10\%$

$$v_i c_i = v_f c_f$$

$$v_i = \frac{v_f c_f}{c_i}$$

$$= \frac{4.00\ \text{L} \times 10\%}{36\%}$$

$$v_i = 1.1\ \text{L}$$

The student should measure out 1.1 L of 36% hydrochloric acid to make 4.00 L of the dilute solution.

The dilution technique is especially useful when you need to manipulate the concentration of a solution. For example, when doing scientific or technological research, you may want to slow down a reaction that proceeds too rapidly or too violently with a concentrated solution. You could do this by lowering the concentration of the solution. In the medical and pharmaceutical industries, prescriptions require not only minute quantities, but also extremely precise measurement. If the solutions are diluted before being sold, it is much easier for a patient to take the correct dose. For example, it's easier to accurately measure out 10 mL (two teaspoons) of a cough medicine than it is to measure one-fifth of a teaspoon, which the patient would have to do if the medicine were 10 times more concentrated.

The preparation of standard solutions by dilution requires a means of transferring precise volumes of solution. You know how to use graduated cylinders to measure volumes of solution, but graduated cylinders are not precise enough when working with small volumes. To deliver a very precise, small volume of solution, a laboratory device called a pipet is used. A 10-mL graduated pipet has graduation marks every tenth of a millilitre (**Figure 7**). This type of pipet can transfer any volume from 0.1 mL to 10.0 mL, and is more precise than a graduated cylinder. A volumetric pipet transfers only one specific volume, but has a very high precision. For example, a 10-mL volumetric (or delivery) pipet is designed to transfer 10.00 mL of solution with a precision of ± 0.02 mL. The volumetric pipet is often inscribed with TD to indicate that it is calibrated *to deliver* a particular volume with a specified precision. Both kinds of pipet come in a range of sizes and are used with a pipet bulb.

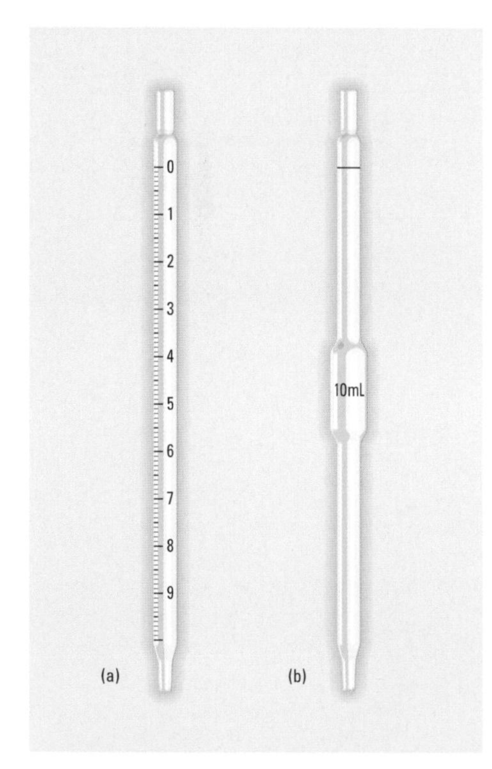

Figure 7
A graduated pipet (a) measures a range of volumes, while a volumetric pipet (b) is calibrated to deliver (TD) a fixed volume.

Activity 6.5.2

A Standard Solution by Dilution

The purpose of this activity is to practise the procedure and skills for precisely diluting the standard solution prepared in the previous activity (Activity 6.5.1).

Materials

lab apron
eye protection
0.5000 mol/L $CuSO_{4(aq)}$ standard solution
150-mL beaker
10-mL graduated or volumetric pipet
pipet bulb
wash bottle of pure water
100-mL volumetric flask with stopper
small funnel
medicine dropper
meniscus finder

Copper(II) sulfate is harmful if swallowed. Wear eye protection and a laboratory apron.

Use of a pipet bulb is essential. Do not pipet by mouth.

Procedure

1. Calculate the volume of a 0.5000 mol/L standard solution of $CuSO_{4(aq)}$ required to prepare 100.0 mL of a 0.05000 mol/L solution.
2. Add 40 mL to 50 mL of pure water to a clean 100-mL volumetric flask.
3. Measure the required volume of the standard solution using a 10-mL pipet.
4. Transfer the required volume of the initial (standard) solution into the 100-mL volumetric flask (**Figure 8(a)**).
5. Add pure water until the final volume is reached (**Figure 8(b)**).
6. Stopper the flask and mix the solution thoroughly (**Figure 8(c)**).

Analysis

(a) Why was 40 mL of water placed in the flask before adding the standard solution?
(b) Why wasn't 100 mL of water placed in the flask?

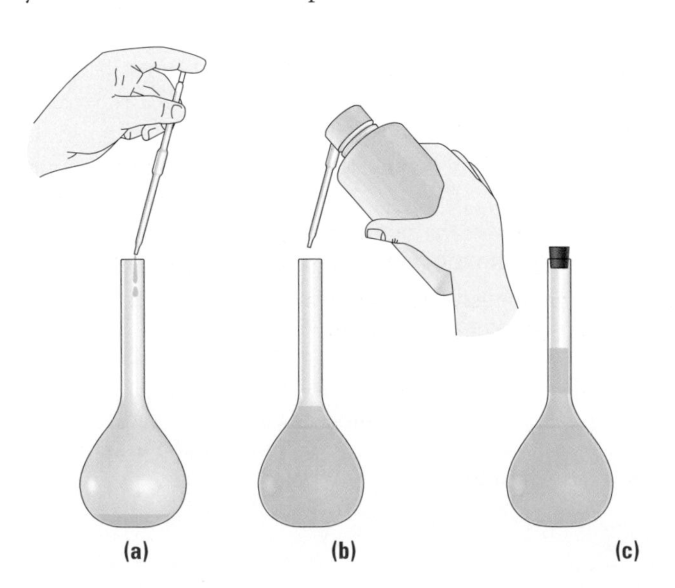

Figure 8
(a) The appropriate volume of $CuSO_{4(aq)}$ is transferred to a volumetric flask.
(b) The initial amount of copper(II) sulfate solute is not changed by adding water to the flask.
(c) In the final dilute solution, the initial amount of copper(II) sulfate is still present, but it is distributed throughout a larger volume; in other words, it is diluted.

Answers

6. 22.5 mL
7. (a) 2.50 mmol/L
   (b) 3.99 mg
8. (a) 1.000:25.00 or 4.000%
   (b) 25.00 times the volume

Practice

Understanding Concepts

6. Many solutions are prepared in the laboratory from purchased concentrated solutions. What volume of concentrated 17.8 mol/L stock solution of sulfuric acid would a laboratory technician need to make 2.00 L of 0.200 mol/L solution by dilution of the original, concentrated solution?
7. In a study of reaction rates, you need to dilute the copper(II) sulfate solution prepared in Activity 6.5.2, A Standard Solution by Dilution. You take 5.00 mL of 0.05000 mol/L $CuSO_{4(aq)}$ and dilute this to a final volume of 100.0 mL.
   (a) What is the final concentration of the dilute solution?
   (b) What mass of $CuSO_{4(s)}$ is present in 10.0 mL of the final dilute solution?
   (c) Can this final dilute solution be prepared directly using the pure solid? Defend your answer.
8. A student tries a reaction and finds that the volume of solution that reacts is too small to be measured precisely. She takes a 10.00-mL volume of the solution with a pipet, transfers it into a clean 250-mL volumetric flask containing some pure water, adds enough pure water to increase the volume to 250.0 mL, and mixes the solution thoroughly.
   (a) Compare the concentration of the dilute solution to that of the original solution.
   (b) Compare the volume that will react now to the volume that reacted initially.
   (c) Predict the speed or rate of the reaction using the diluted solution compared with the rate using the original solution. Explain your answer.
9. List several examples of dilutions you have made from more concentrated solutions in the last week.

Reflecting

10. "Precise dilution can produce solutions of much more accurate concentration compared with dilute solutions prepared directly by measuring the mass of the pure solute." Consider this statement. Under what circumstances would this statement *not* be true?

Section 6.5 Questions

Understanding Concepts

1. List several reasons why scientists make solutions in the course of their work.
2. (a) Briefly describe two different ways of making a solution.
   (b) When would it be most appropriate to use each method?
3. In a quantitative analysis for sulfate ions in a water treatment plant, a technician needs 100 mL of 0.125 mol/L barium nitrate solution. What mass of pure barium nitrate is required?
4. A 1.00-L bottle of purchased acetic acid is labelled with a concentration of 17.4 mol/L. A dyer dilutes this entire bottle of concentrated acid to prepare a 0.400 mol/L solution. What volume of diluted solution is prepared?

5. A 10.00-mL sample of a test solution is diluted in an environmental laboratory to a final volume of 250.0 mL. The concentration of the diluted solution is found to be 0.274 g/L. What was the concentration of the original test solution?

Applying Inquiry Skills

6. A chemical analysis of silver uses 100 mL of a 0.155 mol/L solution of potassium thiocyanate, KSCN. Write a complete, specific procedure for preparing the solution from the solid. Include all necessary calculations and precautions.
7. A laboratory technician needs 1.00 L of 0.125 mol/L sulfuric acid solution for a quantitative analysis experiment. A commercial 5.00 mol/L sulfuric acid solution is available from a chemical supply company. Write a complete, specific procedure for preparing the solution. Include all necessary calculations and safety precautions.
8. As part of a study of rates of reaction, you are to prepare two aqueous solutions of cobalt(II) chloride.
   (a) Calculate the mass of solid cobalt(II) chloride hexahydrate you will need to prepare 100.0 mL of a 0.100 mol/L cobalt(II) chloride solution.
   (b) Calculate how to dilute this solution to make 100.0 mL of a 0.0100 mol/L cobalt(II) chloride solution.
   (c) Write a list of Materials, and a Procedure for the preparation of the two solutions. Be sure to include all necessary safety precautions and disposal steps.
   (d) With your teacher's approval, carry out your Procedure.

Making Connections

9. It has been suggested that it is more environmentally friendly to transport chemicals in a highly concentrated state. List arguments for and against this position.
10. For many years the adage, "The solution to pollution is dilution" was used by individuals, industries, and governments. They did not realize at that time that chemicals, diluted by water or air, could be concentrated in another system later. Identify and describe a system in which pollutants can become concentrated.

Reflecting

11. Recall the work that you did in Grade 10 Science with reaction rates. You compared the reaction rate for a series of solutions with varying concentrations. Write the initial steps of a simple procedure to prepare the solutions for testing the reaction rates of, for example, zinc with stock hydrochloric acid. How have your knowledge and skills progressed since you did this investigation in Grade 10?

Chapter 6 Summary

Key Expectations

Throughout the chapter, you have had the opportunity to do the following:

- Explain the formation of solutions involving various solutes in water and nonpolar solutes in nonpolar solvents. (6.1, 6.2)
- Supply examples from everyday life of solutions involving all three states. (6.1)
- Describe and explain the properties of water, and demonstrate an understanding of its importance as a universal solvent. (6.2)
- Use the terms: solute, solvent, solution, electrolyte, leachate, runoff, concentration, standard solution, stock solution, and dilution. (all sections)
- Solve solution concentration problems using a variety of units. (6.3, 6.5)
- Explain the origins of pollutants in natural waters and identify maximum allowable concentrations of metallic and organic contaminants in drinking water. (6.4)
- Develop and use the technological skills for the preparation of solutions of required concentrations. (6.5)
- Describe consumer and commercial examples of solutions, including those in which the concentration must be precisely known. (6.5)

Key Terms

acid
aquifer
agricultural runoff
aqueous solution
base
concentrated
concentration
dilute
dilution
dissociation
electrolyte
homogeneous mixture
intermolecular force
intramolecular force
landfill leachate
molar concentration
mole
neutral
nonelectrolyte
parts per million
pure water
solute
solution
solvent
standard solution
stock solution

Make a Summary

1. Make four small concept maps to express each of the following central concepts. Draw them in the corners of a page.
   (a) electrolytes and nonelectrolytes
   (b) calculations related to concentration of solution
   (c) preparation of a solution from a solid solute and by dilution of an existing solution
   (d) the sources of water and pollution and the refining of water

   What central concept or substance can be used to link all of these concept maps? Add this central concept or substance to the centre of the page and link it to the four concept maps prepared above.

Reflect on your Learning

Revisit your answers to the Reflect on Your Learning questions at the beginning of this chapter.

- How has your thinking changed?
- What new questions do you have?

Chapter 6 Review

Understanding Concepts

1. What do all concentration units have in common?
2. In general, what type of solvent dissolves
   (a) ionic compounds?
   (b) polar compounds?
   (c) nonpolar compounds?
3. Water is capable of dissolving many things.
   (a) Provide some reasons, based on the theory you have studied, for water being "the universal solvent."
   (b) List some reasons for dissolving substances in water. Give examples for these reasons.
   (c) How does water's property as a powerful solvent affect our drinking water?
4. Write IUPAC names for the solute and solvent in the following household solutions:
   (a) brine, $NaCl_{(aq)}$
   (b) vinegar, $HC_2H_3O_{2(aq)}$
   (c) washing soda, $Na_2CO_{3(aq)}$
   (d) pancake syrup, $C_{12}H_{22}O_{11(aq)}$
   (e) vodka, $C_2H_5OH_{(aq)}$
5. Partly skimmed milk contains 2.0 g of milk fat (MF) per 100 mL of milk. What mass of milk fat is present in 250 mL (one glass) of milk?
6. A shopper has a choice of yogurt with three different concentrations of milk fat: 5.9% MF, 2.0% MF, and 1.2% MF (**Figure 1**). If the shopper wants to limit his milk fat intake to 3.0 g per serving, calculate the mass of the largest serving he could have for each type of yogurt.

Figure 1
The label tells us the concentration of milk fat in yogurt.

7. What volume of vinegar contains 15 mL of pure acetic acid (**Figure 2**)?

Figure 2
The label tells us the concentration of acetic acid in vinegar.

8. Water from a well is found to have a nitrate ion concentration of 55 ppm, a level considered unsafe for drinking. Calculate the mass of nitrate ions in 200 mL of the water.
9. Calculate the molar concentration of the following solutions:
   (a) 0.35 mol copper(II) nitrate is dissolved in water to make 500 mL of solution.
   (b) 10.0 g of sodium hydroxide is dissolved in water to make 2.00 L of solution.
   (c) 25 mL of 11.6 mol/L $HCl_{(aq)}$ is diluted to a volume of 145 mL.
   (d) A sample of tap water contains 16 ppm of magnesium ions.
10. The label on a bottle of "sports drink" indicates that the beverage contains water, glucose, citric acid, potassium citrate, sodium chloride, and potassium phosphate, as well as natural flavours and artificial colours. The label also indicates that the beverage contains 50 mg of sodium ions and 55 mg of potassium ions per 400 mL serving.
    (a) Write chemical formulas for all the compounds named on the label, and classify them as ionic or molecular. Further classify the molecular

compounds as acid or neutral. You may find the list of Common Chemicals in Appendix C useful in answering this question.
(b) Which compound imparts a sweet taste to the beverage, and which imparts a tangy taste?
(c) Calculate the concentration in parts per million of the sodium and potassium ions in the beverage.

11. Standard solutions of sodium oxalate, $Na_2C_2O_{4(aq)}$, are used in a variety of chemical analyses. What mass of sodium oxalate is required to prepare 250.0 mL of a 0.375 mol/L solution?
12. Phosphoric acid is the active ingredient in many commercial rust-removing solutions. Calculate the volume of concentrated phosphoric acid (14.6 mol/L) that must be diluted to prepare 500 mL of a 1.25 mol/L solution.
13. Laboratories order hydrochloric acid as a concentrated solution (e.g., 36% W/V). What initial volume of concentrated laboratory hydrochloric acid should be diluted to prepare 5.00 L of a 0.12 mol/L solution for an experiment?

Applying Inquiry Skills

14. Describe two methods used to prepare standard solutions.
15. What is a standard solution, and why is such a solution necessary?
16. Scientists have developed a classification system to help organize the study of matter. Describe an empirical test that can be used to distinguish between the following classes of matter:
    (a) electrolytes and nonelectrolytes
    (b) acids, bases, and neutral compounds
17. Standard solutions of potassium hydrogen tartrate, $KHC_4H_4O_{6(aq)}$, are used in chemical analyses to determine the concentration of bases such as sodium hydroxide.
    (a) Calculate the mass of potassium hydrogen tartrate that is measured to prepare 100.0 mL of a 0.150 mol/L standard solution.
    (b) Write a complete procedure for the preparation of this standard solution, including specific quantities and equipment.
18. In chemical analysis we often dilute a stock solution to produce a required standard solution.
    (a) What volume of a 0.400 mol/L stock solution is required to produce 100.0 mL of a 0.100 mol/L solution? (Note that the high precision of each of these solutions indicates that they are standard solutions.)
    (b) Write a complete procedure for the preparation of this standard solution, including specific quantities and equipment.
19. A chemistry student was given the task of identifying four colourless solutions. Complete the **Analysis** of the investigation report.

Problem

Which of the solutions labelled A, B, C, and D is calcium hydroxide? Which is glucose? potassium chloride? sulfuric acid?

Experimental Design

Each solution, at the same concentration and temperature, is tested with red and blue litmus paper and conductivity apparatus.

Evidence

Table 1: Litmus and Conductivity Tests

Solution	Red litmus	Blue litmus	Conductivity
A	stays red	blue to red	high
B	stays red	stays blue	none
C	red to blue	stays blue	high
D	stays red	stays blue	high

Analysis

(a) Which solution is which?

Making Connections

20. In March 1989, the *Exxon Valdez* oil tanker ran aground in Prince William Sound off Alaska, spilling 232 000 barrels of oil and causing extensive environmental damage (**Figure 3**). Use the Internet to research this event.
    (a) Do oil and water dissolve in each other? Provide some reasons for your answer.
    (b) What are some of the environmental effects of an oil spill such as the one from the *Exxon Valdez*?
    (c) What are some methods used to clean up spilled oil?
    (d) Should the transportation of oil by large tanker ships be prohibited? Discuss, including some risks and benefits of this practice.

Follow the links for Nelson Chemistry 11, Chapter 6 Review.

GO TO www.science.nelson.com

Figure 3
The oil spill from the *Exxon Valdez* attracted worldwide attention and a demand for more responsible transportation of oil.

21. Collect at least ten examples of common solutions. The solutes should represent the three states of matter, and there should be at least two different solvents. Present your samples in a display, with a description for each sample including, if known, the concentration of each solution. Each solution should be accompanied by a card indicating any necessary safety precautions, along with the reasons for these precautions.
22. It is vitally important that medicines be administered in the correct dosages. In a hospital setting, many substances are given intravenously as solutions. Research at least five intravenous solutions, and create a table showing the name of the solution, a typical concentration, and the medical purpose for which the solution is administered.
23. How do various pollutants get into natural water? Create a flow chart to illustrate the source, route, and potential health effect of each contaminant.
24. Use the Internet to discover the maximum acceptable levels of microorganisms and chemicals in drinking water. Read a summary of Health Canada's *Guidelines for Canadian Drinking Water Quality*.
    (a) Choose one potentially dangerous contaminant from Health Canada's list; find the MAC of that contaminant.
    (b) Find an example of a place where the MAC of your chosen contaminant has been exceeded in drinking water.
    (c) What were some of the problems resulting form this incident?
    (d) What steps were taken to correct the problem?

    Follow the links from Nelson Chemistry 11, Chapter 6 Review.

Exploring

25. Reverse osmosis is a water treatment technique currently widely used for producing many brands of bottled water, and also for purifying seawater. Outdoor supply stores sell reverse osmosis kits for purifying water on wilderness trips, or as emergency equipment for use in lifeboats. Use the Internet to research reverse osmosis. Find the pressures needed to produce pure drinking water from seawater by this process. Find out under what circumstances reverse osmosis is commercially viable, and find an example of a reverse osmosis plant that is operating, desalinating seawater, today. Present your findings as if you were trying to convince a coastal community to install a reverse osmosis facility.

    Follow the links for Nelson Chemistry 11, Chapter 6 Review.

    GO TO www.science.nelson.com

26. When travelling in the wilderness or in another country, Canadians are sometimes advised to take chlorine tablets with them to put in their drinking water. Use the Internet to research the purpose of using these tablets, their ingredients, and their advertised effects.

    Follow the links for Nelson Chemistry 11, Chapter 6 Review.

Chapter

7

Solubility and Reactions

In this chapter, you will be able to

- describe the effect of temperature on the solubility of solids, liquids, and gases in water;
- explain hardness of water, its consequences, and water-softening methods;
- predict combinations of aqueous solutions that produce precipitates and represent these reactions using net ionic equations;
- describe the technology and the major steps involved in the purification of water and the treatment of waste water;
- perform qualitative and quantitative analyses of solutions.

It is easier to handle a great many chemicals when they are in solution, particularly those that are toxic, corrosive, or gaseous. Both in homes and at worksites, transporting, loading, and storing chemicals are more convenient and efficient when the chemicals are in solution rather than in solid or gaseous states. Also, performing a reaction in solution can change the rate (speed), the extent (completeness), and the type (kind of product) of the chemical reaction.

Solutions make it easy to

- handle chemicals—a solid or gas is dissolved in water for ease of use or transportation;
- complete reactions—some chemicals do not react until in a solution where there is increased contact between the reacting entities;
- control reactions—the rate, extent, and type of reactions are much more easily controlled when one or more reactants are in solution.

These three points all apply to the liquid cleaning solution in **Figure 1**. The cleaning solution is easy to handle, and the fact that it is sold in a spray bottle adds to its convenience. Spraying a solution is an effective way of handling a chemical that is dissolved in water. Secondly, the solution allows a reaction to occur between the cleaning chemicals and the dirty deposit, whereas a pure gas or solid would not react well with a solid. Thirdly, the manufacturer can control the rate of the reaction (and thus the safety) by choosing the ideal concentration of the cleaning solution. Having the chemical in solution rather than in its pure state increases our ability to handle and control it.

In this chapter we will examine several concepts, including the extent to which one substance will dissolve in another, and the effect of temperature on the extent of dissolving. In chemical reactions we often wish to produce a product that is more or less soluble than the reactants. In fact, this is one of the most common techniques used for separating chemical substances. This chapter will help you understand how to do this.

Reflect on your Learning

1. A liquid you are using at home or in the laboratory may be a solution or a pure substance. Which is most likely? How would you test the liquid to determine whether it is a solution or a pure substance?
2. When dissolving a chemical in water, the rate of dissolving is often confused with the extent of dissolving. How do you speed up the rate of dissolving of a solute? How do you know if no more solute will dissolve?
3. Reactions in solution are common. If two aqueous solutions are mixed, what evidence would indicate that a reaction has occurred? How would you know if the change was just a physical change rather than a chemical change?

Figure 1
The low solubility of the soap deposit is overcome by a chemical reaction.

Try This **Activity**

Measuring the Dissolving Process

Are there different kinds of salt? How much salt can you get to dissolve? What happens to the volume of a solution when a solute is added to it? This quick activity will help you to think about the answers to these questions.

Materials: distilled or deionized water, table salt, coarse pickling salt (pure $NaCl_{(s)}$), a measuring teaspoon (5 mL), two 250-mL Erlenmeyer flasks, with stoppers, one 100-mL graduated cylinder or measuring cup

- Place a level teaspoonful of table salt into 100 mL of pure water at room temperature in a 250-mL Erlenmeyer flask. Swirl the flask's contents thoroughly for a minute or two. Record your observations.
- Repeat with pickling salt, again recording your observations.

(a) What does the result, with common table salt as a solute, show about the nature of the substance being used? Compare it with the solution in the second flask.
(b) List the ingredients in common table salt, according to the package label, and explain your observations of the contents of the first flask.

- Add a further teaspoon of pickling salt to the second flask, and swirl until the solid is again completely dissolved. Keeping track of how much pickling salt you add, continue to dissolve level teaspoons of salt until no amount of swirling will make all of the solid crystals disappear.

(c) How many level teaspoons of pickling salt (pure $NaCl_{(s)}$) could you get to dissolve in 100 mL of $H_2O_{(l)}$ in the second flask?
(d) What is the final volume of your $NaCl_{(aq)}$ solution in the second flask?
(e) If you dissolve 20.0 mL of $NaCl_{(s)}$ in 100.0 mL of liquid water, what do you suppose the volume of the solution would be? Can you think of a way to test this? The answer is very interesting.

saturated solution: a solution containing the maximum quantity of a solute at specific temperature and pressure conditions

solubility: a property of a solute; the concentration of a saturated solution of a solute in a solvent at a specific temperature and pressure

Figure 1
The excess of solid solute in the mixture is visible evidence for a saturated solution.

INQUIRY SKILLS

- ○ Questioning
- ○ Hypothesizing
- ○ Predicting
- ○ Planning
- ● Conducting
- ● Recording
- ● Analyzing
- ● Evaluating
- ● Communicating

7.1 Solubility

When you add a small amount of pickling salt (pure sodium chloride) to a jar of water and shake the jar, the salt dissolves and disappears completely. What happens if you continue adding salt and shaking? Eventually, some visible solid salt crystals will remain at the bottom of the jar, despite your efforts to make them dissolve. You have formed a **saturated solution**—a solution in which no more solute will dissolve. We say it is at maximum solute concentration. If the container is sealed, and the temperature stays the same, no further changes will ever occur in the concentration of this solution. The quantity (mass) of solute that remains undissolved will also stay the same. The **solubility** of sodium chloride in water is the concentration of your saturated solution. The units for solubility and maximum concentration are therefore the same: usually grams of solute per 100 mL of solvent. You will learn in this chapter that solubility depends on the temperature, so it is a *particular* maximum concentration value. Every solubility value must be accompanied by a temperature value.

When calculating and using solubility values we have to make one assumption: the solute is not reacting with the solvent.

Solubility of Solids

Every pure substance has its own unique solubility. For example, we can find from a reference source, such as the *CRC Handbook of Chemistry and Physics*, that the solubility of sodium sulfate in water at 0°C is 4.76 g/100 mL. Remember that this means 4.76 g of solute can be dissolved in 100 mL of water; *not* that you will have 100 mL of solution after dissolving 4.76 g of solute. If more than 4.76 g of this solute is added to 100 mL of water in the container, the excess will not dissolve under the specified conditions (**Figure 1**). The quickest way to see whether you have a saturated solution is to look for the presence of undissolved solids in the solution. There are several experimental designs that can be used to determine the solubility of a solid. For example, the solvent from a measured volume of saturated solution might be removed by evaporation, leaving the crystallized solid solute behind—which can then be collected and measured.

Investigation 7.1.1

Solubility Curve of a Solid

Because the solubility of a substance changes with the temperature of the saturated solution, it is sometimes useful to plot graphs of the relationship between these two variables. A graph gives an instant visual picture of the relationship and can then be used to determine the solubility at any temperature. A graph of solubility and temperature of the solution is called a *solubility curve.*

In this investigation, you will measure the temperature at which the solute precipitates out of solution, indicating that the solute has reached its maximum concentration. You will do this for several different concentrations. You will then use your evidence to create a solubility curve. Complete the **Analysis** and **Evaluation** sections of the lab report.

Question

What is the relationship between the solubility of potassium nitrate and the temperature of its solution?

Experimental Design

A known mass of potassium nitrate is dissolved in warm water. As the solution cools, the temperature is recorded when the *first* sign of crystal formation occurs. This Procedure is repeated several times with different volumes of water (i.e., different solution concentrations).

(a) Identify dependent, independent, and controlled variables in this design.

Materials

eye protection
laboratory apron
oven mitts
250-mL beaker
thermometer
centigram balance
hot plate
medicine dropper
laboratory scoop
small test tube
10-mL glass graduated cylinder
pure (distilled) water
$KNO_{3(s)}$ (about 3 g)

Be careful with the hot plate; it looks the same whether hot or cold. Use oven mitts when handling hot apparatus.

Wear eye protection and a laboratory apron.

Procedure

1. Set up a hot-water bath by placing the 250-mL beaker, about two-thirds full of water, on the hot plate. Heat the water to about 85°C.
2. Add about 3 g of $KNO_{3(s)}$ to a clean, dry test tube and record the actual mass used.
3. Add 10.0 mL pure water to the graduated cylinder.
4. Use the medicine dropper to remove 2.0 mL of pure water from the graduated cylinder. Transfer the 2.0 mL of water to the test tube.
5. Insert the thermometer into the test tube and then place the test tube in the hot-water bath.
6. Carefully stir the contents of the test tube with the thermometer until the solid has completely dissolved. Do not leave the test tube in the water bath any longer than necessary to dissolve the solid.
7. Remove the test tube from the bath, stir carefully, and record the temperature when the *first crystals appear*. It may help to hold the test tube up to a light.
8. Repeat steps 4 to 7 four more times with the same test tube, adding 1.0 mL more water (instead of 2.0 mL) each time.
9. Empty the contents of the test tube into the labelled beaker provided to recycle the potassium nitrate.

Analysis

(b) For each trial, calculate the solubility in grams of $KNO_{3(s)}$ per 100 mL of water.
(c) Plot a graph of solubility (in g/100 mL) against temperature.
(d) Write a sentence to answer the Question: What is the relationship between the solubility of potassium nitrate and the temperature of its solution?

Evaluation

(e) Evaluate the Evidence by judging the Experimental Design, Materials, Procedure, and your skills. Identify sources of experimental uncertainty or error. Note any flaws and potential improvements.
(f) Suggest a different experimental design that could be used to determine the solubility of a solid such as potassium nitrate.

Synthesis

(g) What property of the solution is illustrated at the moment the crystals start forming? Explain briefly.

(h) According to your graph, which of the following mixtures is a saturated solution and which is an unsaturated solution?

(i) 100 g of $KNO_{3(s)}$ in 100 mL of $H_2O_{(l)}$ at 40°C

(ii) 50 g of $KNO_{3(s)}$ in 100 mL of $H_2O_{(l)}$ at 70°C

(iii) 120 g of $KNO_{3(s)}$ in 200 mL of $H_2O_{(l)}$ at 60°C

(i) According to your graph, what mass of potassium nitrate will dissolve in 100 mL of water at

(i) 20°C? (ii) 50°C?

Solubility Curves

Graphs of solubility (maximum concentration) against temperature allow quick and easy reference, and are very useful for a wide variety of questions involving solution concentrations. **Figure 2** shows some solubility curves for ionic compounds in aqueous solution.

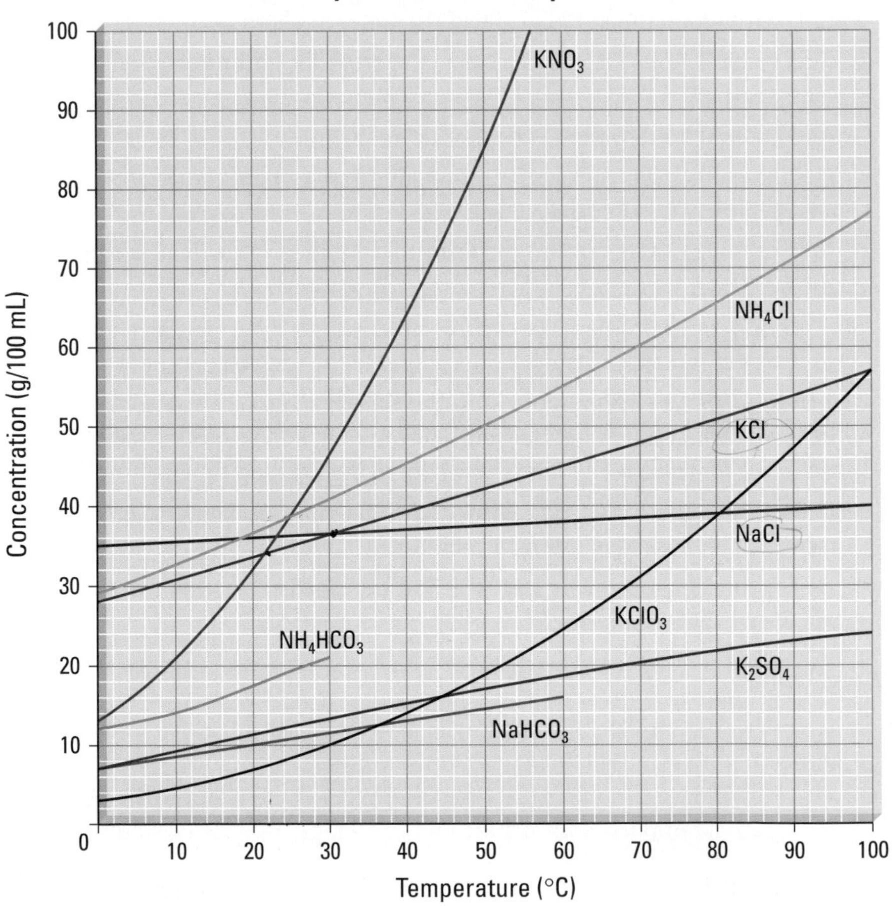

Figure 2
The solubility of ionic compounds in water is related to temperature.

Practice

Understanding Concepts

1. Refer to **Figure 2**.
   (a) What mass of K_2SO_4 can be dissolved in 100 mL of water at 70°C?

(b) At what temperature are the solubilities of KNO_3 and KCl approximately equal?
(c) Why is it not possible to determine the molar concentration of a saturated solution of $KClO_3$ at 25°C from its solubility curve? What additional information would be required?
(d) Which substance forms a saturated solution in which the mass ratio of solute to solvent is 1:1, and at what temperature?

2. A solution containing equal masses of NaCL and KCL is slowly cooled.
   (a) At what temperatures might sodium chloride precipitate first, without precipitating potassium chloride?
   (b) At what temperatures might potassium chloride precipitate first, without precipitating sodium chloride?

Applying Inquiry Skills

3. If you completed Investigation 7.1.1, you generated a solubility curve for $KNO_{3(aq)}$. Compare your curve to the curve in **Figure 2**. Are there differences? How would you account for any differences?

Solubility of Gases

Although we are not so aware of gases dissolving in liquids, they do indeed do so. Swimming pools contain dissolved chlorine, rivers and streams contain dissolved oxygen, and cans of pop contain dissolved carbon dioxide.

Gas Solubility

This activity illustrates the various effects of temperature (and even some effects of pressure) on solubility. It also highlights an interesting point: air dissolves in water, and water dissolves in air!

Materials: clear, colourless container (e.g., beaker or drinking glass), a long match, any clear, light-coloured carbonated beverage

- Open the carbonated beverage, and fill your container half full. Observe the rapidity of formation, source, and size of the bubbles.
- Let the bubbling drink stand a few moments in a draft-free area, then light a match and lower it into the top half of the container. Observe what happens to the flame.

  (a) Describe and explain your observation.

- Empty and rinse your container.
- Run a tap until the water is quite hot and then fill your container. Normally, the water will at first be clouded or milky in appearance. Let it stand for a minute or two as you observe closely what happens. Record your observations. Dispose of the hot water.
- Run a tap until the water is very cold and then fill your container. Normally, the water will at first be nicely clear in appearance. Let it stand for a few minutes as you observe closely what happens to the outside of the container. Record your observations.

  (b) In terms of solubility, explain what causes the condensation here, and describe how you could test qualitatively what the liquid is.

- Let your container of cold water stand for another half-hour. Occasionally observe and record what happens inside of container.

  (c) In terms of solubility, explain your observations of the container of cold water. What causes the bubbles to form? Discuss what might cause the bubbles to be larger than they were in the hot water.

INQUIRY SKILLS

- ○ Questioning
- ○ Hypothesizing
- ○ Predicting
- ○ Planning
- ○ Conducting
- ○ Recording
- ● Analyzing
- ○ Evaluating
- ○ Communicating

Lab Exercise 7.1.1

Solubility of a Gas

We have seen that the solubility of a solid increases with increasing temperature. How does the solubility of a gas change as the temperature changes? Using the evidence provided in this lab exercise, create a generalization for the effect of temperature on the solubility of a gas. Complete the **Experimental Design, Analysis,** and **Synthesis** sections of the report.

Question

What effect does temperature have on the solubility of various gases in water?

Experimental Design

(a) Based on **Table 1**, identify the independent, dependent, and controlled variables.

Evidence

Table 1: Solubility of Gases at Different Temperatures

Gas	Solubility (g/100 mL) at specified temperature		
	0°C	**20°C**	**50°C**
$N_{2(g)}$	0.0029	0.0019	0.0012
$O_{2(g)}$	0.0069	0.0043	0.0027
$CO_{2(g)}$	0.335	0.169	0.076
$NH_{3(g)}$	89.9	51.8	28.4

Analysis

(b) Use the Evidence to answer the Question. Write your response as a generalization.

Synthesis

(c) What assumption must we make about the dissolving of a gas in water for the solubility evidence to be valid?

(d) Use your knowledge of the kinetic molecular theory to explain the generalization for the solubility of gas as the temperature increases.

Practice

Understanding Concepts

4. When you open a can of pop, which is more likely to fizz and spray: a can at room temperature or a cold can from the refrigerator? Explain why using your understanding of gas solubility.

Reflecting

5. (a) The solubility of oxygen in blood is much greater than its solubility in pure water. Suggest a reason for this observation.
   (b) If the solubility of the oxygen in blood were the same as in pure water, how would your life be different?

Solubility in Water—Generalizations and Examples

Scientists have carried out a very large number of experiments as they investigated the effects of temperature on the solubility of various solutes. From the results of their experiments, they have developed several useful generalizations about the solubility of solids, liquids, and gases in water. In all cases, we assume that the solid, liquid, or gas does not react with the solvent, water. The following list outlines how the solubility of various solutes varies with temperature.

- Solids usually have higher solubility in water at higher temperatures. Figure 2 shows the solubility curves of many ionic compounds. This trend is generally the same for soluble molecular compounds. For example, sucrose has a solubility of about 180 g/100 mL at 0°C and 487 g/100 mL at 100°C.
- Gases always have higher solubility in water at lower temperatures. The solubility of gases decreases as the temperature increases. The relationship between solubility and temperature is inverse and approximately linear.
- It is difficult to generalize about the effect of temperature on the solubility of liquids in water. However, for polar liquids in water, the solubility usually increases with temperature. A prediction of the solubility of liquids with temperature will not be as reliable as a prediction for solids and gases.
- Some liquids (mostly nonpolar liquids) do not dissolve in water to any appreciable extent, but form a separate layer. Liquids that behave in this way are said to be **immiscible** with water. For example, benzene, gasoline, and carbon disulfide (which is used in the process of turning wood pulp into rayon or cellophane) are all virtually insoluble in water.
- Some liquids (such as those containing small polar molecules with hydrogen bonding) dissolve completely in water in any proportion. Liquids that behave in this way are said to be **miscible** with water. For example, ethanol (in alcoholic beverages), acetic acid (in vinegar), and ethylene glycol (in antifreeze) all dissolve completely in water.
- Elements generally have low solubility in water. For example, some people put a lump of sulfur in their dog's water bowl, to keep the water "fresh." The same piece of sulfur can last, apparently unchanged, for years. Carbon is used in many water filtration systems to remove organic compounds that cause odours. The carbon does not dissolve in the water passing through it.
- Although the halogens and oxygen dissolve in water to only a very tiny extent, they are so reactive that even in tiny concentrations they are often very important in solution reactions.

immiscible: two liquids that form separate layers instead of dissolving

miscible: liquids that mix in all proportions and have no maximum concentration

Of course, there are exceptions to all generalizations. For example, the solubility of lithium carbonate in water actually decreases as the temperature increases.

Practice

Understanding Concepts

6. For any solute, what important condition must be stated in order to report the solubility?
7. Sketch a solubility graph showing two lines labelled "solids" and "gases." Assume a straight-line relationship and show the generalization for the change in solubility of each type of substance with increasing temperature.

Answer

8. (f) 1.7 g

8. A common diagnostic test for $CO_{2(g)}$ involves the use of a saturated solution of calcium hydroxide, $Ca(OH)_{2(aq)}$, historically called limewater. Calcium hydroxide has low solubility—but is actually soluble enough to have a noticeable effect. The solution is easily prepared by placing a spoonful of solid $Ca(OH)_{2(s)}$ in pure water, and stirring or shaking for a while. A small amount of the solid dissolves. The excess solute settles to the bottom, leaving a clear, saturated solution. Bubbling carbon dioxide gas through this saturated solution causes reactions that finally precipitate calcium carbonate, which is roughly 10 times less soluble than calcium hydroxide.
   (a) The first reaction is the apparent "dissolving" of a tiny amount of carbon dioxide in water. What we believe really occurs is that some carbon dioxide gas reacts with water to form carbonic acid in solution. Write a balanced chemical equation to represent this reaction.
   (b) The acid reacts with dissolved calcium hydroxide to form calcium carbonate precipitate. Write a balanced chemical equation to represent this reaction.
   (c) Use the reaction equations to explain why this is a test for the presence of $CO_{2(g)}$.
   (d) If the calcium hydroxide solution were not saturated, or nearly so, would this diagnostic test still work? Explain the reasoning behind your answer.
   (e) Plot a solubility curve for $Ca(OH)_2$, using the data in **Table 2**.
   (f) Assuming you wished to prepare 1.0 L of saturated limewater solution to be stored at a normal room temperature of 22°C, what minimum mass of solid calcium hydroxide would you require? Why should you actually use considerably more than the minimum required?
   (g) Does the solubility curve of this compound fit the generalization about the solubility of ionic solids at different temperatures? Does this mean that the generalization is not valid or useful? Discuss briefly.

Table 2: Solubility of Calcium Hydroxide at Various Temperatures

Temperature (°C)	Solubility (g /100 mL)
0	0.18
20	0.17
40	0.14
60	0.11
80	0.09
100	0.08

Applying Inquiry Skills

9. An experiment is conducted to test the generalization for the temperature dependence of the solubility of solids. Complete the **Prediction** and **Analysis** sections of the following report.

   Question

   How does temperature affect the solubility of potassium chlorate?

   Prediction

   (a) Use the generalization for the solubility of solids to answer the Question and state your reasons.

   Evidence

Table 3: Solubility of Potassium Chlorate

Temperature (°C)	Solubility (g /100 mL)
0	5.0
20	8.5
40	16.3
60	27.5
80	42.5
100	59.5

Analysis

(b) Plot a solubility curve for potassium chlorate.

(c) What is the answer to the Question, according to the Evidence?

Making Connections

10. Some industries, particularly electric power generating stations, get rid of waste hot water by releasing it into a nearby lake or river. Use your knowledge of solubility and temperature to describe why this thermal pollution is detrimental to most fish.

Crystallization

Have you ever opened a jar of liquid honey only to find that the clear, golden syrup is mixed with hard, white particles (**Figure 3**)? This is an example of what happens to a saturated sugar solution (honey is mostly sugar and water) when some of the water evaporates. The white particles are sugar crystals that have come out of solution in a process called crystallization. This crystallization process also occurs naturally (and extremely slowly) during the formation of stalactites and stalagmites in caves (**Figure 4**).

Crystallization can be artificially speeded up by heating a solution to evaporate off the solvent. With less solvent present, the concentration of the solute quickly exceeds the solubility, so the excess solute crystallizes out. This evaporation method is used industrially to isolate the solid solute from many solutions. For example, table salt and table sugar are produced industrially by heating saturated solutions of the substances. Rapid evaporation forms lots of small crystals, which are then screened for uniformity of size. Crystals of unwanted sizes are just redissolved and cycled through the process again. Large crystals can be formed by slowly evaporating the solvent (**Figure 5**).

There are several experimental designs that we can use to determine the solubility (maximum concentration) of a solid. In Unit 2 you used precipitation and filtration as an experimental design to determine the quantity of a solute present in a solution. Later in this unit you will use a titration procedure to determine the concentration of a solution. At this point you are going to use a procedure called crystallization. Crystallization involves removing the solvent from a solution by evaporation, leaving behind the solid solute—which can then be collected and measured.

Figure 3
Liquid honey normally crystallizes over time. The label instructions usually suggest that you just place the container in warm water for a while if you want the sugar crystals to redissolve.

Figure 4
Stalactites and stalagmites form in caves when calcium carbonate crystallizes from ground water solutions. The solvent evaporates.

Figure 5
You can create your own stalactite–stalagmite system with a saturated solution of Epsom salts, two glasses, a plate, and some heavy string—and it won't take centuries to form!

INQUIRY SKILLS

- ○ Questioning
- ○ Hypothesizing
- ● Predicting
- ○ Planning
- ● Conducting
- ● Recording
- ● Analyzing
- ● Evaluating
- ● Communicating

Wear eye protection and a laboratory apron.

When using a laboratory burner, keep long hair tied back and loose clothing secured. If using a hot plate, take all necessary precautions.

When heating anything in a container, reduce the danger of splattering by keeping the container loosely covered.

Use oven mitts or heat-proof gloves to handle hot apparatus.

Table 4: Solubility of Sodium Chloride in Water

Temperature (°C)	Solubility (g /100 mL)
0	35.7
20	35.9
40	36.4
60	37.1
80	38.0
100	39.2

Investigation 7.1.2

The Solubility of Sodium Chloride in Water

A significant part of the work of science is to test existing theories, laws, and generalizations. Your purpose in this investigation is to test the solubility curve for an ionic solid. To do this, you will create a graph from the solubility data (**Table 4**), and use this graph to predict the solubility of sodium chloride in water at a particular temperature. You will then compare the predicted value with a value that you determine experimentally—by crystallization of sodium chloride from a saturated solution.

Complete the **Prediction**, **Evidence**, **Analysis**, and **Evaluation** sections of this report.

Question

What is the solubility of sodium chloride at room temperature?

Prediction

(a) Create a graph of the data in **Table 4**, and use it to predict the solubility of sodium chloride in water at the measured room temperature. (You will be able to improve the precision of your prediction if you start the vertical axis at 35 g/100 mL instead of the usual zero value.)

Experimental Design

A precisely measured volume of a saturated $NaCl_{(aq)}$ solution at room temperature is heated to evaporate the solvent and crystallize the solute. The mass of the dry solute is measured and the concentration of the saturated solution is calculated.

Materials

lab apron
eye protection
oven mitts or heatproof gloves
saturated $NaCl_{(aq)}$ solution
laboratory burner with matches or striker, or hot plate
centigram balance
thermometer
laboratory stand
ring clamp
wire gauze
250-mL beaker
100-mL beaker
10-mL pipet with pipet bulb to fit

Procedure

1. Measure and record the total mass of a clean, dry 250-mL beaker (plus watch glass cover, if used).
2. Obtain about 40 mL to 50 mL of saturated $NaCl_{(aq)}$ in a 100-mL beaker.
3. Measure and record the temperature of the saturated solution to a precision of 0.2°C.
4. Pipet a 10.00-mL sample of the saturated solution into the 250-mL beaker.

5. Using a laboratory burner or hot plate, heat the solution evenly in the beaker until all the water boils away, and dry, crystalline $NaCl_{(s)}$ remains (**Figure 6**).
6. Shut off the burner or hot plate, and allow the beaker (plus any cover) and contents to cool for at least 5 min.
7. Measure and record the total mass of the beaker, cover, and contents.
8. Reheat the beaker and the residue and repeat steps 6 and 7 until two consecutive measurements of the mass give the same value. Record the final mass. (If the mass remains constant, this confirms that the sample is dry.)
9. Dispose of the salt as regular solid waste.

Figure 6
A saturated sodium chloride solution is heated to evaporate the water and crystallize the solute.

Analysis

(b) Using the evidence you collected, answer the Question. Write the solubility of sodium chloride in grams per 100 mL.

Evaluation

(c) Evaluate the Evidence by judging the Experimental Design, Materials, Procedure, and skills used. Note any flaws and suggest improvements. List some sources of experimental uncertainty or error.
(d) Considering your answers to (c), how certain are you about the experimental answer that you obtained?
(e) Determine the accuracy of your result by calculating the percentage difference.
(f) Use the accuracy you calculated and your answer to (c) to evaluate the Prediction you made based on the solubility curve.
(g) Based on your Evaluation of the Prediction, is the authority you used (the solubility data and curve) acceptable?

Solubility Categories

People have been using the property of solubility for thousands, perhaps millions, of years. Over the last couple of hundred years, experimenters have been investigating solubility in a more quantitative fashion, and have developed tables of solubilities of various substances under a variety of conditions. Measurements of extremely high precision show that all substances are soluble in water to some extent—so the question then just becomes, to what extent? The solubilities of various simple ionic compounds range from very **high solubility**, like that of ammonium chloride, to extremely **low solubility**, like that of silver chloride. Strictly speaking, nothing is absolutely insoluble in water, but we use the term **insoluble** to mean negligible solubility—where the effect of the quantity that will dissolve is not easily detectable. This means that, although your drinking glass may dissolve to the extent of a few hundred molecules every time you wash it, you won't have to worry about the effect for a few million years.

high solubility: with a maximum concentration at SATP (standard ambient temperature and pressure) of greater than or equal to 0.1 mol/L

low solubility: with a maximum concentration at SATP of less than 0.1 mol/L

insoluble: a substance that has a negligible solubility at SATP

As you probably found in Chapter 3, reading solubility tables takes a little practice. You will notice that **Table 5**, page 324, shows only the anions as headings. We can safely assume that most compounds containing those anions have similar solubilities. It is useful to classify simple ionic compounds into categories of high and low aqueous solubility. The classification allows you to predict the state of many compounds formed in single and double displacement reactions. Of course, any defined cutoff point between high and low solubility is entirely arbitrary. We usually assign a solubility of 0.1 mol/L as the cutoff point, because most ionic compounds have solubilities either significantly greater or significantly less than

Table 5: Solubility of Ionic Compounds at SATP

		Anions						
		Cl^-, Br^-, I^-	S^{2-}	OH^-	SO_4^{2-}	CO_3^{2-}, PO_4^{3-}, SO_3^{2-}	$C_2H_3O_2^-$	NO_3^-
Cations	High solubility (aq) ≥0.1 mol/L (at SATP)	most	Group 1, NH_4^+ Group 2	Group 1, NH_4^+ Sr^{2+}, Ba^{2+}, Tl^+	most	Group 1, NH_4^+	most	all
		All Group 1 compounds, including acids, and all ammonium compounds are assumed to have high solubility in water.						
	Low solubility (s) <0.1 mol/L (at SATP)	Ag^+, Pb^{2+}, Tl^+, Hg_2^{2+}, (Hg^+), Cu^+	most	most	Ag^+, Pb^{2+}, Ca^{2+}, Ba^{2+}, Sr^{2+}, Ra^{2+}	most	Ag^+	none

this value. As well, we often work with solution concentrations of 0.01 mol/L to 1.0 mol/L in school laboratories. In other words, we choose this 0.1 mol/L cutoff because it usually works for our purposes.

As with all simple generalizations that *usually* work, there are exceptions. There are some compounds with solubilities that are low, but close enough to our cutoff to cause problems in investigation. Calcium sulfate has a low solubility by our definition (about 0.02 mol/L at SATP), and when a spoonful is stirred into water it doesn't appear to dissolve. However, a conductivity test shows that enough does dissolve for the saturated solution to noticeably conduct electric current (**Figure 7**). This also means that, if you washed a sample of $CaSO_{4(s)}$ in a filter paper with water, you would be losing a little bit of your sample every time you rinsed it, slowly ruining your experiment's accuracy.

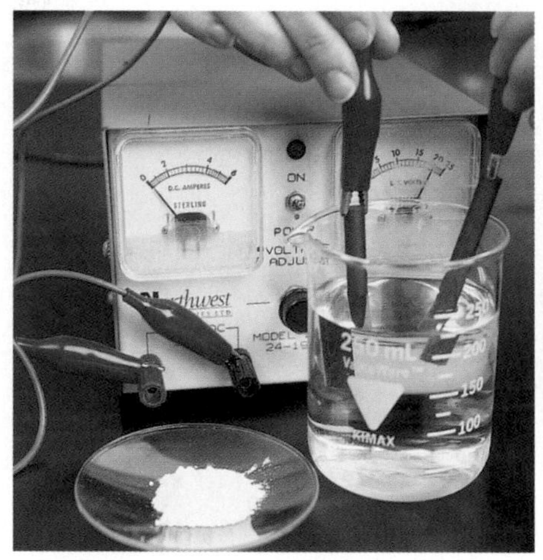

Figure 7
If we rely on visual evidence, calcium sulfate appears insoluble, but other evidence (i.e., conductivity) indicates that the solubility is very low but not negligible.

precipitate: (*verb*) form a low solubility solid from a solution; (*noun*) the solid formed in a chemical reaction or by decreased solubility

There is another familiar compound that is only slightly soluble in water: calcium hydroxide, $Ca(OH)_{2(s)}$, used to make limewater, which is used in a common diagnostic test for $CO_{2(g)}$. This compound is defined as having low solubility by our arbitrary cutoff point, but it is actually soluble enough to have a noticeable effect. Bubbling carbon dioxide gas through this saturated solution causes reactions that finally **precipitate** calcium carbonate, which has *very* low solubility. A precipitate is a pure, solid substance that appears in a solution, either as a result of a reaction or because of decreasing solubility.

One final example of a substance that is defined as having low solubility, but that in fact does dissolve very slightly, is oxygen. The solubility of oxygen gas in home aquariums is, for example, approximately 0.002 mol/L at 20°C—about 50 times lower than our "low solubility" cutoff—but obviously the fish don't care. This is a sufficient concentration for their purposes.

Practice

Understanding Concepts

11. The solubility table (**Table 5**) summarizes many solubility generalizations for common ionic compounds in water at SATP.
    (a) What cutoff point is used to separate low and high solubility substances in water?
    (b) State two reasons for the choice of this concentration value.
    (c) Describe, in your own words, how to use this table to find out whether an ionic compound has low or high solubility.
    (d) In a chemistry class, how do the terms "soluble" and "insoluble" relate to this table?
12. Classify the following compounds as having high or low solubilities:
    (a) $NaOH_{(s)}$ (oven cleaner)
    (b) $MnCl_{2(s)}$ (used in the dyeing industry)
    (c) $Al(OH)_{3(s)}$ (water purification)
    (d) $Ca_3(PO_4)_{2(s)}$ (raw matertial in production of fertilizers)
    (e) $CuSO_4{\cdot}5H_2O_{(s)}$ (bluestone, a fungicide)
13. Specialty sugar, made up of crystals as large as 5 mm long, is sold for sweetening coffee.
    (a) Suggest how this sugar might have been produced.
    (b) Predict the solubility of this sugar, compared with that of regular white sugar.
    (c) Predict the rate of dissolving of this sugar, compared with that of regular white sugar.

Applying Inquiry Skills

14. Describe two experimental designs that can be used to determine the solubility of a solid in water.

Making Connections

15. Some pollutants in natural waters, such as heavy metals and organic compounds, would be classified as having low solubility. What are some origins of pollutants in natural waters? If some are low solubility compounds, why are these a problem?

The Solvay Process—An Effect of Solubility

In Chapter 5 you studied the sequence of chemical reactions called the Solvay process. This commercial process was developed to produce sodium carbonate, $Na_2CO_{3(s)}$, a compound historically called soda ash and now commonly known as washing soda. The overall reaction for the Solvay process is

$$CaCO_{3(s)} + 2\ NaCl_{(aq)} \rightarrow Na_2CO_{3(aq)} + CaCl_{2(aq)}$$

The key to the process, however, wasn't really a chemical change. Ernest Solvay used a solubility effect to separate chemicals in solution, and found a way to produce the product at one-third of the previous cost. This made Solvay both wealthy and famous, as you can imagine.

The reactants in the crucial step of the overall process are $NH_4HCO_{3(aq)}$ and $NaCl_{(aq)}$. Using our solubility chart, we find this reaction improbable.

$$NH_4HCO_{3(aq)} + NaCl_{(aq)} \rightarrow NH_4Cl_{(aq)} + NaHCO_{3(s)}$$

From the generalizations you have learned, all the possible products are high solubility compounds, and should not precipitate in water solution. According to our solubility table (**Table 5**), none of the possible ion combinations has low solubility.

However, the solubility table gives solubilities at SATP. Solvay's effect was made possible by Solvay's realization that, in cold water, the solubility of $NaHCO_3$ (7 g/100 mL at 0°C) is lower than the solubility of NH_4HCO_3 (12 g/100 mL at 0°C). It is also much lower than the solubilities of either NH_4Cl or NaCl (**Figure 2**). This meant that the solution, at room temperature, would already be nearly saturated with sodium hydrogen carbonate. When he decreased the temperature, the $Na^+_{(aq)}$ and $HCO_3^-{}_{(aq)}$ combination reached its solubility limit and crystallized out of the solution as $NaHCO_{3(s)}$, leaving the other more soluble ions behind.

Practice

Understanding Concepts

16. Sketch a solubility graph (qualitative only) to illustrate the crucial step for the Solvay process in which sodium hydrogen carbonate was precipitated from an aqueous mixture of sodium hydrogen carbonate and ammonium chloride.

Making Connections

17. How is the development of the Solvay process an example of science leading to a new technology?

Section 7.1 Questions

Understanding Concepts

1. In a chemical analysis experiment, a student notices that a precipitate has formed, and separates this precipitate by filtration. The collected liquid filtrate, which contains aqueous sodium bromide, is set aside in an open beaker. Several days later, some white solid is visible along the top edges of the liquid and at the bottom of the beaker.
   (a) What does the presence of the solid indicate about the nature of the solution?
   (b) What interpretation can be made about the concentration of the sodium bromide in the remaining solution? What is the term used for this concentration?
   (c) State two different ways to convert the mixture of the solid and solution into a homogeneous mixture.
2. Describe how the solubilities of solids and gases in water depend on temperature.

Applying Inquiry Skills

3. The following investigation is carried out to test the generalization about the effect of temperature on the solubility of an ionic

compound classified as having low solubility. Complete the **Prediction, Experimental Design, Analysis,** and **Evaluation** sections of the investigation report.

Question
What is the relationship between temperature and the solubility of barium sulfate?

Prediction
(a) Answer the Question, including your reasoning.

Experimental Design
Pure barium sulfate is added to three flasks of pure water until no more will dissolve and there is excess solid in each beaker. The flasks are sealed, and each is stirred at a different temperature until no further changes occur. The same volume of each solution is removed and evaporated to crystallize the solid.
(b) Identify the independent, dependent, and controlled variables.

Evidence

Solubility of Barium Sulfate

Temperature (°C)	Mass of $BaSO_{4(s)}$ (mg/100 mL)
0	0.19
20	0.25
50	0.34

Analysis
(c) Plot a solubility curve for barium sulfate.
(d) Write a sentence to answer the Question, based on the Evidence collected.

Evaluation
(e) Suggest a significant improvement that could be made in this experiment.
(f) Was the prediction verified? State your reasons.
(g) Does the solubility generalization appear acceptable for low solubility ionic compounds? Is one example enough of a test? Discuss briefly.

Making Connections

4. In "dry" cleaning, a non-aqueous solvent is used to remove grease stains from clothing. Suggest a reason (hypothesis) why a non-aqueous solvent is used, rather than water and detergent.
5. Different species of fish are adapted to live in different habitats. Some, such as carp, can survive perfectly well in relatively warm, still water. Others, such as brook trout, need cold, fast-flowing streams, and will die if moved to the carp's habitat.
   (a) Describe and explain the oxygen conditions in the two habitats.
   (b) Hypothesize about the oxygen requirements of the two species of fish.
   (c) Predict the effect of thermal pollution on trout in their streams.

7.2 Hard Water Treatment

When using soap in the bath, do you find that a floating scum forms on the surface of the bath water? Is there a hard, greyish-white scale inside your kettle and pipes (**Figure 1**)? When you wash your hands, does your soap fail to produce a good lather? This is all evidence that you have **hard water**. When detergents were initially developed they were hailed as a laundry revolution. They were chemicals formulated to work even in hard water, without producing the precipitates common when using soap.

hard water: water containing an appreciable concentration of calcium and magnesium ions

Table 1: Water Hardness

Hardness index (mg/L or ppm)	Water classification
< 50	soft
50–200	slightly hard
200–400	moderately hard
400–600	hard
> 600	very hard

Figure 1
Scale deposits caused by hard water can reduce the efficiency of some kitchen appliances.

When water travels through soil and rock it dissolves some of the minerals from the rock. Some minerals will always dissolve in ground water: The amount that dissolves depends on the rock type. Some rocks, such as granite, contain mostly low-solubility minerals. Other rocks, such as limestone, contain large quantities of minerals that dissolve relatively easily in water. The ground water that flows through or over such rock will then contain metal cations, such as calcium, $Ca^{2+}_{(aq)}$, magnesium, $Mg^{2+}_{(aq)}$, iron(II), $Fe^{2+}_{(aq)}$, iron(III), $Fe^{3+}_{(aq)}$, and manganese, $Mn^{2+}_{(aq)}$.

Water hardness is primarily due to calcium and magnesium ions. The hardness index used by chemists is the total concentration of these two ions, expressed in milligrams per litre (which is equivalent to parts per million). There is no agreed-upon standard system for grading water hardness, but **Table 1** shows an approximate classification.

Figure 2
The use of home water-softening units is increasing steadily.

Water Softening

Water is "softened" by removing the calcium and magnesium ions. This may be done during large-scale water treatment, or within a home or business by various kinds of commercial water-softening devices (**Figure 2**).

In some parts of Ontario the hardness of the water is a problem. As you learned earlier in your study of freshwater treatment, municipalities in these areas often use the **soda-lime process** to soften the water. Both sodium carbonate (the soda) and calcium hydroxide (called hydrated lime or slaked lime) are added to the hard water. After a series of reactions, the calcium and magnesium ions in the hard water are precipitated out as the respective carbonates, softening the water. This process depends entirely on differences in solubility, as both calcium and magnesium carbonates are much less soluble than the respective hydrogen carbonate (bicarbonate) compounds found in water.

soda-lime process: a water-softening process involving sodium carbonate and calcium hydroxide, in which calcium carbonate and magnesium carbonate are precipitated out

$$Na_2CO_{3(aq)} + Ca(HCO_3)_{2(aq)} \rightarrow CaCO_{3(s)} + 2\ NaHCO_{3(aq)}$$

Home water softening usually uses an ion exchange process. In the softener tank is a bed of small resin (plastic) grains, through which the water is passed. You can think of the molecules of resin as extremely long chains, along which are attached many charged sulfonate groups: SO_3^- (**Figure 3**). The resin begins with sodium ions attached to each sulfonate group. When hard water moves through the resin, calcium and magnesium ions attach to the sulfonate groups, displacing the sodium ions into the water (**Figure 3**). Water emerging from the resin is thus "softened": Sodium ions have been exchanged for the water's calcium and magnesium ions.

Eventually, there will be no more sodium ions left on the resin. The water softener must then be regenerated or the disposable cartridge replaced. Regeneration involves washing the resin with a saturated brine (sodium chloride) solution from a salt tank attached to the water softener. This causes sodium ions to replace the calcium and magnesium ions on the resin. The hard water ions are flushed down the drain, and the softener is ready to work for another cycle.

Home softeners may regenerate by clock setting, usually during the night. More advanced models measure water usage, and regenerate only when the resin's capacity to exchange ions is nearly exhausted. This means that less salt is wasted, but these models are more expensive. Water softened in this way becomes higher in sodium ions, which may be a concern for people on low-sodium diets. Potassium chloride is sometimes used in place of sodium chloride in the brine tank. Marketers claim that since plants require potassium, this may be better for the environment, as well as for people on low-sodium diets. Potassium ions do not work quite as well as sodium ions, though, and $KCl_{(s)}$ is more expensive to buy than $NaCl_{(s)}$.

(a)

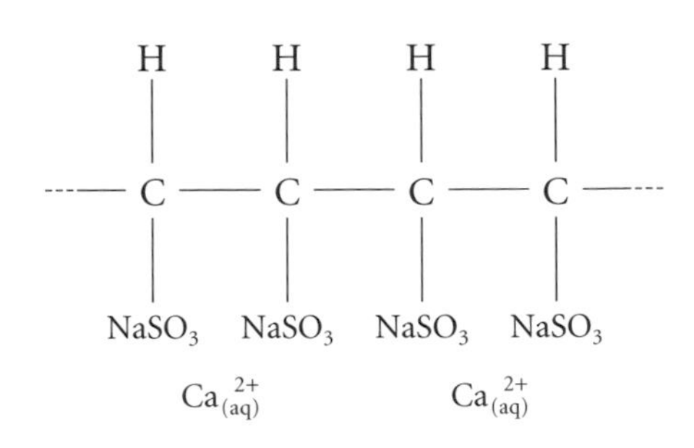

(b)

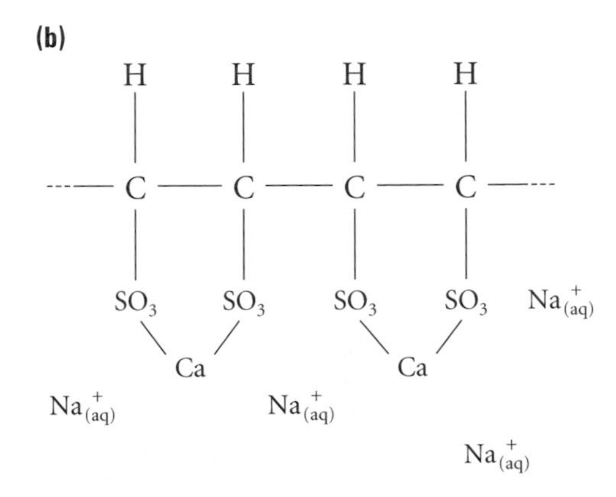

Figure 3

(a) Each resin molecule is actually hundreds of thousands of these units long. Each sulfonate group acts as a site where positive (metal) ions may be attached and held.

(b) Two aqueous calcium ions have been attached and held by the resin surface—exchanged for four aqueous sodium ions.

Practice

Understanding Concepts

1. Hard water is not a problem when cleaning laundry with modern laundry detergents, but it can still cause other problems.
   (a) State two observations that are evidence of hard water.
   (b) Explain why ground water is hard in some areas.
2. The soda-lime process uses two common chemicals to react with and remove certain metal ions from water.
   (a) Write the chemical formulas for the chemicals used to remove the ions.
   (b) Explain how this reaction solves the hard-water problem.
3. In municipal water treatment to soften water, calcium ions are mostly removed from the water by precipitation of calcium carbonate, which has a low solubility of 7.1×10^{-5} mol/L.
   (a) Calculate the concentration of calcium carbonate in units of milligrams per litre in the treated water.
   (b) What is the concentration of calcium carbonate in units of parts per million?
   (c) Ignoring other ions that may be present, classify the hardness of the treated water.
4. Why does a home water softener unit have to be regenerated?

Making Connections

5. Suggest reasons why an ion exchange water softener would not be appropriate for large-scale municipal water softening.

Answer

3. (a) 7.1 mg/L

6. "When detergents were initially developed they were hailed as a laundry revolution." Comment on this statement, giving examples of their positive and negative effects.
7. When homes in rural areas have water softeners, does the soft water go to all of the taps in the house? Use the Internet to find out. Report on what you find. What are some health concerns of drinking water with high sodium ion content?

   Follow the links for Nelson Chemistry 11, 7.2.

GO TO www.science.nelson.com

Section 7.2 Questions

Understanding Concepts

1. Why is water in some parts of Ontario soft, and in other parts hard?
2. What are the two metal ions most responsible for the hardness of water?
3. If you have hard water at home, what are the consequences
   (a) that you would likely notice first?
   (b) that are not easily visible? What serious problems may result?
4. In municipal water softening, calcium ions in water are precipitated as solid calcium carbonate. Assuming that the hard water contains aqueous calcium hydrogen carbonate, write the complete balanced chemical equation for the reaction of the hard water with aqueous sodium carbonate.
5. Explain the following features of ion-exchange resins in a home water softener.
   (a) Negatively charged sulfonate "sites" on resin molecules attract $Ca^{2+}_{(aq)}$ ions better than $Na^{+}_{(aq)}$ ions. (Consider the ion charges in your answer.)
   (b) Sodium ions from saturated $NaCl_{(aq)}$ replace calcium ions during regeneration. (Consider the concentration and collision theory in your answer.)

Applying Inquiry Skills

6. According to many news media reports, a researcher in the United Kingdom claimed that magnets could be used to soften water flowing through a pipe.
   (a) Write a specific question for a possible investigation and design a controlled experiment to answer the question.
   (b) The claim of water-softening magnets was, in fact, erroneously reported by most news media. The researcher actually claimed that a magnet attached to a water pipe containing flowing water makes it easier to remove the hard-water scale once it has formed. Evaluate your design in (a). What improvements can you now suggest?

Making Connections

7. Summarize the stages of ion-exchange water softening as a circular flow chart. Note the benefits inside the circle. Note the environmental or health concerns of each stage outside the circle.

7.3 Reactions in Solution

At the beginning of this chapter, we mentioned the removal of soap deposits in, for example, a bathtub or shower enclosure. The deposits (precipitates) form when soluble soap chemicals react with hard-water ions. The resulting calcium stearate, being insoluble, precipitates out on the walls of the shower.

$$\text{sodium stearate}_{(aq)} + \text{calcium hydrogen carbonate}_{(aq)} \rightarrow \text{calcium stearate}_{(s)} + \text{sodium hydrogen carbonate}_{(aq)}$$

To remove these deposits, we could initiate a chemical reaction between the calcium stearate and the cleaning agent to produce soluble products that would rinse away.

$$\text{calcium stearate}_{(s)} + \text{cleaning agent}_{(aq)} \rightarrow \text{products}_{(aq)}$$

Your solubility table is useful in predicting solubilities, but there are many more low- and high-solubility ionic compounds than are presented in the table. From a scientific perspective, choosing the right chemical for the cleaning agent is a matter of choosing the right cation and anion. In this section, we look at chemical reactions that will either produce precipitates in the laboratory or remove precipitates at home.

According to the collision theory, all chemical reactions must involve collisions between atoms, ions, or molecules. If there are more collisions, there will be more reactions. Imagine a reaction between a solid and a gas. Collisions can occur only on the relatively small surface area of the solid. If the same quantity of solid is broken up, it will have far more surface area, so more collisions will occur and the substances will react more quickly. Chopping firewood into kindling to make it easy to light is a simple example of this. Dissolving a solid in a solution is the ultimate breaking-up of that solid, since the substance is reduced to the smallest possible particles: separate atoms, ions, or molecules. This creates the largest possible surface area and, therefore, the greatest number of collisions. This situation is only possible in gas- or liquid-phase solutions, but is very common.

Investigation 7.3.1

Precipitation Reactions in Solution

INQUIRY SKILLS

- ○ Questioning
- ○ Hypothesizing
- ● Predicting
- ○ Planning
- ● Conducting
- ● Recording
- ● Analyzing
- ● Evaluating
- ● Communicating

The purpose of this investigation is to test the solubility generalizations contained in a solubility table. To do this, you will use the solubility table (**Table 5**, page 324, or Appendix C) to predict possible precipitates.

Make a **Prediction**, carry out the **Procedure**, then complete the **Analysis** and **Evaluation** sections of the investigation report.

Question

Which aqueous solutions of cobalt(II) chloride, silver nitrate, and lead(II) nitrate react with each of sodium hydroxide, potassium iodide, and sodium carbonate?

Prediction

(a) Use the solubility table (**Table 5**) to predict the precipitates formed from each of the nine combinations of solutions.

The silver compound solution is an irritant, will stain skin, and is mildly toxic by ingestion.

The lead(II) compounds are toxic.

Wear eye protection, gloves, and a laboratory apron.

Reaction mixtures with either silver or lead should not be put down the drain. Dispose of these products carefully in the containers labelled "silver recycle" or "heavy metal waste."

Experimental Design

A few drops of each pair of solutions are combined and evidence of a precipitate is noted. The volumes, concentrations, and temperatures of the solutions are all controlled.

Materials

lab apron
eye protection
rubber gloves
spot or well plate
2 medicine droppers
dropper bottles containing 0.10 mol/L aqueous solutions of
- cobalt(II) chloride
- silver nitrate
- lead(II) nitrate
- sodium hydroxide
- potassium iodide
- sodium carbonate

wash bottle of distilled water
waste containers for waste silver, lead, and cobalt compounds

Procedure

1. Place the spot or well plate on a sheet of paper and label, on the paper, the first three columns with the chemical formulas for sodium hydroxide, potassium iodide, and sodium carbonate.
2. Label the first three rows with the chemical formulas for cobalt(II) chloride, silver nitrate, and lead(II) nitrate.
3. At the intersection of each row and column, combine two drops of the solutions listed. Record specific observations for each combination of solutions.
4. Deposit all waste solutions in the containers provided by your teacher labelled "silver recycle" and "heavy metal waste."

Analysis

(b) Answer the Question by identifying which combinations did and did not produce precipitates, according to your evidence.

Evaluation

(c) Evaluate the evidence gathered by considering the Experimental Design, Materials, and Procedure. Did you gather the kind of evidence necessary to test your Prediction? If not, what evidence would be necessary?

(d) If you are confident in your evidence, compare it with your Prediction. Does your evidence support or cast doubt on the generalizations that form the basis of the solubility table?

Practice

Understanding Concepts

1. Classify each of the following substances as having high or low solubility:

(a) silver sulfide
(b) magnesium nitrate
(c) zinc carbonate

2. Which combinations of the reactants listed below will produce a precipitate? For each precipitate predicted, write the chemical formula.
(a) aqueous strontium nitrate and aqueous potassium sulfate
(b) aqueous sodium acetate and aqueous strontium chloride
(c) aqueous copper(II) bromide and aqueous sodium sulfite

Net Ionic Equations

Two students, performing Investigation 7.3.1, were curious about the yellow precipitate that formed when they mixed lead(II) nitrate with potassium iodide. They looked in the solubility table, and predicted that the precipitate is lead(II) iodide. They set out to design an experiment to provide evidence in support of their prediction. They decided to mix aqueous lead(II) nitrate with aqueous sodium iodide, and aqueous lead(II) acetate with aqueous magnesium iodide to see if they got the same yellow precipitate as they observed for lead(II) nitrate with potassium iodide.

The first student mixed aqueous solutions of lead(II) nitrate and sodium iodide and observed the instant formation of a bright yellow precipitate (**Figure 1**). The student's lab partner recorded the same observation after mixing aqueous solutions of lead(II) acetate and magnesium iodide. The balanced chemical equations for these reactions show some similarities and some differences.

Figure 1
A solution of sodium iodide reacts immediately, when mixed with a solution of lead(II) nitrate, to form a bright yellow precipitate.

$$(1)\ Pb(NO_3)_{2(aq)} + 2\ KI_{(aq)} \rightarrow PbI_{2(s)} + 2\ KNO_{3(aq)}$$

$$(2)\ Pb(NO_3)_{2(aq)} + 2\ NaI_{(aq)} \rightarrow PbI_{2(s)} + 2\ NaNO_{3(aq)}$$

$$(3)\ Pb(C_2H_3O_2)_{2(aq)} + MgI_{2(aq)} \rightarrow PbI_{2(s)} + Mg(C_2H_3O_2)_{2(aq)}$$

The evidence gathered from these reactions is identical: yellow precipitates. This supports the students' prediction that the yellow precipitate is lead(II) iodide. The results of this test increase our confidence in both the double displacement reaction generalization and the solubility generalizations.

Let's now see if we can develop a theoretical description for reactions 1, 2, and 3 above.

Using the Arrhenius theory of dissociation, we can describe these reactions in more detail. We believe that each of the high-solubility ionic reactants dissociates in aqueous solution to form separate cations and anions. For example, reaction (1) could be written as follows:

$$Pb^{2+}_{(aq)} + 2\ NO^{-}_{3(aq)} + 2\ K^{+}_{(aq)} + 2\ I^{-}_{(aq)} \rightarrow PbI_{2(s)} + 2\ K^{+}_{(aq)} + 2\ NO^{-}_{3(aq)}$$

Notice that the precipitate, $PbI_{2(s)}$, is not dissociated because lead(II) iodide has a low solubility. If an ionic compound does not dissolve, it cannot dissociate into individual ions. The chemical equation above is called a **total ionic equation**. It shows what entities are present in the reaction. In this total ionic equation, the nitrate and potassium ions present on the reactant side also appear unchanged among the products.

total ionic equation: a chemical equation that shows all high-solubility ionic compounds in their dissociated form

Chemical reaction equations normally do not show chemical substances that do not change in any way. We can extend this practice to entities in solution—in this case, the ions that do not change. Any ion (or other entity, such as molecules or atoms) present in a reaction system that does not change during the course of

spectator: an entity such as an ion, molecule, or ionic solid that does not change or take part in a chemical reaction

net ionic equation: a way of representing a reaction by writing only those ions or neutral substances specifically involved in an overall chemical reaction

the chemical reaction is called a **spectator.** A spectator ion is like a spectator at a hockey game—present for the action but not taking part in the game. By ignoring these ions and rewriting this reaction equation showing only the entities that change, you write a **net ionic equation** for the reaction. When modifying a total ionic equation to give a net ionic equation, be careful to cancel only identical amounts of identical entities appearing on both reactant and product sides.

$$(1)\ Pb^{2+}_{(aq)} + 2\ \cancel{NO^{-}_{3(aq)}} + 2\ \cancel{K^{+}_{(aq)}} + 2\ I^{-}_{(aq)} \rightarrow PbI_{2(s)} + 2\ \cancel{K^{+}_{(aq)}} + 2\ \cancel{NO^{-}_{3(aq)}}$$

can be shortened to

$$Pb^{2+}_{(aq)} + 2\ I^{-}_{(aq)} \rightarrow PbI_{2(s)}\ \text{(net ionic equation)}$$

Similarly, for reaction (2),

$$(2)\ Pb^{2+}_{(aq)} + 2\ \cancel{NO^{-}_{3(aq)}} + 2\ \cancel{Na^{+}_{(aq)}} + 2\ I^{-}_{(aq)} \rightarrow PbI_{2(s)} + 2\ \cancel{Na^{+}_{(aq)}} + 2\ \cancel{NO^{-}_{3(aq)}}$$

becomes

$$Pb^{2+}_{(aq)} + 2\ I^{-}_{(aq)} \rightarrow PbI_{2(s)}\ \text{(net ionic equation)}$$

If we apply the same procedure to reaction (3),

$$(3)\ Pb^{2+}_{(aq)} + 2\ \cancel{C_2H_3O^{-}_{2(aq)}} + \cancel{Mg^{2+}_{(aq)}} + 2\ I^{-}_{(aq)} \rightarrow PbI_{2(s)} + \cancel{Mg^{2+}_{(aq)}} + 2\ \cancel{C_2H_3O^{-}_{2(aq)}}$$

is written

$$Pb^{2+}_{(aq)} + 2\ I^{-}_{(aq)} \rightarrow PbI_{2(s)}\ \text{(net ionic equation)}$$

Notice that the same observation made from three apparently different chemical reactions gives the same net ionic equation. Although solutions of different substances were combined in these two cases, the net ionic equation makes it clear that the same chemical reaction occurs. Apparently, bright yellow lead(II) iodide will precipitate when we mix any two solutions containing both lead(II) ions and iodide ions.

SUMMARY Writing Net Ionic Equations

1. Write the balanced chemical equation with full chemical formulas for all reactants and products.
2. Using solubility information, such as the table in Appendix C, rewrite the formulas for all high-solubility ionic compounds as dissociated ions, to show the total ionic equation.
3. Cancel identical amounts of identical entities appearing on both reactant and product sides.
4. Write the net ionic equation, reducing coefficients if necessary.

Sample Problem 1

Write the net ionic equation for the reaction of aqueous barium chloride and aqueous sodium sulfate.

Solution

$$BaCl_{2(aq)} + Na_2SO_{4(aq)} \rightarrow BaSO_{4(s)} + 2\ NaCl_{(aq)}$$
$$Ba^{2+}_{(aq)} + 2\ \cancel{Cl^{-}_{(aq)}} + 2\ \cancel{Na^{+}_{(aq)}} + SO_{4(aq)}^{2-} \rightarrow BaSO_{4(s)} + 2\ \cancel{Na^{+}_{(aq)}} + 2\ \cancel{Cl^{-}_{(aq)}}$$
$$Ba^{2+}_{(aq)} + SO_{4(aq)}^{2-} \rightarrow BaSO_{4(s)} \text{ (net ionic equation)}$$

Net ionic equations are useful for more than double displacement reactions that produce precipitates. We can also use them for communicating other reactions. The following single displacement reaction is a good example.

Sample Problem 2

Write the net ionic equation for the reaction of zinc metal and aqueous copper(II) sulfate, and then write a statement to communicate the meaning of the net ionic equation.

Solution

$$Zn_{(s)} + CuSO_{4(aq)} \rightarrow Cu_{(s)} + ZnSO_{4(aq)}$$
$$Zn_{(s)} + Cu^{2+}_{(aq)} + \cancel{SO_{4(aq)}^{2-}} \rightarrow Cu_{(s)} + Zn^{2+}_{(aq)} + \cancel{SO_{4(aq)}^{2-}}$$
$$Zn_{(s)} + Cu^{2+}_{(aq)} \rightarrow Cu_{(s)} + Zn^{2+}_{(aq)} \text{ (net ionic equation)}$$

Placing solid zinc in any aqueous solution containing copper(II) ions will produce solid copper and aqueous zinc ions.

Practice

Understanding Concepts

3. Strontium compounds are often used in flares because their flame colour is bright red. One industrial process to produce low-solubility strontium compounds (that are less affected by getting wet) involves the reaction of aqueous solutions of strontium nitrate and sodium carbonate. Write the balanced chemical equation, total ionic equation, and net ionic equation for this reaction.
4. Placing aluminum foil in any solution containing aqueous copper(II) ions will result in a reaction. The reaction is slow to begin with, but then proceeds rapidly.
   (a) Referring to the solubility table (Appendix C), name at least four ionic compounds that could be dissolved in water to make a solution containing aqueous copper(II) ions.
   (b) Write a balanced chemical equation for the reaction of aluminum with one of the compounds you suggested in (a).
   (c) Write the total ionic equation for the reaction.
   (d) Write the net ionic equation for the reaction.
5. One industrial method of producing bromine is to react seawater, containing a low concentration of sodium bromide, with chlorine gas. The chlorine gas is bubbled through the seawater in a specially designed vessel. Write the net ionic equation for this reaction.

6. In a hard-water analysis, sodium oxalate solution reacts with calcium hydrogen carbonate (in the hard water) to precipitate a calcium compound. Write the net ionic equation for this reaction.
7. In a laboratory test of the metal activity series, a student places a strip of lead metal into aqueous silver nitrate. Write the net ionic equation for this reaction.

Making Connections

8. Some natural waters contain iron ions that affect the taste of the water and cause rust stains. Aeration converts any iron(II) ions into iron(III) ions. A basic solution (containing hydroxide ions) is added to produce a precipitate.
   (a) Write the net ionic equation for the reaction of aqueous iron(III) ions and aqueous hydroxide ions.
   (b) What separation method is most likely to be used during this water treatment process?

Explore an Issue

Debate: Producing Photographs

DECISION-MAKING SKILLS

- Define the Issue
- Identify Alternatives
- Research
- Analyze the Issue
- Defend a Decision
- Evaluate

Photographers and the photographic developing industry use a great many resources. You are probably aware of some of the drawbacks of using film. There are disposable cameras. Film is bought in plastic canisters. Every day developers and printers use huge quantities of chemicals in solution, some of which are toxic, and others of which are costly to produce. All of these aspects of film photography have an environmental impact. But what about digital photography? Think about the resources required to create and distribute digital photographs. Is this technology as environmentally clean as it first appears?

(a) In small groups, prepare for a debate on the proposition, "Digital cameras are more environmentally friendly than film cameras."

(b) Your group will be either defending or opposing the proposition. Brainstorm and research arguments in support of your position. Collect scientific evidence wherever possible. Try to go to primary sources as much as possible for your information.

(c) Assemble your evidence into separate, logical subtopics, and prepare to debate. (In the debate, each subtopic could perhaps be presented by a different member of the group.)

(d) After the debate, discuss in your group how you could have improved on your group's performance. Which (if any) aspects of your preparation could have been done better?

Follow the links for Nelson Chemistry 11, 7.3.

GO TO www.science.nelson.com

Section 7.3 Questions

Understanding Concepts

1. A common method for the disposal of soluble lead waste is to precipitate the lead as the low-solubility lead(II) silicate. Write the net ionic equation for the reaction of aqueous lead(II) nitrate and aqueous sodium silicate.

2. In a water treatment plant, sodium phosphate is added to remove calcium ions from the water. Write the net ionic equation for the reaction of aqueous calcium chloride and aqueous sodium phosphate.
3. As part of a recycling process, silver metal is recovered from a silver nitrate solution by reacting it with copper metal. Write the net ionic equation for this reaction.

Making Connections

4. To reduce the amount of poisonous substances dumped in our landfill sites, many municipalities have hazardous waste disposal depots. Find out whether there is one in your area, and research the skills and qualifications of the people who work there, and the types of substances they handle.

   Follow the links for Nelson Chemistry 11, 7.3.

GO TO www.science.nelson.com

7.4 Waste Water Treatment

In Chapter 6 we looked at the treatment of drinking water, and the ways in which we make sure that it is safe. Then, earlier in this chapter we explored the technologies used to reduce the effects of hard water. In this section we will look at another aspect of water treatment: the treatment of waste water (**Figure 1**).

Ideally, all used water returned to the environment would be clean enough to be drinkable, whatever kind of treatment was necessary. In reality, though, some municipalities still permit raw sewage to be discharged into our rivers, lakes, and coastal waters.

Sewage is anything that gets flushed out of our homes through the sewer pipes, and includes a considerable amount of organic waste. There are three major problems with organic waste.

First is the issue of disease transmission. Untreated sewage may contain bodily wastes from people with a wide range of infectious diseases, many of which can be transmitted by contact with feces. If the disease-causing organisms are not killed, they could easily cause infection in anyone who drinks, swims in, or washes with the contaminated water.

The second problem is that the decomposition (rotting) of organic wastes uses up a lot of oxygen. Biologists measure the use of dissolved oxygen by bacteria, and call it the BOD—the biological oxygen demand. This is defined as the quantity of oxygen used up by bacteria over five days at 20°C to decompose any organic matter that is present in (or added to) water. A high BOD reading indicates that there is a lot of organic matter in the water, and that a considerable quantity of oxygen is being removed from the water in order to decompose those organics. Oxygen is really only very slightly soluble in water. Under natural conditions, oxygen is present in water to about a 0.010 g/L (10 ppm) level at a temperature of 10°C, and about 8 ppm at 25°C. Many fish and other aquatic organisms require a minimum of about 4–5 ppm of oxygen to survive. Bacteria, growing as a result of raw sewage discharge into a river, can therefore cause a critical lowering in oxygen levels for a long way downstream. This effect is even more serious if the river flows slowly and so the water is not naturally aerated. The low oxygen level can be a serious problem for anything living in the water.

Figure 1
Modern sewage treatment technology allows waste treatment plants to be less offensive to neighbours. The resulting sludge has a wide range of uses, and is even useful in some industrial processes.

DID YOU KNOW ?

Natural Oxygen Levels

The dissolved oxygen content in rivers and streams is considerably less than the maximum solubility of oxygen in water at the same temperature. There are many reasons for this, including the small surface area in contact with the air (in deep, smoothly flowing water) and the use by organisms in the water of the dissolved oxygen during respiration.

The third problem is that the decomposition of organic matter releases chemicals, including nitrates and phosphates, that stimulate the rapid growth of aquatic plants and algae. As the surface plants grow in much greater abundance than is usual, they block out light to deeper water, killing the bottom-dwelling plants. Later, as the artificially fertilized "bloom" dies away, it becomes another load of organic material to be decomposed—and to use up dissolved oxygen.

Along with organic waste, many other materials are delivered into the sewers: cleaning solutions (e.g., phosphates), paint (e.g., oil-based paints), garden chemicals (e.g., fertilizers and pesticides), insoluble materials (e.g., calcium and magnesium carbonates), minerals from water softeners (e.g., sodium and chloride ions), and so on. All of these would be harmful if discharged

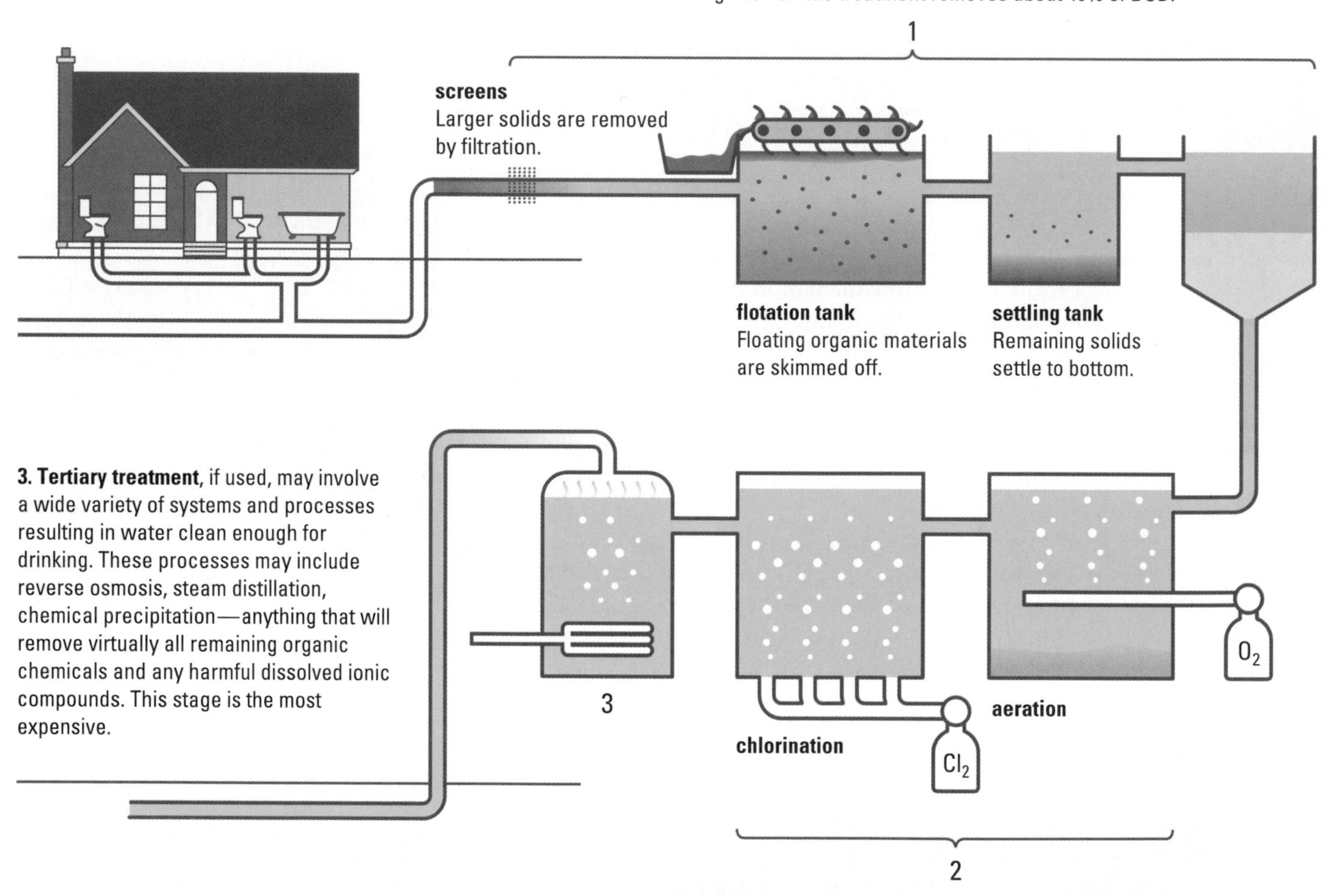

Figure 2
Three stages of a waste water treatment system

directly into the natural environment. Fortunately, provided that they are only present in small quantities, the waste water treatment system can deal with them.

Treatment of waste water is becoming more and more important as urban populations grow, and high-quality drinking water becomes more scarce. The general sequence for treatment of municipal sewage is shown in **Figure 2**.

In Canada there is public pressure to place the responsibility for waste water treatment on the user. The idea is that waste water should be cleaned and treated to a point that is not polluting. Not only is the treatment beneficial to life forms in the river, but also to anyone downstream who wants to use that river's water as a municipal supply of drinking water.

There are many parts of Ontario where homes and businesses are not connected to municipal sewage treatment facilities. In most cases, this is not a problem. The sewage is either stored in a holding tank until it is periodically pumped out and trucked away, or it is piped directly into a septic system (**Figure 3**). Here the solid matter is decomposed by bacteria, and the liquid is piped off to a leaching bed, where it gradually trickles through layers of sand and gravel and eventually runs down into the water table. Although septic systems generally remove most of the organic material and bacteria from the sewage, they cannot remove the other substances, such as phosphates (from cleaning products) and nonaqueous liquids.

Figure 3
Many rural homes and cottages have septic systems to purify their waste water.

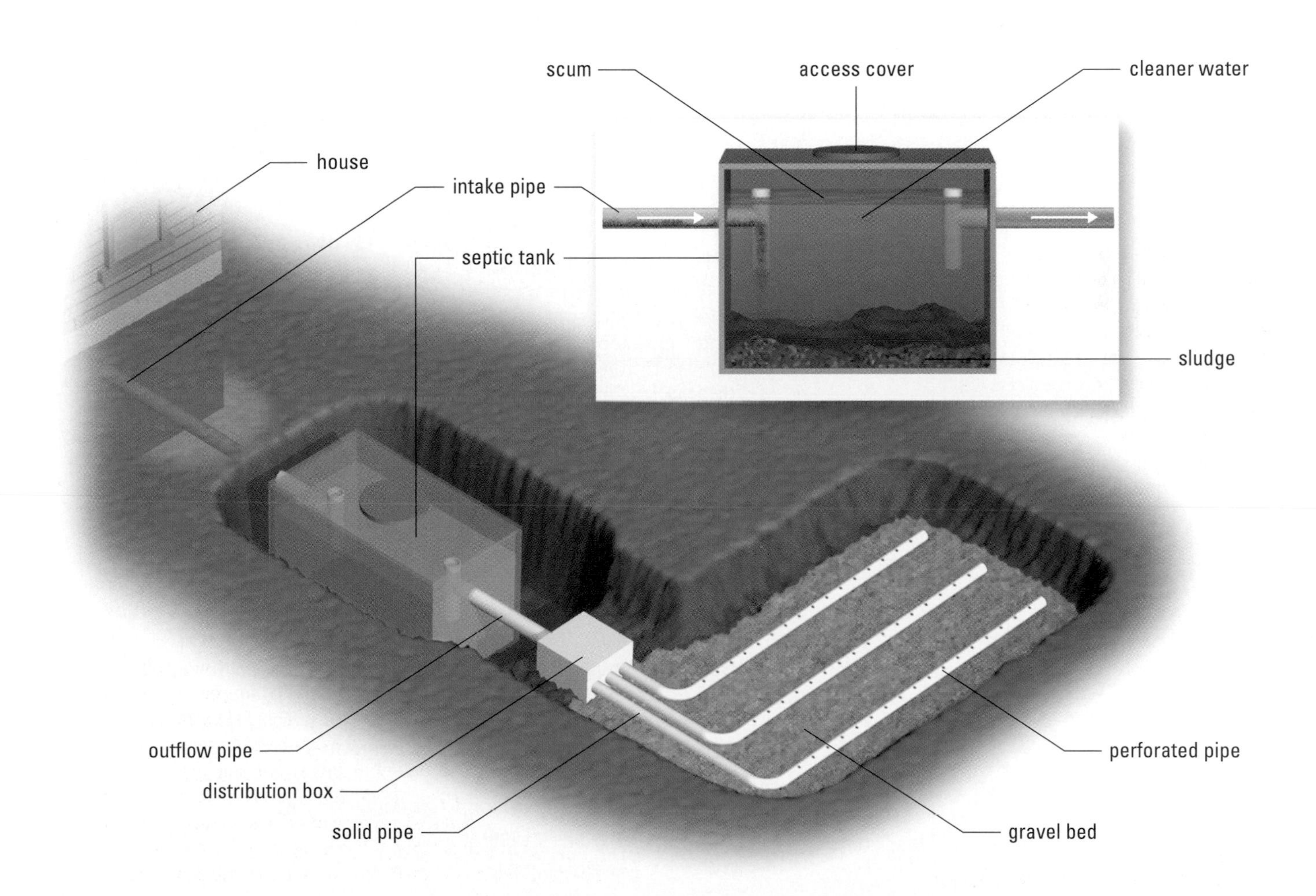

The natural environment can tolerate a certain amount of pollution. For example, the combination of sunlight with a healthy community of organisms living in large bodies of fresh water is able to break down some of the pollutants from domestic sewage, agricultural fertilizers, and industrial waste. Small quantities of these chemicals—nitrates, phosphates, and organic compounds—can be broken down by chemical reactions or by natural bacteria into simple, non-polluting substances such as carbon dioxide and nitrogen.

Practice

Understanding Concepts

1. Outline the problems that are likely to result from releasing untreated sewage into the environment.
2. Briefly outline the causes and effects of a high BOD.
3. Create a diagram indicating the stages of waste water treatment.
4. Explain why rural homeowners should be especially careful about what they flush down their drains.

Reflecting

5. Will the information in this section affect your attitude toward the materials that you pour down the sink at home? at school?

Section 7.4 Questions

Understanding Concepts

1. Create a flow chart diagram indicating the ways in which water is treated during its cycle from raw, untreated water to drinkable water in our homes, and then returned to the environment as treated waste water.

Applying Inquiry Skills

2. A team of environmental scientists discovers many dead fish in a river downstream from an industrial town. The scientists need to find the reason for this observation.
   (a) Write two possible hypotheses that they might test.
   (b) Write a prediction and experimental design for each of the hypotheses.

Making Connections

3. To reduce the quantity of garbage going to landfill sites, some people have suggested that food waste be disposed of in a garburator. These household devices grind up the waste and flush it, with lots of water, into the sewer. What are some advantages and disadvantages of this suggestion? What would be the effect of this material on the sewage treatment system?
4. How is sewage treated after it leaves your home? Research to find out, and draw your own conclusions (with further research if necessary) about whether this treatment is adequate. Outline your position (supporting it with evidence) in a communication to the person or organization responsible for your water treatment.

   Follow the links for Nelson Chemistry 11, 7.4.

   GO TO www.science.nelson.com

7.5 Qualitative Chemical Analysis

Chemical analysis of an unknown chemical sample can include both

- **qualitative** analysis—the identification of the specific substances present,

and

- **quantitative** analysis—the measurement of the quantity of a substance present.

qualitative: describes a quality or change in matter that has no numerical value expressed
quantitative: describes a quantity of matter or degree of change of matter

You have probably heard about at least one widely used chemical analysis: the blood test for alcohol. Technicians test a blood sample to determine whether ethanol (from alcoholic beverages) is present, and if so, how much is present. As another example, water quality analysts are continually checking our drinking water for a wide variety of dissolved substances, some potentially harmful, some beneficial.

You have already seen many examples of qualitative analysis, such as the colour reaction of litmus paper to identify the presence of an acid or a base. The conductivity test for electrolytes, the limewater test for carbon dioxide, the explosion ("pop") test for hydrogen gas, and the glowing splint test for oxygen are all qualitative analyses. As you can see, qualitative analysis is often used for diagnostic testing. We can plan diagnostic tests using the format, "If [procedure], and [evidence], then [analysis]." For example, if cobalt(II) chloride paper is exposed to a liquid or vapour [procedure], and the paper turns from blue to pink [evidence], then water is likely present [analysis].

Flame tests, such as those that you may have conducted in Investigation 1.4.2, are also qualitative analyses.

Table 1: Colours of Solutions

Ion	Solution colour
Groups 1, 2, 17	colourless
$Cr^{2+}_{(aq)}$	blue
$Cr^{3+}_{(aq)}$	green
$Co^{2+}_{(aq)}$	pink
$Cu^{+}_{(aq)}$	green
$Cu^{2+}_{(aq)}$	blue
$Fe^{2+}_{(aq)}$	pale green
$Fe^{3+}_{(aq)}$	yellow-brown
$Mn^{2+}_{(aq)}$	pale pink
$Ni^{2+}_{(aq)}$	green
$CrO^{2-}_{4(aq)}$	yellow
$Cr_2O^{2-}_{7(aq)}$	orange
$MnO^{-}_{4(aq)}$	purple

Qualitative Analysis by Colour

You have already learned in Unit 1 that some ions have a characteristic colour in a solution, a flame, or a gas discharge tube.

First, let's look at the colours of aqueous solutions. Most aqueous solutions are colourless, as **Table 1** shows: ions of elements in Group 1, 2, and 17 have no colour at all. Other ions, not listed in the table, are also colourless. However, many solutions containing monatomic and polyatomic ions of the transition elements do have a colour in solution.

Sample Problem 1

According to the evidence in **Table 2** and **Figure 1**, which of the numbered solutions is potassium dichromate, sodium chloride, sodium chromate, potassium permanganate, nickel(II) nitrate, and copper(II) sulfate? (Refer to **Table 1**.)

Table 2: Colours of the Unknown Solutions

Solution	1	2	3	4	5	6
Colour	purple	colourless	green	yellow	blue	orange

Solution

According to **Table 2** and **Table 1**, the solutions are (1) potassium permanganate, (2) sodium chloride, (3) nickel(II) nitrate, (4) sodium chromate, (5) copper(II) sulfate, and (6) potassium dichromate.

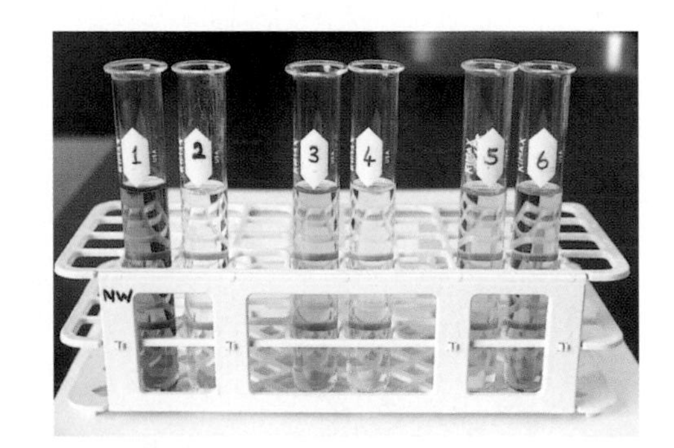

Figure 1
Which solution is which?

Table 3: Colours of Flames

Ion	Flame colour
H^+	colourless
Li^+	bright red
Na^+	yellow
K^+	violet
Ca^{2+}	yellow-red
Sr^{2+}	bright red
Ba^{2+}	yellow-green
Cu^{2+}	blue (halides) green (others)
Pb^{2+}	light blue-grey
Zn^{2+}	whitish green

You may be familiar with the blue colour of solutions containing copper(II) ions, such as copper(II) sulfate, but did you know that copper(II) ions produce a green flame (**Figure 2**)?

We can use flame tests to indicate the presence of several metal ions, such as copper(II), calcium, and sodium (**Table 3**). In a flame test, a clean platinum or nichrome wire is dipped into a test solution and then held in a nearly colourless flame (**Figure 2**). The wire is cleaned by dipping it alternately into hydrochloric acid and then into the flame, until very little flame colour is produced. There are other ways to conduct flame tests: you could dip a wood splint in the aqueous solution and then hold it close to a flame; you could hold a tiny solid sample of a substance in the flame; or you could spray the aqueous solution into the flame.

Robert Wilhelm Bunsen and Gustav Robert Kirchhoff took the idea of the flame test and developed it into a technique called spectroscopy (**Figure 3**). Bunsen had previously invented an efficient gas laboratory burner that produced an easily adjustible, hot, nearly colourless flame. Bunsen's burner made better research possible—a classic example of technology leading science—and made his name famous. Bunsen and Kirchhoff soon discovered two new elements, cesium and rubidium, by examining the spectra produced by passing the light from flame tests through a prism.

Flame tests are still used for identification today. Of course, the technology has become much more sophisticated. The equipment used is called a spectrophotometer (**Figure 4**). It analyzes the light produced by samples vaporized in a flame. It can even detect light not visible to humans. The spectrophotometer can detect minute quantities of substances, in concentrations as tiny as parts per billion. By measuring the quantity of light emitted, this device can also do quantitative analysis—measuring the concentrations of various elements precisely and accurately. Forensic scientists may use this technique of high-tech qualitative analysis when gathering evidence for criminal investigations. Similar technology is used in a completely different branch of science: astronomy. Astronomers study the light spectra from distant stars to find out what elements the stars are composed of.

Figure 2
Copper(II) ions usually impart a green colour to a flame. This green flame, and the characteristic blue colour in aqueous solution, can be used as diagnostic tests for copper(II) ions.

Practice

Understanding Concepts

1. What is the expected colour of solutions that contain the following? (Refer to **Table 1**.)
   (a) $Na^+_{(aq)}$
   (b) $Cu^{2+}_{(aq)}$
   (c) $Fe^{3+}_{(aq)}$
   (d) $Cr_2O_7{}^{2-}_{(aq)}$
   (e) $Cl^-_{(aq)}$
   (f) $Ni^{2+}_{(aq)}$
2. What colour is imparted to a flame by the following ions in a flame test?
   (a) calcium
   (b) copper(II) (halide)
   (c) Na^+
   (d) K^+
   (e) H^+
3. Flame tests on solids produce the same results as flame tests on solutions. These tests may be used as additional evidence to support

the identification of precipitates. What colour would the following precipitates give to a flame?
(a) $CaCO_{3(s)}$
(b) $PbCl_{2(s)}$
(c) $SrSO_{4(s)}$
(d) $Cu(OH)_{2(s)}$

Applying Inquiry Skills

4. Complete the **Analysis** and **Evaluation** sections of the following report.

Question

What ions are present in the solutions provided?

Experimental Design

The solution colour is noted and a flame test is conducted on each solution.

Evidence

Table 4: Solution and Flame Colours

Solution	Solution colour	Flame colour
A	colourless	violet
B	blue	green
C	colourless	yellow
D	colourless	yellow-red
E	colourless	bright red

Analysis

(a) Which ions are present in solutions A–E?

Evaluation

(b) Critique the Experimental Design.

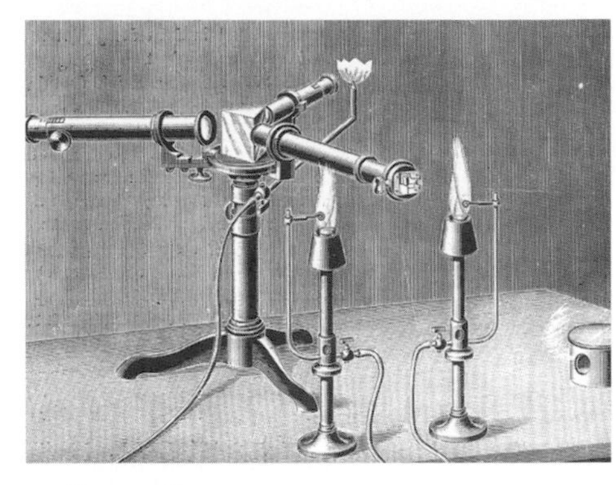

Figure 3
In 1860 Bunsen and Kirchhoff developed techniques of spectroscopy for analysis.

Sequential Qualitative Chemical Analysis

In addition to colours of solutions and flames, analytical chemists have created very specific qualitative tests for ions in aqueous solution. These specific tests use the precipitation of low-solubility products. The chemist plans a double displacement reaction involving one unknown solution and one known solution, and predicts that, if a precipitate forms, then a certain ion must have been present in the unknown solution. Solubility tables help the chemist (and us) to choose reactants that will produce precipitates as evidence of the presence of specific ions. This type of test is called a **qualitative chemical analysis**.

Suppose you were given a solution that might contain either lead(II) ions or strontium ions, or possibly both or neither. How could you determine which ions, if any, were present? You could perform a qualitative analysis. For this, you must design an experiment that involves two diagnostic tests, one for each ion—that is, tests that definitely identify a particular substance. Of course, it is hard to identify an ion in a solution unless it forms a compound with low solubility. In other words, unless it forms a precipitate. So you choose reactants that, if lead(II) ions or strontium ions were present, would form a precipitate.

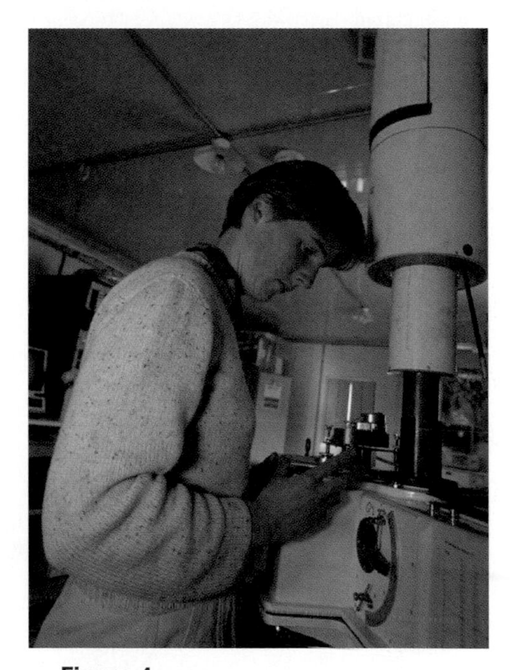

Figure 4
The spectrophotometer is a valuable tool, essential for precise qualitative and quantitative analyses in many areas of science.

qualitative chemical analysis: the identification of substances present in a sample; may involve several diagnostic tests

Test for the Presence of Lead(II) Ions

Let's test for the possible presence of lead(II) ions in a solution, before going on to test for the presence of strontium ions (**Figure 5**, page 344). Chloride ions form a low-solubility compound with lead(II) ions, so if lead(II) ions are present

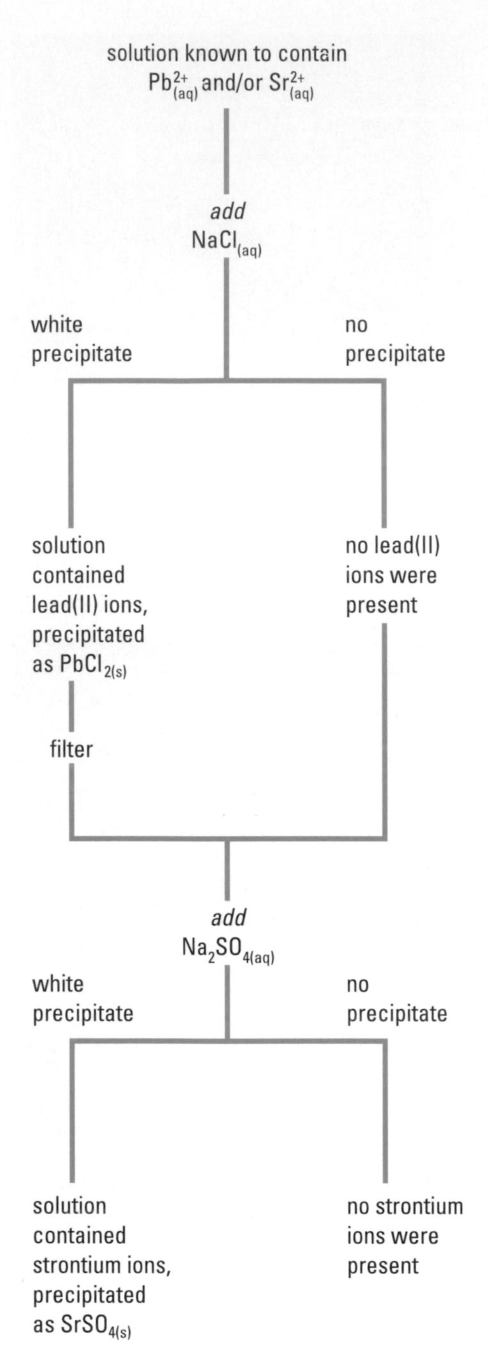

Figure 5
Reading down, we see that this is one experimental design for analyzing a solution for lead(II) and/or strontium ions. In this example, the two tests could not be done in reverse order, because both lead(II) and strontium ions precipitate with sulfate ions.

limiting reagent: a reactant that is completely consumed in a chemical reaction

excess reagent: the reactant that is present in more than the required amount for complete reaction

in a solution, a precipitate forms when chloride ions are added. An appropriate source of chloride ions would be a solution of sodium chloride—a high-solubility chloride compound. Refer to the solubility table (Appendix C) to confirm that strontium ions do not precipitate with chloride ions, but that lead(II) ions do.

If sodium chloride solution is added to a sample of the test solution and a precipitate forms, then lead(II) ions are likely present and the following reaction has taken place.

$$Pb^{2+}_{(aq)} + 2\ Cl^{-}_{(aq)} \rightarrow PbCl_{2(s)}$$

We add an excess of sodium chloride so that there are sufficient chloride ions to precipitate all of the lead(II) ions, leaving none in solution. This makes certain that there are no leftover lead(II) ions to interfere with a subsequent test for strontium ions.

When a sample under investigation (e.g., our unknown solution) is combined with an excess quantity of another reactant (e.g., sodium chloride), all of the sample reacts. The reactant that is completely used up (in this case, lead(II) ions) is called the **limiting reagent**. The reactant that is present in more than the required quantity (the chloride ions) is called the **excess reagent**.

Our precipitate indicates that lead(II) ions were present in the solution. Now we need to test our solution for strontium ions. First, however, we should remove the lead(II) chloride precipitate by filtering the solution. The remaining filtrate can then be tested for strontium ions.

Test for the Presence of Strontium Ions

What strontium-containing compound would form a precipitate? A look at the solubility table shows us that a compound containing sulfate ions and strontium ions is relatively insoluble. If a highly soluble sulfate solution (e.g., sodium sulfate) is added to the filtrate, and a precipitate forms, then strontium ions are likely to be present. (In the unlikely event that there are still some lead(II) ions in solution, they will react with the sulfate ions and would precipitate also.) If $Sr^{2+}_{(aq)}$ is present, then

$$Sr^{2+}_{(aq)} + SO^{2-}_{4(aq)} \rightarrow SrSO_{4(s)}$$

If we see a precipitate form, it is likely to be strontium sulfate. This indicates the presence of strontium ions in the original solution. We could also confirm the presence of strontium ions with a flame test.

In our example, both the unknown solution and the diagnostic-test solutions contained dissociated ions. If a precipitate forms, collisions must have occurred between two kinds of ions to form a low-solubility solid. The chloride and sodium ions present in the diagnostic-test solutions are spectator ions.

By carefully planning an experimental design for qualitative chemical analysis, you can do a sequence of diagnostic tests to detect many different ions, beginning with only one sample of the unknown solution.

We can use our experimental design as a guide for planning diagnostic tests for many ions (**Figure 6**, page 346).

SUMMARY

To complete a sequential analysis involving solubility, follow these steps. (These steps are written for cation (metal ion) analysis. For anion analysis, reverse the words cations and anions.)

1. Locate the possible cations on the solubility table.
2. Determine which anions precipitate the possible cations.
3. Plan a sequence of precipitation reactions that uses anions to precipitate a single cation at a time.
4. Use filtration between steps to remove cation precipitates that might interfere with subsequent additions of anions.
5. Draw a flow chart to assist your testing and communication.

Recognize that the absence of a precipitate is an indicator that the ion being tested for is not present. Realize that sometimes an experimental design can be created where parallel tests rather than sequential steps are used. For example, you can test for the presence or absence of calcium and mercury(I) ions by adding sulfate and chloride ions to separate samples of a solution.

Investigation 7.5.1

Sequential Chemical Analysis in Solution

INQUIRY SKILLS

- ○ Questioning
- ○ Hypothesizing
- ○ Predicting
- ● Planning
- ● Conducting
- ● Recording
- ● Analyzing
- ○ Evaluating
- ● Communicating

A security guard noticed a trickle of clear liquid dripping from a drainage channel in the warehouse of a chemical plant. The guard alerted the chemists. To find out which of the containers of chemical solutions was leaking, the chemists needed to find out what ions were in the leaked solution. They decided to test for the most likely ions first.

In this investigation you are one of the chemists identifying the spill. Your purpose is to test a sample of the spilled solution for the presence of silver, barium, and zinc ions. You have to plan and carry out a qualitative chemical analysis using precipitation reactions, and analyze the evidence gathered.

Complete the **Materials**, **Experimental Design**, **Procedure**, and **Analysis** sections of the investigation report.

See Handling Chemicals Safely in Appendix B.

Barium solutions are toxic and should be treated with care. Avoid swallowing and avoid contact with the skin. Wear gloves.

Silver solutions are corrosive and must be kept away from the eyes and skin, and must not be swallowed. Wear eye protection and do not rub your eyes.

Zinc nitrate is harmful if swallowed.

Dispose of all waste substances in a special container labelled "Heavy Metal Waste."

Question

Are there any silver, barium, and/or zinc ions present in the leaked solution?

Experimental Design

(a) Plan a sequential qualitative analysis involving precipitation and filtration at each step. Use sodium chloride, sodium sulfate, and sodium carbonate to test for silver, barium, and zinc ions. Draw a flow chart to communicate the experimental design and procedure.

(b) Suggest flame tests on the precipitates that would increase your confidence in the qualitative analysis.

(c) Create a list of Materials and write out your Procedure. Include any necessary safety precautions and special disposal arrangements.

Procedure

1. With your teacher's approval, carry out your investigation. Record your observations at every step.

Analysis

(d) Analyze your evidence to answer the Question.

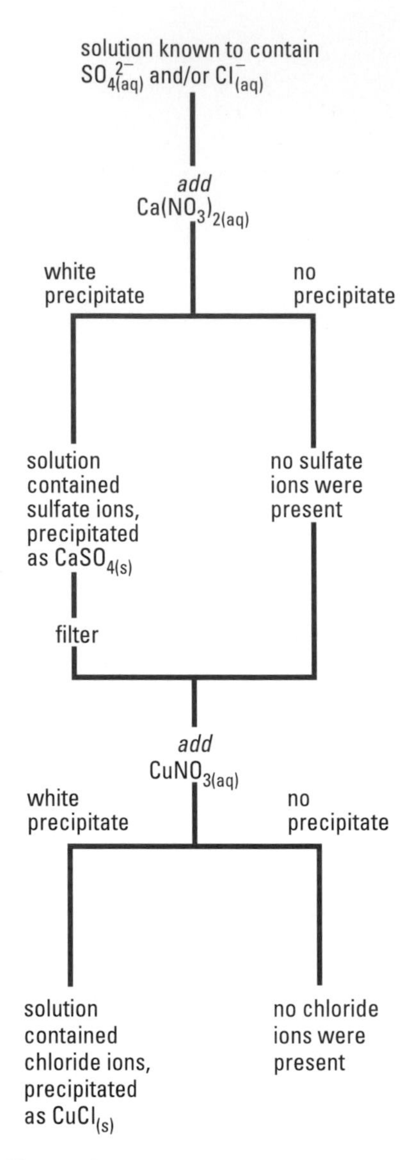

Figure 6
Reading down, we see that this is one experimental design for analyzing a solution for sulfate and/or chloride ions. In this particular example, the calcium nitrate and copper(II) nitrate tests could be done in reverse order, but this is usually not true in sequential analyses.

Practice

Understanding Concepts

5. Distinguish between qualitative and quantitative analysis.
6. What three parts are included in the "If..., and...,then..." statement of a diagnostic test? Provide an example to illustrate this.
7. Use **Table 5:** Solubility of Ionic Compounds at SATP (page 324) to suggest ion(s) that could be used to precipitate the ion listed.
   (a) $Mg^{2+}_{(aq)}$
   (b) $Ba^{2+}_{(aq)}$
   (c) $I^{-}_{(aq)}$
   (d) $SO_{4(aq)}^{2-}$

Applying Inquiry Skills

8. (a) Design a qualitative analyis for carbonate ions, using a reactant that would not precipitate sulfide ions in the sample. Refer to **Table 5:** Solubility of Ionic Compounds at SATP (page 324).
   (b) Write a net ionic equation for the precipitation reaction.
9. You are given a solution that may contain acetate and carbonate ions. Design an experiment to analyze a single sample of this solution to find out which of the ions (if any) it contains.

Making Connections

10. Provide some examples of qualitative chemical analyses that are important in our society. Use the Internet in your research.
    Follow the links for Nelson Chemistry 11, 7.5.

GO TO **www.science.nelson.com**

Section 7.5 Questions

Understanding Concepts

1. Predict which of the following combinations of aqueous chemicals produce a precipitate. Write a net ionic equation (including states of matter) for the formation of any precipitate.
   (a) lead(II) nitrate and calcium chloride
   (b) ammonium sulfide and zinc bromide
   (c) potassium iodide and sodium nitrate
   (d) silver sulfate and ammonium acetate
   (e) barium nitrate and ammonium phosphate
   (f) sodium hydroxide and calcium nitrate
2. Predict the colour of aqueous solutions containing the following ions.
   (a) iron(III)
   (b) sodium
   (c) $Cu^{2+}_{(aq)}$
   (d) $Ni^{2+}_{(aq)}$
   (e) $Cl^{-}_{(aq)}$
3. (a) Which of the ions, lithium, calcium, and strontium, in separate unlabelled solutions, can be distinguished using a simple flame test with a laboratory burner?

(b) Suggest a different method to distinguish experimentally between the remaining ions.

Applying Inquiry Skills

4. (a) Design an experiment to determine whether potassium and/or strontium ions are present in a solution.
   (b) Evaluate your design.
5. Design an experiment to analyze a single sample of a solution for any or all of the $Tl^{+}_{(aq)}$, $Ba^{2+}_{(aq)}$, and $Ca^{2+}_{(aq)}$ ions.
6. An investigation is planned to test for the presence of copper(II) and/or calcium ions in a single sample of a solution.

   Experimental Design

   A sample of the solution is examined for colour, and is tested using a flame test.

   (a) Critique the Experimental Design.
   (b) Write an alternative Experimental Design to test for the presence of the two ions in one solution.
7. Provide an example of a household solution that contains a gas as a solute. Identify the gas and state a diagnostic test for this gas.
8. Provide an example of a household solution that contains a solid as a solute. State diagnostic tests for the presence of the ions of this solid in the solution.
9. To be transported and used in your body, minerals must exist in solution. Oxalic acid in foods such as rhubarb, Swiss chard, spinach, cocoa, and tea may remove the minerals through precipitation. Design an investigation to determine which metal-ion minerals are precipitated by oxalic acid.
10. You have been told, by an unreliable source, that a product marketed as a water-softening agent is sodium carbonate.
    (a) Design an investigation to test this hypothesis.
    (b) With your teacher's approval, conduct your investigation.

Making Connections

11. Using the Internet, research the career of forensic chemist. Find out how a spectrophotometer is used in this line of work, and give examples of the kinds of substances analyzed. Why must the amounts of substances be so precisely known?
    Follow the links for Nelson Chemistry 11, 7.5.

GO TO www.science.nelson.com

DID YOU KNOW ?

Northern Lights

The northern lights (**Figure 7**), a phenomenon that occurs high in the atmosphere, usually near Earth's poles, is an example of a natural event where the colours of light observed depend on the chemicals present. The brightest colours are caused by $N_{2(g)}$ (violet and red), $N^{+}_{2(g)}$ (blue-violet), and free oxygen atoms (green and red).

Figure 7

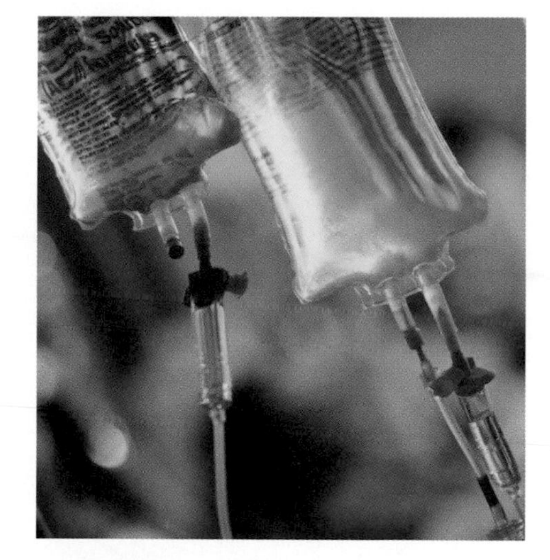

Figure 1
Human blood plasma contains the following ions: $Na^{+}_{(aq)}$, $K^{+}_{(aq)}$, $Ca^{2+}_{(aq)}$, $Mg^{2+}_{(aq)}$, $HCO_3{}^{-}{}_{(aq)}$, $Cl^{-}_{(aq)}$, $HPO_{4(aq)}^{2-}$, and $SO_{4(aq)}^{2-}$, as well as many complex acid and protein molecules. The concentrations of these various substances can indicate, to a trained medical professional, the state of your health.

7.6 Quantitative Analysis

Quantitative analyses are carried out for many reasons. They can help chemical engineers find out whether an industrial chemical process is operating optimally. They can also be used by water quality analysts to establish whether a contaminant is concentrated enough to be toxic. And medical laboratory technicians use quantitative analyses to decide whether the concentration of a substance in human blood is healthy or legal (**Figure 1**).

Quantitative analysis quite often follows qualitative analysis. Many of us can detect the presence of alcohol by its odour. This is a crude (but direct) form of qualitative analysis. If the police suspect someone of drinking and driving, they may decide to use a breathalyzer. Breathalyzers can indirectly determine the level of blood alcohol quantitatively. Exhaled air (which may contain alcohol) is blown through a bright orange solution containing dichromate ions. If there is alcohol in the breath, it reacts with the dichromate ions to produce chromium(III) ions. The new solution is pale green in colour. This change in colour is detected when light of a certain wavelength is shone through the solution. The light that passes through is converted into an electric current by a photocell. The size of the electric current is indicated on a current meter, which is labelled in terms of blood alcohol content.

There is a more direct (and therefore more certain) way of measuring blood alcohol content. This method involves taking a blood sample and reacting it for several hours with the same kind of solution as in the breathalyzer. A laboratory technologist then performs a procedure called titration to determine the concentration of alcohol in the blood. The legal limit for blood alcohol varies from 0 to 0.080 g/100 mL (800 ppm), depending on what kind of driving licence the suspect holds. If the suspect's blood alcohol concentration is greater than the legal limit, the police may decide to press impaired driving charges.

There are various experimental designs used to determine blood alcohol content. Most of the current designs require chemical reactions and stoichiometry for quantitative analysis. In this section you will learn how to complete stoichiometric calculations involving reactions in solution.

Solution Stoichiometry

In Unit 2 you saw the usefulness of gravimetric (mass-to-mass) stoichiometry for both prediction and analysis. However, most reactions in research and industry take place in aqueous solutions. This is because chemicals in solution are easy to manipulate, and reaction rates and extents in solution are relatively easy to control—not to mention the fact that many chemical reactions only occur if the reagents are in solution. We therefore need to learn how to find the concentration of reactants in solutions. Scientists have developed a concept to help them do this, a concept known as **solution stoichiometry**.

solution stoichiometry: a method of calculating the concentration of substances in a chemical reaction by measuring the volumes of solutions that react completely; sometimes called volumetric stoichiometry

When using the concept of solution stoichiometry, we often know some of the information, such as the mass or concentration of one of the products of a reaction. We can then use this information to determine another quantity, such as the molar concentration of a reactant.

SUMMARY Stoichiometry Calculations

When performing any kind of stoichiometry calculation by any method, follow these general steps.

1. Write a balanced equation for the reaction, to obtain the mole ratios.
2. Convert the given value to an amount in moles using the appropriate conversion factor.
3. Convert the given amount in moles to the required amount in moles, using the mole ratio from the balanced equation.
4. Convert the required amount in moles to the required value using the appropriate conversion factor.

Lab Exercise 7.6.1

Quantitative Analysis in Solution

INQUIRY SKILLS

- ○ Questioning
- ○ Hypothesizing
- ○ Predicting
- ○ Planning
- ○ Conducting
- ○ Recording
- ● Analyzing
- ○ Evaluating
- ○ Communicating

It is more financially viable to recycle metals if they are in fairly concentrated solutions, so recycling companies will pay more for those solutions than for more dilute solutions. How do companies find out how much silver, for example, is in a solution? Technicians carry out a reaction that involves removing all the silver from a known volume of the solution, drying it, and measuring its mass. Knowing the mass of silver and the volume of solution, they can calculate the molar concentration of silver in the solution.

The purpose of this investigation is to use the stoichiometric concept to find the unknown concentration of a known volume of aqueous silver nitrate: to extend gravimetric stoichiometry into solution stoichiometry.

Although you will not be conducting this investigation, the investigation report models the type of report that you will be expected to produce in later investigations. Here the evidence is provided. You are to complete the **Analysis** section.

Question

What is the molar concentration of silver nitrate in the aqueous solution provided?

Experimental Design

A precisely measured volume of aqueous silver nitrate solution, $AgNO_{3(aq)}$, is completely reacted with excess copper metal, $Cu_{(s)}$. The silver metal product, $Ag_{(s)}$, is separated by filtration and dried, and the mass of silver measured to the precision of the balance. The concentration of the initial solution is calculated from the mass of product by the stoichiometric method.

Materials

>100 mL $AgNO_{3(aq)}$ of unknown concentration
centigram balance
#16–#20 gauge solid (not braided) copper wire
fine steel wool
wash bottle of pure water
wash bottle of pure acetone, $CH_3COCH_{3(l)}$
filtration apparatus
filter paper
250-mL beaker, with watch glass to fit
400-mL waste beaker for acetone
400-mL waste beaker for filtrate
100-mL graduated cylinder
stirring rod

Procedure

Day 1

1. Using a graduated cylinder, obtain 100 mL of silver nitrate solution and pour it into a 250-mL beaker.
2. Clean about 30 cm of solid copper wire with fine steel wool, and form about 20 cm of it into a coil with a 10-cm handle, so the coiled section will

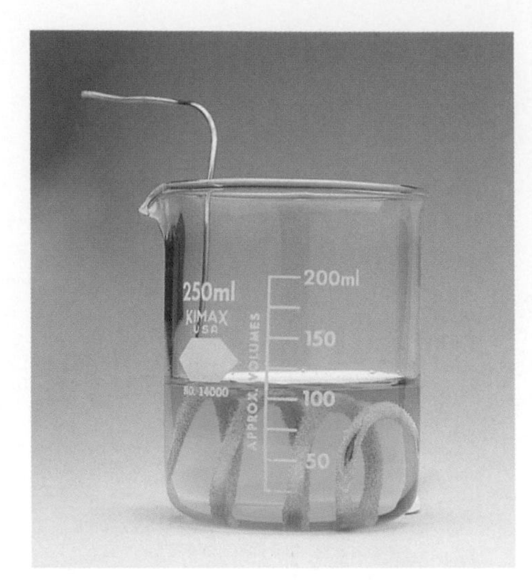

Figure 2
The blue solution that appears verifies that the most common ion of copper, $Cu^{2+}_{(aq)}$, is being slowly formed in this reaction. Silver crystals and excess copper are also visible.

be submerged when placed in the beaker (**Figure 2**) of silver nitrate solution.

3. Record any immediate evidence of chemical reaction, cover with a watch glass, and set aside until the next day.

Day 2

4. Check for completeness of reaction. If the coil is intact, with unreacted (excess) copper remaining, the reaction is complete; proceed to step 6. If all of the copper from Day 1 has reacted away, proceed to step 5.
5. Add another coil of copper wire, cover with a watch glass, and set aside until the next day.
6. When the reaction is complete, remove the wire coil, using shaking, washing, and rubbing with a stirring rod to make sure that all solid silver particles remain in the beaker.
7. Measure and record the mass of a piece of filter paper.
8. Filter the beaker contents to separate the solid silver from the filtrate.
9. Do the final three washes of the solid silver and filter paper with acetone from a wash bottle. Catch the rinsing acetone in a waste beaker.
10. Place the unfolded filter paper and contents on a paper towel to dry for a few minutes.
11. Measure and record the mass of the dry silver plus filter paper.
12. Dispose of the remaining substances by placing solids in the garbage (or recycling them), and rinsing aqueous solutions (not acetone) down the drain with plenty of water. Transfer the acetone to a recycle container.

Evidence

The following observations were recorded:
volume of silver nitrate solution = 100.0 mL
mass of filter paper = 0.93 g
mass of filter paper plus silver = 2.73 g
The copper wire in the silver nitrate solution turned black, grey, and then silver, with spikes of silvery crystals growing from the copper wire.
The solution gradually went from colourless to blue.
There was an excess of copper wire at the end of the reaction time.
After cleaning and washing, the copper wire was noticeably thinner.

Analysis

(a) Analyze the Evidence and use it to answer the Question.

Evaluation

(b) Why was the copper wire cleaned with steel wool before starting the reaction?
(c) Why did the Procedure specify to use acetone for the final three rinses?
(d) What alternative is there to the acetone rinses?
(e) What other test could be done to determine whether all of the silver nitrate in solution had reacted?
(f) If some copper were to break off the wire and join the silver product, how could you separate the two metals?

Using Molar Concentration

In Lab Exercise 7.6.1 you calculated the molar concentration of silver nitrate from the mass of silver, the molar mass of silver, and the volume of the solution. This was a stoichiometric calculation.

Some solution stoichiometry problems are a little more complicated. Sometimes a problem may involve determining the amount (in moles) of a substance by considering how much of that substance reacts with, or is produced by, a known quantity of another substance. In other words, solving the problem is dependent upon writing a balanced chemical equation for the reaction, and converting from concentration (or volume) to amount in moles, and back again.

When you did gravimetric stoichiometry calculations in Chapter 5, you used molar mass as the conversion factor to convert from mass to amount in moles, and back to mass. In solution (volumetric) stoichiometry you use molar concentration as a conversion factor.

However, the general stoichiometric method remains the same. Whatever value you initially measure, you have to first convert it to an amount in moles by using a conversion factor. Second, use the mole ratio between the chemicals with given and required values. Finally, use an appropriate conversion factor to convert the required amount in moles into the requested value with appropriate units.

Let us now work through a typical stoichiometry calculation involving molar concentrations of aqueous solutions.

Many popular chemical fertilizers include ammonium hydrogen phosphate (**Figure 3**). This compound is made commercially by reacting concentrated aqueous solutions of ammonia and phosphoric acid. The reaction is most efficient if the right proportions of reactants are used. Of course, the amounts required will depend on the concentrations of the reactant solutions. What volume of 14.8 mol/L $NH_{3(aq)}$ would be needed to react completely with each 1.00 kL (1.00 m^3) of 12.9 mol/L $H_3PO_{4(aq)}$ to produce fertilizer in a commercial operation?

Figure 3
Fertilizers can have a dramatic effect on plant growth. The plants on the left were fertilized with ammonium hydrogen phosphate. Compare them with the unfertilized plants on the right.

Step 1: Write a balanced chemical equation to find the relationship between the amount of ammonia (in moles) and the amount of phosphoric acid (in moles).

Beneath the equation, list both the given and the required measurements, and the conversion factors, in that order.

$2\ NH_{3(aq)}$	$+\ H_3PO_{4(aq)}$	$\rightarrow (NH_4)_2HPO_{4(aq)}$
v	1.00 kL	
14.8 mol/L	12.9 mol/L	

m is mass in, for example, grams (g).
v is volume in, for example, millilitres (mL).
n is amount in, for example, moles (mol).
C is molar concentration in moles per litre (mol/L).
M is molar mass in, for example, grams per mole (g/mol).

$$v_{H_3PO_4} = 1.00\ \text{kL}$$
$$C_{H_3PO_4} = 12.9\ \text{mol/L}$$
$$C_{NH_3} = 14.8\ \text{mol/L}$$

Step 2: Convert the information given for the reactant of known volume (phosphoric acid) to an amount in moles.

$$n_{H_3PO_4} = 1.00\ \text{k}\cancel{\text{L}} \times \frac{12.9\ \text{mol}}{1\ \cancel{\text{L}}}$$

$$n_{H_3PO_4} = 12.9\ \text{kmol}$$

Step 3: Use the mole ratio to calculate the amount of the required substance (ammonia). According to the balanced chemical equation in Step 1, two moles of ammonia react for every one mole of phosphoric acid.

$$n_{NH_3} = 12.9\ \text{kmol} \times \frac{2}{1}$$

$$n_{NH_3} = 25.8\ \text{kmol}$$

Step 4: Convert the amount of ammonia to the quantity requested in the question; in this example, volume. To obtain the volume of ammonia, the molar concentration (which is the amount/volume ratio) is used to convert the amount in moles to the solution volume. Note that the prefix "kilo" does *not* cancel, while the unit "mol" does.

$$v_{NH_3} = 25.8\ \text{k}\cancel{\text{mol}} \times \frac{1\ \text{L}}{14.8\ \cancel{\text{mol}}}$$

$$v_{NH_3} = 1.74\ \text{kL}$$

According to the stoichiometric method, the required volume of ammonia solution is 1.74 kL.

We can summarize steps 2 to 4 into one step, shown below. Note that the order of use of the conversion factors is the same as above. Again, in the quantity 1.00 kL, note that only the unit cancels.

$$v_{NH_3} = 1.00\ \text{k}\cancel{\text{L}}\ \cancel{H_3PO_4} \times \frac{12.9\cancel{\text{mol}}\ \cancel{H_3PO_4}}{1\ \cancel{\text{L}}\ \cancel{H_3PO_4}} \times \frac{2\ \cancel{\text{mol}}\ \cancel{NH_3}}{1\ \cancel{\text{mol}}\ \cancel{H_3PO_4}} \times \frac{1\ \text{L}\ NH_3}{14.8\cancel{\text{mol}}\ \cancel{NH_3}}$$

$$v_{NH_3} = 1.74\ \text{kL}$$

The following sample problem shows how to solve and communicate a solution to another stoichiometric problem involving solutions.

Sample Problem 1

Chemical technologists work in the laboratories of chemical industries. One of their jobs is to monitor the concentrations of solutions in the process stream. For example, sulfuric acid is a reactant in the production of sulfates (e.g., ammonium sulfate) in fertilizer plants. A technician needs to determine the concentration of the sulfuric acid solution. In the experiment, a 10.00-mL sample of sulfuric acid reacts completely with 15.9 mL of 0.150 mol/L potassium hydroxide solution. Calculate the molar concentration of the sulfuric acid.

Solution

$H_2SO_{4(aq)}$ +	$2\ KOH_{(aq)} \rightarrow 2\ H_2O_{(l)} + K_2SO_{4(aq)}$
10.00 mL	15.9 mL
C	0.150 mol/L

$$n_{KOH} = 15.9\ \text{mL} \times \frac{0.150\ \text{mol}}{1\ \text{L}} = 2.39\ \text{mmol}$$

$$n_{H_2SO_4} = 2.39\ \text{mmol} \times \frac{1}{2} = 1.19\ \text{mmol}$$

$$C_{H_2SO_4} = \frac{1.19\ \text{mmol}}{10.00\ \text{mL}}$$

$$C_{H_2SO_4} = 0.119\ \text{mol/L}$$

or $$C_{H_2SO_4} = 15.9\ \text{mL KOH} \times \frac{0.150\ \text{mol KOH}}{1\ \text{L KOH}} \times \frac{1\ \text{mol}\ H_2SO_4}{2\ \text{mol KOH}} \times \frac{1}{10.00\ \text{mL}}$$

$$C_{H_2SO_4} = 0.119\ \text{mol/L}$$

According to the stoichiometric method, the concentration of the sulfuric acid solution is 0.119 mol/L.

Practice

Understanding Concepts

1. Ammonium sulfate is a "high-nitrogen" fertilizer. It is manufactured by reacting sulfuric acid with ammonia. In a laboratory study of this process, 50.0 mL of sulfuric acid reacts with 24.4 mL of a 2.20 mol/L ammonia solution to yield the product ammonium sulfate in solution. From this evidence, calculate the molar concentration of the sulfuric acid at this stage in the process.
2. Slaked lime is sometimes used in water treatment plants to clarify water for residential use. The lime is added to an aluminum sulfate solution in the water. Fine particles in the water stick to the floc precipitate produced, and settle out with it. Calculate the volume of 0.0250 mol/L calcium hydroxide solution that can be completely reacted with 25.0 mL of 0.125 mol/L aluminum sulfate solution.
3. In designing a solution stoichiometry experiment for her class to perform, a chemistry teacher wants 75.0 mL of 0.200 mol/L iron(III) chloride solution to react completely with an excess of 0.250 mol/L sodium carbonate solution.
   (a) What is the *minimum* volume of this sodium carbonate solution needed?
   (b) What would be a *reasonable* volume of this sodium carbonate solution to use in this experiment? Provide your reasoning.

Applying Inquiry Skills

4. Every concept introduced in science goes through a create-test-use cycle. Even after a scientific concept has been used for some time, scientists continue to test it in new conditions, using new technologies and new experimental designs. This is why scientific knowledge is considered trustworthy. Concepts that have been around for a long time have withstood the tests of time. Their certainty is high.

   The purpose of this investigation is to test solution stoichiometry. Complete the **Prediction**, the **Analysis**, and the **Evaluation**.

   Question
   What is the mass of precipitate produced by the reaction of 20.0 mL of a 2.50 mol/L stock solution of sodium hydroxide with an excess of zinc chloride solution?

Answers

1. 0.537 mol/L
2. 375 mL
3. (a) 90.0 mL
   (b) 100 mL

Answers

4. (a) 2.48 g
   (b) 2.39 g
   (c) 3.6 % error

Prediction

(a) Use the stoichiometric method to predict the mass of precipitate.

Evidence

volume of $NaOH_{(aq)}$ = 20.0 mL
mass of filter paper = 0.91 g
mass of filter paper plus precipitate = 3.30 g
Litmus tests of the filtrate showed no change in colour.

Analysis

(b) Use the Evidence to answer the Question.

Evaluation

(c) Use your answer in (b) to evaluate your Prediction.
(d) Use your evaluation of the Prediction to evaluate the stoichiometry concept.

INQUIRY SKILLS

- Questioning
- Hypothesizing
- Predicting
- Planning
- Conducting
- Recording
- Analyzing
- Evaluating
- Communicating

Soluble barium compounds, such as barium chloride, are toxic and must not be swallowed. Wear gloves and wash hands thoroughly after handling the barium ion solution.

Wear eye protection and a laboratory apron.

Investigation 7.6.1

Percentage Yield of Barium Sulfate

Barium sulfate is a white, odourless, tasteless powder that has a variety of different uses: as a weighting mud in oil drilling; in the manufacture of paper, paints, and inks; and taken internally for gastrointestinal X-ray analysis. It is so insoluble that it is non-toxic, and is therefore very safe to handle.

The reaction studied in this investigation is similar to the one used in the industrial manufacture of barium sulfate. The purpose of this investigation is to evaluate this procedure for producing barium sulfate by comparing the experimental to the expected yield in a percentage yield calculation. To do this, you will first have to decide which is the limiting and which the excess reagent, and then assuming 100% yield, predict the expected mass of product by the stoichiometric method.

Complete the **Question**, **Prediction**, **Materials**, **Procedure**, **Analysis**, and **Evaluation** sections of the investigation report.

Question

(a) Write a Question for this investigation.

Prediction

(b) From the information in the Experimental Design, decide which is the limiting and which the excess reagent. Use solution stoichiomety to predict the expected yield.

Experimental Design

A 40.0-mL sample of 0.15 mol/L sodium sulfate solution is mixed with 50.0 mL of 0.100 mol/L barium chloride solution. A diagnostic test is used to test the filtrate for excess reagent (**Figure 4**). The actual mass produced is compared with the predicted mass to assess the completeness of this reaction.

(c) Write a numbered step-by-step Procedure, including all necessary safety and disposal precautions.
(d) Write a list of Materials.

Figure 4
Once the precipitate settles and the top layer becomes clear, you can test for the completeness of a reaction. Carefully run a drop or two of the excess reagent down the side of the beaker and watch for any additional precipitation, which would indicate that some of the limiting reagent remains in the filtrate solution. No precipitate formation indicates that the reaction of the limiting reagent is complete.

Procedure

1. With your teacher's approval, carry out your experiment and record your observations.

Analysis

(e) Analyze your evidence to calculate the experimental yield.

Evaluation

(f) Evaluate the method of production by comparing the experimental yield with the expected yield by calculating a percentage yield.

Section 7.6 Questions

Applying Inquiry Skills

1. A student wishes to precipitate all the lead(II) ions from 2.0 L of solution containing, among other substances, 0.34 mol/L $Pb(NO_3)_{2(aq)}$. The purpose of this reaction is to make the filtrate solution non-toxic. If the student intends to precipitate lead(II) sulfate, suggest and calculate an appropriate solute, and calculate the required mass of this solute.
2. When designing an experiment to determine the concentration of an aqueous solution, two students decide to carry out a precipitation reaction, followed by a crystallization. They precipitate one of the ions of the solution, boil the water away, and then measure the mass of solid remaining. Critique this experimental design.
3. The purpose of the following investigation is to test solution stoichiometry. Complete the Prediction, Analysis, and Evaluation of the investigation report.

   Question

   What mass of precipitate is produced by the reaction of 20.0 mL of 0.210 mol/L sodium sulfide with an excess quantity of aluminum nitrate solution?

   Prediction

   (a) Use the stoichiometric concept to predict an expected yield.

   Experimental Design

   The two solutions provided react with each other and the resulting precipitate is separated by filtration and dried. The mass of the dried precipitate is determined.

   Evidence

   A precipitate was formed very rapidly when the solutions were mixed.
   mass of filter paper = 0.97 g
   mass of dried filter paper plus precipitate = 1.17 g
   A few additional drops of the aluminum nitrate solution added to the filtrate produced no precipitate.
   A few additional drops of the sodium sulfide solution added to the filtrate produced a yellow precipitate.

   Analysis

   (b) Use the Evidence to answer the Question.

(continued)

Evaluation

(c) Evaluate the experimental design, the prediction, and the stoichiometric method.

4. Once a scientific concept has passed several tests, it can be used in industry. Many industries recycle valuable byproducts, such as silver nitrate solution. Suppose you are a technician in an industry that needs to determine the molar concentration of a solution of silver nitrate. Complete the **Analysis** of the investigation report.

Question

What is the molar concentration of silver nitrate in the solution to be recycled?

Experimental Design

A sample of the silver nitrate solution to be recycled reacts with an excess quantity of sodium sulfate in solution. The precipitate formed is filtered and the mass of dried precipitate is measured.

Evidence

A white precipitate was formed in the reaction.
A similar precipitate formed when a few drops of silver nitrate were added to the filtrate.
No precipitate formed when a few drops of sodium sulfate were added to the filtrate.
volume of silver nitrate solution = 100 mL
mass of filter paper = 1.27 g
mass of dried filter paper plus precipitate = 6.74 g

Analysis

(a) Analyze the Evidence to answer the Question.

Making Connections

5. Antifreeze is used in car cooling systems. It is usually an aqueous solution of ethylene glycol, $C_2H_4(OH)_{2(aq)}$. The concentration of this solution is usually 50% by volume. What is the most common way to determine the concentration of antifreeze solutions at home and in the garage? What other solutions in an automobile have their concentrations determined in a similar way? Describe how the quantitative test works and what units of measure are used to express the concentration of the solutions.

   Follow the links for Nelson Chemistry 11, 7.6.

GO TO www.science.nelson.com

Chapter 7 Summary

Key Expectations

Throughout this chapter, you have had the opportunity to do the following:

- Describe, in words and by graphs, the dependence on temperature of the solubility of solids, liquids, and gases in water. (7.1)
- Explain hardness of water, its consequences, and water-softening methods. (7.2)
- Predict common combinations of aqueous solutions that produce precipitates and represent these reactions using net ionic equations. (7.3, 7.5)
- Describe the technology and the major steps involved in the treatment of waste water. (7.4)
- Perform qualitative and quantitative analyses of solutions. (7.5, 7.6)
- Solve solution stoichiometry problems. (7.6)
- Use the Key Terms for this chapter to communicate clearly. (all sections)

Key Terms

excess reagent
hard water
high solubility
immiscible
insoluble
limiting reagent
low solubility
miscible
net ionic equation
precipitate
qualitative
qualitive chemical analysis
quantitative
saturated solution
soda-lime process
solubility
solution stoichiometry
spectator
total ionic equation

Make a Summary

Make a concept map showing how an understanding of solubility is connected to, and is central to, each of the other topics discussed in this chapter. Use as many of the Key Terms as you can.

Reflect on your Learning

Revisit your answers to the Reflect on Your Learning questions at the begining of this chapter.

- How has your thinking changed?
- What new questions do you have?

Chapter 7 Review

Understanding Concepts

1. Equal volumes of 1.0 mol/L solutions of each of the following pairs of solutions are mixed. Predict which combinations will form a precipitate and write net ionic equations for the predicted reactions.
   (a) $CuSO_{4(aq)}$ and $NaOH_{(aq)}$
   (b) $H_2SO_{4(aq)}$ and $NaOH_{(aq)}$
   (c) $Na_3PO_{4(aq)}$ and $CaCl_{2(aq)}$
   (d) $AgNO_{3(aq)}$ and $KCl_{(aq)}$
   (e) $MgSO_{4(aq)}$ and $LiBr_{(aq)}$
   (f) $CuNO_{3(aq)}$ and $NaCl_{(aq)}$

2. A lab technician uses 1.0 mol/L $Na_2CO_{3(aq)}$ to precipitate metal ions from waste solutions. The resulting filtered solids can be disposed of more easily than large volumes of solution. Write net ionic equations for the reaction between the $Na_2CO_{3(aq)}$ and each of the following waste solutions.
   (a) $Zn(NO_3)_{2(aq)}$
   (b) $Pb(NO_3)_{2(aq)}$
   (c) $Fe(NO_3)_{3(aq)}$
   (d) $CuSO_{4(aq)}$
   (e) $AgNO_{3(aq)}$
   (f) $NiCl_{2(aq)}$
   (g) Defend the technician's choice of $Na_2CO_{3(aq)}$ as the excess reagent.

3. The purification of water can involve several precipitation reactions. Write balanced net ionic equations to represent the reactions described below.
   (a) aqueous aluminum sulfate reacts with aqueous calcium hydroxide
   (b) aqueous sodium phosphate reacts with dissolved calcium bicarbonate
   (c) dissolved magnesium bicarbonate reacts with aqueous calcium hydroxide
   (d) aqueous calcium hydroxide reacts with dissolved iron(III) sulfate

4. When a flame test is done on a blue solution, a green flame is produced (**Figure 1**). Which aqueous cation could account for these observations?

5. List the ions whose compounds are assumed to have high solubility in water. (Refer to the solubility table on the inside back cover.)

6. A flame test on an unknown colourless solution produces a violet flame. When $HgNO_{3(aq)}$ is added to the unknown solution, a precipitate is formed. Name one possible solute for the unknown solution.

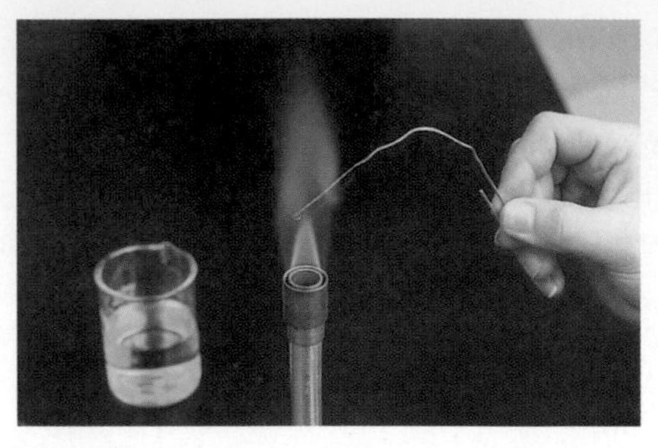

Figure 1

7. How can colour be used to distinguish between the following pairs of ions?
   (a) $Cu^{+}_{(aq)}$ and $Cu^{2+}_{(aq)}$
   (b) $Fe^{2+}_{(aq)}$ and $Fe^{3+}_{(aq)}$
   (c) $CrO_4{}^{2-}{}_{(aq)}$ and $Cr_2O_7{}^{2-}{}_{(aq)}$

8. Which ions are primarily responsible for "hard" water?

9. Copper(II) ions can be precipitated from waste solutions by adding aqueous sodium carbonate.
   (a) What is the minimum volume of 1.25 mol/L $Na_2CO_{3(aq)}$ needed to precipitate all the copper(II) ions in 4.54 L of 0.0875 mol/L $CuSO_{4(aq)}$?
   (b) Suggest a suitable volume to use for this reaction.

10. A 24.89-g piece of zinc is placed into a beaker containing 350 mL of hydrochloric acid. The next day the remaining zinc is removed, dried, weighed, and found to have a mass of 21.62 g. Determine the concentration of zinc chloride in the beaker.

Applying Inquiry Skills

11. A sample of drinking water from a well turns cloudy when $Ba(NO_3)_{2(aq)}$ is added to it. List the anions that could produce the cloudiness. Suggest further tests that could narrow down the possibilities.

12. A solution is known to contain sodium sulfate and/or lithium bromide. Design an experiment to test for the presence of these two compounds.

13. Nitrogen and its compounds are an essential part of the growth and decomposition of plants, and they find their way into soil and surface water. Design an experiment to distinguish among dilute aqueous solutions of the following nitrogen-containing substances: ammonium hydroxide, nitrogen gas, nitrous acid, and potassium nitrate.

14. Seawater contains traces of every chemical compound found on land. A student uses a filtration design to determine the concentration of sodium chloride, the most abundant solute in seawater. Complete the

Materials and **Analysis** sections, and evaluate the **Experimental Design** in the **Evaluation** section.

Question

What is the molar concentration of sodium chloride in a sample of seawater?

Experimental Design

The sodium chloride in a test sample of deep seawater reacts with an excess volume of a 1.00 mol/L $Pb(NO_3)_{2(aq)}$ solution to form a precipitate, which is then filtered and dried.

Materials

(a) What materials would be needed to conduct this investigation?

Evidence

A white precipitate formed when the solutions were mixed.

volume of seawater = 50.0 mL
mass of filter paper = 0.91 g
mass of dried filter paper plus precipitate = 4.58 g

Several drops of potassium iodide solution added to the filtrate produced a yellow precipitate.

Analysis

(b) According to the Evidence, what is the molar concentration of sodium chloride in seawater?

Evaluation

(c) Critique the Experimental Design. What flaws can you see in the student's plan?

15. Copper(II) sulfate is very toxic to algae (tiny water plants). Sometimes copper(II) sulfate is added to swimming pools and water reservoirs to kill algae. A lab technician uses a filtration design to determine the concentration of copper(II) sulfate in a solution prepared for use in a water reservoir. Complete the **Materials, Analysis,** and **Evaluation** sections of the report.

Question

What is the molar concentration of copper(II) sulfate in a solution?

Experimental Design

The copper(II) sulfate solution reacts with an excess volume of a 0.750 mol/L $NaOH_{(aq)}$ solution to form a precipitate, which is then filtered and dried.

Materials

(a) What materials are needed to conduct this investigation?

Evidence

A gelatinous precipitate formed when the solutions were mixed.

volume of copper(II) sulfate solution = 25.0 mL
mass of filter paper = 0.88 g
mass of dried filter paper plus precipitate = 2.83 g

The filtrate is colourless; the dried precipitate is blue.

Analysis

(b) According to the Evidence, what is the molar concentration of copper(II) sulfate in the solution?

Evaluation

(c) Critique the Experimental Design. What flaws can you see in the lab technician's plan?

Making Connections

16. What is the level of hardness in the water source for your community? Find out whether the water is treated for hardness. If so, what treatment is used?

Follow the links for Nelson Chemistry 11, Chapter 7 Review.

GO TO www.science.nelson.com

17. What waste treatment process is used in your community or in a cottage that you know? What are the stages in the process? How old are the process and the equipment—is this of concern?

18. The maximum quantity of oxygen that dissolves in water at 0°C is 14.7 ppm, and at 25°C is 8.7 ppm.
    (a) Calculate the difference in the mass of oxygen that can be dissolved in 50 L of water at the two temperatures.
    (b) If you were a fish, which temperature might you prefer? Explain your answer.

Chapter

8

Acids and Bases

In this chapter, you will be able to

- experimentally determine the empirical properties, including pH, of acids and bases;
- design and conduct an experiment to determine the effect of dilution on pH;
- explain acids and bases, using both Arrhenius and Brønsted-Lowry theories;
- describe and explain the difference between strong and weak acids and bases;
- use the terms: ionization, dissociation, strong acid/base, weak acid/base, hydronium ion, proton transfer, conjugate acid/base, titration, titrant, and endpoint;
- write balanced chemical equations for reactions involving acids and bases;
- develop the skills involved in titration and solve stoichiometry problems using titration evidence;
- describe examples of solutions for which the concentration must be known and exact.

Acid indigestion, commercial antacid remedies for indigestion, pH-balanced shampoos—you don't have to look far in a drugstore to find labels referring to acids or acidity. Many people think that all acids are corrosive, and therefore dangerous, because solutions of acids react with many substances. Yet boric acid is used as an eyewash. Can this be as dangerous as it sounds?

References in the popular media offer no insight into what acids and bases are, or what they do. In fact, such references usually emphasize only one perspective, such as the environmental damage caused by an acid or the cleaning power of a base. As a result, popular ideas are often confusing. An amateur gardener who has just read an article describing the destruction of conifer forests by acid rain may be puzzled by instructions on a package of evergreen fertilizer stating that evergreens are acid-loving plants (**Figure 1**).

This chapter takes a historical approach, presenting evidence and following the development of theories about the substances we call acids and bases. These theories are used to explain and predict the behaviour of acids and bases.

Reflect on your Learning

1. What are some properties of acids?
2. How can you explain these properties of acids?
3. What are some properties of bases?
4. How can you explain these properties of bases?
5. How do your explanations in questions 2 and 4 account for the evidence that acids react with bases?

Try This **Activity**

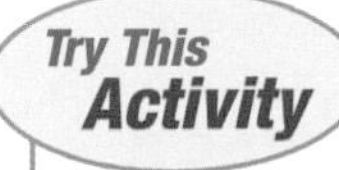

Consumer Products

Look at home or in a store and read the labels on a variety of cleaning products such as drain, oven, wall, floor, window, and toilet bowl cleaners. Find the lists of ingredients as well as any caution notes.

For each product:

(a) Record the product name and the list of ingredients.
(b) Underline the ingredients on the list that you think are active. Give reasons for your choices.
(c) Classify as many of the active ingredients as you can as acids or bases.
(d) Record any warnings about mixing the product with other substances.
(e) Referring to your list in (d), state which combinations represent mixtures of acids and bases.

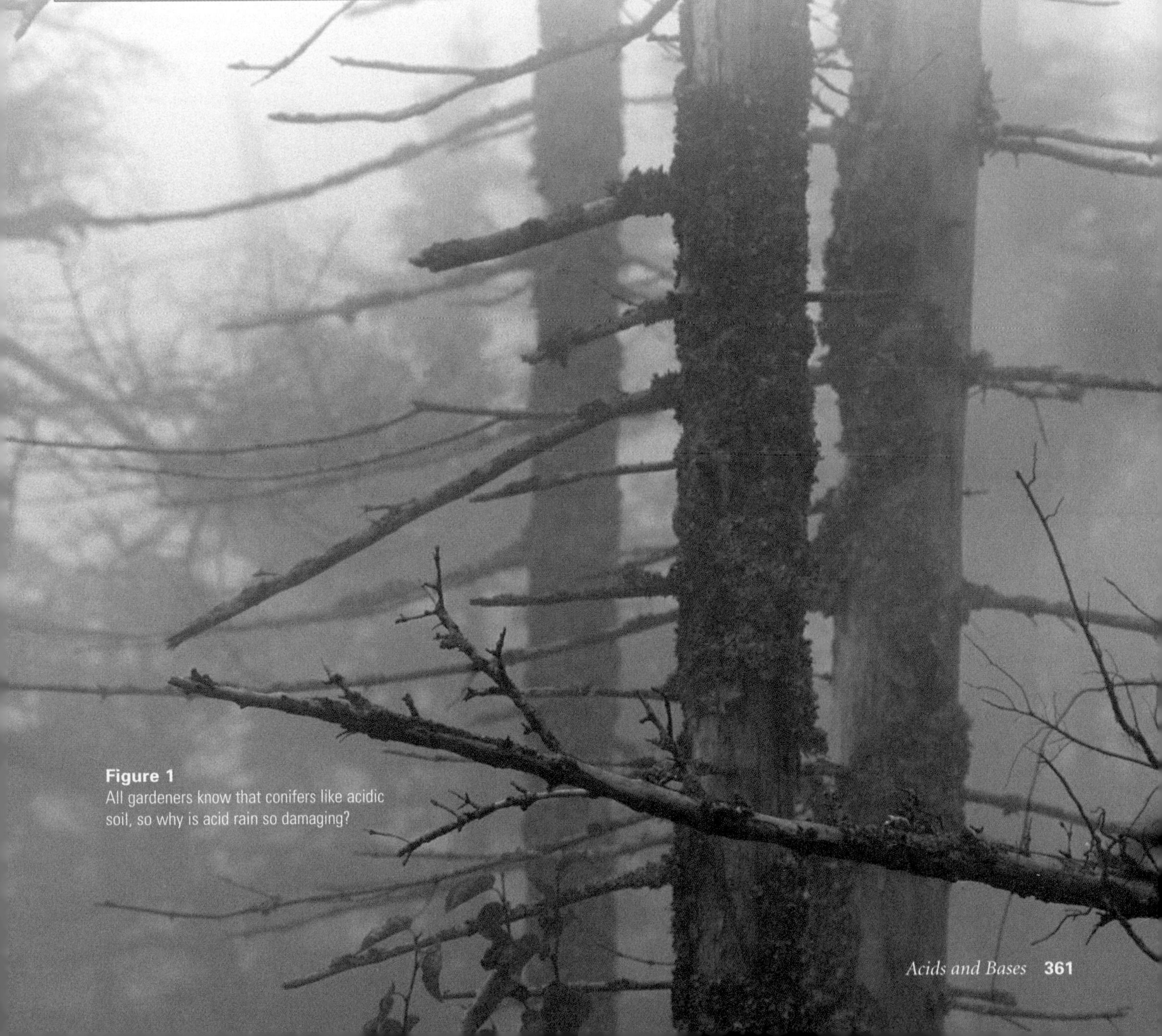

Figure 1
All gardeners know that conifers like acidic soil, so why is acid rain so damaging?

8.1 Understanding Acids and Bases

The story of acids and bases is progressive: It is unfolding like a detective story. Our understanding of acids and bases has changed over time as we have extended our concepts to include more and more acids and bases. Early scientists described acids, for example, as compounds that produce hydrogen gas when reacted with an active metal. These scientists realized that acids (at least, some acids) contain hydrogen atoms. To other investigators, acids were substances that contain oxygen and that react with limestone to produce carbon dioxide.

Later, investigators discovered that acids in aqueous solution change blue litmus to red and conduct electricity. These observations did not fit with the earlier definitions of acids. A new explanation was needed.

Acids and bases share some properties with molecular and ionic substances, and have some properties that are unique.

Table 1 shows that pure liquid samples of both ionic compounds and bases conduct electricity. Similarly, aqueous solutions of both ionic compounds and bases conduct electricity. In some way, bases seem to be similar to ionic compounds. What is it about ionic compounds that enables them to conduct electricity? According to Arrhenius, ionic compounds separate into ions when they are liquid or in solution. Can we assume that the same is true of bases? Perhaps they, too, separate into ions. The presence of mobile ions would explain the conductivity.

Table 1: Properties of Pure and Aqueous Substances

	Conductivity			Litmus
Substance	**Solid**	**Liquid**	**Aqueous**	**Aqueous**
most molecular compounds	no	no	no	no effect
most ionic compounds	no	yes	yes	no effect
acids	no	no	yes	blue to red
bases	no	yes	yes	red to blue

Do all bases release the same ion in aqueous solution? Many bases are ionic compounds that contain a hydroxide, OH^-, which could be released in solution. It seems likely that this ion gives a base its characteristics. After all, sodium chloride, NaCl, is not a base but sodium hydroxide, NaOH, is. Going by the evidence we have at this stage, we can conclude that bases are ionic hydroxides that release mobile hydroxide ions in solution.

$$NaOH_{(s)} \rightarrow Na^+_{(aq)} + OH^-_{(aq)}$$
$$Ca(OH)_{2(s)} \rightarrow Ca^{2+}_{(aq)} + 2\ OH^-_{(aq)}$$

Table 1 also shows that molecular substances, including acids, do not conduct electricity in their pure states. However, acids (unlike other molecular substances) become conductors when dissolved in water. Arrhenius explained that molecular substances do not conduct electricity because they contain only electrically neutral particles called molecules. Can we conclude, therefore, that acids in their pure forms contain neutral molecules and not ions, but that acids in solution contain ions? This is certainly what the evidence suggests. It seems that acids are somehow different in structure and/or composition from other molecular substances.

DID YOU KNOW?

Liquid Ionic Compounds

How hot does an ionic compound have to be, before we can test it for conductivity? Because of the strong ionic bonds, it has to be heated to extremely high temperatures (700 – 1000°C). When the ionic bonds are broken, the ions are free to move around and the substance can conduct electricity.

By studying the composition of substances that turn blue litmus red in an aqueous solution, scientists found that acids seem to contain hydrogen atoms. This led scientists to write the chemical formula for acids as $HA_{(aq)}$ (**Table 2**). The electrical conductivity of these acidic solutions led to the theory that acids ionize in water to release hydrogen ions, $H^+_{(aq)}$. Acids, then, according to the evidence, are hydrogen-containing compounds that ionize in water to produce hydrogen ions.

Table 2: Properties of Hydrogen Compounds

Empirical formula	Litmus test	Molecular formula
$CH_2O_{3(aq)}$	blue to red	H_2CO_3
$CH_{4(aq)}$	no change	CH_4
$SH_{2(aq)}$	blue to red	H_2S
$PH_{3(aq)}$	no change	PH_3
$NHO_{3(aq)}$	blue to red	HNO_3

$$HCl_{(g)} \rightarrow H^+_{(aq)} + Cl^-_{(aq)}$$
$$HNO_{3(aq)} \rightarrow H^+_{(aq)} + NO^-_{3(aq)}$$

Arrhenius extended his theory of ions to explain some of the properties of acids and bases. According to Arrhenius, we can write an equation showing that **bases** dissociate into individual positive and negative ions in solution. He proposed that aqueous hydroxide ions were responsible for the properties of basic solutions, such as turning red litmus paper blue. The **dissociation** of bases is similar to that of any other ionic compound, as shown in the following dissociation equation for barium hydroxide.

$$Ba(OH)_{2(s)} \rightarrow Ba^{2+}_{(aq)} + 2\ OH^-_{(aq)}$$

According to the evidence in **Table 1**, acids are electrolytes in solution even though as pure substances, they are molecular compounds. Acids, such as $HCl_{(g)}$ and $H_2SO_{4(l)}$, do not show their acidic properties until they dissolve in water. Since acids in solution are electrolytes, Arrhenius' theory suggests that acid solutions must contain ions. However, the pure solute is molecular; it is made up only of neutral molecules. How, then, can its solution contain ions? Obviously, **acids** do not simply dissolve to form a solution of molecules. According to Arrhenius, after acids dissolve as individual molecules, they then ionize into hydrogen ions and negative ions in solution.

In the case of acids, Arrhenius assumed that the water somehow causes the acid molecules to ionize, but he didn't propose an explanation for this. (We now believe that water molecules help to pull the molecules apart—to ionize the acid.) A typical example of an acid is hydrogen chloride gas dissolving in water to form hydrochloric acid. We can describe this process with an **ionization** equation.

$$HCl_{(g)} \rightarrow H^+_{(aq)} + Cl^-_{(aq)}$$

So, although HCl is a molecular compound, it appears to behave in solution as if it were ionic. It ionizes into ions, which are capable of conducting electricity in solution. We explain the properties of acids by saying that all acids produce hydrogen ions in solution, and define acids as substances that ionize in water to increase the hydrogen ion concentration.

DID YOU **KNOW ?**

Formulas of Acids

The 19th century idea was that acids are salts (compounds) of hydrogen. For example, scientists would say that $HCl_{(aq)}$ is the hydrogen salt of $NaCl_{(aq)}$. This led to the practice of writing hydrogen first in the formulas of substances known to form acidic solutions, such as $H_2SO_{4(aq)}$ and $HCl_{(aq)}$.

base: (according to the Arrhenius theory) an ionic hydroxide that dissociates in water to produce hydroxide ions

dissociation: the separation of ions that occurs when an ionic compound dissolves in water

acid: (according to the Arrhenius theory) a compound that ionizes in water to form hydrogen ions

ionization: any process by which a neutral atom or molecule is converted into an ion

The Arrhenius theory was a major advance in understanding chemical substances and solutions. Arrhenius also provided the first comprehensive theory of acids and bases. The empirical and theoretical definitions of acids and bases are summarized in **Table 3**.

Table 3: Acids, Bases, and Neutral Substances

Type of substance	Empirical definition	Theoretical definition
acids	• in solution, turn blue litmus red • are electrolytes • in solution, neutralize bases	• these hydrogen-containing compounds ionize to produce $H^+_{(aq)}$ ions • $H^+_{(aq)}$ ions react with $OH^-_{(aq)}$ ions to produce water
bases	• in solution, turn red litmus blue • are electrolytes • in solution, neutralize acids	• ionic hydroxides dissociate to produce $OH^-_{(aq)}$ ions • $OH^-_{(aq)}$ ions react with $H^+_{(aq)}$ ions to produce water
neutral substances	• in solution, do not affect litmus • some are electrolytes • some are nonelectrolytes	• no $H^+_{(aq)}$ or $OH^-_{(aq)}$ ions are formed • some exist as ions in solution • some exist as molecules in solution

Sample Problem 1

Write dissociation or ionization equations (as appropriate) for the dissolving of the following chemicals in water. Label each equation as either dissociation or ionization.

(a) potassium chloride (a salt substitute)

(b) hydroiodic acid (a strong acid)

Solution

(a) $KCl_{(s)} \rightarrow K^+_{(aq)} + Cl^-_{(aq)}$ (dissociation)

(b) $HI_{(aq)} \rightarrow H^+_{(aq)} + I^-_{(aq)}$ (ionization)

As Sample Problem 1 shows, ionic substances dissociate in water, but acids ionize.

Acids as pure substances are molecular and, as such, may be solids, liquids, or gases at SATP (standard ambient temperature and pressure). When you are writing ionization equations for acids, you may not always know the initial state of matter. If you do know the state, use (s), (l), or (g) subscripts; if you do not know the pure state of the acid, use (aq). This is correct for now, as all the acids you will be using in this course are in aqueous solution. For example,

$HCl_{(g)}$ or $HCl_{(aq)} \rightarrow H^+_{(aq)} + Cl^+_{(aq)}$ (a gaseous acid)

$HC_2H_3O_{2(l)}$ or $HC_2H_3O_{2(aq)} \rightarrow H^+_{(aq)} + C_2H_3O^-_{2(aq)}$ (a liquid acid)

$H_2C_2O_{4(s)}$ or $H_2C_2O_{4(aq)} \rightarrow H^+_{(aq)} + HC_2O^-_{4(aq)}$ (a solid acid)

Recall that chemicals that turn blue litmus red have acid formulas that begin with H.

Strong and Weak Acids

Different acidic solutions have different electrical conductivity. We can see this from laboratory evidence (**Figure 1**). There seem to be two fairly distinctive classes of acids. If we were to test the electrical conductivity of a variety of acids at equal concentration and temperature, we might collect evidence such as that in **Table 4**. If we were then to analyze the evidence, we could classify the acids according to the acid strength, as is shown in the last column.

Figure 1
In solutions of equal concentration, hydrochloric acid is a very good conductor of electricity; acetic acid conducts electricity less well.

Table 4: Electrical Conductivity and Strength of Various Acids

Acid name	Acid formula	Electrical conductivity	Strength
acetic acid	$HC_2H_3O_{2(aq)}$	low	weak
nitrous acid	$HNO_{2(aq)}$	low	weak
carbonic acid	$H_2CO_{3(aq)}$	low	weak
hydrochloric acid	$HCl_{(aq)}$	high	strong
sulfuric acid	$H_2SO_{4(aq)}$	high	strong
nitric acid	$HNO_{3(aq)}$	high	strong

We can classify all acids as either strong or weak. Acids whose solutions have high electrical conductivity are called **strong acids**. Their high electrical conductivity is explained by their high **percentage ionization**. Most (more than 99%) of the acid molecules ionize. Sulfuric acid, nitric acid, and hydrochloric acid are examples of strong acids. (These strong acids are among those identified in the table of Concentrated Reagents in Appendix C, page 636.) Most other common acids, such as carbonic acid, are **weak acids**. Their low conductivity is explained by their very low percentage ionization.

strong acid: (theoretical definition) an acid that ionizes almost completely (>99%) in water to form aqueous hydrogen ions

percentage ionization: the percentage of molecules that form ions in solution

weak acid: (empirical definition) an acid with characteristic properties less than those of a strong acid; (theoretical definition) an acid that ionizes only partially (<50%) in water to form aqueous hydrogen ions, so exists primarily in the form of molecules

You have probably heard that some acids are dangerous. How do you know which ones to treat with particular caution? For safety purposes you need to pay more attention to the strength of the acid than the concentration. A dilute solution of a strong acid can be more dangerous than a concentrated solution of a weak acid. This is because the corrosive property of acids is due to the hydrogen ion. The more hydrogen ions in solution, the more dangerous the solution. As a general rule, show respect for all acids, but especially strong acids.

To communicate the percentage ionization and the strength of an acid undergoing ionization, we can write the percentage ionization over the chemical equation arrow. Strong acids generally ionize >99%, while weak acids generally ionize <50%.

$$HNO_{3(aq)} \xrightarrow{>99\%} H^+_{(aq)} + NO^-_{3(aq)} \qquad \text{(a strong acid)}$$

$$HC_2H_3O_{2(aq)} \xrightarrow{<50\%} H^+_{(aq)} + C_2H_3O^-_{2(aq)} \qquad \text{(a weak acid)}$$

Sample Problem 2

The following acidic solutions were tested (at equal concentration and temperature) for electrical conductivity. Write ionization equations to explain the relative conductivity of each acid.

(a) hydrobromic acid (aqueous hydrogen bromide): high conductivity

(b) hydrofluoric acid (aqueous hydrogen fluoride): low conductivity

Solution

(a) $HBr_{(aq)} \xrightarrow{>99\%} H^+_{(aq)} + Br^-_{(aq)}$

(b) $HF_{(aq)} \xrightarrow{<50\%} H^+_{(aq)} + F^-_{(aq)}$

SUMMARY Acids and Bases

At this point, the only strong acids that you have to know about are hydrochloric acid, nitric acid, and sulfuric acid. You can assume that all other acids are weak, unless you are told otherwise.

You can also assume that all bases are ionic hydroxides, all of which are strong bases.

Strong Acid: $HA_{(s/l/g/aq)} \xrightarrow{>99\%} H^+_{(aq)} + A^-_{(aq)}$ (>99% ionized)

Weak Acid: $HA_{(s/l/g/aq)} \xrightarrow{<50\%} H^+_{(aq)} + A^-_{(aq)}$ (<50% ionized)

Strong Base: $MOH_{(s)} \longrightarrow M^+_{(aq)} + OH^-_{(aq)}$ (100% dissociated as ions)

Practice

Understanding Concepts

1. What evidence is there that ionic compounds exist as ions in their pure state while molecular compounds, including acids, exist as molecules in their pure state?
2. Based upon their chemical formulas, classify the following chemicals as acid, base, or neutral.
   (a) $H_2SO_{3(aq)}$
   (b) $NaOH_{(aq)}$
   (c) $CH_3OH_{(aq)}$
   (d) $HC_3H_5O_{2(aq)}$
   (e) $NaC_2H_3O_{2(aq)}$
   (f) $Ba(OH)_{2(aq)}$
3. Acids are molecular compounds, but they don't behave quite like other molecular compounds. What properties make acids unique?
4. According to the Arrhenius theory, what causes the change in colour of litmus paper in a basic solution? in an acidic solution?
5. Write an empirical and a theoretical definition of an acid.
6. For each of the following compounds, indicate whether they dissociate or ionize in aqueous solution. Write ionic equations to represent the dissociation or the ionization.
   (a) sodium hydroxide (drain cleaner)
   (b) hydrogen acetate (vinegar)
   (c) hydrogen sulfate (battery acid)
   (d) calcium hydroxide (slaked lime)

Applying Inquiry Skills

7. Complete the **Analysis** and **Evaluation** for the following investigation.

   Question
   Which of the chemicals, numbered 1 to 7, is $KCl_{(s)}$, $Ba(OH)_{2(s)}$, $Zn_{(s)}$, $HC_7H_5O_{2(s)}$, $Ca_3(PO_4)_{2(s)}$, $C_{25}H_{52(s)}$ (paraffin wax), and $C_{12}H_{22}O_{11(s)}$?

   Experimental Design
   Equal amounts of each chemical are added to equal volumes of water. The chemicals are tested for solubility, and their aqueous solutions are tested for their conductivity and effect on litmus paper.

   Evidence

Table 5: Properties of Seven Substances

Chemical	Solubility in water	Conductivity of solution	Effect of solution on litmus paper
1	high	none	no change
2	high	high	no change
3	none	none	no change
4	high	high	red to blue
5	none	none	no change
6	none	none	no change
7	low	low	blue to red

   Analysis
   (a) Based on the Evidence (**Table 5**), which chemical is which?

Evaluation

(b) Use your knowledge of chemicals to suggest improvements to the Experimental Design.

Making Connections

8. What kinds of compounds can be used in solution to conduct electricity in batteries? Using the Internet, find examples (or illustrations) of batteries that use each kind of compound.

   Follow the links for Nelson Chemistry 11, 8.1.

   GO TO **www.science.nelson.com**

9. Which of the following acids should be handled particularly carefully? (Use the table of Concentrated Reagents in Appendix C.) Give your reasons.
   (a) hydrochloric acid (concrete etching)
   (b) carbonic acid (carbonated beverages)
   (c) sulfuric acid (car battery acid)
   (d) acetic acid (vinegar)
   (e) nitric acid (copper etching)
   (f) phosphoric acid (rust remover)
10. Pure water is a molecular compound. Can you be electrocuted if you are standing in pure water? Provide your reasoning.

8.2 pH of a Solution

Skin care and hair care products are often advertised as being pH balanced. What does this mean? It sounds like a good thing, but what is pH, and how can it be balanced?

pH is a way of indicating the concentration of hydrogen ions present in a solution. You have just discovered that all acids release H^+ ions when they ionize in water. And you know that we can define concentration as the amount of a substance (in moles) present in a given volume of a solution. Chemists have combined these concepts into a way of communicating the acidity of a solution: a concise code for the concentration of $H^+_{(aq)}$ ions. This code is the pH scale.

The molar concentration of hydrogen ions is extremely important in chemistry. According to Arrhenius's theory, hydrogen ions are responsible for the properties of acids, and the higher the concentration of hydrogen ions, the more acidic a solution will be. Similarly, the higher the concentration of hydroxide ions, the more basic a solution will be. You might not expect a neutral solution or pure water to contain any hydrogen or hydroxide ions at all. However, careful testing yields evidence that even pure water always contains tiny amounts of both hydrogen and hydroxide ions, due to a slight ionization of the water molecules (**Figure 1**). In a sample of pure water, about two of every billion water molecules ionize to form hydrogen and hydroxide ions.

$$H_2O_{(l)} \rightarrow H^+_{(aq)} + OH^-_{(aq)}$$

This gives a hydrogen ion concentration of about 1×10^{-7} mol/L. Most conductivity tests will show no conductivity for pure water (unless the equipment is extremely sensitive). The concentration of hydrogen ions declines when a base is dissolved in the water. The hydroxide ions released by the base react with the hydrogen ions freed by the ionization of water to produce water molecules. The result is a decline in the concentration of hydrogen ions.

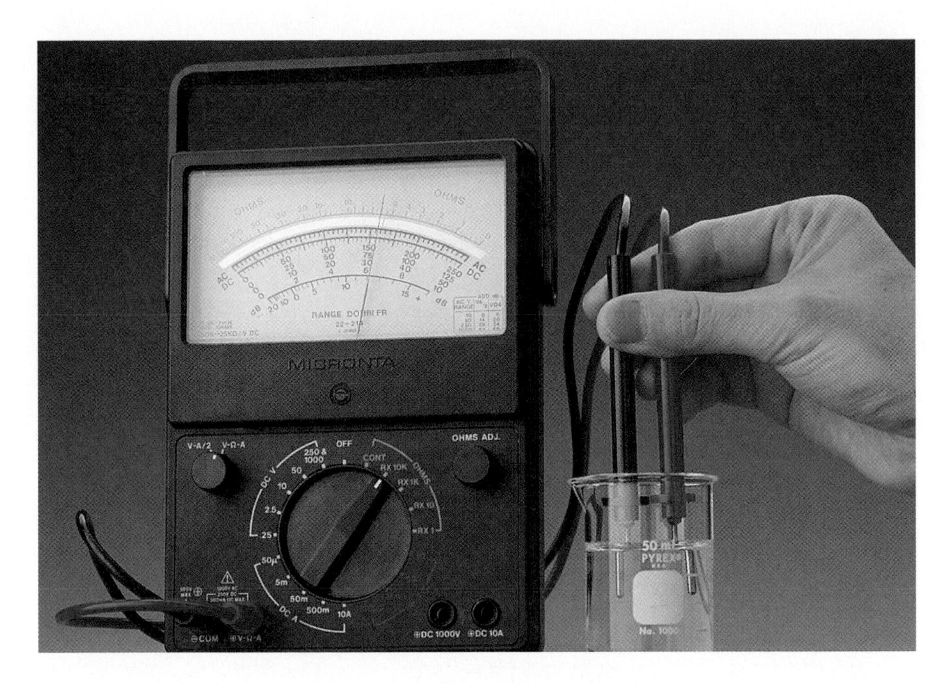

Figure 1
Pure distilled water has a very slight electrical conductivity that is only noticeable when tested with a very sensitive meter.

Aqueous solutions show a phenomenally wide range of hydrogen ion concentrations—from more than 10 mol/L for a concentrated hydrochloric acid solution, to less than 1×10^{-15} mol/L for a concentrated sodium hydroxide solution. Any aqueous solution can be classified as acidic, neutral, or basic using a scale based on the hydrogen ion concentration. Note that the square brackets around the $H^+_{(aq)}$ ion indicate "molar concentration."

- In a neutral solution, $[H^+_{(aq)}] = 1 \times 10^{-7}$ mol/L.
- In an acidic solution, $[H^+_{(aq)}] > 1 \times 10^{-7}$ mol/L.
- In a basic solution, $[H^+_{(aq)}] < 1 \times 10^{-7}$ mol/L.

The extremely wide range of hydrogen ion concentrations led to a convenient shorthand method of communicating these concentrations. This method, called pH, was invented in 1909 by Danish chemist Søren Sørenson. The pH of a solution is defined as the negative of the exponent to the base 10 of the hydrogen ion concentration (expressed as moles per litre). This is not quite as complicated as it sounds. For example, a solution with a hydrogen concentration of 10^{-7} mol/L has a pH of 7 (neutral). Similarly, a pH of 2 corresponds to a much higher hydrogen ion concentration of 10^{-2} mol/L (acidic).

We can rearrange this relationship to show that pH is the negative of the power of 10 of the hydrogen ion molar concentration.

$$[H^+_{(aq)}] = 10^{-pH}$$

This relationship can be used, without complicated mathematics, to convert between pH and the hydrogen ion concentration. For example, if the hydrogen ion concentration is 1×10^{-5} mol/L, then the pH is 5.0. If the pH is 8.0, then the hydrogen ion concentration is 1×10^{-8} mol/L.

Notice, in **Table 1**, that the certainty (as expressed in the number of significant digits) of the hydrogen ion concentration provides the precision (as expressed in the number of decimal places) of the pH. The number of decimal places for the pH is equal to the number of significant digits in the hydrogen ion concentration. This is because the integer of the pH (e.g., 7) does not count as a significant digit any more than the exponent in 10^{-7} (i.e., 7) does.

DID YOU KNOW ?

The "p" in pH

pH was developed only about 100 years ago, but already the origin of the term has become blurred. Some scientists associate pH with **p**ower of **h**ydrogen H; others with **p**otential of **h**ydrogen. Sørenson was Danish, so perhaps the "p" in pH comes from the Danish word "potenz," meaning "strength," or the French word "potentiel." It is strange that we have so quickly lost the origin of such a familiar term.

Table 1: Sample Conversions Between Hydrogen Ion Concentration and pH

$[H^+_{(aq)}]$ (mol/L)	pH
10^{-9}	9
1×10^{-2}	2.0
1.0×10^{-7}	7.00
1.00×10^{-11}	11.000

Sample Problem 1

What is the pH of each of the following solutions?

(a) 1×10^{-2} mol/L hydrogen ion concentration in vinegar

(b) $[H^+_{(aq)}] = 1.0 \times 10^{-12}$ mol/L in household ammonia

Solution

(a) pH = 2.0

(b) pH = 12.00

Sample Problem 2

What is the hydrogen ion concentration for the following solutions?

(a) a carbonated beverage with a pH of 3.0

(b) an antacid solution for which pH = 10.00

Solution

(a) $[H^+_{(aq)}] = 1 \times 10^{-3}$ mol/L

(b) $[H^+_{(aq)}] = 1.0 \times 10^{-10}$ mol/L

---

pH is specified on the labels of consumer products such as shampoos, in water-quality tests for pools and aquariums, in environmental studies of acid rain, and in laboratory investigations of acids and bases. Since each pH unit corresponds to a factor of 10 in the concentration, a huge $[H^+_{(aq)}]$ range can now be communicated by a simple set of positive numbers (**Figure 2**). In these applications, dilution is often an important consideration. What happens to the pH when a solution is diluted? When the pH changes by one unit (e.g., from 5 to 6), the hydrogen ion concentration has been decreased by a factor of ten (i.e., from 10^{-5} mol/L to 10^{-6} mol/L). When the hydrogen ion concentration changes by a factor of 100 (e.g., from 10^{-2} mol/L to 10^{-4} mol/L), the pH changes by 2 units (i.e., from 2 to 4).

neutralization: a reaction between an acid and a base that results in a pH closer to 7

Another way of changing the pH is by a **neutralization** reaction, in which hydrogen ions react with hydroxide ions to move the pH closer to 7. Removing hydrogen ions from the solution is a more effective method of changing the pH than diluting the solution. For this reason, spills of acids and bases are more often neutralized than diluted. Diluting the solution helps to make the solution less hazardous, but neutralizing the solution is more effective.

$$H^+_{(aq)} + OH^-_{(aq)} \rightarrow H_2O_{(l)}$$

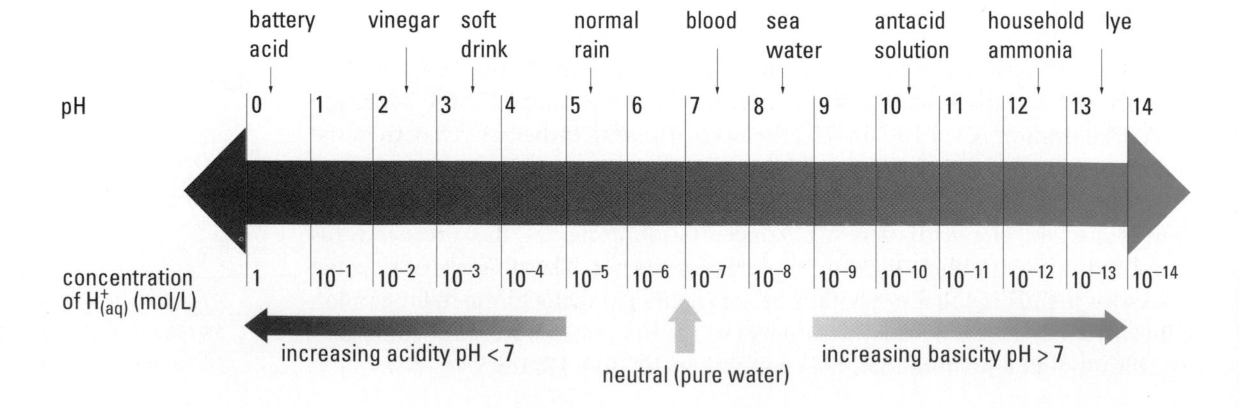

Figure 2
The pH scale can communicate a broad range of hydrogen ion concentrations in a wide variety of substances.

Practice

Understanding Concepts

1. State four examples of products for which the pH may be specified.
2. What is the hydrogen ion concentration in the following household solutions?
   (a) household ammonia: pH = 11.00
   (b) vinegar: pH = 2.0
   (c) soda pop: pH = 4.00
   (d) drain cleaner: pH = 14.00
3. Express the following typical concentrations as pH values.
   (a) grapefruit juice: $[H^+_{(aq)}] = 1.0 \times 10^{-3}$ mol/L
   (b) rainwater: $[H^+_{(aq)}] = 1 \times 10^{-5}$ mol/L
   (c) milk: $[H^+_{(aq)}] = 1 \times 10^{-7}$ mol/L
   (d) liquid soap: $[H^+_{(aq)}] = 1.0 \times 10^{-10}$ mol/L
4. If a water sample test shows a pH of 5, by what factor would the hydrogen ion concentration have to be changed to make the sample neutral? Is this an increase or a decrease in hydrogen ion concentration?
5. Explain why, if the hydrogen ion concentration is 1 mol/L, pH = 0.
6. What amount of hydrogen ions, in moles, is present in 100 L of the following solutions?
   (a) wine: $[H^+_{(aq)}] = 1 \times 10^{-3}$ mol/L
   (b) seawater: pH = 8.00
   (c) stomach acid: $[H^+_{(aq)}] = 10.0$ mmol/L

Reflecting

7. Many chemicals that are potentially toxic or harmful to the environment have maximum allowable concentration levels set by government legislation.
   (a) If the chemical is dangerous, should the limit be zero?
   (b) Is a zero level theoretically possible?
   (c) Is a zero level measurable?
   (d) If a nonzero limit is set, in your opinion, how should this limit be chosen?

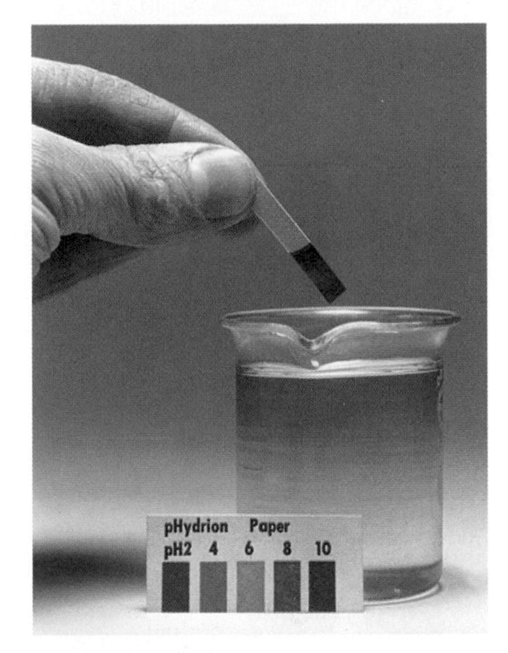

Figure 3
pH paper has a range of possible colours. Each colour corresponds to a particular pH.

Measuring pH of a Solution

You may have already measured the pH of solutions using pH indicators or pH paper (**Figure 3**) to estimate the approximate value. In some situations, such as testing garden soil or aquarium water, this is appropriate. However, in scientific analysis we usually require more precise measurement. For example, when environmental scientists and technicians study the effects of acid rain in waterways, small pH changes in a large body of water can be significant. Precise pH measurements are normally made using a pH meter. All pH meters operate like small electric cells in which the electricity produced by the cell depends on the acidity of the solution. With modern electronics, the tiny electrical signals are detected, converted, and displayed on a screen or dial as the pH. Some pH meters, particularly older ones, are small desktop units (**Figure 4**); some newer meters can be just as precise, yet are as small as a pencil.

Even more recent pH meters consist of a pH probe connected to a computer. The main advantage of this technology is that the pH can be sampled, recorded, and even graphed continuously and automatically. However, if you are just going to take a few pH readings, a system without a computer is perfectly adequate.

Figure 4
Arnold Beckman invented the pH meter in 1935, 26 years after Søren Sørenson had developed the concept of pH for communicating hydrogen ion concentration.

INQUIRY SKILLS

- ○ Questioning
- ○ Hypothesizing
- ○ Predicting
- ● Planning
- ● Conducting
- ● Recording
- ● Analyzing
- ○ Evaluating
- ● Communicating

Figure 5
Unfortunately, it is all too common in Canada for industrial chemicals to flow directly into local waterways.

Figure 6
High concentrations of PCBs have been found in the fat of the beluga whales in the St. Lawrence.

Acidic solutions are corrosive. Wear eye protection and a laboratory apron. Keep your hands away from your eyes and wash your hands when finished.

Investigation 8.2.1

Dilution and pH

For many years it has been common practice for some municipalities, industries, businesses, and consumers to dispose of hazardous wastes in lakes and rivers (**Figure 5**). This is often justified by an argument based on the dilution of the wastes by a relatively large volume of water. Some critics refer to this argument as "the solution to pollution is dilution." There are many reasons why this reasoning is not valid. One reason is the environmental effect of bioamplification, in which initially minute quantities of contaminants are concentrated to dangerous quantities higher up the food chain (**Figure 6**). But there may also be other effects that are not immediately obvious. The purpose of this investigation is to study the effect of dilution on the pH of a solution and develop a generalization for any trends observed.

In previous work (see Section 6.5), you practised preparing a solution by a careful dilution of a starting solution. You can use the same skills and equipment to plan and carry out an experiment to answer the following Question. Complete the **Experimental Design**, **Materials**, **Procedure**, and **Analysis** sections of the report.

Question

What effect does the dilution of an acidic solution have on the pH of the solution?

Experimental Design

The pH of a 0.10 mol/L $HCl_{(aq)}$ solution is measured using a pH meter or pH paper. A sample of this solution is precisely diluted and the pH (dependent variable) is measured. This process is repeated several times to obtain measures of pH for a range of concentrations (independent variable). The temperature of the solution is a controlled variable.

(a) Write a detailed Procedure for this investigation, including any safety precautions.

(b) Draw up a table to record your observations.

Materials

(c) Create a list of Materials.

Procedure

1. Obtain your teacher's approval, and then conduct your investigation.

Analysis

(d) Answer the Question. Based on your Evidence, what relationship exists between concentration and pH?

Calculating pH and Hydrogen Ion Concentration

Earlier in this section you saw that a hydrogen ion concentration can be expressed as a simple pH. The relationship between $[H^+_{(aq)}]$ and pH is easy to calculate if the concentration is only a power of ten, such as 10^{-3} mol/L where the pH = 3, or $[H^+_{(aq)}] = 1 \times 10^{-7}$ mol/L where the pH is 7.0. This is clearly illustrated in **Figure 2**, page 370. What if the concentration of hydrogen ions is not so simple, such as 2.7×10^{-3} mol/L? To answer this question you need to know that, in mathematics, the logarithm of a number is the exponent when the number is written in exponential form. For example, ignoring certainty in significant digits,

$$100 = 10^2 \qquad \log_{10}(10^2) = 2$$
$$0.001 = 10^{-3} \qquad \log_{10}(10^{-3}) = -3$$

In general, if $y = 10^x$, then $\log_{10}(y) = x$. Fortunately, all scientific calculators have a function key, labelled "log," that will find the logarithm (exponent) for any number in the display of the calculator (**Figure 7**). Because hydrogen ion concentrations are usually less than 1 mol/L, **pH** is defined as the negative logarithm to avoid having almost all pHs as negative numbers. This is simply a convenience agreed to by all scientists—a convention of communication. On your calculator and in general usage, log is normally understood to mean $\log_{10}$. Therefore, the mathematical definition of pH becomes,

pH: a measure of the acidity of a solution; the negative logarithm, to the base ten, of the hydrogen ion molar concentration

$$pH = -\log[H^+_{(aq)}]$$

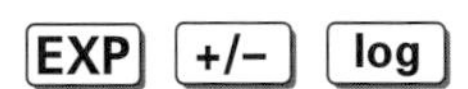

Figure 7
You will become familiar with these keys when converting from hydrogen ion concentration to pH.

Values of pH can be calculated from the hydrogen ion concentration, as shown in the following example. The digits preceding the decimal point in a pH value are determined by the digits in the exponent of the hydrogen ion concentration. These digits locate the position of the decimal point in the concentration value and do not indicate the certainty of the value. However, *the number of digits following the decimal point in the pH value is equal to the number of significant digits expressing the certainty of the hydrogen ion concentration.* For example, a hydrogen ion concentration of 2.7×10^{-3} mol/L (note the two significant digits) corresponds to a pH of 2.57 (note the two decimal places). The 2.7 in the 2.7×10^{-3} mol/L value indicates the certainty as two significant digits. The 3 in the 2.7×10^{-3} mol/L value only indicates where the decimal point goes. The .57 in the pH value of 2.57 communicates a certainty of two significant digits. The 2 in the 2.57 pH value does not count as measured digits; it only indicates where the decimal place in the value goes, i.e., the power of 10 of the value. The examples below will help to clarify this for you.

Sample Problem 3

An antacid solution has a hydrogen ion concentration of 4.7×10^{-11} mol/L. What is its pH? (See **Figure 8**.)

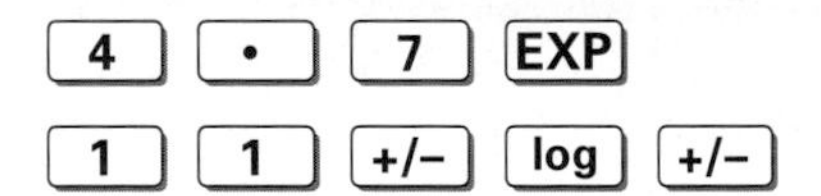

Figure 8
On many calculators, $-\log(4.7 \times 10^{-11})$ may be entered by pushing the above sequence of keys.

Solution

$$pH = -\log[H^+_{(aq)}]$$
$$= -\log(4.7 \times 10^{-11})$$
$$pH = 10.33$$

(Note that the certainty of both values is expressed as two significant digits.)

If pH is measured in an acid–base analysis, you may have to convert a pH reading to the molar concentration of hydrogen ions. This conversion is based on the mathematical concept that a base ten logarithm represents an exponent. Therefore, the pH becomes the exponent.

$$[H^+_{(aq)}] = 10^{-pH}$$

The method of calculating the hydrogen ion concentration from the pH value is shown in Sample Problem 4.

1 0 • 3 3
+/− either INV log
or 2nd log

Figure 9
On many calculators, $10^{-10.33}$ may be entered by pushing the sequence of keys shown above.

Sample Problem 4

The pH reading of a solution is 10.33. What is its hydrogen ion concentration? Be sure to indicate your answer with the correct certainty (**Figure 9**).

$$[H^+_{(aq)}] = 10^{-pH}$$
$$= 10^{-10.33} \text{ mol/L}$$
$$[H^+_{(aq)}] = 4.7 \times 10^{-11} \text{ mol/L}$$

(Note that the two decimal places in the pH yield two significant digits in the hydrogen ion concentration. The 10 is not counted in the pH for the same reason that the 11 is not counted in the hydrogen ion concentration: Both the 10 and the 11 only tell you where the decimal place goes—the power of ten.)

SUMMARY Hydrogen Ion Concentration and pH

pH is the negative power of ten of the hydrogen ion concentration.

$$pH = -\log[H^+_{(aq)}] \quad \text{or} \quad [H^+_{(aq)}] = 10^{-pH}$$

solution:	acidic	neutral	basic
$[H^+_{(aq)}]$:	$>10^{-7}$	10^{-7}	$<10^{-7}$
pH:	<7	7	>7

Note the inverse relationship between $[H^+_{(aq)}]$ and pH. The higher the hydrogen ion molar concentration, the lower the pH.

Practice

Understanding Concepts

8. What are two ways used to measure the pH of a solution?
9. (a) What is the pH of pure water?
   (b) What is the hydrogen ion concentration of pure water?
10. Food scientists and dietitians measure the pH of foods when they devise recipes and special diets.
    (a) Copy and complete **Table 2**.
    (b) Based on pH only, which of the foods should taste the most sour?

Making Connections

11. What are some benefits and risks of using acidic and basic substances in your home? When do you consider the benefits to exceed the risks? When do the risks exceed the benefits? Provide some examples of each.

Answers

10. (a) 2.26; 4×10^{-9} mol/L; 4.6×10^{-4} mol/L; 3.4

Table 2: Acidity of Foods

Food	$[H^+_{aq}]$ (mol/L)	pH
oranges	5.5×10^{-3}	
asparagus		8.4
olives		3.34
blackberries	4×10^{-4}	

Sections 8.1–8.2 Questions

Understanding Concepts

1. A household cleaner has a pH of 12 and some fruit juice has a pH of 3.
   (a) What is the hydrogen ion concentration in each solution?
   (b) Compare the concentration of hydrogen ions in the fruit juice to that of the hydrogen ions in the cleaner. How many times more concentrated is the hydrogen ion in the juice than the cleaner?
2. What is the pH of each of the following water samples?
   (a) tap water: $[H^+_{aq}] = 1 \times 10^{-8}$ mol/L
   (b) pure water: $[H^+_{aq}] = 1 \times 10^{-7}$ mol/L
   (c) normal rainwater: $[H^+_{aq}] = 2.5 \times 10^{-6}$ mol/L
   (d) acid rain: $[H^+_{aq}] = 1.3 \times 10^{-4}$ mol/L
3. Common household vinegar has a pH of 2.4 and some pickling vinegar has a pH of 2.2.
   (a) Which vinegar solution is more acidic?
   (b) Which has a greater hydrogen ion concentration?
   (c) Calculate the hydrogen ion concentration in each solution.
4. A student tested the pH of 0.1 mol/L solutions of hydrochloric acid and acetic acid. The pH of the hydrochloric acid solution was 1.1 and the pH of the acetic acid was 2.9.
   (a) Explain the difference in pH.
   (b) Communicate the difference by writing ionization equations for each of the acids.
   (c) Which of the solutions deserves greater caution when being used? Why?
5. The pH of a cleaning solution was determined using a variety of technologies. Convert the pH into a molar concentration of hydrogen ions, with the correct certainty in the answer. Suggest what technology might have been used in each of these measurements.
   (a) pH = 10
   (b) pH = 9.8
   (c) pH = 9.84
   (d) pH = 9.836
6. Hydrangeas are garden shrubs that may produce blue, purple, or pink flowers. Research has indicated that the colour is dependent on the pH of the soil: blue at pH 5.0–5.5, purple at pH 5.5–6.0, and pink at pH 6.0–6.5. Convert the following expressions of acidity from pH to $[H^+_{(aq)}]$.
   (a) pH = 5.4 (blue)
   (b) pH = 5.72 (purple)

   Convert these concentrations to pH and predict the colour of the flower.
   (c) $[H^+_{(aq)}] = 5 \times 10^{-7}$ mol/L
   (d) $[H^+_{(aq)}] = 7.9 \times 10^{-6}$ mol/L

Applying Inquiry Skills

7. Write an Experimental Design to determine which of six provided acids are strong and which are weak.

(continued)

8. Design and conduct an investigation to test the concept that the pH of a weak acid changes predictably by dilution. Complete every section of the report: Prediction, Experimental Design, Materials, Procedure, Evidence, Analysis, and Evaluation. Before conducting the investigation, have your chemistry teacher authorize your design, materials, and procedure.

   Question
   What will be the increase in the pH of vinegar when a 1.0 mL sample is diluted 100 times?

9. A scientist wants to determine the pH of several toothpastes but must add water to the pastes in order to measure the pH with a pH meter probe. Critique this experimental design by supporting and defending the design and by suggesting an alternative design.

Making Connections

10. Use the Internet or other resources to find the effect of soil pH on plants. Report this information in a short Did You Know? column for a gardening magazine.

    Follow the links for Nelson Chemistry 11, 8.2.

GO TO www.science.nelson.com

11. Large restaurants give vegetables such as lettuce an acid wash before serving. Why would they follow this practice and which acid are they most likely to use? Provide your reasoning.

Reflecting

12. The term "weak" is sometimes applied to dilute solutions. Why do we have to be careful not to use "weak" in this context, when discussing acids?

8.3 Working with Solutions

Everyone deals with solutions on a daily basis. Tap water, soft drinks, air, gasoline, and alloys such as the "gold" used in jewellery are just a few common examples of homogeneous mixtures called solutions. Many foods contain acidic solutions. Cleaning solutions often contain bases. Sometimes we prepare solutions, for example, when we dissolve flavour crystals and sugar in water to make an inexpensive drink, or make a cup of tea, or dilute a concentrated cleaner in water. In most cases we pay some attention to concentration, although it may be measured only roughly, and adjust it to our requirements.

Some people regularly handle solutions as a part of their work. The kinds of solutions vary tremendously, depending on the occupation. However, the preparation and use of solutions is generally more technologically advanced and precise than we would encounter at home (**Figure 1**).

Figure 1
Working with solutions requires a variety of specialized glassware.

Practice

Applying Inquiry Skills

1. State the name and use of each piece of apparatus in **Figure 1**.

CAREER

Career Solutions

The training requirements for careers that involve solutions vary from high school chemistry for a job as a tree-planter to a Ph.D. degree in chemistry for a career in pure research chemistry.

Water-Quality Analyst

A water-quality analyst or technician in a water treatment plant works with aqueous solutions every day. Physical and chemical tests are routinely done to determine the treatment of the raw water and to monitor the quality of the final treated water. Many chemical tests (such as the analysis of dissolved iron ions, calcium ions, and chlorine) require the preparation of other reagent (reactant) solutions to conduct the tests. Both solution preparation and reactions in solution are important parts of the job of a water-quality analyst.

Chemistry Teacher

A chemistry or physical science teacher must have a knowledge of solutions, and be able to handle them safely. In this course and many of your previous science courses, you will have seen and used solutions on many occasions. At most schools, the teachers prepare the solutions that you use, plan the reactions that you do, and sometimes need to be very resourceful in cleaning some stains from glassware by reacting the stain with other chemicals such as acids and bases. Chemistry and other physical science teachers need a good understanding of solutions in order to prepare for lab activities. These teachers are in great demand in schools and universities.

Environmental Chemist

Environmental chemists often specialize in particular aspects of the air, water, or soil. Many of them use solutions as either reactants or samples in chemical analysis. The concentrations of these samples are usually critically important. This career requires a higher degree of chemistry training than technicians and teachers. To be an environmental researcher requires considerable perseverance and optimism as well as an ability to ask questions and design experiments. Some of the research involves understanding the components and processes in the environment and some research may focus on the nature and effects of pollution.

Practice

Making Connections

2. Choose one of the careers discussed that you might be interested in and use the Internet to research this career. What are the specific educational requirements? Does this occupation require certification by some organization? If so, state the organization. What are the job prospects in this area?
   Follow the links for Nelson Chemistry 11, 8.3.

GO TO www.science.nelson.com

8.4 Acid–Base Theories

Acids and bases can be distinguished by means of a variety of properties (**Table 1**). Some properties of acids and bases are more useful than others to a chemist, especially those that can be used as diagnostic tests, such as the litmus test.

Table 1: Empirical Properties of Acids and Bases

Acids	Bases
sour taste*	taste bitter and feel slippery*
turn blue litmus red	turn red litmus blue
have pH less than 7	have pH greater than 7
neutralize bases	neutralize acids
react with active metals to produce hydrogen gas	
react with carbonates to produce carbon dioxide	

*Note that for reasons of safety it is not appropriate to use taste or touch as diagnostic tests in the laboratory.

Many acids and bases are sold under common or traditional names. As you have learned, concentrated hydrochloric acid is sometimes sold as muriatic acid. Sodium hydroxide, called lye as a pure solid, has a variety of brand names when sold as a concentrated solution for cleaning plugged drains. Generic or "no-name" products often contain the same kind and quantity of active ingredients as brand name products. You can save time, trouble, and money by knowing that, in most cases, the chemical names of compounds used in home products must be given on the label. If you discover that your favourite brand of scale remover is an acetic acid solution, you may be able to substitute vinegar to do the same job less expensively, assuming that the concentrations are similar.

Strong and Weak Acids

Are all acids similar in their reactivity and their pH? Do acidic solutions at the same concentration and temperature possess acidic properties to the same degree? Not surprisingly, the answer to this question is that each acid is unique. The pH of an acid may be only slightly less than 7, or it may be as low as –1. We discussed in Section 8.1 that solutions of strong acids have a much greater conductivity than those of weak acids and the difference can be explained by percentage ionization. As you might suspect, the percentage ionization has an effect on the pH of acid solutions.

We can explain the differences in properties between strong and weak acids using the Arrhenius theory and percentage ionization. For example, hydrogen chloride is a strong acid because it is believed to ionize completely (more than 99%) in water. The high concentration of $H^+_{(aq)}$ ions gives the solution strong acid properties and a low pH.

$$HCl_{(aq)} \xrightarrow{>99\%} H^+_{(aq)} + Cl^-_{(aq)}$$

This means that for each mole of hydrogen chloride dissolved, about one mole of hydrogen ions is produced.

There are relatively few strong acids: Hydrochloric, sulfuric, and nitric acids are the most common.

A weak acid is an acid that ionizes partially in water. Measurements of pH indicate that most weak acids ionize less than 50%. Acetic acid, a common weak acid, is only 1.3% ionized in solution at 25°C and 0.10 mol/L concentration. The relatively low concentration of $H^+_{(aq)}$ ions gives the solution weaker acid properties and a pH closer to 7.

$$HC_2H_3O_{2(aq)} \xrightarrow{1.3\%} H^+_{(aq)} + C_2H_3O^-_{2(aq)}$$

For each mole of acetic acid dissolved, only 0.013 mol of $H^+_{(aq)}$ ions is produced.

When we observe chemical reactions involving acids we can see that some acids (such as acetic acid), although they react in the same manner and amount as other acids (such as hydrochloric acid), do not react as quickly. This is why weak acids are generally so much safer to handle, and even to eat or drink, than strong acids. Most of the acids you are likely to encounter are classed as weak acids (**Figure 1**).

The concepts of strong and weak acids were developed to describe, explain, and predict these differences in properties of acids.

Figure 1
Many naturally occurring acids are weak organic (carbon chain) acids. Methanoic (formic) acid is found in the stingers of certain ants, butanoic acid in rancid butter, citric acid in citrus fruits such as lemons and oranges, oxalic acid in tomatoes, and long-chain fatty acids, such as stearic acid, in animal fats.

SUMMARY Properties of Strong and Weak Acids of Equal Concentration

Property	Strong Acid	Weak Acid
pH	<<7	<7
Ionization	>99%	<50%
Rate of reaction	fast	slow
Corrosion	fast	slow

Practice

Understanding Concepts

1. Which empirical property listed in **Table 1** is not a diagnostic test used in a chemistry laboratory? Is there a situation where knowledge of this property might be useful?
2. Strong and weak acids can be differentiated by their rates of reaction. Complete and balance the following chemical equations. Predict whether each reaction will be fast or slow.
   (a) $Mg_{(s)} + HCl_{(aq)} \rightarrow$
   (b) $Mg_{(s)} + HC_2H_3O_{2(aq)} \rightarrow$
   (c) $HCl_{(aq)} + CaCO_{3(s)} \rightarrow$
   (d) $HC_2H_3O_{2(aq)} + CaCO_{3(s)} \rightarrow$
3. Which property listed in **Table 1** would be the best to distinguish between strong and weak acids? Justify your choice.
4. What is the theoretical distinction between strong and weak acids?
5. Suppose 100 molecules of a strong acid are dissolved to make a litre of solution and 100 molecules of a weak acid are also dissolved to make a litre of solution in a different container. Assume a 2% ionization for the weak acid.
   (a) What is the concentration of hydrogen ions for each solution, expressed as the number of hydrogen ions per litre?
   (b) How does your answer to (a) explain the difference in properties of strong and weak acids?

Answers

5. (a) 100 H^+/L
   2 H^+/L

Applying Inquiry Skills

6. Bases can also be classified as strong or weak. Predict some differences that you might expect to observe between strong and weak bases. Outline an Experimental Design (including controlled variables) to test your Predictions.
7. Complete the **Experimental Design** and the **Analysis** of the report for the following investigation that determines the relative strength of some acids.

   Question

   What is the order of several common acids in terms of decreasing strength?

   Experimental Design

   (a) Write a description of a design that would produce the Evidence listed in **Table 2**. Include the independent, dependent, and controlled variables in your description.

   Evidence

   Table 2: Acidity of 0.10 mol/L Acids

Acid solution	Formula	pH
hydrochloric acid	$HCl_{(aq)}$	1.00
acetic (ethanoic) acid	$HC_2H_3O_{2(aq)}$	2.89
hydrofluoric acid	$HF_{(aq)}$	2.11
methanoic acid	$HCHO_{2(aq)}$	2.38
nitric acid	$HNO_{3(aq)}$	1.00
hydrocyanic acid	$HCN_{(aq)}$	5.15

   Analysis

   (b) List the acids in decreasing strength.
8. You are given six unlabelled solutions, each containing the same concentration of one of the following six substances: $HCl_{(aq)}$, $HC_2H_3O_{2(aq)}$, $NaCl_{(aq)}$, $C_{12}H_{22}O_{11(aq)}$, $Ba(OH)_{2(aq)}$, and $KOH_{(aq)}$. Your job is to identify each solution. Write an Experimental Design including the specific tests that you would use in your qualitative analysis. You may present your answer as a paragraph, a table of expected evidence, or a flow chart.

Making Connections

9. In the media, especially movies, acids are often portrayed as dangerous, with the ability to "burn through" or "eat away" almost anything. Is this accurate? Justify your answer with personal experience, examples, and explanations. What acids are the most dangerous? How should the media more accurately portray the degree of reactivity of acids?
10. What is acid deposition? What are the typical acids that may be present? Which ones are strong and which are weak? Is it possible to predict which acids have a greater effect in the environment? Why or why not?

    Follow the links for Nelson Chemistry 11, 8.4.

GO TO www.science.nelson.com

Reflecting

11. In a 0.01 mol/L solution of an acid, what is the maximum concentration of H^+ ions? What further information would allow you to give a more accurate answer?

The Arrhenius Concept of Acids and Bases

In 1887 Svante Arrhenius created a theory of ions to explain the electrical conductivity of solutions. Arrhenius explained that ionic compounds form solutions that conduct electricity because these compounds dissociate as they dissolve to release an anion and a cation. For example, potassium hydrogen sulfate forms an electrically conductive solution because it dissolves as two ions.

$$KHSO_{4(s)} \rightarrow K^{+}_{(aq)} \text{ or } HSO^{-}_{4(aq)}$$

Scientists had previously agreed that acids were hydrogen compounds. Arrhenius added to this theory by suggesting that acids are hydrogen compounds that ionize to increase the hydrogen ion concentration of a solution. For example, hydrogen chloride gas dissolves in water and ionizes almost completely to increase the hydrogen ion concentration.

$$HCl_{(g)} \rightarrow H^{+}_{(aq)} + Cl^{-}_{(aq)}$$

Arrhenius was also able to explain in a theoretical way why bases have their characteristic properties. He suggested that bases are ionic hydroxides that dissolve in water to increase the hydroxide ion concentration of the solution. For example, potassium hydroxide dissociates in water to increase the hydroxide ion concentration.

$$KOH_{(s)} \rightarrow K^{+}_{(aq)} + OH^{-}_{(aq)}$$

SUMMARY The Arrhenius Theoretical Definitions

acid $\rightarrow H^{+}_{(aq)}$ + anion

base $\rightarrow$ cation + $OH^{-}_{(aq)}$

other ionic compounds $\rightarrow$ ions but no $H^{+}_{(aq)}$ or $OH^{-}_{(aq)}$

Investigation 8.4.1

Testing Arrhenius' Acid–Base Definitions

INQUIRY SKILLS

- ○ Questioning
- ○ Hypothesizing
- ● Predicting
- ● Planning
- ● Conducting
- ● Recording
- ● Analyzing
- ● Evaluating
- ● Communicating

The purpose of this investigation is to test the Arrhenius definitions of acids and bases. As part of this investigation, you are asked to make a Prediction. Do not base your Prediction on experience or a personal hypothesis—use only the Arrhenius theoretical definitions. Assume that the Arrhenius concept restricts dissociation and ionization: Bases dissociate to produce $OH^{-}_{(aq)}$ and a cation; and acids ionize to produce $H^{+}_{(aq)}$ and an anion.

You are expected to design an experiment to classify a number of common substances in solution (see **Materials**) as acidic, basic, or neutral. (Refer to Chapter 2 if you need help writing the formulas for each of the substances.) Complete a report, including the **Prediction**, **Experimental Design**, **Analysis**, and **Evaluation**.

Question

Which of the chemicals tested may be classified as an acid, a base, or neutral?

Prediction

(a) Based on Arrhenius' definitions, predict which of the chemicals in the Materials list will test as an acid, which as a base, and which as neutral.

Experimental Design

(b) Create an Experimental Design. Be sure to identify all variables, including any controls. Your experiment should involve qualitative analysis, incorporating one or more diagnostic tests. You should also note any necessary safety or disposal precautions.

(c) Write up your Procedure. Obtain your teacher's approval before conducting your experiment.

Materials

Hydrochloric acid and sodium hydroxide are irritants. Wear eye protection and a laboratory apron.

lab apron
eye protection
aqueous 0.10 mol/L solutions of:
- hydrogen chloride (a gas in solution)
- hydrogen acetate (vinegar)
- sodium hydroxide (lye, caustic soda)
- calcium hydroxide (slaked lime)
- ammonia (cleaning agent)
- sodium carbonate (washing soda, soda ash)
- sodium hydrogen carbonate (baking soda)
- sodium hydrogen sulfate (toilet bowl cleaner)
- calcium oxide (lime)
- carbon dioxide (carbonated beverage)
- aluminum nitrate (salt solution)
- sodium nitrate (fertilizer)

conductivity apparatus
blue litmus paper
red litmus paper
any other materials necessary for diagnostic tests

Procedure

1. Conduct your experiment.

Analysis

(d) Answer the Question: Which of the substances tested may be classified as an acid, a base, or neutral?

Evaluation

(e) Evaluate the validity of your Experimental Design, your Prediction, and the Arrhenius definition it was based on.

Revision of Arrhenius' Definitions

Arrhenius' definitions cannot always predict whether a substance is an acid or a base. Using Arrhenius' definitions, you would probably correctly predict that $HCl_{(aq)}$ and $HC_2H_3O_{2(aq)}$ are acids; that $NaOH_{(aq)}$ and $Ca(OH)_{2(aq)}$ are bases;

and that $NaNO_{3(aq)}$ is neutral. However, by Arrhenius's definitions, we would predict all of the following compounds to be neutral, but they are not:

- compounds of **hydrogen polyatomic ions** ($NaHCO_{3(aq)}$ and $NaHSO_{4(aq)}$)
- oxides of metals and nonmetals ($CaO_{(aq)}$ and $CO_{2(g)}$)
- compounds that are neither oxides nor hydroxides (e.g., $NH_{3(aq)}$ and $Na_2CO_{3(aq)}$), but yet are bases
- compounds that contain no hydrogen (e.g., $Al(NO_3)_{3(aq)}$), but yet are acids

hydrogen polyatomic ion: a bi-ion; a polyatomic ion with an available hydrogen (e.g., hydrogen carbonate (bicarbonate) ion, hydrogen sulfite (bisulfite) ion)

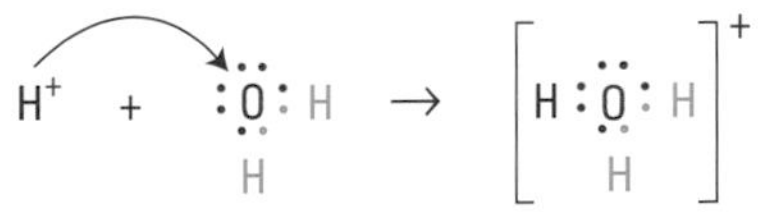

Clearly, the Arrhenius theoretical definitions of acid and base need to be revised or replaced.

Figure 3
By passing infrared light through solutions of acids, Paul Giguère of the Université Laval, Quebec City, obtained clear evidence for the existence of hydronium ions in solution.

A theoretical concept has two major purposes: to explain current evidence and to predict the results of new experiments. While the first purpose is useful, it is not valued as much as the ability to predict results. Theoretical progress is made when theories not only explain what is known but also allow valid predictions to be made about new situations.

We need to revise Arrhenius' acid–base definitions to explain the exceptions listed above. The new theory involves two key ideas: collisions with water molecules and the nature of the hydrogen ion. Since all substances tested are in aqueous solution, then particles will constantly be colliding with, and may also react with, the water molecules present.

It is highly unlikely that the particle we call an aqueous hydrogen ion, $H^+_{(aq)}$, actually exists in an acidic solution. If such a particle were to come near polar water molecules, it would bond strongly to one or more of the molecules (**Figure 2**), that is, it would be hydrated. There is no evidence for the existence of unhydrated hydrogen ions in aqueous solution. However, the Canadian scientist Paul Giguère has done experiments that provide clear evidence for the existence of hydrated protons (**Figure 3**). The simplest representation of a hydrated proton is $H_3O^+_{(aq)}$, commonly called the **hydronium ion** (**Figure 4**).

$$H^+ + :\ddot{O}:H \rightarrow [H:\ddot{O}:H]^+$$

Figure 2
The Lewis (electron dot) diagram for a hydrogen ion has no electrons. A water molecule is believed to have two lone pairs of electrons, as shown in its Lewis diagram. The hydrogen ion (proton) is believed to bond to one of these lone pairs of electrons to produce the H_3O^+ ion.

hydronium ion: a hydrated hydrogen ion (proton), conventionally represented as $H_3O^+_{(aq)}$

Figure 4
The hydronium ion is represented as a pyramidal structure. The oxygen atom is the apex and the three identical hydrogen atoms form the base of the pyramid.

We can now explain the formation of acidic solutions by strong acids such as $HCl_{(aq)}$ as a reaction with water, forming hydronium ions (**Figure 5**).

$$HCl_{(g)} + H_2O_{(l)} \xrightarrow{>99\%} H_3O^+_{(aq)} + Cl^-_{(aq)}$$

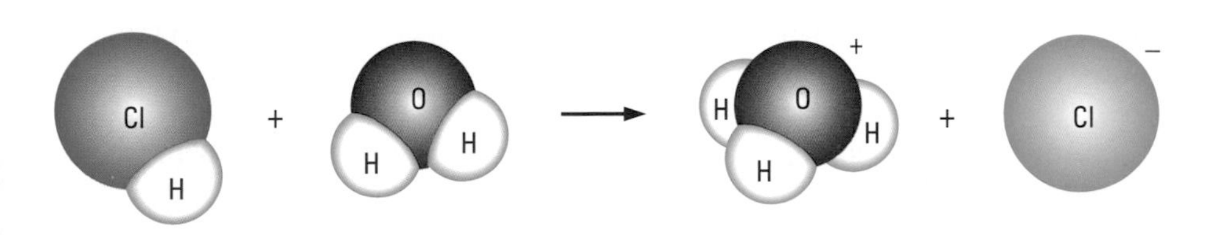

Figure 5
When gaseous hydrogen chloride dissolves in water, the HCl molecules are thought to collide and react with water molecules to form hydronium and chloride ions.

A strong acid, such as HCl, is considered to react completely with water. In other words, the collisions with water molecules are very successful, producing a 100% reaction. What about weak acids? Because they have lesser acidic properties,

DID YOU KNOW ?

pH

How does this new theory about acids affect our concept of pH? Fortunately, the mathematical definition of pH works equally well with hydronium ions as it did with hydrogen ions.

there must be fewer hydronium ions produced from the same volume and concentration of solution compared with strong acids. Therefore, the collisions of weak acid molecules with water cannot be very successful. Based on pH measurements, 0.10 mol/L acetic acid, a common weak acid, is only successful in forming hydronium ions in 1.3% of its collisions with water molecules.

$$HC_2H_3O_{2(aq)} + H_2O_{(l)} \xrightarrow{1.3\%} H_3O^+_{(aq)} + C_2H_3O^-_{2(aq)}$$

In general, *acidic solutions form when substances react with water to form hydronium ions.*

The concept of acids reacting with water to produce hydronium ions is a small adjustment in thinking when explaining or predicting the behaviour of typical acids. In most contexts, we can think of acids as either ionizing to produce hydrogen ions or reacting with water to produce hydronium ions.

Strong and Weak Bases

Evidence indicates that there are both strong bases (e.g., sodium hydroxide) and weak bases (e.g., ammonia). For equal concentrations of solutions, strong bases have high electrical conductivity and very high pH (>>7), whereas weak bases have low electrical conductivity and pH closer to 7. We can explain the behaviour of strong bases: They dissociate to increase the hydroxide ion concentration in an aqueous solution. Further evidence indicates that all ionic hydroxides are **strong bases:** 100% of the dissolved ionic hydroxides dissociates to release hydroxide ions.

strong base: (according to the Arrhenius theory and the "reaction-with-water" theory) an ionic hydroxide that dissociates 100% in water to produce hydroxide ions

weak base: (according to the "reaction-with-water" theory) a chemical that reacts less than 50% with water to produce hydroxide ions

What about **weak bases**? How can we explain their properties? The pure compounds (e.g., $NH_{3(g)}$) do not contain hydroxide ions, so they cannot dissociate to release hydroxide ions. Nevertheless, solutions of weak bases appear to contain hydroxide ions in a higher concentration than does pure water. Where do they come from? Clearly, this question cannot be answered by the Arrhenius definition of bases. We need to revise his theory to include a new concept: that weak base molecules or ions react with water to produce hydroxide ions. This remains consistent with the explanation for strong bases and for strong and weak acids. Weak bases do not react 100% with water. Evidence indicates that they commonly react less than 10%. This means that they produce fewer hydroxide ions than a similar amount of a strong base, which accounts for the weaker basic properties of weak bases.

Recall from Section 8.1 that ionic hydroxides produce basic solutions by simple dissociation. We know that ionic hydroxides, such as barium hydroxide, are strong bases.

$$Ba(OH)_{2(s)} \rightarrow Ba^{2+}_{(aq)} + 2\ OH^-_{(aq)} \qquad \text{(a strong base)}$$

Here, there is no need to consider a reaction with water because we know that ionic hydroxides, such as $Ba(OH)_2$, dissociate to produce hydroxide ions. However, there are many common examples of bases that are not ionic hydroxides, such as ammonia (window cleaner) and sodium carbonate (washing soda). Most bases, other than soluble ionic hydroxides, are weak bases. Weak bases may be either ionic or molecular compounds in their pure state.

Ammonia and sodium carbonate each form basic aqueous solutions as demonstrated by a litmus paper test. This equation for ammonia shows the theory to explain the evidence:

$$NH_{3(aq)} + H_2O_{(l)} \xrightarrow{<50\%} OH^-_{(aq)} + NH^+_{4(aq)} \qquad \text{(a weak base)}$$

The presence of hydroxide ions explains the basic solution, and the less than 100% reaction explains the weak base properties. (Note that both atoms and charge are conserved in the balanced equation.)

Sodium carbonate is an ionic compound with high solubility. According to the Arrhenius theory, sodium carbonate dissociates in water to produce aqueous ions of sodium and carbonate.

$$Na_2CO_{3(s)} \rightarrow 2\ Na^+_{(aq)} + CO^{2-}_{3(aq)}$$

The sodium ion cannot be responsible for the basic properties of the solution, because many sodium compounds (e.g., $NaCl_{(aq)}$) form neutral solutions. The basic character of carbonate solutions can be explained as resulting from their reaction with water.

$$CO^{2-}_{3(aq)} + H_2O_{(l)} \xrightarrow{<50\%} OH^-_{(aq)} + HCO^-_{3(aq)}$$

Note again that hydroxide ions explain the basic properties and that atoms and charge are conserved in the balanced equation.

We now have explanations for the production of hydroxide ions by bases: Strong bases (commonly ionic hydroxides) dissociate to produce hydroxide ions; and weak bases react with water to increase the hydroxide ion concentration. This theory is sometimes called the revised Arrhenius theory.

Sample Problem 1

A forensic technician tested the pH of a sodium cyanide solution and found that it had a pH greater than 7. Explain this evidence using chemical equations.

Solution

$$NaCN_{(s)} \rightarrow Na^+_{(aq)} + CN^-_{(aq)}$$

$$CN^-_{(aq)} + H_2O_{(l)} \xrightarrow{<50\%} OH^-_{(aq)} + HCN_{(aq)}$$

The cyanide ion produces hydroxide ions in reaction with water, so is a weak base. As a weak base, we expect the reaction to be less than 50% and the solution to have a pH greater than 7 but not, say, greater than 13.

SUMMARY Strong and Weak Acids and Bases

	Strong acids	Weak acids	Strong bases	Weak bases
empirical	very low pH (<<7)	medium to low pH (<7)	very high pH (>>7)	medium to high pH (>7)
	high conductivity	low conductivity	high conductivity	low conductivity
	fast reaction rate	slow reaction rate	fast reaction rate	slow reaction rate
solute	molecular	molecular and polyatomic ion	ionic hydroxide	molecular and polyatomic ion*
theoretical (Arrhenius)	completely ionized to form $H^+_{(aq)}$	partially ionized to form $H^+_{(aq)}$	completely dissociated into $OH^-_{(aq)}$	—
theoretical (revised Arrhenius)	completely reacted with water to form $H_3O^+_{(aq)}$	partially reacted with water to form $H_3O^+_{(aq)}$	completely dissociated to form form $OH^-_{(aq)}$	partially reacted with water to form $OH^-_{(aq)}$

* Except the hydroxide ion, $OH^-_{(aq)}$

Figure 6
Johannes Brønsted created new theoretical definitions for acids and bases based upon proton transfer.

Practice

Understanding Concepts

12. What were some of the early ideas about the chemistry of acids? What evidence eventually showed that these ideas were false?
13. How well does the original Arrhenius theory predict and explain acids and bases?
14. What is the more recent replacement for the idea of a hydrogen ion causing acidic properties? State its name and formula.
15. Write chemical equations to explain the pH of a 0.1 mol/L solution of each substance.
    (a) $HCN_{(aq)}$; pH = 5
    (b) $HNO_{3(aq)}$; pH = 1
    (c) $Na_2SO_{4(aq)}$; pH = 8
    (d) $Sr(OH)_{2(aq)}$; pH = 13
16. In the previous question, identify the strong and weak acids and bases.

The Brønsted–Lowry Concept

Acid and base definitions, revised to include the ideas of the hydronium ion and reaction with water, are more effective in describing, explaining, and predicting the behaviour of acids and bases than are Arrhenius' original definitions. However, chemical research has shown that even these revised definitions are still too restrictive. Reactions of acids and bases do not always involve water. Also, evidence indicates that some entities that form basic solutions (such as $HCO^-_{3(aq)}$) can actually neutralize the solution of a stronger base. A broader concept is needed to describe, explain, and predict these properties of acids and bases.

New theories in science usually result from looking at the evidence in a way that has not occurred to other observers. A new approach to acids and bases was developed in 1923 by Johannes Brønsted (1879–1947) of Denmark (**Figure 6**) and independently by Thomas Lowry (1874–1936) of England. These scientists focused on the role of an acid and a base in a reaction rather than on the acidic or basic properties of their aqueous solutions. An acid, such as aqueous hydrogen chloride, functions in a way opposite to a base, such as aqueous ammonia. According to the Brønsted-Lowry concept, hydrogen chloride donates a proton (H^+) to a water molecule,

$$HCl_{(aq)} + H_2O_{(l)} \rightarrow H_3O^+_{(aq)} + Cl^-_{(aq)}$$

(H^+ transferred from HCl to H_2O)

and ammonia accepts a proton from a water molecule.

$$NH_{3(aq)} + H_2O_{(l)} \rightarrow OH^-_{(aq)} + NH^+_{4(aq)}$$

(H^+ transferred from H_2O to NH_3)

Water does not have to be one of the reactants. For example, the hydronium ions present in a hydrochloric acid solution can react directly with dissolved ammonia molecules.

$$\underset{\text{acid}}{H_3O^+_{(aq)}} + \underset{\text{base}}{NH_{3(aq)}} \rightarrow H_2O_{(l)} + NH^+_{4(aq)}$$

(H^+ transferred from H_3O^+ to NH_3)

We can describe this reaction as NH_3 molecules removing protons from H_3O^+ ions. Hydronium ions act as the **acid**, and ammonia molecules act as the **base**. Water is present as the solvent, but not as a primary reactant. In fact, water does not even have to be present, as evidenced by the reaction of hydrogen chloride and ammonia gases (**Figure 7**).

Figure 7
One hazard of handling concentrated solutions of ammonia and hydrochloric acid is gas fumes. The photograph shows ammonia gas and hydrogen chloride gas escaping from their open bottles, and reacting to form a white cloud of very tiny crystals of $NH_4Cl_{(s)}$.

$$\underset{\text{acid}}{HCl_{(g)}} + \underset{\text{base}}{NH_{3(g)}} \rightarrow NH_4Cl_{(s)} \quad (H^+ \text{ transferred from } HCl \text{ to } NH_3)$$

A substance can be classified as a Brønsted-Lowry acid or base only for a specific reaction. It is not a general property of a substance. This point is important—a substance may gain protons in one reaction, but lose them in another reaction with another substance. (For example, in the reaction of HCl with water shown above, water acts as the Brønsted-Lowry base; whereas, in the reaction of NH_3 with water, water acts as the Brønsted-Lowry acid.) A substance that appears to act as a Brønsted-Lowry acid in some reactions and as a Brønsted-Lowry base in other reactions is called **amphiprotic**. The hydrogen carbonate ion ($HCO^-_{3(aq)}$) in baking soda (**Figure 8**) is amphiprotic, like every other hydrogen polyatomic ion. Hydrogen polyatomic ions, as their name suggests, are polyatomic ions containing hydrogen. Examples of amphiprotic substances include $HCO^-_{3(aq)}$, $H_2O_{(l)}$, $HSO^-_{3(aq)}$, $H_2PO^-_{4(aq)}$, and $HPO^{2-}_{4(aq)}$.

Note that amphiprotic entities can either gain or lose a proton, as shown by the following reactions. First let's see what happens when the bicarbonate ion is added to the solution of a strong acid, which will contain hydronium ions.

acid: (according to the Brønsted-Lowry concept) a proton donor

base: (according to the Brønsted-Lowry concept) a proton acceptor

amphiprotic: a substance capable of acting as an acid or a base in different chemical reactions; an entity that can gain or lose a proton (sometimes called amphoteric)

$$\underset{\text{base}}{HCO^-_{3(aq)}} + \underset{\text{acid}}{H_3O^+_{(aq)}} \rightarrow H_2CO_{3(aq)} + H_2O_{(l)} \quad (H^+ \text{ transferred from } H_3O^+ \text{ to } HCO_3^-)$$

(neutralizes a strong acid)

Now let's look at the reaction of the bicarbonate ion with the solution of a strong base, which will contain hydroxide ions.

$$\underset{\text{acid}}{HCO^-_{3(aq)}} + \underset{\text{base}}{OH^-_{(aq)}} \rightarrow CO^{2-}_{3(aq)} + H_2O_{(l)} \quad (H^+ \text{ transferred from } HCO_3^- \text{ to } OH^-)$$

(neutralizes a strong base)

In both cases, the bicarbonate ion moves the pH of the solutions toward 7.

According to the Brønsted-Lowry concept, acid–base reactions involve the transfer of a proton. Therefore, the products formed in these reactions must differ from the reactants by a proton (H^+). If you look again at the equations above, you can see that this is true.

As another example, when acetic acid reacts with water, an acidic solution (containing hydronium ions) is formed.

Figure 8
Baking soda (sodium hydrogen carbonate, $NaHCO_3$) is a common household substance that is useful for many purposes other than baking. You can use it to neutralize both spilled acids and bases, and it can also be used as an extinguisher for small fires.

$$\underset{\text{acid}}{HC_2H_3O_{2(aq)}} + \underset{\text{base}}{H_2O_{(l)}} \rightarrow C_2H_3O^-_{2(aq)} + H_3O^+_{(aq)} \quad (H^+ \text{ transferred from } HC_2H_3O_2 \text{ to } H_2O)$$

conjugate base: the base formed by removing a proton (H^+) from an acid

conjugate acid: the acid formed by adding a proton (H^+) to a base

The acetate ion product is simply what is left after an acetic acid molecule loses its proton. The hydronium ion product is what is formed as a result of a water molecule gaining a proton. Any proton that is lost can, in principle, be regained and any proton that is gained can be lost in some other reaction. Therefore, we can consider the acetate ion to be a potential Brønsted-Lowry base: It could act as a base in another reaction. A product formed as a result of an acid losing a proton is called a **conjugate base**. Similarly, the hydronium ion is a potential Brønsted-Lowry acid: It could act as an acid in another reaction. A product resulting from a base gaining a proton is called a **conjugate acid**.

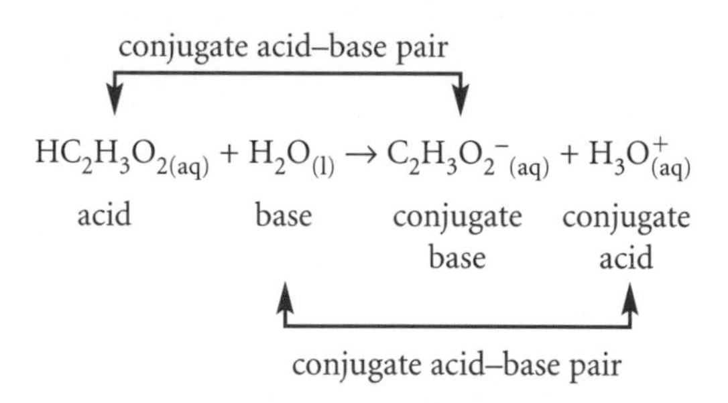

conjugate acid–base pair: an acid–base pair that differs by one proton (H^+)

A pair of substances that differ only by a proton is called a **conjugate acid–base pair** (Table 3).

Table 3: Some Examples of Conjugate Acid–Base Pairs

Conjugate acid		Conjugate base
$H_2O_{(l)}$	and	$OH^-_{(aq)}$
$H_3O^+_{(aq)}$	and	$H_2O_{(l)}$
$HC_2H_3O_{2(aq)}$	and	$C_2H_3O^-_{2(aq)}$

SUMMARY Definitions of Acids and Bases

Arrhenius Definitions

- An acid ionizes in water to increase the hydrogen ion concentration.
- A base dissociates in water to increase the hydroxide ion concentration.
- A neutralization reaction involves the reaction of a hydrogen ion with a hydroxide ion.

Revised Arrhenius Definitions

- An acid reacts with water to increase the hydronium ion concentration.
- A base reacts with water to increase the hydroxide ion concentration.
- A neutralization reaction involves the reaction of a hydronium ion with a hydroxide ion.

Brønsted–Lowry Definitions

- An acid is a proton donor.
- A base is a proton acceptor.
- An acid–base neutralization reaction involves the transfer of one proton from the strongest acid present to the strongest base present.
- An amphiprotic substance is one that appears to act as a Brønsted-Lowry acid in some reactions and as a Brønsted-Lowry base in other reactions.
- A conjugate acid–base pair consists of two substances that differ only by one proton.

Practice

Understanding Concepts

17. According to the Brønsted-Lowry definitions, how are acids and bases different?
18. Classify each reactant in the following equations as a Brønsted-Lowry acid or base.
    (a) $HF_{(aq)} + SO_{3(aq)}^{2-} \rightarrow F_{(aq)}^{-} + HSO_{3(aq)}^{-}$
    (b) $CO_{3(aq)}^{2-} + HC_2H_3O_{2(aq)} \rightarrow C_2H_3O_{2(aq)}^{-} + HCO_{3(aq)}^{-}$
    (c) $H_3PO_{4(aq)} + OCl_{(aq)}^{-} \rightarrow H_2PO_{4(aq)}^{-} + HOCl_{(aq)}$
19. An aqueous hydrogen sulfate ion acts as the Brønsted-Lowry acid in the neutralization of a solution of hydrogen carbonate ions.
    (a) Write the chemical equation.
    (b) Identify two conjugate acid–base pairs.
20. What restrictions to acid–base reactions do the Brønsted-Lowry definitions remove?

Figure 9
Antoine Lavoisier

Changing Ideas on Acids and Bases

Usually chemists discover the empirical properties of substances long before a theory is developed to describe, explain, and predict their behaviour. For example, several of the distinguishing properties of acids and bases were known by the middle of the 17th century. Additional properties, such as pH and the nature of acid–base reactions, were discovered by the early 20th century.

Let's take an overview of the developing theory of acids and bases. It is a story that took place over several hundred years, thanks to the innovative thoughts and painstaking laboratory investigations of many great scientists.

Antoine Lavoisier (1743–1794) (**Figure 9**) assumed that oxygen was responsible for acid properties and that acids were combinations of oxides and water. For example, sulfuric acid, H_2SO_4, was described as hydrated sulfur trioxide, $SO_3{\cdot}H_2O$. There were immediate problems with this theory because some oxide solutions, such as CaO, are basic, and several acids, such as HCl, are not formed from oxides. This evidence led to the rejection of the oxygen theory, although we still use the generalization that nonmetallic oxides (e.g., SO_3) form acidic solutions.

Figure 10
Sir Humphry Davy

Figure 11
Justus von Liebig

Sir Humphry Davy (1778–1829) (**Figure 10**) advanced a theory that the presence of hydrogen gave a compound acidic properties. Justus von Liebig (1803–1873) (**Figure 11**) later expanded this theory to include the idea that acids are salts (compounds) of hydrogen. This meant that acids could be thought of as ionic compounds in which hydrogen had replaced the metal ion. However, this theory did not explain why many compounds containing hydrogen have neutral properties (e.g., CH_4) or basic properties (e.g., NH_3).

Svante Arrhenius (1859–1927) (**Figure 12**) developed a theory in 1887 that provided the first useful theoretical definition of acids and bases. He described acids as substances that ionize in aqueous solution to form hydrogen ions, and bases as substances that dissociate to form hydroxide ions in solution. This theory explained the process of neutralization by assuming that $H_{(aq)}^{+}$ and $OH_{(aq)}^{-}$ ions combine to form $H_2O_{(l)}$. The various strengths of acids were explained in terms of the degree (percentage) of ionization, but Arrhenius's theory is limited to aqueous solutions and cannot explain the properties of many common substances.

Figure 12
Svante Arrhenius

DID YOU KNOW ?

Prove Me Wrong!

In science, no theory can be proven. Well-established, accepted theories have a substantial quantity of supporting evidence. On the other hand, a theory can be disproven by a single, significant, reproducible observation. In Einstein's words: "No amount of experimentation can ever prove me right; a single experiment can prove me wrong."

Paul Giguère of the Université Laval, Quebec, found evidence that, in a solution, hydrogen ions are bonded to water molecules. The simplest representation of an aqueous hydrogen ion is $H_3O^+_{(aq)}$, commonly known as a hydronium ion. The concept of the hydronium ion was used to revise the Arrhenius definitions of acids and bases. Now acids could be described as substances that react with water to form hydronium ions, and bases could be described as substances that dissociate or react with water to form hydroxide ions. This revised Arrhenius theory is still limited to aqueous solutions but it does provide an explanation for the properties of aqueous solutions of nonmetal oxides, metal oxides, and polyatomic anions.

Johannes Brønsted (1879–1947) of Denmark and Thomas Lowry (1874–1936) of England independently developed a theory that focused on the role of acids and bases in a reaction rather than on the properties of their aqueous solutions. They defined acids as substances that donate protons (H^+) and bases as substances that accept protons in a chemical reaction. In the Brønsted-Lowry concept, a substance can only be defined as an acid or a base for a specific reaction. Ions such as hydrogen carbonate, $HCO^-_{3(aq)}$, or hydrogen sulfite, $HSO^-_{3(aq)}$, can act as acids in one reaction and as bases in another.

Changes in Knowledge

History indicates that it is unwise to assume that any scientific concept is the final word. Whenever scientists assume that they understand a subject, two things usually happen. Conceptual knowledge tends to remain static for a while, because little falsifying evidence exists or because any falsifying evidence is ignored. Then, when enough falsifying evidence accumulates, a revolution in thinking occurs within the scientific community in which the current concept is drastically revised or entirely replaced. The experience of the Swedish chemist Svante Arrhenius gives some insight into the difficulty scientists have in getting new ideas accepted.

While Arrhenius was attending the University of Uppsala near his home, he became intrigued by the question of why some aqueous solutions conduct electricity, but others do not. This problem had puzzled chemists ever since Sir Humphry Davy and Michael Faraday experimented over half a century earlier by passing electric currents through chemical substances.

Faraday believed that an electric current produces particles of electricity, which he called ions, in some solutions. He could not explain what ions were, or why they did not form in aqueous sugar or alcohol solutions.

As a university student, Arrhenius noticed that conducting solutions differed from non-conducting solutions in terms of another important property. The freezing point of any aqueous solution is lower than the freezing point of pure water; the more solute that is dissolved in the water, the more the freezing point is lowered (depressed). Arrhenius found that the freezing point depression of electrolytes in solution was always two or three times lower than that of non-electrolytes, in solutions of the same concentration. He concluded that when a solution such as pure table salt, NaCl, dissolves, it does not separate into NaCl molecules in solution but rather into two types of particles. Since the NaCl solution also conducts electricity, he reasoned that the particles must be electrically charged. In Arrhenius' view, the conductivity and freezing point evidence indicated that pure substances that form electrolytes are composed of charged ions, not neutral atoms. The stage was now set for a scientific controversy. Faraday was an established, respected scientist and his explanation agreed with Dalton's model of indivisible, neutral atoms. Arrhenius was an unknown university student and his theory contradicted Dalton's widely accepted model.

Despite strong supporting evidence, Arrhenius's creative idea was rejected by most of the scientific community, including his teachers. When Arrhenius presented his theory and its supporting evidence as part of his doctoral thesis, the examiners questioned him for four gruelling hours. They grudgingly passed him, but with the lowest possible mark.

For over a decade, only a few people supported Arrhenius' theory. Gradually, more supporting evidence accumulated, including J.J. Thomson's discovery of the electron in 1897. Soon, Arrhenius's theory of ions became widely accepted as the simplest and most logical explanation of the nature of electroytes. In 1903 Arrhenius won the Nobel Prize for the same thesis that had nearly failed him in his PhD examination years earlier.

Arrhenius' struggle to have his ideas accepted is not so unusual. Ideally, scientists are completely open-minded, but they are people, and many people resist change. We are sometimes reluctant to accept new ideas that conflict radically with familiar ones.

DID YOU KNOW ?

Scientific Concepts

"Creating a new theory is not like destroying an old barn and erecting a skyscraper in its place. It is rather like climbing a mountain, gaining new and wider views, discovering unexpected connections between our starting point and its rich environment. But the point from which we started out still exists and can be seen, although it appears smaller and forms a tiny part of our broad view gained by the mastery of the obstacles on our adventurous way up."

Albert Einstein (1879–1955)
German-born American theoretical physicist

Explore an Issue

Role Play: Evaluating New Ideas in Science

You are a member of the PhD examination committee for Svante Arrhenius. Is his experimental work reliable? Were his experiments well designed and carefully repeated? Is his interpretation of the experimental results valid? Is his reasoning logical and based on the evidence?

(a) Read the short summary of Arrhenius' work above, and research more detailed information in other references.

(b) Choose a role and prepare to question Arrhenius, played by your teacher, on his PhD thesis. You might consider the following roles:
- a senior professor at the university who firmly believes in Dalton's theory that atoms are indivisible, neutral particles;
- a scientist who frequently corresponded with Michael Faraday during his long, distinguished career;
- the professor who supervised Arrhenius' research and who frequently discussed the experimental results with him;
- a scientist who is dissatisfied with the current theories of electricity; or
- a young scientist who wants to know how Arrhenius' ideas would explain the acid–base properties of solutions.

DECISION-MAKING SKILLS

- ○ Define the Issue
- ● Identify Alternatives
- ● Research
- ● Analyze the Issue
- ● Defend a Decision
- ○ Evaluate

Sections 8.3–8.4 Questions

Understanding Concepts

1. Distinguish between a strong and weak acid using the concept of reaction with water.
2. What class of substances are strong bases? Explain their properties.
3. What are the properties of a weak base? Explain these properties.
4. Write appropriate chemical equations to explain the acidic or basic properties of each of the following substances added to water.
   (a) hydrogen bromide (acidic)
   (b) potassium hydroxide (basic)
   (c) benzoic acid, $HC_7H_5O_{2(aq)}$ (acidic)
   (d) sodium sulfide (basic)
5. Theories in science develop over a period of time. Illustrate this development by writing theoretical definitions of an acid, using the following concepts. Begin your answer with, "According to [name of concept], acids are substances that..."
   (a) the Arrhenius concept
   (b) the revised Arrhenius concept
   (c) the Brønsted-Lowry concept
6. Repeat question 5, defining bases. Refer to both strong and weak bases in your answer.
7. According to the Brønsted-Lowry concept, what happens in an acid–base reaction?
8. Use the Brønsted-Lowry definitions to identify each of the reactants in the following equations as acids or bases.
   (a) $HCO^{-}_{3(aq)} + S^{2-}_{(aq)} \rightarrow HS^{-}_{(aq)} + CO^{2-}_{3(aq)}$
   (b) $H_2CO_{3(aq)} + OH^{-}_{(aq)} \rightarrow HCO^{-}_{3(aq)} + H_2O_{(l)}$
9. Complete the following chemical equations to predict the acid–base reaction products.
   (a) $HSO^{-}_{4(aq)} + PO^{3-}_{4(aq)} \rightarrow$
   (b) $H_3O^{+}_{(aq)} + HPO^{2-}_{4(aq)} \rightarrow$
10. Some ions can form more than one conjugate acid–base pair. List the two conjugate acid–base pairs involving a hydrogen carbonate ion.
11. Identify the two acid–base conjugate pairs in each of the following reactions.
    (a) $H_3O^{+}_{(aq)} + HSO^{-}_{3(aq)} \rightarrow H_2O_{(l)} + H_2SO_{3(aq)}$
    (b) $OH^{-}_{(aq)} + HSO^{-}_{3(aq)} \rightarrow H_2O_{(l)} + SO^{2-}_{3(aq)}$

Applying Inquiry Skills

12. Baking soda is a common chemical but its chemical properties are difficult for chemists to explain and predict. Baking soda is amphiprotic and forms a basic solution. List some of the chemical properties of baking soda and indicate why some of these properties are difficult to explain and predict.
    Follow the links for Nelson Chemistry 11, 8.4.

GO TO www.science.nelson.com

Making Connections

13. Common kitchen-variety baking soda has so many uses that it has entire books written about it. Use references to gather a list of uses for baking soda. Identify the uses that involve acid–base reactions.

8.5 Acid–Base Reactions

Figure 1
The first thing the emergency crew must do is identify the spilled chemical.

Acids take part in several characteristic reactions, including their reaction with bases. This means that we can sometimes get clues about an unknown substance by observing how it reacts, and what the products of the reaction are. For example, imagine an emergency response team arriving at the site of a traffic accident involving a chemical spill (**Figure 1**). At first team members are not sure what the spilled chemical is, but then they notice that it seems to be reacting with the magnesium/aluminum wheels of a car. They know that all acids react with active metals (e.g., $Mg_{(s)}$, $Zn_{(s)}$, and $Al_{(s)}$) to produce hydrogen gas, so it is likely that the spilled chemical is an acid. They could confirm their suspicions with a strip of litmus paper or by doing a quick test of a sample of the gas being produced. The acid/metal reaction is a single displacement reaction in which the hydrogen in the acid behaves like a metal.

$$\text{active metal} + \text{acid} \rightarrow H_{2(g)} + \text{ionic compound}$$

For example:

$$Mg_{(s)} + 2\ HCl_{(aq)} \rightarrow H_{2(g)} + MgCl_{2(aq)}$$

All acids also react at various rates with carbonates (e.g., $Na_2CO_{3(aq)}$ and $CaCO_{3(s)}$) in a double displacement reaction producing carbonic acid, $H_2CO_{3(aq)}$. To clean up spills, emergency response teams often use a mixture of chemicals that includes sodium carbonate. The carbonic acid produced is unstable and decomposes quickly to form carbon dioxide and water.

$$\begin{aligned}\text{acid} + \text{carbonate} &\rightarrow H_2CO_{3(aq)} + \text{ionic compound}\\ &\rightarrow CO_{2(g)} + H_2O_{(l)} + \text{ionic compound}\end{aligned}$$

For example:

$$2\ HNO_{3(aq)} + Na_2CO_{3(s)} \rightarrow CO_{2(g)} + H_2O_{(l)} + 2\ NaNO_{3(aq)}$$

Finally, acids react with bases in another double displacement reaction often called a neutralization reaction. The result of this reaction is that the pH of the solution moves toward seven. Emergency response teams might test the pH of a spill with pH paper and then monitor the neutralization of the spill by repeated tests with pH paper.

$$\text{strong acid} + \text{strong base} \rightarrow \text{water} + \text{ionic compound}$$

For example:

$$HCl_{(aq)} + KOH_{(aq)} \rightarrow H_2O_{(l)} + KCl_{(aq)}$$
$$H^+_{(aq)} + Cl^-_{(aq)} + K^+_{(aq)} + OH^-_{(aq)} \rightarrow H_2O_{(l)} + K^+_{(aq)} + Cl^-_{(aq)} \text{ (total ionic equation)}$$

$$H^+_{(aq)} + OH^-_{(aq)} \rightarrow H_2O_{(l)} \text{ (net ionic equation)}$$
or $$H_3O^+_{(aq)} + OH^-_{(aq)} \rightarrow 2\ H_2O_{(l)} \text{ (net ionic equation)}$$

Although not a defining property of acids, acids also undergo precipitation reactions with some ionic compounds. Instead of a double displacement neutralization reaction, a double displacement precipitation reaction occurs. For example, we can determine the concentration of hydroiodic acid by reacting the solution with aqueous lead(II) nitrate and measuring the mass of precipitate formed.

acid + ionic compound → precipitate + acid

For example:

$$2\ HI_{(aq)} + Pb(NO_3)_{2(aq)} \rightarrow PbI_{2(s)} + 2\ HNO_{3(aq)}$$

Practice

Understanding Concepts

1. List three chemical-reaction properties that are characteristic of acids.
2. (a) A spill of hydrobromic acid can be neutralized by reacting it with zinc, lye, and/or washing soda. Write chemical equations to represent each of these reactions individually. (Common names are given in Appendix C.)
   (b) List advantages and/or disadvantages for each of the neutralization methods above.
3. Oxalic acid, $H_2C_2O_{4(aq)}$, found in foods such as rhubarb, undergoes typical acid reactions. Write chemical equations to represent the following reactions.
   (a) Oxalic acid reacts with an aluminum cooking pot to make it look shinier. Assume that one product is an oxalate (low solubility).
   (b) Oxalic acid reacts with aqueous calcium, say, calcium chloride in your blood, to produce insoluble crystals of calcium oxalate that can grow to become bladder or kidney stones.
   (c) The high oxalic acid content in spinach reacts with the high iron content in spinach to precipitate iron (III) ions out during digestion as, for example, iron(III) oxalate. Assume that the iron in the stomach appears as iron(III) chloride. How does this destroy the myth of Popeye getting his strength from iron in spinach?

Figure 2
An initial volume reading is taken before any titrant is added to the sample solution. Then titrant is added until the reaction is complete, that is, when a drop of titrant changes the colour of the sample. The final buret reading is then taken. The difference in buret readings is the volume of titrant added.

Acid–Base Titration

titration: a laboratory procedure involving the carefully measured and controlled addition of a solution from a buret into a measured volume of a sample solution

titrant: the solution in the buret during a titration

Titration is a common laboratory technique used to determine the concentration of substances in solution. It is a reliable, efficient, and economical technology that is simple to use. A known volume of the sample to be analyzed is usually transferred into an Erlenmeyer flask (**Figure 2**). The solution in the buret (called the **titrant**) is added, drop by drop, to the sample. Alternatively, the standard solution could be in the flask, so the solution of unknown concentration would be the titrant. The titrant is added drop by drop until the reaction is judged to be complete. To help us identify this point, we select an indicator that changes colour when the reaction is complete (**Table 1**). The point at which the indicator

Table 1: Indicator Colour Change as the Endpoint of Titration

Indicator	Acidic	Basic
litmus	red	blue
methyl orange	red	yellow
bromothymol blue	yellow	blue
phenolphthalein	colourless	red

changes colour is called the **endpoint.** This is at, or close to, the point at which the titrant and sample have completely reacted.

endpoint: the point in a titration at which a sharp change in a property occurs (e.g., a colour change)

Reactions between aqueous reactants are generally fast. If this reaction involves acids and bases, and at least one of these is strong, then the reaction will normally proceed as in a balanced chemical equation (be stoichiometric), require no special conditions (be spontaneous), and be complete (quantitative). These are the necessary requirements for the use of titration for chemical analysis. Typical chemical analyses include analysis of acids in the environment (acid deposition studies), quality control in industrial and commercial operations, and scientific research. A typical practice titration is the chemical analysis of acetic acid in a sample of vinegar, using a sodium hydroxide solution in a buret as the titrant.

When you perform a chemical analysis by titration, you will use a number of volumetric techniques such as using a pipet to transfer portions of the sample for analysis, the titration using a buret, and measuring solution volumes. In order to obtain precise and reliable results, you must know the concentration of one of the reactants; that is, you must use a **standard solution.**

standard solution: a solution of precisely and accurately known concentration

When doing a titration, you come to a point when the reaction is complete and the indicator suddenly changes colour: the endpoint. At the endpoint you stop the titration and record the volume of titrant used. Chemically equivalent amounts of reactants, as determined by the mole ratio in the balanced chemical equation, have now been combined.

A titration procedure should involve several trials, using different samples of the unknown solution to improve the reliability of the answer. A typical requirement is to repeat measurements until three trials result in volumes within 0.1 mL to 0.2 mL of each other. These three results are then averaged before carrying out further calculations.

SUMMARY Titration Requirements

For titration, a chemical reaction must be

- spontaneous—chemicals react on their own without a continuous addition of energy
- fast—chemicals react instantaneously when mixed
- quantitative—the reaction is more than 99% complete
- stoichiometric—there is a single, whole number mole ratio of amounts of reactants and products

Example: Titration of Hydrochloric Acid

Manufacturers of commercial chemicals must ensure that their products meet certain standards. Quality control technicians are responsible for checking samples of product to ensure that they are acceptable. For aqueous solutions of acids such as muriatic acid (hydrochloric acid), the concentration of the product must be within certain limits. Titration is an excellent technique to test concentration. A sodium carbonate solution can be used as the reagent to analyze the hydrochloric acid. Suppose 1.59 g of anhydrous sodium carbonate, $Na_2CO_{3(s)}$, is dissolved to make 100.0 mL of a standard solution. Samples (10.00 mL) of this standard solution are then taken and titrated with the $HCl_{(aq)}$ product, which has been diluted by a factor of 10. The titration evidence collected is shown in **Table 2** (page 396).

Table 2: The Titration of $Na_2CO_{3(aq)}$ with $HCl_{(aq)}$

Trial	1	2	3	4
final buret reading (mL)	13.3	26.0	38.8	13.4
initial buret reading (mL)	0.2	13.3	26.0	0.6
volume of $HCl_{(aq)}$ added (mL)	13.1	12.7	12.8	12.8

To analyze this evidence you first need to calculate the molar concentration of the sodium carbonate solution.

$v_{Na_2CO_3} = 100\ \text{mL}$

$M_{Na_2CO_3} = 105.99\ \text{g/mol}$

$m_{Na_2CO_3} = 1.59\ \text{g}$

$$n_{Na_2CO_3} = 1.59\ \cancel{\text{g}} \times \frac{1\ \text{mol}}{105.99\ \cancel{\text{g}}}$$

$$= 0.0150\ \text{mol}$$

$$c_{Na_2CO_3} = \frac{0.0150\ \text{mol}}{0.10000\ \text{L}}$$

$$c_{Na_2CO_3} = 0.150\ \text{mol/L}$$

Now you can start the stoichiometry procedure by writing the balanced chemical equation. Notice in **Table 2** that four trials were done, and the volume in the first trial is significantly higher than in the others. The volume you use for $HCl_{(aq)}$ should be your best average, typically three results within ±0.1 mL of each other. The value, 12.8 mL, is the average of trials 2, 3, and 4. (Keep the unrounded value in your calculator as usual.) The rest of the stoichiometry procedure follows the usual steps.

$2\ HCl_{(aq)}$	+	$Na_2CO_{3(aq)}$	→	$H_2CO_{3(aq)}$	+	$2\ NaCl_{(aq)}$
12.8 mL		10.00 mL				
		0.150 mol/L				

Remember that we are using only 10.00 mL of the Na_2CO_3 solution for each trial, not 100 mL.

$$n_{Na_2CO_3} = 10.00\ \text{m}\cancel{\text{L}} \times \frac{0.150\ \text{mol}}{1\ \cancel{\text{L}}}$$

$$= 1.50\ \text{mmol}$$

$$n_{HCl} = 1.50\ \text{mmol} \times \frac{2}{1}$$

$$= 3.00\ \text{mmol}$$

$$c_{HCl} = \frac{3.00\ \cancel{\text{m}}\text{mol}}{12.8\ \cancel{\text{m}}\text{L}}$$

$$c_{HCl} = 0.234\ \text{mol/L}$$

Alternatively, we could combine these steps into one calculation as shown below.

$$c_{HCl} = 10.00\ \cancel{\text{mL}}\ \cancel{Na_2CO_3} \times \frac{0.150\ \cancel{\text{mol}}\ \cancel{Na_2CO_3}}{1\ \text{L}\ \cancel{Na_2CO_3}} \times \frac{2\ \text{mol HCl}}{1\ \cancel{\text{mol}}\ \cancel{Na_2CO_3}} \times \frac{1}{12.8\ \cancel{\text{mL}}}$$

$$c_{HCl} = 0.234\ \text{mol/L}$$

Since the sample of muriatic acid had been diluted by a factor of 10, the original concentration of hydrochloric acid must be 10 times greater, or 2.35 mol/L.

Sample Problem 1

An acid rain sample containing sulfurous acid was analyzed in a laboratory using a titration with a standard solution of sodium hydroxide. Use the evidence given in **Table 3** to determine the concentration of the sulfurous acid.

Table 3: Titration of 25.0 mL of $H_2SO_{3(aq)}$ with 0.105 mol/L $NaOH_{(aq)}$

Trial	**1**	**2**	**3**
final buret reading (mL)	11.1	21.7	32.4
initial buret reading (mL)	0.3	11.1	21.7
volume of $NaOH_{(aq)}$ added	10.8	10.6	10.7

Solution

$$H_2SO_{3(aq)} + 2\ NaOH_{(aq)} \rightarrow Na_2SO_{3(aq)} + 2\ H_2O_{(l)}$$

$H_2SO_{3(aq)}$	$2\ NaOH_{(aq)}$
25.0 mL	10.7 mL
C	0.105 mol/L

$$n_{NaOH} = 10.7\ \text{m}\not{\text{L}} \times \frac{0.105\ \text{mol}}{1\ \not{\text{L}}}$$

$$= 1.12\ \text{mmol}$$

$$n_{H_2SO_3} = 1.12\ \text{mmol} \times \frac{1}{2}$$

$$= 0.562\ \text{mmol}$$

$$C_{H_2SO_3} = \frac{0.562\ \not{\text{m}}\text{mol}}{25.0\ \not{\text{m}}\text{L}}$$

$$C_{H_2SO_3} = 0.0225\ \text{mol/L}$$

or

$$C_{H_2SO_3} = 10.7\ \not{\text{mL}}\ \not{\text{NaOH}} \times \frac{0.105\ \not{\text{mol}}\ \not{\text{NaOH}}}{1\ \text{L}\ \not{\text{NaOH}}} \times \frac{1\ \text{mol}\ H_2SO_3}{2\ \not{\text{mol}}\ \not{\text{NaOH}}} \times \frac{1}{25.0\ \not{\text{mL}}}$$

$$C_{H_2SO_3} = 0.0225\ \text{mol/L}$$

The concentration of the sulfurous acid in the sample is 0.0225 mol/L or 22.5 mmol/L.

INQUIRY SKILLS

- ○ Questioning
- ○ Hypothesizing
- ○ Predicting
- ○ Planning
- ● Conducting
- ● Recording
- ● Analyzing
- ● Evaluating
- ● Communicating

Investigation 8.5.1

Titration Analysis of Vinegar

Consumer products are required by law to have the minimum quantity of the active ingredient listed on the product label. Companies that produce chemical products usually employ analytical chemists and technicians to monitor the final product in a process known as quality control. Government consumer affairs departments also use chemists and technicians to check products, particularly in response to consumer complaints.

In this investigation, you will be the quality control chemist. You have received a report that a local high-school cafeteria has been serving watered-down vinegar to the students. Your purpose is to test the acetic acid concentration of the vinegar to discover whether it has been diluted (i.e., below the 5.0% W/V acetic acid indicated on the purchased container). Complete the **Analysis** and **Evaluation** sections of the report.

Question

What is the molar concentration of acetic acid in a sample of vinegar?

Prediction

The manufacturer claims on the label that the vinegar contains 5.0% acetic acid, which translates into a 0.87 mol/L concentration of acetic acid. The concentration of acetic acid in the vinegar sample should be the same.

Experimental Design

A sample of vinegar from the school cafeteria is diluted by a factor of 10 to make a 100.0-mL solution. The diluted solution is titrated with a standard sodium hydroxide solution using phenolphthalein as the indicator.

Materials

Wear eye protection and a lab apron.

At these dilutions, the chemicals are fairly safe and can be disposed of down the drain.

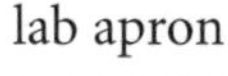

lab apron
eye protection
$NaOH_{(aq)}$
vinegar
phenolphthalein
wash bottle of pure water
two 100-mL or 150-mL beakers
250-mL beaker.
100-mL volumetric flask with stopper
50-mL buret
10-mL volumetric pipet
pipet bulb
ring stand
buret clamp
stirring rod
small funnel
two 250-mL Erlenmeyer flasks
meniscus finder

Procedure

1. Obtain about 30 mL of vinegar in a clean, dry 100-mL beaker.
2. Pipet one 10.00-mL portion into a clean 100-mL volumetric flask and dilute to the mark.
3. Stopper and invert several times to mix thoroughly.
4. Obtain about 70 mL of $NaOH_{(aq)}$ in a clean, dry, labelled 100-mL beaker.
5. Set up the buret with $NaOH_{(aq)}$, following the accepted procedure for rinsing and clearing the air bubble.
6. Pipet a 10.00-mL sample of diluted vinegar into a clean Erlenmeyer flask.
7. Add 1 or 2 drops of phenolphthalein indicator.
8. Record the initial buret reading to the nearest 0.1 mL.
9. Titrate the sample with $NaOH_{(aq)}$ until a single drop produces a permanent change from colourless to faint pink.
10. Record the final buret reading to the nearest 0.1 mL.
11. Repeat steps 6 to 10 until three consistent results are obtained.

Analysis

(a) Answer the Question: What is the molar concentration of acetic acid in a sample of vinegar?

Evaluation

(b) Evaluate your evidence: How confident are you that your techniques and measurements resulted in good evidence?
(c) Evaluate the Prediction: Assuming the manufacturer's claim is accurate, is someone in the cafeteria diluting the vinegar? Include an accuracy calculation (percentage difference) in your evaluation.

Practice

Understanding Concepts

4. Briefly describe three types of characteristic reactions of acids.
5. What are the four reaction requirements in order to use a reaction in a titration in a chemical analysis?
6. What are the two reactants in a titration, and what equipment is used to contain them?
7. What is a standard solution?
8. Why are several trials usually done in a titration?

Applying Inquiry Skills

9. Analysis shows that 9.44 mL of 0.0506 mol/L $KOH_{(aq)}$ is needed for the titration of 10.00 mL of a water sample taken from an acidic lake. Determine the molar concentration of acid in the lake water, assuming that the acid is sulfuric acid.

Answer

9. 0.0239 mol/L or 23.9 mmol/L

Answers

10. (b) 1.08 mol/L

11. (b) 2.66 mol/L

12. 0.278 mol/L

10. Solutions of oxalic acid, $H_2C_2O_{4(aq)}$, have many applications. Like $H_2SO_{4(aq)}$, oxalic acid reacts in a 2:1 mole ratio with sodium hydroxide. Complete the **Evidence**, **Analysis**, and **Evaluation** sections of the following investigation report.

Question
What is the concentration of oxalic acid in a rust-removing solution?

Prediction
The oxalic acid solution is labelled as 10% W/V, or 1.11 mol/L.

Experimental Design
The original oxalic acid solution (rust remover) is diluted by a factor of 100, that is, 10.00 mL to 1000 mL. The concentration of dilute oxalic acid solution is determined by titration with a sodium hydroxide solution.

Evidence
(a) Copy and complete **Table 4**.

Table 4: Volume of 0.0161 mol/L Sodium Hydroxide Required to Neutralize 10.00 mL of Diluted Oxalic Acid

Trial	1	2	3	4
Final buret reading (mL)	14.3	27.8	41.1	13.8
Initial buret reading (mL)	0.2	14.3	27.8	0.4
Volume of $NaOH_{(aq)}$ used (mL)				

Analysis
(b) Using the Evidence in **Table 5**, calculate the concentration of oxalic acid in the rust remover.

Evaluation
(c) Evaluate the Prediction: Is the manufacturer's label accurate?

11. Complete the **Evidence** and **Analysis** for the following titration.

Question
What is the molar concentration of the hydrochloric acid in a solution of kettle-scale remover?

Experimental Design
The hydrochloric acid in a solution of kettle-scale remover is titrated with a standardized solution of barium hydroxide. The colour change of bromothymol blue indicator (from blue to green) is the endpoint.

Evidence
(a) Copy and complete **Table 5**.

Table 5: Titration of 10.00-mL Samples of $HCl_{(aq)}$ with 0.974 mol/L $Ba(OH)_{2(aq)}$

Trial	1	2	3	4
final buret reading (mL)	15.6	29.3	43.0	14.8
initial buret reading (mL)	0.6	15.6	29.3	1.2
volume of $Ba(OH)_{2(aq)}$ added (mL)				
colour at endpoint	blue	green	green	green

Analysis
(b) Using the Evidence in **Table 5**, calculate the concentration of the hydrochloric acid in the kettle-scale remover.

12. Samples of sulfuric acid were titrated with 0.484 mol/L sodium hydroxide. The evidence is shown in **Figure 3**. Calculate the concentration of the sulfuric acid solution.

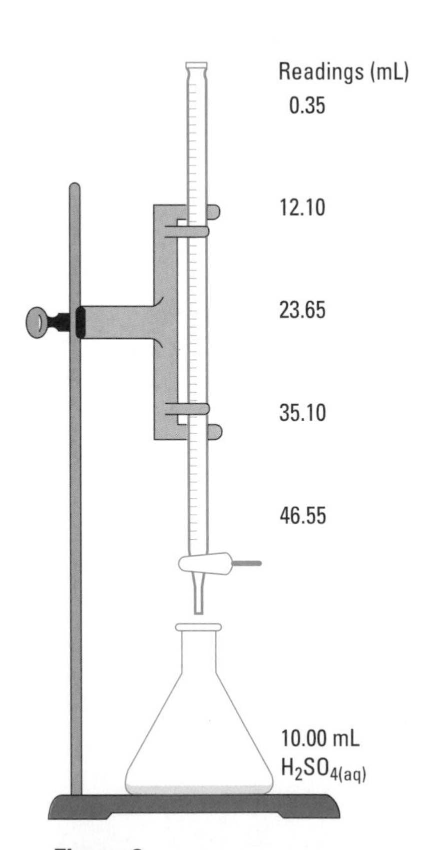

Figure 3
Sodium hydroxide titrant is added to samples of sulfuric acid in successive trials.

Section 8.5 Questions

Understanding Concepts

1. Write specific balanced chemical equations to illustrate each of the three types of characteristic reactions of acids.
2. An antacid tablet contains 0.912 g aluminum hydroxide to neutralize excess stomach acid. What volume of 0.10 mol/L stomach acid (assume $HCl_{(aq)}$) can one tablet neutralize?
3. Slaked lime, $Ca(OH)_{2(s)}$, can be used to neutralize the water in lakes that have been "killed" by acid rain. Ecologists hope that the original plants and animals will become reestablished. If the concentration of sulfuric acid in an acid lake is 1.2×10^{-3} mol/L, and 1.0 t of slaked lime is added to the lake, then what is the volume of the lake that could be neutralized?

Applying Inquiry Skills

4. Write two different Experimental Designs, involving different kinds of chemical reactions, to answer this Question: "What is the concentration of sodium hydroxide in an unknown solution?" List the Materials that you intend to use in your designs.
5. Laboratory technicians perform quantitative chemical analyses to determine the concentration of oxalic acid in, for example, foods and blood. Using the typical reactions of acids, and the technique of titration, write an Experimental Design to answer the Question: What is the concentration of oxalic acid, $H_2C_2O_{4(aq)}$, in a provided solution? Include a list of Materials and any necessary safety and disposal precautions. Then write out a detailed Procedure. With your teacher's approval, carry out your Procedure, record your Evidence, and perform an Analysis. Your Evaluation should critique your Design, Materials, Procedure, skills, and therefore, your Evidence.

DID YOU KNOW ?

Oxalic Acid

Oxalic acid is a solid in its pure form and is relatively toxic to humans and animals. In nature, it is found in a number of green-leaved plants, including rhubarb (especially the leaves). Daily consumption of oxalic acid in an ordinary diet averages around 150 mg; the lethal dose is around 1500 mg.

Chapter 8 Summary

Key Expectations

Throughout this chapter, you have had the opportunity to do the following:

- Demonstrate an understanding of the Arrhenius and Brønsted-Lowry theories of acids and bases. (8.1, 8.4, 8.5)
- Explain qualitatively the difference between strong and weak acids and bases. (8.1, 8.2, 8.4)
- Demonstrate an understanding of the operational definition of pH. (8.2)
- Design and conduct an experiment to determine the effect of dilution on pH. (8.2)
- Identify and describe science- and technology-based careers that work with acidic and basic solutions. (8.3)
- Describe examples of solutions for which the concentration must be known and exact. (8.3, 8.5)
- Use the terms: ionization, dissociation, strong acid/base, weak acid/base, hydronium ion, proton transfer, conjugate acid/base, titration, titrant, and endpoint. (8.4, 8.5)
- Write balanced chemical equations for reactions involving acids and bases. (8.4, 8.5)
- Use titration procedure to determine the concentration of an acid or base in solution. (8.5)

Key Terms

acid (2)
amphiprotic
base (2)
conjugate acid
conjugate acid–base pair
conjugate base
dissociation
endpoint
hydrogen polyatomic ion
hydronium ion
ionization
neutralization (2)
percentage ionization
pH
standard solution
strong acid
strong base
titrant
titration
weak acid
weak base

Make a Summary

In this chapter, you have studied acids and bases. To summarize your learning, sketch the equipment for a titration.

- Use, for example, 0.12 mol/L hydrochloric acid as the titrant to determine the concentration of sodium bicarbonate in "gripe water," an antacid for babies.
- Label the equipment and use key expectations and key terms to accompany your sketch.
- Try to display all of your learning from this chapter in and around this sketch.
- Use the chemicals, their concentrations, and their chemical formulas and chemical reactions as much as possible in your summary.

Reflect on your Learning

Revisit your answers to the Reflect on Your Learning questions at the begining of this chapter.

- How has your thinking changed?
- What new questions do you have?

Chapter 8 Review

Understanding Concepts

1. Provide an empirical definition for
   (a) an acid
   (b) a base
2. Many familiar household substances are acids or bases. Write IUPAC names for the following compounds and classify them as acids or bases.
   (a) lye, $NaOH_{(s)}$
   (b) vinegar, $HC_2H_3O_{2(aq)}$
   (c) milk of magnesia, $Mg(OH)_{2(s)}$
   (d) muriatic acid, $HCl_{(aq)}$
   (e) slaked lime, $Ca(OH)_{2(s)}$
   (f) window cleaner, $NH_{3(aq)}$
3. For each of the substances listed in the previous question, write a Prediction for whether the substance would form an acidic or a basic solution; and a chemical equation to explain your Prediction.
4. Solutions of hydrochloric acid and acetic acid are prepared, both with the same volume and concentration.
   (a) Compare the concentrations of hydrogen ions in the two solutions.
   (b) Explain the difference in hydrogen ion concentrations in the two solutions.
   (c) Compare the volumes of the sodium hydroxide solution required to neutralize each solution.
5. A student prepared 0.10 mol/L solutions of acetic acid, ammonia, hydrochloric acid, sodium chloride, and sodium hydroxide. Rank the solutions in order of increasing pH.
6. Calculate the pH of the following solutions.
   (a) lemon juice with $[H^+_{(aq)}] = 7.5 \times 10^{-3}$ mol/L
   (b) 2.5×10^{-3} mol/L nitric acid
7. Calculate the hydrogen ion concentration in each of the following solutions.
   (a) cleaning solution with a pH = 11.56
   (b) fruit juice with a pH = 3.50
8. List three empirical properties that may be used to rank acids in order of strength.
9. Aqueous solutions of nitric acid ($HNO_{3(aq)}$, a strong acid) and nitrous acid ($HNO_{2(aq)}$, a weak acid) of the same concentration are prepared.
   (a) How do their pH values compare?
   (b) Explain your answer, using chemical equations and the Brønsted-Lowry concept.
10. One sample of rainwater has a pH = 5, while another has a pH = 6. How do the hydrogen ion concentrations in the two samples compare?
11. Formal concepts of acids have existed since the 18th century. State the main idea and the limitations of each of the following: the oxygen concept; the hydrogen concept; Arrhenius's concept.
12. Use the Brønsted-Lowry concept to identify the reactants as acids or bases in the following reactions. In your answers, indicate the conjugate acid–base pairs.
   (a) $HSO^-_{4(aq)} + HCO^-_{3(aq)} \rightarrow SO^{2-}_{4(aq)} + H_2CO_{3(aq)}$
   (b) $HPO^{2-}_{4(aq)} + HSO^-_{4(aq)} \rightarrow H_2PO^-_{3(aq)} + SO^{2-}_{4(aq)}$
   (c) $H_2PO^-_{4(aq)} + H_2BO^-_{3(aq)} \rightarrow HPO^{2-}_{4(aq)} + H_3BO_{3(aq)}$
   (d) $HS^-_{(aq)} + HCO^-_{3(aq)} \rightarrow CO^{2-}_{3(aq)} + H_2S_{(aq)}$
   (e) $HSO^-_{3(aq)} + NH_{3(aq)} \rightarrow SO^{2-}_{3(aq)} + NH^+_{4(aq)}$
13. Some sources of drinking water contain low concentrations of nitrogen compounds. Because many nitrogen compounds are harmful to human health, especially that of infants, Health Canada has established maximum allowable levels for nitrogen compounds in drinking water. Write IUPAC names for the following nitrogen-containing entities and classify them as potential Brønsted-Lowry acids and/or bases.
   (a) $NH_{3(aq)}$
   (b) $NH^+_{4(aq)}$
   (c) $NO^-_{2(aq)}$
   (d) $NO^-_{3(aq)}$
14. The household cleaner TSP contains sodium phosphate as the active ingredient. In solution the phosphate ion reacts with water as shown in the following equation.

$$PO^{3-}_{4(aq)} + H_2O_{(aq)} \rightarrow HPO^{2-}_{4(aq)} + OH^-_{(aq)}$$

   (a) Identify two conjugate acid/base pairs in the above equation.
   (b) Predict whether a TSP solution would be acidic, basic, or neutral, and explain your prediction.
15. One component of acid rain is sulfurous acid. It forms when sulfur dioxide in the atmosphere dissolves in rainwater. The neutralization of sulfurous acid with a strong base proceeds in a two-step reaction. Complete the reaction equations.
   (a) $H_2SO_{3(aq)} + OH^-_{(aq)} \rightarrow$
   (b) $HSO^-_{3(aq)} + OH^-_{(aq)} \rightarrow$
16. Baking soda, $NaHCO_{3(s)}$, is a versatile substance as it reacts with both strong acids and strong bases.
   (a) Write a balanced chemical equation and a net ionic equation for the reaction of baking soda with hydrochloric acid.
   (b) Write a balanced chemical equation and a net ionic equation for the reaction of baking soda with sodium hydroxide.

(c) Classify each of the entities in the net ionic equations as Brønsted-Lowry acids or bases. Identify the two conjugate acid–base pairs in each reaction.
(d) What do the reactions in (a) and (b) indicate about the nature of the hydrogen carbonate ion? What term is used for this characteristic?

17. List the requirements of a chemical reaction used in a titration analysis to determine the concentration of a solute.
18. Why is it necessary to start a titration with at least one standard solution?
19. Define the following terms:
(a) titration
(b) titrant
(c) endpoint
20. The chemical reactions that define acids are illustrated by the following reactions for sulfuric acid accidentally spilled from a car battery during an upset and a cleanup. Write chemical equations for each combination and indicate what it is about each reaction that helps define sulfuric acid as an acid.
(a) Sulfuric acid reacts with zinc on the galvanized fenders of the car.
(b) Sulfuric acid reacts with washing soda used to neutralize the acid.
(c) Sulfuric acid reacts with slaked lime used to neutralize the acid.

Applying Inquiry Skills

21. What specific volumetric equipment is required to
(a) contain the solution in the final steps of preparing a standard solution?
(b) deliver precisely 7.8 mL of a solution in a dilution procedure?
(c) deliver precisely 10.00 mL of a sample to be analyzed in a titration?
22. Design an experiment to test the generalization that diluting an acidic and a basic solution by a factor of 10 changes the pH by 1. Provide a list of Materials necessary to carry out your Experimental Design.
23. Complete the **Analysis** and **Evaluation** sections of the following investigation report.

Question

Which of the 0.1 mol/L solutions labelled 1, 2, 3, 4, and 5 is $KNO_{3(aq)}$, $NaHCO_{3(aq)}$, $HCl_{(aq)}$, $H_2SO_{3(aq)}$, and $NaOH_{(aq)}$?

Experimental Design

The solutions are all at the same concentration and temperature. A sample of each solution is tested to show its effect on litmus, and tested for conductivity and pH. Each solution is tested in an identical way.

Evidence

Table 1: Properties of the Unknown Solutions

Unknown	Litmus paper	Conductivity	pH
1	red to blue	high	10
2	no change	high	7
3	blue to red	low	2
4	blue to red	high	1
5	red to blue	high	13

Analysis

(a) Using the Evidence in **Table 1**, identify the unknown solutions.

Evaluation

(b) Critique the Experimental Design.

24. A chemistry student was given the task of determining the concentration of a hydrochloric acid solution so it can be used as a standard solution. Complete the **Evidence** and **Analysis** sections of the following investigation report.

Question

What is the molar concentration of the hydrochloric acid solution?

Experimental Design

Samples of a standard sodium carbonate solution are titrated with the unknown solution of hydrochloric acid. The colour change of a methyl orange indicator from yellow to orange is used as the endpoint. (Methyl orange is yellow in a basic solution and red in an acidic solution.)

Evidence

(a) Copy and complete **Table 2**.

Table 2: Titration of 10.00 mL of 0.120 mol/L $Na_2CO_{3(aq)}$ with $HCl_{(aq)}$

Trial	1	2	3	4
final buret reading (mL)	17.9	35.0	22.9	40.1
initial buret reading (mL)	0.3	17.9	5.9	22.9
volume of $HCl_{(aq)}$ added				
colour at endpoint	red	orange	orange	orange

Analysis

(b) Calculate the molar concentration of the hydrochloric acid solution.

25. Phosphoric acid is the active ingredient in many commercial rust removers. A technician in a product testing lab uses titration to determine the concentration of phosphoric acid in a bottle of commercial rust remover. Complete the **Evidence**, **Analysis**, and **Evaluation** sections for the investigation report.

Question

What is the molar concentration of the phosphoric acid in a rust-removing solution?

Prediction

According to the label on the bottle of commercial rust remover, the concentration of phosphoric acid in the product is 42.5% W/V, or 7.30 mol/L.

Experimental Design

The rust-removing solution is diluted by a factor of 10, and then 10.00-mL samples of the diluted product are titrated with a standardized 1.25 mol/L $NaOH_{(aq)}$ solution to the endpoint of the following reaction. Phenolphthalein is used as the indicator.

$$2\ NaOH_{(aq)} + H_3PO_{4(aq)} \rightarrow Na_2HPO_{4(aq)} + 2\ H_2O_{(l)}$$

Evidence

(a) Copy and complete Table 3.

Table 3: Titration of 10.00 mL of $H_3PO_{4(aq)}$ with 1.25 mol/L $NaOH_{(aq)}$

Trial	1	2	3	4
final buret reading (mL)	13.3	25.0	36.8	48.4
initial buret reading (mL)	0.4	13.3	25.0	36.8
volume of $NaOH_{(aq)}$ added				
colour at endpoint	deep red	pale pink	pale pink	pale pink

Analysis

(b) Calculate the molar concentration of the phosphoric acid in the solution.

Evaluation

(c) Evaluate the Prediction. Is the manufacturer's label accurate?

Making Connections

26. Acids are frequently portrayed in movies as being extremely dangerous. For example, in *Dante's Peak* the aluminum boat is quickly eaten away by the acidic lake water. Find a clip from a movie that involves acids, and critique the portrayal of its chemistry.
27. Laboratory safety rules require students to wear eye protection when handling acids such as hydrochloric acid and sulfuric acid, yet dilute boric acid, $H_3BO_{3(aq)}$, is sold in drugstores as a soothing eye wash. Explain, including balanced chemical equations, why many acids are harmful to your eyes, but this solution is not.
28. Hardness of water is directly related to the quantity of calcium and magnesium ions present. At home, the scale that appears inside kettles, in pots, in coffee makers, and on showerheads after prolonged usage is evidence of hard water. Assuming the scale is entirely carbonates, design a process to remove it. Identify common Materials to be used, write appropriate chemical equations, and outline a Procedure, including safety and disposal. If you have this problem at home, obtain permission from a parent or guardian to remove the scale from some household items.
29. Measure the pH of tap water in your community, home, or school. If you live in a community with municipal drinking water treatment, use the Internet to find out how the pH level of your drinking water is regulated.

    Follow the links for Nelson Chemistry 11, Chapter 8 Review.

GO TO www.science.nelson.com

30. All gardeners know that conifers like acidic soil, so why is acid rain so damaging? Find out why acid is good for trees in one case, but not in others.

    Follow the links for Nelson Chemistry 11, Chapter 8 Review.

GO TO www.science.nelson.com

Exploring

31. The Brønsted-Lowry concept is not the last acid–base theory. Research the Lewis theory of acids and bases. What are its advantages and disadvantages, compared with the Brønsted-Lowry concept?

    Follow the links for Nelson Chemistry 11, Chapter 8 Review.

GO TO www.science.nelson.com

Unit 3 Performance Task

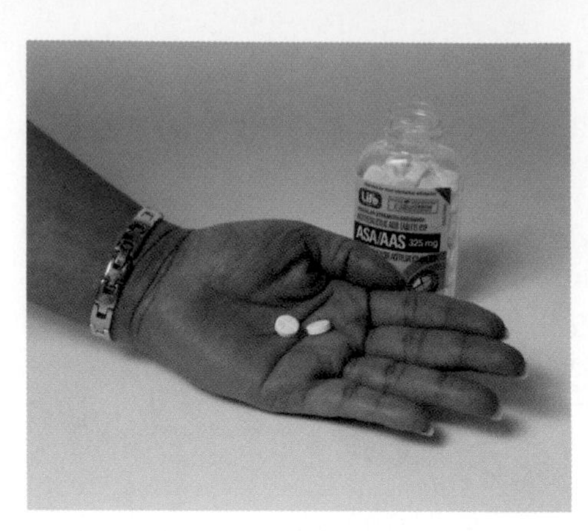

Figure 1

Analysis of ASA

Acetylsalicylic acid, widely known as ASA or Aspirin (**Figure 1**), is the world's most commonly used pain-relieving drug—over ten thousand tonnes are manufactured every year in North America alone. ASA, $HC_9H_7O_{4(s)}$, is an organic (carbon chain) acid like acetic acid, $HC_2H_3O_{2(l)}$, and reacts with strong bases such as sodium hydroxide in the same way. You will be provided with a standard solution of $NaOH_{(aq)}$—one for which the concentration is known with a fair degree of certainty.

The main purpose of this task is to perform an investigation to accurately determine the ASA content of an over-the-counter pain reliever, and to compare the result with the quantity stated on the product label. There are several complicating factors that arise from the nature of ASA tablets:

- All tablet medications contain other "inert" ingredients besides the "medicinal" ones—and some of these other ingredients are not soluble.
- ASA is a large molecule with only one OH group to provide hydrogen bonding, so it is not very soluble in water. *The CRC Handbook* lists this compound as soluble in alcohols, ethers, and chloroform.
- Some tablets of this type have special coatings or are "buffered," meaning that other compounds are added that can and do interfere with ASA's reaction as an acid.

The use of a solvent other than water to dissolve your ASA tablet presents some more complications: There are three alcohols commonly available to the public. They are methanol, $CH_3OH_{(l)}$, ethanol, $C_2H_5OH_{(l)}$, and isopropanol, $CH_3CHOHCH_{3(l)}$. Methanol is sold as fondue fuel and gas line or windshield washer antifreeze, often called methyl hydrate. It is quite toxic if ingested, so it is not usually named as an alcohol for public sale. The idea is not to tempt or confuse consumers into thinking that it is drinkable. The other two alcohols are sold as "rubbing" alcohols in pharmacies. Note that while ethanol is not toxic in small quantities, ethanol used in anything other than liquors (alcoholic beverages) has other additives to make it toxic and undrinkable.

Investigation

Titration Analysis of an ASA Tablet

For this task you will use a standardized solution of sodium hydroxide to perform a titration to test the manufacturer's claim for the mass of ASA in a commercial tablet. The equation for the reaction of sodium hydroxide and ASA is

$$HC_9H_7O_{4(al)} + NaOH_{(aq)} \rightarrow NaC_9H_7O_{4(aq)} + H_2O_{(l)}$$

Question

What is the quantity of "active ingredient" in a commercial ASA tablet?

Prediction

(a) Write a prediction, based on the manufacturer's claim on the label of the bottle of ASA tablets.

Experimental Design

(b) Describe the design of your experiment.
(c) Explain why it would not be appropriate to use ether or chloroform as the solvent for your ASA tablet titration.
(d) Write a complete Procedure, including the precautions you will take to avoid safety hazards, and have it approved by your teacher.

Materials

ASA tablets
(e) Complete the Materials list.

Analysis

(f) Use stoichiometric calculations to determine the mass of ASA in a tablet.

Evaluation

(g) Evaluate the evidence gathered by evaluating the Experimental Design, the Materials, the Procedure, and the skills of experimenters. Indicate sources of error and suggest ways to improve the investigation.
(h) Calculate the percentage difference between the predicted and experimental mass.
(i) Evaluate the Prediction and the manufacturer's claim.

Synthesis

(j) Why are some ASA tablets coated or buffered?
(k) What other ingredients are included in medication tablets, and why?

Assessment

Your completed task will be assessed according to the following criteria:

Process

- Develop an appropriate Experimental Design.
- Choose and safely use appropriate Materials.
- Carry out the approved investigation.
- Record observations with appropriate precision.
- Analyze the results.
- Evaluate the Evidence.

Product

- Prepare a suitable lab report, including a discussion of the ingredients in this typical medication.
- Demonstrate an understanding of the relevant scientific concepts.
- Use terms, symbols, equations, and SI metric units correctly.

Unit 3 Review

Understanding Concepts

1. Using examples, explain why solutions are important to the study of chemistry.
2. A water testing lab uses samples of different types of water for reference. Indicate which of the following samples are solutions. Give reasons for your answers.
   (a) cloudy water
   (b) hard water
   (c) pure water
   (d) soft water
   (e) water with high iron content
3. What classes of compounds are
   (a) electrolytes?
   (b) nonelectrolytes?
4. The label on a bottle of a popular soft drink indicates that it contains the following solutes. Write the chemical formulas for these solutes, and classify them as ionic, molecular, or acid. (You may find the list of Common Chemicals in Appendix C useful.)
   (a) glucose
   (b) citric acid
   (c) sodium citrate
   (d) sodium benzoate
   (e) carbon dioxide
5. Modern solution theory is based on experimental work done by the Swedish chemist Svante Arrhenius.
   (a) What two properties of solutions did Arrhenius study to develop his theory of dissociation of electrolytes?
   (b) According to Arrhenius' theory, which ions are responsible for the acidic and the basic properties of a solution?
   (c) Use Arrhenius' theory to explain why acids and bases are always electrolytes, but compounds that form neutral solutions can be electrolytes or nonelectrolytes.
6. What happens when scientists find a theory to be inadequate? Use the Arrhenius theory of acids as an example.
7. The theoretical definition of a base changed from the Arrhenius concept, to the revised Arrhenius concept, to the Brønsted-Lowry concept. Describe these changes.
8. Most of the common household chemicals used in cooking, cleaning, and gardening are aqueous solutions. Copy and complete **Table 1**, classifying the active compound in each of the household products as acid, base, or neutral, and then as electrolyte or nonelectrolyte. Also provide the IUPAC name of each compound.

Table 1: Properties of Household Chemicals

Household chemical	Chemical formula	Acid, base, or neutral	Electrolyte or nonelectrolyte	IUPAC name
(a) syrup	$C_{12}H_{22}O_{11(aq)}$			
(b) kettle scale remover	$HCl_{(aq)}$			
(c) windshield washer fluid	$CH_3OH_{(aq)}$			
(d) oven cleaner	$NaOH_{(aq)}$			
(e) plant fertilizer	$KNO_{3(aq)}$			

9. A chemistry student conducts qualitative analysis tests on some common household products. Write net ionic equations for the predicted reactions.
   (a) Silver nitrate solution is added to a dilute solution of table salt, sodium chloride.
   (b) Aqueous sodium hydroxide is added to a solution of bluestone, copper(II) sulfate.
   (c) Calcium chloride solution is added to a solution of washing soda, sodium carbonate.
   (d) Aqueous lead(II) nitrate is added to a solution of potassium chloride lawn fertilizer.
   (e) Silver nitrate solution is added to a dilute solution of road salt, calcium chloride.
10. Give one example in which a high concentration of a solute is beneficial and one example in which it is harmful.
11. Ethane, $C_2H_{6(g)}$, from natural gas is used to synthesize hundreds of different compounds. Predict the solubility in water of each of the following ethane derivatives as high or low, based on molecular polarity.
   (a) $C_2H_3Cl_{(l)}$
   (b) $CH_3CHO_{(l)}$
   (c) $C_2H_5OH_{(l)}$
   (d) $C_2H_{4(g)}$
   (e) $HC_2H_3O_{2(l)}$
   (f) $C_2H_{2(g)}$
   (g) $C_2H_5NH_{3(l)}$
12. Predict from theory which compound in each pair is more soluble in water. Provide a theoretical explanation for your choice.
   (a) $(CH_3)_2CO_{(l)}$, nail polish remover, or $C_3H_7OH_{(l)}$, rubbing alcohol
   (b) $C_2H_4(OH)_{2(l)}$, radiator antifreeze, or $C_2H_4Cl_{2(l)}$, industrial solvent
   (c) two different industrial solvents, $CCl_{4(l)}$ or $CHCl_{3(l)}$

13. According to the label on a package of tea, a typical cup of green tea contains 20 mg of caffeine compared with 100 mg in a typical cup of black tea. Assuming a typical cup holds 225 mL, calculate the caffeine concentration for each beverage.
14. A 25.0-mL sample of saturated potassium chlorate solution at 22°C is evaporated to dryness, forming 2.16 g of crystals. What is the concentration of saturated potassium chlorate in g/100 mL at 22°C?
15. Water from Lake Ontario contains 162 ppm of dissolved minerals. If 2.5 L of this water is boiled to dryness in a kettle, what mass of minerals would remain?
16. Water is added to a 40.0-mL sample of 2.50 mol/L aqueous sodium hydroxide solution until the volume becomes 5.00 L. Calculate the concentration of the final solution.
17. One brand of bottled water contains 150 mg of calcium in a 2.00-L bottle. Calculate the concentration of calcium in
    (a) parts per million
    (b) moles per litre
18. A bottle of household vinegar is labelled 5% acetic acid (by volume). What minimum volume of vinegar contains 60 mL of acetic acid?
19. Copper(II) sulfate pentahydrate, $CuSO_4 \cdot 5H_2O_{(s)}$, is a fungicide used to treat fenceposts (before they are put in the ground) to delay the rotting process. The maximum concentration of copper(II) sulfate pentahydrate at 0°C is 14.3 g/100 mL of solution.
    (a) Calculate the molar solubility of copper(II) sulfate pentahydrate at 0°C.
    (b) Find the maximum mass of copper(II) sulfate pentahydrate that can be dissolved to make 4.54 L of solution.
20. Precipitation reactions are often used in "chemistry magic shows" to produce sudden colour changes. Write net ionic equations to represent the following reactions.
    (a) Solutions of silver nitrate and potassium iodide are mixed to produce a pale yellow precipitate.
    (b) Solutions of copper(II) nitrate and potassium hydroxide produce a blue precipitate when mixed.
    (c) Lead(II) nitrate and sodium sulfide solutions are mixed to produce a black precipitate.
    (d) Solutions of sodium carbonate and calcium chloride are mixed to produce a white precipitate.
    (e) When solutions of lead(II) nitrate and sodium iodide are mixed, a bright yellow precipitate is formed.
21. Dilution of aqueous solutions is an essential laboratory skill. Use the values for concentrated reagents in Appendix C to answer the following questions.
    (a) A student needs to prepare 500 mL of 1.00 mol/L $HCl_{(aq)}$. Calculate the volume of concentrated reagent required.
    (b) A lab technician finds that she has only 250 mL of concentrated phosphoric acid in the storage cabinet. What is the maximum volume of 2.00 mol/L $H_3PO_{4(aq)}$ that she will be able to prepare?
    (c) A student dilutes 25 mL of saturated sulfurous acid to a final volume of 750 mL. What is the concentration of sulfurous acid in the diluted solution?
    (d) A lab technician needs to prepare 4.54 L of 2.50 mol/L ammonia solution. What volume of concentrated ammonia will he require?
22. On what factors does the hydrogen ion concentration depend in any solution of a weak acid?
23. Copy and complete **Table 2**.

Table 2: pH and Hydrogen Ion Concentration of Some Common Beverages

Beverage	pH	$[H^+_{(aq)}]$ (mol/L)
antacid	10.46	
apple juice	3.14	
beer		3.12×10^{-5}
cider		7.18×10^{-4}
tap water	7.86	

24. The following chemicals are sometimes used in water treatment plants. Write dissociation equations for the solutes in each of the following solutions.
    (a) $Al_2(SO_4)_{3(aq)}$
    (b) $Ca(OH)_{2(aq)}$
    (c) $NaOCl_{(aq)}$
25. The pesticide malathion has a solubility of 145 ppm at room temperature. Calculate the volume of room-temperature water needed to completely dissolve 1.00 g of malathion.
26. Canadians use an average of 200 L of water per person per day. The sodium carbonate concentration in a sample of well water is tested and found to be 225 ppm. Determine the mass of sodium carbonate in 200 L of well water.
27. Arsenic commonly occurs in the ores of gold, lead, copper, and nickel. The unintended release of compounds containing arsenic into ground water is a concern in areas where these metals are mined and smelted.

Because of its toxicity, Health Canada has set the maximum acceptable concentration (MAC) of arsenic in drinking water at 25 ppb.

(a) If a community's water supply contained the MAC of arsenic, what mass of arsenic would a person consume if he/she drank an average of 1.5 L of this water per day for one year (365 days)?

(b) A 25.0-mL sample of water taken from a stream near an abandoned gold mine is found to contain 1.2 μg of arsenic. Does this exceed the Health Canada MAC for arsenic?

(c) Arsenic in the compound sodium arsenate, $Na_3AsO_{4(aq)}$, can be removed from drinking water by treatment with iron(III) chloride solution, to precipitate the toxic anion. Write a net ionic equation for the reaction of iron(III) chloride solution with the aqueous sodium arsenate.

28. A 250-mL sample of polluted, acidic pond water is titrated with 0.0085 mol/L $NaOH_{(aq)}$. If 9.3 mL of the base was required to reach the endpoint, what was the molar concentration of acid (assume $HCl_{(aq)}$) in the pond water? (Assume the acid in the pond water is dilute $HCl_{(aq)}$.)

29. When 0.10 mol/L solutions of potassium benzoate and sodium hydrogen sulfate are mixed, a precipitate of white crystalline flakes is formed. The net ionic equation for the reaction is

$$C_7H_5O_{2(aq)}^{-} + HSO_{4(aq)}^{-} \rightarrow HC_7H_5O_{2(s)} + SO_{4(aq)}^{2-}$$

(a) Identify each reactant as a Brønsted-Lowry acid or base.

(b) Provide a theoretical explanation for the observation that the $C_7H_5O_2^-$ ion is more soluble in water than the $HC_7H_5O_2$ molecule.

30. Does a Brønsted-Lowry acid in solution always form an acidic solution? Provide an example to explain your answer.

Applying Inquiry Skills

31. Design an experiment to identify five unlabelled solutions of equal concentration that are known to contain the following compounds: road salt, $CaCl_{2(aq)}$; table sugar, $C_{12}H_{22}O_{11(aq)}$; vinegar, $HC_2H_3O_{2(aq)}$; oven cleaner, $NaOH_{(aq)}$; and battery acid, $H_2SO_{4(aq)}$. Your materials can include litmus paper and a conductivity meter.

32. A solution is believed to contain chloride ions and sulfide ions. Assuming that the solution does not contain any interfering ions, design an experiment to test for the presence of these two ions.

33. An unknown solution conducts electricity, turns red litmus paper blue, and forms a precipitate when sodium sulfate solution is added. What is one possible chemical formula for the solute present in the original solution?

34. A lab technician wants to prepare 250 mL of 0.50 mol/L nitric acid by diluting a stock solution.

(a) Calculate the volume of 15.4 mol/L nitric acid required to prepare this solution.

(b) Write a complete Procedure for preparing this solution.

35. Complete the Analysis of the following investigation report, and write an Evaluation of the Experimental Design.

Question

Which of the 0.1 mol/L solutions labelled 1, 2, 3, 4, 5, 6, and 7 is $KCl_{(aq)}$, $CuBr_{2(aq)}$, $HCl_{(aq)}$, $HC_2H_3O_{2(aq)}$, $NH_{3(aq)}$, $CH_3OH_{(aq)}$, and $NaOH_{(aq)}$?

Experimental Design

The solutions are prepared so that they all have the same concentration and temperature. A sample of each solution is observed to determine its colour, and is then tested for pH, conductivity, and effect on blue and red litmus papers.

Evidence

Table 3: Properties of the Unknown Solutions

Unknown	Colour	pH	Conductivity	Litmus paper
1	colourless	13	high	red to blue
2	blue	7	high	no change
3	colourless	7	low	no change
4	colourless	1	high	blue to red
5	colourless	7	high	no change
6	colourless	11	high	red to blue
7	colourless	3	high	blue to red

Analysis

(a) Using the Evidence in **Table 3**, identify each of the unknown solutions.

Evaluation

(b) Was the Experimental Design effective?

Making Connections

36. Several solutions, such as CLR, are sold in retail stores as household rust and lime (hard-water scale) removers. The labels all indicate that the active ingredient is glycolic acid, but do not provide the formula for this com-

pound. A student wonders if another acid could be substituted for the commercial solution, and checks out a library reference. The student finds that the percentage ionization of a 0.10 mol/L solution of glycolic acid is 3.9%. In Chapter 8, the percentage ionization of a 0.10 mol/L solution of acetic acid (the acid in vinegar) is given as 1.3%.

(a) Use any library or Internet resource to find the correct chemical formula for glycolic acid (also known by its IUPAC name: hydroxyacetic acid).
(b) Is glycolic acid a strong acid or a weak acid?
(c) Would a solution of glycolic acid react more or less rapidly with rust than a vinegar solution of the same concentration? Explain your reasoning.
(d) Would a solution of glycolic acid react with more or less iron(III) hydroxide (rust) than a vinegar solution of the same concentration, or with the same amount? Explain your reasoning.
(e) Identify and briefly discuss several factors that might affect the student's decision on whether to substitute vinegar for a commercial glycolic acid solution for rust and hard-water scale removal tasks around the home.

37. A shopper finds a 700-mL spray bottle of "bleach cleaner" with a label claiming that the product has a "fresh" scent, contains a powerful cleaner, and that it kills 99.9% of germs. The fine print states that it contains 1.84 % W/V sodium hypochlorite, $NaOCl_{(aq)}$, bleach when packed. The "powerful" cleaner is not identified, so it is probably a non-controlled product like liquid detergent. Many cleaning solutions have perfumes added, so that likely explains the "fresh" scent. A 3.6-L jug of regular household bleach with another brand name, found on the same shelf, is labelled 5.25% W/V sodium hypochlorite, when packed.
(a) When bleach bottles are opened, an easily identified odour of chlorine tells you that some chlorine has escaped. Explain what this observation tells you about the tendency of aqueous hypochlorite ions to react, and why the bottle labels must use the phrase, "when packed," when listing the minimum concentration value of sodium hypochlorite.
(b) What minimum mass of $NaOCl_{(s)}$ was required to make the bleach solution contained in the large 3.6-L jug?
(c) If your main purpose is to kill microorganisms, you could just dilute some regular bleach and wipe your cutting board, countertop or stove surface. How many times more NaOCl solute, by mass, is in the large jug than is in the spray bottle?
(d) Find some bleach solution containers at home or in the supermarket. Try to find large and small ones with the same brand name, such as Clorox or Javex. Record the volumes, concentration, and prices; compare the two by calculating the cost per gram of NaOCl in each container.
(e) The shopper notes that both bottles are labelled as corrosive, and marked with the Hazardous Household Product Symbol (HHPS) in a triangular frame. What is the symbol used, and what does the triangular frame mean?
(f) The labels also say "Dangerous Gas Formed When Mixed with Acid." Write the net ionic equation for hydrogen ions reacting with hypochlorite ions, $OCl^-_{(aq)}$, and chloride ions, to form chlorine gas and water. Explain why this reaction as written cannot be considered a simple Brønsted-Lowry acid–base reaction.

Exploring

38. On a large jug of household bleach, the words CAUTION and ATTENTION, and a HHPS corrosive logo are in large print, in black, on a white background. On a smaller spray bottle of a more dilute solution, the same words and logo are used, but in small print, in light green, on a slightly darker green background. Both containers also state "Dangerous Gas Formed When Mixed with Acid"—which is a *very* important caution—so that people do not try to clean their sinks and toilets by mixing acids (such as rust-removing cleaners) with bleach.
(a) Use the Internet to determine what the legal limit is for exposure to chlorine gas in air, and what effect breathing chlorine has on humans.
(b) Use the Internet to determine what the Canadian legal standards are (if any) for hazard warnings on product labels. Suggest some changes in the standards that would improve safety.

Follow the links for Nelson Chemisty 11, Unit 3 Review.

GO TO www.science.nelson.com

Unit 4

Gases and Atmospheric Chemistry

"During recent years, health specialists have shown that ground-level ozone (smog) and fine particulate matter present in our atmosphere are linked to respiratory problems, and possibly premature death. However, after eight years working as an air quality consultant, I realized that current knowledge of these important atmospheric issues is limited, and that much more research was necessary. I wanted to help close the information gap. Now I do research focusing on understanding the interactions between gases and particles in the atmospheres. By comparing our model results to actual measurements made in the field, we can understand the data and test our theories."

Diane Michelangeli, Associate Professor, Atmospheric Science, York University, Toronto

Overall Expectations

In this unit, you will be able to

- demonstrate an understanding of the laws that govern the behaviour of gases;
- investigate through experimentation the relationships among the pressure, volume, and temperature of a gas and solve problems involving amount of substance in moles, molar masses and volumes, and the gas laws;
- describe how knowledge of gases has helped to advance technology and how such technological advances have led to a better understanding of environmental phenomena and issues.

Unit

4

Gases and Atmospheric Chemistry

Are You Ready?

Knowledge and Understanding

1. This unit is about gases and the atmosphere. The state of a substance (solid, liquid, or gas) is determined by temperature and pressure. Complete **Table 1** by indicating the state of the following substances at the given temperature and standard pressure. The melting and boiling points of the elements can be found in the periodic table.

Table 1: States of Matter of Elements

Substance	State at –150°C	State at SATP	State at 150°C
argon			
bromine			
chlorine			
sulfur			
water			

2. Provide the names for the intermolecular forces described below:
   (a) The intermolecular bond caused by the attractions of oppositely charged ends of polar molecules is called ____________________.
   (b) The intermolecular bond caused by attractions of electrons in one molecule by positive nuclei of atoms in nearby molecules is called ____________________.
   (c) The relatively strong intermolecular force between molecules containing F—H, O—H, or N—H bonds is called ____________________.

Inquiry and Communication

3. Sodium azide, $NaN_{3(s)}$, is the solid that rapidly decomposes when an air bag is activated during a collision. In North American cars, about 250 g of sodium azide is a typical quantity used. The equation for the decomposition of sodium azide is

$$2\ NaN_{3(s)} \rightarrow 2\ Na_{(s)} + 3\ N_{2(g)}$$

Complete **Table 2.**

Table 2

type of reaction	?
mass of $NaN_{3\,(s)}$ in typical air bag	250g
maximum amount of $Na_{(s)}$ formed	?
maximum amount of $N_{2(g)}$ formed	?
maximum mass of $Na_{(s)}$ formed	?

4. Sulfur dioxide is a significant factor in the formation of acid rain. The sulfur dioxide is oxidized in air (oxygen) to form sulfur trioxide. The sulfur trioxide readily combines with water to form sulfuric acid.

(a) Write the balanced equations for the conversion of
   (i) sulfur dioxide to sulfur trioxide
   (ii) sulfur trioxide to sulfuric acid
(b) What amount, in moles, of sulfur dioxide is contained in 1.0 t of gas emitted from a smelter?
(c) What is the maximum amount of sulfur trioxide that can be formed?
(d) If all of the sulfur trioxide from (c) combines with water, what mass of sulfuric acid is formed?

Making Connections

5. Based on a local weather forecast, complete **Table 3**.

Table 3

	Value/Description	Instrument used to make measurement
date	?	?
location	?	?
local atmospheric pressure (kPa)	?	?
local relative humidity (%)	?	?

Math Skills

6. In many jobs, the salary (without deductions) is directly related to the time spent on the job.
   (a) Using axes like those in **Figure 1**, draw a graph for the direct variation between these two variables.
   (b) Like many relationships in science, the slope of the graph has a specific meaning. What does the slope of your graph in (a) represent?
   (c) Suppose the salary varied inversely with the time spent. Using axes like those in (a), draw the graph for this inverse relationship. Would you want a job that pays this way?

Figure 1

Technical Skills and Safety

7. Flammable and combustible materials require special attention in a laboratory.
   (a) What do the WHMIS symbols in **Figure 2** represent?
   (b) Complete **Table 4** by using check marks to indicate the types of fire extinguishers suitable for the various classes of fire.

Figure 2

Table 4

Class of fire	Water	Carbon dioxide	Dry chemical
Class A (wood, paper, cloth)			
Class B (flammable liquids)			
Class C (live electrical equipment)			

Chapter

9

The Gas State

In this chapter, you will be able to

- explain different states of matter in terms of the forces among atoms, molecules, and ions;
- use the kinetic molecular theory to describe and explain the behaviour of gases;
- determine through experimentation the algebraic and graphical relationships among the pressure, volume, and temperature of an ideal gas;
- describe the mathematical relationships among the pressure, volume, temperature, and amount of an ideal gas;
- solve quantitative problems involving laws that describe the properties and behaviour of gases;
- use the terms standard temperature and pressure (including STP and SATP), absolute or Kelvin temperature, and ideal gas;
- convert between various units of pressure and between Celsius and Kelvin temperatures;
- describe various natural events and technological products and processes associated with gases;
- identify technological uses and safety concerns of compressed gases;
- identify the components of the atmosphere and describe Canadian initiatives to improve air quality.

The photograph in **Figure 1** is a dramatic depiction of how a gas can save a human life. In a car crash, an air bag, especially in combination with a seat belt, can protect the driver from serious injury. Upon collision, sensors in the steering column and in the bumper initiate the decomposition of sodium azide into sodium metal and nitrogen gas. This reaction is extremely fast: Nitrogen gas is produced and expands into the bag in less than 0.04 s. After cushioning the impact, the air bag gradually deflates as the nitrogen gas escapes through the permeable bag. Instead of taking a trip to the hospital, the driver takes a trip to the automobile body shop to have the air bag mechanism recharged and the triggering devices reset.

Air bags are not the only use of gases in the operation of automobiles: Tires and shock absorbers are inflated with pressurized air to provide a safe and comfortable ride. Air enters through the car's vents and is cooled by the air conditioner to keep us comfortable on hot summer days or is heated by the car engine to keep us warm in winter. Inside the combustion cylinders of the engine, a gasoline and oxygen explosion produces a large amount of gas at high temperature, which moves a piston. This is an example of converting chemical energy into motion. Finally, the gases emitted by automobile exhausts, such as carbon oxides and nitrogen oxides, diffuse into the atmosphere as pollutants.

As you can see, gases play an important role in both technology and our natural environment. In this chapter, you will learn more about the properties and uses of gases.

Reflect on your Learning

1. Since many gases are invisible, how do you think we can study them?
2. All around us, we see examples of all three states of matter. Why are some substances solid, liquid, or gas? How is this explained by the forces between the molecules?
3. Weather reports often refer to low- and high-pressure systems. What does pressure of a gas mean?
4. What determines the quality of the air in our atmosphere?

Throughout this chapter, note any changes in your ideas as you learn new concepts and develop your skills.

Try This **Activity**

Creating and Testing Ideas About Gases

Coming up with ideas to explain experiments with gases requires some imagination, because most gases are invisible. Can you figure out what is happening to the aluminum cans and water vapour (a gas in our atmosphere) in this activity?

Materials: water, 5 aluminum pop cans, 5 large beakers or containers, hot plate, beaker tongs, ice cubes, eye protection

Care is required handling hot items. Steam can scald skin. Switch off hot plate immediately after use.

- Place about 20 mL of water in an empty aluminum pop can.
- Heat the can on a hot plate until steam rises steadily out of the top for a couple of minutes.
- Fill a large beaker to near the top with cold water.
- Using the tongs, lift the can and move it quickly to the beaker of cold water.
- Invert the can, and dip the top rim of the can just under the surface of the water.
- Record your observations.

  (a) Create a Hypothesis for what happens.

- Repeat the Procedure without placing any water in the can. (Heat the can for a few minutes.)

  (b) What happens now? Does this support or refute your Hypothesis?

  (c) Using your original or revised Hypothesis, predict the results if you repeat the Procedure inverting the steaming can into ice water and warm water.

  (d) Try each of these, then judge your Prediction and Hypothesis.

- Recycle the cans.

Figure 1
Air bags are a good example of how knowledge of gas reactions and gas properties can be used in life-saving technology.

9.1 States of Matter

The empirical properties (**Table 1**) of the three states of matter—solid, liquid, gas—provide important clues about the nature and structure of matter. At the same temperature some substances are solids, some are liquids, and some are gases. This suggests that the forces between the particles that make up the substance vary in strength. Unlike liquids and gases, solids maintain their shape and volume (**Figure 1**). This evidence suggests that the forces between the particles in the solid are strong; in fact, these forces are the strongest of the three states. At room temperature, all ionic compounds and all metals (except mercury) are solid. Some molecular substances, including both molecular elements and compounds, can also be in the solid state. The ionic bonds in ionic compounds, intermolecular forces in some molecular substances, covalent bonds in covalent crystals, and metallic bonds in metallic substances all provide strong attractions that bond the particles tightly together. Strong bonds are also believed to prevent solids from flowing easily, since this would require particles to be able to slip past one another. Strong bonds would also suggest that there are few empty spaces between the particles, and this would explain why solids are difficult to compress.

Table 1: Empirical Properties of States of Matter

State	Properties	Example
solid	• solids have definite shape and volume • are virtually incompressible • do not flow easily	**Figure 1** A crystal of table salt
liquid	• liquids assume the shape of the container but have a definite volume • are virtually incompressible • flow readily	**Figure 2** Coloured water
gas	• gases assume the shape and volume of the container • are highly compressible • flow readily	**Figure 3** Bicycle air pump

At room temperature, most liquids are molecular compounds. This evidence plus their generally low boiling points suggest that the intermolecular forces (dipole–dipole, London, and hydrogen bonding) are not as strong as ionic, covalent, or metallic bonds. However, in these liquids, the intermolecular forces must be sufficiently strong to hold the molecules closely together but not locked in place, allowing them to move past one another. Therefore, the molecules in a liquid can spread out to take the shape of the container while keeping their volume constant (**Figure 2**). Because gases have no definite shape or volume (**Figure 3**), there appears to be an absence of forces between the molecules in a gas.

Atomic theory predicts that noble gases are composed of monatomic molecules (e.g., $Ne_{(g)}$). Since these molecules are nonpolar, intermolecular force theory suggests that any attraction between noble gas molecules must be explained by London (dispersion) forces. The larger the noble gas molecule, the larger the London forces, due to the larger number of electrons per molecule. Similarly, the intermolecular attractions between diatomic molecules of elements, for example, $H_{2(g)}$ and $Cl_{2(g)}$, can be explained by London forces. The van der Waals attraction between diatomic molecules of compounds, for example, $HCl_{(g)}$ and $CO_{(g)}$, is explained as being a combination of London and dipole–dipole forces. In general, because of large intermolecular distances in a gas, hydrogen bonding is not possible between gas molecules.

An attempt to explain the states of matter based only on the strengths of forces fails when considering how solids can be changed to liquids and liquids changed to gases by increasing the temperature. The reverse changes in state also occur when the temperature decreases. The strengths of bonds between molecules cannot be the complete explanation for the different states. To more completely understand the states of matter, we must also consider the motion of the molecules.

DID YOU KNOW ?

Atmospheres

The Earth has an atmosphere, Saturn's moon Titan has an atmosphere, but the Moon does not. Why? This can be explained, at least partially, by kinetic molecular theory and gravitational attraction. The higher the temperature, the faster the particles in an atmosphere would move. The faster they move, the more likely they are to escape the pull of the body's gravity. (Molecules, like rockets, must reach escape velocity to leave an atmosphere.) Cool bodies are more likely to have an atmosphere, as are massive bodies, which pull more strongly on atmospheric particles. The Earth and Moon are the same distance from the Sun. Without an atmosphere their temperatures would be roughly the same. However, Earth is much more massive than the Moon, and so can retain an atmosphere. Saturn's moon Titan, although its mass is less than that of Earth, has a denser atmosphere. Temperatures are much lower so far from the Sun.

The Kinetic Molecular Theory

How would you explain why a drop of food colouring added to a glass of cold water slowly spreads out, or diffuses, throughout the water? How would you explain why the amount of water in an open container slowly decreases as some of the water evaporates? For the first question, scientists would say that the molecules of food colouring and the molecules of water are moving and colliding with each other, and this causes them to mix. The answer to the second question is also molecular motion: Some of the water molecules in the open container obtain sufficient energy from collisions to escape from the liquid. The idea of molecular motion that is used to explain these observations has led to the **kinetic molecular theory,** which has become a cornerstone of modern science.

The fundamental idea of the kinetic molecular theory is that solids, liquids, and gases are composed of particles that are continually moving and colliding with other particles. These particles may be atoms, ions, or molecules. As they move about, they collide with each other and with objects in their path. Very tiny objects, such as pollen grains or specks of smoke, are buffeted by these particles in air and move erratically, as shown in **Figure 4**.

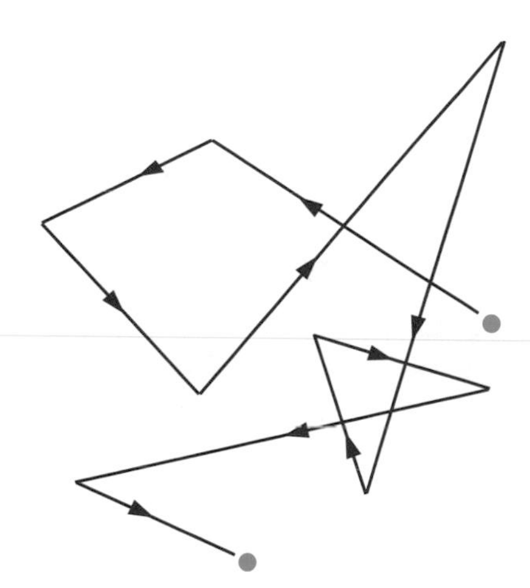

Figure 4
Observation of microscopic particles such as pollen grains or specks of smoke shows a continuous, random motion known as Brownian motion, named for Scottish scientist Robert Brown (1773–1858), who first described it. Scientists' interpretations of this evidence led to the formation of the kinetic molecular theory.

There are three types of motion that any particle can exhibit: translational (straight-line), rotational (spinning), and vibrational (back-and-forth motion of atoms within the molecule). Which type of motion predominates depends on the freedom of movement of the particles; this, in turn, depends on the strengths of the forces between the particles. If particles are restricted to mainly vibrational motion, as in solids, then particles stay together in a relatively ordered state

kinetic molecular theory: the idea that all substances contain particles that are in constant, random motion

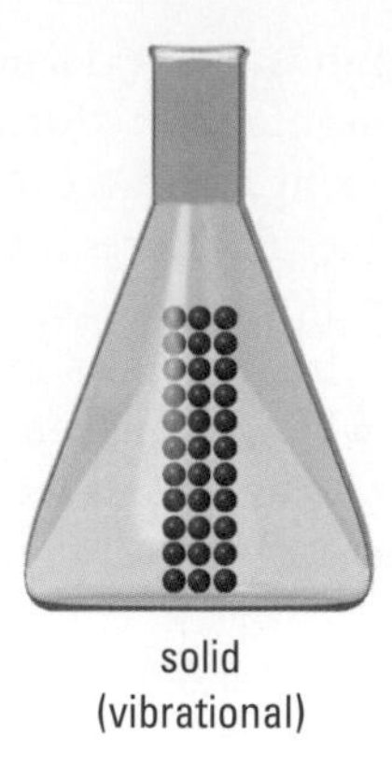

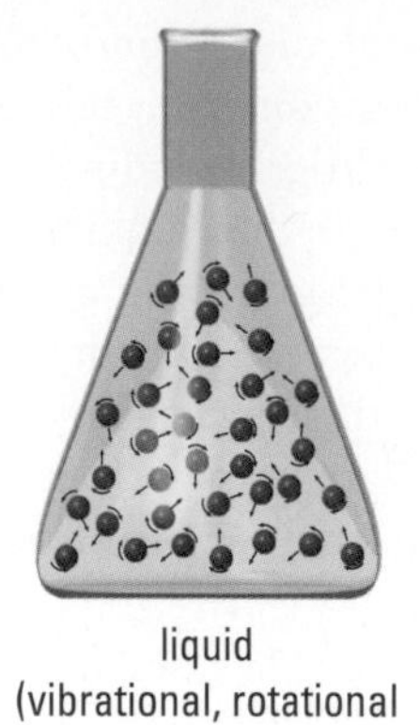

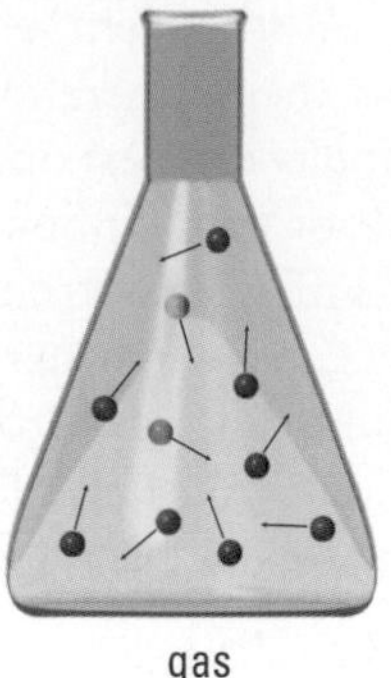

Figure 5
According to the kinetic molecular theory, the motion of molecules is different in solids, liquids, and gases. Particles in solids have primarily vibrational motion; particles in liquids have vibrational, rotational, and translational motion; and the most important form of motion in gases is translational.

(**Figure 5**). Liquids with some of each type of motion remain largely together but in a more jumbled, less orderly state. Gas molecules rotate and vibrate but their translational (straight-line) motion is most significant. This produces random collisions and the most disordered state with no organization.

Any moving object has energy called kinetic energy. A moving car, bird, and molecule all have kinetic energy. The faster the motion of an object, the greater its kinetic energy. Because the molecules of a substance are always colliding, at any instant some molecules are moving faster than others. Therefore, in a large group of molecules, there will be a range of kinetic energies from very low to very high values. The temperature of a substance is a measure of the average kinetic energy of its particles. If a substance is heated, its temperature rises until a change of state occurs. As the temperature increases, the average kinetic energy of the particles increases and, on average, each particle moves faster and has more kinetic energy. When the average kinetic energy exceeds a particular (threshold) energy, a substance changes state; for example, melts or boils. The high-energy particles escape, and with a constant energy input from the outside of the system, the temperature remains constant until the change of state is complete.

Explaining the Gas State

Theoretical chemists have created theories to explain the empirical properties of the gas state. They explain that for solids and liquids, the intermolecular forces hold the particles together. These forces of attraction are opposed by forces of repulsion when the particles get too close together. We say that the particles collide. If the intermolecular forces are relatively strong and the kinetic energy is relatively low, then a condensed state is favoured. The properties of gases (**Table 1**) suggest that the intermolecular forces are virtually nonexistent. Consider a closed bottle of perfume. Initially, the perfume molecules are contained within the bottle. If the bottle is opened in a closed room, some of the perfume molecules that are continually escaping from the solution through evaporation can now leave the open bottle. The perfume gas molecules will slowly diffuse throughout their new container—the room. Opening the door allows the perfume molecules to eventually occupy the whole building. And opening the front door of the building means that the perfume gas molecules can diffuse into the atmosphere, a very large container. There does not appear to be a limit to the diffusion of gas molecules as long as they do not react with other substances. Based on this evidence and other properties of gases (e.g., lack of a definite shape), we will assume that there are no attractive forces between gas molecules and that the molecules move in straight lines independently of each other. This is a very good approximation when starting the study of gases.

DID YOU KNOW ?

Expansion upon Heating

Most solids, liquids, and gases expand (their density decreases) when the temperature increases. This is explained as an increase in the average kinetic energy of the particles, which carries the particles farther apart before the intermolecular forces can pull them back. Of course, the increase in density when ice melts is explained differently. Do you know the explanation?

Activity 9.1.1

Molecular Motion

Models in science are used to describe abstract ideas and make them easier to understand. Some models are mathematical (e.g., mathematical models of motion describing falling bodies), some are mechanical (e.g., model cars and trains), and some are physical (e.g., physical models of the biological cell). It is important to remember that all models have limitations. The model is good if the descriptions, explanations, and predictions it gives for natural phenomena outweigh its limitations.

In this activity, you will use a molecular motion demonstrator to better visualize and understand the models of solids, liquids, and gases.

Figure 6
A molecular motion demonstrator is a mechanical device that uses small spheres to represent molecules and agitation to simulate molecular motion.

Materials

molecular motion demonstrator (**Figure 6**)
watch glass
plastic spheres
overhead projector

Procedure

1. Set the molecular motion demonstrator on the overhead projector and place the watch glass inside the metal square. Add some plastic spheres to the watch glass.
2. Turn on the power to the demonstrator and set it to low speed. Observe and record the shape and volume of the whole sample and the motion of the individual spheres.
3. Slowly increase the speed until the spheres start to leave the main group. Observe and record the characteristics of the sample and the individual spheres.
4. Remove the watch glass and place the plastic spheres inside the main compartment. Slowly increase the speed until it is at the maximum setting. Record your observations.

Analysis

(a) How does the molecular model demonstrator simulate the attractive forces between particles of a solid or liquid?

(b) Based on your observations, describe the degree of order or organization of the particles in each model of the three states of matter.

Synthesis

(c) Evaluate the ability of this model to represent the molecular motion within solids, liquids, and gases.

(d) How could you use this model to show diffusion within a liquid or gas? the compressibility of gases?

DID YOU KNOW ?

Distances Between Molecules

In a molecular solid, the distance between molecules is about the same size as the molecules themselves; in a liquid it is slightly greater; and in a gas the distance between molecules is about 20 to 30 times the size of the molecules. To picture yourself as a particle in a solid, imagine yourself seated in a regular classroom. For a liquid, picture a school dance. For a gas, imagine yourself and three friends skating randomly at the Air Canada Centre.

Practice

Understanding Concepts

1. Arrange the three states of matter in order of increasing strengths of forces between the particles. Explain your order using the evidence of shapes and volumes.
2. Which state of matter has the highest degree of order and which has the lowest? How is this related to the forces between the particles in each state?
3. The approach of starting with a single observation, such as "gases are compressible," and asking a series of "why" questions illustrates the limits of our current theories.
   (a) Why are gases compressible? Answer this question in terms of the model of a gas.
   (b) Use your answer to (a) to ask and answer a further "why" question.
   (c) How far can you extend this series of questions? Try this and stop when you cannot answer a "why" question.

Making Connections

4. Hydraulic devices, such as the brake system of a car, have a piston at one end that pushes on a liquid connected by a hose to a piston at the other end of the closed system.
   (a) What property of a liquid allows this system to work?
   (b) Why is it dangerous if the liquid (brake fluid) leaks out and is replaced by an air bubble?

Reflecting

5. Models are useful to help visualize and understand abstract ideas. Different kinds of models, such as models that use physical objects, diagrams, and mathematical equations, often appeal to different people. What kind of model works best for you? If presented with one kind of model, do you sometimes try to switch to another kind?

Section 9.1 Questions

Understanding Concepts

1. (a) List the four classes of chemical substances that are or can be solids at room temperature.
   (b) Identify the types of bonds for each class.
   (c) Explain briefly why these substances are in the solid state.
2. Solids and liquids are often referred to as condensed states.
   (a) What empirical property is the same for both these states? Explain briefly.
   (b) What properties are different? Explain the differences in terms of forces and motion.
3. For substances that are gases at room temperature,
   (a) what interpretation can be made about the forces between the molecules?
   (b) what is the predominant type of motion for gas molecules?
4. Using appropriate theoretical concepts, explain each of the empirical properties of gases listed in **Table 1**, page 418.

Making Connections

5. What properties of gases make the air bag in an automobile useful as a safety device? For each property chosen, show how this applies to air bags.

9.2 Gas Laws

Gases have always been important to us—we need them to breathe after all—but as our society has advanced technologically, the importance of gases has been expanding. We use gases in our daily lives—natural gas as fuel, gases as refrigerants, anesthetic gases for surgery—and we generate gases for special investigation, for example we create artificial atmospheres for deep-sea diving and for the exploration of outer space. It is not surprising then that the study of gases has a long history in chemistry. Many experimental properties of gases were studied long before the development of our modern understanding of the composition and molecular motion of substances. In fact, some important ideas such as atomic theory, kinetic molecular theory, and the mole concept were made possible to a large extent by the large body of empirical knowledge about gases. Let's now look more closely at the empirical properties of gases.

A Simulation of Gas Properties

Suppose five nitrogen gas cylinders are assembled using the conditions listed in **Table 1**. Each cylinder contains the same mass of nitrogen gas.

Table 1: Comparison of Nitrogen Gas Cylinders

Cylinder number	Volume (L)	Temperature (°C)
1	1.0	800
2	2.0	200
3	2.0	300
4	4.0	200
5	4.0	800

(a) What is the order of gas cylinders from most likely to least likely to explode? Write your Prediction and provide your reasoning for the order you choose.

(b) If you were designing and testing cylinders for the safe and efficient transportation of various gases, which variables would you need to consider?

Pressure and Volume: Boyle's Law

pressure: force per unit area

We live at the bottom of an ocean of air. That air has many different properties that can be altered experimentally in a laboratory, including temperature and **pressure** (properties familiar to us from weather reports), volume, and amount of gas. In any controlled experiment, the plan is to manipulate one variable and observe its effect on another variable while keeping all other properties constant. We begin our study of gases by looking at the relationship between pressure and volume at a constant temperature and amount of gas.

Earth's gravity exerts a downward force on you, and you, in turn, exert an equal force on the ground. However, the force you exert can be distributed over a large or a small area. The area is large when you lie down, and small when you stand on the tips of your toes. The greater the area, the lower the pressure. For example, when you wear snowshoes, the force is distributed over the surface area of the snowshoe, so you exert less pressure on the ground directly below your feet than you would if you were wearing regular shoes. This allows you to walk over snow instead of sinking into it. Pressure of a gas is also force per unit area, but in this case the force is exerted by the moving molecules as they collide with objects in their path, particularly the walls of a container.

DID YOU KNOW ?

Standard Pressure

Another way to think about standard pressure is as the equivalent to the weight of one kilogram resting on every square centimetre of your body.

Scientists have agreed, internationally, on units, symbols, and standard values for pressure. The SI unit for pressure, pascal (Pa), represents a force of 1 N (newton) on an area of 1 m^2; 1 Pa = 1 N/m^2. Atmospheric pressure and the pressure of many gases are often more conveniently measured in kilopascals (kPa); 1 kPa = 1000 Pa = 1 kN/m^2 (exactly).

atmospheric pressure: the force per unit area exerted by air on all objects

At sea level, average **atmospheric pressure** is about 101 kPa. Scientists used this value as a basis to define one standard atmosphere (1 atm), or *standard pressure,* as exactly 101.325 kPa. For convenience, *standard ambient pressure* has been more recently defined as exactly 100 kPa.

For many years, standard conditions for work with gases were a temperature of 0°C and a pressure of 1 atm (101.325 kPa); these conditions are known as standard temperature and pressure (STP). However, 0°C is not a convenient temperature, because laboratory temperatures are not close to 0°C. Scientists have since agreed to use another set of standard conditions, not only for gases but also for reporting the properties of other substances. The new standard is called standard ambient temperature and pressure (SATP), defined as 25°C and 100 kPa. The new standard is much closer to laboratory conditions.

Since the empirical properties of gases were measured long before the development of SI, pressure of gas has been expressed in a bewildering variety of units over the years. In 1643, Evangelista Torricelli (1608–1647), following up on a suggestion from Galileo, accidentally invented a way of measuring atmospheric pressure. He was investigating Aristotle's notion that nature abhors a vacuum. His experimental design involved inverting a glass tube filled with mercury and placing it into a tub also containing mercury (**Figure 1**). Noticing that the mercury level changed from day to day, he realized that his device, which came to be called a mercury barometer, was a means of measuring atmospheric pressure. In Torricelli's honour, standard pressure was at one time defined as 760 torr, or 760 mm Hg. (Mercury vapour is toxic; in modern mercury barometers, a thin film of water or oil is added to prevent the evaporation of mercury from the open reservoir.)

Many areas of study that employ gases, such as medicine and meteorology, and several technological applications, such as deep-sea diving, still use non-SI units (**Table 2**). Using the definitions in **Table 2**, it is possible to easily convert between SI and non-SI units.

Table 2: SI and Non-SI Units of Gas Pressure

Unit name	Unit symbol	Definition/Conversion
pascal	Pa	1 Pa = 1 N/m^2
atmosphere	atm	1 atm = 101.325 kPa (exactly)
millimetres of mercury	mm Hg	760 mm Hg = 1 atm = 101.325 kPa
torr	torr	1 torr = 1 mm Hg

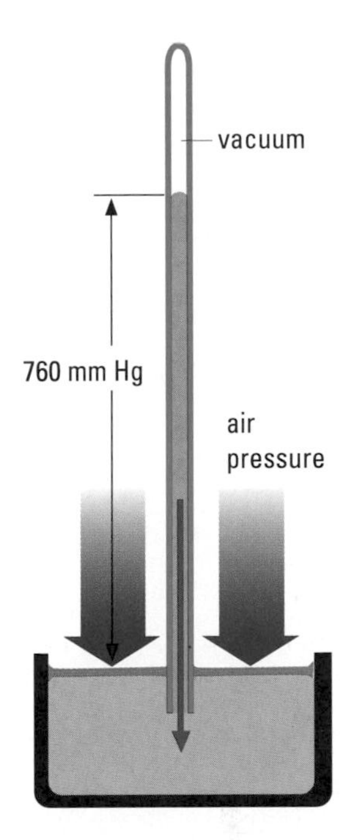

Figure 1
When a tube filled with mercury is inverted, the weight of the column of mercury pulls it toward Earth. However, the weight of air directly above the open dish pushes down on the surface of the mercury and prevents all of the mercury from falling out of the tube. The two opposing forces—weight of mercury and weight of air—balance each other when the height of mercury is about 760 mm. If the tube of mercury is longer than 760 mm, the mercury drops, leaving a vacuum above. Why is mercury used in most barometers, and not other liquids such as water, which is plentiful and nontoxic? The answer is density: Mercury is much denser than water. The weight of air in the atmosphere will support a column of water about 10 m high, which might be difficult to fit in a room!

Sample Problem 1

Convert standard ambient pressure, defined as 100 kPa, to the corresponding values in atmospheres and millimetres of mercury.

Solution

$$100\ \cancel{\text{kPa}} \times \frac{1\ \text{atm}}{101.325\ \cancel{\text{kPa}}} = 0.987\ \text{atm}$$

$$100\ \cancel{\text{kPa}} \times \frac{760\ \text{mm Hg}}{101.325\ \cancel{\text{kPa}}} = 750\ \text{mm Hg}$$

Practice

Understanding Concepts

1. Define STP and SATP.
2. Copy and complete **Table 3**. Show your work using appropriate conversion factors.

Table 3: Converting Pressure Units

	Pressure (kPa)	Pressure (atm)	Pressure (mm Hg)
(a)	0.50		
(b)	96.5		
(c)			825
(d)		2.50	

3. What are the advantages of having only one unit for pressure?

Making Connections

4. When using a medicine dropper or a meat baster, you squeeze the rubber bulb and insert the end of the tube into a liquid. Why does the liquid rise inside the dropper or baster when you release the bulb?

Answers

2. (a) 0.0049 atm, 3.8 mm Hg
   (b) 0.952 atm, 724 mm Hg
   (c) 110 kPa, 1.09 atm
   (d) 253 kPa, 1.90×10^3 mm Hg

INQUIRY SKILLS

- ○ Questioning
- ○ Hypothesizing
- ○ Predicting
- ● Planning
- ● Conducting
- ● Recording
- ● Analyzing
- ● Evaluating
- ● Communicating

Investigation 9.2.1

Pressure and Volume of a Gas

The purpose of this investigation is to determine the general relationship between the pressure and volume of a gas. Complete the **Experimental Design, Analysis,** and **Evaluation** sections of the lab report.

Question

What effect does increasing the pressure have on the volume of a gas?

Experimental Design

(a) Using the Procedure and **Figure 2**, write a brief plan to summarize this experiment.
(b) Identify the independent, dependent, and two controlled variables.
(c) Design a table to record your observations.

Materials

Boyle's law apparatus or 35-mL plastic syringe
large rubber stopper
cork borer
5 textbooks or equal masses (1 kg)
utility stand
buret clamp
mass balance

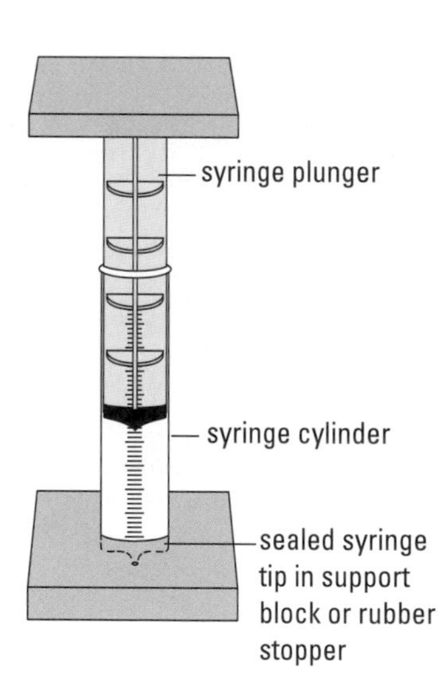

Figure 2
Setup of Boyle's law apparatus

Procedure

1. Pull out the syringe plunger so that 30 mL of air is inside the cylinder.
2. If a syringe cap is not provided, bore a small hole deep enough in the rubber stopper so that the tip of the syringe is inside the stopper. This should be a tight fit. Make sure the tip of the syringe does not leak.
3. Hold the syringe barrel vertical and measure the initial volume. Clamp the syringe on a retort stand.
4. While holding the syringe securely, carefully place one textbook or mass on the end of the plunger (**Figure 2**). (Your partner should balance the mass and be prepared to catch it if it starts to tilt.) Record the mass and new volume of air.
5. Repeat step 4 for a total of 4 or 5 books or masses.
6. If time permits, repeat steps 3 to 5 for an additional one or two trials.

Analysis

(d) Plot a graph of gas volume (or average volume from trials) versus mass added and draw a best-fit line.
(e) How does changing the mass on the syringe plunger affect the pressure on the air inside the syringe?
(f) According to the evidence you have collected, what effect does increasing pressure have on the volume of a gas?

Evaluation

(g) What are some sources of experimental error or uncertainty in this experiment? In your judgment, are these major or minor problems?
(h) How does your graph provide some indication of experimental errors or uncertainties in this experiment?
(i) Suggest some improvements that might raise the quality of the Evidence. Be as specific as possible.

The Relationship Between Pressure and Volume

Analysis of the evidence produced in an investigation similar to Investigation 9.2.1 suggests an inverse variation between the pressure and the volume of a gas; that is, as the pressure increases, the volume decreases (**Figure 3**). Using the evidence given in SI units in **Table 4**, you can see that when the pressure is doubled (100 kPa to 200 kPa), the volume is halved (3.00 L to 1.50 L). If the pressure is tripled, the volume is reduced to one-third. Check the other values to see similar results.

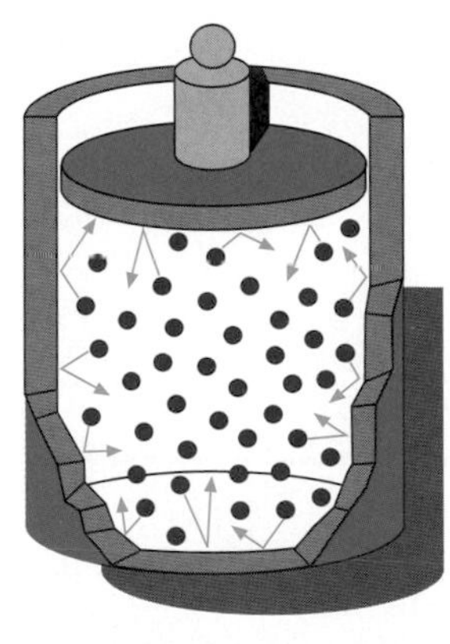

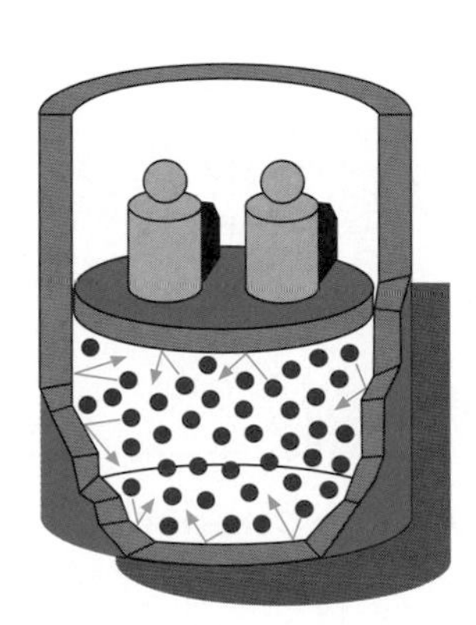

Figure 3
As the pressure on a gas increases, the volume of the gas decreases.

Table 4: Pressure and Volume of Gas Samples

Pressure (kPa)	Volume (L)	*pv* (kPa·L)
100	3.00	300
200	1.52	304
300	1.01	303
400	0.74	296
500	0.60	300

If p_1 and v_1 represent the initial conditions, the other values of pressure and volume from **Table 4** may be stated as follows:

$$(p_1, v_1) \quad (2p_1, \tfrac{1}{2}v_1) \quad (3p_1, \tfrac{1}{3}v_1) \quad (4p_1, \tfrac{1}{4}v_1) \quad (5p_1, \tfrac{1}{5}v_1)$$

For all the conditions listed above, the product of the pressure and volume is equal to p_1v_1. Mathematically, the relationship is represented as $pv = k$, where k is a constant. This simple relationship was first determined by Robert Boyle in 1662 (**Figure 4, page 428**). **Boyle's law** states that *as the pressure on a gas increases, the volume of the gas decreases proportionally, provided that the temperature and amount of gas remain constant.* In other words, the volume of a gas is inversely proportional to the pressure of the gas, providing that the temperature and amount of gas are held constant. Boyle's law can be conveniently written comparing any two sets of pressure and volume measurements:

Boyle's law: as the pressure on a gas increases, the volume of the gas decreases proportionally, provided that the temperature and amount of gas remain constant; the volume and pressure of a gas are inversely proportional

$$p_1v_1 = p_2v_2 \quad \textit{(Boyle's law)}$$

This can also be expressed as a calculation of a new pressure inversely related to the volumes ratio:

$$p_2 = \frac{p_1v_1}{v_2}$$

Figure 4
Anglo-Irish chemist Robert Boyle (1627–1691) determined the effect of pressure on the volume of a gas in quantitative terms. He was a founding member of the Royal Society of London and is reported to have coined its anti-Aristotelian motto: "Nothing by Authority." In the early 1660s, Boyle worked with Robert Hooke, the able inventor, who helped him construct an air pump. Using this necessary technology, Boyle demonstrated the physical characteristics of air and the necessary role of air in combustion, respiration, and the transmission of sound. In 1661, he reported to the Royal Society on the relationship now known as Boyle's law. Boyle became so famous that foreign academics wouldn't consider a trip to England complete until they had met him. He was elected president of the Royal Society in 1680, but declined the honour.

Answers

6. 263 kPa
7. 137 L
8. 0.16 L
9. 21 kL

Sample Problem 2

A 2.0-L party balloon at 98 kPa is taken to the top of a mountain where the pressure is 75 kPa. Assume the temperature is the same. What is the new volume of the balloon?

Solution

$v_1 = 2.0 \text{ L}$
$p_1 = 98 \text{ kPa}$
$p_2 = 75 \text{ kPa}$
$v_2 = ?$

$$p_1 v_1 = p_2 v_2$$

$$v_2 = \frac{p_1 v_1}{p_2}$$

$$= \frac{98 \text{ kPa} \times 2.0 \text{ L}}{75 \text{ kPa}}$$

$$v_2 = 2.6 \text{ L}$$

or $$v_{\text{balloon}} = 2.0 \text{ L} \times \frac{98 \text{ kPa}}{75 \text{ kPa}} = 2.6 \text{ L}$$

The new volume of the balloon is 2.6 L.

Practice

Understanding Concepts

5. Define atmospheric pressure.
6. A bicycle pump contains 0.650 L of air at 101 kPa. If the pump is closed, what pressure is required to change the volume to 0.250 L?
7. A weather balloon containing 35.0 L of helium at 98.0 kPa is released and rises. Assuming the temperature is constant, find out the volume of the balloon when the atmospheric pressure is 25.0 kPa.
8. A small oxygen canister contains 110 mL of oxygen gas at a pressure of 3.0 atm. This oxygen is released into a balloon with a final pressure of 2.0 atm. What is the final volume of the balloon?
9. A diving bell contains 32 kL of air at a pressure of 98 kPa at the surface. About 5 m below the surface, the pressure on the air trapped inside the bell is 150 kPa (**Figure 5**). What is the volume of air in the bell, if you assume the temperature remains the same?

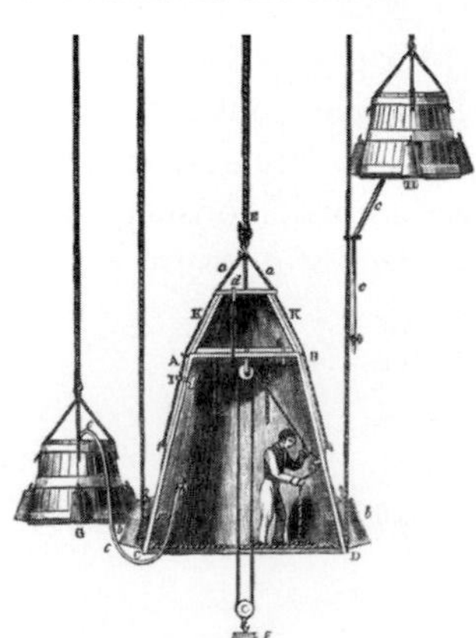

Figure 5
Before underwater diving apparatus became common, divers used a diving bell to explore underwater.

10. Why does atmospheric pressure depend on your location or vary over time at your location?

Making Connections

11. Use the Internet to investigate the invention and refinement of barometers and manometers as technologies to measure the pressure of a gas. Create a chronological flow chart or multimedia presentation including technologies, dates, names, and diagrams.

Follow the links for Nelson Chemistry 11, 9.2.

GO TO www.science.nelson.com

Volume and Temperature: Charles's Law

More than a century after Boyle had determined the relationship between the volume and pressure of a gas, French physicist Jacques Charles (**Figure 6**) determined the relationship between the volume and temperature of a gas. Charles became interested in the effect of temperature on gas volume after observing the hot-air balloons that had become popular as flying machines.

Figure 6
Jacques Charles (1746–1823) designed and flew the first hydrogen balloon in 1783. His invention was based on Archimedes' concept of buoyancy, Henry Cavendish's calculations for the density of hydrogen, and his own observations. Later, his experiences and experiments led to the formulation of Charles's law.

Investigation 9.2.2

Temperature and Volume of a Gas

INQUIRY SKILLS

- ○ Questioning
- ○ Hypothesizing
- ○ Predicting
- ○ Planning
- ● Conducting
- ● Recording
- ● Analyzing
- ● Evaluating
- ● Communicating

The purpose of this investigation is to determine how the temperature and volume of a gas are related. Complete the **Analysis** and **Evaluation** sections of the lab report.

Question

What effect does increasing the temperature have on the volume of a gas?

Experimental Design

A volume of air is sealed inside a syringe, which is then placed in a water bath. As the temperature of the water (independent variable) is changed, the volume of air (dependent variable) is measured. The amount of gas inside the syringe and the pressure on the gas are two controlled variables.

(a) Read the Procedure and design a table to record your observations.

Materials

lab apron
eye protection
600-mL beaker
water, 600 mL, room temperature
plastic syringe (35–60 mL)
cap or stopper for the syringe tip
buret clamp
thermometer and clamp
ring stand
plastic stirring rod
hot plate

Heat the water slowly and ensure that the tested gas in the syringe does not eject the syringe plunger. Wear eye protection.

Procedure

1. Set the syringe plunger to about 15–20 mL of air.
2. Seal the tip of the syringe with a cap or stopper.
3. Set up the ring stand with the 600-mL beaker on the hot plate (**Figure 7**).
4. Use the buret clamp to hold the syringe as far as possible into the beaker without touching the sides or bottom.
5. Clamp the thermometer so that the bulb is beside the end of the plunger but not touching the syringe.
6. Add room-temperature water to about 1 cm from the top of the beaker. After a few minutes, record the temperature and volume of air.
7. Turn on the hot plate.
8. Heat the water slowly, stirring occasionally.
9. Record the gas volume and temperature about every 10°C until about 90°C. (It may be necessary to tap or twist the plunger occasionally to make sure it is not stuck.)

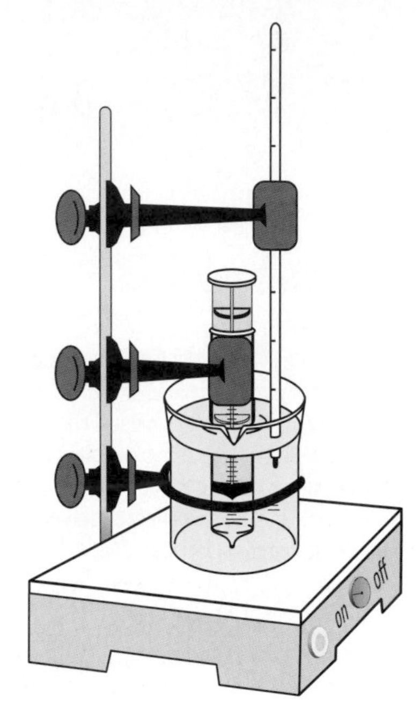

Figure 7
Setup of apparatus

Analysis

(b) Plot a graph of gas volume versus temperature and draw a best-fit line.
(c) According to the evidence you collected, what effect does increasing the temperature have on the volume of a gas?

Evaluation

(d) Within your lab group, discuss the Experimental Design, Materials, and Procedure used in this experiment. Decide whether these were adequate, and state your reasons in your report. List some sources of error or uncertainty and possible improvements that would raise the quality of the evidence collected.
(e) Write a summary of your discussions and indicate how certain you are about the results of this experiment.

Kelvin Temperature Scale

The mathematical equation describing the relationship between temperature and volume may not be apparent from the graph you created in Investigation 9.2.2; however, if the two variables are graphed as in **Figure 8(a)**, a straight line is obtained, so a simple relationship does exist. When the line is extrapolated downward, it meets the horizontal axis at –273°C. It appears that, if the gas did not liquefy, its volume would become zero at –273°C. If this experiment is repeated with different quantities of gas or with samples of different gases, straight-line relationships between temperature and volume are also observed. When the lines are extrapolated, they all meet at –273°C, as shown in **Figure 8(b)**. This temperature, called **absolute zero**, is the lowest possible temperature. Scientists with sophisticated technology are coming within an increasingly smaller fraction of a degree from absolute zero.

absolute zero: believed to be the lowest possible temperature

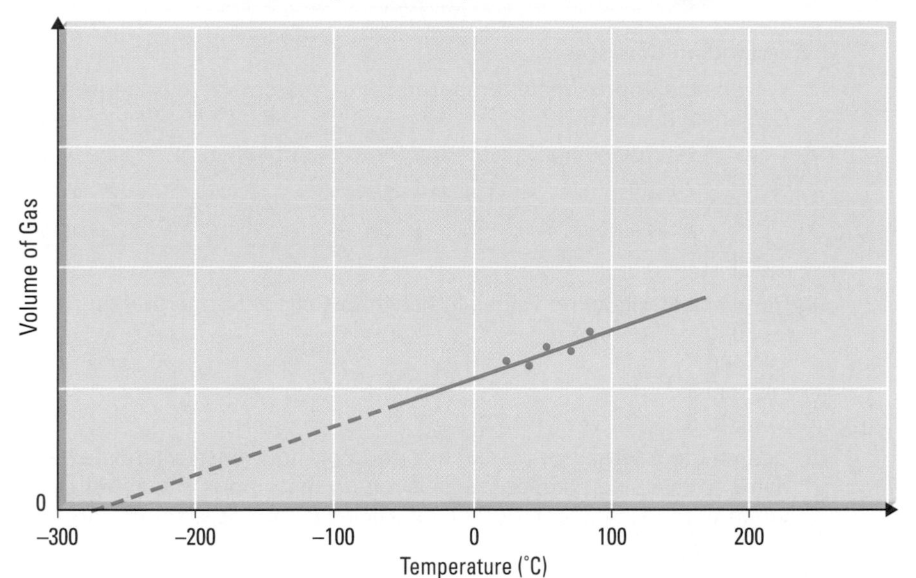

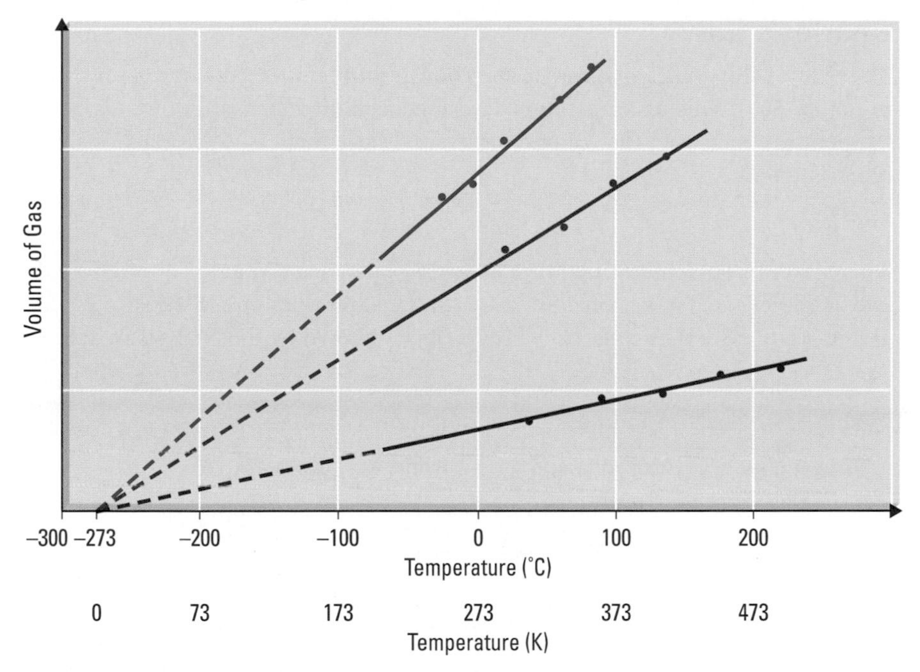

Figure 8
When the graphs of several careful volume–temperature experiments are extrapolated, all the lines meet at absolute zero, –273°C or 0 K.

DID YOU KNOW?

Lord Kelvin

Sir William Thomson (1824–1907), also known as Lord Kelvin, was a Scottish engineer, mathematician, and physicist who profoundly influenced the scientific thought of his generation. His contributions to science included the absolute temperature scale (measured in kelvin). The style and character of Thomson's scientific and engineering work reflected his active personality. While a student at the University of Cambridge, he was awarded silver sculls for winning the university championship in racing single-seater rowing shells. He was a traveller all of his life; he spent much time in Europe and made several trips to the United States. Thomson risked his life several times during the laying of the first transatlantic cable.

Absolute zero is the basis of another temperature scale, called the absolute or **Kelvin temperature scale**. On the Kelvin scale, absolute zero (–273°C) is zero kelvin (0 K), as shown in **Figure 8(b).** (Note that no degree symbol is used for kelvin.) To convert degrees Celsius to kelvin, add 273 (**Figure 9**, page 432). STP and SATP are each defined by two exact values with infinite significant digits (i.e., STP is 273.15 K and 101.325 kPa; SATP is 298.15 K and 100 kPa). For convenience, however, use STP as 273 K and 101 kPa and SATP as 298 K and 100 kPa. Several other values are commonly rounded off to three significant digits for calculation purposes; for example, Avogadro's number, 6.02×10^{23}.

Kelvin temperature scale: a temperature scale with zero kelvin (0 K) at absolute zero and the same size divisions as the Celsius temperature scale

Answers

13. (a) 273 K
    (b) 373 K
    (c) 243 K
    (d) 298 K
14. (a) –273°C
    (b) –173°C
    (c) 27°C
    (d) 100°C

Practice

Understanding Concepts

12. What is the approximate temperature for absolute zero in degrees Celsius and kelvin?
13. Convert the following Celsius temperatures to kelvin:
    (a) 0°C
    (b) 100°C
    (c) –30°C
    (d) 25°C
14. Convert the following values in kelvin to Celsius temperatures:
    (a) 0 K
    (b) 100 K
    (c) 300 K
    (d) 373 K
15. Search the Internet for research reports on how close scientists have come to reaching absolute zero. What do the reports say about whether the kinetic energy of all particles is zero at absolute zero?
    Follow the links for Nelson Chemistry 11, 9.2.

www.science.nelson.com

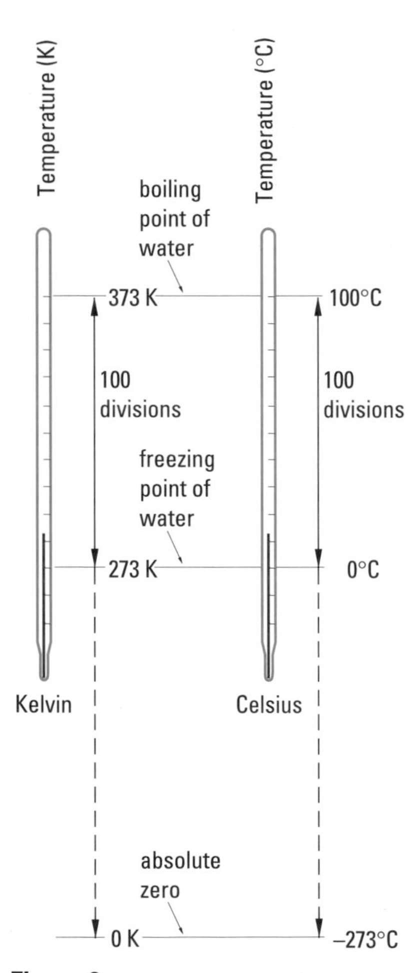

Figure 9
Jacques Charles predicted –273°C to be the temperature at which the volume of a gas would become zero, if the gas did not liquefy before reaching that temperature. Lord Kelvin considered –273°C to be the temperature at which the kinetic energy of all particles of solids, liquids, or gases would become zero. The debate continues.

Charles's Law

The relationship in kelvin between the volume and temperature of a gas is shown in **Figure 8(b)**. This relationship is described as a direct variation; that is, as the temperature increases, the volume increases. Mathematically, this relationship is represented as

$$v = kT$$

where T represents the temperature in kelvin. This means that the quotient of the two variables (v/T) has a constant value (k), which is the slope of the straight-line graph (**Figure 8(b)**). A constant value is clearly shown by the analysis in **Table 5**.

Table 5: Analysis of Temperature and Volume of a Gas Sample

Temperature, *t* (°C)	Temperature, *T* (K)	Volume, *v* (L)	Constant, *v/T* (L/K)
25	298	5.00	0.0168
50	323	5.42	0.0168
75	348	5.84	0.0168
100	373	6.26	0.0168
125	398	6.68	0.0168

The relationship between volume and absolute temperature is known as **Charles's law**, which states that, *as the temperature of a gas increases, the volume increases proportionally, provided that the pressure and amount of gas remain constant* (**Figure 10**). Charles's law can be conveniently written comparing any two sets of volume and temperature measurements:

$$\frac{v_1}{T_1} = k \text{ and } \frac{v_2}{T_2} = k$$

$$\text{therefore, } \frac{v_1}{T_1} = \frac{v_2}{T_2} \text{ (Charles's law)}$$

This can also be expressed as a calculation of a new volume directly related to the temperatures ratio:

$$v_2 = v_1\frac{T_2}{T_1} \text{ or } v_2 = \frac{v_1T_2}{T_1}$$

Charles's law: the volume of a gas varies directly with its temperature in kelvin, if the pressure and amount of gas are constant

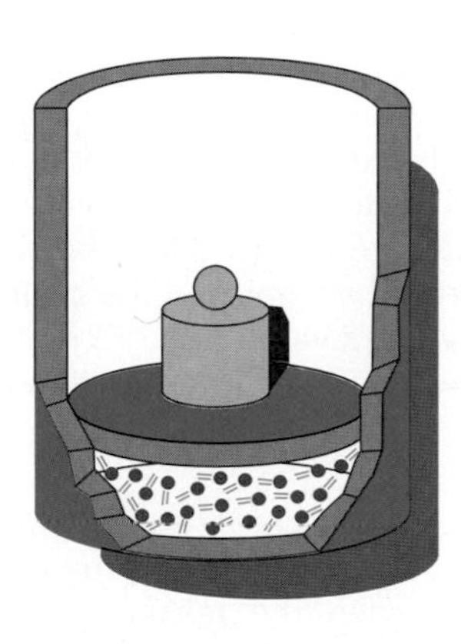

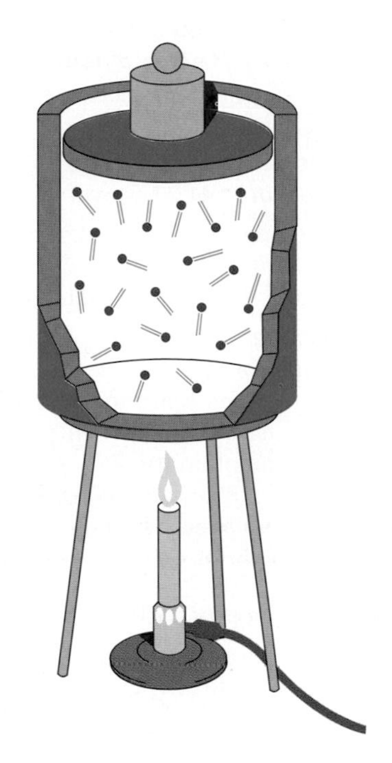

Figure 10
The volume of a gas—in this case, in a container with a movable piston—increases as the temperature of the gas increases. The pressure, equal to the pressure exerted by the mass, the piston, and the atmosphere, remains constant.

Sample Problem 3

A gas inside a cylinder with a movable piston (**Figure 10**) is to be heated to 315°C. The volume of gas in the cylinder is 0.30 L at 25°C. What is the final volume when the temperature is 315°C?

Solution

$v_1 = 0.30$ L

$T_1 = 25°C = 298$ K

$v_2 = ?$

$T_2 = 315°C = 588$ K

$$\frac{v_1}{T_1} = \frac{v_2}{T_2}$$

$$v_2 = \frac{v_1T_2}{T_1}$$

$$= \frac{0.30 \text{ L} \times 588 \not{\text{K}}}{298 \not{\text{K}}}$$

$$v_2 = 0.59 \text{ L}$$

or $v_{N_2} = 0.30 \text{ L} \times \frac{588 \not{\text{K}}}{298 \not{\text{K}}}$

$$v_{N_2} = 0.59 \text{ L}$$

The final volume of the nitrogen gas at 315°C is 0.59 L.

Answers

16. 11.3 mL
17. 0.12 L
18. 26% increase
19. 1.20 L
21. (a) 3.82:1.00

Practice

Understanding Concepts

16. Butane lighters work very poorly outdoors in very cold weather. If 12.7 mL of butane gas is released from a lighter at 22°C, what volume would this same amount of butane occupy at –11°C?
17. An open, "empty" 2-L plastic pop container, which has an actual inside volume of 2.05 L, is removed from a refrigerator at 5°C and allowed to warm up to 21°C on a kitchen counter. What volume of air, measured at 21°C, will leave the container as it warms?
18. Cooking pots have loose-fitting lids to allow air to escape while food is being heated. If a 1.5-L saucepan is heated from 22°C to 100°C, any gas in the pan will increase in volume by what percentage?
19. Jacques Charles became interested in temperature–volume relationships for gases because of his curiosity about hot-air balloon flight, at a time when burning straw, with all its hazards and inconveniences, was used to heat the interior air. Hot-air balloons are open containers that maintain the air inside at (very nearly) atmospheric pressure. When a modern balloon's propane burner has warmed the air inside from an average value of 20°C to an average value of 80°C, find the final volume of each 1.00 L of air that was initially in the balloon.

Applying Inquiry Skills

20. A student decides to make a gas expansion thermometer by trapping some air (about 50–70 mL) inside an inverted 100-mL graduated cylinder, the open end of which is submerged in a beaker of water. The student reasons that she should be able to calculate the temperature of the surrounding air by measuring the volume of air inside the cylinder using the graduated scale on the cylinder walls.

    Evaluate the design of this technology, using your knowledge of gas behaviour, and predict whether this design would provide accurate values. Suggest possible improvements.

Making Connections

21. Jet aircraft engines use energy from burning fuel to power the process of taking in cold air and releasing hot gases. The 78% of the intake air that is nitrogen reacts only in negligible amounts, so basically four-fifths of the air is just heated strongly. The expanding gas mixture escapes backward, and the reaction of this force drives the engine forward.
    (a) Assuming the $N_{2(g)}$ in the air is heated from –60°C to 540°C in the engine, express the volume increase as a ratio of final volume to initial volume, to three significant digits.
    (b) Describe what other work the expanding gases in a jet engine must do, besides providing forward thrust. (Hint: An older term is "turbojet" engine.)

Pressure and Temperature Law

If you read the warning on any aerosol can, such as a can of spray paint, you will see a caution about the danger of the can exploding if heated, for example, in a fire. As you might expect from this warning, raising the temperature increases the pressure of the gas inside the can until it can no longer contain the pressure and the can ruptures. Mathematically, the direct variation is represented as

$$p = kT \text{ or } \frac{p}{T} = k$$

If we assume the volume and amount of the gas remain the same, the quotient of the two variables (p/T) has a constant value (k). This means that two sets of pressure and temperature measurements can easily be compared. This is the **pressure and temperature law** (sometimes called Gay-Lussac's law):

pressure and temperature law: The pressure exerted by a gas varies directly with the absolute temperature if the volume and amount of gas remain constant.

$$\frac{p_1}{T_1} = \frac{p_2}{T_2} \quad \textit{(pressure and temperature law)}$$

This can also be expressed as a calculation of a new pressure directly related to the temperatures ratio:

$$p_2 = p_1 \frac{T_2}{T_1}$$

Note the similarity between this law and Charles's law comparing the volume and temperature of a gas. Both laws represent direct relationships and both require the use of absolute temperature in kelvin.

Sample Problem 4

A sealed storage tank contains argon gas at 18°C and a pressure of 875 kPa at night. What is the new pressure if the tank and its contents warm to 32°C during the day?

Solution

$T_1 = 18°C = 291\ K$
$p_1 = 875\ kPa$
$p_2 = ?$
$T_2 = 32°C = 305\ K$

$$\frac{p_1}{T_1} = \frac{p_2}{T_2}$$

$$p_2 = \frac{p_1 T_2}{T_1}$$

$$= \frac{875\ kPa \times 305\ \cancel{K}}{291\ \cancel{K}}$$

$$p_2 = 917\ kPa$$

or $p_{Ar} = 875\ kPa \times \frac{305\ \cancel{K}}{291\ \cancel{K}}$

$$p_{Ar} = 917\ kPa$$

The new pressure is 917 kPa.

DID YOU KNOW ?

Gay-Lussac's Law?

According to some books, the French chemist Joseph Gay-Lussac discovered the direct relationship between the temperature and pressure of a gas in the early 1800s. Therefore, this relationship is sometimes called Gay-Lussac's law. However, history of science references say that Charles, Dalton, and Gay-Lussac were all involved in investigating this relationship, with Charles and Dalton doing their work before Gay-Lussac.

Practice

Understanding Concepts

22. A closed, "empty" tank containing air at 97 kPa and 22°C survives intact in a fire. If the tank is able to withstand a maximum internal pressure of 350 kPa, what is the maximum temperature it could have reached during the fire?
23. Use the kinetic molecular theory to provide an explanation for the increase in pressure of a constant volume of gas when the temperature rises.

Answer

22. 791°C

Answers

24. (a) 167 kPa
    (b) 2.80%
25. (a) 328 kPa
    (b) 228 kPa

Applying Inquiry Skills

24. A sample of neon gas in a gas cylinder has a pressure of 125 kPa at 300 K. The cylinder is slowly heated to a temperature of 400 K.
    (a) Using the pressure and temperature law, predict the pressure at 400 K.
    (b) If the measured pressure in the cylinder at 400 K is 162 kPa, what is the percent difference?

Making Connections

25. Car manufacturers suggest that you check and adjust the air in the tires of a car when the tires are cold. To do this, you use a pressure gauge, which must read zero before you attach it to the valve stem of the tire. Therefore, the pressure reading on the gauge is actually the amount by which the tire pressure exceeds atmospheric pressure.
    (a) Suppose the pressure gauge shows a tire pressure of 210 kPa (for a total pressure of 310 kPa) at 21°C. After driving for a period of time, the tires reach a temperature of 38°C. What is the new total pressure?
    (b) What would the pressure gauge read at the tire temperature of 38°C?
    (c) What problems may be created if you set the recommended tire pressure when the tires are hot from a long period of highway driving?

The Combined Gas Law

combined gas law: The product of the pressure and volume of a gas sample is proportional to its absolute temperature in kelvin; $pv = kT$.

When Boyle's, Charles's, and the pressure–temperature laws are combined, the resulting **combined gas law** states the relationship among the volume, temperature, and pressure of any fixed amount of gas:

Boyle's law: $pv =$ a constant (T and n are controlled variables)

Charles's law: $\frac{v}{T} =$ a constant (p and n are controlled variables)

pressure–temperature law: $\frac{p}{T} =$ a constant (v and n are controlled variables)

If the product pv is constant at a fixed temperature, then $p(\frac{v}{T})$ should also be a constant because v divided by a constant temperature is also constant. If the temperature changes, then Charles's law tells us that the ratio $\frac{v}{T}$ is constant at a fixed pressure. Therefore, multiplying a constant pressure by a constant ratio of volume to temperature certainly produces a number that is a constant. Alternatively, you could employ the same reasoning using Boyle's and the pressure and temperature laws to again show that $(\frac{p}{T})v$ must also be constant. Using this reasoning or a mathematical method of joint variation, we can conclude that the product of the pressure and volume of a gas divided by its absolute temperature is a constant as long as the amount of gas is controlled, that is, does not change:

$$\frac{pv}{T} = k$$

The relationship can be expressed in a convenient form for calculations involving changes in volume, temperature, or pressure for a particular gas sample:

$$\frac{p_1v_1}{T_1} = \frac{p_2v_2}{T_2} \quad \textit{(combined gas law)}$$

The combined gas law is a useful starting point for all cases involving pressure, volume, and temperature, even if one of these variables is constant (as in Boyle's, Charles's, and the pressure–temperature laws). A variable that is constant can easily be eliminated from the combined gas law equation. For example, a steel cylinder with a fixed volume contains a gas at a pressure of 652 kPa and a temperature of 25°C. If the cylinder is heated to 150°C, what will be the new pressure? Because the volume is constant, we can cancel v_1 and v_2 from the combined gas law equation because $v_2 = v_1$:

$$\frac{p_1 \cancel{v_1}}{T_1} = \frac{p_2 \cancel{v_2}}{T_2}$$

We can now solve for p_2 and then substitute the pressures and temperatures (after converting to kelvin):

$$p_2 = \frac{p_1 T_2}{T_1}$$

$$= \frac{652 \text{ kPa} \times 423 \cancel{\text{K}}}{298 \cancel{\text{K}}}$$

$$P_2 = 925 \text{ kPa}$$

This can also be expressed as a calculation of a new pressure directly related to the temperatures ratio and inversely related to the volumes ratio (which cancels to one in this case):

$$p_{gas} = p_1 \frac{T_2}{T_1} \frac{\cancel{v_1}}{\cancel{v_2}}$$

$$= 652 \text{ kPa} \times \frac{423 \cancel{\text{K}}}{298 \cancel{\text{K}}}$$

$$p_{gas} = 925 \text{ kPa}$$

If we assume that the steel walls are sufficiently strong, the gas will have a pressure of 925 kPa inside the cylinder.

Sample Problem 5

A balloon containing hydrogen gas at 20°C and a pressure of 100 kPa has a volume of 7.50 L. Calculate the volume of the balloon after it rises 10 km into the upper atmosphere, where the temperature is –36°C and the outside air pressure is 28 kPa. Assume that no hydrogen gas escapes and that balloons are free to expand so that the gas pressure within them remains equal to the air pressure outside.

Solution

$T_1 = 20°\text{C} = 293 \text{ K}$

$p_1 = 100 \text{ kPa}$

$v_1 = 7.50 \text{ L}$

$v_2 = ?$

$T_2 = -36°\text{C} = 237 \text{ K}$

$p_2 = 28 \text{ kPa}$

$$\frac{p_1 v_1}{T_1} = \frac{p_2 v_2}{T_2}$$

$$v_2 = \frac{p_1 v_1 T_2}{p_2 T_1}$$

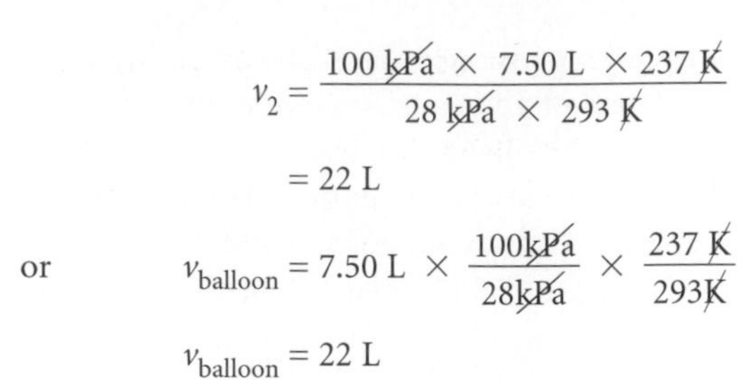

$$v_2 = \frac{100\ \text{kPa} \times 7.50\ \text{L} \times 237\ \text{K}}{28\ \text{kPa} \times 293\ \text{K}}$$

$$= 22\ \text{L}$$

or

$$v_{\text{balloon}} = 7.50\ \text{L} \times \frac{100\text{kPa}}{28\text{kPa}} \times \frac{237\ \text{K}}{293\text{K}}$$

$$v_{\text{balloon}} = 22\ \text{L}$$

The volume of the balloon is 22 L.

Figure 11
The lightness of baked goods such as bread and cakes is a result of gas bubbles trapped in the dough or batter when it is heated. The leavening, or production of gas bubbles, can be due to vaporization of water, expansion of gases already in the dough or batter, or leavening agents such as yeast and baking powder. Yeasts are living organisms that feed on sugar, producing carbon dioxide and either water or ethanol; baking powder is a mixture of sodium hydrogen carbonate and a solid acid that react together to produce carbon dioxide; the bubbles of gas are part of the light and delectable baked goods that result from kitchen chemistry.

SUMMARY Gas Laws

STP: 0°C and 101.325 kPa (exact values)
SATP: 25°C and 100 kPa (exact values)

101.325 kPa = 1 atm = 760 mm Hg (exact values) or 101 kPa (for calculation)
absolute zero = 0 K or –273.15°C, or –273°C (for calculation)
T (K) = t (°C) + 273 (for calculation)

Boyle's law: $p_1v_1 = p_2v_2$ (for constant temperature and amount of gas)

Charles's law: $\frac{v_1}{T_1} = \frac{v_2}{T_2}$ (for constant pressure and amount of gas)

pressure–temperature law: $\frac{p_1}{T_1} = \frac{p_2}{T_2}$ (for constant volume and amount of gas)

combined gas law: $\frac{p_1v_1}{T_1} = \frac{p_2v_2}{T_2}$ (for constant amount of gas)

Practice

Understanding Concepts

26. A syringe contains 50.0 mL of a gas at a pressure of 101 kPa. The end is sealed and the plunger is pushed to compress the gas to a volume of 12.5 mL. What is the new pressure, assuming constant temperature?
27. Carbon dioxide produced by yeast in bread dough causes the dough to rise, even before it is baked (**Figure 11**). During baking, the carbon dioxide gas expands. Predict the final volume of 0.10 L of carbon dioxide in bread dough that is heated from 25°C to 190°C at a constant pressure.
28. A storage tank is designed to hold a fixed volume of butane gas at 150 kPa and 35°C. To prevent dangerous pressure buildup, the tank has a relief valve that opens at 250 kPa. At what (Celsius) temperature does the valve open?
29. A balloon has a volume of 5.00 L at 20°C and 100 kPa. What is its volume at 35°C and 90 kPa?
30. A cylinder of helium gas has a volume of 1.0 L. The gas in the cylinder exerts a pressure of 800 kPa at 30°C. What volume would this gas occupy at SATP?
31. For any of the calculations in the previous questions, does the result depend on the identity of the gas? Explain briefly.

Answers

26. 404 kPa
27. 0.16 L
28. 240°C
29. 5.8 L
30. 7.9 L

32. A 2.0-mL bubble of gas is released at the bottom of a lake where the pressure is 6.5 atm and the temperature is 10°C. What is the volume of the gas bubble when it reaches the surface, where the pressure is 0.95 atm and the temperature is 24°C?
33. What assumption was made in all of the previous calculations?

Answer

32. 14 mL

Making Connections

34. Popcorn is a favourite snack food for many people (**Figure 12**). As you learned in Chapter 4, the corn kernel is heated, and some of the moisture inside the kernel vaporizes, starting a chain of events that leads to the tasty popped corn.
    (a) If we assume a constant volume kernel (before popping), what happens to the pressure inside the kernel as the temperature increases? Justify your answer using appropriate mathematical equations or relations.
    (b) The pressure inside the kernel forces some superheated water and steam to penetrate into the starch granules, making them soft and gelatinous. When the hull of the kernel breaks at about 900 kPa, what happens to the volume of water vapour when the pressure quickly drops to about 100 kPa? Justify your answer using appropriate mathematical equations or relations.

Figure 12
Popcorn was used by the original Native peoples in North America long before the arrival of Europeans. The popping method used very hot clay pots, which is a method similar to today's hot-air poppers.

Reflecting

35. When you are solving questions, the gas laws are like tools. How do you know which gas law is the appropriate one to use? How can the combined gas law be used instead of either Boyle's or Charles's law?

Section 9.2 Questions

Understanding Concepts

1. Copy and complete **Table 6**. Show your work using appropriate conversion factors.

Table 6: Using Pressure Units

	Pressure (kPa)	Pressure (atm)	Pressure (mm Hg)
(a)		0.875	
(b)	25.0		
(c)			842

2. Copy and complete **Table 7**.

Table 7: Celsius and Kelvin

	t (°C)	*T* (K)
(a)	25	
(b)	–35	
(c)		312
(d)		208

(continued)

3. The gas laws described in this section involve the properties of volume, pressure, and temperature. Some of these variables have a direct relationship (as one increases, so does the other), and some have an inverse relationship (as one increases, the other decreases). For each pair of the following variables, state whether the relationship is direct or inverse:
   (a) pressure and volume at a constant temperature
   (b) temperature and volume at a constant pressure
   (c) temperature and pressure at a constant volume
   (d) What other property of a gas must also be constant for all of the above?
4. An automobile tire has an internal volume of 27 L at 225 kPa and 18°C.
   (a) What volume would the air inside the tire occupy if it escaped? (Atmospheric pressure at the time is 98 kPa and the temperature remains the same.)
   (b) How many times larger is the new volume compared with the original volume? How does this compare with the change in pressure?
5. In a cylinder of a diesel engine, 500 mL of air at 40.0°C and 1.00 atm is powerfully compressed just before the diesel fuel is injected. The resulting pressure is 35.0 atm. If the final volume is 23.0 mL, what is the final temperature in the cylinder?

Applying Inquiry Skills

6. The purpose of the following investigation is to test the combined gas law for the relationship between the pressure and the temperature of a gas. Complete a report, including the **Hypothesis, Experimental Design, Analysis,** and **Evaluation.**

   Question
   What effect does the temperature of nitrogen gas have on the pressure it exerts (**Figure 13**)?

   Hypothesis
   (a) State the hypothesis used to answer the Question and provide your reasoning.

   Experimental Design
   (b) Briefly describe an experiment, using the apparatus in **Figure 13**, that would allow you to answer the Question.

   Analysis
   (c) Analyze the Evidence in **Table 8**. Include in your analysis a graph and final word statement answering the Question.

   Evaluation
   (d) Evaluate the Evidence and the Hypothesis.

Table 8: Evidence

Temperature (°C)	Pressure (kPa)
0	100
20	106
40	115
60	123
80	129
100	135

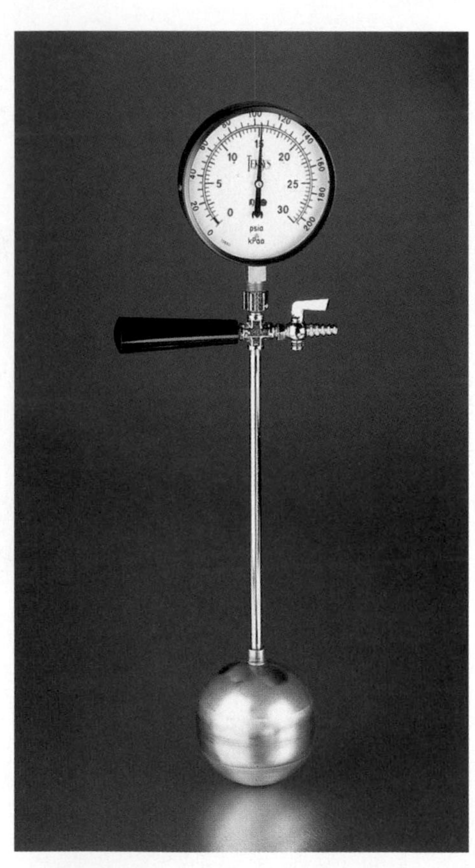

Figure 13
This apparatus consists of a hollow metal sphere to which a pressure gauge is attached. Because the gas inside the sphere cannot expand, the relationship between temperature and pressure of a gas can be determined.

Making Connections

7. For a typical geyser (**Figure 14**), underground water seeps into a deep narrow shaft in the ground and is heated from below. Because of the depth, the pressure on the water is high so the water at the bottom of the shaft boils at a much higher temperature than normal.
   (a) What happens to the volume of a 1.0-L bubble of water vapour at 130°C and 305 kPa when it reaches the surface, where the conditions are 93°C and 101 kPa?
   (b) Why is a narrow shaft necessary to produce the geyser effect?

Figure 14
Geysers are unusual and dramatic examples of geothermal energy used to heat water in a confined space.

9.3 Compressed Gases

Not only are gases a major part of our lives, but compressed gases, that is, gases at pressures above atmospheric pressure, are particularly useful:

- The tires of vehicles contain pressurized air.
- Many people use gas barbecues with a pressurized propane fuel tank.
- Aerosol cans contain a propellant that carries the contents of the can out the nozzle; the propellant is a pressurized gas.
- Major surgery usually involves oxygen administered from a pressurized oxygen tank and is often accompanied by an anesthetic, which may also be a pressurized gas, such as dinitrogen monoxide.

Certain occupations require some work with pressurized gases. In the medical field, paramedics and doctors use oxygen tanks. Firefighters use compressed air tanks like those used by underwater divers. Some welders use oxyacetylene torches (**Figure 1**). This form of welding requires both a pressurized oxygen tank and a pressurized acetylene tank. Many scientists and their graduate students routinely use pressurized gases for research because the gas is part of the reaction system or because it provides an inert (nonreactive) environment. Noble gases, such as argon, are also used to provide an inert environment in the computer chip industry, where oxygen would cause undesirable reactions.

The chemical safety hazards of some gases are similar to those of many other chemicals, which may be corrosive, toxic, flammable, dangerously reactive, or oxidizing agents. What makes compressed gases much more dangerous is the physical hazard of a potential rocket. In commercial gas cylinders, gas pressures can be as high as 15 MPa (about 150 atm). The hole in the tank, to which the valve stem and valve are connected, is the diameter of a pencil. If the gas is suddenly released through such a small opening, the very great pressure propels the tank, making it a formidable projectile. If the tank is mishandled, dropped, or falls over and the valve stem breaks, the tank can fly through solid brick walls and cause considerable damage.

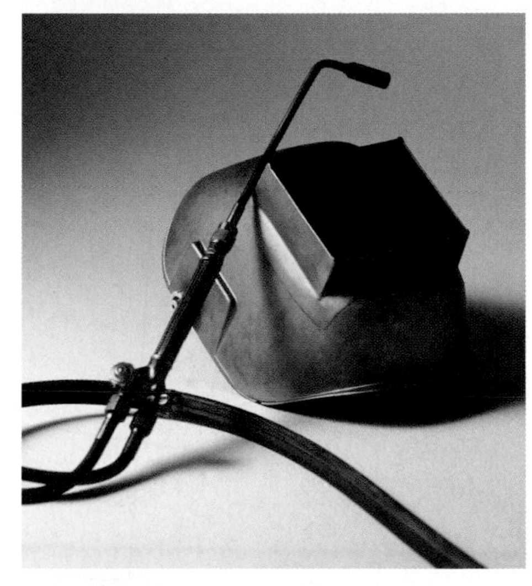

Figure 1
The use of a controlled mixture of oxygen and acetylene provides the best combustion and very high temperatures necessary for cutting or welding metal. Note the two hoses leading to the torch.

Figure 2
Every 10 m of depth adds about 1 atm (100 kPa) of pressure to the normal air pressure. At a depth of 20 m, the total pressure is about 300 kPa. In order to breathe, the air pressure then must be about 300 kPa.

Professional and recreational underwater divers face risks associated with the use of a pressurized air tank or scuba (self-contained underwater breathing apparatus). The tank containing compressed air is attached to a regulator that releases the air at the same pressure as the underwater surroundings (**Figure 2**). The pressure underwater can be quite substantial. Breathing pressurized air is necessary to balance the internal and external pressures on the chest to allow divers to inflate their lungs. However, this creates problems if divers ascend to normal pressure too quickly or while holding their breath. According to Boyle's law, if the pressure is decreased, the volume of air increases. When the volume of air is contained in the lungs, the lungs would expand to accommodate the increased volume, but this is very dangerous because the lungs could rupture. This is one reason why a person needs an understanding of gases and gas laws in order to obtain a scuba-diving licence and to dive safely.

Another problem of breathing air under pressure is that it forces more air to dissolve in the diver's bloodstream. If the diver comes up too quickly, the solubility of air (mostly nitrogen) decreases as the pressure decreases, and nitrogen bubbles form in the blood vessels. These nitrogen bubbles are the cause of a diving danger called "the bends" (so named because divers typically bend over in agony as they try to relieve the pain). Nitrogen bubbles are especially dangerous if they form in the brain or spinal cord. The bends may be avoided by ascending very slowly or corrected by using a decompression chamber.

Practice

Understanding Concepts

1. A scuba diver with a total lung capacity of 6.0 L breathes air at a pressure of 300 kPa and returns to the surface (100 kPa) while trying to hold her breath.
   (a) Calculate the new volume of air in her lungs at surface conditions.
   (b) Why is this a dangerous situation?
   (c) What assumption is made for the calculation in (a)?
2. Compressed gases, including those in aerosol cans, should not be heated. Use your knowledge of the properties of gases to describe the potential safety hazard.

Answer

1. (a) 18 L

Making Connections

3. Identify several careers that involve work with pressurized gases. How does knowledge of gas properties help those people in their jobs?
4. Suggest a chemical or physical reason why pressure regulators, which control the release of a gas from a pressurized tank, are labelled "for use only with certain gases." For example, only a regulator marked "oxygen" should be used on oxygen tanks.
5. Helium has many uses (**Figure 3**) but one that is familiar to many people is its use in party balloons. Sometimes people inhale helium because it produces an unusual change in a person's voice. In a paragraph based on your Internet research, describe this effect as well as the dangers of inhaling helium.

   Follow the links for Nelson Chemistry 11, 9.3.

   GO TO **www.science.nelson.com**

Figure 3
Modern airships are filled with the light noble gas, helium. Helium is also used in party balloons.

Section 9.3 Questions

Making Connections

1. Identify two consumer or commercial products that use or contain compressed air and two that involve other compressed gases.
2. How is the safety hazard of compressed gases different from that of other chemicals? State the potential hazard, how it may be created, and some possible consequences.
3. Even if a compressed gas is slowly released at a normal pressure, the gas may still be dangerous. What are some of the possible dangers?

DID YOU KNOW?

First to Commercially Produce Helium

John McLennan (1867–1935), University of Toronto research scientist and inventor, was the first to produce commercial quantities of helium (by extracting it from Alberta natural gas). The process dropped the price from \$20 000/m^3 to \$3/m^3. Helium is a safe alternative to hydrogen for airships.

9.4 The Ideal Gas Law

In this book, all real gases are dealt with in calculations as if they were ideal. An **ideal gas** is a hypothetical gas that obeys all the gas laws perfectly under all conditions; that is, it does not condense into a liquid when cooled, and graphs of its volume and temperature and of its pressure and temperature relationships are perfectly straight lines. A single, ideal gas equation describes the relationship among pressure, temperature, volume, and amount of matter, the four variables that define an ideal gaseous system.

ideal gas: a hypothetical gas composed of particles that have zero size, travel in straight lines, and have no attraction to each other (zero intermolecular force)

- According to Boyle's law, the volume of a gas is inversely proportional to the pressure: $v \propto 1/p$.
- According to Charles's law, the volume of a gas is directly proportional to the absolute temperature: $v \propto T$.
- According to the pressure–temperature law, the pressure of a gas is directly proportional to the absolute kelvin temperature: $p \propto T$.
- As anyone knows who has blown up a balloon, the more air you blow into the balloon, the bigger it gets; in other words, the greater the amount of air, in moles, at the same temperature and pressure, the greater the volume. Therefore, $v \propto n$.

Combining the three mathematical relationships for the volume of a gas produces the following relationship:

$$v \propto \frac{1}{p} \times T \times n$$

Another way of stating this is

$$v = (\text{constant}, R) \times \frac{1}{p} \times T \times n$$

$$v = \frac{nRT}{p}$$

$$pv = nRT$$

This last equation is known as the **ideal gas law**, and the constant R is known as the **gas constant**. The value for the gas constant is obtained by substituting known values into the ideal gas law and solving for R. For example, references state that at STP, 1.00 mol of an ideal gas would occupy a volume of 22.414 L.

ideal gas law: The product of the pressure and volume of a gas is directly proportional to the amount and the kelvin temperature of the gas; $pv = nRT$.

gas constant: the constant of variation, R, that relates the pressure in kilopascals, volume in litres, amount in moles, and temperature in kelvin of an ideal gas

DID YOU KNOW ?

van der Waals Forces

In 1873, Johannes Diderik van der Waals (1837–1923) hypothesized the existence of attraction between gas molecules to explain deviations from the ideal gas law. The general attraction, called van der Waals forces, was later explained as being London forces by Fritz London and dipole–dipole forces by, for example, Peter Debye.

$v = 22.414$ L

$n = 1.00$ mol (to give a certainty for R of three significant digits)

$p = 101.325$ kPa

$T = 0°C = 273.15$ K

$$pv = nRT$$

$$R = \frac{pv}{nT}$$

$$= \frac{101.325 \text{ kPa} \times 22.414 \text{ L}}{1.00 \text{ mol} \times 273.15 \text{ K}}$$

$$= \frac{8.31 \text{ kPa·L}}{\text{mol·K}}$$

As with several other constant values used for calculations in this book, a certainty of three significant digits for R is normally used. The value of the gas constant depends on the units chosen to measure volume, pressure, and temperature. If any three of the four variables in the ideal gas law are known, the fourth can be calculated by means of this equation. However, often the mass of a gas is a known quantity. In this case, a two-step calculation is required.

For example, 0.78 g of hydrogen at 22°C and 125 kPa is produced. What volume of hydrogen would be expected? To use the ideal gas law, you first need to convert the mass into an amount in moles of hydrogen:

$$n_{H_2} = 0.78 \not{g} \times \frac{1 \text{ mol}}{2.02 \not{g}} = 0.39 \text{ mol}$$

(Retain the unrounded value for subsequent calculation.) Now you can use the ideal gas law to determine the volume of hydrogen at the conditions specified:

$$pv = nRT$$

$$v_{H_2} = \frac{nRT}{p}$$

$$= \frac{0.39 \text{ mol} \times \frac{8.31 \text{ kPa} \cdot \text{L}}{1 \text{ mol} \cdot \text{K}} \times 295 \text{ K}}{125 \text{ kPa}}$$

$$v_{H_2} = 7.6 \text{ L}$$

Sample Problem 1

What mass of neon gas should be introduced into an evacuated 0.88-L tube to produce a pressure of 90 kPa at 30°C?

Solution

$v = 0.88$ L

$p = 90$ kPa

$m_{Ne} = ?$

$T = 30°C = 303$ K

$$pv = nRT$$

$$n_{Ne} = \frac{pv}{RT}$$

$$= \frac{90 \text{ kPa} \times 0.88 \text{ L}}{\frac{8.31 \text{ kPa} \cdot \text{L}}{\text{mol} \cdot \text{K}} \times 303 \text{ K}}$$

$$= 0.031 \text{ mol}$$

$$m_{Ne} = 0.031 \text{ mol} \times \frac{20.18 \text{ g}}{1 \text{ mol}}$$

$$m_{Ne} = 0.63 \text{ g}$$

The mass of neon gas required to produce a pressure of 90 kPa at 30°C is 0.63 g.

Practice

Understanding Concepts

1. List three ways of reducing the volume of gas in a shock absorber (cylinder and piston) of an automobile.
2. Under what conditions is a gas closest to the properties of an ideal gas? Why?
3. What amount of methane gas is present in a sample that has a volume of 500 mL at 35.0°C and 210 kPa?
4. What volume does 50 kg of oxygen gas occupy at a pressure of 150 kPa and a temperature of 125°C?
5. The density of a gas is the mass per unit volume of the gas in units of, for example, grams per litre. By finding the mass of one litre (assume 1.00 L) of gas, you can then calculate the density of the gas.
   (a) What is the density of propane, $C_3H_{8(g)}$, at 22°C and 96.7 kPa?
   (b) If the density of air at this temperature is 1.2 g/L, what happens to propane gas that may leak from a propane cylinder in a basement or from the tank of an automobile in an underground parkade? Why is this a problem?
6. Determining the molar mass of gases is an important experiment for qualitative analysis. Starting with the ideal gas law, derive a formula to calculate the molar mass, M, of a gas, given the mass and volume of the gas at a specific pressure and temperature, and that $n = \frac{m}{M}$.

Answers

3. 41.0 mmol
4. 34 kL (or 34 m^3)
5. (a) 1.74 g/L

Applying Inquiry Skills

7. The purpose of the following investigation is to use the ideal gas law to determine an important property of a substance, its molar mass. Complete the **Analysis** section of the following investigation report.

   Question

   What is the molar mass of an unknown gas?

   Experimental Design

   A measured mass of the gas is collected and the volume, temperature, and pressure of the gas are measured.

Answers

7. (a) 28.4 g/mol
9. (a) 1.78 g/L
   (c) 1.1 g/L

Evidence

mass = 1.25 g   pressure = 100 kPa

volume = 1.00 L   temperature = 0°C

Analysis

(a) What is the molar mass of the gas?

Making Connections

8. Hot-air balloons (**Figure 1**) rise up through the air because the density of the air inside the balloon is less than the density of the outside air. Using the ideal gas law, describe how this occurs.
9. Knowledge of the densities of gases compared to air (at 1.2 g/L) can save your life.
   (a) What is the density of carbon dioxide gas at SATP?
   (b) If you are caught in a fire, is the suffocating carbon dioxide gas found closer to the floor or to the ceiling?
   (c) What is the density of carbon monoxide gas at 20°C and 98 kPa in a home?
   (d) Where should a carbon monoxide detector be located, close to the floor or close to the ceiling?
   (e) If potentially lethal carbon dioxide comes from a fire and the carbon monoxide comes from the furnace, what other variable might affect the density of these gases released within a home?

Figure 1
A particular hot-air balloon has a fixed size when fully inflated and is open at the bottom. The air is usually heated with a propane burner.

Reflecting

10. In conversations and discussions, it is not unusual for people to say, "In an ideal world, ..." followed by some statement like, "the buses would always run on time." How does the use of the word "ideal" compare with the use of this word in the "ideal gas law"? What other concepts have you learned in science that are also ideal?

INQUIRY SKILLS

- ○ Questioning
- ○ Hypothesizing
- ● Predicting
- ○ Planning
- ● Conducting
- ● Recording
- ● Analyzing
- ● Evaluating
- ● Communicating

Investigation 9.4.1

Determining the Molar Mass of a Gas

The molar mass of a compound is an important constant that, in some cases, can help to identify a substance. Molar masses of known compounds are obtained using accepted atomic molar masses from the periodic table. The purpose of this investigation is to use the ideal gas law to evaluate the Experimental Design by determining the molar mass of a well-known substance. Butane, $C_4H_{10(g)}$, is suggested, but you may substitute another gas. Complete the **Prediction, Analysis, Evaluation** and **Synthesis** sections of the lab report.

Butane is highly flammable. Do not conduct this experiment near an open flame. Flints must be removed from butane lighters.

Good ventilation in the laboratory is essential.

Eye protection and a lab apron are required.

Question

What is the molar mass of butane?

Prediction

(a) According to the chemical formula and atomic molar masses in the periodic table, what is the molar mass of butane?

Experimental Design

A sample of butane gas from a lighter is collected in a graduated cylinder by downward displacement of water. The volume, temperature, and pressure of the gas are measured, along with the change in mass of the butane lighter. The Experimental Design is evaluated on the basis of the accuracy of the experimental value for the molar mass of butane, which is compared with the accepted value.

Materials

lab apron
eye protection
butane lighter (with flint removed) or butane cylinder with tubing
plastic bucket
100–500-mL graduated cylinder
balance
thermometer
barometer

Procedure

1. Determine the initial mass of the butane lighter.
2. Pour water into the bucket until it is two-thirds full. Then completely fill the graduated cylinder with water; invert the graduated cylinder in the bucket. Ensure that no air has been trapped in the cylinder.
3. Hold the butane lighter in the water under the graduated cylinder (**Figure 2**) and release the gas until you have collected half to three-quarters of the cylinder. Make sure all the bubbles enter the cylinder.
4. Equalize the pressures inside and outside the cylinder by adjusting the position of the cylinder until the water levels inside and outside the cylinder are the same.
5. Read the measurement on the cylinder and record the volume of gas collected. Record the ambient (room) temperature and pressure. Record your measurements as precisely as possible with the equipment available.
6. Thoroughly dry the butane lighter and determine its final mass.
7. Release the butane gas from the cylinder in a fume hood or outdoors.

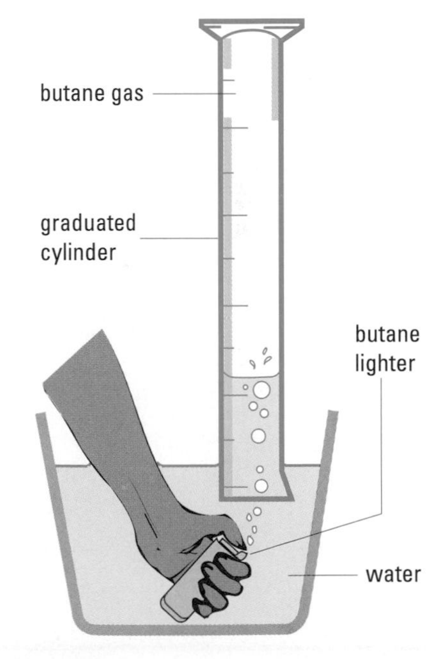

Figure 2
The gas from a butane lighter can be collected by downward displacement of water. This apparatus can be used to determine the molar mass of butane.

Analysis

(b) Using the difference in the masses of the butane container before and after the experiment, calculate the mass of butane that was released from the container.
(c) Combine the $pv = nRT$ and $n = \frac{m}{M}$ to solve for molar mass, M.
(d) Using your answers to (b) and (c), answer the Question by calculating the molar mass of butane according to your Evidence.

Evaluation

(e) Calculate the percentage difference. How accurate was your experimental answer for the molar mass of butane?
(f) Every experiment has some experimental errors or uncertainties. In order to judge whether your percentage difference was reasonable for this particular experiment, you need to consider possible flaws, experimental errors or uncertainties, and any other limitations that may affect the accuracy of your result. Within your lab group, make a list and rank the items from most to least significant.

(g) In your opinion, can the percentage difference you calculated in (e) be reasonably explained by your list of experimental errors or uncertainties? If not, explain briefly.
(h) Are the Experimental Design, Materials, and Procedure adequate for this experiment? Provide reasons.

Synthesis

(i) Evaluating the evidence from an experiment is not an easy task, and it is possible to omit a source of experimental error. In this experiment, you may or may not have realized that the gas you collected is not completely pure butane. It also contains a small quantity of water vapour. For a typical room temperature, this would correspond to about 2.6 kPa of the pressure you recorded. Subtract this value from the pressure you used, and use the new pressure to recalculate the molar mass and the percentage difference. Is your answer significantly more accurate?
(j) Are there any other possible sources of error? For example, what assumption must be made about the butane from the lighter or canister you used?

SUMMARY Properties of an Ideal Gas

- v-T and p-T graphs are perfectly straight lines
- gas does not condense to a liquid when cooled
- gas volume = 0 at absolute zero
- $pv = nRT$
- gas particles are point size (volume of particle = 0)
- gas particles do not attract each other

Section 9.4 Questions

Understanding Concepts

1. Unlike an ideal gas, a real gas condenses to a liquid when the temperature is low enough. What does this indicate about the interaction between the particles and why the gas is real versus ideal?
2. What are three variables that can determine the pressure of a gas? How is each of these variables related to the pressure?
3. Determine the pressure in a 50-L compressed air cylinder if 30 mol of air is present in the container, which is heated to 40°C.
4. At what temperature does 10.5 g of ammonia gas exert a pressure of 85.0 kPa in a 30.0-L container?
5. Using atmospheres as the pressure unit, determine the value of the gas constant, R, at SATP for 1.00 mol of a gas that occupies 24.8 L.
6. A 1.49-g sample of a pure gas occupies a volume of 981 mL at 42.0°C and 117 kPa.
   (a) Determine the molar mass of the compound.
   (b) If the chemical formula is known to be XH_3, identify the element "X."

9.5 Air Quality

Air quality, like water quality, has often been taken for granted. Increasingly, both have come under public scrutiny as the quality of these two requirements of life has deteriorated. The Ontario Medical Association estimates that 1900 people die prematurely each year in Ontario due to poor air quality, and almost $10 billion a year is spent in health-care costs, lost work time, and other quantifiable expenses resulting from poor air quality. In southern Ontario, the number of "smog alert days" has increased dramatically over the past few years, and the recorded asthma rate in children has increased by 60% in the past decade. Young children, the elderly, and people with respiratory illnesses are particularly at risk. Public opinion polls commissioned by the Ontario Clean Air Alliance in March 1999 in each of the Hamilton, London, and Kitchener–Waterloo regions showed that approximately 70% (of the 400 adult respondents) were concerned about air pollution, and 80% thought that the problem would continue to get worse in the future. Less than 10% gave the Ontario government a positive rating for its efforts in controlling air pollution, and more than 85% were willing to pay higher utility bills to lower emissions from coal-fired electric generating plants. Although industry, particularly the electric power plants, is a major source of air pollution, the transportation choices made by citizens are also a significant part of the air pollution problem.

Although the composition of air is well known (**Table 1**), its chemistry is very complex and not completely understood. Many of the chemical processes taking place in the atmosphere are induced by solar radiation and are intertwined with chemical processes—both naturally occurring and as a result of human activities. However, it is becoming apparent that the human activities are producing dangerous levels of air pollutants, particularly in large urban areas. Primary air pollutants include

- gases such as various nitrogen oxides, sulfur dioxide, and carbon dioxide;
- vapours from volatile organic compounds (VOCs);
- heavy metals and other toxic and carcinogenic substances emitted primarily during the combustion of coal.

An important secondary air pollutant is ground-level ozone formed when nitrogen oxides and VOCs react in sunlight. According to the Canadian Council of Ministers of the Environment (CCME), the Windsor–Quebec City corridor has one of the worst ozone problems in the country. This section focuses on some of the chemistry of nitrogen and VOCs in the atmosphere and their role in the formation of ground-level ozone.

Table 1: Components of Dry Air at Sea Level

	Composition	
Gas*	**Volume (%)**	**Volume (ppm)**
$N_{2(g)}$	78.08	7.808×10^5
$O_{2(g)}$	20.95	2.095×10^5
$Ar_{(g)}$	0.934	9.34×10^3
$CO_{2(g)}$	0.036	3.6×10^2
$Ne_{(g)}$	0.001 818	18.18
$He_{(g)}$	0.000 524	5.24
$CH_{4(g)}$	0.000 2	2
$Kr_{(g)}$	0.000 114	1.14

* Water is excluded from the table because its concentration can vary widely from one place to another.

The Nitrogen Cycle in Nature

Nitrogen is an important element in living systems (**Figure 1**, page 450). Before atmospheric nitrogen can be absorbed by plants, it must be converted into compounds suitable for assimilation. The conversion of atmospheric nitrogen into nitrates by various microorganisms in the soil (nitrogen-fixing bacteria) and in the roots of certain plants (legumes, e.g., beans and peas) is called nitrogen fixation. Lightning is another natural source of nitrogen compounds for plants. At the high temperature of a lightning strike, nitrogen and oxygen react to produce nitrogen oxides, which react with water to form nitrous and nitric acids. These acids are converted to nitrites and nitrates in the soil. Decaying plants and animals and animal waste products are in turn acted on by bacteria, and the nitrogen in them is again made available for circulation as nitrogen in the atmosphere.

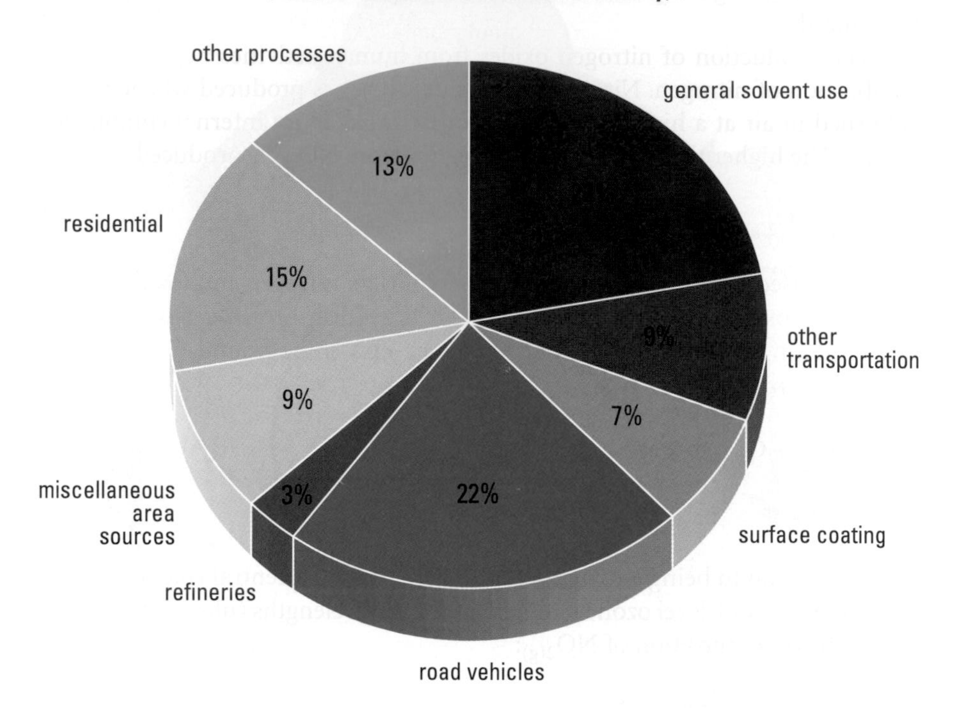

Figure 3
Transportation modes account for approximately 31% of the total VOC emissions in Ontario.

Source: Ontario Ministry of the Environment, 2000

Ozone irritates the respiratory tract and eyes, and people with heart or lung problems are at risk of lung damage. Ground-level ozone is also undesirable because it damages plants, bleaches colour from fabrics, and hardens rubber, thereby shortening the life span of tires.

Plants are damaged in several ways. According to current research, ozone causes injury to foliage, increases susceptibility to diseases in plants and trees, reduces yields in sensitive crops, and increases the mortality rates of individual trees. As a result of ozone pollution from the American Midwest, certain crops, such as white beans, can no longer be grown successfully in southwestern Ontario. The maximum allowable ozone concentration in Canada is 82 ppb averaged over a one-hour period, but this level is often exceeded in urban centres (**Figure 4**). The federal and provincial governments have recently agreed to a target of 65 ppb, averaged over an eight-hour period, by 2010.

In order to reduce the formation of ground-level ozone, it is necessary to reduce the concentration of the reactants in the air. Since it is usually the concentration of NO_x that controls the overall rate of the ozone-producing reactions, most research has focused on developing technologies to reduce the amounts of NO_x released into the atmosphere. Catalytic converters significantly reduce the level of NO_x emitted by automobile engines. In a catalytic converter the hot exhaust gases pass over a surface containing beads of rhodium, platinum, and palladium, which reduces $NO_{(g)}$ to $N_{2(g)}$ and oxidizes $CO_{(g)}$ to $CO_{2(g)}$. The air in Los Angeles used to reach ozone levels of 680 ppb but has been reduced to around 300 ppb, thanks in large part to the development of catalytic converters.

Number of Days per Year with Ozone Levels in Excess of the One-Hour Air Quality Objective of 82 ppb Average of Three Highest Years (1983–90)

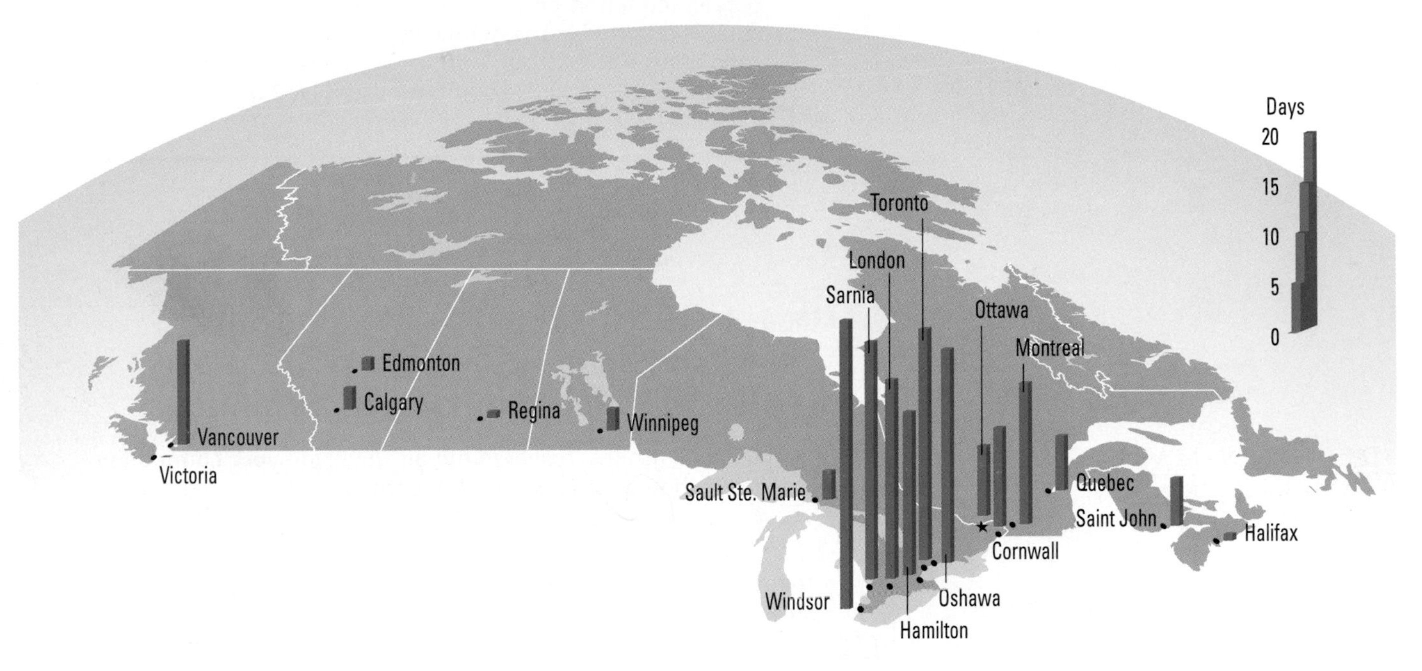

Figure 4
This 1994 Environment Canada map shows the number of days per year with ozone levels in excess of the 82 ppb current standard.

Source: Environment Canada, 1994

Practice

Understanding Concepts

1. Why is water vapour not listed in **Table 1**?
2. Describe the nitrogen cycle.
3. Which human activities contribute the most nitrogen oxides to the atmosphere?
4. Describe the role of nitrogen oxides in producing ground-level ozone.
5. An empty classroom contains 240 kL of air at SATP. Assume the air is dry.
   (a) Calculate the amount (in moles) of nitrogen, oxygen, argon, and carbon dioxide in the room.
   (b) What is the mass of each of the gases in the room? If the total mass of air was liquefied, could you carry this mass of air?
   (c) In what way will the amount of each gas change after a class has been in the room for an hour? State any assumption(s) that you make when answering this question.
   (d) At SAPT, respiration in a classroom converts about 400 L of oxygen per student per school day to carbon dioxide. If there are 30 students in the classroom for this period of time and the air is stagnant, what percentage of the oxygen is consumed?
6. How is it possible for an industry to claim a reduction of a pollutant when its actual emissions have increased?

Answers

5. (a) 7.57 kmol $N_{2(g)}$, 2.03 kmol $O_{2(g)}$, 90.5 mol $Ar_{(g)}$, and 3.5 mol $CO_{2(g)}$
   (b) 212 kg $N_{2(g)}$, 65.0 kg $O_{2(g)}$, 3.62 kg $Ar_{(g)}$, and 0.15 kg $CO_{2(g)}$
   (d) 23.9%

Making Connections

7. Weather stations and government agencies often report air quality using an air quality index (AQI). Use the Internet to discover and report on what this index includes. What is the scale that is reported in the media, including possible health warnings and effects?
Follow the links for Nelson Chemistry 11, 9.5.

 www.science.nelson.com

Explore an Issue

Take a Stand: How Can We Improve the Air Quality in Our Communities?

DECISION-MAKING SKILLS

- Define the Issue
- Identify Alternatives
- Research
- Analyze the Issue
- Defend a Decision
- Evaluate

Working in small groups, research the air quality in your community. Your research should include the following:

- the major air pollutants;
- the sources of these pollutants;
- the measures currently in place to reduce the pollutants.

Follow the links for Nelson Chemistry 11, 9.5.

GO TO www.science.nelson.com

(a) Draft a letter to the appropriate department in your municipality, recommending improvements where you think they are needed, or

(b) Draft a pamphlet for the public that outlines the concerns of your group.

Section 9.5 Questions

Understanding Concepts

1. Identify the four most abundant gases in the atmosphere.
2. Identify the three most abundant noble gases in the atmosphere.
3. Water in the air is so variable that it is often not listed as a component or considered as affecting the properties of air. For example, is moist air denser than dry air? Provide your reasoning. (No calculations are necessary.)
4. List and describe the three main categories of primary air pollutants.
5. Why is ozone not classified as a primary air pollutant?
6. Describe how catalytic converters in automobiles improve air quality.
7. Air quality is the responsibility of everyone on our planet. What are some consumer, commercial, and/or industrial strategies to reduce $NO_{x(g)}$ and VOC emissions?

Chapter 9 Summary

Key Expectations

Throughout this chapter, you have had the opportunity to do the following:

- Explain different states of matter in terms of the forces among atoms, molecules, and ions. (9.1)
- Describe the gaseous state, using the kinetic molecular theory, in the terms of degree of disorder and the types of motion of atoms and molecules. (9.1)
- Describe natural phenomena and technological products associated with gases. (9.1, 9.2, 9.3, 9.4)
- Determine through experimentation the quantitative and graphical relationships among the pressure, volume, and temperature of an ideal gas. (9.2)
- Solve quantitative problems involving the following gas laws: Charles's law, Boyle's law, the combined gas law, the pressure and temperature law (Gay-Lussac's law), and the ideal gas law. (9.2, 9.4)
- Describe the quantitative relationships that exist among the following variables for an ideal gas: pressure, volume, temperature, and amount of substance. (9.2, 9.4)
- Use and interconvert appropriate units to express pressure and temperature. (9.2, 9.3, 9.4)
- Use appropriate scientific vocabulary to communicate ideas related to gases. (9.2, 9.3, 9.4)
- Identify technological products and safety concerns associated with compressed gases. (9.3)
- Identify the major and minor components of the atmosphere and describe Canadian initiatives to improve air quality. (9.5)

Key Terms

absolute zero
atmospheric pressure
Boyle's law
Charles's law
combined gas law
gas constant
ideal gas
ideal gas law
Kelvin temperature scale
kinetic molecular theory
pressure
pressure and temperature law

Make a Summary

1. Draw a simple model of a solid, a liquid, and a gas. With each model, list the
   (a) empirical properties
   (b) possible forces present
   (c) type of motion of the particles
   (d) degree of order
2. Sketch a series of graphs showing the relationship between volume and each of the following variables: pressure, temperature, and amount of gas. For each graph, include a mathematical equation and indicate the variables that are controlled.
3. Write the ideal gas law. Using arrows, label each symbol in the equation with the name of the variable and the typical units.

Reflect on your Learning

Revisit your answers to the Reflect on Your Learning questions, at the beginning of this chapter.

- How has your thinking changed?
- What new questions do you have?

Chapter 9 Review

Understanding Concepts

1. Chlorine, bromine, and iodine are all members of the halogen family and have similar chemical properties. What does the state (solid, liquid, or gas) of these elements at SATP reveal about the strength of their intermolecular forces?
2. All three states of matter for H_2O are always present on Earth: Most of the surface of Earth is covered by water and ice (**Figure 1**), and the atmosphere contains water vapour.
   (a) Which physical properties of water and ice are similar?
   (b) Which physical properties of water and ice are different?
   (c) Why is a gas like water vapour highly compressible?
   (d) In general terms, describe the degree of disorder of the H_2O molecules in all three states.
   (e) Briefly describe all three states of matter, using forces and motion of H_2O molecules.
   (f) Water vapour is a small and variable part of the composition of the atmosphere. List the two major components and two minor components of the atmosphere.
   (g) Water is a small component of the atmosphere, but it is a very important one. List some effects that water has as a component of Earth's atmosphere.

Figure 1

3. State each of the following laws in a sentence beginning, "The volume of a gas sample ..."
   (a) Boyle's law
   (b) Charles's law
   (c) the ideal gas law
4. Predict how the volume of a given mass of gas will differ when the following changes in the temperature and pressure are made:
   (a) The pressure is tripled while the absolute temperature is doubled.
   (b) The absolute temperature is doubled while the pressure is reduced to half.
   (c) The pressure and the absolute temperature are both doubled.
5. What is the major difference between a scientific law and a theory? Use Charles's law and the kinetic molecular theory to support your answer.
6. Convert each of the following gas pressures to units of kilopascals:
   (a) A mercury barometer gives the atmospheric pressure as 745 mm Hg.
   (b) A vacuum pump reduces the pressure in a container to 150 Pa.
   (c) An industrial process maintains a pressure of 2.50 atm in the reaction chamber.
7. Convert each of the following temperatures into kelvin temperatures:
   (a) freezing point of water
   (b) 21°C room temperature
   (c) 37°C body temperature
   (d) absolute zero
8. Pressurized hydrogen gas is used to fuel some prototype automobiles. What is the new volume of a 28.8-L sample of hydrogen for which the pressure is increased from 100 kPa to 350 kPa? State the assumptions that you need to make in order to answer this question.
9. One of the most common uses of carbon dioxide gas is carbonating beverages such as soft drinks.
   (a) Squeezing a plastic bottle increases pressure inside the bottle. What is the new volume of a 300-mL sample of carbon dioxide gas when the pressure doubles?
   (b) A 2-L bottle of pop, containing 300 mL of carbon dioxide gas at 125 kPa, is removed from the refrigerator. The temperature of the bottle and its contents increases from 7°C to 30°C. What is the new pressure of the carbon dioxide gas?
   (c) Explain why a can of carbonated pop sometimes overflows when opened.
10. A glass container can hold an internal pressure of only 195 kPa before breaking. The container is filled with a gas at 19.5°C and 96.7 kPa and then heated. Predict the temperature at which the container will break.
11. Electrical power plants and ships commonly use steam to drive turbines, producing mechanical energy from the pressure of the steam. The rotating turbine is connected to a generator that produces electricity. Steam enters a turbine at a high temperature and pressure and exits, still a gas, at a lower temperature and pressure.

Determine the final pressure of steam that is converted from 10.0 kL at 600 kPa and 150°C to 18.0 kL at 110°C.

12. Weather folklore tells of many signs that warn of approaching storms, such as wind rushing out of caves and vegetation floating to the surface of ponds. These phenomena are most likely due to the drop in atmospheric pressure associated with an advancing low-pressure area, which often accompanies a storm.
    (a) A cave holds 3000 m^3 of air when the atmospheric pressure is 103 kPa. Calculate the increase in volume of the air when the atmospheric pressure drops to 97 kPa.
    (b) Decaying vegetation at the bottom of a pond contains thousands of bubbles of trapped methane gas. If one piece of vegetation contains a total volume of 100.0 mL of methane gas at 103 kPa, calculate the volume of gas when the pressure drops to 97 kPa.
    (c) In light of your answers to (a) and (b), explain why wind rushing out of caves and decaying vegetation floating to the surface of ponds might be evidence of an advancing storm.
13. Many campers use propane as a fuel for cooking. If a tank contains 4.54 kg of propane, what volume of propane gas could be supplied at 12°C and 96.5 kPa?
14. An advertiser needs 5000 helium-filled balloons for a special promotion. If each balloon has a volume of 7.5 L on a day when the temperature is 18°C and the atmospheric pressure is 102.7 kPa, calculate the mass of helium required.
15. What amount of air (in moles) is present in an empty room with the dimensions 2.95 m × 3.50 m × 2.45 m at 21°C and 99.5 kPa?

Applying Inquiry Skills

16. For the purpose of testing Boyle's law, a student performs an experiment similar to Investigation 9.2.1, which studied the relationship between the volume and pressure of a gas. Complete the **Hypothesis, Analysis,** and **Evaluation** sections of the following lab report.

Question

What is the relationship between the volume and pressure of a gas?

Hypothesis

(a) Answer the Question and provide your reasoning.

Experimental Design

A syringe is sealed at the tip and then placed inside a 2-L plastic bottle. The pressure on the syringe is gradually increased by pumping air into the bottle. Gas pressure in the bottle is the independent variable, and gas volume in the syringe is the dependent variable.

Evidence

Air Pressure and Volume in a Syringe

Pressure (kPa)	Volume (mL)
100	50
150	33
200	25
250	20
300	17

Analysis

(b) What is the answer to the Question?

Evaluation

(c) Evaluate your Hypothesis. Is Boyle's law supported by the Evidence gathered in this investigation?

17. Design an experiment using a syringe that is sealed at the tip to determine the value of absolute zero.

Making Connections

18. Describe two natural phenomena and two technological products or processes using the gas laws.
19. Propane tanks for barbecues have a limited life. Propane refill stations can refuse to fill a tank and consumers may be stuck with the tank because recycling and disposal sites often do not accept old tanks. Suggest some reasons for the problems with refilling and disposal.
20. Describe briefly how air bags work and identify some risks and benefits of their use.
21. Air has many components.
    (a) List some categories that can be used to classify the components of air. Provide some examples for each category of gas.
    (b) Choose one of the components of air and tell its story from a variety of perspectives, for example, from scientific, technological, economic, and environmental perspectives.

Exploring

22. Use the Internet to research one of the noble gases to determine its abundance in the atmosphere, its commercial uses, and how commercial quantities are obtained.

    Follow the links for Nelson Chemistry 11, Chapter 9 Review.

GO TO www.science.nelson.com

Chapter

10

Gas Mixtures and Reactions

In this chapter, you will be able to

- describe the total pressure of a gas mixture as the sum of the partial pressures;
- state natural and technological examples in which partial pressure plays a role;
- explain Dalton's law of partial pressures using the kinetic molecular theory;
- use the law of combining volumes to solve for reacting volumes of gases;
- state Avogadro's theory and describe its importance in understanding gas reactions;
- define and use molar volumes at STP and SATP to convert between volumes and amounts of gases;
- determine, experimentally, the molar volume of a gas;
- perform stoichiometric calculations for reactions involving gases;
- describe how knowledge of gases relates to natural phenomena, technological products, and other areas of study.

Seen from space, our atmosphere does not look very impressive—only a thin blue layer around the planet—yet it is essential to our existence (**Figure 1**). A complex, variable mixture of gases, it is also the medium for a great variety of both useful and harmful gas reactions. Many familiar gases in the atmosphere have a dual personality: Oxygen gas is essential for us to breathe, but too little or too much is deadly. Nitrogen is an important nutrient source for plants (and indirectly for us), but its reactions with oxygen can produce dangerous substances such as those found in smog and acid rain. Carbon dioxide is necessary for green plants to produce food and oxygen by photosynthesis, but too much may have long-term, undesirable climate effects or can be suffocating (**Figure 2**). Ozone is another good example of an atmospheric gas that is both useful and dangerous. High in the atmosphere, a low-density ozone layer protects us from harmful UV radiation, but in the lower atmosphere ozone is a very destructive and lethal chemical.

These gases are only some of the constituents of the atmosphere. There are many others that are formed by natural or technological processes. An understanding of the properties of gas mixtures and reactions is necessary to begin to understand the complex atmosphere in which we live.

Reflect on your Learning

The concentrations of the gases in our atmosphere change with time. Sometimes these changes are local; sometimes they are global. Sometimes the concentrations vary as a result of natural causes, such as a volcano, and sometimes they are due to activities by humans, such as our production of car exhaust pollution.

1. How do the properties of gas mixtures, like the atmosphere, compare with those of other homogeneous mixtures, such as aqueous solutions?
2. How is the study of the reactions of gases similar to and different from other reactions you have studied?
3. What are some beneficial applications of gases and some problems associated with gases?

Throughout this chapter, note any changes in your ideas as you learn new concepts and develop your skills.

Figure 2
The release of carbon dioxide from underground gas pockets and the decomposition of limestone around Horseshoe Lake in California produces about 50 t to 150 t of carbon dioxide per day. This has killed many trees, particularly at the north end of the lake; it can also easily asphyxiate people and animals.

Producing a "Natural" Gas

If you do this activity at home and do not have baking powder, you can use Alka-Seltzer tablets with tap water or baking soda (with some diluted 5% V/V vinegar instead of tap water).

Materials: a clear drinking glass, water, measuring spoon, baking powder, piece of cardboard to cover the glass, matches

- Fill the glass about two-thirds full with tap water.
- Add 10 mL (around 2 level teaspoons) of baking powder to the water and cover the glass with a piece of cardboard.
- Record your observations, noting in particular the activity above the surface of the liquid.
- When the reaction subsides, light a match, lift the cardboard momentarily, and insert the lit match into the top of the glass. Remove the match and replace the cardboard. Record your observations.
- Repeat the reaction. Light another match, remove the paper, and gently "pour" the gas in the glass over the lit match. Record your observations.
- Dispose of the liquid down the sink. Ensure the used matches are extinguished by dipping them in water before disposal in the regular garbage.

(a) What gases are present in the glass before adding the baking powder?
(b) What gas is present after the reaction? How do you know?
(c) What additional test could be done to test your answer to (b)?
(d) The gas produced in this reaction can be poured out of a glass. What does this suggest about the properties of this gas?
(e) What is a common technological application of this gas?
(f) What natural occurrence mentioned earlier exhibits the same physical events you observed above the liquid during the reaction?

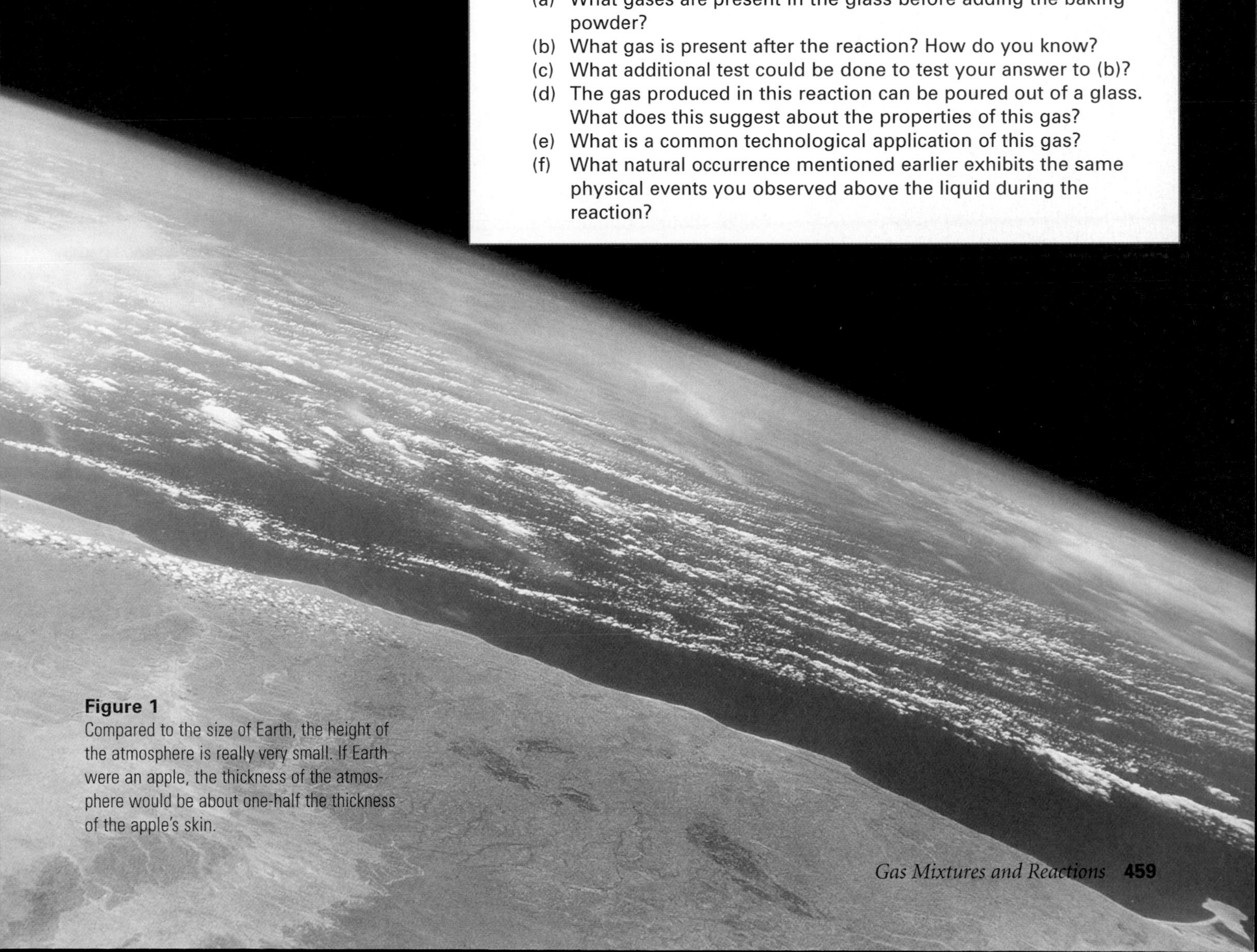

Figure 1
Compared to the size of Earth, the height of the atmosphere is really very small. If Earth were an apple, the thickness of the atmosphere would be about one-half the thickness of the apple's skin.

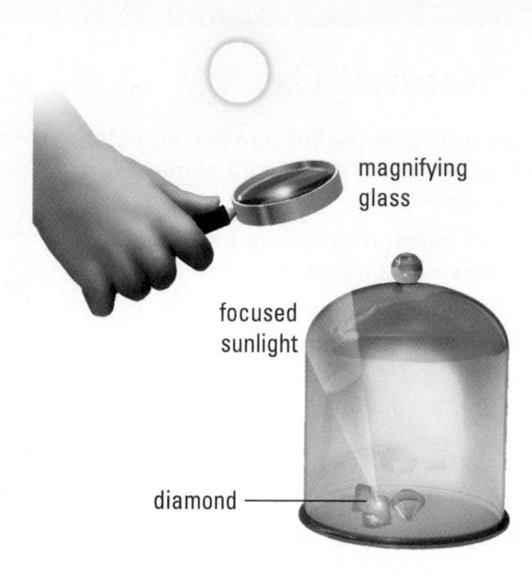

Figure 1
A prominent Parisian jeweller supplied the diamonds that Lavoisier used to show that diamonds will burn in air (to produce carbon dioxide) but would not burn in the absence of air.

10.1 Mixtures of Gases

In the 18th century, many scientists believed that the atmosphere was a single chemical compound, not a mixture of gaseous elements and compounds as we know today. In France in the 1770s, Antoine Lavoisier (1743–1794), acknowledged as the father of modern chemistry, began a series of experiments involving burning substances, for example, diamonds, in air (**Figure 1**). After numerous quantitative experiments over several years, he concluded that air consisted of at least two different gases: one that supported combustion and one that did not. These experiments not only led to the discovery of the presence of oxygen in air, but also established the law of conservation of mass for chemical reactions.

In some of his combustion experiments Lavoisier noticed that when various metal oxides were formed from a heated metal and air sealed in a flask, room air rushed into the flask when it was opened. This indicated a decrease in pressure inside the flask because the room air, at a higher pressure, was immediately forced into the flask. Gases naturally move from a higher-pressure to a lower-pressure region. The reduction of the air pressure inside the flask clearly supported his idea that part of the air (the oxygen) was reacted and therefore removed. There is no indication that Lavoisier himself measured any changes in pressure. However, his experimental designs and methods for careful quantitative observations were continued by many chemists and much was learned about chemical compounds, the composition of air, and the properties of gases.

Of particular note on the properties of gases is the work by John Dalton (1766–1844). In the early 1800s, Dalton hypothesized that gas particles behaved independently and that the pressure exerted by a particular gas is the same whether it exists by itself or in a gas mixture (if we assume the same temperature). Dalton proceeded to test this hypothesis by conducting a series of experiments to determine the pressure that each component of air contributes to the total air pressure (**Table 1**). The pressure of an individual gas in a gas mixture can be thought of like the concentration of a solute in a solution. For example, a household vinegar solution contains 5% acetic acid; the remaining 95% is water. The partial pressure of oxygen in air can be thought of as a concentration, or fraction, of air. Notice that the pressure (in kilopascals) of each component in air (**Table 1**) is very close to its actual percentage composition. This is largely a consequence of the unit of measure—normal air pressure in SI units is about 100 kPa. If other units of pressure, such as millimetres of mercury, are used, this direct similarity does not exist. However, the fraction is still the same no matter what pressure units are used.

Dalton also determined that the water vapour pressure was quite variable and depended on the temperature of the air. Furthermore, the pressure of water vapour is the same whether by itself (with no other gases present) or in a gas mixture such as air.

Table 1: Dalton's Analysis of Gases in the Atmosphere

Chemical name of air component		Pressure		Composition
Dalton	**IUPAC**	**(mm Hg)**	**(kPa)**	**(%)**
azotic gas	nitrogen	593.3	79.11	78.08
oxygenous gas	oxygen	157	20.9	20.95
aqueous vapour	water	11	1.5	varies
carbonic acid gas	carbon dioxide	0.5	0.07	0.04

Dalton's hypothesis was verified by his extensive experiments as well as those of other chemists and is now known as **Dalton's law of partial pressures:** *The total pressure of a mixture of nonreacting gases is equal to the sum of the partial pressures of the individual gases* (**Figure 2**). In symbols, this law is represented as

$$p_{total} = p_1 + p_2 + p_3 + \ldots$$

Any pressure unit may be used with this definition as long as the same unit is used for all partial pressures and the total pressure of the mixture.

partial pressure: the pressure, *p*, a gas in a mixture would exert if it were the only gas present in the same volume and at the same temperature

Dalton's law of partial pressures: The total pressure of a mixture of nonreacting gases is equal to the sum of the partial pressures of the individual gases.

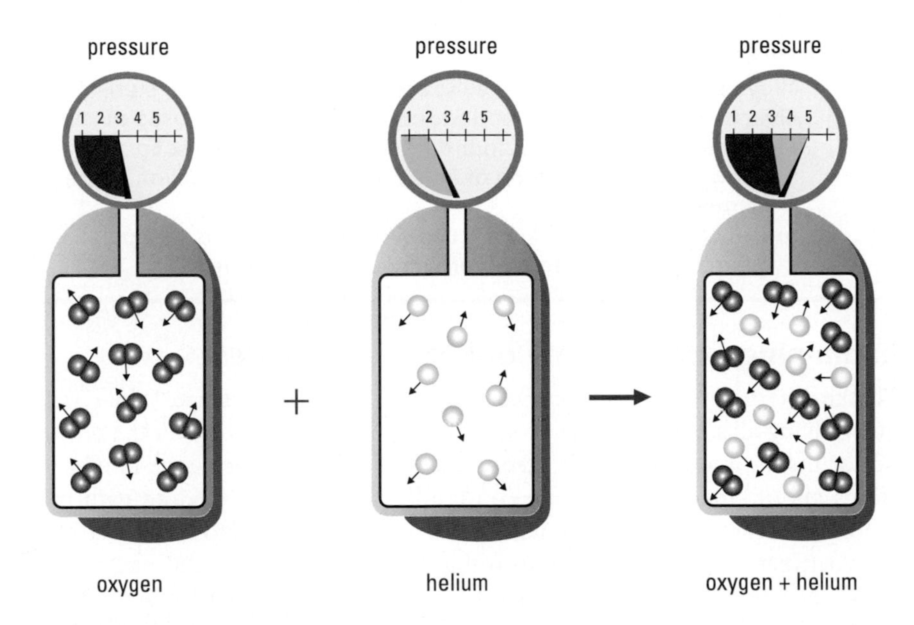

Figure 2
Dalton's law of partial pressures. The amount, temperature, and volume of gas is held constant.

Sample Problem 1

In a compressed air tank for scuba diving to a depth of 30 m, a mixture with an oxygen partial pressure of 28 atm and a nitrogen partial pressure of 110 atm is used. What is the total pressure in the tank?

Solution

$p_{O_2} = 28$ atm
$p_{N_2} = 110$ atm
$p_{total} = ?$

$$p_{total} = p_{O_2} + p_{N_2}$$
$$= 28 \text{ atm} + 110 \text{ atm}$$
$$p_{total} = 138 \text{ atm}$$

The total pressure in the tank is 138 atm.

Practice

Understanding Concepts

1. A 1-L flask of air at SATP contains carbon dioxide with a partial pressure of 2 kPa. If all other gases were removed from the flask and the temperature remained the same, what would be the pressure of carbon dioxide by itself in the flask?

Answers

2. (a) 762 mm Hg, 101.6 kPa
3. 12.9 atm

2. Using Dalton's values in **Table 1**, page 460,
   (a) find the total pressure for the four gases listed, using both pressure units given in the table;
   (b) show that both totals are the same by converting millimetres of mercury into kilopascals.
3. Beyond 60 m, compressed air is not used for underwater diving because of the toxicity of oxygen at high partial pressures. Suppose a commercial diver needs to work at a pressure of 14.0 atm using a helium–oxygen breathing mixture (known as heliox) containing 1.1 atm of oxygen. What is the partial pressure of helium in this mixture?

Reflecting

4. Alloys used in jewellery are solid solutions. For example, sterling silver contains 92% silver and 8% copper. Some alloys used in jewellery can contain three or four different metals in one solid solution.
   (a) How is the composition of an alloy like the composition of the atmosphere?
   (b) What quantity and law are used for solid solutions that parallel partial pressures and Dalton's law for gas mixtures?

Explaining Dalton's Law of Partial Pressures

The kinetic molecular theory is a cornerstone of modern science. While useful for studying all states of matter, it was initially developed to explain the properties of gases. According to this theory, gases consist of a very large number of atoms and/or molecules that are constantly moving and colliding with each other and the walls of the container. Gas particles are believed to move in straight lines between collisions, which means that the molecules do not attract or repel each other. If we assume a nonreacting mixture, the gas particles in a mixture behave like any other collection of gas molecules. It makes no difference if a container holds only one kind of molecule or several kinds of gas molecules (**Figure 3**).

DID YOU KNOW?

Laws and Theories

Laws tend to precede theories. For example, the gas laws you are studying in this unit were all solidly established before the kinetic molecular theory was created to explain them. Theories explain laws in terms of unobservables, such as molecules and intermolecular forces. Laws are based on observables (measurements).

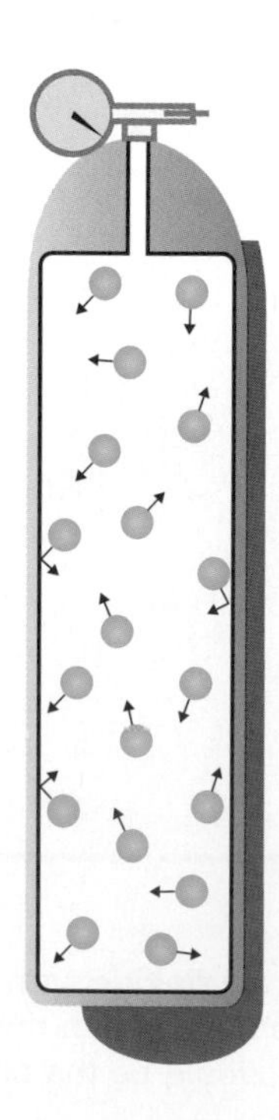
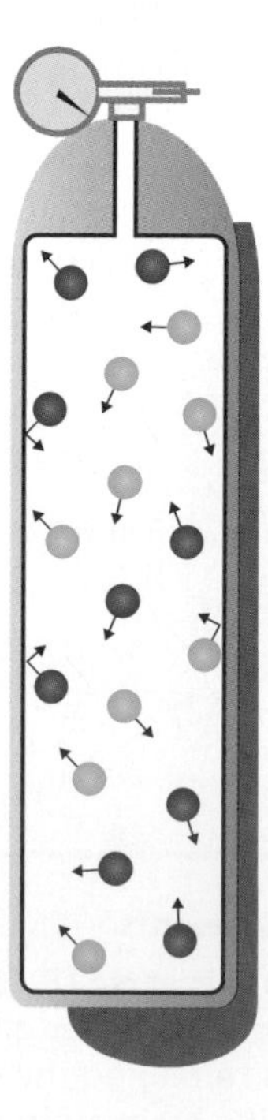

Figure 3
In both cylinders, the frequency of the collisions of particles with the walls of the container is the same. The collisions may involve only one kind of gas molecule or several kinds of gas molecules.

Dalton's law of partial pressures can be explained by two concepts from the kinetic molecular theory:

- The pressure of a gas is caused by the collisions of molecules with the walls of the container.
- Gas molecules act independently of each other.

Therefore, the total pressure (total of the collisions with the walls) is the sum of the individual pressures (collisions of only one kind of particle) of each gas present.

Practice

Understanding Concepts

5. According to the kinetic molecular theory,
   (a) what is the cause of the pressure exerted by a gas?
   (b) what is the reason gas molecules travel in straight lines?
6. Suppose you have 100 molecules of gas X in a container at a certain temperature.
   (a) If you added another 100 molecules of X while keeping the volume and temperature constant, what happens to the pressure? Briefly explain your answer.
   (b) If instead you added 100 molecules of gas Y to the 100 molecules of gas X that are present, what happens to the pressure?
   (c) How does your answer to (b) explain Dalton's law of partial pressures?
7. Why is Dalton's law of partial pressures restricted to nonreacting gases? For example, consider what might happen to the pressure if the two gases in 6 (b) reacted to form a gaseous compound.

Applications of Partial Pressure

Gas exchange between living organisms and the environment depends on the properties of gases, in particular, partial pressure and solubility. Respiration is one of the most important processes because we need to breathe in oxygen and breathe out carbon dioxide in order to live. Results from human respiration studies at normal atmospheric pressure clearly show the changes in partial pressures of these two gases (**Table 2**). Notice from **Table 2** that the air we exhale contains less oxygen and considerably more carbon dioxide compared with the inhaled air. Because the total pressure of both inhaled and exhaled air must be the same as atmospheric pressure, the partial pressure of a component such as oxygen represents its fraction of the total pressure. In other words, inhaled air contains about 21% oxygen, but exhaled air contains only about 15% oxygen. The difference is what has been absorbed by blood in the capillaries surrounding the lungs of a person.

Table 2: Partial Pressure Changes During Respiration

	Partial Pressure (kPa)	
Gas	**Inhaled air**	**Exhaled air**
$N_{2(g)}$	79.3	75.9
$O_{2(g)}$	21.3	15.5
$CO_{2(g)}$	0.040	3.7
$H_2O_{(g)}$*	0.67	6.2

* The quantity of water in air varies. The value used in this table is based on a relatively low humidity.

DID YOU KNOW ?

Absorbing Oxygen

You may have about 3.5×10^8 tiny alveoli (air sacs) in each lung. The surface area of contact with capillaries for absorbing oxygen is about 75 m^2, about four times as much area as an average classroom floor!

Figure 4
In December 1991, eight people were sealed inside a glass-walled ecosystem to test the ability of humans to survive in a space station. The plants were to supply the oxygen and use up the exhaled carbon dioxide. Unfortunately, about 27 t of oxygen unexpectedly disappeared, probably into the soil, and the project ended after about one year.

Oxygen is the most important gas in the atmosphere; it makes up approximately 21% of the volume of dry air. Partial pressure of a gas is more useful than a percentage composition because it is the pressure (force per unit area) of oxygen that determines how much oxygen is absorbed by the lungs of a person. The Biosphere project in Arizona (**Figure 4**) came to an abrupt end because the people sealed inside the dome were not able to breathe very well when the oxygen levels dropped to about 17 kPa (17% at normal atmospheric pressure). At the top of a high mountain the percentage oxygen may still be a normal 21%, but the partial pressure of oxygen may be sufficiently low that the human system cannot function very well. For example, 21% of 101 kPa (one atmosphere) is about 21 kPa for the partial pressure of oxygen, p_{O_2}, but 21% of 33 kPa (at the top of Mount Everest) is only about 7 kPa for p_{O_2}. Most people, the exceptions being extremely fit people adapted to high altitudes, require about 10 kPa in order to survive.

Another application of partial pressures that you will encounter in chemistry is the collection of gases by the displacement of water. Hydrogen and oxygen gases are often generated in the laboratory and collected by bubbling the gases into a container filled with water (**Figure 5**). Both of these gases have a relatively low solubility in water. However, water evaporates relatively easily and the gas collected will be mixed with some water vapour. Water vapour is a gas like any other gas and the pressure exerted by a gas above its liquid is called its vapour pressure. The vapour pressure of water at different temperatures is well known (**Table 3**). Dalton's law of partial pressures and a table of known vapour pressures of water can be used to determine the pressure of dry gas that has been collected.

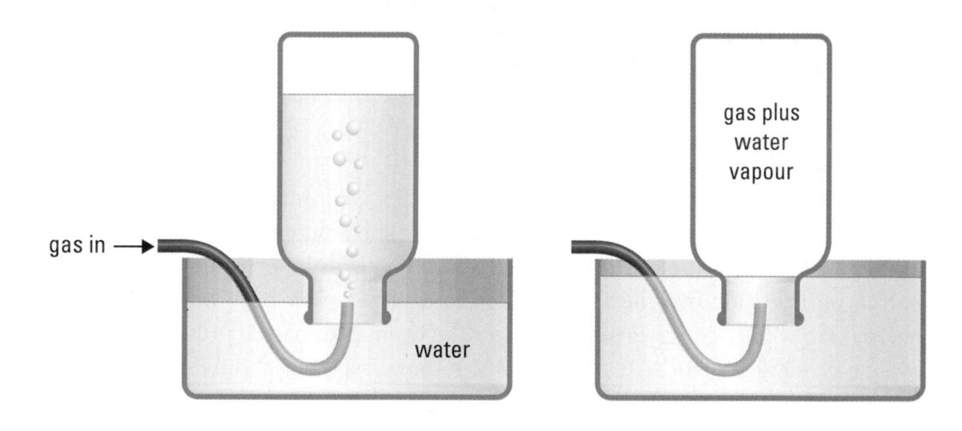

Figure 5
As the gas bubbles enter the jar, water is forced out the bottom. The space occupied by the gas also contains some water vapour. The final result is a bottle filled with a gas mixture, mostly the gas collected plus a small quantity of water vapour.

Table 3: Vapour Pressure of Water at Various Temperatures

Temperature (°C)	Vapour pressure (kPa)
17.0	1.94
18.0	2.06
19.0	2.20
20.0	2.34
21.0	2.49
22.0	2.64
23.0	2.81
24.0	2.98
25.0	3.17
26.0	3.36
27.0	3.57
28.0	3.78
29.0	4.01
30.0	4.24

Sample Problem 2

In a laboratory, oxygen gas was collected by water displacement at an atmospheric pressure of 96.8 kPa and a temperature of 22°C. Using **Table 3,** calculate the partial pressure of dry oxygen.

Solution

$p_{total} = 96.8\ \text{kPa}$
$p_{H_2O} = 2.64\ \text{kPa}$
$p_{O_2} = ?$

$$p_{total} = p_{O_2} + p_{H_2O}$$
$$p_{O_2} = p_{total} - p_{H_2O}$$
$$= 96.8\ \text{kPa} - 2.64\ \text{kPa}$$
$$p_{O_2} = 94.2\ \text{kPa}$$

The partial pressure of dry oxygen is 94.2 kPa.

Practice

Understanding Concepts

8. A sealed container of bottled water sits on a store shelf at a temperature of 23°C. What is the partial pressure of water vapour in the air space inside the container?
9. Nitrogen gas is collected at 20°C and a total ambient pressure of 98.1 kPa using the method of water displacement. What is the partial pressure of dry nitrogen?
10. In an experiment, a student collected a 275-mL sample of hydrogen at 92.4 kPa and 25°C using the water displacement method.
    (a) What is the partial pressure of hydrogen?
    (b) What volume would this hydrogen occupy at standard ambient pressure of 100 kPa?

Applying Inquiry Skills

11. Ammonia is a gas at SATP and is very soluble in water. Suggest an Experimental Design to collect bottles of ammonia with no air present.

Making Connections

12. Carbonated drinks are sold in pressurized bottles and cans. List the gases whose partial pressures contribute to the total pressure in the space between the liquid and the top of the container. What can you deduce about the magnitude of this total pressure?

Answers

8. 2.81 kPa
9. 95.8 kPa
10. (a) 89.2 kPa
    (b) 245 mL

Section 10.1 Questions

Understanding Concepts

1. A tank of compressed air for an underwater diver holds nitrogen at a partial pressure of 240 kPa and has a total pressure of 385 kPa.
   (a) What is the partial pressure of oxygen in the tank?
   (b) What assumptions must be made to predict the partial pressure of oxygen?
2. **Table 2** provides the partial pressures for the main gases present in a sample of air before breathing in and after breathing the air out.
   (a) Calculate the total pressure of the inhaled air and the total pressure of the exhaled air.
   (b) Convert your answers into units of atmospheres (atm).
   (c) Compare and explain your calculated answers for inhaled and exhaled air. Answer from both a biology and a chemistry perspective.
3. Can you use Dalton's law for any mixture of gases? Explain briefly.
4. What two theoretical ideas can be used to explain Dalton's law?
5. When applying Dalton's law of partial pressures, are you assuming ideal or real (nonideal) gases? Explain briefly.

Making Connections

6. A large quantity of various gases such as carbon dioxide and sulfur dioxide is added to the atmosphere every year from natural sources such as volcanoes and from industrial sources such

(continued)

as power plants. Suggest some reasons why the average total pressure of the atmosphere has not been steadily increasing.

Reflecting

7. In this and previous courses, you have often been asked to make summaries using a visual tool such as a diagram. Draw a diagram to illustrate Dalton's law. What are some advantages of visual representations compared to mathematical ones?

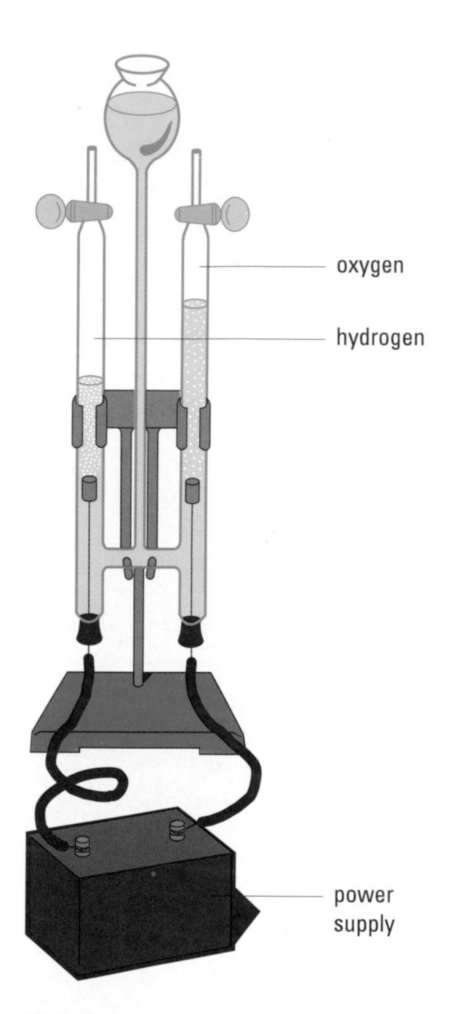

Figure 1
Water decomposes to hydrogen and oxygen gases. Notice that about twice the volume of hydrogen is collected compared to the volume of oxygen. Gay-Lussac is often credited with discovering this ratio in 1805.

10.2 Reactions of Gases

In 1809, Joseph Gay-Lussac, a French scientist and a colleague of Jacques Charles, measured the relative volumes of gases involved in chemical reactions. For example, when he combined hydrogen and chlorine gases at the same temperature and pressure, he noticed that every one litre of hydrogen gas, $H_{2(g)}$, reacted with one litre of chlorine gas, $Cl_{2(g)}$, to produce two litres of hydrogen chloride gas, $HCl_{(g)}$.

	hydrogen	+	chlorine	→	hydrogen chloride
	1.0 L		1.0 L		2.0 L
ratio	1		1		2

His observations of several gas reactions led to the **law of combining volumes,** which states that *when measured at the same temperature and pressure, volumes of gaseous reactants and products of chemical reactions are always in simple ratios of whole numbers.* This law is also known as Gay-Lussac's law of combining volumes. Not all reactants and products need be gases, but the law deals only with the gases consumed or produced. An example of this is the simple decomposition of liquid water, in which the volumes of hydrogen gas and oxygen gas produced are always in the ratio of 2:1 (**Figure 1**).

law of combining volumes: When measured at the same temperature and pressure, volumes of gaseous reactants and products of chemical reactions are always in simple, whole-number ratios; also called Gay-Lussac's law of combining volumes.

Two years after this law was formulated, Amedeo Avogadro proposed an explanation which, unfortunately, was largely ignored for about half a century. Avogadro initially got his idea from the simple relationship between volume and temperature (Charles's law, Chapter 9) but did not follow up on it. When Gay-Lussac published his work about combining volumes, Avogadro was intrigued by the fact that reacting volumes of gases were in whole-number ratios, just like the coefficients in a balanced equation. (Remember that this was only about eight years after Dalton had presented his atomic theory of matter.) Suggesting an explanation for the relationship between the volume ratios and coefficient ratios, he proposed that *equal volumes of gases at the same temperature and pressure contain equal numbers of molecules,* a statement that is best called **Avogadro's theory.** Avogadro's initial idea was a hypothesis. Although it is still sometimes referred to as a hypothesis, the idea is firmly established. Therefore, Avogadro's idea has the status of a theory.

Avogadro's theory: Equal volumes of gases at the same temperature and pressure contain equal numbers of molecules.

This theoretical concept explains the law of combining volumes. For example, if a reaction occurs between two volumes of one gas and one volume of another at the same temperature and pressure, the theory says that two molecules of the first substance react with one molecule of the second. Another example is the reaction of nitrogen and hydrogen, in which ammonia is produced (**Figure 2**).

When all gases are at the same temperature and pressure, the law of combining gas volumes provides an efficient way of predicting the volumes of gases

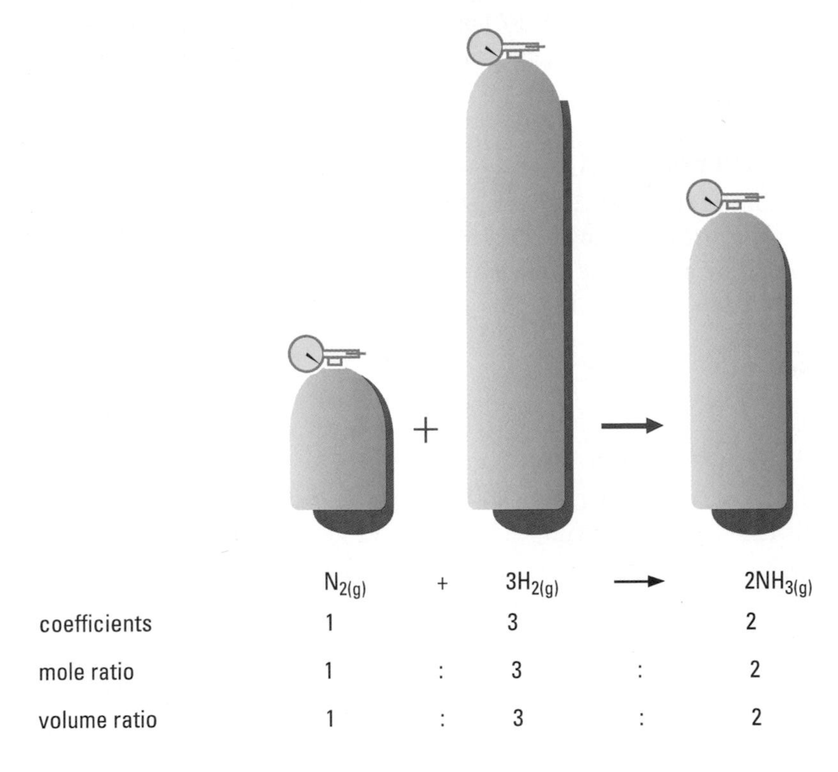

Figure 2
One volume of nitrogen reacts with three volumes of hydrogen, producing two volumes of ammonia.

involved in a chemical reaction. As explained by Avogadro's theory, the mole ratios provided by the balanced equation are also the volume ratios. For example,

coefficients	$2\ C_4H_{10(g)}$ +	$13\ O_{2(g)}$ →	$8\ CO_{2(g)}$ +	$10\ H_2O_{(g)}$
amounts	2 mol	13 mol	8 mol	10 mol
volumes	2 L	13 L	8 L	10 L
example	4 L	26 L	16 L	20 L

For this reaction, suppose you want to predict the volume of oxygen required for the complete combustion of 120 mL of butane gas from a lighter. The first step is to write the balanced chemical equation including the information you are given and what you need to find:

$$2\ C_4H_{10(g)} + 13\ O_{2(g)} \rightarrow 8\ CO_{2(g)} + 10\ H_2O_{(g)}$$

120 mL $\qquad v$

From this chemical equation you can see that 13 mol of oxygen are required for every 2 mol of butane. Therefore, the volume of oxygen has to be greater than 120 mL by a factor of 13/2:

$$v_{O_2} = 120\ \text{mL} \times \frac{13}{2}$$

$$v_{O_2} = 780\ \text{mL}$$

To make sure that the ratio is used in the correct order, include the chemical formula with each quantity as shown following:

DID YOU KNOW ?

Law of Combining Volumes

The law of combining volumes should probably be called Gay-Lussac's law. It is clear from historical documents that Gay-Lussac originally did this work. The pressure–temperature relationship is sometimes called Gay-Lussac's law even though most historians of science do not credit him with the pressure–temperature law. The latter is attributed to earlier work done by Charles and Dalton, but their names have not been directly associated with this law.

Figure 3
More sulfuric acid is manufactured in North America than any other chemical. The industrial chemical reaction used is

$SO_{3(g)} + H_2O_{(l)} \rightarrow H_2SO_{4(aq)}$

$$v_{O_2} = 120 \text{ mL } \cancel{C_4H_{10}} \times \frac{13 \cancel{\text{mol}} O_2}{2 \cancel{\text{mol}} \cancel{C_4H_{10}}} = 780 \text{ mL}$$

Note the cancellation of the units and chemical formulas. Therefore, the volume of oxygen required is 780 mL.

Sample Problem 1

A catalytic converter in the exhaust system of a car uses oxygen (from the air) to convert carbon monoxide to carbon dioxide, which is released through the tailpipe. If we assume the same temperature and pressure, what volume of oxygen is required to react with 125 L of carbon monoxide produced during a 100-km trip?

Solution

$$2\,CO_{(g)} + O_{2(g)} \rightarrow 2\,CO_{2(g)}$$
$$125 \text{ L} \qquad v$$

$$v_{O_2} = 125 \text{ L } \cancel{CO} \times \frac{1 \cancel{\text{mol}} O_2}{2 \cancel{\text{mol}} \cancel{CO}}$$

$$v_{O_2} = 62.5 \text{ L}$$

or $$v_{O_2} = 125 \text{ L} \times \frac{1}{2}$$

$$v_{O_2} = 62.5 \text{ L}$$

The volume of oxygen required is 62.5 L.

Practice

Understanding Concepts

1. Gay-Lussac was the first to notice and publish evidence of simple volume ratios of reacting gases. What important idea was necessary to relate this observation to chemical equations?
2. State two empirical observations that led Avogadro to his theory.
3. If 5.00 L of propane is burned in a gas barbecue, what volume of oxygen, at the same temperature and pressure, is required for complete combustion?
4. In modern automobile catalytic converters, nitrogen monoxide (a pollutant) reacts with hydrogen to produce nitrogen and water vapour (part of the exhaust). The catalytic converter of a car meeting current emission standards removes about 1.2 L of nitrogen monoxide at SATP for every kilometre of driving. What volume of nitrogen gas is formed from 1.2 L of nitrogen monoxide at the same temperature and pressure?
5. The production of sulfuric acid is a very important chemical industry in any developing or developed country (**Figure 3**). The main reactant required to produce sulfuric acid is sulfur, which can be obtained from a variety of sources.
   (a) One technology for removing hydrogen sulfide from sour natural gas involves converting part of the hydrogen sulfide to sulfur dioxide, which then reacts with the remaining hydrogen sulfide as shown in the reaction equation below. Predict the volume of sulfur dioxide needed to react completely with 248 kL of hydrogen sulfide. The gases are measured at 350°C and 250 kPa.

$$16\,H_2S_{(g)} + 8\,SO_{2(g)} \rightarrow 3\,S_{8(s)} + 16\,H_2O_{(g)}$$

Answers

3. 25.0 L
4. 0.60 L
5. (a) 124 kL
   (b) 250 kL
   (c) 325 kL $SO_{2(g)}$, 163 kL $O_{2(g)}$

(b) Solid sulfur is the starting material in the typical manufacture of sulfuric acid. First it is burned to form sulfur dioxide. Predict the volume of oxygen required to produce 250 kL of sulfur dioxide with all gases at 450°C and 200 kPa.

(c) In the presence of the catalyst $V_2O_{5(s)}$, sulfur dioxide—from the burning of sulfur or directly from the output of a smelter—reacts with oxygen to form sulfur trioxide.

$$2\ SO_{2(g)} + O_{2(g)} \rightarrow 2\ SO_{3(g)}$$

Predict the volumes of sulfur dioxide and oxygen needed to produce 325 kL of sulfur trioxide when all gases are measured at the same temperature and pressure.

Reflecting

6. Which law studied in Unit 2 is similar to the law of combining volumes? Describe the similarity and any differences.

Molar Volume of Gases

The evolution of scientific knowledge often involves integrating two or more concepts. For example, Avogadro's idea and the mole concept (Chapter 4) can be integrated. According to Avogadro's theory, equal volumes of any gas at the same temperature and pressure contain an equal number of particles. The mole concept indicates that a mole is a specific number of particles—Avogadro's number of particles. Therefore, for all gases at each specific pressure and temperature, there must be a certain volume that contains one mole of particles. Logically, this **molar volume** is the same for all gases at the same temperature and pressure. (This is not true for liquids and solids.) For scientific work, the most useful specific pressure and temperature conditions are either SATP or STP. It has been determined empirically that the molar volume of a gas at SATP is 24.8 L/mol. The molar volume of a gas at STP is 22.4 L/mol (**Figure 4**).

Knowing the molar volume of gases allows scientists to work with easily measured volumes of gases when specific amounts of gases are needed. Measuring the volume of a gas is much more convenient than measuring its mass. Imagine trapping a gas in a container and trying to measure its mass on a balance—and then making corrections for the buoyant force of the surrounding air. Also, working with gas volumes is more precise because the process involves measuring relatively large volumes rather than relatively small masses. Molar volume can be used as a conversion factor to convert amount in moles to volume, as shown in the following example.

In SI symbols, the relationship of amount (n), volume (v), and molar volume (V) is expressed as

$$n = \frac{v}{V} \quad \text{or} \quad v = nV$$

DID YOU KNOW ?

Real Molar Volumes

When performing calculations, we will assume in this text that real gases behave like ideal gases. You can make your own judgment on just how valid this assumption is by considering the empirical values for molar volumes of selected gases.

Gas	Molar volume (L/mol at STP)
O_2	22.397
N_2	22.402
CO_2	22.260
H_2	22.433
NH_3	22.097

Of these gases, you can suggest why ammonia behaves least like an ideal gas?

molar volume: The volume that one mole, in this case a gas, occupies at a specified temperature and pressure.

Figure 4
At STP, one mole of gas has a volume of 22.4 L, which is approximately the volume of 11 "empty" 2-L pop bottles.

Sample Problem 2

What volume is occupied by 0.024 mol of carbon dioxide gas at SATP?

Solution

$n_{CO_2} = 0.024\ \text{mol}$

$V_{SATP} = 24.8\ \text{L/mol}$

$$v_{CO_2} = 0.024 \text{ mol} \times \frac{24.8 \text{ L}}{1 \text{ mol}}$$

$$v_{CO_2} = 0.60 \text{ L}$$

The carbon dioxide gas occupies the volume 0.60 L.

Notice how the units cancel in the example above. A molar volume can also be used to convert from a volume to an amount in moles. In this case, the molar volume ratio must be inverted to allow for the correct cancellation of units.

Sample Problem 3

What amount of oxygen, in moles, is available for a combustion reaction in a volume of 5.6 L at STP?

Solution

$v_{O_2} = 5.6 \text{ L}$

$V_{STP} = 22.4 \text{ L/mol}$

$$n_{O_2} = 5.6 \text{ L} \times \frac{1 \text{ mol}}{22.4 \text{ L}}$$

$$n_{O_2} = 0.25 \text{ mol}$$

The amount of oxygen available is 0.25 mol.

Gases such as oxygen and nitrogen are often liquefied for storage and transportation, then allowed to vaporize for use in a technological application. Helium is stored and transported as a compressed gas. Both liquefied and compressed gases are sold by mass. Molar volume and molar mass can be combined to calculate the volume of gas that is available from a known mass of a substance.

For example, helium-filled balloons (**Figure 5**), often used for party decorations, are less dense than air, so they stay aloft and will rise unless tied down by a string. What volume does 3.50 g of helium gas occupy at SATP? To answer this question, we first need to convert the mass into an amount in moles:

$$n_{He} = 3.50 \text{ g} \times \frac{1 \text{ mol}}{4.00 \text{ g}}$$

$$n_{He} = 0.875 \text{ mol}$$

Now we can convert this amount into a volume at SATP, using the molar volume constant:

$$v_{He} = 0.875 \text{ mol} \times \frac{24.8 \text{ L}}{1 \text{ mol}}$$

$$v_{He} = 21.7 \text{ L}$$

Once these two steps are clearly understood, they can be combined into a single calculation, as shown below. Notice how you can plan use of the conversion factors by planning the cancellation of the units. All units except the final unit cancel.

$$v_{He} = 3.50 \text{ g} \times \frac{1 \text{ mol}}{4.00 \text{ g}} \times \frac{24.8 \text{ L}}{1 \text{ mol}}$$

$$v_{He} = 21.7 \text{ L}$$

Figure 5
Helium-filled balloons are popular items for parties and store promotions.

Sample Problem 4

A propane tank for a barbecue contains liquefied propane. If the tank mass drops by 9.1 kg after a month's use, what volume of propane gas at SATP was used for cooking?

Solution

$m_{C_3H_8} = 9.1\ kg$

$M_{C_3H_8} = 44.11\ g/mol$

$V_{SATP} = 24.8\ L/mol$

$$v_{C_3H_8} = 9.1\ \text{kg} \times \frac{1\ \text{mol}}{44.11\ \text{g}} \times \frac{24.8\ \text{L}}{1\ \text{mol}}$$

$$v_{C_3H_8} = 5.2\ \text{kL}$$

The volume of propane gas used was 5.2 kL, or 5.2 m^3.

SUMMARY Concepts Useful in Gas Reactions

Law of combining volumes: when measured at the same temperature and pressure, volumes of gaseous reactants and products of chemical reactions are always in simple ratios of whole numbers

Avogadro's theory: equal volumes of gases at the same temperature and pressure contain equal numbers of molecules

Molar volume: the volume that one mole of a gas occupies at a specified temperature and pressure

$V_{STP} = 22.4\ L/mol$ $\qquad$ $V_{SATP} = 24.8\ L/mol$

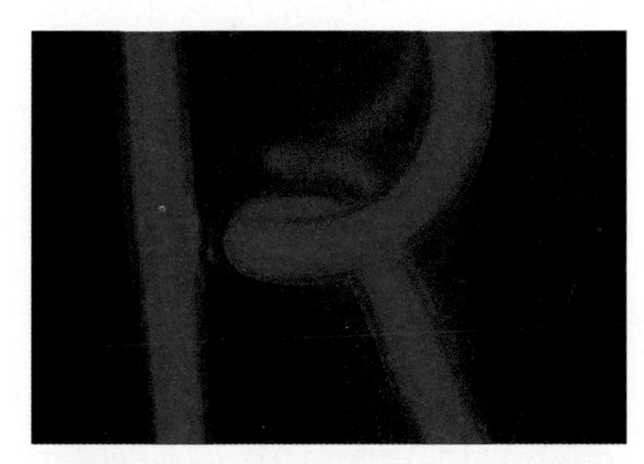

Figure 6
When an electric current is passed through a glass tube containing the noble gas neon, the gas glows a characteristic red colour.

Practice

Understanding Concepts

7. Sulfur dioxide gas is emitted from marshes, volcanoes, and refineries that process crude oil and natural gas. What amount in moles of sulfur dioxide is contained in 50 mL of the gas at SATP?
8. Neon gas under low pressure emits the red light that glows in advertising signs (**Figure 6**).
   (a) What volume does 2.25 mol of neon gas occupy at STP before being added to neon tubes in a sign?
   (b) What pressure is reached when the gas is heated to 35°C by the Sun?
   (c) When designing the tube for this application, what specifications for the quality of the tube are necessary? Provide your reasoning.
9. One gram of baking powder produces about 0.13 g of carbon dioxide. What volume is occupied by 0.13 g of carbon dioxide gas at SATP?
10. Volatile liquids vaporize rapidly from opened containers or if spilled. Some vapours, such as those from gasoline, contribute to the formation of smog. What volume at STP is occupied by gasoline vapours from 50.0 g of spilled gasoline (assume octane, $C_8H_{18(l)}$)?
11. Millions of tonnes of nitrogen dioxide are dumped into the atmosphere each year by automobiles and are a major cause of smog formation. What is the volume of 1.00 t (1.00 Mg) of nitrogen dioxide at SATP?

Answers

7. 2.0 mmol
8. (a) 50.4 L
   (b) 114 kPa
9. 73 mL
10. 9.80 L
11. 539 kL

Answers

12. 0.727 g
13. (a) 22.4 L/mol
    (b) 21.9 L/mol
    (c) 2% difference
14. (a) 1.8 g/L

12. Water vapour plays an important role in the weather patterns on Earth. What mass of water must vaporize to produce 1.00 L of water vapour at SATP?

Applying Inquiry Skills

13. Complete the **Prediction, Analysis,** and **Evaluation** sections for the following investigation report.

    Question

    What is the molar volume of oxygen at STP?

    Prediction

    (a) Predict the answer to the Question. Include the basis (the authority) for your Prediction.

    Experimental Design

    A small oxygen cylinder (**Figure 7**) sold for mini-welding applications provides the oxygen, which is collected by water displacement.

    Evidence

    initial mass of oxygen cylinder = 46.84 g

    final mass of oxygen cylinder = 45.79 g

    volume of gas collected = 848 mL

    temperature = 22°C

    atmospheric pressure = 95.2 kPa

    water vapour pressure (see **Table 3,** page 464) = 2.64 kPa

    Analysis

    (b) Based on the Evidence, what is the answer to the Question? (Adjust the pressure to correct for the vapour pressure of water.)

    Evaluation

    (c) Does the answer based on the Evidence agree with your Prediction? Calculate the percent difference.
    (d) Based on the Evidence, evaluate the authority you used to make your Prediction.

Making Connections

14. Carbon dioxide is commonly used in fire extinguishers.
    (a) What is the density (in grams per litre) of carbon dioxide at SATP (two significant digits)?
    (b) If the density of air at SATP is 1.2 g/L, use this value and the answer to (a) to suggest one reason for the use of carbon dioxide as a fire-extinguishing agent.
    (c) What is another important characteristic of carbon dioxide that makes it suitable for use in a fire extinguisher?

Figure 7
Mini cylinders of oxygen are commercially available.

INQUIRY SKILLS

- ○ Questioning
- ● Hypothesizing
- ● Predicting
- ○ Planning
- ● Conducting
- ● Recording
- ● Analyzing
- ● Evaluating
- ● Communicating

Investigation 10.2.1

Molar Volume of a Gas

The purpose of this investigation is to test the accepted value for the molar volume of a gas at STP. To do this, determine the molar volume of a gas (obtained from a liquid that is easily vaporized) from the measured volume of the vapour (converted to STP conditions), the molar mass of the substance, and the mass of the condensed vapour. Complete the **Prediction, Analysis,** and **Evaluation** sections of the lab report.

Question

What is the molar volume of a gas at STP?

Prediction

(a) According to the accepted constant, what is the molar volume of any gas at STP?
(b) What assumption is made when stating this value?

Experimental Design

Excess liquid is slowly vaporized inside a flask with a small opening. The pressure, volume, and temperature of the vapour remaining in the flask are measured and the mass of the vapour is determined by condensing the vapour back into its liquid state.

Materials

Eye protection and a lab apron are needed.
Assume that the liquid is flammable. Do not conduct this activity near an open flame. Good ventilation is important.

lab apron
eye protection
mass balance
aluminum foil, 5 cm × 5 cm
elastic band
125-mL Erlenmeyer flask
10-mL graduated cylinder
volatile liquid, 3 mL
pin
ring stand
buret clamp
600-mL beaker
hot plate
water
boiling chips (optional)
plastic stirring rod
thermometer
barometer
100-mL or 250-mL graduated cylinder
paper towel

Procedure

1. Measure the total mass of the aluminum foil, elastic band, and a dry 125-mL Erlenmeyer flask.
2. Record the name and formula of the volatile liquid.
3. Using the 10-mL graduated cylinder, measure 3 mL of the volatile liquid and pour it into the flask.
4. Fold the aluminum foil tightly over the flask opening and secure, just under the lip, with the elastic band. Use a pin to make a *tiny* hole, as small as possible, in the centre of the foil.
5. Place the beaker on the hot plate, and clamp the flask to the stand so that the flask is slightly tilted and as far into the beaker as possible (**Figure 8**).
6. Add water to the beaker so that the flask is almost completely surrounded by water. Add a few boiling chips if directed by your teacher.
7. Heat the beaker of water slowly, with occasional stirring, until the last traces of the liquid in the flask have evaporated. Monitor the temperature of the water. Do not exceed the maximum temperature specified by your teacher.

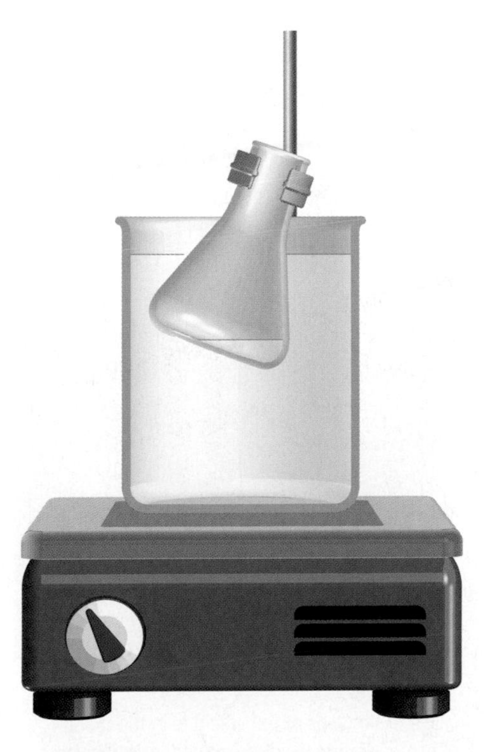

Figure 8
The flask should be almost completely surrounded by water and the water heated slowly to the maximum temperature specified by your teacher.

8. When all of the liquid in the flask has evaporated, immediately remove the flask from the beaker and measure the temperature of the water.
9. Record the atmospheric pressure.
10. After the flask has cooled to room temperature, wipe the flask and cap carefully with a paper towel to remove all the water. Be sure to remove any water trapped between the ends of the aluminum foil and the flask.
11. Determine the total final mass of the flask, cap, and condensed liquid.
12. Remove the cap from the flask and fill the flask completely with tap water. Use a large graduated cylinder to measure the volume of water in the flask.
13. Dispose of the water mixture down the drain with lots of water.

Analysis

(c) Using the final mass and the masses of the individual components, calculate the mass of the condensed liquid remaining in the flask.
(d) Convert the mass to an amount, in moles, of the liquid used.
(e) Convert the volume of the vapour used in this experiment to the volume it would occupy at STP conditions.
(f) According to your Evidence and calculations, what is the molar volume of the vapour at STP?

Evaluation

(g) Evaluate the Evidence collected by considering the quality of the Experimental Design, Materials, and Procedure. Identify possible sources of experimental error or uncertainty.
(h) What improvements could be made to this experiment? Would these significantly affect the final outcome?
(i) Calculate the accuracy of your experimental result by determining the percentage difference between your value and the predicted value.
(j) Was your Prediction verified? Justify your answer.
(k) Does the assumption about the nature of the gas appear to be valid? Discuss briefly.

Figure 9
One of the uses of nitric acid is in the manufacture of nitroglycerine, a potent explosive. The explosive characteristic results from the production of gases that expand rapidly.

Section 10.2 Questions

Understanding Concepts

1. Avogadro's theory was not immediately recognized as an important development in chemistry.
   (a) State Avogadro's theory.
   (b) In a paragraph, explain why this theory is so important in the chemistry of gases.
2. The production of nitric acid is important to the fertilizers and explosives industries **(Figure 9)**.
   (a) The production of nitric acid by the Ostwald process begins with the combustion of ammonia:

   $$4\ NH_{3(g)} + 5\ O_{2(g)} \rightarrow 4\ NO_{(g)} + 6\ H_2O_{(g)}$$

   Predict the volume of oxygen required to react with 100 L of ammonia as well as the volumes of nitrogen oxide and water vapour produced. All gases are measured at 800°C and 200 kPa.

(b) In another step of the Ostwald process, nitrogen monoxide reacts with oxygen to form nitrogen dioxide. Predict the volume of oxygen at 800°C and 200 kPa required to produce 750 L of nitrogen dioxide at the same temperature and pressure.

(c) Nitric acid is produced by reacting nitrogen dioxide with water:

$$3\ NO_{2(g)} + H_2O_{(l)} \rightarrow 2\ HNO_{3(aq)} + NO_{(g)}$$

Predict the volume of nitrogen monoxide produced by the reaction of 100 L of nitrogen dioxide with excess water. Both gases are measured at the same temperature and pressure.

(d) A high-nitrogen fertilizer is made by reacting ammonia gas with nitric acid to produce aqueous ammonium nitrate. Can the law of combining volumes be used to predict the volume of ammonia gas required to react with 100 L of nitric acid? Justify your answer.

3. Weather balloons filled with hydrogen gas are occasionally reported as UFOs. They can reach altitudes of about 40 km.
   (a) Assume SATP and that balloons expand and contract without resistance. What volume does 7.50 mol of hydrogen gas occupy when a small weather balloon is released at ground level?
   (b) What volume does the gas occupy at –47°C and 1.2 kPa (the average conditions at 30 km altitude)?
4. Oxygen is released by plants during photosynthesis and is used by plants and animals during respiration. What amount, in moles, of oxygen is present in 20.0 L of air at STP? Assume that air is 20% oxygen (by volume).
5. Human beings exhale millions of tonnes of carbon dioxide into the atmosphere each year. Determine the volume occupied by 1.00 t of carbon dioxide at STP.
6. To completely burn 1.0 L of gasoline in an automobile engine requires about 1.9 kL of oxygen at SATP (**Figure 10**). What mass of oxygen gas is consumed by burning 1.0 L of gasoline?

Figure 10
For years, the fuel of choice for automobiles has been gasoline, but now consumers can buy a hybrid vehicle that runs on gasoline and battery-supplied electricity; the small gasoline engine is used only when extra power is required for acceleration.

10.3 The Ozone Layer

As you learned in Chapter 9, ground-level ozone is the main harmful ingredient in smog. However, ozone in the stratosphere (10–50 km above Earth's surface) is beneficial because it protects us and the environment by preventing most of the short-wavelength, high-energy ultraviolet (UV) radiation from reaching Earth's surface (**Figure 1**, page 476). If all of the UV radiation in sunlight were to reach Earth's surface, serious damage would occur to life on Earth. The higher-energy UV radiation has enough energy to break covalent bonds; the resulting chemical changes cause sunburn and are linked to the development of cataracts and skin cancer. It can also induce genetic mutations and destroy vegetation. Canada has been active in ozone research since the 1930s and has operated the World Ozone Atmospheric Data Centre since the 1950s. Monitoring ozone levels began in Canada in the late 1950s. In addition to atmospheric studies, Canadian scientists have conducted extensive studies of the impact of increased UV radiation on important species in agriculture and forestry, and on freshwater and marine organisms.

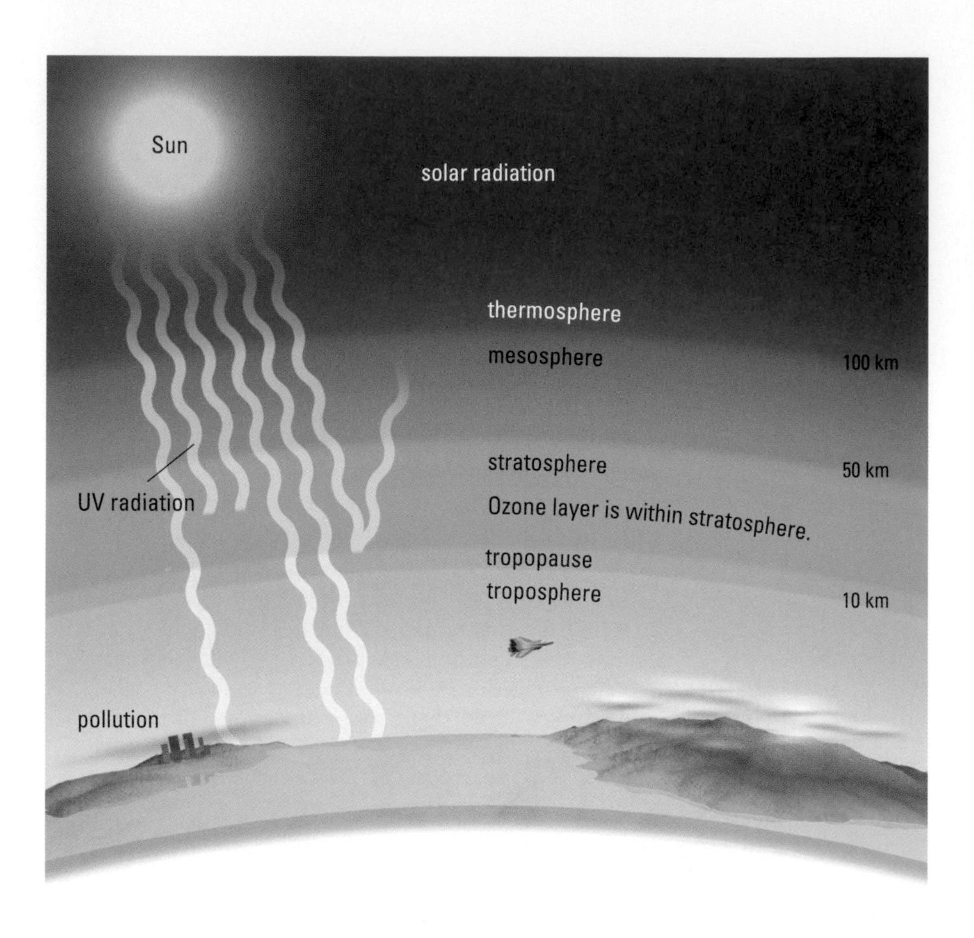

Figure 1
Ozone is formed when UV radiation from the Sun is absorbed by oxygen gas in the stratosphere.

Ultraviolet radiation is involved in producing the ozone layer, as shown in the following reactions:

$$O_{2(g)} \xrightarrow{UV} O_{(g)} + O_{(g)}$$

$$O_{(g)} + O_{2(g)} \rightarrow O_{3(g)}$$

Molecular oxygen in the upper atmosphere absorbs only the higher-energy UV light, and by absorbing this, splits into two oxygen atoms. Oxygen atoms are extremely reactive, and most of them react with an oxygen molecule to form ozone. This reaction is exothermic, that is, releases heat, and is believed to be responsible for the higher temperatures in the stratosphere compared to the atmospheric levels immediately above and below it. Ozone molecules also absorb certain parts of UV light and decompose back to oxygen molecules and oxygen atoms. Through the processes of formation and decomposition of ozone, almost all of the harmful UV light is absorbed. The ozone layer in the stratosphere actually contains very little ozone, rarely exceeding 10 ppm in concentration. If all of the ozone were to be collected at ground level at SATP, the layer would only be about 0.3 mm thick.

For the past 30 years, scientists have been concerned about the effect of certain chlorofluorocarbons (CFCs) on the ozone reaction. The CFCs, which are generally known by their trade name, Freon, were first synthesized in the 1930s. Some of the common ones are $CFCl_{3(g)}$ (Freon-11), $CF_2Cl_{2(g)}$ (Freon-12), $C_2F_3Cl_{3(g)}$ (Freon-113), and $C_2F_4Cl_{2(g)}$ (Freon-114). Because these compounds are readily liquefied, relatively inert, nontoxic, noncombustible, and volatile, they have been used as coolants in refrigerators and air conditioners. Large quantities

DID YOU KNOW?

Saving Your Skin

UV radiation damages DNA. Because the radiation cannot penetrate far into tissue, damage is confined to skin and eyes. Ozone, and sunscreen, absorbs this dangerous radiation.

of CFCs were also used in the manufacture of disposable foam products, such as cups and plates, and as aerosol propellants in spray cans. Most of the CFCs produced for commercial and industrial use are eventually released to the atmosphere. Because they are relatively inert, CFCs remain unchanged in the atmosphere and eventually diffuse to the stratosphere, where UV radiation causes them to decompose. Scientific theory and evidence indicate that, once emitted to the atmosphere, these compounds significantly deplete the ozone layer that shields the planet from damaging UV radiation. For example, the following equations show the reactions started by a molecule of Freon-11:

$$CFCl_{3(g)} \xrightarrow{UV} CFCl_{2(g)} + Cl_{(g)}$$

The reactive chlorine atoms undergo the following reactions:

$$\cancel{Cl}_{(g)} + O_{3(g)} \rightarrow \cancel{ClO}_{(g)} + O_{2(g)}$$

$$\cancel{ClO}_{(g)} + O_{(g)} \rightarrow \cancel{Cl}_{(g)} + O_{2(g)}$$

Looking at the net reaction by cancelling the particles that are the same on both sides of the reaction equations, you can see that the net reaction is the conversion of ozone to oxygen molecules:

$$O_{3(g)} + O_{(g)} \rightarrow 2\ O_{2(g)}$$

Note that the chlorine atom plays the role of a catalyst, since it is not used up and can, therefore, take part in many such reactions. Research has shown that a single chlorine atom can cause the destruction of thousands of ozone molecules.

Scientists have been measuring Antarctic ozone levels since the early 1970s. The first "hole" in the ozone layer was discovered in 1985. "Holes" in the ozone layer are more accurately referred to as areas of ozone thinning, because they are much like the worn-out places in an old sweater where there are still threads covering the worn-out area but the fabric is so thin you can see right through it. In September 2000, scientists at the National Aeronautics and Space Administration (NASA) detected the largest-ever area of ozone thinning over Antarctica. The area spanned 28.3 million square kilometres, nearly three times the size of Canada. Ozone thinning has also been found in the stratosphere over the Arctic regions of Canada (**Figure 2**).

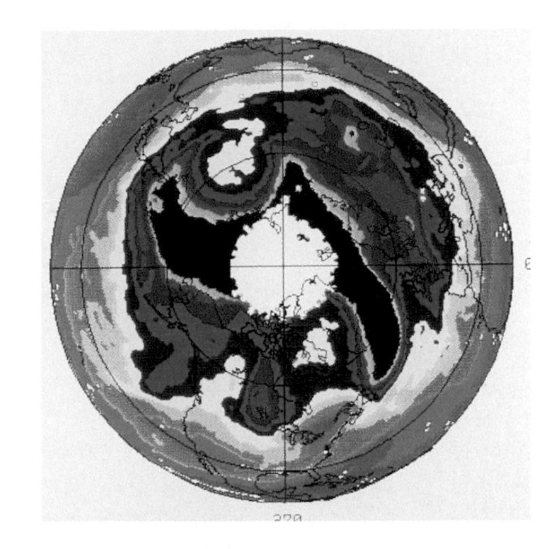

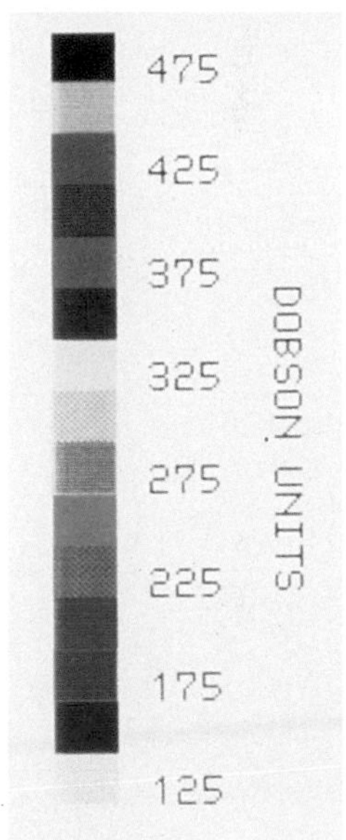

Figure 2
Satellite image of ozone concentrations in the Arctic

Why is ozone thinning occurring only over the polar regions, and particularly Antarctica? According to current theory and evidence, the long, extremely cold polar winter results in the formation of polar stratospheric clouds. During the dark winter, the particles in these clouds tie up water and nitrogen compounds and promote reactions at the particles' surface, which release chlorine. In early spring, the returning sunlight triggers the photochemical reaction catalyzed by the free chlorine. The freezing and chemical reactions in the clouds take time and very low temperatures. In the Arctic, the temperatures are not as low and the air is not as stable as in the Antarctic. Therefore, ozone thinning is not as extensive in Canada's North.

Significant ozone depletion over Antarctica has effects well beyond this isolated region. The hole itself contributes to a general dilution of ozone concentration, because winds in the spring move large quantities of air very low in ozone out of the polar regions and into the general circulation in the hemispheres. Evidence from satellite measurements shows a general, continuous decline in ozone levels across the globe. A thinned-out ozone layer could result

DID YOU **KNOW?**

UV-B

UV radiation is classified by wavelength.

UV-C 200–280 nm
UV-B 280–320 nm
UV-A 320–400 nm

UV-C (the shortest wavelengths) is most dangerous but also makes up less of solar radiation. Ozone has its greatest effect at UV-B wavelengths.

in more skin cancers and cataracts. Scientists are looking into possible harm to agriculture, since there is some evidence of damage to plant life in Antarctic seas. Research and ozone data gathering continue. Canada continues its contribution through its Arctic Observatory and the development of sophisticated ozone-measuring instruments. In addition, the National Research Council is actively engaged in researching and developing alternatives to CFCs.

The Montreal Protocol

In the late 1970s, the use of CFCs in spray cans in North America and Europe was declining mainly by voluntary agreements. However, other uses of CFCs continued to grow. There were several attempts by the United Nations Environment Program to call on governments to reduce all uses of CFCs but most governments ignored the plea. In the early 1980s, Canada, along with a few other countries, began pressing for a ban on all spray-can uses. After much political debate and accusations among countries, a convention was held in Geneva that laid the groundwork for the final round of meetings in Montreal. The Montreal Protocol on substances that deplete the ozone layer is a landmark international agreement designed to protect the ozone layer. The treaty was signed by 27 countries in 1987, now expanded to 175 countries. The agreement was substantially amended in 1990, 1992, and, more recently, in January 2001. The original Montreal Protocol stipulated that the production and consumption of compounds that deplete ozone in the stratosphere were to be phased out by 2000. **Table 1** lists these compounds and their uses.

Table 1: Ozone-Destroying Substances Identified in the Montreal Protocol

Name	Uses
CFCs (chlorofluorocarbons)	air conditioning, refrigeration, aerosol sprays, solvents, foaming agents in the manufacture of plastic
halons (CF_3Br, CF_2BrCl, $C_2F_4Br_2$)	fire extinguishers
carbon tetrachloride (CCl_4)	solvent for oils, fats, lacquers, varnishes, and resins; starting material for the manufacture of organic compounds
methylchloroform (1,1,1-trichloroethane)	widely used solvent for cleaning plastic moulds; cold type metal cleaning

Scientists estimate that the current use of the ozone-destroying substances (ODSs) listed in **Table 1** is about 40% of what it was when the Montreal Protocol was signed—well short of the complete phase-out intended by 2000. Even when production of these chemicals stops, it will be some time before the ozone layer is repaired. With the ozone-destroying chemicals that are already in the stratosphere and those that will arrive within the next few years, ozone destruction will likely continue for another 50 years.

Most of the countries that signed the Montreal Protocol also developed national policies to maximize recycling, ban nonessential uses, develop labelling requirements, and examine safe alternatives for ODSs. In Canada, the production and import of CFCs have been reduced by 85%. The provinces have developed regulations for recycling Freon-12, widely used in refrigerators and automobile air conditioners. Canada has done a good job in controlling ODSs and helping developing countries meet the commitments to the Protocol. However, some environmental groups think that Canada could do much more to completely eliminate, rather than control, ODSs and prohibit questionable replacements for CFCs.

Unfortunately, most of the CFCs in equipment before the Montreal Protocol still exist, and there is no plan to dispose of them. According to the Sierra Club of Canada, about 130 Mt of CFCs are still in use in Canada, and 40% of these are in our homes and automobiles. What are the alternatives? Some industries, such as the automobile manufacturers, have switched to hydrochlorofluorocarbons (HCFCs) or hydrofluorocarbons (HFCs) in their air-conditioning units. Although they may claim to be "ozone friendly," because HCFCs and HFCs have less impact on the ozone layer, this is only an interim solution. Both are potent greenhouse gases, so we are partially solving one problem only to create another potential environmental problem.

A promising alternative for many CFC uses is hydrofluoroether ($C_4F_9OCH_{3(l)}$), also known as HFE. Research indicates that this substance has no effect on ozone, is not a greenhouse gas, and is nontoxic. Another alternative is the use of hydrocarbons such as propane, pentane, isobutane, and cyclopentane in refrigerators (**Figure 3**) and foam containers and insulation. Carefully chosen hydrocarbons have zero ozone depletion potential and little impact as greenhouse gases. They are also very efficient refrigerants and widely available. Unfortunately for the chemical industry giants such as DuPont and ICI, these hydrocarbons cannot be licensed like CFCs and the related compounds in which they have invested billions of dollars. It is therefore not surprising that hydrocarbon-based refrigerators are widely available on all continents except North America. Are the Canadian and U.S. governments protecting the large chemical industries who have a vested interest in CFC-related compounds?

Figure 3
Greenfreeze technology developed when Greenpeace International brought together scientists with experience in the use of hydrocarbons as refrigerants and an East German refrigerator company. The new Greenfreeze refrigerators use propane-isobutane (propane, $C_3H_{8(g)}$, isobutane, $C_4H_{10(g)}$) mixtures or pure isobutane for the refrigerant and cyclopentane-blown polyurethane foam insulation (cyclopentane, $C_5H_{10(l)}$).

Practice

Understanding Concepts

1. What kind of UV radiation is prevented by ozone from reaching Earth's surface? Describe briefly how this happens.
2. Why were CFCs developed?
3. What environmental problem is caused by the release of CFCs into the atmosphere?
4. Is the ozone hole an actual hole or opening (like an open window)? Explain.
5. The halons listed in **Table 1** are classed as ozone-destroying substances because they undergo reactions similar to the sequence shown earlier for Freon-11, with a bromine atom taking the place of a chlorine atom. Write the reaction sequence initiated by UV radiation splitting a bromine atom from a molecule of CF_3Br.
6. Why is the ozone depletion less in the Arctic compared with the Antarctic?
7. What implications does the thinning of the ozone layer have for people who like to suntan?

DECISION-MAKING SKILLS

- Define the Issue
- Identify Alternatives
- Research
- Analyze the Issue
- Defend a Decision
- Evaluate

Explore an Issue

Take a Stand: Out with the Old?

As the substances listed in the Montreal Protocol are phased out, new technologies are being developed, but this raises questions about the environmental impact of the replacement technologies. Obviously, we should replace technologies that produce ODSs with technologies that do not harm our environment.

Working in small groups, research a replacement substance for a CFC technology, for example,

- hydrochlorofluorocarbons (HCFCs)
- hydrofluorocarbons (HFCs)
- hydrofluoroethers (HFEs)
- hydrocarbons
- hydrofluoroalkanes (HFAs)
- carbon dioxide

(a) Prepare a poster on the risks and benefits of the new technology. Conclude with a recommendation for adopting, restricting, or banning the new technology.

Follow the links for Nelson Chemistry 11, 10.3.

www.science.nelson.com

Section 10.3 Questions

Understanding Concepts

1. Describe the nature of the Montreal Protocol and explain its purpose.
2. Give an example of an ozone-depleting substance that is recycled.
3. List at least two alternatives to CFCs and state where they might be used.
4. List some of Canada's contributions in ozone research and solutions to ozone depletion.
5. Explain the following phrase about ozone: "good up high, bad nearby."

10.4 Gas Stoichiometry

Many chemical reactions involve gases. One common consumer example is the combustion of propane in a home gas barbecue (**Figure 1**). The reaction of chlorine in a water treatment plant is a commercial example. An important industrial application of a chemical reaction involving gases is the production of the fertilizer ammonia from nitrogen and hydrogen gases. These technological examples feature gases as either valuable products, such as ammonia, or as part of an essential process, such as water treatment.

Studies of chemical reactions involving gases (e.g., the law of combining volumes) have helped scientists develop ideas about molecules and explanations for chemical reactions, such as the collision–reaction theory (Chapter 3). In both

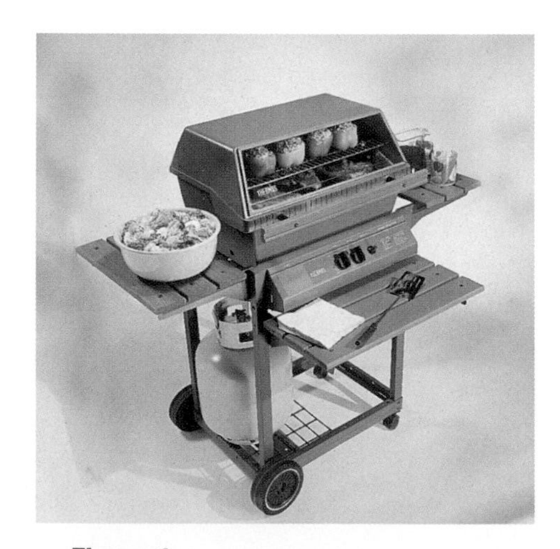

Figure 1
Propane gas barbecues have become very popular. Charcoal barbecues are now banned in parts of California because they produce five times as much pollution (nitrogen oxides, hydrocarbons, and particulates) as gas barbecues.

technological applications and scientific studies of gases, it is necessary to accurately calculate quantities of gaseous reactants and products.

The method of stoichiometry applies to all chemical reactions. This section extends stoichiometry to gases, using gas volume, molar volume, and the ideal gas law. For example, if 275 g of propane burns in a gas barbecue, what volume of oxygen measured at STP is required for the reaction? To answer this question, write a balanced chemical equation to relate the amount propane to the amount oxygen. List the given and the required measurements and the conversion factors for each chemical, just as you did in the previous stoichiometry questions:

$C_3H_{8(g)}$	+	$5\,O_{2(g)}$	$\rightarrow$	$3\,CO_{2(g)}$	+	$4\,H_2O_{(g)}$
275 g		v				
44.11 g/mol		22.4 L/mol				

Since propane and oxygen are related by their mole ratio, you must convert the mass of propane to an amount in moles:

$$n_{C_3H_8} = 275\ \cancel{g} \times \frac{1\ \text{mol}}{44.11\ \cancel{g}}$$

$$n_{C_3H_8} = 6.23\ \text{mol}$$

The balanced equation indicates that one mole of propane reacts with five moles of oxygen. Use this mole ratio to calculate the amount of oxygen required, in moles. (This step is common to all stoichiometry calculations.)

$$n_{O_2} = 6.23\ \cancel{\text{mol}\ C_3H_8} \times \frac{5\ \text{mol}\ O_2}{1\ \cancel{\text{mol}\ C_3H_8}}$$

$$n_{O_2} = 31.2\ \text{mol}$$

The final step involves converting the amount of oxygen to the required quantity, in this case, volume:

$$v_{O_2} = 31.2\ \cancel{\text{mol}} \times \frac{22.4\ \text{L}}{1\ \cancel{\text{mol}}}$$

$$v_{O_2} = 698\ \text{L}$$

Note that the final step uses the molar volume at STP as a conversion factor, in the same way that molar mass is used in gravimetric stoichiometry.

Once you understand the logic of the individual steps, the calculation may be done as a single step:

$$v_{O_2} = 275\ \cancel{\text{g}\ C_3H_8} \times \frac{1\ \cancel{\text{mol}\ C_3H_8}}{44.11\ \cancel{\text{g}\ C_3H_8}} \times \frac{5\ \cancel{\text{mol}\ O_2}}{1\ \cancel{\text{mol}\ C_3H_8}} \times \frac{22.4\ \text{L}\ O_2}{1\ \cancel{\text{mol}\ O_2}}$$

$$v_{O_2} = 698\ \text{L}$$

The following example illustrates how to communicate solutions to stoichiometric problems involving gases.

Sample Problem 1

Hydrogen gas is produced when sodium metal is added to water. What mass of sodium is necessary to produce 20.0 L of hydrogen at SATP?

Solution

$$2\ Na_{(s)} + 2\ H_2O_{(l)} \rightarrow H_{2(g)} + 2\ NaOH_{(aq)}$$

m 20.0 L

22.99 g/mol   24.8 L/mol

$$n_{H_2} = 20.0\ \text{L} \times \frac{1\ \text{mol}}{24.8\ \text{L}}$$

$$= 0.806\ \text{mol}$$

$$n_{Na} = 0.806\ \text{mol}\ H_2 \times \frac{2\ \text{mol Na}}{1\ \text{mol}\ H_2}$$

$$= 1.61\ \text{mol}$$

$$m_{Na} = 1.61\ \text{mol} \times \frac{22.99\ \text{g}}{1\ \text{mol}}$$

$$m_{Na} = 37.1\ \text{g}$$

or

$$m_{Na} = 20.0\ \text{L}\ H_2 \times \frac{1\ \text{mol}\ H_2}{24.8\ \text{L}\ H_2} \times \frac{2\ \text{mol Na}}{1\ \text{mol}\ H_2} \times \frac{22.99\ \text{g Na}}{1\ \text{mol Na}}$$

$$m_{Na} = 37.1\ \text{g}$$

The mass of sodium necessary is 37.1 g.

Note that the general steps of the stoichiometry calculation are the same for both solids and gases. Changes from mass to amount or from volume to amount, or vice versa, are done using the molar mass or the molar volume, respectively, of the substance. Although the molar mass depends on the chemical involved, the molar volume of a gas depends only on temperature and pressure. If the conditions are not standard (i.e., STP or SATP), then the ideal gas law ($pv = nRT$), rather than the molar volume, is used to find the amount or volume of a gas, as in the following example.

Sample Problem 2

Recall the Haber process from Chapter 5 in which ammonia to be used as fertilizer is produced from the reaction of nitrogen and hydrogen. What volume of ammonia at 450 kPa pressure and 80°C can be obtained from the complete reaction of 7.5 kg of hydrogen?

Solution

$$N_{2(g)} + 3\ H_{2(g)} \rightarrow 2\ NH_{3(g)}$$

7.5 kg   450 kPa/353 K

2.02 g/mol   8.31 kPa•L/(mol•K)

$$n_{H_2} = 7.5\ \text{kg} \times \frac{1\ \text{mol}}{2.02\ \text{g}}$$

$$= 3.7\ \text{kmol}$$

$$n_{NH_3} = 3.7\ \text{kmol}\ H_2 \times \frac{2\ \text{mol}\ NH_3}{3\ \text{mol}\ H_2}$$

$$= 2.5\ \text{kmol}$$

$$v_{NH_3} = \frac{nRT}{p}$$

$$= \frac{2.5 \text{ kmol} \times \frac{8.31 \text{ kPa} \cdot \text{L}}{1 \text{ mol} \cdot \text{K}} \times 353 \text{ K}}{450 \text{ kPa}}$$

$$v_{NH_3} = 16 \text{ kL}$$

or $$v_{NH_3} = 7.5 \text{ kg } H_2 \times \frac{1 \text{ mol } H_2}{2.02 \text{ g } H_2} \times \frac{2 \text{ mol } NH_3}{3 \text{ mol } H_2} \times \frac{8.31 \text{ kPa} \cdot \text{L}}{1 \text{ mol } NH_3 \cdot \text{K}} \times \frac{353 \text{ K}}{450 \text{ kPa}}$$

$$v_{NH_3} = 16 \text{ kL}$$

The volume of ammonia obtained is 16 kL.

SUMMARY Gravimetric, Gas, and Solution Stoichiometry

1. Write a balanced chemical equation and list the measurements and conversion factors for the given substance and the one to be calculated.
2. Convert the measurement to an amount in moles using the appropriate conversion factor.
3. Calculate the amount of the other substance by using the mole ratio from the balanced equation.
4. Convert the calculated amount to the final quantity requested by using the appropriate conversion factor.

Stoichiometry Calculations

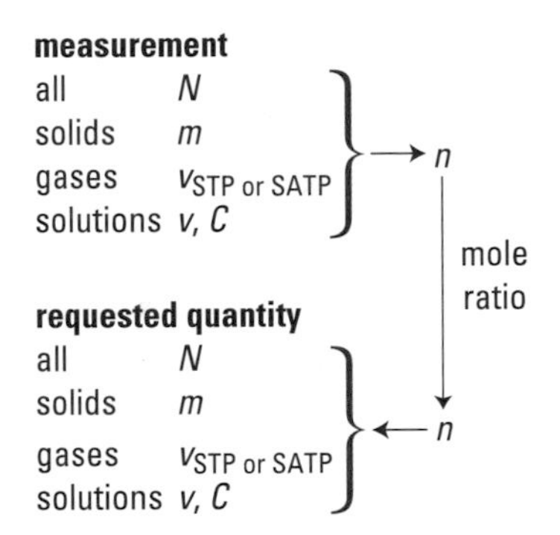

Practice

Understanding Concepts

1. What volume of oxygen at STP is needed to completely burn 15 g of methanol in a fondue burner?
2. A Down's Cell is used in the industrial production of sodium from the decomposition of molten sodium chloride. A major advantage of this process compared with earlier technologies is the production of the valuable byproduct chlorine. What volume of chlorine gas is produced (measured at SATP), along with 105 kg of sodium metal, from the decomposition of sodium chloride?
3. Most combustion reactions use oxygen from the air (assume 20% oxygen). What mass of propane from a tank can be burned using 125 L of air at SATP?

Applying Inquiry Skills

4. The purpose of this investigation is to test the stoichiometric method for gas reactions. In this test, aqueous hydrogen peroxide is decomposed to water and oxygen gas. Complete the **Prediction, Analysis,** and **Evaluation** sections of the following report.

Question
What volume of oxygen at room conditions can be obtained from the decomposition of 50.0 mL of 0.88 mol/L aqueous hydrogen peroxide?

Prediction
(a) Predict the answer to the Question. Include the basis (the authority) for your Prediction.

Experimental Design
A measured volume of a hydrogen peroxide solution (3%, 0.88 mol/L) is decomposed using manganese dioxide as a catalyst. The oxygen produced is collected by water displacement.

Answers

1. 16 L
2. 56.6 kL or 56.6 m^3
3. 8.9 g

   (a) 0.58 L

Answers

4. (b) 556 mL
   (c) 4% difference

Evidence

volume of 0.88 mol/L $H_2O_{2(aq)}$ = 50.0 mL

volume of $O_{2(g)}$ = 556 mL

temperature = 21°C

atmospheric pressure = 94.6 kPa

water vapour pressure = 2.49 kPa

Analysis

(b) Based on the Evidence, what is the answer to the Question?

Evaluation

(c) Calculate the percentage difference. Does the answer based on the Evidence agree with your Prediction?
(d) Based on your evaluation of your Prediction, evaluate the authority you used to make your Prediction.

Making Connections

5. (a) Hydrogen gas is burned in "pollution-free" vehicles in which hydrogen and oxygen gases react to produce water vapour. What volume of oxygen at 40°C and 1.50 atm is necessary to burn 300 L of hydrogen gas measured at the same conditions? (Recall the law of combining volumes.)
   (b) What pollutants might still be formed in such a vehicle?

INQUIRY SKILLS

- ○ Questioning
- ● Hypothesizing
- ● Predicting
- ○ Planning
- ● Conducting
- ● Recording
- ● Analyzing
- ● Evaluating
- ● Communicating

Investigation 10.4.1

Magnesium and Hydrochloric Acid: Testing the Gas Stoichiometry Method

The purpose of this investigation is to test the stoichiometry method applied to reactions that involve gases. Complete the **Prediction, Analysis,** and **Evaluation** sections of the lab report. In the **Prediction** and **Analysis** sections, there are several possible approaches that can be used. The suggested method is to predict the volume of gas at STP and, in your Analysis, convert the measured volume to STP conditions using the combined gas law or the ideal gas law. (For more accurate results, it will be necessary to correct the measured pressure for the vapour pressure of water. See **Table 3**, page 464.)

Question

What is the yield at STP of hydrogen gas from the reaction of magnesium with excess hydrochloric acid?

Prediction

(a) According to the stoichiometric method of the balanced chemical equation and the mass of magnesium used, predict the volume of hydrogen expected at STP conditions.

Experimental Design

A known mass of magnesium ribbon reacts with excess hydrochloric acid. The temperature, pressure, and volume of the hydrogen gas that is produced are measured.

Materials

lab apron
eye protection
disposable plastic gloves
magnesium ribbon, 60–70 mm
centigram or analytical balance
piece of fine copper wire, 100–150 mm
100-mL graduated cylinder
hydrochloric acid (6 mol/L), 15 mL
250-mL beaker
water
large beaker, 600 mL or 1000 mL
2-hole stopper to fit cylinder
thermometer
barometer

Eye protection, a lab apron, and disposable gloves must be worn.

Hydrochloric acid in 6 mol/L concentration is very corrosive. If acid is splashed into your eyes, rinse them immediately with water for 15 to 20 min. Acid splashed onto the skin should be rinsed immediately with plenty of water. If acid is splashed onto your clothes, neutralize with baking soda, then wash thoroughly with plenty of water. Notify your teacher.

Rinse your hands well after step 7 in case you got any dilute acid on your skin.

Procedure

1. Measure and record the mass of the strip of magnesium.
2. Fold the magnesium ribbon to make a small compact bundle that can be held by a copper cage (**Figure 2**).
3. Wrap the fine copper wire all around the magnesium, making a cage to hold it but leaving 30 mm to 50 mm of the wire free for a handle.
4. Carefully pour 10 mL to 15 mL of the hydrochloric acid into the graduated cylinder.
5. Slowly fill the graduated cylinder to the brim with water from a beaker. As you fill the cylinder, pour slowly down the side of the cylinder to minimize mixing of the water with the acid at the bottom. In this way, the liquid at the top of the cylinder is relatively pure water and the acid remains at the bottom.
6. Half-fill the large beaker with water.
7. Bend the copper wire handle through the holes in the stopper so that the cage holding the magnesium is positioned about 10 mm below the bottom of the stopper (**Figure 2**).
8. Insert the stopper into the graduated cylinder; the liquid in the cylinder will overflow a little. Cover the holes in the stopper with your finger. Working quickly, invert the cylinder, and immediately lower it into the large beaker so that the stopper is below the surface of the water before you remove your finger from the stopper holes (**Figure 3** page 486).
9. Observe the reaction, then wait about 5 min after the bubbling stops to allow the contents of the graduated cylinder to reach room temperature.
10. Raise or lower the graduated cylinder so that the level of liquid inside the beaker is the same as the level of liquid in the graduated cylinder. (This equalizes the gas pressure in the cylinder with the pressure of the air in the room.)
11. Measure and record the volume of gas in the graduated cylinder.
12. Record the laboratory (ambient) temperature and pressure.
13. The liquids in this investigation may be poured down the sink, but rinse the sink with lots of water.

Hydrogen gas, produced in the reaction of hydrochloric acid and magnesium, is flammable. Ensure that there is adequate ventilation and that there are no open flames in the classroom.

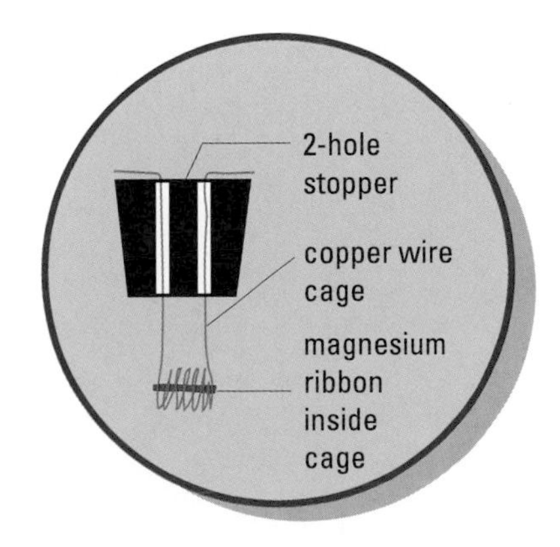

Figure 2
Fasten the copper wire handle to the stopper.

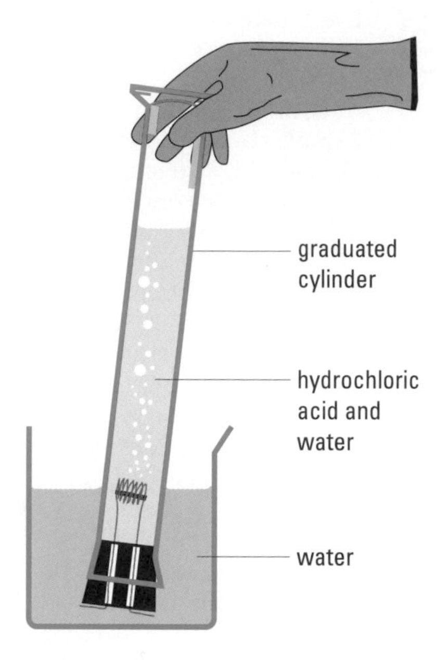

Figure 3
While holding the cylinder so it does not tip, rest it on the bottom of the beaker. The acid, which is denser than water, will flow down toward the stopper and react with the magnesium. The hydrogen produced should remain trapped in the graduated cylinder.

Analysis

(b) Calculate the partial pressure of hydrogen from the total pressure and the partial pressure of water vapour at the temperature measured.

(c) Convert the volume of hydrogen at the conditions measured to the volume it would occupy at STP. According to your Evidence and gas laws, what is the volume (yield) of hydrogen at STP?

Evaluation

(d) Are there any obvious flaws in any part of the Experimental Design, Materials, or Procedure? Look carefully at all measurements made and evaluate each. What are some improvements that could be made to obtain more accurate and more precise evidence?

(e) How would you judge the quality of the Evidence collected? Provide reasons for your judgment.

(f) How accurate was your Prediction? Include a percentage difference calculation with your answer. Consider if the difference obtained can be reasonably explained by normal experimental errors or uncertainties.

(g) Using your answer to (f), how would you judge the stoichiometric method used in the Prediction? Does it appear to be acceptable? Provide reasons.

Section 10.4 Questions

Understanding Concepts

1. The first recorded observation of hydrogen gas was made by the famous alchemist Paracelsus (1493–1541) when he added iron to sulfuric acid. Predict the volume of hydrogen gas at STP produced by adding 10 g of iron to an excess of sulfuric acid.
2. A typical Canadian home heated with natural gas (assume methane, $CH_{4(g)}$) consumes 2.00 ML of natural gas during the month of December. What volume of oxygen at SATP is required to burn 2.00 ML of methane measured at 0°C and 120 kPa?
3. Ammonia reacts with sulfuric acid to form the important fertilizer ammonium sulfate. What mass of ammonium sulfate can be produced from 75.0 kL of ammonia at 10°C and 110 kPa?
4. Methane hydrate, a possible energy resource, looks like ice but is an unusual substance with the approximate chemical formula $CH_4{\cdot}6H_2O_{(s)}$. It occurs in permafrost regions and in large quantities on the ocean floor. Current, rough estimates of the quantity of methane hydrate suggest that it is at least twice the total known reserves of coal, oil, and natural gas combined. Considerable research is now under way to find ways to tap this huge energy resource. If 1.0 kg of solid methane hydrate decomposes to methane gas and water, what volume of methane is produced at 20°C and 95 kPa?

Making Connections

5. Describe briefly one consumer, one industrial, and one laboratory application of gases that involve a chemical reaction that uses or produces gases. For each example, include a complete balanced chemical equation.

Reflecting

6. In this and previous units, you have seen many examples of stoichiometry used in laboratory, industry, and consumer applications.
   (a) What do all kinds of stoichiometry have in common?
   (b) Using an example, describe how knowledge of stoichiometry is important in chemical industries.
   (c) How do consumers use stoichiometry? Describe one household example.

10.5 Applications of Gases

Many people have some direct experience with and practical knowledge about gases. We may be familiar with pressure in car tires and why this should be adjusted as the seasons change. Some of us use aerosol cans containing pressurized gases and have flown in pressurized cabins on airplanes. Some people with joint problems claim that they can tell when a weather system is moving in because the air pressure starts to change and they "feel it in their bones." Ideally, none of us will experience an inflated air bag as a result of an automobile collision. All of these examples involve gases and their properties. A more detailed and scientific understanding of gases is required in a wide variety of areas, such as meteorology, medicine, and deep-sea diving.

Meteorology is the study of the atmosphere and weather forecasting. You have probably seen weather reports citing high- and low-pressure systems (**Figure 1**, page 488). However, this is only one aspect of how gas properties are important in meteorology. The atmosphere is a complex, variable mixture of gases. The composition and partial pressures of these gases determine the density of an air mass, which is an important characteristic for understanding and predicting weather. There are many other variables such as Earth's tilt and the variations in the heating of Earth's surface that make describing the empirical properties of the atmosphere, explaining these properties, and then predicting changes in weather extremely complicated. Meteorologists use sophisticated computer models to simulate weather changes based on vast quantities of data collected around the world. These models rely, in part, on data gathered by helium weather balloons. Some weather balloons may go as high as 30 km, where the air pressure is 1.2 kPa and the air temperature is −47°C, so designing them requires not only knowledge of gas properties but also knowledge of properties of the materials used to make them.

In the area of medicine, most gas applications are in artificial ventilation and the use of anaesthetics. Ventilation (breathing) is the process by which oxygen and carbon dioxide are transported to and from the lungs. Breathing is something we do naturally, without thinking, but occasionally the respiratory system can fail and we need help, sometimes immediately. This could be provided by another person who applies artificial respiration, a portable source of oxygen, or best, by a ventilator that controls both oxygen and carbon dioxide. Artificial ventilation becomes particularly important when anaesthetics are used to make a person unconscious during surgery. Many drugs and techniques used in anaesthesia

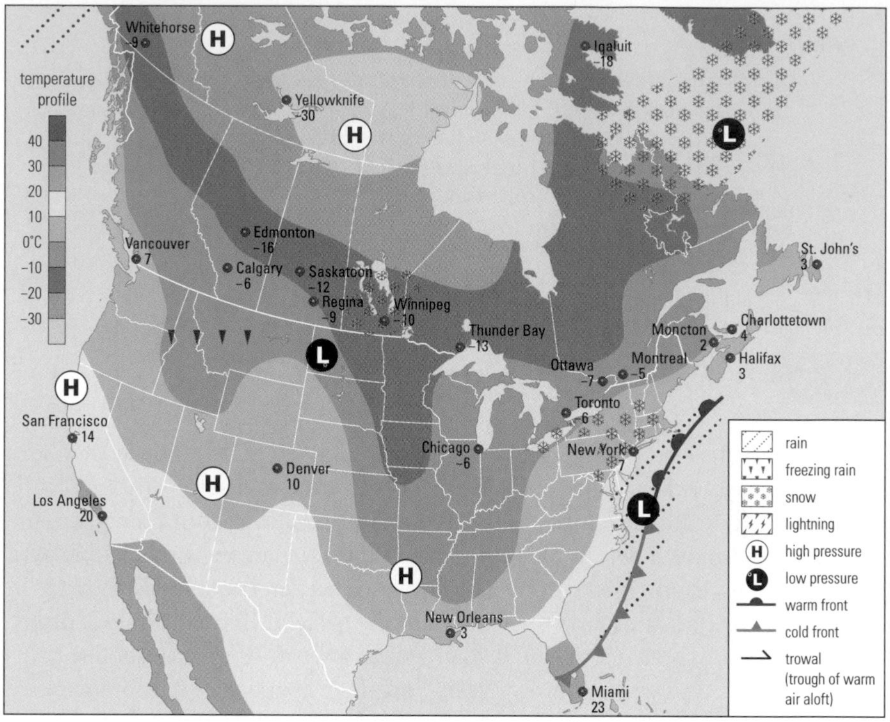

Figure 1
Other than temperature, the most commonly used gas property in weather reports is pressure, often referred to as "highs" and "lows."

interfere with our natural ventilation. An anaesthetic machine, such as Boyle's machine (**Figure 2**), controls not only the ventilation of the patient but also the administration of gaseous anaesthetics like dinitrogen oxide (nitrous oxide) or volatile liquid anaesthetics. As you can see in **Figure 2,** gas cylinders, pressure gauges, and flow meters are important components of the apparatus. To work in any area of medicine as a medical doctor, nurse, or technician dealing with respiration requires a good understanding of gas laws and lung mechanics.

Practice

Understanding Concepts

1. A helium balloon with a volume of 4.0 kL, a pressure of 100 kPa, and temperature of 20°C is released to study ozone concentrations in the stratosphere. What would be its volume when it is 30 km above Earth's surface? (Assume the balloon expands and contracts without resistance.)
2. Humidity is an important property of an air mass. Is an air mass with a high humidity (i.e., high partial pressure of water vapour) denser or less dense than a very dry air mass at the same volume, temperature, and pressure? Justify your answer, using your knowledge of gas laws, molar masses, and density.
3. Dinitrogen oxide (nitrous oxide) is a colourless, sweet-tasting gas first discovered by Joseph Priestley in 1793. It is commonly known as "laughing gas" and for almost 50 years it was used solely for recreational enjoyment and public shows or carnivals. Its use as an anaesthetic for dentistry and medicine began in the early 1840s and continues today (**Figure 2**).
   (a) Dinitrogen oxide is commercially prepared by decomposing pure ammonium nitrate. Water vapour is the only other product. What mass of ammonium nitrate is required to produce 1.0 L of $N_2O_{(g)}$ at 100 kPa and 20°C?

Answers

1. 0.26 ML
3. (a) 3.3 g
   (b) 300 kPa

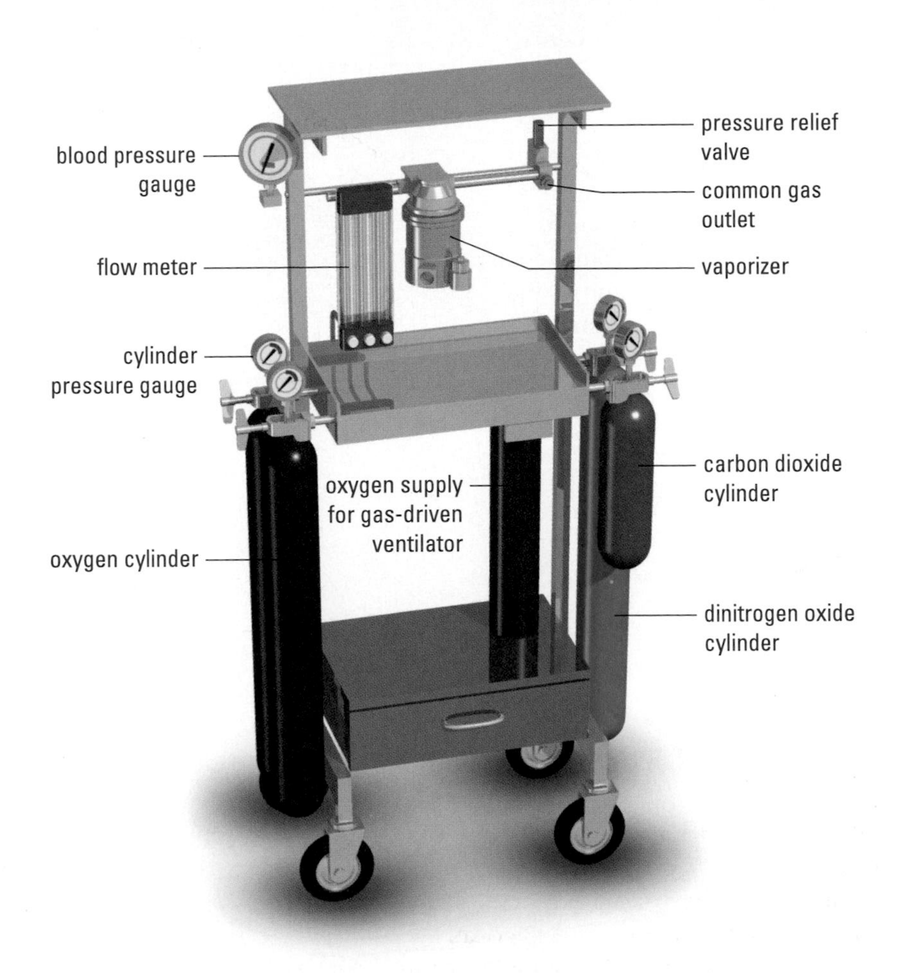

Figure 2
Boyle's machine is a continuous-flow anaesthetic apparatus. The basic design of today's apparatus is the same as the original machine invented by the famous British anaesthetist H.E.G. Boyle (1875–1941).

(b) During surgery, an oxygen–dinitrogen oxide gas mixture with a total pressure of 400 kPa is administered to the patient. If the mixture contains a partial pressure of 100 kPa of oxygen, what is the partial pressure of dinitrogen oxide?

Making Connections

4. What gas properties are most often used by meteorologists when communicating weather information?
5. When any form of ventilation is used for medical purposes, why is it important to know about the partial pressure of oxygen and carbon dioxide?
6. List some gas properties that would be important for an underwater diver to understand.

Section 10.5 Questions

Making Connections

1. Identify three technological uses of gases and compressed gases; for each example, identify the gas.
2. Compressed gases create possible chemical and physical hazards. Identify one specific chemical and one physical hazard.
3. Choose one area of study that involves gases.
   (a) Briefly describe how knowledge of gases is used in that area.
   (b) Identify and describe a career associated with it.

CAREER

Careers with Gases

There are many different types of careers that involve the study and use of gases. Some careers require many years of university preparation, some require training at a technical school, and some require individual courses and on-the-job training. Have a look at the careers described in this activity and find out more about one of them or another career with gases that interests you.

Commercial Diver
A commercial diver requires a good working knowledge of gas properties, laws, and some physiology. Underwater divers undergo various levels of training and require certification for different underwater breathing systems.

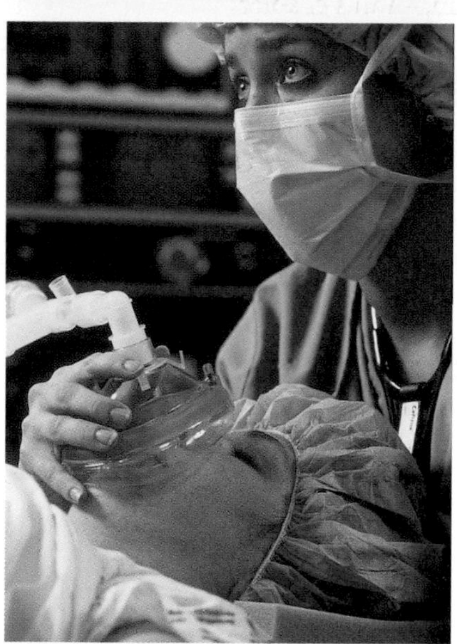

Anaesthetic Technician
An anaesthetic technician is the person who actually controls the ventilation equipment under the direction of medical doctors. Every part of this job requires knowledge of the properties of gases.

Meteorologist
Some meteorologists work for television stations, but most are employed in various capacities with government environment agencies and universities. Meteorologists use advanced knowledge of the physical and chemical properties of gases in their work.

Practice

7. (a) Identify and describe a career that requires a knowledge of gases from this page or by searching the Internet. Be sure to include in your description details on how knowledge of gases learned in this chapter is used in the career.
   (b) Which courses would you have to take at high school to get into this career?
   (c) What training beyond high school is required and how long would this take?
   (d) Survey newspapers in your area or career Web sites for job opportunities in this career. Where would you work and what might be your salary?
   Follow the links for Nelson Chemistry 11, 10.5.

www.science.nelson.com

Chapter 10 Summary

Key Expectations

Throughout this chapter, you have had the opportunity to do the following:

- Explain Dalton's law of partial pressures. (10.1)
- Use appropriate scientific vocabulary to communicate ideas related to gases. (10.1, 10.2, 10.4)
- Use and interconvert appropriate units to express pressure. (10.1, 10.2, 10.4)
- Solve quantitative problems involving the following gas laws: the combined gas law; the law of combining volumes (Gay-Lussac's law), Dalton's law of partial pressures, the ideal gas law. (10.1, 10.4)
- Describe natural phenomena and technological products associated with gases. (10.1, 10.2, 10.4, 10.5)
- State Avogadro's theory and describe his contributions to our understanding of reactions of gases. (10.2)
- Perform stoichiometric calculations involving the quantitative relationships among the quantity of substances in moles, the mass, and the volume of the substances in a balanced chemical equation. (10.2, 10.4)
- Determine the molar volume of a gas through experimentation. (10.2)
- Describe Canadian initiatives to improve air quality. (10.3)
- Identify technological products associated with compressed gases. (10.5)
- Describe how knowledge of gases is applied in other areas of study. (10.5)
- Identify and describe science- and technology-based careers. (10.5)

Key Terms

Avogadro's theory
Dalton's law of partial pressures
law of combining volumes (Gay-Lussac's law of combining volumes)
molar volume
partial pressure

Make a Summary

In this chapter, you studied two important gas laws: Dalton's law of partial pressures and the law of combining volumes. On a piece of paper, write the law in the centre of each half-page. Draw lines to connect each law to related empirical properties, theoretical concepts, applications, and careers. Some connections may cross from one diagram to another and some previous knowledge from Chapter 9 may be used to make your diagrams more complete.

Reflect on your Learning

Revisit your answers to the Reflect on Your Learning questions at the begining of this chapter.

- How has your thinking changed?
- What new questions do you have?

Chapter 10 Review

Understanding Concepts

1. From his experimental work with gases, John Dalton developed the law of partial pressures. State Dalton's law of partial pressures
   (a) in a sentence;
   (b) as a mathematical equation.
2. Using the kinetic molecular theory, explain Dalton's law of partial pressures.
3. A sample of sour natural gas contains gases with the following partial pressures: 230 kPa methane, 13 kPa ethane, and 7 kPa hydrogen sulfide. Calculate the total pressure of the sour natural gas mixture.
4. Oxygen gas generated by the decomposition of potassium chlorate is collected at SATP by the downward displacement of water. Calculate the partial pressure of the oxygen, assuming it is saturated with water vapour. (The vapour pressure of water is given in **Table 3** of Section 10.1.)
5. From his experimental work with chemical reactions involving gases, Joseph Gay-Lussac developed the law of combining volumes.
   (a) Using your own words, state the law of combining volumes in a sentence.
   (b) Use the law of combining volumes to predict the volume of each gas produced when 5.00 L of oxygen gas is consumed in the complete combustion of trinitrotoluene, TNT. Assume all gases are measured at the same temperature and pressure.

$$4\ C_7H_5(NO_2)_{3(s)} + 21\ O_{2(g)} \rightarrow 28\ CO_{2(g)} + 6\ N_{2(g)} + 10\ H_2O_{(g)}$$

6. Amedeo Avogadro proposed an idea to explain the law of combining volumes. Avogadro's idea is sometimes called a principle, a hypothesis, a law, and a theory. Is Avogadro's idea empirical or theoretical? Explain your answer.
7. Disastrous explosions have resulted from the unsafe storage and handling of the fertilizer ammonium nitrate, which can decompose rapidly.

$$2\ NH_4NO_{3(s)} \rightarrow 2\ N_{2(g)} + 4\ H_2O_{(g)} + O_{2(g)}$$

   What volume of gases measured at SATP is produced by the decomposition of 1.00 mol of ammonium nitrate?
8. Argon is used in incandescant light bulbs and fluorescent tubes at a pressure of 400 Pa above atmospheric pressure.
   (a) What amount of argon is required to fill a 125-mL light bulb when the temperature is 20°C and atmospheric pressure is 100.0 kPa?
   (b) Bulbs heat up as they generate light. What will the pressure be inside a light bulb when the temperature rises to 200°C?
   (c) A steel tank contains 50.0 kg of compressed argon. If one fluorescent tube has an internal volume of 0.915 L, how many tubes can be filled from the tank? Assume a pressure of 100.4 kPa inside each tube when the temperature is 20°C.
9. One of the major sources of atmospheric sulfur dioxide, $SO_{2(g)}$, is the extraction of metals from their sulfide ores. Metal extraction often involves heating the metal sulfide in air to form the metal oxide and sulfur dioxide, for example,

$$2\ ZnS_{(s)} + 3\ O_{2(g)} \rightarrow 2\ ZnO_{(s)} + 2\ SO_{2(g)}$$

   What volume of sulfur dioxide at SATP is produced by the reaction of 1.00 t of zinc sulfide with an excess of oxygen?
10. Sulfur dioxide released to the atmosphere forms acid rain, but it can also attack calcium carbonate directly (as seen in the sculpture in **Figure 1**).

$$2\ CaCO_{3(s)} + 2\ SO_{2(g)} + O_{2(g)} \rightarrow 2\ CaSO_{4(s)} + 2\ CO_{2(g)}$$

   If 500 kL of sulfur dioxide reacts at STP,
   (a) What volume of carbon dioxide gas at STP is produced?
   (b) What mass of calcium carbonate is consumed?

Figure 1

11. Sulfur trioxide in the atmosphere reacts with rainwater to form sulfuric acid.

$$SO_{3(g)} + H_2O_{(l)} \rightarrow H_2SO_{4(aq)}$$

If 1.00 t of atmospheric sulfur trioxide dissolves in rainwater, what volume of 0.12 mmol/L sulfuric acid could be formed?

12. In a demonstration of the decomposition of water by electrolysis, 50.0 mL of hydrogen gas is produced at 23°C and 103 kPa.
    (a) What volume of oxygen gas is produced at the same temperature and pressure?
    (b) What would be the volume of hydrogen produced if it was measured at STP?
    (c) What mass of water is decomposed?
13. After 2.00 mol of $N_{2(g)}$ and 3.00 mol of $H_{2(g)}$ are added to a reaction vessel, the total pressure is measured at 200 kPa. Calculate the partial pressure of each gas in the vessel.
14. How does the molar volume of a gas change
    (a) when temperature increases?
    (b) when pressure increases?

Applying Inquiry Skills

15. Complete the **Prediction, Analysis,** and **Evaluation** sections of the following investigation report.

Question
What is the molar volume of propane, $C_3H_{8(g)}$, at STP?

Prediction
(a) Predict the answer to the Question and state the basis for your Prediction.

Experimental Design
A small propane cylinder sold for home maintenance applications provides the propane gas that is collected at ambient pressure by water displacement.

Evidence
initial mass of propane cylinder = 426.79 g
final mass of propane cylinder = 424.92 g
volume of gas collected = 1065 mL
ambient pressure = 98.23 kPa
ambient temperature = 21.0°C

Analysis
(b) Based on the Evidence, what is the answer to the Question? (The pressure must be adjusted to correct for the vapour pressure of water; see **Table 3,** Section 10.1.)

Evaluation
(c) Calculate the percentage difference. Does the answer based on the Evidence agree with your predicted value?

(d) Based on your evaluation of the Prediction, evaluate the authority you used to make your Prediction.

16. A cylinder of compressed gas is known to contain a noble gas. Design an experiment to identify the gas by its molar mass.

Making Connections

17. List one natural and one technological use or source for each of the following gases:
    (a) oxygen
    (b) methane
    (c) helium
    (d) air
    (e) water vapour
    (f) carbon dioxide
18. In the 1930s, chemists produced a series of synthetic chemicals called Freons. Some common Freons are $CFCl_{3(g)}$, $CF_2Cl_{2(g)}$, $C_2F_3Cl_{3(g)}$, and $C_2F_4Cl_{2(g)}$.
    (a) What were some of the initial uses of Freons?
    (b) Why has the production of Freons been banned in many countries, including Canada?
    (c) What volume does 1.00 kg of escaped $CF_2Cl_{2(g)}$ occupy at SATP?
19. Ethanethiol, $C_2H_5SH_{(g)}$, is a very smelly compound that is added to natural gas so that gas leaks can be detected easily. When the natural gas is burned, the ethanethiol undergoes combustion as well:

$$2\ C_2H_5SH_{(g)} + 9\ O_{2(g)} \rightarrow 4\ CO_{2(g)} + 6\ H_2O_{(g)} + 2\ SO_{2(g)}$$

    (a) Calculate the volume at SATP of each gas produced when 1.00 g of ethanethiol is burned.
    (b) Discuss some of the risks and benefits of adding ethanethiol to natural gas.
20. Ontario's Air Quality Index (AQI) network includes 33 state-of-the-art air quality monitoring stations across the province. Each AQI site monitors some or all of the six most common air pollutants: sulfur dioxide, ozone, nitrogen dioxide, total reduced sulfur compounds, carbon monoxide, and suspended particles. Locate the AQI station closest to your school, and research the source of the most common pollutant and the health risks associated with it. Prepare a one-page summary of your findings.

    Follow the links for Nelson Chemistry 11, Chapter 10 Review.

    GO TO www.science.nelson.com

A Study of a Technological System

Gases play an important part in many technological systems. Now that you have finished Unit 4, you will have the opportunity to demonstrate your understanding of gases in the context of a specific technology. There are three technological systems from which to choose: hot-air balloons (**Figure 1**), underwater submersibles (**Figure 2**), and the atmosphere of a spacecraft (**Figure 3**). Read the task description and guidelines below before choosing a technology.

Task

Prepare a report based on one of the three technological systems, using the guidelines listed on the following page. Although you may cooperate with other students in your research, your report should be done individually. You may also be asked to do a brief group presentation using a poster, a model, overhead transparencies, a video, or electronic slides.

Figure 1
Hot-air balloons

Report Guidelines

Your report should include the following scientific, technological, societal, and environmental perspectives:

1. A brief history of the technological development leading to the current application (including some quantitative details);
2. A description of the use of scientific principles including:
   - gas reactions (desirable and/or undesirable) with a quantitative example
   - kinetic molecular theory explanation related to the operation of some part of the system
   - typical values for pressure, volume, amount, and temperature, with an example of the use of the ideal gas law
   - role of partial pressures, including an example
3. An outline of the advantages and disadvantages to society of the technological system;
4. The identification of environmental concerns, including safety hazards;
5. An outline of the current trends in the technology;
6. A brief discussion of a related career, including educational qualification and working conditions;
7. A final paragraph with an overview of how scientific knowledge of gases explains the technological system or has advanced the technology.

Assessment

Your completed task will be assessed according to the following criteria:

Process

- Use a variety of sources in your research.

Product

- Prepare a report that follows the stated guidelines in written or electronic format.
- Demonstrate an understanding of the concepts developed in this unit.
- Use terms, symbols, equations, and SI metric units correctly.

Figure 2
Underwater submersible

Figure 3
Spacecraft interior

Unit 4 Review

Understanding Concepts

1. The substance known as "dry ice" is solid carbon dioxide. At normal atmospheric pressure and –78.5°C, solid carbon dioxide sublimes to the gas phase, that is, does not pass through a liquid phase—hence the term "dry ice."
   (a) What does the sublimation point reveal about the strength of the intermolecular forces in solid carbon dioxide?
   (b) Why is carbon dioxide gas highly compressible while solid carbon dioxide is not?
   (c) In general terms, describe the degree of disorder in solid and gaseous carbon dioxide.
   (d) Carbon dioxide is the fourth most abundant gas in the atmosphere. Which three gases are more abundant?
2. The pressure of a gas in a sealed, rigid container depends on the temperature of the gas. Express the relationship between temperature and pressure of a gas under these conditions
   (a) in a sentence;
   (b) as a mathematical equation;
   (c) in a graph.
3. Use the kinetic molecular theory to explain the relationship between the temperature and pressure of a gas in a sealed, rigid container.
4. In your own words, define the following terms:
   (a) standard temperature
   (b) standard pressure
   (c) molar volume
   (d) ideal gas
5. Convert the following gas pressures to units of pascals:
   (a) A bike tire is pumped up to a pressure of 4.0 atm.
   (b) During manufacture, light bulbs are filled with argon to a pressure of 763 mm Hg.
   (c) Ammonia is produced in a reaction vessel at a pressure of 450 atm.
6. Convert each of the following gas volumes into an amount in moles:
   (a) 5.1 L of carbon monoxide gas at SATP
   (b) 20.7 mL of fluorine gas at STP
   (c) 90 kL of nitrogen dioxide gas at SATP
7. Calculate the volumes at SATP of the following amounts of gas:
   (a) 500 mol of hydrogen (most common element in the universe)
   (b) 56 kmol of hydrogen sulfide (a toxin found in sour natural gas)
   (c) 45.6 mmol of neon (used in neon lights)
8. Large quantities of chlorine gas are produced from salt to make bleach and for water treatment. What is the volume of 26.5 kmol of chlorine gas at 400 kPa and 35°C?
9. Bromine is produced by reacting chlorine with bromide ions in seawater. What amount of bromine is present in an 18.8-L sample of bromine gas at 60 kPa and 140°C?
10. Argon gas is an inert carrier gas that moves other gases through a research or industrial system. What is the volume occupied by 4.2 kg of argon gas at SATP?
11. Uranium hexafluoride is a very dense gas used to separate isotopes of uranium for nuclear applications. What is the density of this gas at 200°C and 100 kPa?
12. Suppose you were trapped in a room in which there was a slow natural gas leak (assume pure methane). In order to breathe as little natural gas as possible, should you be near the ceiling or the floor? Justify your answer.
13. A typical passenger hot-air balloon contains 5.7 ML of air at an average temperature of 100°C. Show that the density of the air in the balloon is noticeably less than the surrounding air at SATP.
14. Helium is used in a spacecraft breathing system. A steel cylinder is filled with helium to a pressure of 200 atm at 23°C. If the temperature drops to –17°C after the cylinder is transported to a launch site, what is the pressure of the helium inside the cylinder?
15. A 5.00-L balloon contains helium at SATP at ground level. What is the balloon's volume when it floats to an altitude where the temperature is –15°C and the atmospheric pressure is 91.5 kPa?
16. Chinooks—warm, dry winter winds occurring in Canada's southern Prairies—cause rapid changes in weather. Calculate the final volume of a cubic metre (1.00 m^3) of air at –23°C and 102 kPa when the temperature and pressure change to 12°C and 96 kPa.
17. On summer afternoons, warm air masses often rise rapidly through the atmosphere, creating cumulus clouds or, sometimes, cumulonimbus (thunderstorm) clouds. Use the kinetic molecular theory to explain this rising of the warm air mass.
18. In an experiment, 125 mL of hydrogen gas is collected over water. If the atmospheric pressure is 99.6 kPa and the temperature is 22°C, what is the partial pressure of

the hydrogen? (Values for the vapour pressure of water are given in **Table 3**, page 464.)

19. A piece of clean sodium metal is dropped into a flask filled with air at SATP; the flask is then closed with a stopper. The excess sodium reacts with the oxygen in the air, forming solid sodium oxide. The initial composition of the air was 78% nitrogen, 21% oxygen, and 1% argon. Calculate the partial pressure of the gases remaining after all the oxygen has reacted and the temperature has returned to 25°C.

20. At one stage in the production of nitric acid, ammonia is reacted with oxygen to produce nitrogen monoxide:

$$4\ NH_{3(g)} + 5\ O_{2(g)} \rightarrow 4\ NO_{(g)} + 6\ H_2O_{(g)}$$

   (a) Calculate the volumes of ammonia and oxygen required to produce 1.00 L of nitrogen monoxide. All gases are measured at the same temperature and pressure.
   (b) Use Avogadro's theory to explain the relationship used to calculate the volumes in (a).

21. The explosion of dynamite can be represented by the following reaction equation:

$$4\ C_3H_5(NO_3)_{3(l)} \rightarrow 12\ CO_{2(g)} + 6\ N_{2(g)} + 10\ H_2O_{(g)} + O_{2(g)}$$

   (a) Calculate the volume at SATP of each gaseous product formed by the reaction of 100 g of $C_3H_5(NO_3)_{3(l)}$
   (b) In the reaction of $C_3H_5(NO_3)_{3(l)}$, the temperature of the gases produced is much higher than 25°C. How does this change the volumes of gases produced?
   (c) When blasting rock with dynamite, a deep hole is drilled into the rock, dynamite is placed into the hole, and then discharged from a safe distance (**Figure 1**). If 3.50 mol of gaseous product at 900°C is formed in a 2.00-L cavity, calculate the gas pressure in the cavity.

22. Ammonia gas reacts with sulfuric acid to form the important fertilizer ammonium sulfate. What mass of ammonium sulfate can be produced from 84 kL of ammonia at 12°C and 115 kPa?

23. Standard ambient temperature and pressure (SATP) is a convention established by scientists to suit conditions on Earth. Suppose scientists were to establish standard ambient conditions on the planet Venus as 800°C and 7500 kPa. What would be the molar volume of Venus's mainly carbon dioxide atmosphere under these standard ambient conditions?

Figure 1

24. Yeast cells in bread dough convert glucose into either carbon dioxide and water or carbon dioxide and ethanol, as shown in the following chemical equations:

$$C_6H_{12}O_{6(s)} + 6\ O_{2(g)} \rightarrow 6\ CO_{2(g)} + 6\ H_2O_{(g)}$$

$$C_6H_{12}O_{6(s)} \rightarrow 2\ CO_{2(g)} + 2\ C_2H_5OH_{(g)}$$

   (a) Use the law of combining volumes to predict the volume of carbon dioxide produced when 50 mL of oxygen gas reacts with excess glucose.
   (b) For equal amounts of glucose reacted, which of the two reactions will produce the greater degree of leavening? Justify your answer.
   (c) What is the volume occupied by 1.0 g of carbon dioxide gas trapped in bread dough at SATP?
   (d) Steam production during baking is a secondary reason why bread and cakes rise. What volume of water vapour is produced inside a cake when 1.0 g of water is vaporized at 190°C and 103 kPa?

Applying Inquiry Skills

25. An investigation was conducted to determine the relationship between the pressure and the solubility of nitrogen in water.

Question

What effect does the pressure of nitrogen gas have on its solubility in water at a fixed temperature?

Evidence

Table 1: Solubility of Nitrogen Gas in Water

Pressure (kPa)	Solubility (mmol/L)
50	0.33
100	0.67
150	1.04
200	1.35
250	1.61
300	1.98

All values were measured at 25°C.

Analysis

(a) Graph the Evidence.
(b) Answer the Question.

Synthesis

(c) From the graph, find the amount of nitrogen gas that could dissolve at 300 kPa in 5.00 L of blood (assume mostly water) of a scuba diver.
(d) Calculate the volume of nitrogen gas that would come out of solution at 100 kPa if the diver had been submerged at 300 kPa and surfaced too quickly.

26. Hydrochlorofluorocarbons (HCFCs) are being employed as replacements for CFCs because HCFCs are believed to do less damage to the ozone layer. The purpose of the following investigation is to use molar mass to identify a HCFC gas. Complete the **Analysis** section of the following report.

Question

Is the HCFC sample tested $CHF_2Cl_{(g)}$, $C_2H_3FCl_{2(g)}$, or $C_2H_3F_2Cl_{(g)}$?

Experimental Design

A sample of a HCFC from a canister of the compressed gas is collected in a graduated cylinder by the downward displacement of water. The volume, temperature, and pressure of the gas are measured, along with the change in mass of the gas canister. Assume the HCFC is not soluble in water.

Evidence

initial mass of canister = 457.64 g
atmospheric pressure = 100.1 kPa
final mass of canister = 454.26 g
ambient temperature = 22.0°C
volume = 845 mL

Analysis

(a) Calculate the molar mass of the gas, and identify which HCFC is in the sample.

27. The purpose of the following investigation is to determine the yield of a reaction that involves gases. Complete the **Analysis** section of the report.

Question

What is the yield of carbon dioxide from the reaction of calcium carbonate with excess hydrochloric acid?

Experimental Design

A known mass of calcium carbonate reacts with excess hydrochloric acid. The carbon dioxide produced is captured by the downward displacement of water, and the temperature, pressure, and volume of the gas are measured. Assume that carbon dioxide does not dissolve appreciably in water.

Evidence

mass of calcium carbonate = 3.02 g
atmospheric pressure = 98.5 kPa
volume = 748 mL
ambient temperature = 23.0°C

Analysis

(a) What is the yield of carbon dioxide?
(b) What is the percentage yield of carbon dioxide?

28. The purpose of the following investigation is to determine the percentage yield of a gas in a chemical reaction. Complete the **Analysis** and **Evaluation** sections of the report.

Question

What is the percentage yield of hydrogen from the reaction of zinc with excess hydrochloric acid?

Experimental Design

A known mass of zinc reacts with excess hydrochloric acid. The hydrogen gas produced is collected by the downward displacement of water. The temperature, pressure, and volume of the hydrogen are measured.

Evidence

mass of zinc = 0.29 g
atmospheric pressure = 98.7 kPa
volume = 94.5 mL
ambient temperature = 19.8°C

Analysis

(a) Calculate the percentage yield of hydrogen.

Evaluation

(b) Evaluate the Experimental Design.

29. Ammonia can be prepared by mixing equal volumes of solid sodium hydroxide and ammonium chloride and then adding a few drops of water. A student is assigned the task of designing an investigation to determine the molar volume of ammonia. The student proposed the following design.

    "Ammonia gas is produced from the reaction of sodium hydroxide and ammonium chloride in a small flask with tubing leading to an inverted graduated cylinder filled with water. After the ammonia is collected by method of water displacement, the volume, pressure, and temperature are measured. The ideal gas law is used to calculate the molar volume."

    (a) What are some criteria used to evaluate an experimental design?

    (b) Using these criteria, evaluate the student's design.

    (c) Suggest an improvement or replacement for this design.

Making Connections

30. On December 6, 1917, two ships collided in the narrowest part of Halifax harbour (**Figure 2**). One ship was carrying 2766 t of explosives, which were ignited by sparks from the collision. In the resulting explosion over 1600 people were killed, 9000 injured, and much of Halifax was destroyed. The main explosive on the munitions ship was picric acid, $C_6H_3N_3O_{7(s)}$, the complete combustion of which is shown by the following equation:

    $$4\ C_6H_3N_3O_{7(s)} + 25\ O_{2(g)} \rightarrow 24\ CO_{2(g)} + 6\ H_2O_{(g)} + 12\ NO_{2(g)}$$

    (a) Calculate the volume at 100 kPa and 1000°C of each gas produced in the complete combustion of 2000 t of picric acid.

Figure 2

    (b) What can we learn from the Halifax explosion about the transportation of dangerous goods?

31. The following gases are commonly considered air pollutants: sulfur dioxide, ozone, nitrogen dioxide, and carbon monoxide.

    (a) Describe one harmful effect of each gas.

    (b) Identify one beneficial use of each gas.

32. Weather reports usually include the local barometric (air) pressure.

    (a) Check and record the barometric pressure for several days, using your local newspaper or your local radio or TV station as a source.

    (b) What is the significance of a rising or falling barometric pressure?

    (c) Actual barometric pressure varies with altitude. How does the air pressure change with increasing altitude? Why?

    (d) Meteorologists measure but do not report actual barometric pressures. The pressures you obtain from all media outlets are all adjusted pressures. Find out how the pressure is adjusted and why this is done.

    Follow the links for Nelson Chemistry 11, Unit 4 Review.

    GO TO www.science.nelson.com

33. The Bermuda Triangle has claimed many boats, planes, and people. One hypothesis is that large volumes of natural gas released from the ocean floor may sink the boats or drop the planes.

    (a) Explain why boats might sink.

    (b) Explain why planes might drop.

    (c) Use the Internet to research the Bermuda Triangle. Which explanation for the losses seems most reasonable to you? Defend your choice in a brief report.

    Follow the links for Nelson Chemistry 11, Unit 4 Review.

    GO TO www.science.nelson.com

Exploring

34. Research how gases are used in

    (a) medical anaesthetics

    (b) undersea exploration

    Choose one application for each example and prepare a two-paragraph description for each one.

    Follow the links for Nelson Chemistry 11, Unit 4 Review.

    GO TO www.science.nelson.com

Unit 5

Hydrocarbons and Energy

"Syncrude Canada was just about to begin operating its oil sands plant when I was graduating. I had attended several talks about Syncrude, so my primary reason for interviewing with the company was to see their pilot plants. However, when they offered me a job, I accepted. Syncrude now produces about 13% of Canada's crude oil requirements. One of the advantages of working in such a large company is that I have had the opportunity to carry out research in numerous areas. I like the variety of the work.

Jean Cooley, Research Associate, Syncrude Canada

"When I joined Syncrude, I was concerned whether I had the correct qualifications for the job. I have since learned that my formal training was really just the beginning of my learning. What I learned in school is how to approach a problem. On the job, I've expanded my knowledge and learned how to work in diverse teams to solve our problems."

Overall Expectations

In this unit, you will be able to

- demonstrate an understanding of the structure and properties of hydrocarbons, especially with respect to energy changes that occur during their combustion;
- describe and investigate the properties of hydrocarbons and apply calorimetric techniques to the calculation of energy changes; and
- evaluate the impact of hydrocarbons on our quality of life and the environment through an examination of some of their uses.

Unit

5

Hydrocarbons and Energy

Are You Ready?

Knowledge and Understanding

1. According to Lewis theory, the number of bonding electrons in an atom determines the number of bonds the atom will form. Complete **Table 1** for elements involved with hydrocarbons and chemicals derived from hydrocarbons.

Table 1: Bonding Electrons in Common Nonmetals

Element	Lewis symbol	Number of bonding electrons
?	$\cdot\overset{\cdot}{\underset{\cdot}{X}}\cdot$	?
?	$:\overset{\cdot\cdot}{\underset{\cdot\cdot}{X}}\cdot$	?
?	$:\overset{\cdot\cdot}{\underset{\cdot\cdot}{X}}\cdot$	?
?	$X\cdot$	?
?	$\cdot\overset{\cdot\cdot}{\underset{\cdot}{X}}\cdot$	?
?	$:\overset{\cdot\cdot}{\underset{\cdot}{X}}\cdot$	?

2. Carbon forms many useful compounds. Complete **Table 2**.

Table 2: Some Common Carbon Compounds

IUPAC name	Formula	Lewis structure	Use
methane	?	$\begin{matrix} & H & \\ H: & \overset{\cdot\cdot}{\underset{\cdot\cdot}{C}} & :H \\ & H & \end{matrix}$	in natural gas
dichlorodifluoromethane	?	$\begin{matrix} & Cl & \\ F: & \overset{\cdot\cdot}{\underset{\cdot\cdot}{C}} & :F \\ & Cl & \end{matrix}$	dangerous refrigerant
methanol (methylalcohol)	$CH_3OH_{(l)}$	?	gas line antifreeze
ethene (ethylene)	$C_2H_{4(g)}$	?	making polymers
ethyne (acetylene)	$C_2H_{2(g)}$	?	cutting torches
?	$CO_{2(s)}$	?	dry ice

3. Hexane, C_6H_{14}, is a hydrocarbon in gasoline that is a liquid at SATP. Describe the intermolecular forces of attraction acting between molecules of hexane (**Figure 1**).
4. Chloromethane, $CH_3Cl_{(g)}$, a substance used in "love meters," has a structure similar to that of methane, $CH_{4(g)}$. However, these two substances have very different boiling points. Chloromethane boils at –24°C, whereas methane boils at –161°C. Provide an explanation for this difference in physical properties.
5. Draw structural diagrams to explain the following empirically determined formulas: $C_2H_{6(g)}$, $C_2H_{4(g)}$, and $C_2H_{2(g)}$.
6. In **Figure 2**, a camper is using a propane stove.
   (a) Write a balanced chemical equation for the combustion of propane, $C_3H_{8(g)}$, in air to produce carbon dioxide and water vapour.
   (b) Butane is often used in lighters. Write a balanced chemical equation for the combustion of butane, $C_4H_{10(g)}$, in air to produce carbon dioxide and water vapour.

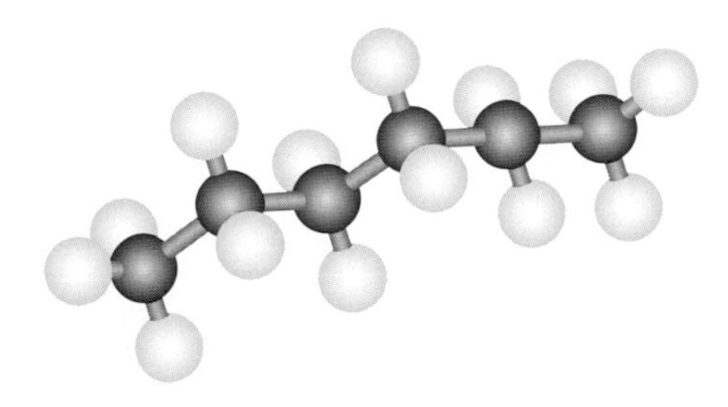

Figure 1
Ball-and-stick diagram of hexane

Inquiry and Communication

7. Complete **Table 3** by providing the required SI names and symbols.

Table 3: Some SI Names and Symbols

Quantity name	Quantity symbol	Unit name	Unit symbol
mass	?	?	?
?	*n*	?	?
?	?	?	°C
?	?	joule	?

8. Determine the molecular formula of the unknown chemical from the following evidence gathered from a combustion analyzer (percentage composition) and a mass spectrometer (molar mass). Explain the molecular formula by drawing a structural diagram.

   percentage of carbon = 79.85%

   percentage of hydrogen = 20.15%

   molar mass of compound = 30.08 g/mol

Figure 2
Hydrocarbons are used everywhere in our society. There is almost nothing in this picture, including the tent, the food, the camper, and the trees, that does not contain some hydrocarbons or that does not involve hydrocarbons in its production.

Technical Skills and Safety

9. The smaller-molecule hydrocarbons commonly used at home and in the laboratory are readily flammable and often evaporate easily to form explosive mixtures with air.
   (a) What is the emergency procedure to follow if there is a fire?
   (b) Suggest some measures for dealing with flammable materials that might prevent a fire.

Chapter

11

Hydrocarbons

In this chapter, you will be able to

- describe the origins and major sources of organic compounds, particularly hydrocarbons;
- describe and explain the characteristics of different structures of hydrocarbons, using Lewis bonding theory;
- describe the process of petroleum refining and the uses society has for the products of this process;
- describe and determine by experimentation some physical and chemical properties of single-, double-, and triple-bonded hydrocarbon compounds;
- name and draw structural formulas for a wide variety of hydrocarbon compounds;
- represent a wide variety of hydrocarbon compounds by using molecular models to show their atom arrangement;
- describe science- and technology-based careers in the petrochemical industry;
- discuss the risks and benefits of society's dependence on petrochemicals, particularly hydrocarbons.

Organic chemicals include most chemicals that compose living material—both plants and animals. All carbohydrates, proteins, and fats (lipids) are organic compounds. Pesticides and antibiotics are organic compounds. DNA, RNA, chromosomes, and genes are composed of organic compounds. Fossil fuels, lubricants, and gasoline are organic. Polyethylene, polypropylene, nylon, and polyester are organic polymers. In this unit you will study hydrocarbons, a class of organic chemicals made up of only carbon and hydrogen atoms.

Hydrocarbons serve a dual purpose in society: They are fuels that may be used to produce electricity, run automobiles, and cook meals; and they are used to produce petrochemicals and thus may be the source of almost anything manufactured, including furniture, processed foods, computers, stereos, plastics, synthetic fabrics, cosmetics, synthetic sweeteners, soaps, solvents, and refrigerants. The list grows longer and longer each year.

This dual purpose of hydrocarbons is integrated into the carbon cycle, as illustrated in **Figure 1**. Knowledge of the chemistry of hydrocarbons will help you make increasingly important decisions about our society and our planet.

Reflect on your Learning

1. Recall the products of incomplete combustion of hydrocarbons. Which one is visible?
2. What are fossil fuels? Name one fuel that is not a fossil fuel.

Throughout this chapter, note any changes in your ideas as you learn new concepts and develop your skills.

Butane Behaviour

The hydrocarbon butane, $C_4H_{10(g)}$, is used in a wide variety of small personal devices, such as lighters. Butane is a gas at SATP, but liquefies under only moderate pressure, a property that makes it a very convenient fuel.

Butane is flammable. Do not heat the cylinder directly.

Conduct this activity in a fume hood.

Materials: a small (sandwich-size) resealable plastic bag, a butane cylinder (for refilling butane lighters), a 250-mL beaker, eye protection, matches

- Hold the cylinder with the nozzle end down, inside the open plastic bag, and pull the nozzle back so that a small amount (approximately 1–2 mL) of liquid butane sprays into the bag.
- Seal the plastic bag and hold it so that the liquid butane inside is warmed by your hand. Don't overdo this and freeze your skin; boiling butane is very cold.
- Observe the physical properties (volume, temperature, colour) of butane as it changes physical state. Then open the bag upside down

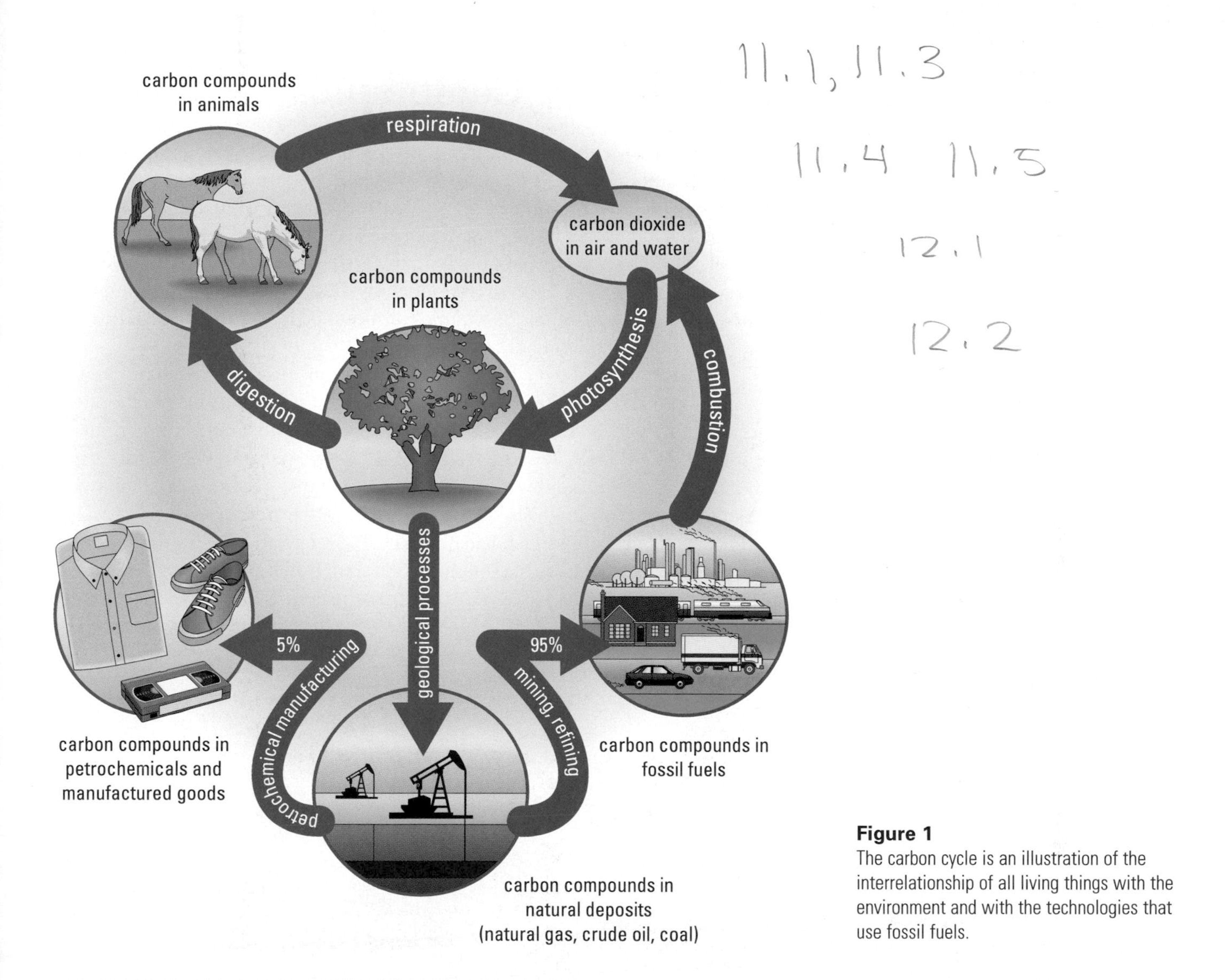

Figure 1
The carbon cycle is an illustration of the interrelationship of all living things with the environment and with the technologies that use fossil fuels.

in a running fume hood (or outdoors, out of an open window) to release the butane from the bag.

- Put on your eye protection and tie back any loose hair. On a level table or bench top in a relatively draft-free area, hold the cylinder with the nozzle end down and spray approximately 1–2 mL of liquid butane into the beaker. Allow the liquid butane to completely vaporize.
- After 5–10 s light a match, and use it to ignite the butane gas in the beaker from the side, being careful not to position your hand directly above the beaker when you light the gas.
- Observe the combustion; when the flame stops, pick up the beaker and feel the bottom. Now wipe the inside top half of the beaker with a clean white tissue, and observe the tissue.

(a) Compare the volumes of the fixed amount of butane in the bag, in its liquid and gaseous forms. How do you know the amount of butane is constant?

(b) Where does the energy come from to boil the butane in the bag? What observation supports your statement?

(c) What does releasing the butane gas from the bag by holding it upside down suggest about the density of butane gas compared to air? Does this explain why butane gas in a beaker will mostly stay in the beaker? What observation tells you this must be true, even though the gas itself is invisible?

(d) Where does the energy come from to boil the butane in the beaker? What observation supports your statement?

(e) What is the dark substance on the tissue after wiping the beaker?

11.1 Organic Compounds

In the early 19th century, Swedish chemist Jöns Jakob Berzelius (1779–1848) classified compounds into two primary categories: those obtained from living organisms, which he called organic, and those obtained from mineral sources, which he called inorganic. At that time, most chemists believed that organic chemicals could be synthesized only in living systems. A theory called "vitalism" proposed that the laws of nature were somehow different for living and nonliving systems and that the synthesis of organic compounds necessarily involved a "vital force," present only in living organisms.

This theory was shown to be unacceptable in 1828 by German chemist Friedrich Wöhler (1800–1882). Wöhler performed a revolutionary laboratory experiment in which he used the inorganic compound ammonium cyanate, $NH_4OCN_{(s)}$, to synthesize urea, $H_2NCONH_{2(s)}$, a well-known organic compound that is produced by many living organisms. In the years following Wöhler's experiment, chemists synthesized many other organic compounds. For example, acetic acid (ethanoic acid), $HC_2H_3O_{2(l)}$, a relatively simple molecule, was synthesized in 1845 in Germany by Adolph Kolbe (1818–1884). Sucrose, $C_{12}H_{22}O_{11(s)}$, (**Figure 1**), has a more complex structure and was long thought impossible to create in a laboratory—until 1953, when it was first synthesized by Canadian chemist Raymond Lemieux (1920–2001).

Figure 1
Sucrose (table sugar, $C_{12}H_{22}O_{11(s)}$) occurs naturally in plants and in fairly high concentrations in sugar beets and sugar cane (shown here). No sugar cane is grown in Canada, although it is refined in New Brunswick, Quebec, and Ontario. Sugar beets are grown and refined in Quebec, Manitoba, and Alberta. High-fructose (fruit sugar, $C_6H_{12}O_{6(s)}$) corn syrup is refined in Ontario.

organic compounds: compounds that contain carbon, except $CO_{(g)}$, $CO_{2(g)}$, and ionic compounds with carbon

organic chemistry: the study of organic compounds

Today, **organic compounds** are defined as compounds that contain carbon, and **organic chemistry** is defined as the study of organic compounds. This can seem like a limitless task, since more than nine million such compounds have been identified. We simplify our organization by grouping organic compounds according to their properties and molecular structures, which, we explain, are a result of the structure of the covalent bonds within organic molecules. By common convention, the two oxides of carbon, $CO_{2(g)}$ and $CO_{(g)}$, are not normally considered to be organic compounds. Compounds of carbonate, bicarbonate, cyanide, cyanate, and thiocyanate ions all contain carbon atoms, but these are also not considered to be organic compounds because their properties are explained by ionic bonds.

The simplest way to begin a detailed study of organic compounds is with **hydrocarbons,** those compounds containing only carbon and hydrogen atoms. Based on previous generalizations (Unit 1), hydrocarbons are all nonpolar substances. Therefore, they all have a low solubility in water, and their physical properties—such as states, densities, and melting and boiling points—form clear trends largely explained by London (dispersion) forces. We consider all organic compounds to be derivatives of hydrocarbons for classification purposes; that is, we classify all organic compounds as though they were hydrocarbons that were then changed by atom rearrangement, addition, and substitution into their present chemical structure. Because we use this system, understanding hydrocarbons is essential for the study of any area within the field of organic chemistry.

hydrocarbons: organic compounds that contain only carbon and hydrogen atoms in their molecular structure

Sources and Uses

Coal, crude oil, oil sands, heavy oil, and natural gas are the nonrenewable sources of fuels that power our society. All of these substances are formed over millions of years from plant and animal material that, over geologic time periods, is subjected to heat and pressure. The original compounds gradually convert to a very complex mix of organic compounds we generally name according to the physical state of the mixture, for example, sand, oil, and gas. We call all of these compound mixtures "fossil" fuels because of the way they are formed. These fuels are primarily hydrocarbons. Over 95% of our society's hydrocarbon use is for combustion reactions to provide heat and electrical energy (primary uses). Hydrocarbons are also the starting materials in the industrial chemical synthesis of thousands of products such as fuels, plastics, solvents, medicines, and synthetic fibres (secondary uses). Some hydrocarbons are obtained directly by physical separation from petroleum and natural gas, whereas others come from oil and gas refining (**Figure 2**).

Figure 2
The first oil company in North America was created in 1854 by Charles Tripp, and the first oil well was dug in 1859 by James Williams at Oil Springs, Ontario. Williams constructed the first oil refinery in Canada at Hamilton in 1860. A replica of one of the oil wells is found near the Oil Museum of Canada in Oil Springs, Ontario.

Refining is the technology that includes separating complex mixtures into simpler purified components. The refining of coal and natural gas involves physical as well as chemical processes; for example, coal may be crushed and treated with solvents. Components of natural gas are separated either by selectively dissolving them in chosen solvents or by condensation and distillation. Petroleum refining, that is, refining crude oil, is much more complex than coal or gas refining, but many more products are obtained from crude oil.

refining: the physical and/or chemical process that converts complex organic mixtures into simpler mixtures or purified substances

Sources of hydrocarbons other than fossil fuels are all living things, and, of course, we believe fossil fuels were living things very long ago. Plant crops and animal decomposition products can be used to produce hydrocarbons and other organic compounds. For example, in old landfill waste disposal sites, methane, $CH_{4(g)}$, is produced underground by bacterial decomposition of organic waste material. Many cities tap into these sites to capture the gas, which can then be burned as fuel in power-generating plants.

Practice

Understanding Concepts

1. Describe the difference between the original and the current definitions of organic compounds.
2. What are the major sources of hydrocarbons used by our society?
3. How do scientists explain how these sources of hydrocarbons were originally formed?
4. List the primary and some secondary uses of hydrocarbons in our society.

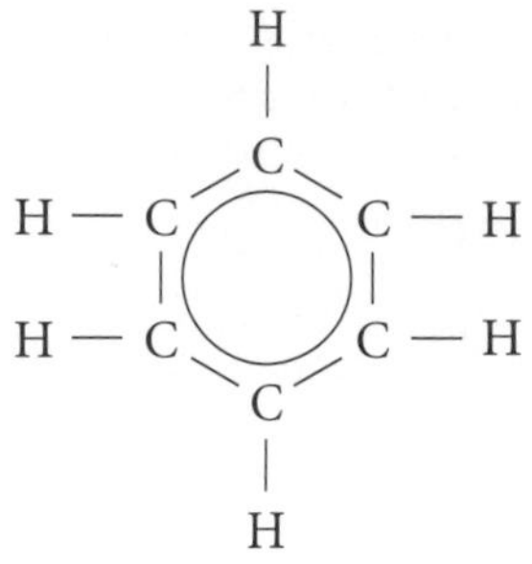

Making Connections

5. Considering the percentages of primary and secondary uses of hydrocarbons in our technological society, write a paragraph on the need to conserve this nonrenewable resource.
6. Some politicians from developed countries are calling for a significant international reduction of hydrocarbon combustion. What might this mean for the future development of developing countries on our planet?

Reflecting

7. What other high-school science course heavily involves organic chemistry, and why?

Classification of Hydrocarbons

The molecular structures of hydrocarbons determine how they are classified (**Figure 3**). Organic compounds with molecular structures that are straight or branched chains or rings of carbon-to-carbon bonds are called **aliphatic compounds**. As you can see in **Figure 3**, this subgroup of hydrocarbons is further classified into alkanes, alkenes, and alkynes. The main purpose of this chapter is to study the properties, molecular structure, and naming of these three groups of aliphatic hydrocarbons, including both open-chain and ring compounds. Hydrocarbons are either aliphatic or aromatic; aromatic compounds are organic compounds that contain a benzene-like ring structure (**Figure 4**). For now, the only aromatic compound that you need to know about is benzene, $C_6H_{6(l)}$. You may study aromatic hydrocarbons and the hydrocarbon derivatives in future chemistry courses.

benzene

Figure 4
Benzene, $C_6H_{6(l)}$, is the simplest example of an aromatic compound. It is an important component of gasoline. A benzene molecule has some unique carbon–carbon bonds, indicated by the circle in the ring.

aliphatic compound: one that has a structure based on straight or branched chains or rings of carbon atoms; not including aromatic compounds, for example, benzene

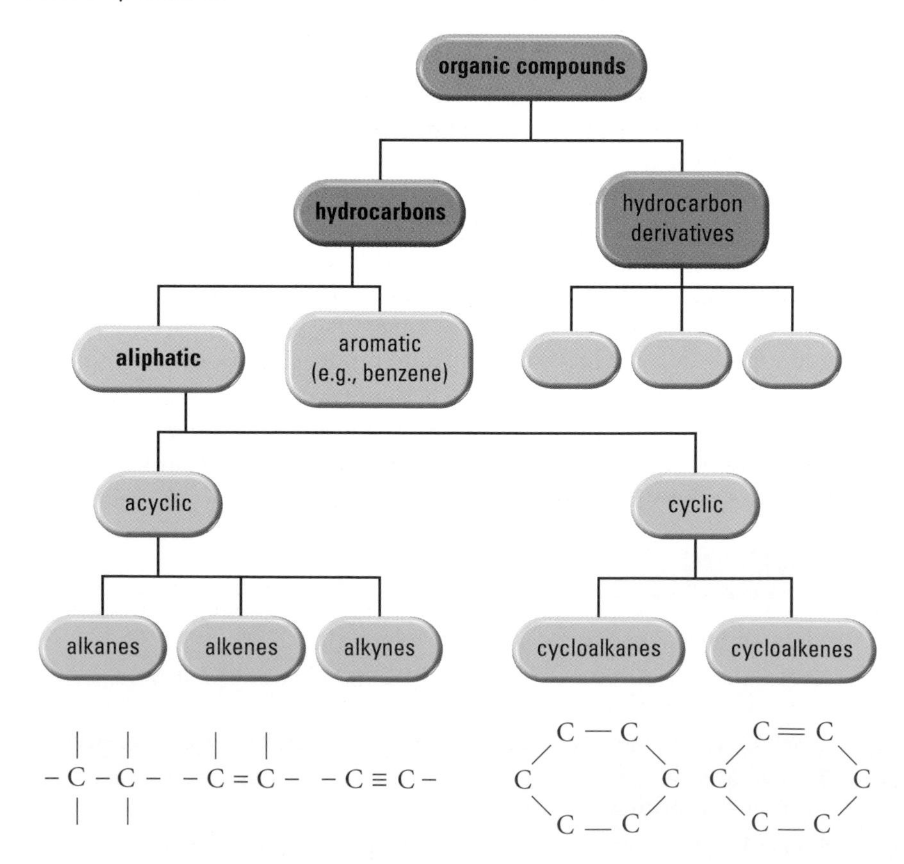

Figure 3
This classification system helps scientists organize their knowledge of organic compounds. You start your study of organic compounds with simple alkanes.

Simple Alkanes

alkane: a hydrocarbon with only single bonds between carbon atoms

Aliphatic hydrocarbons with molecules containing only single carbon-to-carbon bonds with an open-chain (noncyclic) structure are called **alkanes**. The classification system in **Figure 3** shows where the study of alkanes fits into the big picture of organic compounds. The simplest member of the alkane series is methane, $CH_{4(g)}$, which is the main constituent of natural gas used for home heating. Other simple alkanes containing up to 10 carbon atoms in a continuous chain are listed in **Table 1**. Each molecular formula in the series has one more CH_2 group than the one before it. Such a series of compounds, where each differs from the one before it by the same increment (CH_2 in this case), is called a homologous series. Comparing the formulas, you can see that the general formula for these alkanes is always C_nH_{2n+2}; that is, they are a series of CH_2 units plus two terminal hydrogen atoms.

The empirical formulas for the alkanes, such as the ones listed in **Table 1**, can be determined by combustion analysis and mass spectrometry. Empirically determined chemical formulas are useful for communicating the numbers of atoms present in a molecule. For simple molecules, the structure is often evident. As the number of atoms in a molecule increases, the molecular formula must be expanded in order to communicate the structure. One simple alternative is to cluster groups of atoms, for example, writing $CH_3CH_2CH_2CH_2CH_3$ to represent the way in which the atoms bond in C_5H_{12} molecules.

The *expanded* molecular formula, $CH_3CH_2CH_2CH_2CH_3$, is one alternative to the usual molecular formula for pentane, $C_5H_{12(l)}$. Chemists have created other ways to communicate the structures of these compounds. People are very visual beings and often find models to be an effective way of understanding how we believe something exists and operates. In chemistry, ball-and-stick models and space-filling models help us visualize the structures and shapes of molecules. Ball-and-stick models, like the pentane model in **Figure 5**, are more effective in showing types of covalent bonds and the angles between the bonds, but not as good at showing molecular shape and atomic size. Space-filling models (**Figure 6**) do this well.

Structural diagrams are another way to communicate molecular structure, but less completely than 3-D models. A complete structural diagram, as in **Figure 7(a)**, shows all atoms and bonds; a condensed structural diagram, as in **Figure 7(b)**, omits showing the C—H bonds but shows the carbon–carbon bonds. A line structural diagram, as in **Figure 7(c)**, is an efficient way to represent long chains of carbon atoms; the ends of each line segment represent carbon atoms, and hydrogen atoms are not shown. Since information is left out of any diagram, it must be replaced by your knowledge. For example, in a line structural diagram, wherever you see fewer than four lines at an intersection, it is assumed that H atoms are bonded there to make four bonds to each carbon.

Table 1: The Alkane Family of Organic Compounds

IUPAC name	Molecular formula
methane	$CH_{4(g)}$
ethane	$C_2H_{6(g)}$
propane	$C_3H_{8(g)}$
butane	$C_4H_{10(g)}$
pentane	$C_5H_{12(l)}$
hexane	$C_6H_{14(l)}$
heptane	$C_7H_{16(l)}$
octane	$C_8H_{18(l)}$
nonane	$C_9H_{20(l)}$
decane	$C_{10}H_{22(l)}$
-ane	C_nH_{2n+2}

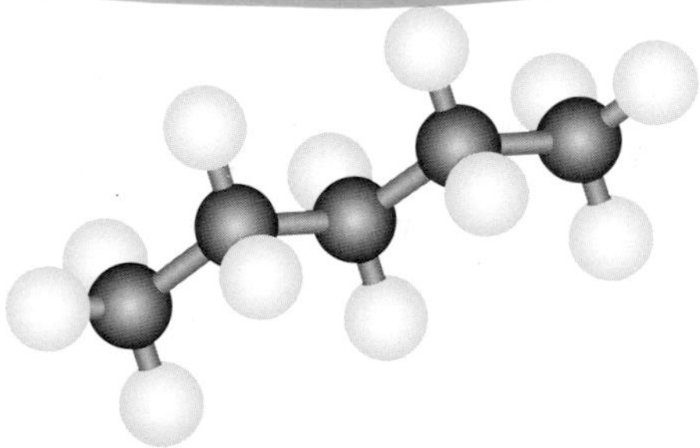

Figure 5
A ball-and-stick model of pentane, $C_5H_{12(l)}$, helps us visualize this theoretical structure.

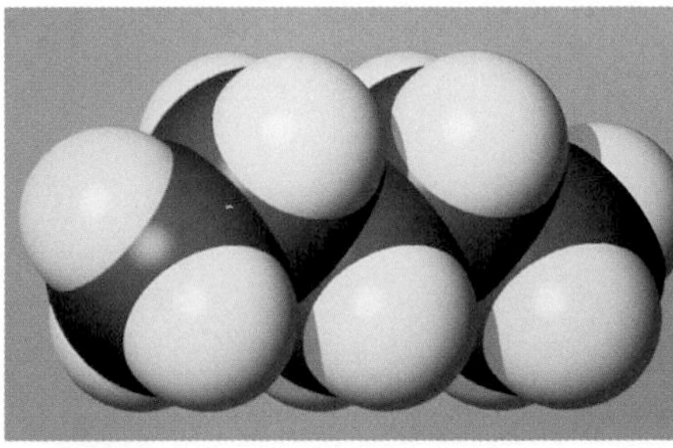

Figure 6
This is a space-filling model, also of pentane, $C_5H_{12(l)}$, used to show the shape of the molecule and the relative size of the atoms.

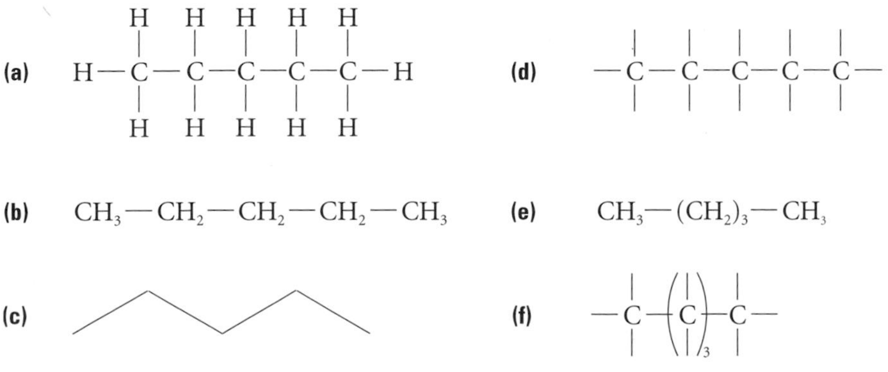

Figure 7
These structural diagrams all represent the same pentane molecule, C_5H_{12}.

Other variations of structural diagrams, such as **Figure** 7(d), (e), and (f) page 509, are also used by some people.

Structural diagrams, in general, do not communicate the shape of the molecule. Knowledge of the three-dimensional structure of the molecule is sacrificed to make the communication simpler. When knowledge of the three-dimensional character of a molecule is required, a structural diagram is replaced by a more sophisticated diagram.

Sample Problem 1

Octane is a component of gasoline. Draw a complete structural diagram and a condensed structural diagram for octane.

Solution

```
    H   H   H   H   H   H   H   H
    |   |   |   |   |   |   |   |
H — C — C — C — C — C — C — C — C — H
    |   |   |   |   |   |   |   |
    H   H   H   H   H   H   H   H
```

$CH_3 - CH_2 - CH_2 - CH_2 - CH_2 - CH_2 - CH_2 - CH_3$

Practice

Understanding Concepts

8. Why do scientists create classification systems?
9. Using your own words, define the class of organic compounds called hydrocarbons.
10. Name the two major classes of hydrocarbons as created by scientists.
11. Are hydrocarbon molecules polar or nonpolar? What does this suggest about the type of intermolecular forces present between hydrocarbon molecules?
12. Draw a Lewis diagram for a carbon atom and a hydrogen atom, and indicate how many covalent bonds each of these atoms can form.
13. Draw a Lewis diagram and a complete structural diagram for methane, the main component of natural gas.
14. Crude oil is a complex mixture of hydrocarbons that includes most of the simple alkanes. Draw complete structural diagrams for the straight-chain alkanes, from ethane through decane. Label each diagram.
15. What is the molecular formula for an alkane containing 30 carbon atoms?
16. What is the common feature in the names of the simple alkanes in **Table 1**? Where do you think this common feature came from?

Section 11.1 Questions

Understanding Concepts

1. Classify each of the following compounds as inorganic or organic:
   (a) $CaCO_{3(s)}$
   (b) $C_6H_{6(l)}$
   (c) $CO_{2(g)}$
   (d) $C_4H_{10(g)}$
   (e) $CH_3(CH_2)_6CH_{3(l)}$
2. What is believed to be the origin of most hydrocarbons on Earth?
3. Identify the sources of most organic compounds.
4. List three common fuels that are hydrocarbon compounds.
5. Draw a complete structural diagram to explain each of the following empirical formulas:
   (a) $C_3H_{8(g)}$
   (b) $C_5H_{12(l)}$
   (c) $C_7H_{16(l)}$
6. Name the following hydrocarbons, which are found in a sample of crude oil:
   (a) $C_2H_{6(g)}$
   (b) $C_4H_{10(g)}$
   (c) $C_6H_{14(l)}$
   (d) $C_9H_{20(l)}$
7. Can the hydrocarbon $C_{45}H_{92(s)}$ be classified as an alkane? Justify your answer.

Applying Inquiry Skills

8. Complete the Analysis section of the following lab report.

   Question

   What is the chemical formula, molecular structure, and name of an unknown gas?

   Experimental Design

   A sample of a gas is analyzed with a combustion analyzer and a mass spectrometer.

   Evidence

   percent by mass of carbon = 81.68%

   percent by mass of hydrogen = 18.32%

   molar mass by analysis = 44.01 g/mol

   Analysis

   (a) Determine the empirical molecular formula of the hydrocarbon, name it, and draw a structural diagram.

Making Connections

9. Are fossil fuels a finite source of hydrocarbons? Provide your reasoning.

Reflecting

10. What will we use for an energy source and raw material for making plastics, fabric, detergents, and so on if sources of fossil fuels are depleted?

11.2 Refining Petroleum

petroleum: a complex gas and/or liquid mixture composed mostly of hydrocarbons, obtained by drilling into underground deposits

natural gas: the fraction of petroleum that vaporizes at normal temperature and pressure, consisting of low molar mass hydrocarbons; also primarily methane sold as heating fuel

Petroleum is a complex liquid mixture of hundreds of thousands of compounds, most of which are hydrocarbons with 1 to 40 carbon atoms per molecule. Over millions of years, heat and pressure deep within the Earth have transformed prehistoric plant and bacterial material into this highly complex mixture. Those compounds that emerge from a drilled well in gaseous form are called **natural gas** and include only those compounds with low molar masses, that is, those with one to five carbon atoms in their molecular structures. The proportions of the components of natural gas vary from well to well. The gas that emerges from wells usually contains carbon dioxide, $CO_{2(g)}$, as well, and often the extremely toxic hydrogen sulfide, $H_2S_{(g)}$. Natural gas that contains hydrogen sulfide is called sour gas; this is a dangerous substance, so workers treat it with great respect. The term natural gas is also used commercially for the fuel sold for home and industrial heating; this fuel is refined natural gas with a high proportion of methane, $CH_{4(g)}$, to which a trace of a very smelly compound has been added (methane is odourless so ethanethiol, $C_2H_5SH_{(l)}$, is added in order to detect a leak). Thus, the meaning of the term natural gas depends on the context in which it is used.

crude oil: the fraction of petroleum that is liquid at normal pressure, consisting of higher molar mass hydrocarbons than natural gas; it is refined to separate it into gas, liquid, and solid hydrocarbon components.

Crude oil is also a complex mixture consisting primarily of hydrocarbons. This mixture contains dissolved gases and solids but is composed of mostly liquid components. The oil varies in consistency from heavy to light, from thick to thin, depending on the region in the country or the world. Crude oil is often forced from the well by the pressure of surrounding natural gas. Heavier crude may even be in the form called heavy oil or oil (tar) sands. Heavy oil is removed by pumping steam into the well to heat the oil so that it flows more easily, not unlike heating honey or molasses. Oil sands are most often mined with gigantic draglines (huge scoops on a conveyer belt). The oil film is then separated from the sand in large rotating cylinders where hot water and sodium hydroxide are added. The oil (bitumen) is so viscous that it needs to be semi-refined before it can be shipped by pipeline to a conventional oil refinery. This semi-refined oil is called synthetic crude.

fractional distillation: the separation of components of petroleum by distillation, using differences in boiling points; also called fractionation

Some of the liquid compounds in crude oil boil at temperatures lower than 30°C. At the other end of the scale, the least volatile components of crude oil boil at temperatures above 400°C. The differences in boiling points of the compounds making up petroleum allow us to physically separate these compounds in a refining process called **fractional distillation**, or fractionation. The most desired fraction of the original mixture is of hydrocarbon substances with 5 to 12 carbon atoms in their molecular structure because this is the basis of gasoline, the most marketable product.

When crude oil is heated to about 500°C in the absence of air, most of its constituent compounds vaporize. The compounds with boiling points higher than 500°C remain unvaporized as complex hydrocarbon mixtures called asphalts and tars, familiar materials used in road and roof surfacing. The vaporized components of the petroleum rise and gradually cool as they move higher in a metal fractionating tower (**Figures 1** and **2**). Where the temperature in the higher parts of the tower is below the boiling points of the vaporized compounds moving up the tower, the substances in the vapour begin to condense into liquids. The substances with high boiling points condense in the lower, hotter parts of the tower, and the substances with lower boiling points condense near the top of the tower, where it's cooler. At various levels in the tower, trays collect mixtures of liquid substances as they condense; each mixture contains compounds with roughly similar boiling points. These mixtures are called petroleum fractions.

The fractions with the lowest boiling points contain the smallest molecules. The low boiling points can be explained by intermolecular forces: Small molecules have fewer electrons and weaker London forces compared with large molecules. The fractions with higher boiling points contain much larger hydrocarbon molecules. Some typical fractions are shown in **Table 1**, page 514. The physical process of fractionation is then followed by chemical processes in which the fractions are converted into more valuable products.

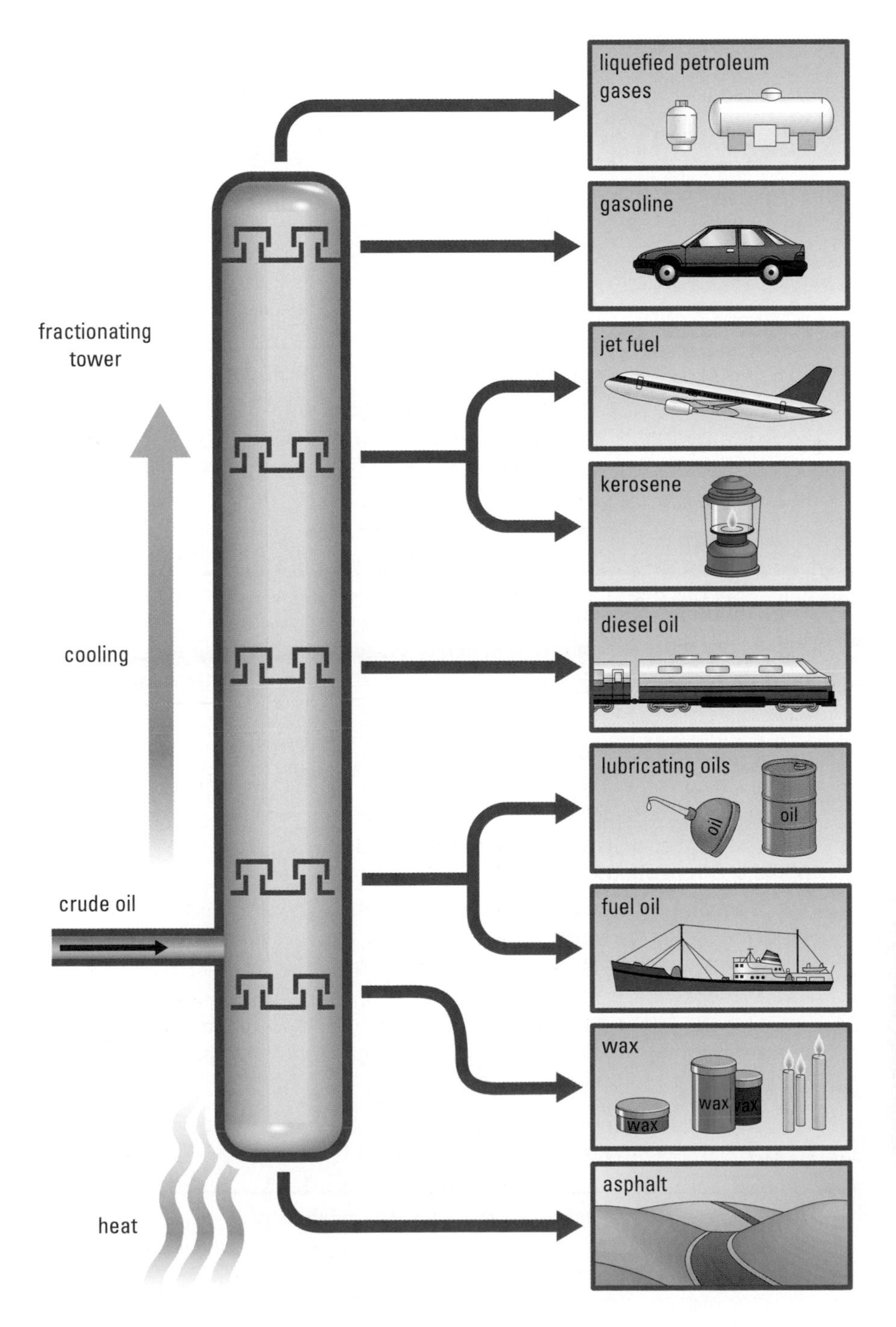

Figure 1
A fractional distillation tower contains liquid collection trays positioned at various levels. Heated crude oil enters near the bottom of the tower. The bottom of the tower is kept very hot; toward the top of the tower, the temperature gradually decreases. The lower the boiling point of a hydrocarbon, the higher the tray on which its vapour condenses.

Figure 2
Fractionating towers look complicated, but most of the structure you see here is to allow access for maintenance.

Table 1: Fractional Distillation of Petroleum

Boiling point range of fraction (°C)	Carbon atoms per molecule	Fraction (intermediate product)	Some common applications
below 30	1 to 5	gases	gaseous fuels for cooking and for heating homes
30 to 90	5 to 6	petroleum, ether	dry-cleaning solvents, naphtha gas, camping fuel
30 to 200	5 to 12	straight-run gasoline	automotive gasoline
175 to 275	12 to 16	kerosene	fuel for diesel and jet engines and kerosene heaters, cracking stock (raw materials for fuel and petrochemical industries)
250 to 375	15 to 18	light gas oil or fuel oils	furnace oil, home heating, cracking stock
over 350	16 to 22	heavy gas oil	lubricating oils, cracking stock
over 400	18 and up	greases	lubricating greases, cracking stock
over 450	20 and up	paraffin, candle waxes	waxed paper, cosmetics, polishes, cracking stock
over 500	26 and up	unvaporized solid residues	asphalts and tars for roofing and paving

Practice

Understanding Concepts

1. $C_{15}H_{27}$ hydrocarbon molecules in petroleum condense on their way up a fractionating tower. What might the fraction containing these molecules be called?
2. Candle wax must first melt, then be drawn up a wick and vaporize before it will mix with air, ignite, and burn. According to **Table 1**, what minimum temperature must you create with your match or lighter to get the wax to vaporize?

Making Connections

3. Gasoline for cars must vaporize and mix with air before the mixture can be burned in the engine. Gasoline is blended differently in summer and winter for retailers selling it in Canada. How and why would refineries change the mixture of substances for winter use?
4. Use the Internet to research the toxicity of $H_2S_{(g)}$, and prepare a short report that explains how it affects humans, how lethal it is, and the issues raised by its presence in crude oil and natural gas.

   Follow the links for Nelson Chemistry 11, 11.2.

GO TO www.science.nelson.com

Lab Exercise 11.2.1

Fractional Distillation

INQUIRY SKILLS

- ○ Questioning
- ○ Hypothesizing
- ● Predicting
- ○ Planning
- ○ Conducting
- ○ Recording
- ● Analyzing
- ● Evaluating
- ● Communicating

In the refining of petroleum, fractional distillation separates a complex mixture into a series of mixtures, or fractions, containing similar-sized hydrocarbons. Fractional distillation is also a very common laboratory technique for separating mixtures so that some or all of the fractions are pure substances. The purpose of this lab exercise is to use fractional distillation to test our knowledge of intermolecular forces by separating a simple liquid mixture of organic compounds. Complete the **Prediction**, read the **Procedure**, and complete the **Analysis**, **Evaluation**, and **Synthesis** sections of the lab report.

Question

What is the order of boiling points, from lowest to highest, of pentane, $C_5H_{12(l)}$, 2-methyl-2-propanol, $C_4H_9OH_{(l)}$, and pentacosane (paraffin wax), $C_{25}H_{52(s)}$, in a mixture (solution) of these compounds?

Prediction

(a) Predict an answer to the Question and provide your reasoning.

Experimental Design

A sample of the mixture is slowly heated in a fractional distillation apparatus to separate the components. Separated samples are identified by a comparison of odour and state with pure reference samples of the original constituent compounds. A temperature–time graph is used to analyze the separation.

Materials

eye protection
fractional distillation apparatus (**Figure 3**, page 516)
apron
100-mL graduated cylinder
pentane, $C_5H_{12(l)}$
2-methyl-2-propanol, $C_4H_9OH_{(l)}$
pentacosane, $C_{25}H_{52(s)}$
round-bottom flask with fittings
heating mantle

Procedure

1. Measure 15 mL of pentane, 20 mL of 2-methyl-2-propanol, and approximately 2 g of pentacosane into the round-bottom flask. Stir to mix.
2. Assemble the fractional distillation apparatus (**Figure 3**) in the fume hood, including the flask with the mixture, making sure that all fittings are secure.
3. Turn up the heating mantle slowly; heat the flask and mixture, taking the vapour temperature at 30 s intervals.
4. After most of the first fraction has boiled off, as indicated by the end of a constant temperature period, change the collection flask at the outflow of the condenser.
5. When the second fraction has boiled off, turn the heating mantle off.
6. Compare the state and odour of the collected fractions with the pure substances used to make the mixture.
7. Dispose of all substances in a container marked "nonhalogenated organic compound waste."

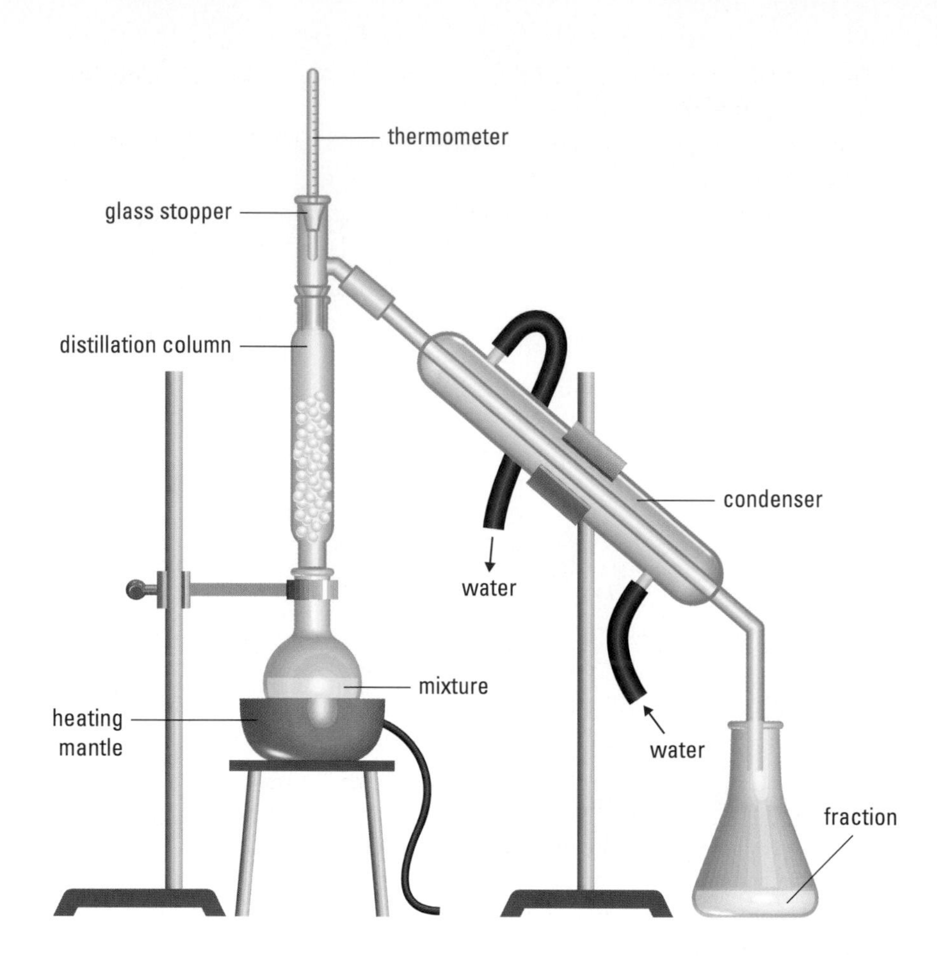

Figure 3
A typical fractional distillation apparatus in the laboratory

Evidence

Table 2: Vapour Temperature in the Distillation Apparatus

Time (min)	Temperature (°C)	Time (min)	Temperature (°C)
0.0	25.5	8.0	63.5
0.5	27.0	8.5	68.5
1.0	30.0	9.0	71.0
1.5	31.0	9.5	72.5
2.0	32.0	10.0	73.5
2.5	32.0	10.5	74.0
3.0	32.1	11.0	74.2
3.5	32.1	11.5	74.2
4.0	32.1	12.0	74.2
4.5	32.5	12.5	74.2
5.0	33.8	13.0	74.2
5.5	35.5	13.5	74.0
6.0	38.0	14.0	74.0
6.5	42.0	14.5	73.2
7.0	49.0	15.0	73.0
7.5	56.8		

- The first fraction collected was a clear, colourless liquid with an odour similar to a sample of pentane.
- The second fraction collected was a clear, colourless liquid with an odour similar to 2-methyl-2-propanol.
- A small quantity of clear, colourless liquid was left in the flask, and the liquid solidified into a white, waxy solid when the flask cooled.

Analysis

(b) Graph the vapour temperature versus time from the Evidence in **Table 2**.
(c) Use the observations and graph to identify the fractions, including boiling points for the first two fractions.
(d) Use the Evidence to answer the Question.

Evaluation

(e) Identify some sources of experimental error or uncertainty.
(f) Suggest some improvements that might be made in this fractional distillation.
(g) Evaluate the Prediction. Was it verified, falsified, or inconclusive?
(h) Evaluate the reasoning used to make the Prediction. Can the concepts and rules of intermolecular forces be used to make predictions of boiling points?

Synthesis

(i) What are some limitations of your knowledge of intermolecular forces applied to boiling points? For example, can the actual boiling points of substances, rather than just the order, be predicted?

Figure 4
The octane rating you see on gas pumps is based on the scale that arbitrarily assigns heptane a rating of zero and isooctane (a branched form of octane) a rating of 100. The higher the number, the better the combustion characteristics of the fuel.

Cracking and Reforming Reactions

Straight fractional distillation of petroleum does not produce enough of the gasoline fraction (called straight-run gasoline) to meet the demand for gasoline. Other fractions, from light gas oil to paraffin (**Table 1**, page 514), are chemically altered to produce more gasoline hydrocarbons with 5 to 12 carbon atoms per molecule. Hydrocarbons in these fractions are broken into smaller fragments in a technological process called **cracking**, which occurs in the absence of air. Notice how many of the fractions in **Table 1** are used as cracking stock, primarily to make gasoline. For example, hydrocarbons of large molar mass (C_{15} to C_{18}) are converted by cracking into gasoline component hydrocarbons (C_5 to C_{12}):

cracking: reaction of a larger hydrocarbon molecule that breaks it into two or more smaller hydrocarbon molecules

$$C_{17}H_{36(l)} + H_{2(g)} \rightarrow C_9H_{20(l)} + C_8H_{18(l)}$$

Originally, only high temperatures in the absence of air would cause these reactions in an industrial process called thermal cracking. Today, catalysts are used to speed up the reactions and allow them to happen at lower temperatures in a process called catalytic cracking.

The opposite of cracking is a **reforming** reaction. Catalysts are also used in reforming reactions. In catalytic reforming, large molecules are formed from smaller ones. For example, pentane can be reformed to decane:

reforming: reaction of two or more smaller hydrocarbon molecules that combines them into a larger or more branched hydrocarbon molecule

$$C_5H_{12(l)} + C_5H_{12(l)} \rightarrow C_{10}H_{22(l)} + H_{2(g)}$$

Reforming may also produce molecules with more branching, which is a desirable characteristic for gasoline. Reforming reactions most commonly convert low-grade (low-octane, inefficient) gasolines into higher-grades gasolines with a higher octane rating, which is necessary for today's highly tuned automobile engines (**Figure 4**). Larger hydrocarbon molecules for synthetic lubricants and petrochemicals can also be made by reforming smaller hydrocarbons.

Practice

Understanding Concepts

5. Write a chemical equation showing the cracking of a $C_{14}H_{30}$ molecule into octane and one other hydrocarbon molecule.
6. Write the reaction equation showing the reforming of a C_4H_8 molecule and a C_5H_{12} molecule into a single, larger hydrocarbon molecule that is the only product. Name this product.

Making Connections

7. What are some alternatives to gasoline-powered vehicles? For one of these alternatives, state an advantage and a possible disadvantage.

INQUIRY SKILLS

- ○ Questioning
- ○ Hypothesizing
- ○ Predicting
- ○ Planning
- ● Conducting
- ● Recording
- ● Analyzing
- ○ Evaluating
- ● Communicating

Most organic compound vapours are flammable. Perform the investigation in a fume hood or well-ventilated area. Tie back long hair and roll up sleeves when using the burner.

Investigation 11.2.1

Destructive Distillation

Destructive distillation is the process of heating substances to a high temperature in the absence of oxygen. Volatile components are driven off from the substance, which also decomposes in the process. Canadian scientist Abraham Gesner (**Figure 5**) used this process extensively in his study of fossil fuels and charcoal. In comparison, cracking is a relatively modern, controlled version of destructive distillation applied to hydrocarbon mixtures. The purpose of this investigation is to study changes in the destructive distillation of sawdust. Complete the **Analysis** section of the lab report.

Figure 5
Abraham Gesner (1797–1864) lived and worked in Nova Scotia. He was a physician, geologist, chemist, author, and the inventor of kerosene oil. For his extensive original work with fossil fuels, Gesner is considered to be the founder of the modern petroleum industry.

Question

What are the products of the destructive distillation of sawdust?

Materials

eye protection
apron
large test tube
glass stirring rod
matches or burner lighter
laboratory stand
buret clamp
laboratory burner
sawdust

Procedure

1. Add sawdust to a clean, dry test tube and pack it down with the end of a stirring rod until you have about 3 cm of packed sawdust.
2. Attach the test tube to the clamp and stand.
3. Light the burner; heat the bottom of the test tube gently at first and then more strongly. Record all observations of changes.
4. When the sawdust starts to darken, try lighting the vapour coming out of the end of the test tube. Record your observations.
5. After heating strongly for several minutes and observing no further changes, shut off the burner.

6. When the test tube has cooled, remove it from the clamp and pour out the contents onto a piece of paper. Observe and describe the final solid product.

Analysis

(a) Interpret your Evidence, including hypotheses about possible products.

Section 11.2 Questions

Understanding Concepts

1. What is petroleum and why is it an important resource?
2. Is petroleum renewable? Provide your reasoning.
3. State and briefly describe the three main processes used in the refining of petroleum to produce gasoline.
4. Are the fractions obtained in the fractional distillation of petroleum pure substances? Provide an example to support your answer.
5. What are some useful products, other than gasoline, of the refining operation?
6. How does the size of a hydrocarbon molecule relate to its boiling point? What is the explanation for this relationship?

Making Connections

7. We depend on hydrocarbons as a fuel for a wide range of technological applications. Imagine that our supply of hydrocarbons was suddenly removed. What would be some of the consequences? List at least 10.

11.3 Combustion of Hydrocarbons

You get up in the morning to reasonably warm surroundings, have a hot shower, cook breakfast, and catch a bus to school. All of these activities directly or indirectly are made possible by the combustion of fossil fuels. Less obvious to us is our dependence on the combustion of fossil fuels for the production of most things that we possess: our clothes, shelter, books, and entertainment. Our way of life depends on the combustion of hydrocarbons. If the sources of hydrocarbons were suddenly, or even gradually, cut off (as they eventually will be), our way of life would dramatically change.

Complete Combustion of Hydrocarbons

Combustion is a very common hydrocarbon reaction, the one that is most familiar to us. In ideal or **complete combustion**, a hydrocarbon reacts with oxygen to produce carbon dioxide gas and water vapour. All hydrocarbons burn in air provided they are warm enough to vaporize. If the vaporization temperature (boiling point) is low and sufficient oxygen is available, the hydrocarbon burns so rapidly it forms an explosive mixture with air. Ninety-five percent of petroleum ends up being used as fuels in combustion reactions to produce energy. For example,

complete combustion: the reaction of an element or compound with oxygen to produce the most common oxides, for example, carbon dioxide, sulfur dioxide, nitrogen dioxide, and water

$$2\ C_8H_{18(l)} + 25\ O_{2(g)} \rightarrow 16\ CO_{2(g)} + 18\ H_2O_{(g)} + \text{energy}$$

DID YOU **KNOW?**

Cold Vapour

The white trails behind jet aircraft are formed from condensed water vapour generated by the combustion of jet fuel. These are not unlike the vapour trails left by automobiles on a cold day. Sometimes vapour trails will abruptly end as the jet leaves a cold air mass and enters a warm air mass.

Recall that in hydrocarbon combustion reactions, the physical states of reactants are given for SATP conditions (before any reaction), ignoring the fact that a hydrocarbon must be vaporized to begin the reaction. The water produced is always shown as water vapour, as it normally remains gaseous even after cooling.

So much carbon dioxide is produced by burning fossil fuels that human activities are measurably increasing the concentration of this gas in our atmosphere. Many scientists are concerned that this may be contributing to an increase in the average temperature of Earth. Of course, all the energy produced by hydrocarbon combustion eventually transfers to the atmosphere. For example, most Canadians know that winter temperatures are always several degrees higher in and around large cities than they are in rural areas, due to emitted heat from buildings, people, and vehicles.

The exploitation of fossil fuels has its consequences. On the one hand, inexpensive fossil fuels have contributed to our high standard of living. On the other hand, we may be paying dearly for this good fortune. Environmental problems such as global warming, rising costs of scarce resources, and shortages of raw materials for the petrochemical industry are some of the disadvantages of our dependency.

Practice

Understanding Concepts

1. List a major use and some minor uses of hydrocarbons.
2. Write the reaction equation showing the complete combustion in a jet engine of tetradecane, $C_{14}H_{30(l)}$, a hydrocarbon substance found in jet fuel.
3. When the combustion reaction for one mole of a hydrocarbon is balanced with whole numbers, four moles of carbon dioxide and five moles of water vapour have formed. Write the molecular formula for this hydrocarbon substance.
4. State some risks and benefits of our reliance on the combustion of fossil fuels.

Making Connections

5. Airplanes used by major airlines carry a large load of jet fuel. In the event of a crash, especially on takeoff, airports need to be equipped to deal with a jet fuel fire. Research the Internet to answer the following questions.
   (a) What substances are used to fight jet fuel fires outside of the airplane?
   (b) Fire spreading inside the aircraft is another major concern of firefighters. What is done if this occurs?
   (c) Some consumer groups are concerned about the availability of suitable firefighting equipment and personnel trained to use it at Canadian airports. Does the airport near your home or one that you have used have firefighting equipment?

   Follow the links for Nelson Chemistry 11, 11.3.

GO TO **www.science.nelson.com**

Reflecting

6. How would your life be different without hydrocarbons? What would you notice first, and what are some longer-term effects?

Global Warming

The chemical products of the combustion of hydrocarbons can be variable. With sufficient oxygen present, the combustion of hydrocarbons such as methane produces carbon dioxide and water vapour:

$$CH_{4(g)} + 2\ O_{2(g)} \rightarrow CO_{2(g)} + 2\ H_2O_{(g)}$$

Carbon dioxide and water vapour are the two most important greenhouse gases. These gases absorb heat radiation (long wavelength infrared) from Earth and emit some of this radiation back to Earth (**Figure 1**). The term *greenhouse effect* refers to the trapping of the thermal energy in the atmosphere, producing the same effect as the glass or plastic panels of a greenhouse. The greenhouse effect is an important natural process that has made Earth habitable. Without this process, Earth would be cooler, and life, if there were any, would be confined to the warm rock beneath the surface. An important current issue is the concern about an enhanced greenhouse effect in which additional carbon dioxide and other greenhouse gases, such as methane, accumulate in the atmosphere and increase the heating effect. The resulting increase in average global temperature, called **global warming**, is considered by many scientists to be the most crucial environmental problem in the world today.

global warming: the increase in the average temperature of Earth's atmosphere

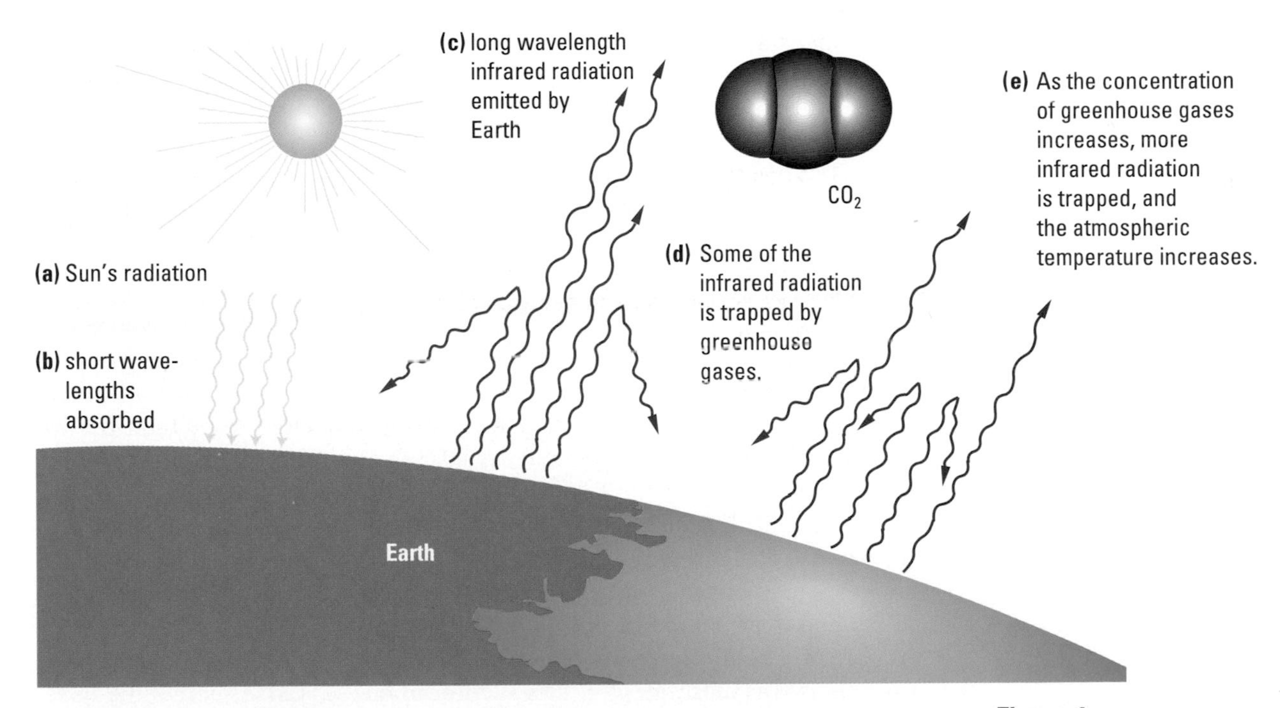

Figure 1
Earth's surface absorbs sunlight, then heats up and radiates longer wavelength infrared radiation into the atmosphere. Greenhouses gases, such as carbon dioxide, absorb this longer wavelength radiation, thereby increasing the temperature of the atmosphere.

Although scientists do not completely understand the cycle of carbon dioxide on a global scale, there is little doubt that the concentration of carbon dioxide in the atmosphere is steadily increasing (**Figure 2**, page 522). This increase in $CO_{2(g)}$ concentration is largely attributed to the combustion of fossil fuels: coal, oil, and natural gas. Can an increasing concentration of $CO_{2(g)}$ be absorbed in the biosphere? If not, what effect will this have on the global temperature? Temperature increases of only a few degrees can cause a melting of polar ice caps, a rise in sea levels, and significant climate changes. The lack of complete understanding of the natural carbon cycle means that we cannot be very certain about any predictions about global warming. Many people, including some scientists, believe that the threat of the greenhouse effect is minimal. Earth's temperature has fluctuated in the past, for example, during the ice ages. It is possible that, independent of human interference, the temperature of Earth is increasing naturally. It is also possible Earth is in the midst of a cooling trend and that the human-generated greenhouse effect is preventing another ice age. Models of the atmosphere are complex but inadequate for making precise predictions. The capacity of the oceans to absorb higher levels of carbon dioxide is not known, and the effect of an increased concentration of atmospheric carbon dioxide on plant growth is not predictable.

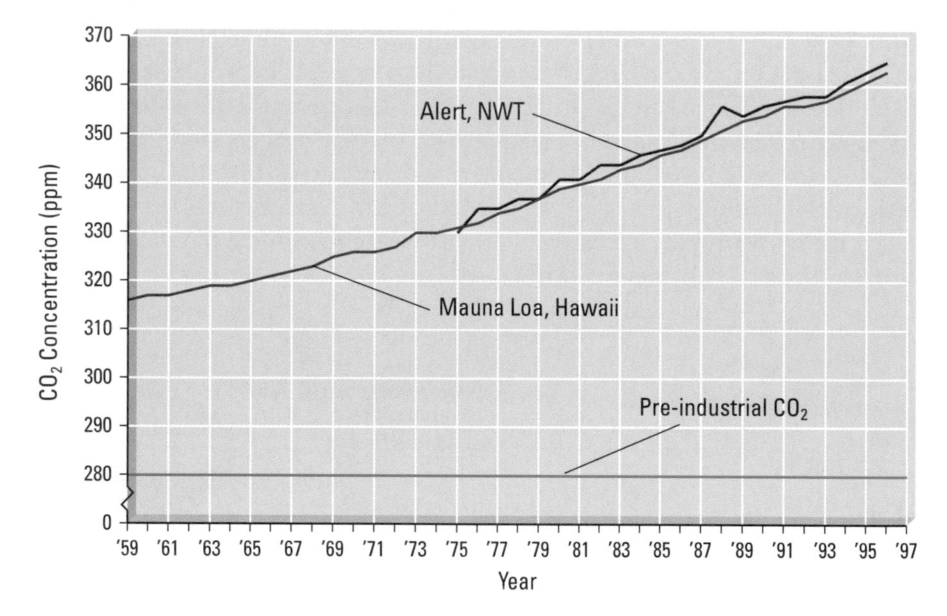

Figure 2
Annual concentrations of CO_2, Mauna Loa, Hawaii (1959–1996), and Alert, Northwest Territories (1975–1996)

Source: Carbon Dioxide Information Analysis Center, Environment Canada

Although the rate and extent of global warming are difficult to predict, it seems reasonable to reduce the production of greenhouse gases to avoid upsetting the delicate balance of the biosphere. We can create technologies to switch from high-carbon fuels to low-carbon fuels and use conventional fuels more efficiently. We can also practise energy conservation and exploit energy sources that do not produce carbon dioxide, such as solar energy, wind power, fuel cells, and photovoltaic cells.

Canada, with about 0.5% of the world's population, produces about 2% of the world's total greenhouse gas emissions. In 1997, Canada was one of over 160 countries to adopt the Kyoto Protocol, which commits industrialized countries to binding targets for reduction in greenhouse gas emissions. The emissions of six gases will be affected: carbon dioxide, methane, nitrous oxide, and three halocarbons used as substitutes for chlorofluorocarbons (CFCs). As part of the agreement, Canada has committed to reducing its 1990 level of greenhouse gas emissions by 6% by the year 2012. Countries that do not meet their own emission targets can strike deals with nations that do better than required and buy their excess "quota," a process known as "carbon trading." In 1997, Canada emitted about 682 Mt of carbon dioxide, making us the eighth largest greenhouse gas emitter in the world, and the fourth largest on a per capita basis. Several factors contribute to Canada's greenhouse gas emissions:

- Canada's population is distributed over an enormous area, creating significant transportation needs for people and goods.
- Canada's northern climate places a heavy demand on energy consumption.
- A large part of Canada's economy is resource based, which means a high demand for energy, for example, in smelting ore.

If Canada continues with "business as usual," it will fall far short of its target. Canada will need innovative technological solutions to slow the trend of climate change and to adapt to changes we cannot avoid. Equally important are changes in the attitudes that lead us to squander energy resources.

Practice

Understanding Concepts

7. Is the greenhouse effect useful or harmful? Explain briefly.
8. How is carbon dioxide recycled in the biosphere?
9. How might Canada be affected if global warming effects become significant?
10. Explain how a nation could benefit from "carbon trading."
11. Coal (assume pure carbon) and natural gas (assume pure methane) are two fossil fuels widely used in electric power-generating stations.
    (a) Write the chemical equation for the complete combustion of natural gas.
    (b) Calculate the mass of carbon dioxide produced for 1.00 g of methane consumed.
    (c) Write the chemical equation for the complete combustion of coal.
    (d) To obtain the same energy as 1.00 g of natural gas, you need about 2.00 g of coal. Calculate the mass of carbon dioxide produced for 2.00 g of carbon consumed.
    (e) Which fuel, coal or natural gas, is better in terms of carbon dioxide produced? Justify your answer.
    (f) Suggest a reason why coal is still widely used in electric power-generating stations, even in places where natural gas is available.

Answers

11. (b) 2.74 g
    (d) 7.33 g

Making Connections

12. The Kyoto Protocol takes effect during the period 2008–2012. One way in which countries like the United States and Canada can meet their targets is the use of "emissions trading." What does this term mean and how does this trading work?

    Follow the links for Nelson Chemistry 11, 11.3.

GO TO www.science.nelson.com

Explore an Issue

Take a Stand: How Can Technology Help Canadians Reduce CO_2 Gas Emissions?

As Canada strives to meet its target under the Kyoto Protocol, various technologies are being considered to decrease our emission of greenhouse gases. Technological fixes are not the only answer; changes in lifestyle can have a huge impact on how efficiently we use energy. Working in small groups, research one of the technologies being

DECISION-MAKING SKILLS

- ● Define the Issue
- ● Identify Alternatives
- ● Research
- ○ Analyze the Issue
- ● Defend a Decision
- ● Evaluate

recommended as a means of reducing the production of greenhouse gases, for example:

- more fuel-efficient cars
- high-efficiency natural gas furnaces
- improved insulation in buildings
- improved mass transportation
- home energy-saving technologies (e.g., set-back thermostats)

(a) Use library and Internet resources to collect information about one of the technologies.

Follow the links from Nelson Chemistry 11, 11.3.

GO TO **www.science.nelson.com**

(b) Before people will change their lifestyles or switch to a new technology, they have to be convinced that there is a benefit. Identify a target group and prepare an advertisement (poster, video, audio, or Web ad) designed to promote the technology you researched.

(c) In a separate report, identify your target group, explain why you think your ad will be effective, and summarize any risks and benefits of the technology that do not appear in the ad.

INQUIRY SKILLS

- ○ Questioning
- ○ Hypothesizing
- ○ Predicting
- ○ Planning
- ● Conducting
- ● Recording
- ● Analyzing
- ○ Evaluating
- ● Communicating

Investigation 11.3.1

Combustion of a Hydrocarbon

All hydrocarbons burn, but the characteristics of the combustion depend on many variables. The purpose of this investigation is to determine some characteristics of the combustion of hydrocarbons. Complete the **Analysis** section of the lab report.

Question

What are the characteristics of the combustion of paraffin wax and methane?

Experimental Design

Two hydrocarbons—paraffin wax, $C_{25}H_{52(s)}$, and methane—are burned. Observations of the burning are made to determine the characteristics of the combustion.

Materials

eye protection
apron
paraffin candle on dish or tin lid
matches or lighter
large beaker (e.g., 1000 mL)
2 smaller beakers (e.g., 400 mL)
timer or stopwatch
index card
bent glass tubing
laboratory burner (connected to natural gas supply)
beaker tongs
paper towels

Procedure

Part 1: Paraffin Candle

Long hair should be tied back and loose sleeves should be contained by sleeve protectors or a lab coat.

Be careful when handling hot glassware.

Wash and dry the glass tubing before use. Do not share with others.

1. Light the candle in an area with little or no draft.
2. Record as many observations as possible about the candle flame.
3. Quickly move an index card into the middle of the candle flame and remove it before it ignites. Record observations of the underside of the card.
4. Invert the large beaker over the candle and determine the time the candle continues to burn.
5. Examine the inside of the beaker and note any deposits.
6. Repeat steps 4 and 5 using the smaller beaker.
7. Remove the beaker and relight the candle. Place the short end of the glass tubing close to the wick and gently blow air up into the candle flame (**Figure 3**). Note any changes in the flame. Relight the candle if it goes out and repeat as many times as necessary to be sure of the changes observed.
8. Place the short end of the glass tubing in the dark part of the flame just above the wick (**Figure 4**). Hold it there for several minutes and describe the vapour coming out the other end of the tubing.
9. When it looks like the maximum quantity of vapour is coming out of the glass tubing, try lighting the vapour and observe. You may have to try this several times.
10. Move the end of the tubing that is in the flame up near the tip of the bright yellow region of the flame. Observe the appearance of the vapour, and again try lighting it.
11. Remove the glass tubing from the flame and let it cool for a few minutes. Record your observations of the outside and inside of the tube.

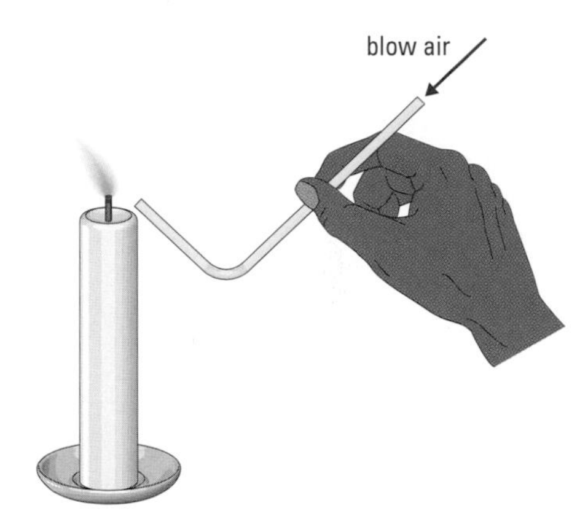

Figure 3
With a little practice, you should be able to gently blow air into the candle flame without extinguishing the flame.

Part 2: Laboratory Burner

12. Connect the lab burner to the natural gas supply and make sure that the air vents are closed.
13. Light the burner and record your observations of the flame.
14. Clean the smaller beaker with paper towel and fill it to the half to three-quarters level with cold water.
15. Using beaker tongs, hold the beaker over the flame. Remove the beaker and observe and record any deposits on the outside of the beaker.
16. Slowly open the air vents on the burner and record changes in the flame.
17. With the burner adjusted to obtain the usual pale blue flame, repeat steps 14 and 15.

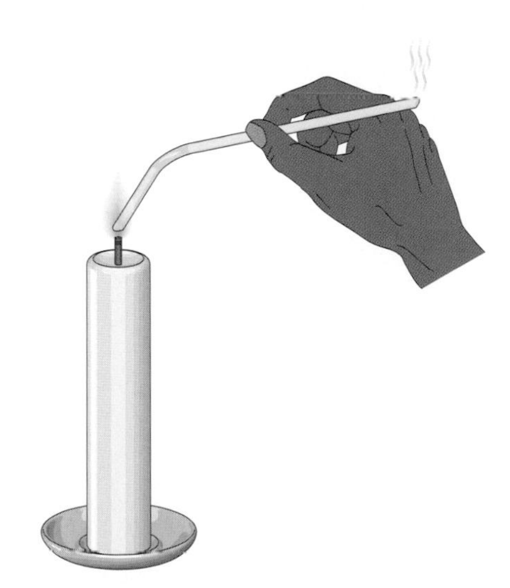

Figure 4
Hold the glass tubing by the end, away from the candle, and keep the short end in the darkest region of the flame, just above the wick.

Analysis

(a) Describe the characteristics of the candle flame, including colour and temperature of different parts of the flame.

(b) What interpretation can be made about the times to extinguish the candle in steps 4 and 6?

(c) For a normal yellow candle flame, what is one obvious product of the combustion? State any Evidence for other products.

(d) When extra air is supplied to the candle flame (step 7), the colour of the flame changes. Provide an explanation.

(e) Based on your Evidence from steps 8 to 11, what is actually burning in a candle flame?
(f) How does the combustion in a laboratory burner compare with that in a candle?
(g) In one or two sentences, summarize the combustion characteristics of the paraffin wax in a candle.
(h) In one or two sentences, summarize the combustion characteristics of natural gas in a laboratory burner.

Incomplete Combustion of Hydrocarbons

incomplete combustion: the reaction of an element or compound with oxygen to produce some oxides with less oxygen content than the most common oxides, for example, carbon monoxide and nitrogen monoxide

Ideally, the combustion of hydrocarbons produces only carbon dioxide and water. If, however, there is insufficient oxygen available, the combustion of hydrocarbons can also produce carbon monoxide and/or carbon. This is called **incomplete combustion.**

The larger the hydrocarbon molecule, the more likely there is to be incomplete combustion. This is why incomplete combustion of gasoline (e.g., octane) is more of a technological and environmental problem than for natural gas (methane). In each of the following chemical equations, the amount of oxygen reacting with one mole of octane is different. Note that less oxygen means less carbon dioxide and more carbon monoxide or carbon produced.

Complete combustion:

$$C_8H_{18(l)} + \frac{25}{2} O_{2(g)} \rightarrow 8\ CO_{2(g)} + 9\ H_2O_{(g)}$$

Some incomplete combustions:

$$C_8H_{18(l)} + 12\ O_{2(g)} \rightarrow 7\ CO_{2(g)} + CO_{(g)} + 9\ H_2O_{(g)}$$

$$C_8H_{18(l)} + \frac{23}{2} O_{2(g)} \rightarrow 6\ CO_{2(g)} + 2\ CO_{(g)} + 9\ H_2O_{(g)}$$

$$C_8H_{18(l)} + 5\ O_{2(g)} \rightarrow CO_{(g)} + 7\ C_{(s)} + 9\ H_2O_{(g)}$$

Incomplete combustion of a fuel such as octane is undesirable for several reasons. It decreases the fuel economy of the automobile, because less energy is produced, and the spark plugs may become fouled with carbon. Incomplete combustion also produces toxic carbon monoxide. Let's review the properties of carbon monoxide: It is a colourless, odourless gas that is usually found in concentrations of a few parts per million in indoor and outdoor air. Parking garages may have levels in the range of 10–20 ppm. Up to 30 ppm is considered a safe level of continuous exposure. Automobile emissions in large cities during rush hours cause health problems. If there is no wind and/or there is a temperature inversion, then headaches (a symptom of mild carbon monoxide poisoning) become common.

Automobile emissions of carbon monoxide have been reduced in newer models; however, older models that are not kept tuned emit carbon monoxide at a toxic level. Parking in a car with the motor running for extended periods of time can result in carbon monoxide poisoning if there are leaks in the exhaust system. Faulty exhaust systems have been the cause of deaths of children and other passengers in the back seat.

Carbon monoxide poisoning can happen in our homes too: Backdrafts down the furnace chimney when you are using a fireplace can result in toxic levels of carbon monoxide in a house. Using a briquette barbecue inside the home can also result in toxic levels of carbon monoxide. Emergency space-heating of homes with kerosene heaters can also be lethal. Many homes now have carbon monoxide detectors. Outside the home, every year there are reports of campers dying after using barbecues to heat their tents.

The symptoms of carbon monoxide poisoning include headache, fatigue, and unconsciousness. Be aware of situations where incomplete combustion may occur and watch for the symptoms. Some chemical knowledge could save your life or the lives of your family and friends.

Practice

Understanding Concepts

13. If you were looking for evidence of incomplete combustion, describe what you would expect to find while examining the flame of a furnace, the exhaust pipe of a car, and emissions from a smokestack.
14. How do the products of complete combustion and incomplete combustion differ?
15. Without additional information, why is it difficult to write a balanced chemical equation for the incomplete combustion of a hydrocarbon?
16. Heptane is a component of gasoline. Write a balanced chemical equation for the
    (a) complete combustion of heptane;
    (b) incomplete combustion of heptane, assuming equal amounts of the two carbon oxides as the only carbon-containing products.
17. If the concentration of carbon monoxide from an untuned car engine is 15 ppt, what is the percentage (parts per hundred) of carbon monoxide in the gases at the exhaust pipe?

Answer

17. 1.5×10^{-6}%

Making Connections

18. In some Canadian cities a bylaw requires homes to be equipped with carbon monoxide detectors (**Figure 5**). How do these detectors work? Do they have to be checked periodically? In a brief report, explain whether you would be in favour or against such a bylaw if you owned a home.

    Follow the links for Nelson Chemistry 11, 11.3.

    GO TO **www.science.nelson.com**

19. A carbon tax has been suggested as a way of reducing carbon dioxide emissions. Its proponents believe that people who pay more for a resource will be more likely to conserve that resource. If a carbon tax were added differentially to various fuels, in your opinion, which hydrocarbons should be taxed the most?

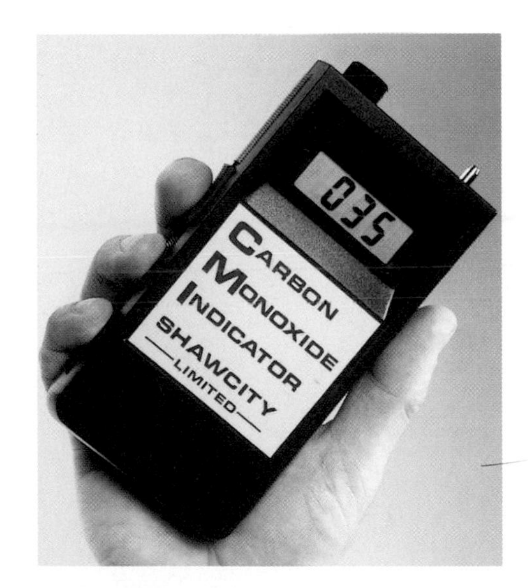

Figure 5
A typical home carbon monoxide detector

Reflecting

20. List as many ways as you can think of for use of the word "burn" in our daily lives. Do all of these uses convey the same meaning as the term "combustion"?

Section 11.3 Questions

Understanding Concepts

1. What are some of the consequences of incomplete combustion? Write a sentence each from scientific, technological, economic, and environmental perspectives.
2. Write the complete and one possible incomplete combustion equation for the following reactions:
   (a) butane in a barbecue lighter
   (b) nonane in gasoline
   (c) $C_{24}H_{50(s)}$ in candle wax

Applying Inquiry Skills

3. Complete the Analysis and Evaluation sections of the following report.

   Question

   What are the products of combustion of natural gas?

   Evidence

   A beaker of cold water moved through the flame condenses a clear, colourless liquid, which when tested with cobalt chloride paper turns the paper from blue to pink. No black deposit is observed.

   An Erlenmeyer flask inverted over the flame is then stoppered. When some limewater is added to the flask and shaken, the limewater turns cloudy.

   Analysis

   (a) According to the Evidence collected, what is the answer to the Question?

   Evaluation

   (b) Evaluate the Evidence collected. For example, is it possible to determine whether the combustion was complete?
   (c) Suggest an improvement to this experiment to be more certain about the type of combustion.

11.4 Alkanes and Cycloalkanes

Why is octane rating based on a chemical called 2,2,4-trimethylpentane when octane is just a simple string of eight carbon atoms? Why is the chemical in a butane lighter called methylpropane or isobutane rather than just butane? What is the proliferation of names all about? The answer to these questions is that, so far, your study of hydrocarbons has been restricted to an emphasis on the first 10 straight-chain alkanes: methane through decane. In order to look at the real impact of hydrocarbons on our everyday life, we now look at branched and cyclic alkanes. However, the restriction to aliphatic hydrocarbons remains; aromatic hydrocarbons other than the parent aromatic, benzene, are left for your next course.

An analysis of hydrocarbon samples from crude oil and natural gas reveals evidence that needs to be explained. Careful fractional distillation of the samples in an organic chemistry laboratory reveals precise fractions that have identical empirical formulas but different physical and chemical properties. For example, two fractions with the same empirical formula of $C_4H_{10(g)}$ have different boiling points: –0.50°C and –11.7°C. To explain this behaviour for $C_4H_{10(g)}$, chemists created the concept of isomers. Isomers are chemicals with the same molecular formula but with different structural diagrams; in this text, you will study **structural isomers** (Figure 1) and geometric isomers. (Geometric isomers will be discussed later in this chapter.) Before we study this area further, we will see how these hydrocarbons are identified, both by name and by their structures.

structural isomers: chemicals with the same molecular formula, but with different structures and different names

alkyl group: a hydrocarbon group derived from an alkane by the removal of a hydrogen atom; often a substitution group or branch of an organic molecule

H H H H
| | | |
$H — {}_1C — {}_2C — {}_3C — {}_4C — H$
| | | |
H H H H

butane

H CH_3 H
| | |
$H — {}_1C — {}_2C — {}_3C — H$
| | |
H H H

isobutane or 2-methylpropane

Figure 1
Butane and 2-methylpropane have the same molecular formula but different structural diagrams. They are structural isomers.

Naming Straight-Chain Alkanes

Since an important way to differentiate between isomers is to write names for the structural formulas, it is important to be able to name straight- and branch-chain alkanes correctly. IUPAC has created a series of internationally accepted rules for naming alkanes.

The first syllable in the name of an alkane is a prefix that indicates the number of carbon atoms in the longest chain in the molecule (**Figure 2**). The prefixes shown in **Table 1** are used in naming all organic compounds. The same prefixes identify groups of atoms that form branches on the chains of organic molecules.

Table 1: The Alkane Family of Organic Compounds

IUPAC name	Carbon atoms	Molecular formula
methane	1	$CH_{4(g)}$
ethane	2	$C_2H_{6(g)}$
propane	3	$C_3H_{8(g)}$
butane	4	$C_4H_{10(g)}$
pentane	5	$C_5H_{12(l)}$
hexane	6	$C_6H_{14(l)}$
heptane	7	$C_7H_{16(l)}$
octane	8	$C_8H_{18(l)}$
nonane	9	$C_9H_{20(l)}$
decane	10	$C_{10}H_{22(l)}$
–ane	n	C_nH_{2n+2}

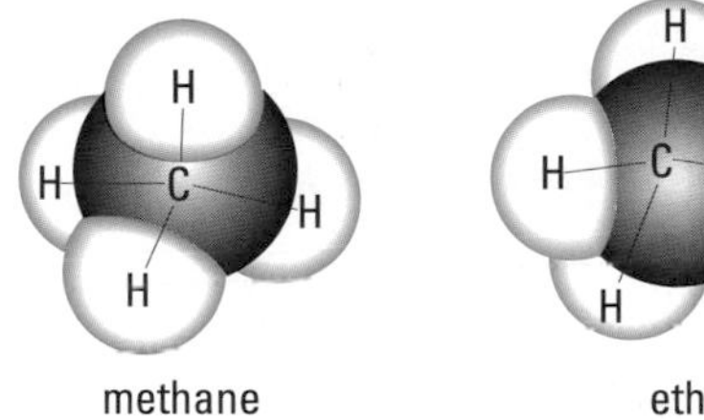

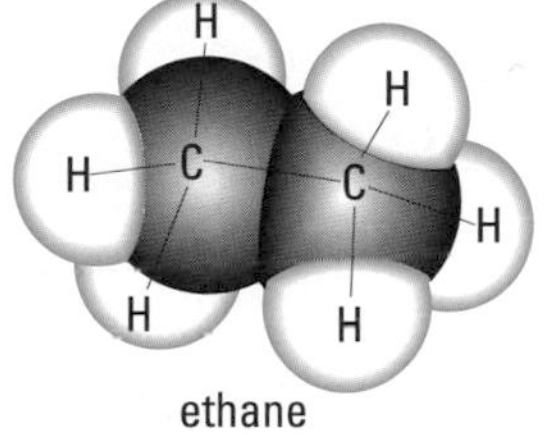

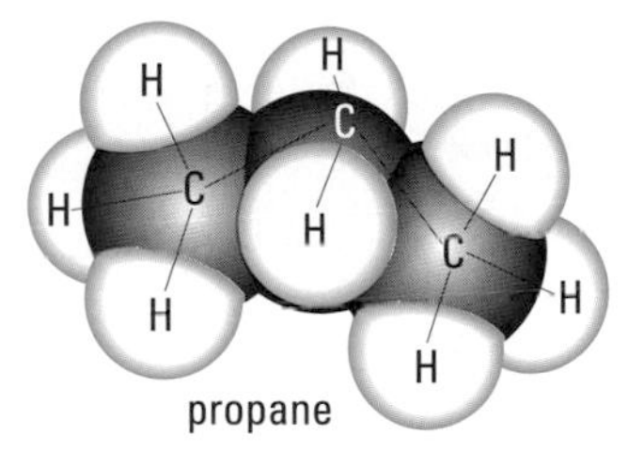

Figure 2
The prefix *meth* indicates one carbon atom, *eth* signifies two carbon atoms, and *prop* signifies three carbon atoms. The suffix *ane* indicates a chain of carbon atoms with single bonds only.

A branch is any group of atoms that is not part of the longest continuous carbon chain. A hydrocarbon branch is called an **alkyl group.** In the names of alkyl groups, the prefixes that indicate the number of carbon atoms in the branch group are followed by the suffix *-yl* (**Table 2**). It is very useful to learn these prefixes for the first 10 alkanes in the alkane series because they occur frequently in organic nomenclature.

After the removal of a hydrogen atom, the general formula for an alkane, C_nH_{2n+2}, becomes C_nH_{2n+1} in the general formula for alkyl groups. For example, the alkane methane, CH_4, becomes the alkyl group methyl, $–CH_3$.

Table 2: Examples of Alkyl Groups

Group	Name
$–CH_3$	methyl
$–CH_2CH_3$ ($–C_2H_5$)	ethyl
$–CH_2CH_2CH_3$ ($–C_3H_7$)	propyl

Practice

Understanding Concepts

1. Write molecular formulas for alkanes where the molecules have
   (a) 11 carbon atoms
   (b) 15 carbon atoms
   (c) 22 carbon atoms
   (d) 77 carbon atoms
2. Which of the following straight-chain molecules could be an alkane?
   (a) $C_{12}H_{22}$
   (b) $C_{12}H_{26}$
   (c) C_6H_{10}
   (d) C_2H_6O
3. Write the alkyl group chemical formula and name for
   (a) three carbon atoms
   (b) four carbon atoms
4. What is the molecular formula for an alkane with a molar mass of 72.17 g/mol?

Naming Branch-Chain Alkanes

When there are side branches on a carbon atom chain, the name of the compound changes to indicate this. For example, consider the three isomers of C_5H_{12} shown in **Figure 3**. The unbranched isomer in **Figure 3(a)** is named pentane. The numbers on the following structural diagram show how the carbon atoms are identified.

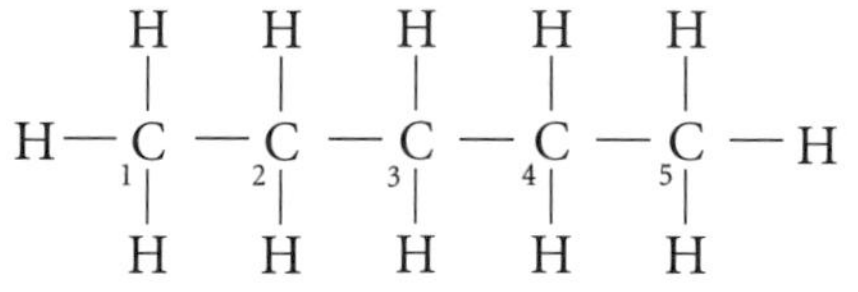

```
                                         H
                                         |
                                     H — C1 — H
    H    H    H    H    H                |
    |    |    |    |    |                |    H    H
H — C1 — C2 — C3 — C4 — C5 — H       H — C2 — C3 — C4 — H
    |    |    |    |    |                |    |    |
    H    H    H    H    H                H    H    |
                                              H — C5 — H
           pentane                                |
                                                  H
```

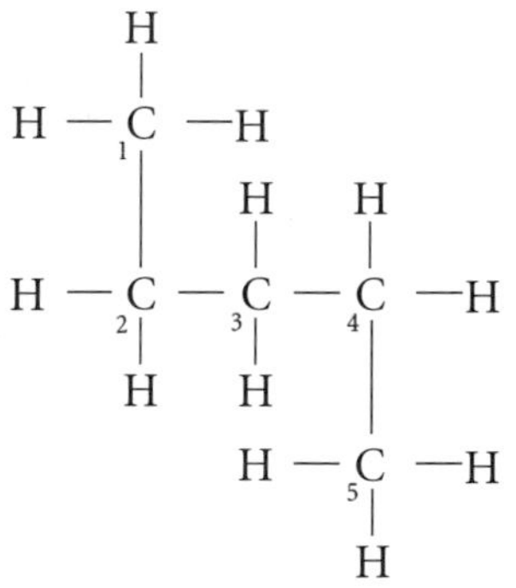

Figure 3(a)
Both structural diagrams show the same molecule, pentane, an isomer of $C_5H_{12(l)}$.

In the second isomer in **Figure 3(b)**, there is a continuous chain of four carbon atoms with a methyl group branching from the second carbon atom. To name this structure, identify and name the parent chain, the longest continuous

Figure 3(b)
Again, all three structural diagrams show the same molecule, 2-methylbutane, an isomer of $C_5H_{12(l)}$.

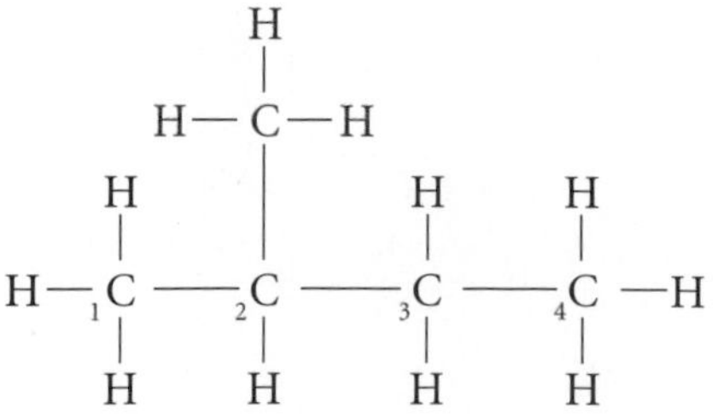

```
          H                                  H                   H    H    H    H
          |                                  |                   |    |    |    |
      H — C — H                          H — C — H           H — C4 — C3 — C2 — C1 — H
 H        |     H    H              H    H   |    H              |    |    |    |
 |        |     |    |              |    |   |    |              H    H    |    H
H — C1 — C2 — C3 — C4 — H       H — C4 — C3 — C2 — C1 — H                H — C — H
 |        |     |    |              |    |   |    |                         |
 H        H     H    H              H    H   H    H                         H

   2-methylbutane                   2-methylbutane                  2-methylbutane
```

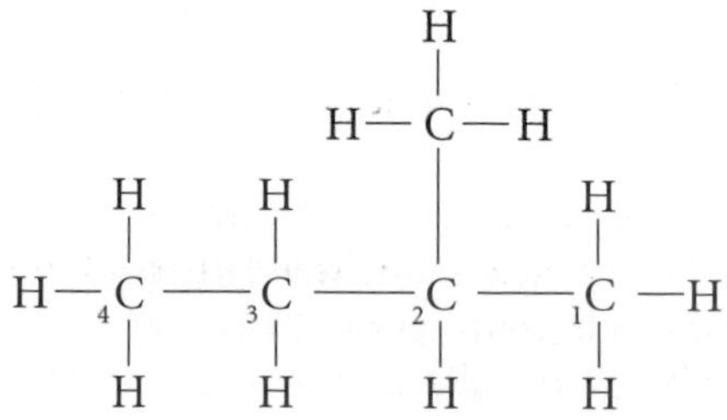

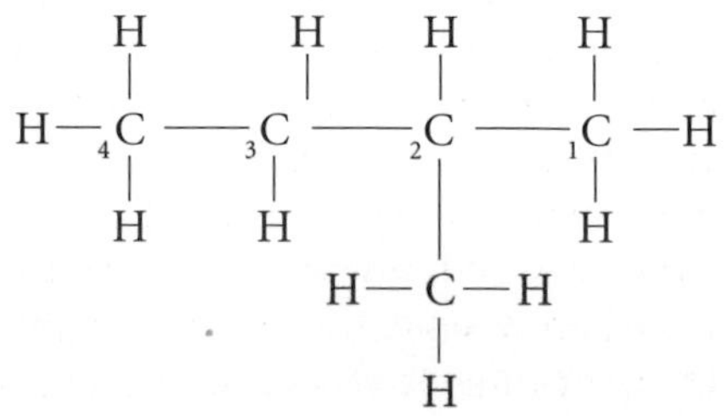

chain of carbon atoms. Here, the four carbon atoms tell us that the parent chain should be called butane. The carbon atoms of a parent chain are always numbered from the end closest to the branch, so this isomer is called 2-methylbutane (a butane chain with one methyl branch on the second carbon atom of the parent chain).

In the third isomer (**Figure 3(c)**), two methyl groups are attached to a three-carbon (propane) parent chain. The methyl groups are both attached to the second carbon atom of the parent chain. Also note that every side branch must have a separate number to identify its location. This third C_5H_{12} isomer is named 2,2-dimethylpropane (the *di* means that there are two methyl branches at the locations given by the numbers). Note the convention for the use of a comma, hyphen, numeral, and single (combined) final word when naming branched alkanes. Commas are used between numbers; hyphens are used between numbers and words; numbers are Arabic; and no spaces are used between the names for prefixes, alkyl groups, and parent chain.

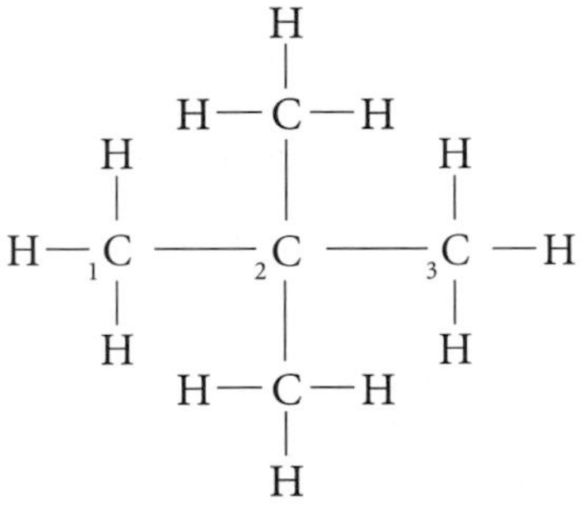

```
            H
            |
        H — C — H
   H        |        H
   |        |        |
H — C₁ ——— C₂ ——— C₃ — H
   |        |        |
   H        |        H
        H — C — H
            |
            H
```

2,2-dimethylpropane

Figure 3(c)
The third isomer of $C_5H_{12(l)}$, 2,2-dimethlypropane.

To name a branched hydrocarbon, we can organize, in steps, the guidelines outlined in the above text. For example, write the IUPAC name corresponding to the following structural diagram:

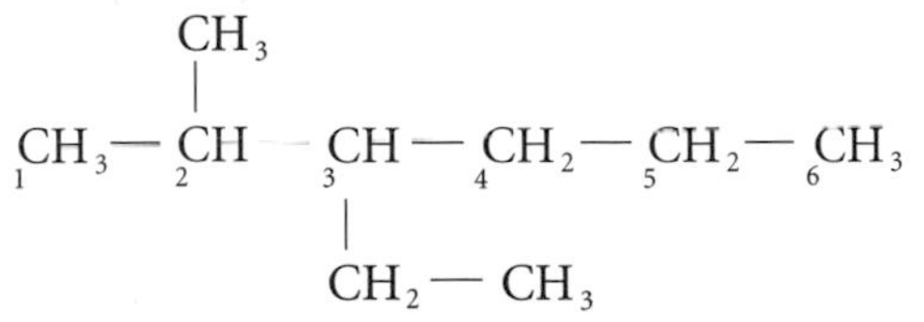

```
        CH₃
        |
CH₃ — CH — CH — CH₂ — CH₂ — CH₃
 1     2    3     4     5     6
            |
            CH₂ — CH₃
```

Step 1: The longest continuous chain has six carbon atoms. Therefore, the name of the parent chain is hexane.
Step 2: Note which end of the parent chain is closest to the first branch. There is a methyl branch at the second carbon atom and an ethyl branch at the third carbon atom of the parent chain. (Also note that a branch at either end of the chain is an extension of the parent chain.)
Step 3: With the branches named in alphabetical order, the compound is 3-ethyl-2-methylhexane.

When writing the name of a branched alkane in a correct order, two systems for listing the branches are commonly used. The branches are listed either (a) in alphabetical order of branch names (ignoring the prefixes) or (b) in order of complexity (numbers of carbon atoms in the branch). All examples and answers in this textbook follow the alphabetical order system.

Sample Problem 1

Write the IUPAC name for the chemical with the following structural diagram:

$$
\begin{array}{cccccc}
 & & & & CH_3 & \\
 & & & & | & \\
CH_3 - & CH_2 - & CH - & CH_2 - & CH - & CH_3 \\
 & & | & & & \\
 & & CH_2 - CH_3 & & &
\end{array}
$$

Solution

According to the IUPAC nomenclature rules, the name for this alkane is 4-ethyl-2-methylhexane.

The longest continuous chain must be the parent chain in an IUPAC name. However, this parent chain need not be drawn as a straight, horizontal chain. It is important that you follow the carbon–carbon bonds (without "backtracking") within a given structure. The longest, continuous chain of carbon atoms can change directions within a structure. The following sample problem illustrates this point.

Sample Problem 2

Write the IUPAC name for the alkane with the following structural diagram:

$$
\begin{array}{ccccc}
 & CH_2 - CH_3 & & & \\
 & | & & & \\
CH_3 - & C - & CH_2 - & CH - & CH_3 \\
 & | & & | & \\
 & CH_3 & & CH_2 - CH_3 &
\end{array}
$$

Solution

According to the IUPAC nomenclature rules, the name for this alkane is 3,3,5-trimethylheptane.

SUMMARY **Naming Alkane Structures**

1. Identify the longest continuous chain of carbon atoms—the parent chain—in the structural diagram. Number the carbon atoms in the parent chain, always starting from the end closest to any branch(es), so the name will have the lowest possible branch numbers.
   - Remember: A branch at either end of the chain is not a branch; it is an extension of the parent chain.
   - If both ends of the parent chain are equidistant from a branch, choose the end that gives the lowest set of numbers; for example, 2,2,4 is lower than 2,4,4.
2. Identify any branches and their location number on the parent chain.
   - Each branch must have a location even if the locations of some branches are the same.

3. Write the complete IUPAC name, following this format:

(location)-(branch name)(parent chain)

- Branch names are listed in alphabetical order of branch names.
- Multiple occurrences of the same branch are grouped together. Use the prefixes *di, tri,* etc., in front of the branch name to indicate the number of branches.

Structural Diagrams from Alkane Names

A structural diagram can illustrate an IUPAC name. For example, 3-ethyl-2,4-dimethylpentane is a substance used as a standard for gasoline performance. This molecule has a pentane parent chain consisting of five carbon atoms joined by single covalent bonds:

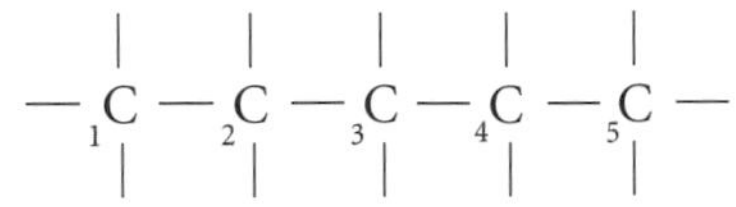

```
   |    |    |    |    |
 — C1 — C2 — C3 — C4 — C5 —
   |    |    |    |    |
```

Numbering this straight chain from left to right establishes the location of the branches. An ethyl branch is attached to the third carbon atom and two methyl branches are attached, one to each of the second and fourth carbon atoms.

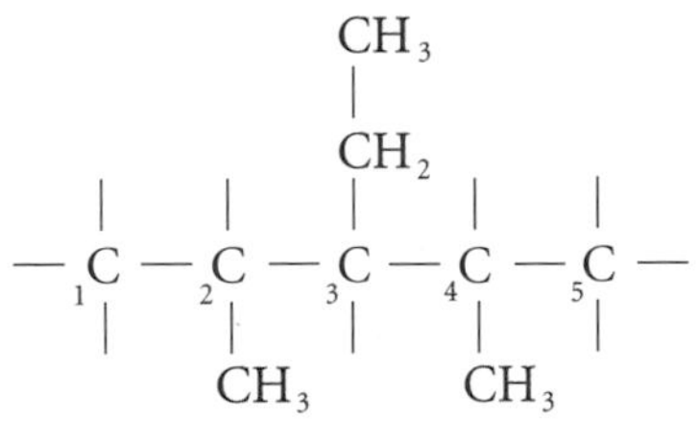

```
             CH3
             |
             CH2
   |    |    |    |    |
 — C1 — C2 — C3 — C4 — C5 —
   |    |    |    |    |
        CH3       CH3
```

In the following structural diagrams, hydrogen atoms are shown at the four bonds around each carbon atom that is left after the branches have been located.

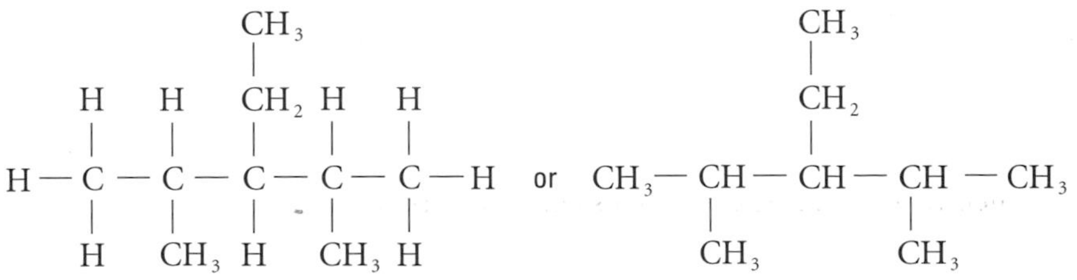

```
              CH3                                    CH3
              |                                      |
    H    H    CH2  H    H                            CH2
    |    |    |    |    |                            |
H — C —  C —  C —  C —  C — H   or   CH3 — CH —  CH —  CH — CH3
    |    |    |    |    |                  |           |
    H    CH3  H    CH3  H                  CH3         CH3
```

3-ethyl-2,4-dimethylpentane

When drawing structural diagrams, it is important to always start with the parent chain given at the end of the name. You can then add all necessary branches at the locations given in the name and finish the structure with hydrogen atoms where required to complete the bonding on each carbon atom.

Sample Problem 3

Synthetic crude is oil (bitumen) from oil sands that has been cracked into smaller molecules for transportation to a standard oil refinery. What is the structural diagram for the synthetic crude constituent 4-ethyl-2,3,5-trimethylheptane?

Solution

```
                        CH3
                         |
           CH3          CH2
            |            |
CH3 — CH — CH — CH — CH — CH2 — CH3
            |            |
           CH3          CH3
```

```
                          CH3
                           |
        CH3               CH2
         |                 |
 CH3 —  CH  —  CH  —  CH  —  CH  —  CH2 —  CH3
                |              |
               CH3            CH3
```

4-ethyl-2,3,5-trimethylheptane

SUMMARY Drawing Alkane Structural Diagrams

1. Draw a straight chain containing the number of carbon atoms represented by the name of the parent chain, and number the carbon atoms, usually from left to right.
2. Attach all branches (e.g., alkyl groups) to their numbered locations on the parent chain.
3. Add enough hydrogen atoms to show that each carbon atom has four bonds.

Practice

Understanding Concepts

5. Automotive gasoline is largely composed of alkanes that have 5 to 12 carbon atoms per molecule. Write IUPAC names for the following components of straight-run gasoline:

(a)
```
CH3—CH—CH2—CH2—CH2—CH2—CH3
    |
    CH3
```

(b)
```
CH3—CH—CH2—CH2—CH2—CH—CH3
    |                  |
    CH2  CH3           CH2  CH3
```

(c)
```
CH3—CH—CH2—CH—CH2—CH—CH3
    |       |       |
    CH3     CH3     CH3
```

(d)
```
CH3—CH—CH2—CH2—CH—CH2—CH3
    |              |
    CH2  CH3       CH2  CH3
```

6. Are octane and 2,2,4-trimethylpentane isomers? Provide your reasoning.
7. Kerosene, used as a fuel for diesel and jet engines and in portable heaters, contains mostly alkanes with 12 to 16 carbon atoms per molecule. Draw structural diagrams for the following substances found in kerosene fuel. Identify compounds that are isomers of each other.
   (a) 4-ethyl-3,5-dimethyloctane
   (b) 3,3-diethyl-2-methylnonane
   (c) 3,5-diethyl-4-propylheptane
   (d) 5-butyl-4,5,6-trimethyldecane
   (e) 3-ethyl-3,6-dimethyl-5-propylnonane
8. Each of the following names has at least two errors. Draw the structural diagram and provide the correct IUPAC name for each.
   (a) 3-dimethyl-hexane
   (b) 1,2,2-trimethyl butane
   (c) 2,4,4-tri-methylpentane

Cycloalkanes

On the evidence of empirically determined formulas and observed chemical properties, chemists have determined that organic compounds sometimes exist in the form of **alicyclic hydrocarbons**, hydrocarbons with a carbon-to-carbon bond structure forming a closed ring; **acyclic hydrocarbons** are open-chain hydrocarbons without any rings of carbon atoms. Alicyclic hydrocarbons are cyclic aliphatics in which the two ends of the molecule have joined together to form a ring. When all the carbon-to-carbon bonds in an alicyclic hydrocarbon are single, the compound is called a **cycloalkane**. The general formula for cycloalkanes, C_nH_{2n}, has two fewer hydrogen atoms than the general formula for alkanes because the end carbon atoms of an alkane are now bonded to each other in a cycloalkane. For example, cyclopropane, $C_3H_{6(g)}$, and cyclobutane, $C_4H_{8(l)}$ (**Figure 4**), are the two simplest cycloalkanes. Notice that the names of the cycloalkanes are derived by adding the prefix *cyclo* to the normal alkane name.

alicyclic hydrocarbons: hydrocarbons that have a structure based on rings of carbon atoms

acyclic hydrocarbons: open-chain hydrocarbons without any rings of carbon atoms

cycloalkane: a class of alicyclic hydrocarbon with only single bonds between carbon atoms that are formed into a ring structure

(a)

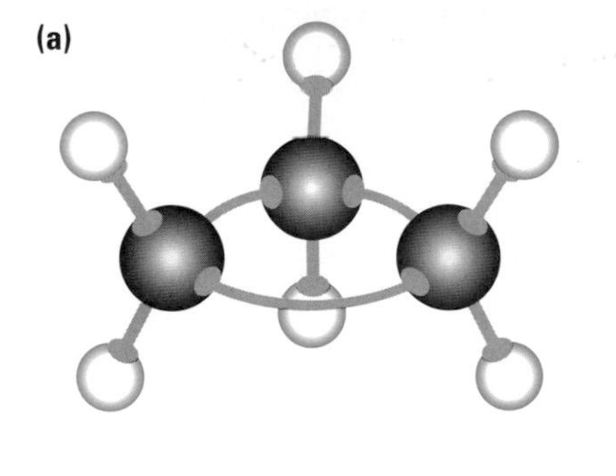

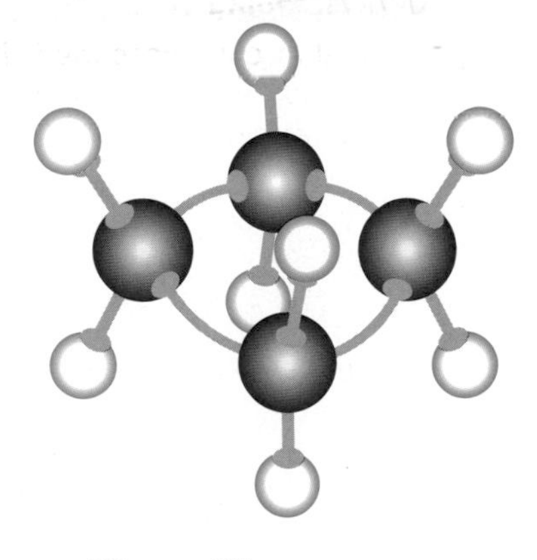

(b)

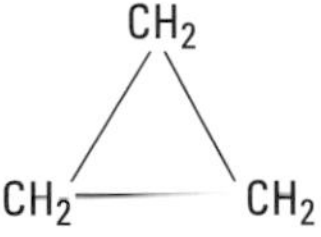

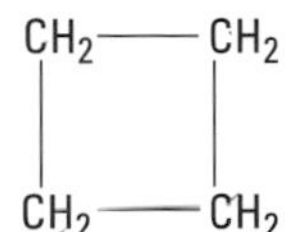

(c)

cyclopropane

cyclobutane

Figure 4
Cycloalkanes such as cyclopropane and cyclobutane are similar to alkanes, except that the two ends of the molecule's carbon chain are joined to form a ring of carbon atoms. The ball-and-spring models, **(a)**, show the approximate molecular shape and orientations of the atoms. Condensed structural diagrams, **(b)**, and the more common line structural diagrams, **(c)**, are drawn in the shape of regular polygons.

Cycloalkanes and other alicyclic hydrocarbons are usually represented by very simple line structural diagrams, drawn as regular polygons, where it is understood that every vertex represents a carbon atom and that two hydrogen atoms are bonded at every vertex. The 3-D ball-and-spring models in **Figure 4** may be similar to physical models you are using in your classroom.

Because you are dealing only with simple cycloalkanes, you should always consider the ring to be the parent chain when naming these hydrocarbons. Cycloalkanes with side branches (e.g., alkyl groups) on the parent ring are named just like open-chain alkanes, including the rule that the ring is always numbered so that the compound name will have the lowest numerals possible for the branches.

Sample Problem 4

Draw structural diagrams for 1,2-dimethylcyclopentane and 1,3-dimethylcyclohexane.

Solution

CH_3
|
$_1CH$
$_5CH_2$ $_2CH-CH_3$
$_4CH_2-{}_3CH_2$

1,2-dimethylcyclopentane

CH_3
|
$_1CH$
$_6CH_2$ $_2CH_2$
$_5CH_2$ $_3CH-CH_3$
$_4CH_2$

1,3-dimethylcyclohexane

Sample Problem 5

Write the correct IUPAC name for the cycloalkanes (a) and (b).

(a)

CH_3
|
$_1CH$
$_5CH_2$ $_2CH_2$
$_4CH_2-{}_3CH-CH_3$

(b)

CH_3
|
$_1CH$
$_6CH_2$ $_2CH_2$
$_5CH_2$ $_3CH_2$
$_4CH$
|
CH_3

Solution

The IUPAC name for (a) is 1,3-dimethylcyclopentane.
The IUPAC name for (b) is 1,4-dimethylcyclohexane.

Practice

Understanding Concepts

9. Cycloalkanes are part of the hydrocarbon mix in fossil fuels. Draw line structural diagrams and write the corresponding molecular formulas, including predicted states of matter, for
   (a) cyclopentane
   (b) cyclohexane
   (c) cyclooctane
   (d) cyclodecane
10. Using the naming rules for alkanes and what you have learned about naming branched cycloalkanes, draw structural diagrams for
   (a) methylcyclobutane
   (b) 1,1-dimethylcyclopentane
   (c) 1,2-dimethylcyclohexane
   (d) 4-ethyl-1-methylcyclohexane
11. Explain why no number is needed for the methyl branch in naming methylcyclobutane.
12. Are cyclopropane and propane isomers? Provide your reasoning.

Properties of Alkanes and Cycloalkanes

The physical properties for all alkanes and cycloalkanes are similar (**Table 3**, page 538), so the generalizations and explanations are also similar. What is said here about the physical properties of alkanes applies to cycloalkanes as well. Initially, when studying the physical properties of alkanes, chemists restricted their studies to creating generalizations relating the molecular formulas of alkane to their physical properties, such as state of matter, solubility, melting point, boiling point, and density. For example, as the size of the chemical formula increases,

- the state of matter at SATP moves from gas to liquid to solid
- the melting point increases
- the boiling point increases
- the density increases

The density of the condensed phases (solid and liquid) increases with the size of the chemical formula but never beyond about 0.8 g/mL. So all alkanes are less dense than water and float on water.

Chemists also discovered very early in their research that alkanes are insoluble (immiscible) in water but are soluble in benzene, $C_6H_{6(l)}$, and carbon tetrachloride, $CCl_{4(l)}$.

Later, the physical properties were explained by theories of intermolecular bonding. For example, chemists now explain that alkanes are composed of nonpolar molecules and, therefore, do not dissolve in water, which is composed of polar molecules. However, alkanes dissolve in benzene, $C_6H_{6(l)}$, and carbon tetrachloride, $CCl_{4(l)}$, because their molecules are all nonpolar. (Recall that polar dissolves in polar and nonpolar dissolves in nonpolar.) When used as solvents themselves, alkanes dissolve nonpolar substances but not polar or ionic substances.

The trends in states of matter and the melting and boiling points are explained by London forces between molecules. The larger the molecule, the larger the number of electrons and, therefore, the larger the London force between molecules. Variations in the melting and boiling points of isomers (which have the same number of electrons and same London forces) are explained by the effect of the shape of the molecule on the close packing of the

DID YOU KNOW?

Dangerous Science

Researchers who experiment with compounds before the medical effects of those compounds are well known are taking chances with their health. The researchers who investigated the solubilities of alkanes in benzene and carbon tetrachloride were not aware that these compounds are toxic and carcinogenic. Solubility research is now done with less hazardous compounds or under a fume hood.

Table 3: Properties of Alkanes and Cycloalkanes

Formula (expanded)	Name (IUPAC)	Melting point (°C)	Boiling point (°C)	Density (g/mL at 20°C)
$CH_{4(g)}$	methane	–183	–162	gaseous
$C_2H_{6(g)}$	ethane	–172	–89	gaseous
$C_3H_{8(g)}$	propane	–187	–42	gaseous
$CH_3(CH_2)_2CH_{3(l)}$	butane	–138	–0.5	gaseous
$CH_3(CH_2)_3CH_{3(l)}$	pentane	–130	36	0.626
$CH_3(CH_2)_4CH_{3(l)}$	hexane	–95	69	0.659
$CH_3(CH_2)_5CH_{3(l)}$	heptane	–91	98	0.684
$CH_3(CH_2)_6CH_{3(l)}$	octane	–57	126	0.703
$CH_3(CH_2)_7CH_{3(l)}$	nonane	–54	151	0.718
$CH_3(CH_2)_8CH_{3(l)}$	decane	–30	174	0.730
$(CH_3)_2CHCH_2CH_{3(g)}$	2-methylbutane	–159	–12	gaseous
$(CH_3)_4C_{(g)}$	2,2-dimethylpropane	–17	10	gaseous
$(CH_3)_2CH(CH_2)_2CH_{3(l)}$	2-methylpentane	–154	60	0.654
$(CH_3CH_2)_2CHCH_{3(l)}$	3-methylpentane	–118	63	0.676
$(CH_3)_3CCH_2CH_{3(l)}$	2,2-dimethylbutane	–98	50	0.649
$(CH_3)_2CHCH(CH_3)_2(l)$	2,3-dimethylbutane	–129	58	0.668
$C_3H_{6(g)}$	cyclopropane	–127	–33	gaseous
$C_4H_{8(g)}$	cyclobutane	–80	13	gaseous
$C_5H_{10(l)}$	cyclopentane	–94	49	0.746
$C_6H_{12(l)}$	cyclohexane	7	81	0.778
$C_7H_{14(l)}$	cycloheptane	–12	118	0.810

molecules. The closer the packing, the larger the net intermolecular force. The density of alkanes is also explained by the size of the molecule and the packing. The larger the molecule, the more carbon atoms packed into a unit volume; therefore, more mass per unit volume (thus, greater density).

From many studies of alkanes, chemists have shown that alkanes are generally not very reactive and when they do react at SATP, their reactions are slow. The exception to this pattern is the complete combustion of alkanes, which, under the right conditions, is usually so fast that it is explosive. Chemists have created theories to explain these chemical properties of alkanes. For example, they explain that alkanes have only single C—C and C—H bonds in their molecular structures, and both kinds of single bonds are stable and not easily broken. A lot of energy is required to break any bond in an alkane, which must happen before any new atom can bond at that point. Alkanes have the maximum possible number of hydrogen atoms bonded to the carbon parent chain; that is, these compounds are said to be **saturated hydrocarbons**. Because atoms or molecules cannot be easily added to a saturated hydrocarbon without removing some existing atoms, most SATP reactions of alkanes are relatively slow.

saturated hydrocarbons: hydrocarbons with only single bonds in their molecules; containing a maximum number of hydrogen atoms

In addition to reacting very slowly at SATP, alkanes react only with very reactive substances like the halogens. For example, when ethane and bromine are mixed in the presence of light, the orange colour of the bromine disappears slowly. A test of the products reveals that moist blue litmus turns red, which indicates the production of $HBr_{(g)}$. As we assume one other product, the balancing species is $C_2H_5Br_{(l)}$. A balanced chemical equation can be written from this laboratory evidence:

$$C_2H_{6(g)} + Br_{2(l)} \xrightarrow{\text{light}} HBr_{(g)} + C_2H_5Br_{(l)}$$

This type of reaction, typical of alkanes, is called a *substitution reaction*. This reaction does not occur in the dark, and in ordinary room light it is still very slow. In the reaction, a hydrogen atom is substituted by a halogen atom as shown more clearly with complete structural diagrams:

```
     H   H                              H   H
     |   |                              |   |
H -- C - C -- H + Br -- Br    →    H -- C - C -- Br + H -- Br
     |   |                              |   |
     H   H                              H   H
```

Note that the hydrogen atom removed from the ethane molecule bonds to one of the bromine atoms from the original bromine molecule, $Br_{2(l)}$. In substitution reactions with alkanes, usually many of the possible isomers are formed and multiple substitutions are possible, so any chemical equation written illustrates only one possible reaction.

The slowness of the reaction of bromine with alkanes is used as a diagnostic test for an alkane. If bromine is added to a hydrocarbon and the reaction is slow, then the hydrocarbon is likely an alkane.

We often think of alkanes (such as butane in lighters) as reactive because they burn rapidly under certain conditions, but this is a special case of reaction with oxygen (which is very reactive). Combustion does not occur unless started by a spark or flame, in other words, at a very high temperature. Once begun, however, combustion produces so much energy that the reaction supports itself; some of the energy produced continually breaks new bonds to provide new places for oxygen atoms to attach to carbons and hydrogens:

$$2\ C_4H_{10(g)} + 13\ O_{2(g)} \rightarrow 8\ CO_{2(g)} + 10\ H_2O_{(g)} + \text{energy}$$

In summary, alkanes have a dual character. They undergo slow substitution reactions or rapid (high-temperature) combustion reactions. In general, alkanes are relatively nonreactive but can undergo explosive combustion reactions if ignited at high temperature. Recall, also, that hydrocarbons undergo cracking and reforming reactions, especially during the secondary refining of crude oil.

SUMMARY Chemical Reactions of Alkanes

Cracking

$$\text{large alkane} + H_{2(g)} \xrightarrow[\text{heat}]{\text{catalyst}} \text{smaller alkane}$$

Reforming

$$\text{small alkane} \xrightarrow[\text{heat}]{\text{catalyst}} \text{larger alkane (or more branched)} + H_{2(g)}$$

Complete Combustion

$$\text{alkane} + \text{oxygen} \rightarrow CO_{2(g)} + H_2O_{(g)} + \text{energy}$$

Diagnostic Test

If its reaction with bromine is slow, then the hydrogen is likely an alkane (orange colour slowly disappears).

$$\text{alkane} + \text{bromine} \rightarrow \text{alkyl bromide} + \text{hydrogen bromide}$$

Practice

Understanding Concepts

13. List three sources of alkanes.
14. Write generalizations concerning the following properties of alkanes:
    (a) the density of alkanes
    (b) the solubility of alkanes in water
    (c) the state of matter of alkanes at SATP
    (d) the position of a liquid alkane that has been mixed with water
15. Write explanations for the generalizations concerning these properties of alkanes:
    (a) the solubility of alkanes in water
    (b) the state of matter of alkanes at SATP
16. Name three classes of chemical reactions for alkanes.
17. Why are alkane reactions generally difficult or slow?
18. Alkanes are saturated hydrocarbons. What does this mean theoretically (molecular structure) and empirically (diagnostic test)?
19. Predict the relative boiling points of octane and 2,2,4-trimethylpentane. Provide your reasoning and then test your prediction and reasoning by obtaining the boiling points from a reference source.
20. Predict the solubility of the gas propane in the liquid hexane. Provide your reasoning.
21. Gases such as methane, ethane, and propane are obtained from the fractional distillation of liquid crude oil. How is this possible?
22. Write chemical equations, using complete structural diagrams, for the diagnostic test reaction of bromine with the following alkanes. Provide only one possible chemical equation for each.
    (a) methane and bromine
    (b) propane and bromine
23. Write chemical equations for the complete combustion of
    (a) methane and oxygen (natural gas combustion);
    (b) 2,2,4-trimethylpentane and oxygen (gasoline combustion).

Making Connections

24. Referring to **Table 3** (page 538), explain why
    (a) skiers and winter hikers don't carry butane lighters in outside pockets;
    (b) propane cylinders must be made of steel, while butane lighters can be plastic;
    (c) converting a truck or taxi to run on liquefied natural gas is initially very expensive, mostly because the tank purchased to hold the liquid methane must be extremely strong and crash-proof.

the vapour pressure of propane is more vp of butane

Section 11.4 Questions

Understanding Concepts

1. Naphtha, commonly used as a camping stove and lantern fuel, is a mixture of alkanes with five or six carbon atoms per molecule. Draw structural diagrams for all of the isomers of C_6H_{14}. Label each diagram with the IUPAC name for the isomer.
2. Cracking reactions are common in the second stage of oil refining. For each of the following word equations, draw a structural diagram of each reactant and product:
   (a) hexane + hydrogen → ethane + butane
   (b) 2-methylpentane + hydrogen → propane + propane
   (c) 2,2-dimethylbutane + hydrogen → ethane + methylpropane
3. Reforming reactions increase the yield of desirable products, such as compounds whose molecules have longer chains or more branches. For each of the following equations, draw structural diagrams when IUPAC names are given and write IUPAC names when structural diagrams are given.
   (a) $CH_3-CH_2-CH_3 + CH_3-CH_2-CH_2-CH_2-CH_3 \rightarrow CH_3-(CH_2)_6-CH_3 + H-H$
   (b) cyclohexane + ethane → ethylcyclohexane + hydrogen
   (c) $CH_3-CH_2-CH_2-CH_2-CH_2-CH_3 \rightarrow CH_3-C(CH_3)_2-CH_2-CH_3$
   (d) Draw structural diagrams and write the IUPAC names for two other structural isomers of the product given in (c).
4. Most of the products of cracking and reforming reactions end up in fuel mixtures such as gasoline. Complete the following equations for complete combustion, including structures and IUPAC names.
   (a) 2,2,4-trimethylpentane + oxygen →
   (b) $CH_3-CH(CH_3)-CH_2-CH_3 + O{=}O \rightarrow$
5. Classify and write structural formula equations for the following organic reactions. Do not balance the equations.
   (a) methane + butane→ pentane + hydrogen
   (b) propane + pentane → octane + hydrogen
   (c) butane + oxygen →
   (d) decane + hydrogen → heptane + propane
   (e) 3-ethyl-5-methylheptane + hydrogen → ethane + propane + methylbutane

(continued)

6. The following hydrocarbons are likely constituents of gasoline. Write the IUPAC name for each the following substances:

(a)

```
      CH2 — CH3
      |
CH3 — C — CH2 — CH3
      |
      CH3
```

(b)

```
      CH3         CH3
      |           |
CH3 — CH — CH  — CH — CH2 — CH3
           |
           CH3
```

(c)

```
                        CH3       CH2 — CH3
                        |         |
CH3 — CH2 — CH2 — CH — C — CH2 — CH3
                              |
                              CH3
```

(d)

```
CH2 — CH2 — CH3
|
(benzene ring)
|
CH3
```

7. Predict the solubility of cyclopentane in water. Provide your reasoning.

8. Referring to molecular structure, explain why methane, CH_4, is a gas, octane, C_8H_{18}, is a liquid, and paraffin, $C_{25}H_{52}$, is a solid at SATP.

9. Describe what you would expect the general properties to be for eicosane, $C_{20}H_{42}$, including physical state at room temperature, density, solubility in water and in gasoline, and combustibility.

10. London forces between molecules depend on surface area and closeness as well as molecular size. The more surface that is close to another molecule, the stronger the attractive force will be. Use this argument and diagrams to explain why pentane has higher melting and boiling points than 2-methylbutane, even though they are isomers.

Applying Inquiry Skills

11. Use values from **Table 3**, page 538, to construct a graph, plotting melting and boiling points of alkanes against the number of carbon atoms per molecule (as the independent variable). Use your graph to predict the melting and boiling points of tetradecane, $C_{14}H_{30}$, and then use a reference to check and report the accuracy (% difference) of your prediction. Note that temperatures must be in kelvins for percentage numerical comparison.

Making Connections

12. Oil spills in ocean water cause a wide variety of environmental problems. Use your knowledge of alkanes to describe and explain what happens physically and chemically when oil is spilled from a tanker.
13. The density of air at SATP is about 1.2 g/L. If a pure gas is assumed to be an ideal gas, then its density can be calculated using its molar mass and the molar volume at SATP (24.8 L/mol).
    (a) What is the density of methane gas at SATP in grams per litre?
    (b) Explain why propane-powered vehicles are prohibited from parking in underground parkades, while natural gas vehicles are allowed.
    (c) Why are gasoline vehicles allowed to park underground?

11.5 Alkenes and Alkynes

Laboratory evidence of hydrocarbon mixtures reveals that there are more kinds of hydrocarbons than just alkanes. Molecular formula determinations reveal the chemical formulas shown in **Table 1**.

Table 1: Series of Empirical Chemical Formulas of Hydrocarbons

Series 1	Series 2	Series 3
$C_2H_{6(g)}$	$C_2H_{4(g)}$	$C_2H_{2(g)}$
$C_3H_{8(g)}$	$C_3H_{6(g)}$	$C_3H_{4(g)}$
$C_4H_{10(g)}$	$C_4H_{8(g)}$	$C_4H_{6(g)}$
$C_5H_{12(l)}$	$C_5H_{10(l)}$	$C_5H_{8(l)}$

Analysis of these series of hydrocarbons reveals a pattern, not only in their formulas, but also in their chemical properties. Series 1, which you will recognize as alkanes, are relatively unreactive compared with the other two series. The molecules in series 1 are alkanes; those in series 2 are called *alkenes*, and the molecules in series 3 are called *alkynes*. Like alkanes, alkenes and alkynes each form a homologous series:

C_nH_{2n+2}	C_nH_{2n}	C_nH_{2n-2}
alkanes	alkenes	alkynes

To explain these series, chemists apply the theory of bonding and bonding capacity to draw structural diagrams. The structural diagrams reveal that the series can be explained using current theory if series 1 contains C—C single bonds, series 2 involves one C═C double bond, and series 3 is explained by one C≡C triple bond. For example, consider the structural diagrams for the three hydrocarbons with two carbon atoms per molecule:

(a) ethane — complete — condensed

$C_2H_{6(g)}$

$$\begin{array}{ccccccc} & & H & & H & & \\ & & | & & | & & \\ H & - & C & - & C & - & H \\ & & | & & | & & \\ & & H & & H & & \end{array}$$

CH_3-CH_3

(b) ethene (ethylene) — complete — condensed

$C_2H_{4(g)}$

$$\begin{array}{ccccccc} & & H & & H & & \\ & & | & & | & & \\ H & - & C & = & C & - & H \end{array}$$

$CH_2=CH_2$

(c) ethyne (acetylene) — complete — condensed

$C_2H_{2(g)}$

$H-C\equiv C-H$

$CH\equiv CH$

The bonding capacity of carbon requires four covalent bonds. With insufficient hydrogen to saturate the molecule with single bonds, a double bond is required to explain the $C_2H_{4(g)}$ formula, and a triple bond is required to explain $C_2H_{2(g)}$. All alkanes, therefore, are explained as having all carbon–carbon single bonds; **alkenes** have one or more double bonds; and **alkynes** have one or more triple bonds.

alkenes: a hydrocarbon family of molecules that contain at least one carbon–carbon double bond; general formula, C_nH_{2n}

alkynes: a hydrocarbon family of molecules that contain at least one carbon–carbon triple bond; general formula, C_nH_{2n-2}

Qualitative and quantitative analysis of petroleum tells us that hydrocarbons containing double or triple covalent bonds are relatively minor constituents. However, these compounds are often formed during cracking reactions during the refining of crude oil and are valuable components of gasoline. Hydrocarbons containing double or triple bonds are vitally important in the petrochemical industry because they are the starting materials for the manufacture of many derivative compounds, including most kinds of plastics.

Hydrocarbons with carbon–carbon double bonds are members of the alkene family (**Figure 1**). The names of alkenes with only one double carbon–carbon bond have the same prefixes as the names of alkanes but take the suffix *-ene* (**Table 2**).

Table 2: The Alkene Family of Organic Compounds

IUPAC name (common name)	Molecular formula
ethene (ethylene)	$C_2H_{4(g)}$
propene (propylene)	$C_3H_{6(g)}$
1-butene (butylene)	$C_4H_{8(g)}$
1-pentene	$C_5H_{10(l)}$
1-hexene	$C_6H_{12(l)}$
–ene	C_nH_{2n}

Figure 1
Ethene and propene are the simplest members of the alkene family. Ethene, called ethylene in the petrochemical industry, is the starting material for an enormous number of consumer, commercial, and industrial petrochemical products. The double covalent bonds are shorter and much more reactive than single carbon–carbon bonds.

The alkyne family has chemical properties that can be explained by assuming the presence of a triple bond between carbon atoms (**Figure 2**). Like alkenes, alkynes are unsaturated and react immediately with small molecules such as hydrogen or bromine in an *addition reaction*; that is, the reaction allows new atoms to add. Alkynes are named like alkenes, except for the *-yne* suffix. The simplest alkyne, ethyne (acetylene), is commonly used as a high-temperature fuel (**Figure 3**). In fact, $C_2H_{2(g)}$ is the only commercially important alkyne, with huge amounts being produced annually as fuel for welding and cutting torches and as starting material for the manufacture of ethanoic acid (acetic acid) and many types of plastics, as well as synthetic rubber for the tire industry. **Table 3** lists the first five members of the alkyne family. Isomers exist for all alkynes larger than propyne.

Table 3: The Alkyne Family of Organic Compounds

IUPAC name (common name)	Molecular formula
ethyne (acetylene)	$C_2H_{2(g)}$
propyne	$C_3H_{4(g)}$
1-butyne	$C_4H_{6(g)}$
1-pentyne	$C_5H_{8(l)}$
1-hexyne	$C_6H_{10(l)}$
–yne	C_nH_{2n-2}

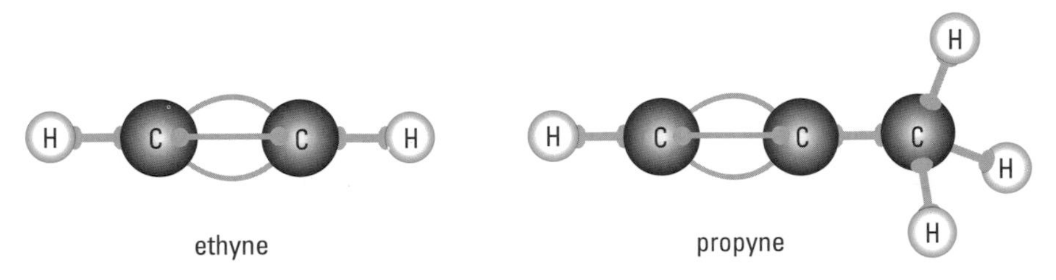

Figure 2
Ethyne and propyne are the simplest members of the alkyne family. Ethyne, called acetylene in industry, is the fuel used in welding torches. Triple bonds are the shortest and most reactive of all carbon–carbon bonds.

Figure 3
The flame of an oxyacetylene torch is hot enough to melt metals easily. Pure oxygen reacts extremely rapidly with the triple bonds of ethyne, releasing large quantities of energy in a very short time.

Naming Alkenes and Alkynes

Since the location of a multiple bond affects the chemical and physical properties of a compound, IUPAC decided that an effective naming system should specify the multiple bond location. Alkenes and alkynes are named much like alkanes, with two additional points to consider:

- The longest or parent chain of carbon atoms must contain the multiple bond, and the chain is numbered from the end closest to the multiple bond.
- The name of the parent chain of the compound is preceded by a number that indicates the position of the multiple bond on the parent chain.

- The name of any branch (e.g., alkyl group) is preceded by a number that indicates the position of the branch on the parent chain. This is the same procedure used with alkanes.

For example, there are two possible butene isomers: 1-butene and 2-butene. (The isomers can also be named but-1-ene and but-2-ene, but we will not use that system in this book.)

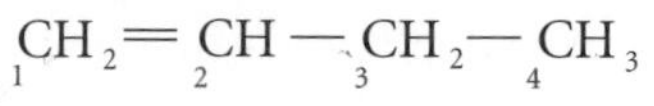

$$\underset{1}{CH_2}{=}\underset{2}{CH}{-}\underset{3}{CH_2}{-}\underset{4}{CH_3} \qquad \underset{1}{CH_3}{-}\underset{2}{CH}{=}\underset{3}{CH}{-}\underset{4}{CH_3}$$

1-butene 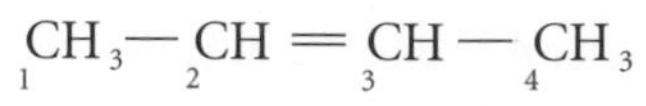 2-butene

Sample Problem 1

Name the hydrocarbon petrochemicals (a) and (b).

(a)

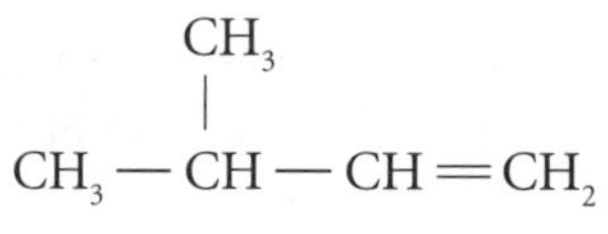

$$CH_3{-}\underset{}{\overset{\overset{\Large CH_3}{|}}{CH}}{-}CH{=}CH_2$$

(b)

$$CH_3{-}CH{=}CH{-}CH_2{-}\overset{\overset{\Large CH_3}{|}}{CH}{-}CH_3$$

Solution

The IUPAC name for (a) is 3-methyl-1-butene.
The IUPAC name for (b) is 5-methyl-2-hexene.

In the following branched alkyne structure, the parent chain is pentyne and there is only one branch, a methyl group:

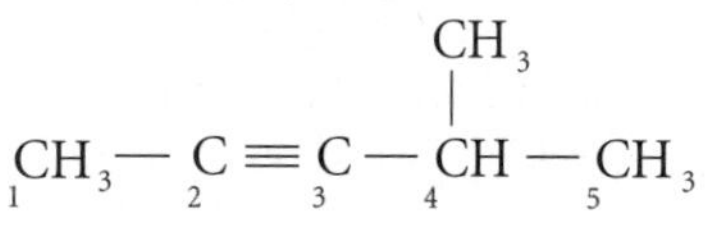

$$\underset{1}{CH_3}{-}\underset{2}{C}{\equiv}\underset{3}{C}{-}\underset{4}{\overset{\overset{\Large CH_3}{|}}{CH}}{-}\underset{5}{CH_3}$$

4-methyl-2-pentyne

The location of the multiple bond in an alkene or alkyne takes precedence over the location of the branches in numbering the carbon atoms of the parent chain. The IUPAC name 4-methyl-2-pentyne follows the same format as that used for alkanes. Branches are listed in alphabetical order. Branched alkynes are rare.

Structural Diagrams from Alkene and Alkyne Names

Whenever you need to draw a structural diagram for any hydrocarbon, you should always look at the end of the name to find the parent chain. You draw the parent alkene or alkyne first and then add the branches listed in the name. Be sure to finish the structure with sufficient hydrogen atoms to complete four bonds of each carbon atom. The following sample problem shows some typical examples of alkenes and alkynes.

Sample Problem 2

Draw structural diagrams for the following alkyne petrochemicals:

(a) 4-methyl-1-pentyne

(b) 3,3-dimethyl-1-butyne

Solutions

(a)

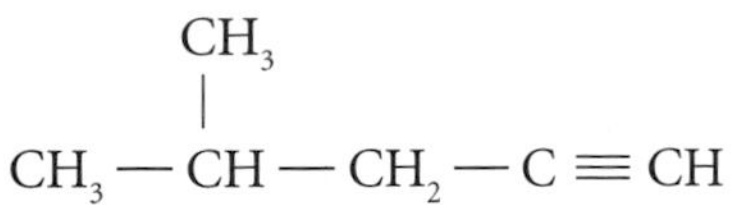

$$CH_3-\underset{\displaystyle CH_3}{\overset{|}{CH}}-CH_2-C\equiv CH$$

(b)

$$CH\equiv C-\overset{\displaystyle CH_3}{\underset{\displaystyle CH_3}{C}}-CH_3$$

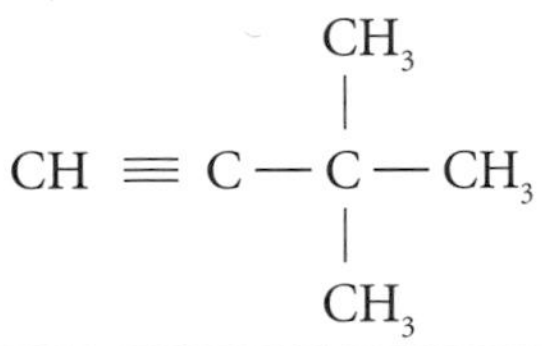

Cycloalkenes and cycloalkynes are classes of hydrocarbons without many members. Chemists explain this low membership by the stress put on the double and triple bonds by creating a cyclic hydrocarbon. However, there are such molecules as cyclohexene, a six-carbon cyclic molecule with one double bond (**Figure 4**). It does not matter where the double bond is shown and no number is necessary. There is only one cyclohexene. Cycloalkanes are isomers of alkenes with the same number of carbon atoms, both with the general formula C_nH_{2n}. Cycloalkenes, similarly, are isomers of alkynes, both with the general formula C_nH_{2n-2}.

Isomers of alkenes and alkynes exist for different locations of the double or triple bond and by changing a straight-chain hydrocarbon into a branched hydrocarbon or into a cyclic hydrocarbon. If you find that several structures have the same formula but different names, then the structures are isomers.

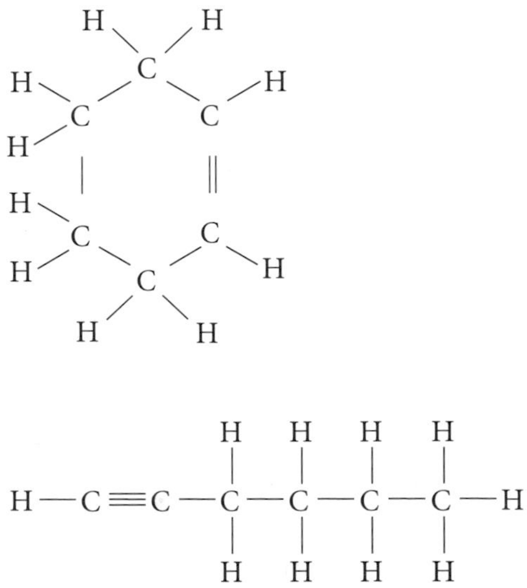

Figure 4
Cyclohexene is a cycloalkene and an isomer of the alkyne hexyne. Both have the formula $C_6H_{10(l)}$.

Practice

Understanding Concepts

1. Classify each of following hydrocarbons as an alkane, alkene, or alkyne and/or as a cycloalkane or cycloalkene.
   (a) $C_2H_{4(g)}$ alkene
   (b) $C_3H_{8(g)}$ alkane
   (c) $C_4H_{6(g)}$ alkyne, cycloalkene
   (d) $C_5H_{10(l)}$

   * need at least 3 carbons to make a ring
2. Draw a structural diagram and write a chemical formula for each of the following.
   (a) propane
   (b) propene
   (c) propyne
   (d) cyclopropane
3. Draw structural diagrams for each of the following petrochemicals.
   (a) propene
   (b) 2-butene
   (c) 2,4-dimethyl-2-pentene
   (d) 1-butyne
4. Why are no numbers required for the location of the multiple bonds in propene or propyne?

5. Write IUPAC names for each of the following structures:

(a)

```
                  CH3
                  |
CH2 = CH — C — CH2 — CH3
                  |
                  CH3
```

(b)

```
          CH3
          |
CH3 —  C — CH = CH2
          |
          CH2 — CH3
```

(c)

```
                    CH3         CH3
                    |           |
CH3 — C≡C — CH — CH — CH2
                          |
                          CH2 — CH3
```

6. Draw structural diagrams and write the IUPAC names for the four structural isomers of $C_4H_{8(g)}$. (Remember alicyclic compounds.)
7. Alkenes and alkynes are the starting materials in the manufacture of a wide variety of organic compounds. Draw structural diagrams for the following starting materials that are used to make the products named in parentheses.
   (a) propene (polypropylene)
   (b) methylpropene (synthetic rubber)

Properties of Alkenes and Alkynes

Table 4: Boiling Points of Alkanes and Alkenes

Alkane name	Boiling point (°C)	Alkane name	Boiling point (°C)
ethane	–88.6	ethene	–103.7
propane	–42.1	propene	–47.4
butane	–0.5	1-butene	–6.3
pentane	36.1	1-pentene	30.0

unsaturated hydrocarbon: a reactive hydrocarbon whose molecules contain double and triple covalent bonds between carbon atoms; for example, alkenes and alkynes

Hydrocarbons with molecules containing one or more carbon–carbon double bonds (alkenes) or triple bonds (alkynes) have very similar physical properties to alkanes of the same molar mass. Melting points, boiling points, solubilities, and densities are all very much like those of comparable alkanes. A change of two or four hydrogen atoms and their electrons is usually a small change in the total number of electrons and, therefore, only a small change in London forces; however, it is sometimes measurable. For example, with two fewer hydrogen atoms, the alkenes have a slightly lower boiling point than the alkanes (**Table 4**).

However, double or triple bonds between carbon atoms in the molecules dramatically affect the chemical properties of the substance. For example, hydrocarbons with double bonds react quickly at room temperature with bromine, compared with alkanes, which react extremely slowly. (**Figure 5**). Organic compounds with carbon–carbon double and triple bonds are said to be **unsaturated** because fewer atoms are attached to the carbon atom framework than the number that *could* be attached if all the bonds were single.

Note that the reaction of a double bond allows two new atoms to add, and the reaction of a triple bond allows up to four new atoms to add. Both of these reactions require only a rearrangement of the electrons involved in the double and triple bonds leaving those forming the single carbon–carbon bond unaffected. These reactions—addition reactions—are generally very fast.

```
  H   H   H   H                       H   H   H   H
  |   |   |   |                       |   |   |   |
H—C = C — C — C—H  +  H—H  →     H— C — C — C — C—H
          |   |                       |   |   |   |
          H   H                       H   H   H   H
```

```
               H                        H   H   H
               |                        |   |   |
H—C ≡ C — C—H  +  2 H—H  →        H— C — C — C—H
               |                        |   |   |
               H                        H   H   H
```

A diagnostic test for the presence of multiple bonds is the bromine water test (**Figure 5**): If bromine water is added to a hydrocarbon and the orange bromine colour disappears instantly, then a multiple bond is likely present. If bromine is added and the orange colour remains, then the hydrocarbon is likely saturated, for example, an alkane. The reaction explaining this diagnostic test is as follows, using ethylene as an example:

$$\underset{\text{colourless}}{CH_2{=}CH_2} + \underset{\text{orange}}{Br{-}Br} \xrightarrow{\text{(fast)}} \underset{\text{colourless}}{CH_2Br{-}CH_2Br}$$

This is a very fast reaction compared to the substitution reaction that saturated hydrocarbons undergo:

$$\underset{\text{colourless}}{CH_3{-}CH_3} + \underset{\text{orange}}{Br{-}Br} \xrightarrow{\text{(slow)}} \underset{\text{colourless}}{CH_3{-}CH_2Br} + \underset{\text{turns moist blue litmus red}}{H{-}Br}$$

The two compounds with the empirical formulas $C_6H_{12(l)}$ and $C_6H_{10(l)}$ have very similar physical properties. Physical properties alone cannot be used to identify separate samples of the two chemicals. A chemical diagnostic test that can be used to differentiate these chemicals is the reaction with bromine water or aqueous potassium permanganate. The slow reaction of $C_6H_{12(l)}$ with either of these reactants indicates the presence of single bonds, that is, a saturated compound. The rapid reaction of $C_6H_{10(l)}$ indicates the presence of multiple (double or triple) bonds, an alkyne or cycloalkene. The simplest interpretation of these results is that $C_6H_{12(l)}$ is cyclohexane and $C_6H_{10(l)}$ is cyclohexene (**Figure 6**).

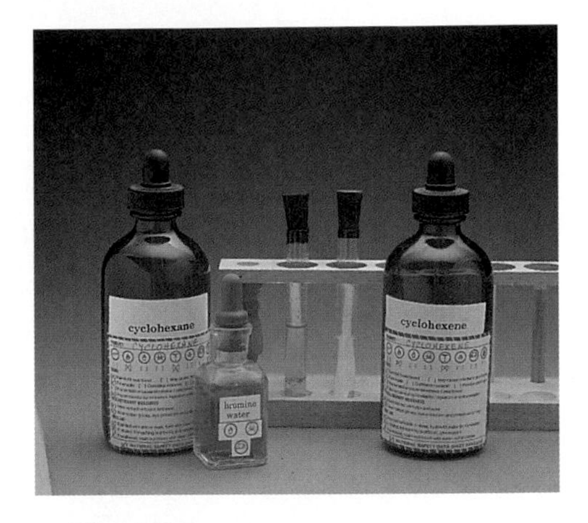

Figure 5
Bromine water (a saturated aqueous solution of bromine) is used in a diagnostic test for unsaturated organic compounds. When an equal amount of bromine water is added simultaneously to cyclohexane and cyclohexene, the unsaturated cyclohexene reacts with the bromine water instantaneously, decolourizing the orange solution. In the saturated cyclohexane, there is no immediate colour change, which is interpreted as no reaction.

(a)

```
        H H
         \/
   H     C     H
    \  /   \  /
  H—C       C—H
    |       |        or   (hexagon)
  H—C       C—H
    /  \   /  \
   H     C     H
         /\
        H H
```

cyclohexane

(b)

```
         H
          \
   H       C
    \   /    \\
  H—C          C—H
    |          |     or   (hexagon with one double bond)
  H—C          C—H
    /  \     /   \
   H     C        H
         /\
        H H
```

cyclohexene

Figure 6
The structural diagram of cyclohexane **(a)** shows that all bonds are single bonds. The cyclohexene structure **(b)** indicates one carbon–carbon double bond. The second structure for diagrams (a) and (b) represents the same molecules with simpler line (polygon) diagrams.

From a theoretical perspective, cyclohexane and cyclohexene are believed to be almost identical, except for the presence of a double bond between two carbon atoms in cyclohexene. These compounds illustrate a relationship between structure and reactivity: Cyclohexene reacts rapidly with bromine water or aqueous potassium permanganate but cyclohexane does not. The reaction is indicated by the disappearance of the orange colour of the bromine or the purple (pink) of the potassium permanganate.

SUMMARY Diagnostic Test Results for Saturated and Unsaturated Hydrocarbons

Hydrocarbon	$Br_{2(aq)}$	$KMnO_{4(aq)}$	Rate
saturated	orange	purple	slow
unsaturated	colourless	brown	fast

Practice

Understanding Concepts

8. Write a generalization describing the trend in boiling points for
   (a) an increasing size of aliphatic hydrocarbon molecules
   (b) alkanes and alkenes with the same number of carbon atoms per molecule
9. Provide theoretical definitions for saturated and unsaturated hydrocarbons.
10. Describe two diagnostic tests for saturated and unsaturated hydrocarbons.
11. Draw condensed structural diagrams for cylcohexane and cyclohexene.

Applying Inquiry Skills

12. Due to the potential hazards of doing diagnostic tests for cyclohexane and cyclohexene with bromine, these tests are available for viewing on the Internet. How does the reaction of cyclohexane with bromine compare with that of cyclohexene?

    Follow the links for Nelson Chemistry 11, 11.5.

www.science.nelson.com

Investigation 11.5.1

Evidence for Multiple Bonds

INQUIRY SKILLS

- ○ Questioning
- ○ Hypothesizing
- ○ Predicting
- ○ Planning
- ● Conducting
- ● Recording
- ● Analyzing
- ○ Evaluating
- ● Communicating

The purpose of this investigation is to use the bromine or potassium permanganate diagnostic test to identify which of the samples provided are saturated and which are unsaturated. Cyclohexane and cyclohexene are provided as optional examples of saturated and unsaturated compounds to model the reaction with bromine water. You will complete the **Analysis** section of the lab report.

Question

Which of the common substances tested are saturated and which are unsaturated?

Experimental Design

The unknown samples and two controls (e.g., cyclohexane and cyclohexene) are tested by adding a few drops of a diagnostic test solution (e.g., potassium permanganate in water). After each sample is mixed with the test solution, evidence of a chemical reaction (a colour change or not) is noted.

Cyclohexane, cyclohexene, acetylene, and propane are highly flammable.

Solid potassium permanganate is an oxidant and a toxin.

Vapours of cyclohexane, cyclohexene, and paint thinner are hazardous because they are flammable and toxic. Avoid inhaling these vapours.

Keep test tubes stoppered and waste containers closed. Work in a fume hood or in a well-ventilated area. Avoid skin contact. Eye protection and gloves must be worn.

Materials

lab apron
eye protection
vinyl gloves
small test tubes with stoppers
test-tube rack
waste container, with lid, for organic substances
potassium permanganate solution in a dropper bottle
cyclohexane or hexane in dropper bottle (or propane gas)
cyclohexene or hexene in dropper bottle (or acetylene gas)
common substances, such as mineral oil, paint thinner, kerosene, liquid paraffin, soybean oil, corn oil, margarine, butter

Procedure

1. Add 10 drops of a known saturated hydrocarbon to a clean test tube.
2. Add 1 drop of the aqueous diagnostic test solution to the test tube. Shake the test tube gently. Repeat this procedure with up to 4 drops of the diagnostic test solution.
3. Dispose of all materials into the labelled waste container.
4. Repeat steps 1 to 3 using a clean test tube and a known unsaturated compound.
5. Repeat steps 1 to 3 using the samples provided. Use a clean test tube each time.

Analysis

(a) Answer the Question.

Geometric Isomers

In alkanes, the rotation of attached groups about the carbon–carbon single bond is quite free. The situation is different for alkenes, where rotation about a carbon–carbon double bond is not possible without breaking the bond. (Molecular models are quite useful in simulating this difference in rotation ability.) Molecular models usually provide a good 3-D representation of a molecule. When you build an alkene model with a ball-and-spring model (**Figure 7**), you can see that the molecule is flat with single bonds approximately 120° from the carbon–carbon double bond. This shape and the lack of rotation about the double bond mean that alkenes can have geometric isomers, which differ from each other only in the position of attached groups relative to the double bond. Unlike structural isomers, the same atoms are bonded to the same parent chain locations, but the molecular shape differs, depending on which side of the double bond they are attached. These are **geometric isomers**; they are also known as **cis-trans isomers.** The term *cis* means on this side, so two groups attached on each side of the double bond are on the same side of the molecule; the term *trans* means across, so two groups attached on each side of the double bond are on opposite sides of the molecule, across from each other. For a simple example, consider the two geometric (cis-trans) isomers of 2-butene, $CH_3CH{=}CHCH_3$:

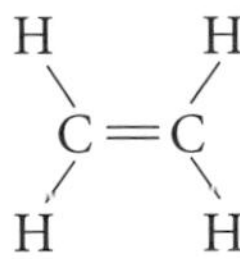

Figure 7
In this molecular model of ethene (ethylene), notice the shape of the molecule and angles between the bonds.

geometric (cis-trans) isomers: Organic molecules that differ in structure only by the position of groups attached on either side of a carbon–carbon double bond. A cis isomer has both groups on the same side of the molecular structure; a trans isomer has the groups on opposite sides of the molecular structure.

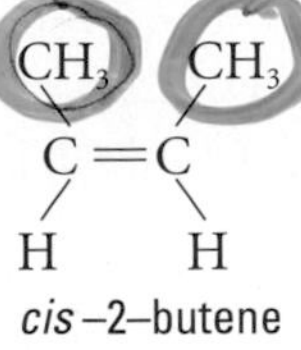

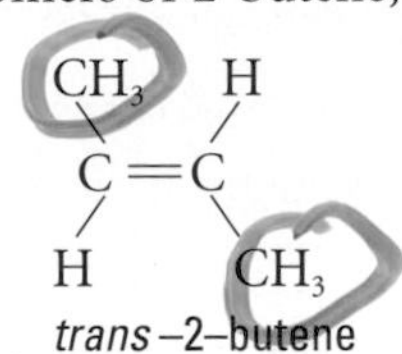

Practice

Understanding Concepts

13. Draw structural diagrams and write IUPAC names for the geometric isomers of $CH_3CHCHCH_2CH_3$.

14. When answering the following questions, use complete structural diagrams to communicate your reasoning.
    (a) Does 1-butene have any geometric isomers?
    (b) Does 3-hexene have any geometric isomers?
    (c) Can an alkene be correctly named 4-hexene?
    (d) Can an alkene be correctly named 4-heptene?
15. Using diagrams, demonstrate whether it is possible to have geometric isomers of
    (a) an alkane
    (b) an alkene
    (c) an alkyne
    (d) a cycloalkane or cycloalkene

Making Connections

16. Physical molecular models are still very useful to a chemist. However, computer molecular models, especially for large biochemically important molecules, are now common and have become an important modelling tool. Using computer models, chemists can construct almost any molecule, rotate it, and even simulate reactions. How do these models appear on a computer screen? How can you manipulate the models? What are some advantages and disadvantages of computer models? (To use computer models, you may need to install a small free program called Chime.)

    Follow the links for Nelson Chemistry 11, 11.5.

    GO TO **www.science.nelson.com**

Activity 11.5.1

Structures and Properties of Isomers

The purpose of this activity is to use molecular models and a chemistry reference to reveal the structures and physical properties of some isomers of unsaturated hydrocarbons. Structures of possible isomers are determined by means of a molecular model kit. Once each structure is named, the boiling and melting points are obtained from a current reference, such as *The CRC Handbook of Chemistry and Physics*, *The Merck Index*, or *Lange's Handbook of Chemistry*.

Materials

molecular model kits
chemical reference

Procedure

1. Use the required atoms to make a model of C_4H_8.
2. Draw a complete structural diagram of the model and write the IUPAC name for the structure.

3. By rearranging bonds, produce models for all isomers of C_4H_8, including cyclic structures. Draw a structural diagram and write the IUPAC name for each structure before disassembling the model.
4. If your model kit creates a C=C bond as two separate bonds, test and note the restricted rotation of groups about the bond axis, and then construct the geometric isomer of your model.
5. Repeat steps 1 to 4 for C_4H_6.
6. In a reference, find the melting and boiling points of each of the compounds you have identified.

Analysis

(a) Prepare a summary table for the molecular structures and relative physical properties of all the substances that are isomers of C_4H_8 and C_4H_6.

Investigation 11.5.2

Preparation and Properties of Ethyne (Acetylene)

INQUIRY SKILLS

- ○ Questioning
- ○ Hypothesizing
- ● Predicting
- ○ Planning
- ● Conducting
- ● Recording
- ● Analyzing
- ● Evaluating
- ○ Communicating

Ethyne (acetylene) is the simplest alkyne, $C_2H_{2(g)}$. The purpose of this investigation is to test the Experimental Design, Materials, and Procedure provided about the production of ethyne. Ethyne can be prepared readily in a laboratory by reacting the ionic compound calcium carbide, $CaC_{2(s)}$, with water, $H_2O_{(l)}$, to produce calcium hydroxide, ethyne gas, and some energy. Complete the **Prediction**, **Analysis**, **Evaluation**, and **Synthesis** sections of the lab report.

Question

What are the products of the reaction of calcium carbide and water?

Prediction

(a) Write and balance an equation for the reaction to synthesize ethyne.

Experimental Design

The expected gaseous product of the reaction of calcium carbide and water is collected by water displacement. The reaction mixture is tested with litmus paper and the expected gaseous product is tested for combustion and for saturation.

Ethyne (acetylene) is very flammable. Work, if possible, in a fume hood, and attempt to ignite only small volumes.

Calcium carbide *must* be kept away from water, unless actually being used to produce ethyne. A beaker of cold water should be handy to slow the reaction if it becomes too vigorous. *All* calcium carbide must be completely reacted before the disposal of any liquids.

Materials

lab apron
eye protection
tongs (for handling the calcium carbide)
water
calcium carbide, $CaC_{2(s)}$, pea size
250-mL beaker
four 18 mm × 150 mm test tubes
bromine water
test-tube rack
stopper for test tube
wooden splint and matches
limewater
red and blue litmus paper

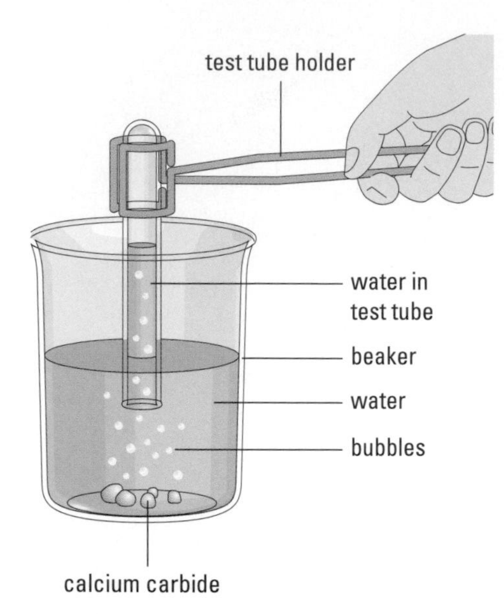

Figure 8
Gases that have low solubility in water, like acetylene, may be collected by downward displacement of water.

Procedure

1. Using the tongs, add one piece of calcium carbide to about 150 mL of water in a 250-mL beaker (**Figure 8**). If the reaction becomes too rapid, cold water should be added to the beaker.
2. Collect two test tubes full of ethyne by downward displacement of water (**Figure 8**). Set the labelled test tubes upside down on the countertop.
3. Collect a third test tube half full of ethyne. Lift the test tube out of the beaker, allowing air to flow in as the water flows out, so that the test tube has an air:ethyne mixture at a roughly 1:1 ratio. Set the labelled test tube upside down in a test-tube rack.
4. Collect in a fourth test tube 1 cm of ethyne. Lift the test tube out of the beaker, allowing air to flow in as the water flows out, so that the test tube has an air:ethyne mixture at a roughly 12:1 ratio. Set the labelled test tube upside down in a test-tube rack.
5. Add 10 drops of bromine water to test tube 1, stopper the test tube, then shake it. Record your observations. Wash any bromine from your hands.
6. Ignite the ethyne in test tubes 2 to 4, one at a time, and record your observations for each.
7. Add a few millilitres of limewater to test tube 4 and shake. Record your observations.
8. Use litmus paper to test the solution in the beaker.
9. Dispose of any extra ethyne in a fume hood and any extra bromine water in the waste container provided. Wash any remaining bromine water and limewater from the test tubes down the sink with lots of water.

Analysis

(b) Answer the Question by listing the products together with the key Evidence that identifies each product.

Evaluation

(c) Is your Prediction supported by the Evidence gathered in this investigation? How certain are you about the evidence collected?
(d) Are the Experimental Design, Materials, and Procedure adequate for the synthesis of ethyne? Include pros and cons, complete with your reasoning.

Synthesis

(e) Why can ethyne be collected by the displacement of water?
(f) In the Procedure, you were asked to keep the test tube with ethyne inverted until ready for use. What does this suggest about the density of ethyne?
(g) How does the Evidence you collected illustrate incomplete and complete combustion of ethyne?
(h) According to the Evidence you collected, what is the best ratio of air:ethyne for complete combustion?
(i) Write a balanced chemical equation for the combustion of ethyne if the products are
   (i) carbon and water vapour;
   (ii) carbon monoxide and water vapour;
   (iii) carbon dioxide and water vapour;
   (iv) carbon, carbon dioxide, carbon monoxide, and water vapour.

DID YOU KNOW ?

Discovery of Acetylene

Acetylene was discovered by notable Canadian scientist Thomas Leopold Willson (1860–1915), who was born in Princeton, Ontario, and attended high school at Hamilton Collegiate Institute. An active inventor and entrepreneur, Willson is credited with more than 60 inventions, from electric arc lighting (which he patented at age 21) to gas navigational buoys and beacons. He is best known for his discovery in 1892 of a more efficient process for making calcium carbide and its byproduct, acetylene gas. Willson's small aluminum smelting furnace in North Carolina produced a slag, which he threw into the nearby stream until a large pile accumulated. One day, upon dumping red-hot slag into the stream, there was a dazzling burst of flame. Willson investigated and discovered that by adding water to the smelting furnace slag he produced a gas that he could ignite with a match. Willson had produced calcium carbide and acetylene gas. Willson's discovery of a method to economically make calcium carbide led to his nickname "Carbide" Willson. The discovery of acetylene helped to establish the automotive industry: The acetylene, with oxygen, is used as fuel in the oxyacetylene torch, an invaluable tool in metal cutting and welding.

(j) How does the ratio you found for complete combustion compare with the ratio of oxygen:ethyne in i(iii) above? (Remember that air is about 20% oxygen.)

The Diversity of Organic Molecules

You have studied relatively small alkanes, alkenes, alkynes, and their corresponding cyclic compounds. You have also seen examples of both structural and geometric isomers. Now consider that there can easily be hydrocarbons with hundreds of thousands of carbon atoms. There are also numerous hydrocarbon derivatives containing carbon, hydrogen, and other nonmetal atoms. And there are also many other kinds of isomers. So you can imagine that there must be a staggering number of organic molecules. Of the more than 10 million compounds known, at least 90% are molecular compounds of the element carbon. The number of known compounds of carbon far exceeds the number of compounds of all other elements combined. This observation is explained by chemists as resulting from the combination of several properties of carbon:

- Carbon is a small atom that can form four bonds, more than atoms of most other elements.
- Carbon atoms have the special property of being able to bond together to form chains, rings, spheres, sheets, and tubes of almost any size (**Figure 9**).
- Carbon can form multiple combinations of single, double, and triple covalent bonds with itself and with atoms of other elements.

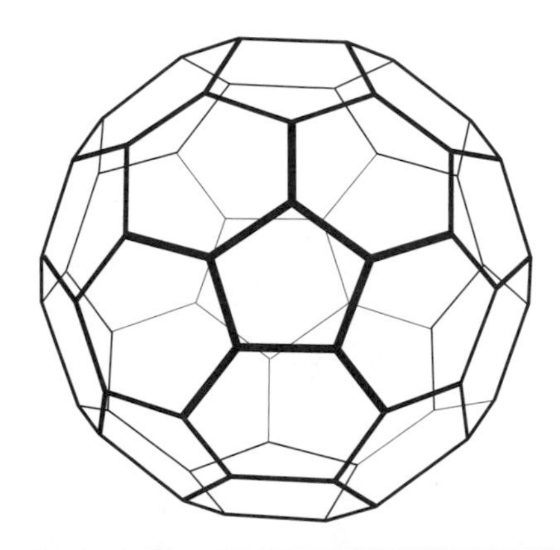

Figure 9
The soccer-ball-shaped C_{60} molecule has pentagons of carbon atoms surrounded by hexagons of carbon atoms. This structure of carbon, called buckminsterfullerene, was discovered in 1985. Common soot contains this molecule.

Polymers

Polymers are substances whose molecules are made up of many similar small molecules (**monomers**) linked together in long chains. **Polymerization** is the formation of polymers from many monomers. These compounds have long existed in nature but were only synthesized by technological processes in the 20th century. They have molar masses up to millions of grams per mole.

Addition Polymers

Many plastics are produced by the polymerization of alkenes. For example, polyethene (polyethylene) is made by polymerizing ethene molecules in a reaction known as **addition polymerization**. Polyethylene is used to make plastic insulation for wires and containers such as plastic milk bottles, refrigerator dishes, and laboratory wash bottles. Addition polymers are formed when monomers join each other in a process that involves the rearranging of electrons in double or triple bonds in the monomer. In addition polymerization, the polymer is the only product formed.

polymers: a long chain molecule made up of many small identical units (monomers)
monomers: the smallest repeating unit of a polymer

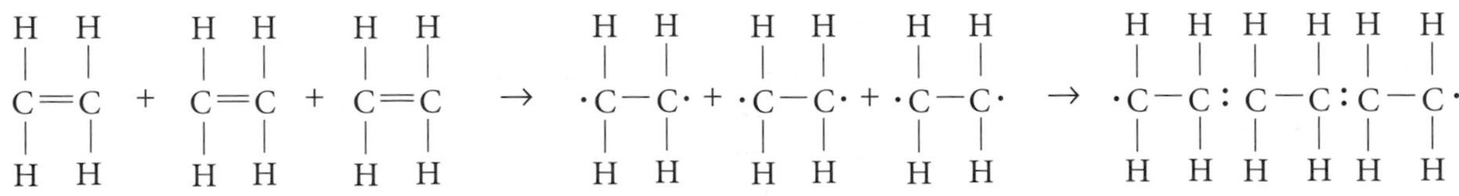

ethylene

part of polyethylene

Using tetrafluoroethene instead of ethene in an addition polymerization reaction produces the substance polytetrafluoroethene, commonly known as Teflon. Teflon has properties similar to polyethylene, such as a slippery surface and a nonreactive nature. But Teflon has a much higher melting point than polyethylene, so it is used to coat cooking utensils. Polypropene (polypropylene), polyvinyl chloride, Plexiglas, polystyrene, and natural rubber are also addition polymers (**Figure 10**, page 556).

polymerization: a type of chemical reaction involving the formation of very large molecules (polymers) from many small molecules (monomers)
addition polymerization: a reaction in which unsaturated monomers combine with each other to form a polymer

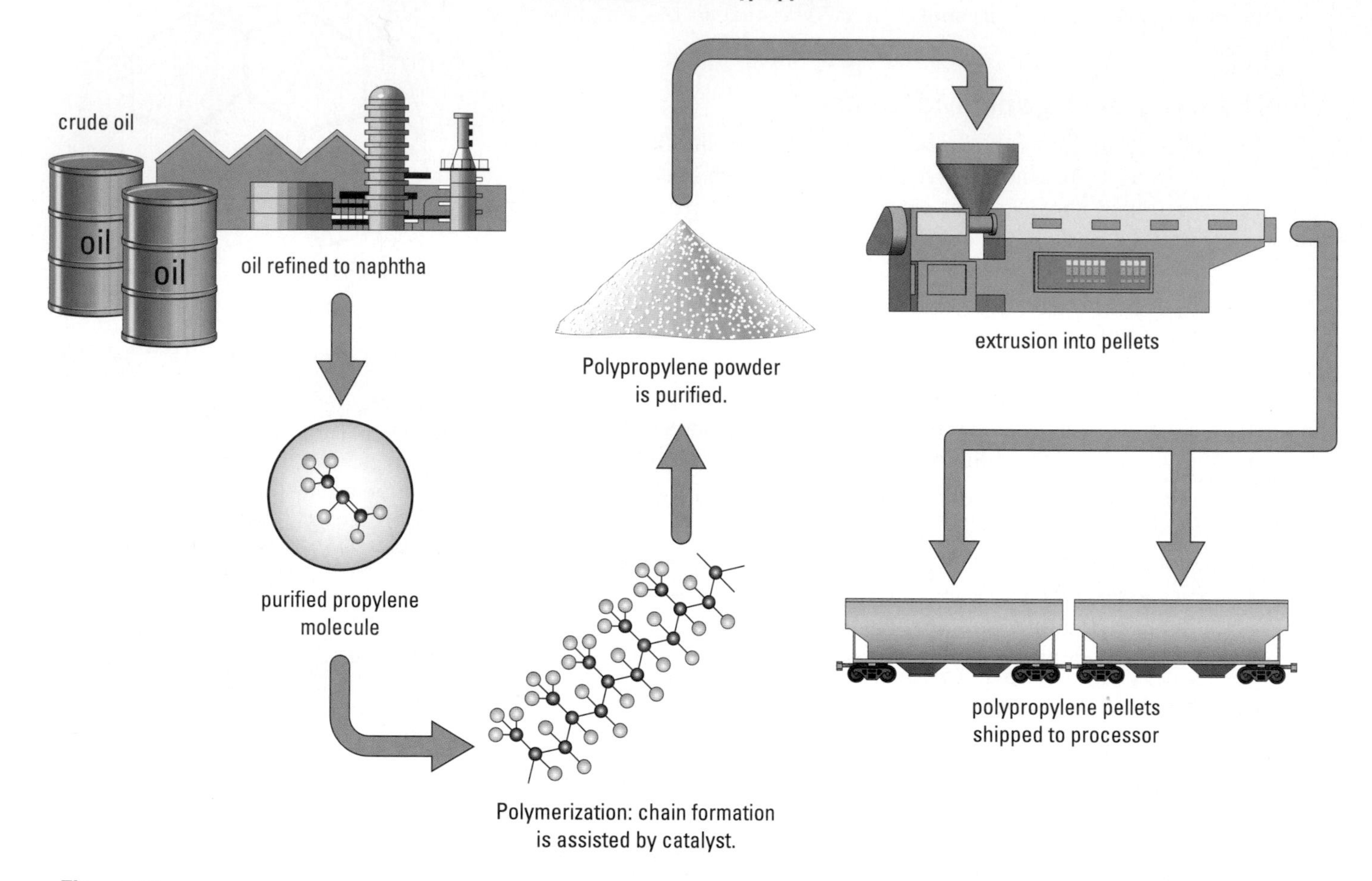

Figure 10
Polypropylene is one of many chemicals derived from crude oil.

Practice

Understanding Concepts

17. Polyethene (polyethylene) is a very common plastic.
    (a) The starting material for polyethylene is ethane, which is obtained from natural gas under high pressure and low temperature. Write a chemical equation for the condensation of ethane gas.
    (b) Write a chemical equation using complete structural diagrams for the synthesis of ethene from ethane.
    (c) Write a chemical equation using complete structural diagrams for the synthesis of polyethylene.
18. List some technological products that are made from polyethylene.

Making Connections

19. Polypropylene and polybutylene are two other common hydrocarbon polymers. Research and list some of the main uses of each of these polymers. Identify some benefits and risks to society and the environment of our use of polymers.

    Follow the links for Nelson Chemistry 11, 11.5.

GO TO www.science.nelson.com

CAREER

Careers in the Petrochemical Industry

Many areas of industry and commerce begin with the production of petrochemicals from petroleum; only a few of the many associated careers are shown here.

Plastics Technologist
Plastics technologists work with a wide range of different materials, matching properties of different plastics to specialized applications. They are concerned with recycling technology, measurement, testing, and fabrication techniques controlled largely by the chemistry of, and bonding structure within, any of the myriad types of plastic materials. Petrochemicals supply the raw materials for the manufacture of most types of plastics.

Refinery Laboratory Technician
Lab technicians are generally concerned with testing and analysis, both to monitor and control refinery processes and to do testing for projects involving research into improvements and changes. Familiarity with reaction and bonding theory is essential, as are the ability to operate a wide variety of technological equipment and the skills to perform analytical techniques.

Petroleum Geologist
Geologists who work in the petrochemical industry require specialized knowledge about the geological formations associated with underground petrochemical deposits. They work extensively with rocks and minerals in test drilling cores and with charts from seismic exploration of underground strata. They must also be familiar with characteristics of fossil fuels under extreme pressure and temperature conditions. Their work may extend into such interesting and exotic areas as examining the increased likelihood of earthquakes in regions where large amounts of materials are removed from underground deposits.

Petrochemical Engineer
Industrial engineering is fundamentally concerned with making processes more efficient and dependable. A petrochemical engineer must understand the chemistry of the reactions, the physical changes that occur, and the technology of the equipment and power requirements for those processes occurring in the industrial workplace. Engineers are expected to operate in a supervisory capacity, directing the efforts of teams of other employees. Engineers are also expected to report findings and procedures in written reports, published papers, and audiovisual presentations to interested groups.

Practice

Making Connections

20. Use the Internet to research any career connected with the petrochemical industry, and write a brief summary that describes
    (a) the type of work and how petrochemicals are involved in it;
    (b) the education required to qualify for employment in this field;
    (c) the current working conditions, opportunities, and salary for an employee in this field.

    Follow the links for Nelson Chemistry 11, 11.5.

GO TO **www.science.nelson.com**

Section 11.5 Questions

Understanding Concepts

1. Alkanes, alkenes, and alkynes are the three main families of aliphatic hydrocarbons.
   (a) What is the general molecular formula for each family?
   (b) What is the main structural feature of each family?
   (c) Why does the number of hydrogen atoms in the molecular formula decrease by 2 as you go from alkanes to alkenes and then to alkynes?
2. Why are there more possible isomers of an alkene than an alkane with the same number of carbon atoms?
3. Compare the physical properties of alkanes, alkenes, and alkynes.
4. Compare the chemical properties of alkanes, alkenes, and alkynes.
5. State one major use of the first member of the alkene and alkyne families.
6. Explain what is meant by the term "unsaturated" as applied to a hydrocarbon.
7. Draw structural diagrams and write IUPAC names for the five acyclic (non-ring) structural isomers of C_5H_{10}.
8. Write IUPAC names for the following hydrocarbons. Draw a structural diagram of and name any geometric isomers formed by these compounds.
   (a) $CH_3—CH_2—CH_2—CH_2—CH_3$
   (b) $CH_3—CH=CH—CH_2—CH_3$
   (c) $CH \equiv C—CH_2—CH_2—CH_3$
   (d) $CH_2=CH—CH_2—CH_3$
   (e) $CH_3—CH(CH_3)—CH=CH—CH_3$ (with the CH_3 branch on the second carbon)
9. Draw a structural diagram and write the IUPAC name for an alicyclic hydrocarbon that is a structural isomer of 1-butyne.
10. Draw structural diagrams, labelled with IUPAC names, for all the acyclic isomers of $C_4H_{6(g)}$.
11. Draw a structural diagram for each of the following hydrocarbons:
    (a) 3-ethyl-4-methyl-2-pentene
    (b) 5-ethyl-2,2,6-trimethyl-3-heptyne
12. List three reasons why there are more molecular compounds of carbon than compounds of all other elements combined.
13. What is the monomer from which polypropene (polypropylene) is made?
14. Polyvinyl chloride, or PVC plastic, has numerous applications. Write a chemical equation to represent the polymerization of chloroethene (vinyl chloride), $C_2H_3Cl_{(g)}$.

Applying Inquiry Skills

15. Fats and oils for cooking and consumption vary in structure. Some edible products are said to be high in polyunsaturated fats. Describe a possible chemical test for multiple bonds in polyunsaturated fats, explaining how to test, what the results might be, and what the possible results would indicate about any substances tested.
16. Using a labelled diagram, describe how the gaseous products of a chemical reaction may be collected. Title the diagram and indicate the kind of gases for which this process is suitable.

Making Connections

17. Ethyne (acetylene) is used in extremely large quantities by industrial processes. Normally, gaseous substances are liquefied under high pressure and stored in steel cylinders in order to provide a reasonably large quantity for use; cylinders of propane are a typical example. Research to find out and report why it is not advisable to highly compress acetylene and how solubility is used to store $C_2H_{2(g)}$ in cylinders.
18. As with most consumer products, the use of polyethylene has benefits and problems. What are some beneficial uses of polyethylene and what problems result from these uses? Suggest alternative substances for each application.
19. Hydrocarbons can be burned (as 95% currently are) or used in the production of petrochemicals such as polymers. Is it right for your generation to be burning a finite, nonrenewable resource? Write one pro and one con statement from economic, social, environmental, and ethical perspectives.

Chapter 11 Summary

Key Expectations

In this chapter, you have had the opportunity to do the following:

- Identify the origins and major sources of organic compounds. (11.1, 11.3)
- Name, using the IUPAC nomenclature system, and draw structural representations for aliphatic (acyclic and cyclic) hydrocarbons containing no more than 10 carbon atoms in the main chain, with or without side chains. (11.1, 11.4, 11.5)
- Use appropriate scientific vocabulary to communicate ideas about hydrocarbons. (11.1, 11.2, 11.3, 11.4, 11.5)
- Describe the steps involved in refining petroleum to obtain gasoline and other useful fractions. (11.2)
- Describe some of the physical and chemical properties of hydrocarbons. (11.2, 11.3, 11.4, 11.5)
- Carry out an experiment involving production or combustion of a hydrocarbon. (11.2, 11.3, 11.5);
- Demonstrate an understanding of the importance of hydrocarbons as fuels and in other applications, such as the manufacture of polymers, and to identify the risks and benefits of these uses to society and the environment. (11.3, 11.5)
- Demonstrate an understanding of the particular characteristics of the carbon atom, especially with respect to bonding in aliphatics and cycloalkanes, including structural isomers. (11.4, 11.5)
- Use molecular models to demonstrate the arrangement of atoms in isomers of hydrocarbons. (11.4, 11.5)
- Determine through experimentation some of the characteristic properties of saturated and unsaturated hydrocarbons. (11.5)
- Describe science- and technology-based careers in the petrochemical industry. (11.5)

Key Terms

acyclic hydrocarbons
addition polymerization
alicyclic hydrocarbons
aliphatic compounds
alkane
alkene
alkyl group
alkyne
complete combustion
cracking
crude oil
cycloalkane
fractional distillation (fractionation)
geometric (cis-trans) isomers
global warming
hydrocarbons
incomplete combustion
monomers
natural gas
organic chemistry
organic compound
petroleum
polymerization
polymers
refining
reforming
saturated hydrocarbons
structural isomer
unsaturated hydrocarbons

Make a Summary

1. Make a large table of hydrocarbons, with columns for alkanes, alkenes, and alkynes and rows for point-form summaries of
   (a) sources and uses
   (b) chemical and physical properties
   (c) typical structure and isomer formation
   (d) general molecular formula
   (e) nomenclature (naming) rules

Reflect on your Learning

Revisit your answers to the Reflect on Your Learning questions at the beginning of this chapter.

- How has your thinking changed?
- What new questions do you have?

Understanding Concepts

1. Using diagrams with labels, describe the geologic process believed to be responsible for the formation of petroleum.
2. What are the two fractions of raw petroleum called, and what property defines the difference between them?
3. What is the principal physical process used in the refining of petroleum?
4. What are the two principal chemical processes used in the refining of petroleum?
5. What are the two principal technological uses of hydrocarbons?
6. When space probes are sent to the other planets or the Moon, soil samples are collected and analyzed for organic compounds. Why are scientists interested in the possibility of the presence of organic compounds?
7. For each hydrocarbon compound formula given, use **Table 1**, page 514, to identify the fraction of refined petroleum in which the compound would be found. Add states of matter to each of the formulas.
   (a) $C_{13}H_{28}$
   (b) $C_{35}H_{72}$
   (c) C_7H_{16}
   (d) C_4H_{10}
   (e) $C_{11}H_{24}$
8. What is the economic purpose of the cracking and reforming reaction processes used in petroleum refining?
9. The main fuels used in Canada are hydrocarbons.
   (a) What are the products of the complete combustion of a hydrocarbon?
   (b) What are the possible products of the incomplete combustion of a hydrocarbon?
   (c) What diagnostic tests are used to test the products of a complete combustion of a hydrocarbon?
10. For each of these reforming or cracking reactions, draw a structural diagram wherever a name is given and a name wherever a structural diagram is given.
   (a) 1-butene + pentane → nonane
   (b) $CH_3-CH=CH-CH_3 + CH_3-CH(CH_3)-CH_3 \rightarrow CH_3-CH_2-CH(CH_3)-CH_2-CH(CH_3)-CH_3$
   (c) tetracosane, $C_{24}H_{50}$, + 2 H_2 → 3 octanes
   (d) $CH_3-CH_2-CH_2-CH_3 \rightarrow 2\ CH_2=CH_2 + H-H$
11. What are the three attributes of carbon atoms that are used to explain why carbon forms so many different compounds?
12. For each of the following hydrocarbon families, provide the general molecular formula and some information about a substance that is a common example of this family.
   (a) alkanes
   (b) alkenes
   (c) alkynes
13. Using examples containing four carbon atoms, draw structural diagrams and write IUPAC names for examples of the following hydrocarbon families. Where isomers are possible, draw and name each of the isomers.
   (a) alkanes
   (b) cycloalkanes
   (c) alkenes
   (d) alkynes
14. Draw and name the nine isomers of heptane, a common constituent of gasoline. Careful naming will automatically check whether any two structures you draw are really different isomers or just appear to be different because of the way they are drawn.
15. Draw all possible structures for 2-hexene and 2-methyl-2-hexene, and use them to explain cis-trans isomerism for the compound that has this feature.
16. Hydrocarbons can be classified as saturated or unsaturated. In the context of organic chemistry, what do the terms *saturated* and *unsaturated* mean?
17. Classify the following acyclic aliphatic hydrocarbons as saturated or unsaturated:
   (a) C_5H_{10}
   (b) C_5H_{12}
   (c) C_6H_{10}
   (d) C_6H_{12}
   (e) C_6H_{14}
   (f) C_7H_{14}

18. Which of the following could represent cyclic hydrocarbons? Draw structural diagrams to explain.
   (a) C_5H_{10}
   (b) C_5H_{12}
   (c) C_6H_{10}
   (d) C_6H_{12}
   (e) C_6H_{14}
   (f) C_7H_{14}

19. Which of the following names cannot be correct according to IUPAC naming rules for hydrocarbons? If the name is incorrect, provide the correct name if possible.
   (a) 2,2,3-trimethylpentane
   (b) 3-pentyne
   (c) 1,2-dimethylpropane
   (d) 3,3-dimethyl-3-hexene
   (e) 2-propene
   (f) 2-butene

20. The incomplete combustion of acetylene in air produces carbon, carbon dioxide, and water vapour. Given that one mole of carbon dioxide is produced for every two moles of acetylene reacted, write and balance a chemical equation for this reaction.

21. Branched alkane isomers have lower boiling points than the straight-chain structures, and the branched compounds' molecules are more nearly spherical in shape and so have smaller surface areas. For example, pentane boils at 36°C, 2-methylbutane at 28°C, and dimethylpropane at 10°C, even though they all have the formula C_5H_{12}.
   (a) Draw complete structural diagrams for these three isomers.
   (b) Use simple sketches (e.g., circles, ovals) and intermolecular forces to explain the boiling point variations.

22. Assuming the generalization for boiling points and branching of isomers is valid, match the following isomers to these boiling points: 69°C, 50°C, and 63°C.
   (a) 3-methylpentane
   (b) hexane
   (c) 2,2-dimethylbutane

23. Draw structural diagrams for the product, and then provide molecular formulas and names for each reactant and product of the following chemical equations:
   (a) $CH_2{=}CH{-}CH_3 + H{-}H \rightarrow$
   (b)

$$\begin{array}{l} CH_3{-}\underset{\displaystyle CH_3}{\underset{|}{CH}}{-}\underset{\displaystyle CH_2{-}CH_3}{\underset{|}{C}}{=}CH{-}CH_2{-}CH_3 + 2\,H{-}H \rightarrow \end{array}$$

   (c) $CH_3{-}CH_2{-}CH_2{-}C{\equiv}CH + 2\,H{-}H \rightarrow$

Applying Inquiry Skills

24. A student is required to use standard tests to identify three unlabelled hydrocarbon liquids. Complete the Analysis section of the investigation report.

Question
Which of the liquids is pentane, 2-methylbutane, and 2-methyl-2-butene?

Experimental Design
The liquids are each tested with a solution containing aqueous bromine. The solutions are then vaporized to measure their boiling points.

Evidence

Compound	Liquid 1	Liquid 2	Liquid 3
$Br_{2(aq)}$ test	no change	turns colourless	no change
boiling point (°C)	36	39	–12

Analysis
(a) Answer the Question.

Making Connections

25. Hydrocarbons are very useful in many applications other than combustion. The manufacture of polymers is only one example of this.
   (a) List some polymers that are used in our society.
   (b) What are some advantages and disadvantages of our use of polymers?
   (c) If we continue to burn about 95% of hydrocarbons, they will eventually become scarce. What effect will this have on the other applications of hydrocarbons? What are some alternatives if we still want the many polymers that we rely on?

26. Increased fuel efficiency in automobiles is beneficial from both an economic and an environmental perspective. Besides the fuel economy rating (L/100 km), what other properties or factors are involved in evaluating automobile fuels from both an economic and an environmental perspective?

27. WD-40 is a common commercial product that, like many other products, is labelled "contains petroleum distillate." It is also labelled with claims to stop squeaks, free rusted parts and sticky mechanisms, and protect metals by displacing moisture. The product can also be used to clean surfaces of greases and adhesives.
    (a) What is petroleum distillate?
    (b) Explain the label claims, using hydrocarbon properties and bonding theory.
    (c) The product "contains no CFCs (chlorofluorocarbons)." Research to determine what the general structure of such molecules would be and why they should not be contained in a commercial spray product.
28. Oxyacetylene torches are used for cutting and welding steel. This process uses both acetylene and oxygen tanks to feed a gas mixture to the torch.
    (a) In terms of combustion characteristics, why is the use of a separate tank of pure oxygen an advantage compared to using the surrounding air as a source of oxygen?
    (b) For the combustion of acetylene to be complete, what minimum volume of oxygen would be required per litre of acetylene measured at the same temperature and pressure?
    (c) If air were used as a source of oxygen, what minimum volume of air (about 20% oxygen) would be required?

Exploring

29. Many gasoline manufacturers advertise a grade of "green gas" that contains ethanol. Find out the composition of these environmentally friendly fuels and how well they are selling. Suggest some reasons for the relative amounts sold of green gas and regular gas.

    Follow the links for Nelson Chemistry 11, Chapter 11 Review.

    GO TO www.science.nelson.com
30. The Swedish chemist Jöns Jakob Berzelius, who is often called the father of modern chemistry, divided compounds into inorganic and organic categories. He believed organic compounds could only be produced by the action of living things. The first experiment suggesting that this theory was unacceptable was performed in 1828 by Berzelius's student and lifelong friend, German chemist Friedrich Wöhler. Wöhler heated the inorganic compound ammonium cyanate, $NH_4OCN_{(s)}$, and discovered that crystals forming from the vapour were urea, $H_2NCONH_{2(s)}$, a well-known organic compound that is a waste product of all mammals. While we now consider Wöhler's experiment a turning point, the scientific community was only finally convinced primarily by the later work of the French chemist Pierre Eugène Marcelin Berthelot (1827–1907). Use any biographical reference for scientists to
    (a) list the major contributions of Berzelius to chemistry theory and practice;
    (b) list a few of the syntheses of Berthelot that made him famous.

    Follow the links for Nelson Chemistry 11, Chapter 11 Review.

    GO TO www.science.nelson.com

Chapter

12

Energy from Hydrocarbons

In this chapter, you will be able to

- explain the role of heat in energy changes in chemical reactions;
- distinguish between heat and temperature;
- identify, measure, and describe the three factors that determine the quantity of heat transferred in an energy change;
- compare the energy changes observed in endothermic and exothermic reactions and explain using changes in chemical bonds;
- describe energy changes in endothermic and exothermic reactions;
- communicate heat transferred in chemical reactions as part of thermochemical equations;
- use the energy terms heat, heat capacity, specific heat, specific heat capacity, heat of reaction, and molar heat of reaction;
- conduct experiments to construct an energy device using the combustion of a hydrocarbon;
- obtain and analyze experimental evidence in calorimetry experiments;
- demonstrate an understanding of the importance of hydrocarbons as fuels and as sources for other products and identify the risks and benefits of these uses to society and the environment.

We use the term "energy" in everyday language in many different ways. Sometimes we say that we have "no energy" to do chores or that an active child is "full of energy." We get "energy" from breakfast cereals. We are mindful of not wasting "energy" by leaving the lights on unnecessarily. Our society is preoccupied by energy—its availability, management, benefits, and future sources. Perhaps that's because we use so much of it. North Americans consume more than one-quarter of the world's energy output, much of it from nonrenewable resources.

Our society depends on heat and on mechanical and electrical energy (**Figure 1**). However, as the old saying goes, "You can't get something for nothing." These valued forms of energy must come from other energy forms. The source of heat and mechanical and electrical energy for technological uses is almost entirely chemical energy from fossil fuels. Burning natural gas in furnaces keeps us warm; we rely on gasoline burned in car engines to travel from one place to another; and electrical utility companies burn natural gas, oil, or coal and convert the released heat into electricity. Our dependence on fossil fuels comes with costs both for ourselves and for others. Rapidly increasing fuel prices may finally force our society to seriously consider our reliance on fossil fuels, and how our demand for energy affects much poorer people in nations that do not have sufficient energy.

In this chapter, you will study energy changes from the combustion of hydrocarbons. You will learn how these energy changes can be calculated from laboratory evidence. You will also gain an understanding of both the usefulness and the impact of the combustion of fossil fuels.

Reflect on your Learning

1. Burning hydrocarbons generates energy (the combustion reaction is exothermic), but why?
2. What is the difference between heat and temperature?
3. What are some of the benefits and risks associated with the combustion of hydrocarbons as a source of heat energy?

Throughout this chapter, note any changes in your ideas as you learn new concepts and develop your skills.

Hot and Cold

What does it mean when something gets hot or cold? If the "something" is a pot on a hot stove element or an ice-cube tray in a freezer, then this seems like a simple process. But what if the "something" is a combination of chemicals that mix and react? What is happening in this case?

Materials: eye protection, a small (sandwich-size) resealable plastic bag, small beaker or graduated cylinder, laboratory scoop or plastic spoon, water, calcium chloride, citric acid, and sodium hydrogen carbonate (baking soda) solids

As a general precaution, wash your hands if you come into contact with any of the substances or solutions.

- Measure about 10 mL of water in the beaker or graduated cylinder.
- Place about 5 mL (approximately 1 tsp.) of calcium chloride solid in one bottom corner of a dry, plastic bag.
- Fold up and hold the corner of the bag containing the solid. Add the water to the opposite corner of the bag so that it is kept separate from the solid. Seal the bag.
- Grasp the top of the bag and allow the water and the solid to mix. Tilting the bag back and forth will ensure mixing.
- Hold the bottom of the bag in your hand.
- Record all observations from the initial mixing until no further changes occur.
- Dispose of the contents into the sink. Rinse and dry the inside of the bag.
- Repeat the above procedure, replacing the calcium chloride solid with about 5 mL (approximately 1 tsp.) each of citric acid and baking soda solids in the same corner of the bag and the water in the opposite corner.

(a) Which mixture felt hot after mixing with water? Where did the heat come from?
(b) Did you notice any change when the dry solids were added together?
(c) Describe the changes when water was added to the second mixture. What role does the water play?
(d) What does it mean when something feels cold? Is "cold" being transferred from the contents of the bag to your hand?
(e) Which of the two combinations you tested showed an endothermic change? an exothermic change?

Figure 1
Fossil fuels account for more than 90% of the total energy use in Canada, and a significant fraction of this is used to generate electricity.

Figure 1
In 2000 the Nanticoke electric power generating station on Lake Erie used 1360 t of coal per hour (at full power) and supplied 14% of Ontario's electrical requirements.

12.1 Classifying Energy Changes

Energy is essential to both life itself and our way of life: Breathing, digestion, the beating of our hearts, and the production of our food, shelter, and clothing all depend on energy. We are more and more dependent on our personal energy devices, such as radios, televisions, computers, and motorized vehicles. And our society demands energy to maintain itself, for example, to run our governments and businesses and to transport people and goods. Where does this energy come from? An *energy resource* is a natural substance or process that provides a useful form of energy from one of four basic energy sources (**Table 1**). For example, coal, oil, and natural gas are natural resources whose **chemical energy** we convert into heat and other forms of energy (**Figure 1**). **Nuclear energy** stored in uranium atoms, for example, is the energy source for nuclear reactors, which produce heat and electricity (**Figure 2**). Radiant energy from the Sun, or **solar energy**, and **geothermal energy** from Earth are used as energy sources to a limited extent. Unfortunately, these two energy sources are often too variable or produce too little energy; in the case of the geothermal sources, these natural resources are often too far from populated areas. A significant solar energy resource is the water in rivers, which is used by hydroelectric power generating stations.

chemical energy: energy derived directly from a chemical reaction; for example, burning of hydrocarbons in a furnace

nuclear energy: energy derived directly from a nuclear reaction; for example, a nuclear reactor for production of electricity

solar energy: energy directly and indirectly from solar radiation; for example, directly through a window or indirectly through the water cycle or winds

geothermal energy: energy from the heat of the Earth; for example, water pumped into and out of hot rock

Table 1: Energy Sources, Resources, and Forms

Energy sources	Natural resources	Technologically useful forms of energy
chemical	fossil fuels, plants	heat electrical energy mechanical energy light, sound
nuclear	uranium, hydrogen	
solar	direct radiant energy from the Sun; indirect energy from wind, water	
geothermal	geysers, hot springs	

As you have learned in previous studies, energy can be converted from one form to another. In fact, most of our scientific understanding of energy comes from studying changes in the forms of energy.

Practice

Understanding Concepts

1. List two energy-consuming devices that you use every day that are essential, two that are practical, efficient, or convenient, and two that are nonessential.
2. For each example in question 1, use **Table 1** to identify
   (a) the technologically useful form of energy;
   (b) the energy source and the natural resource from which the energy was obtained.

Reflecting

3. The nonrenewable nature of fossil fuels and their rising costs will likely force more conservation as well as a search for alternatives. List three examples of energy-conserving strategies or products that you or someone you know could use.

Energy Inventory

- Take a survey of energy consumption at your home, including both indoor and outdoor uses.
- Devise some general categories for the various devices or uses so that you do not list every individual item.
- Prepare a table that includes the following titles: Energy Device/Use, Type of Energy (directly consumed), and Natural Resource (used to supply the energy used).
- List as many general categories of energy devices/uses as you can to cover all energy used at your home.
- Complete the type of energy directly used and the natural resource for each category. You may want to ask other people at home for information—especially the person who pays the bills.

(a) Estimate what fraction of your total energy use depends directly or indirectly on the combustion of fossil fuels. If necessary, state your assumptions.

(b) What are some risks and benefits of a dependence on fossil fuels?

Figure 2
Over 20% of Ontario's electrical needs are supplied by the Pickering nuclear power station, which is situated on Lake Ontario, just east of Toronto.

Heat

Most familiar forms of energy are eventually converted to thermal energy, that is, the temperature of a substance is increased. Let's look at a few examples. The chemical energy from the combustion of gasoline in a car engine is converted mostly to thermal energy and partly to energy of motion of the car, which in turn is also converted to thermal energy by frictional forces.

The electrical energy used to operate a TV set is converted mostly to thermal energy and to some light and sound energy. The light and sound energy allow us to view and listen to the TV show, but this energy is absorbed by materials in the surroundings and converted to thermal energy. People obtain energy from food and lose energy as body heat and in the form of excretions (**Figure 3**). This transfer of energy as heat (whether by conduction, convection, or radiation) results in an increase in the thermal energy of the surroundings. So, **heat** is the energy being transferred between substances. An object possesses thermal energy but cannot possess heat.

heat: the energy being transferred between substances (in units, for example, of kilojoules and represented by the quantity symbol, *q*)

In order to measure the heat transferred, scientists measure the change in temperature with a thermometer to calculate the change in thermal energy. When the thermal energy changes, heat has been transferred. Consider the evidence displayed in **Table 2** (page 568).

Scanning down the heat and mass columns, we can see that the heat transferred and the mass of water and an oil are systematically manipulated to determine their effect on the temperature of the two liquids. According to this evidence, the temperature change, Δt, of a substance in a system or its surroundings varies directly with the quantity of heat, q, flowing into or out of the substance. Increasing the heat, q, gained increases the temperature change, Δt (see trials 1–4). Graphing the evidence shows that the temperature change varies inversely with the mass of the substance (**Figure 4**, page 568). Increasing the mass, m, of a substance decreases the temperature change, Δt (see trials 1, 5, 6, and 7).

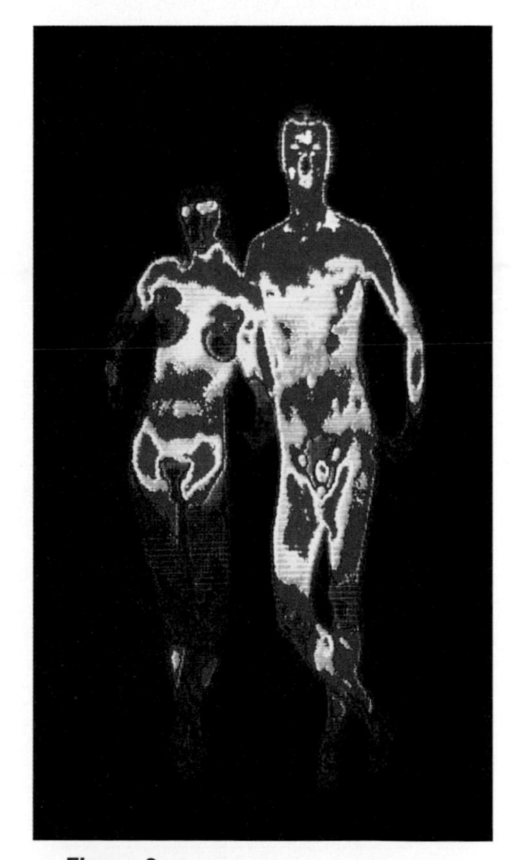

Figure 3
As shown by this thermogram, people radiate heat to the surroundings. The colour purple represents areas with the lowest heat transferred, and white represents areas with the highest heat transferred.

Note also in **Table 2** and **Figure 4** that the change in temperature is dependent on the type of substance involved, in this case, water or an oil. In every instance, the oil increases in temperature more than the water does. The composition of the substance affects the temperature change. Every substance has what

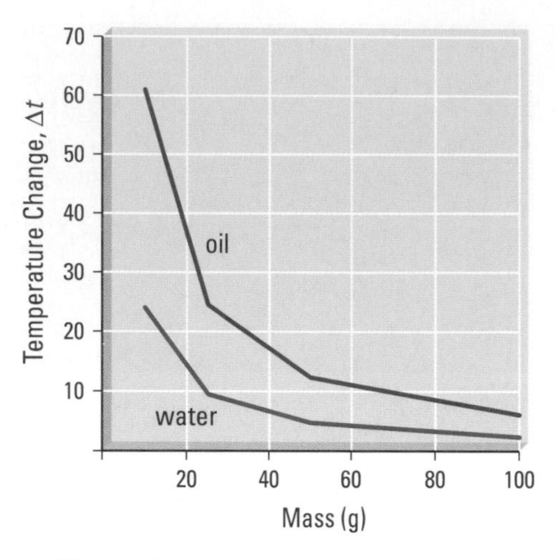

Figure 4
Temperature changes of water and oil at a constant heat input of 1000 J

Table 2: The Effect of Mass and Heat on the Temperature of Water and on Oil

			Temperature change, Δt		Specific heat capacity, c	
Trial	Heat, q (J)	Mass, m (g)	Water (°C)	Oil (°C)	Water (J/(g·°C))	Oil (J/(g·°C))
1	1000	100	2.39	6.10	4.18	1.64
2	2000	100	4.77	12.12	4.19	1.65
3	4000	100	9.58	24.10	4.17	1.66
4	8000	100	19.19	48.78	4.17	1.64
5	1000	50.0	4.77	12.34	4.19	1.62
6	1000	25.0	9.52	24.53	4.20	1.63
7	1000	10.0	24.04	60.98	4.16	1.64
Average					4.18	1.64

is called its *specific heat capacity*, a constant. The larger the temperature change, the smaller the specific heat capacity for a particular substance. A substance is said to have a large specific heat capacity when a relatively large quantity of heat must flow to produce a given temperature change per unit mass of the substance. (Recall that the SI unit for energy is the joule, J.)

specific heat capacity: the quantity of heat required to change the temperature of a unit mass of a substance by one degree Celsius (in units of, for example, J/(g·°C); represented by c)

The **specific heat capacity**, c, is the quantity of heat required to change the temperature of a unit mass (e.g., one gram) of a substance by one degree Celsius. For example, as calculated in the analysis of evidence from **Table 2**, the specific heat capacity of water is 4.18 J/(g·°C). This means that 4.18 J of energy is required to raise the temperature of one gram of water by one degree Celsius. The same quantity of energy is released by one gram of water if its temperature decreases by one degree Celsius. The oil analyzed is a typical component of motor oil, which lubricates and cools car motors. Its specific heat capacity is considerably smaller than that of water. Compared to most other substances, water has a very high specific heat capacity.

Table 3: Specific Heat Capacities (c), for Some Pure Substances (J/(g · °C))

ice	$H_2O_{(s)}$	2.01
water	$H_2O_{(l)}$	4.18
steam	$H_2O_{(g)}$	2.01
aluminum		0.90
calcium		0.65
copper		0.38
gold		0.13
hydrogen		14.27
iron		0.44
lead		0.16
lithium		3.56
magnesium		1.02
mercury		0.14
nickel		0.44
potassium		0.75
silver		0.24
sodium		1.23
sulfur		0.73
tin		0.21
zinc		0.39

From the analysis of the evidence in **Table 2,**

$$\Delta t \propto q$$
$$\Delta t \propto 1/m$$
$$\Delta t \propto 1/c$$
$$\Delta t \propto \frac{q}{mc} \text{ or } \Delta t = \frac{q}{mc}$$

(The units are chosen to make the constant of variations equal to one.) Rearranged, $q = mc\Delta t$.

The quantity of heat, q, that transfers varies directly with the mass of substance m, the specific heat capacity c, and the temperature change Δt. The specific heat capacity of a chemical can be calculated from the heat transferred, the mass, and the change in temperature, as was done in **Table 2:**

$$q = mc\Delta t$$
$$c = \frac{q}{m\Delta t}$$

From trial 1 for water in **Table 2,**

$$c = \frac{1000 \text{ J}}{100 \text{ g} \times 2.39°\text{C}}$$

$$c = \frac{4.18 \text{ J}}{\text{g·°C}}$$

Specific heat capacities vary for different substances and for different states of matter. See, for example, the specific heat capacities for the three states of water and for some other chemicals in **Table 3**. When calculating the quantity of heat that flows into or out of a substance, the value of the specific heat capacity must correspond to both the state of matter of the substance and the measured quantity (mass or volume) of the substance. For example, calculation of the quantity of heat flowing into a measured mass of ice requires the use of the specific heat capacity for $H_2O_{(s)}$, 2.01 J/(g·°C). If necessary, you may have to use the density of the substance to convert from mass to volume or vice versa. Note that in this book, quantities of heat transferred are calculated as absolute values by subtracting the lower temperature from the higher temperature.

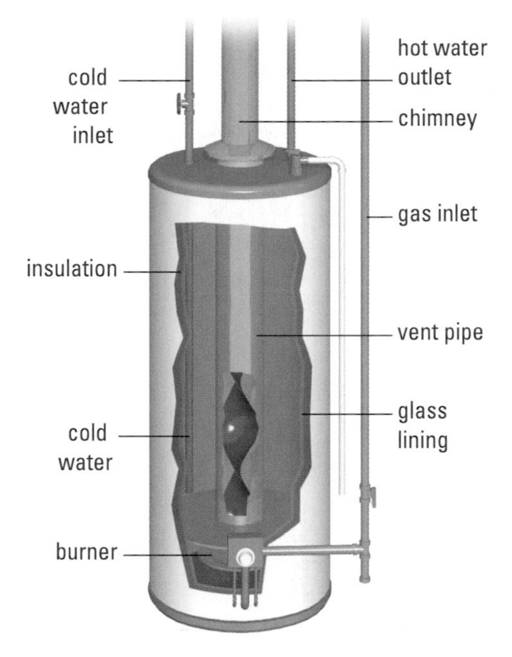

Figure 5
The chemical energy from the combustion of natural gas or fuel oil provides the heat that flows into the water in the insulated, glass-lined tank above the burner.

Sample Problem 1

Many water heaters use the combustion of natural gas (assume methane) to heat the water in the tank (**Figure 5**). When 150 L of water at 10°C is heated to 65°C, how much heat flows into the water?

Solution

$$m = 150\ \text{L} \times \frac{1\ \text{kg}}{1\ \text{L}}$$

$$= 150\ \text{kg}$$

$$c = 4.18\ \text{J/(g·°C)}$$

$$\Delta t = (65 - 10)\text{°C}$$

$$q = mc\Delta t$$

$$= 150\ \cancel{\text{kg}} \times \frac{4.18\ \text{J}}{\cancel{\text{g}}\cdot\cancel{\text{°C}}} \times (65 - 10)\cancel{\text{°C}}$$

$$q = 35\ \text{MJ}$$

The quantity of heat that flows into the water is 35 MJ.

Sample Problem 2

Winter hotels made of ice are becoming popular (**Figure 6**). The engineer at an ice hotel calculates that 10 guests radiate a total of 15 MJ of heat during an average stay. Before the guests arrive, the temperature of the ice in the hotel is −15°C. The engineer calculates that a safe temperature for the ice to rise to is −5°C. What volume of ice will absorb the 15 MJ of heat? Is this significant, or can the hotel easily handle the 10 guests? (The density of ice is 0.917 Mg/m^3, and according to **Table 3**, the specific heat capacity of ice is 2.01 J/(g·°C).)

Figure 6
Canada's first ice hotel near Quebec City was made in 2001 from 4500 t of snow and 250 t of ice.

Solution

$$q = 15\ \text{MJ}$$

$$\Delta t = ((-5) - (-15))\text{°C}$$

$$c = 2.01\ \text{J/(g·°C)}$$

$$q = mc\Delta t$$

$$m = \frac{q}{c\Delta t}$$

$$= \frac{15\ \text{M}\cancel{\text{J}}}{\frac{2.01\ \cancel{\text{J}}}{\text{g}\cdot\cancel{\text{°C}}} \times ((-5) - (-15))\cancel{\text{°C}}}$$

$$= 0.75 \text{ Mg, or } 0.75 \text{ t}$$

$$v_{H_2O_{(s)}} = 0.75 \cancel{\text{Mg}} \times \frac{1 \text{ m}^3}{0.917 \cancel{\text{Mg}}}$$

$$v_{H_2O_{(s)}} = 0.81 \text{ m}^3$$

The volume of ice that is heated is 0.81 m^3. There is more than enough ice to absorb the heat transferred from the people without causing structural damage.

Practice

Understanding Concepts

4. Listing all necessary controlled variables, state what happens to the change in temperature of a substance when it is heated and
   (a) the mass is doubled;
   (b) the heat transferred is doubled;
   (c) the specific heat capacity is doubled.
5. An electric kettle is used to raise the temperature of 1.50 L of water from 18.0°C to 98.7°C. Calculate the quantity of heat that flows into the water.
6. In an industrial plant using combustion of natural gas, 100 kg of steam is heated from 100°C to 210°C. Calculate the quantity of heat that flows into the steam.
7. Some North American Native peoples use rocks heated in fire pits to produce steam in sweat lodges for purification rites (**Figure 7**). If the specific heat capacity of rock is 0.86 J/(g·°C), what quantity of heat is released by a 2.5-kg rock cooled from 350°C to 15°C?
8. A 450-g stainless steel pot is used to warm 1.20 kg of water from 12.0°C to 65.0°C. What is the heat required to warm both the pot and the water? The specific heat capacity of stainless steel is 0.510 J/(g·°C).
9. Assume that the top 10 cm of an ice bed in an ice hotel (**Figure 6**) contains 0.200 m^3 of ice with a density of 0.917 Mg/m^3. Predict the change in temperature of this ice if a sleeping guest loses 1.8 MJ as heat during a night's sleep. Is the ice likely to melt?

Answers

5. 506 kJ
6. 22.1 MJ
7. 0.72 MJ
8. 278 kJ
9. 4.9°C

Applying Inquiry Skills

10. Write an experimental design for the investigation employed to gather the evidence in **Table 2.**
11. Evaluate the experimental design used to gather the evidence displayed in **Table 2.**

Figure 7
A traditional sweat lodge. In the sweat lodges of some North American Native peoples, rocks are heated to high temperatures, then water is sprinkled on them to produce steam.

Making Connections

12. **Table 3** gives the specific heat capacities for ice, water, and steam. For each of these states of matter, describe a simple situation from your everyday life in which its specific heat capacity can be used to calculate the heat transferred.
13. What is the capacity (volume) of the hot-water heater used in your home? Based on measured values of the temperature of the cold water and the hot water in your home, how much heat flows from the heater into the water when the tank is emptied, refilled, and heated?

Reflecting

14. When someone warned you about an object being hot, what did you use to think this meant? What do you think this means now?
15. If you make snowballs with your bare hands, your hands get cold. Does the coldness of the snow transfer to your hands? Explain.

Endothermic and Exothermic Changes

When heat is transferred between a system and its surroundings, evidence obtained from measurements of the temperature of the surroundings is used to classify the change as exothermic or endothermic. **Exothermic** changes usually involve an increase in the temperature of the surroundings (**Figure 8**), that is, the system has transferred heat to the surroundings (*exo* means "outside"). An example of an exothermic change is the burning of hydrocarbons. **Endothermic** changes usually involve a decrease in the temperature of the surroundings (**Figure 8**), that is, the system has absorbed heat from the surroundings (*endo* means "inside"). An example of an endothermic change is the cooking of foods.

exothermic: changes that usually involve an increase in the temperature of the surroundings; energy is transferred as heat from a chemical system to the surroundings.

endothermic: changes that usually involve a decrease in the temperature of the surroundings; energy is transferred as heat from the surroundings to a chemical system.

It is important to have a clear understanding of the relationship among heat, thermal energy, and temperature. Recall the kinetic molecular theory from earlier chapters: Any moving object has kinetic energy (E_k), and, because the molecules of a substance are always colliding, at any instant some molecules are moving faster than others. Therefore, in a large group of molecules there will be a range of kinetic energies from very low to very high values.

The thermal energy available from a substance is related to the total kinetic energy of all of the molecules. As long as there are no changes in state, the transfer of heat to a substance causes faster molecular motion, increasing both the *temperature,* which is related to the average kinetic energy, and the thermal energy, which is related to the total kinetic energy, of the substance. Therefore, a change in the **temperature** of the substance, as measured with a thermometer, is explained theoretically as a change in the average kinetic energy of the particles in the substance. This change in the average kinetic energy is represented by ΔE_k. If the temperature of the substance has decreased and there are no changes in state, then heat has flowed out of the substance.

temperature: a measure of the average kinetic energy of a substance's particles; represented by *t*, when using degrees Celsius

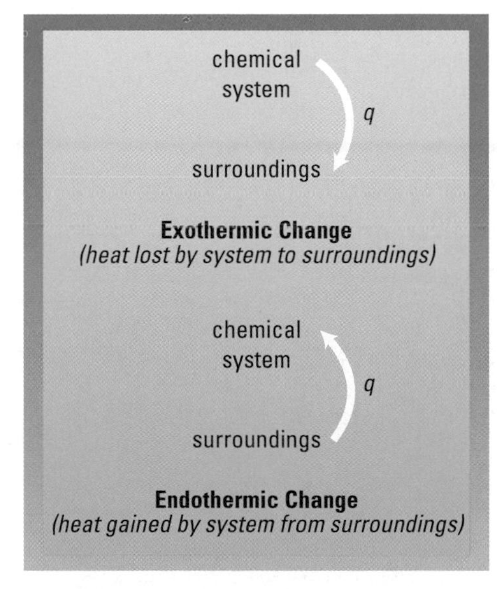

Figure 8
In an exothermic change, energy exits the chemical system, increasing the temperature of the surroundings. In an endothermic change, energy enters the system, decreasing the temperature of the surroundings.

Practice

Understanding Concepts

16. How are the following concepts related?
    (a) temperature *t* and kinetic energy E_k of the particles of a substance
    (b) the quantity of heat *q* transferred to the surroundings and the kinetic energy E_k of the particles of the surroundings
17. What happens to the average kinetic energy of the particles in the surroundings when
    (a) their temperature increases?
    (b) their temperature decreases?
    (c) an exothermic change occurs?
    (d) an endothermic change occurs?
18. Classify the following chemical changes as endothermic or exothermic:
    (a) cooking pasta on an electric element on a stove
    (b) burning methane or propane in a barbecue
19. Describe at the atom/molecular level what is happening to the surroundings during the endothermic and exothermic processes in the previous question.

INQUIRY SKILLS

- ○ Questioning
- ● Hypothesizing
- ○ Predicting
- ● Planning
- ● Conducting
- ● Recording
- ● Analyzing
- ● Evaluating
- ● Communicating

specific heat: the heat transferred per unit mass (in units of, for example, J/kg); represented by *h*

cooling rate: the heat flowing out of a substance per unit time (in units of, for example, J/min)

Investigation 12.1.1

Building a Water Heater

A water heater is an insulated container with an energy source to heat the water. The insulation is not perfect, so the water tends to cool as heat flows from the container to the surroundings. In this investigation you will design, build, and evaluate a water heater. To evaluate your heater, you will use two criteria: **specific heat**, the heat transferred per unit mass, and **cooling rate**, the heat flowing out of a substance per unit time. The better the water heater, the higher the specific heat in joules per kilogram and the lower the cooling rate in joules per minute.

The purpose of this investigation is to practise technological problem-solving skills related to an energy system by designing, building, and evaluating a water heater. As you test your design in the **Procedure**, record all of your observations, including unexpected ones. Record any modifications made to your device. For specific heat, use a temperature change from room temperature to 60°C. For the lab report, complete the **Hypothesis, Experimental Design, Materials, Analysis,** and **Evaluation** sections.

Question

What design for a simple water heater has the highest specific heat and lowest cooling rate?

Hypothesis

(a) Discuss and reach a consensus within your group about the features of your water heater. Record your final decision with reasons for your choices.

Experimental Design

(b) Sketch and label your proposed design, including the energy source you intend to use. Obtain your teacher's approval of the safety of your design.

Do not use flammable materials to construct your water heater.

Materials

(c) Prepare a list of all Materials for the construction and testing of your water heater.

Procedure

1. Construct your water heater.
2. Add a measured volume of room-temperature water.
3. Measure the total mass of the water heater (without the energy source).
4. Measure the initial temperature of the water.
5. Heat the water to 60°C while stirring it constantly.
6. Remove the heat source and let the water heater sit for 10 min.
7. Stir well and measure the final temperature of the water.
8. If time allows, repeat the Procedure using either the same or a modified Experimental Design.

Analysis

(d) Calculate the specific heat and cooling rate of your heater for all trials.
(e) Answer the Question, using your best design.

Evaluation

(f) Evaluate your Experimental Design, Materials, Procedure, and skills, noting any flaws and improvements that could be made in future models.
(g) Identify major sources of experimental error or uncertainty.
(h) Evaluate your Hypothesis. How well did your water heater work? Do your limited results indicate that you had a promising idea or is a better idea required?

Section 12.1 Questions

Understanding Concepts

1. How does specific heat capacity compare with specific heat?
2. Explain how mass, specific heat capacity, and change in temperature determine the quantity of heat transferred to or from a substance. Include the mathematical equation and a brief theoretical description of each term.
3. The combustion of methane in a natural gas stove provides the heat to raise the temperature of 1.1 L of water from 12°C to 98°C. Calculate the heat transferred.
4. What mass of water can be heated from 5°C to 78°C by a small, backpacking propane stove that produces 295 kJ of heat?
5. Greenhouses are easy to keep warm on sunny days, but at night they lose heat to their surroundings. Water, heated by sunlight, can be used to store heat during the day. The stored energy is then transferred to the air in the greenhouse overnight.
   (a) If we assume the same mass and temperature change, would a liquid such as an alcohol with a specific heat capacity of 2.1 J/(g·°C) work better than water for this use? Explain your answer.
   (b) Relative to water, approximately what mass of this alcohol would be required for the same temperature change?

12.2 Calorimetry

When you put a kettle of water on a hot stove element, a pizza in a hot oven, or jump into a cool lake on a hot summer day, heat is transferred. And, according to one of the fundamental laws of energy (known as the laws of thermodynamics), heat always transfers from a hotter object to a cooler object. What about the change in energy of the two objects? According to the law of conservation of energy, energy is neither created nor destroyed in any physical or chemical change. In other words, energy is only converted from one form to another.

To study energy changes, we need carefully designed experiments plus an *isolated system*, that is, one in which no energy can move in or out. If no energy is allowed to move between the inside and the outside of an isolated system, we can more easily study energy changes within the system. An example of such a system is called a **calorimeter** (**Figure 1**). Precise measurements are also needed; the process of measuring energy changes in chemical systems is called **calorimetry.** Inside the calorimeter, the chemical system being studied is surrounded by a known quantity of water (known as the "surroundings"). Energy is transferred between the chemical system and the water. Water is used not only because it is readily available, but also because it has one of the highest specific heat capacities. For example, to determine the energy required to melt ice, the ice may be placed in water inside a calorimeter. As the ice melts, heat transfers from

calorimeter: an isolated system consisting of some chemical system surrounded by a measured quantity of water and other components

calorimetry: the technological process of measuring energy changes in chemical systems

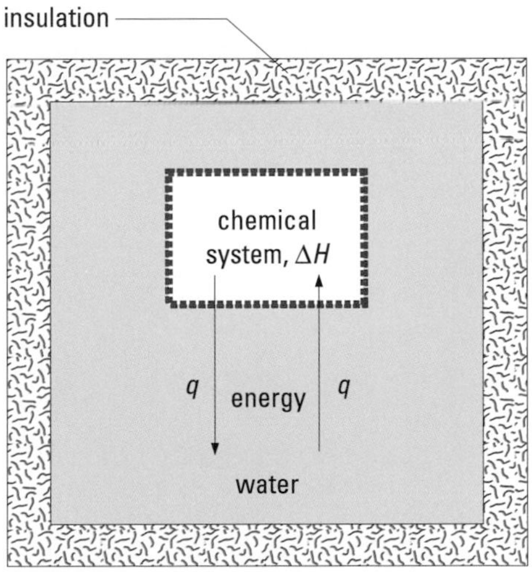

Figure 1
The system inside the calorimeter undergoes either a change in state, such as melting, or a chemical change, such as a double displacement reaction. Energy is either absorbed from the water or released to the water. An increase in the temperature of the water indicates an exothermic change in the system; a decrease in the temperature of the water indicates an endothermic change.

the water to the ice. The total energy gained by the ice is equal to the energy lost by the calorimeter water, as long as both the ice (system) and the water (surroundings) are part of an isolated system. In other words, for measurements to be accurate, no energy may be transferred between the inside of the calorimeter and the environment outside the calorimeter.

The specific heat capacities listed in **Table 3**, page 568, were determined by means of calorimetry. A variety of physical changes, such as changes in state, dissolving, and dilution, can be studied using calorimeters. The dissolving of substances in water may involve noticeable energy changes. The cold packs used in sports to treat sprains and bruises are a practical application of this fact. These packs contain a chemical that absorbs heat from the surroundings when it dissolves in water. The result is that the temperature of the injured part of the body (the surroundings) decreases, which reduces the severity of the injury.

Analysis of calorimetric evidence is based on the law of conservation of energy and on several assumptions. The law of conservation of energy may be expressed in several ways, for instance, "The total energy change of the chemical system is equal to the total energy change of the calorimeter surroundings." Using this approach, we can calculate both the energy change of the chemical system, ΔE, and the quantity of heat, q, as absolute values, without using a positive or a negative sign:

$$\underset{\text{(system)}}{\Delta E} = \underset{\text{(calorimeter)}}{q}$$

The main assumption is that no heat is transferred between the calorimeter and the outside environment. Another assumption is that any heat absorbed or released by the calorimeter materials, such as the container, is negligible. Also, a dilute aqueous solution is assumed to have the same density and specific heat capacity as pure water.

The energy change of the chemical system, ΔE, depends on two quantities: the mass of chemical (directly related to the number of particles that make it up) and the energy constant for the change occurring (related to the particular chemical and the changes in the bonding). One common way to specify the energy constant is as a specific heat, h, the energy absorbed or released per unit mass of a chemical:

$$\Delta E = mh$$

Usually we specify the type of change occurring, such as the specific heat of melting, solution (dissolving), or combustion, by including a subscript with the energy terms. For example, in a calorimetry experiment, 4.2 g of lithium chloride is dissolved in 100 mL (100 g) of water at an initial temperature of 16.3°C. The final temperature of the solution is 25.1°C. To calculate the specific heat of solution, h_s, for lithium chloride, the first step is to use the law of conservation of energy:

$$\underset{\text{(LiCl dissolving)}}{\Delta E_s} = \underset{\text{(calorimeter water)}}{q}$$

In other words, the heat lost by dissolving of the lithium chloride equals the heat gained by the calorimeter water (surroundings). The energy change of lithium chloride dissolving to form a solution depends on the specific heat (in kilojoules per gram) and the mass (in grams); the heat gained by the water is expressed in the usual heat formula, $mc\Delta t$:

$$mh_s = mc\Delta t$$

Assuming that the dilute solution has the same physical properties as pure water, we can now obtain the specific heat of solution by substituting the given information and the appropriate constants into this equation:

$$mh_s = mc\Delta t$$

$$4.2\ \text{g} \times h_s = 100\ \text{g} \times \frac{4.18\ \text{J}}{\text{g} \cdot {}^\circ\text{C}} \times (25.1 - 16.3)^\circ\text{C}$$

$$h_{s_{\text{LiCl}}} = \frac{0.88\ \text{kJ}}{\text{g}}$$

Since the temperature of the water in the calorimeter increases, the dissolving of lithium chloride is exothermic. Therefore, the specific heat of solution for lithium chloride is reported as an exothermic 0.88 kJ/g.

Investigation 12.2.1

Hot and Cold Packs

INQUIRY SKILLS

- ○ Questioning
- ○ Hypothesizing
- ○ Predicting
- ○ Planning
- ● Conducting
- ● Recording
- ● Analyzing
- ○ Evaluating
- ● Communicating

As a general precaution, handle all materials with care and wash your hands if you get any solution on them. Follow the safety precautions in the MSD sheets for the particular chemicals used.

Hot and cold packs usually contain a chemical and water. The hot-pack chemical has an exothermic specific heat of solution, and the cold-pack chemical has an endothermic specific heat of solution. The purpose of this investigation is to investigate the energy changes for two of the chemicals that may be used in these packs. Complete the **Experimental Design** and **Analysis** sections of the laboratory report.

Question

What is the specific heat of solution of some ionic solids?

Experimental Design

A simple calorimeter (**Figure 2**) is used to measure the temperature change when a known mass of a compound is dissolved in the water.

(a) Calculate the mass of each compound required to make 50 mL of a 1.00 mol/L solution.

(b) Use the MSD sheets to determine the hazards associated with the compound and list the necessary precautions.

Materials

lab apron
eye protection
samples of ionic compounds
bottle of distilled water
laboratory scoop
weighing boat or paper
centigram balance
50-mL graduated cylinder
medicine dropper
calorimeter apparatus (**Figure 2**)

Procedure

1. Using a graduated cylinder, measure 50 mL of water and pour it into the calorimeter.
2. Obtain the required mass of one of the compounds in a suitable container.
3. Record the initial temperature of the water.

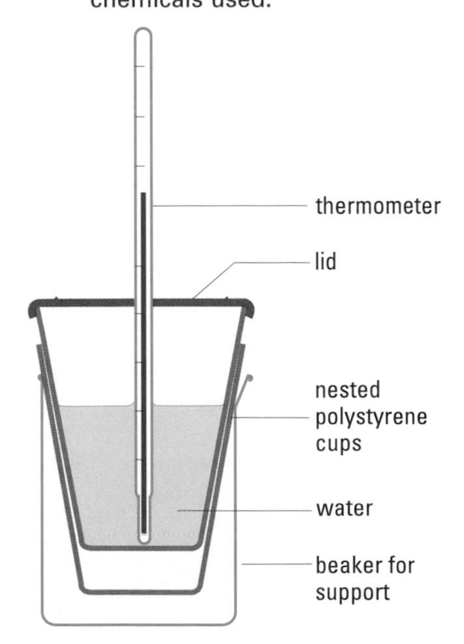

Figure 2
A simple laboratory calorimeter. The chemical system that will undergo an energy change is placed in or dissolved in the water of the calorimeter.

4. Add the compound to the water.
5. Cover the calorimeter and stir with the thermometer.
6. Measure and record the temperature of the water repeatedly. The temperature will either increase or decrease. Record the temperature at which the temperature change is maximum (either the lowest or highest temperature), which is called the final temperature. To ensure you have recorded the best value of the final temperature, continue to measure and record the temperature of the solution for a few minutes as it returns to room temperature.
7. Dispose of the contents of the calorimeter into the sink and flush down the drain with lots of water. Rinse and dry the inside of the calorimeter.
8. Repeat, using other compounds provided.

Analysis

(c) Calculate the specific heat of solution for each compound.

Practice

Understanding Concepts

1. List three assumptions made in student investigations involving simple calorimeters.
2. In a calorimetry experiment, which measurements limit the final certainty (number of significant digits) of the experimental result?
3. Distinguish between specific heat and specific heat capacity.
4. In a chemistry experiment, 10 g of the fertilizer urea, $NH_2CONH_{2(s)}$, is dissolved in 150 mL of water in a simple calorimeter. A temperature change from 20.4°C to 16.7°C is measured. Calculate the specific heat of solution for urea.
5. A commercial hot pack, sold for outdoor use, uses 275 g of calcium chloride and a rubber bladder with a fleece lining. When preparing to use this hot pack, 750 mL of water is placed into the rubber bladder and the calcium chloride is then added. If the calcium chloride has a specific heat of solution of 0.52 kJ/g, what is the maximum temperature change that occurs when it dissolves?
6. A laboratory technician initially adds 43.1 mL of concentrated, 11.6 mol/L hydrochloric acid to water to form 500 mL of dilute solution. The temperature of the solution changes from 19.2°C to 21.8°C. Calculate the specific heat of dilution of hydrochloric acid.
7. A chemistry teacher designs a calorimetry lab in which students prepare a 250-mL solution of ammonium nitrate, which has an endothermic specific heat of solution of 0.31 kJ/g. Predict the mass of ammonium nitrate that must be dissolved to produce a temperature decrease of 5.0°C.

Applying Inquiry Skills

8. It is commonly assumed in calorimetry labs with polystyrene calorimeters that a negligible quantity of heat is absorbed or released by the solid calorimeter materials (the cups, the stirring rod, and the thermometer). Use the data in **Table 1** to answer the following:
   (a) For a temperature change of 5.0°C, calculate the energy change of the water only.
   (b) For a temperature change of 5.0°C, calculate the total energy change of the water, polystyrene cups, stirring rod, and thermometer.

Answers

4. endothermic 0.23 kJ/g
5. 46°C
6. exothermic 0.30 kJ/g
7. 17 g
8. (a) 2.1 kJ
   (b) 2.2 kJ
   (c) 3.6%

DID YOU KNOW ?

Safety Reminder

Note that caution and suitable safety procedures are required when diluting strong, concentrated acids. Small areas within the mixing fluids can become hot, causing the mixture in those local areas to boil and splatter corrosive liquid.

Table 1: Typical Quantities for Materials in a Simple Calorimeter

Material	Specific heat Capacity (J/(g·°C))	Mass (g)
water	4.18	100.00
polystyrene cups	0.30	3.58
glass stirring rod	0.84	9.45
thermometer	0.87	7.67

(c) Calculate the percentage difference introduced by using only the energy change of the water.
(d) Evaluate the assumption of negligible heat transfer to the solid calorimeter materials.

Making Connections

9. Self-heating meals are used by the armed forces and are also available to consumers for outdoor activities such as camping and hiking. Find out how these systems work. What are the advantages and disadvantages of these systems compared with the use of a portable stove or heater?

   Follow the links for Nelson Chemistry 11, 12.2.

GO TO www.science.nelson.com

Reflecting

10. Is the water heater in your home a calorimeter? List some similarities and differences.

Bomb Calorimeters and Heat Capacity

Many chemical reactions that are of interest to chemists and engineers do not take place in aqueous solutions. An important practical example is combustion. Oil companies are always researching better gasolines, and one important characteristic of gasoline is the heat released during combustion. A calorimeter made of polystyrene cannot be used to study the energy changes of combustion reactions, since polystyrene has a low melting point and burns readily. So how do we study combustion reactions?

These types of reactions can be studied by using a bomb calorimeter (**Figure 3**). The inner reaction compartment is called a bomb because in early

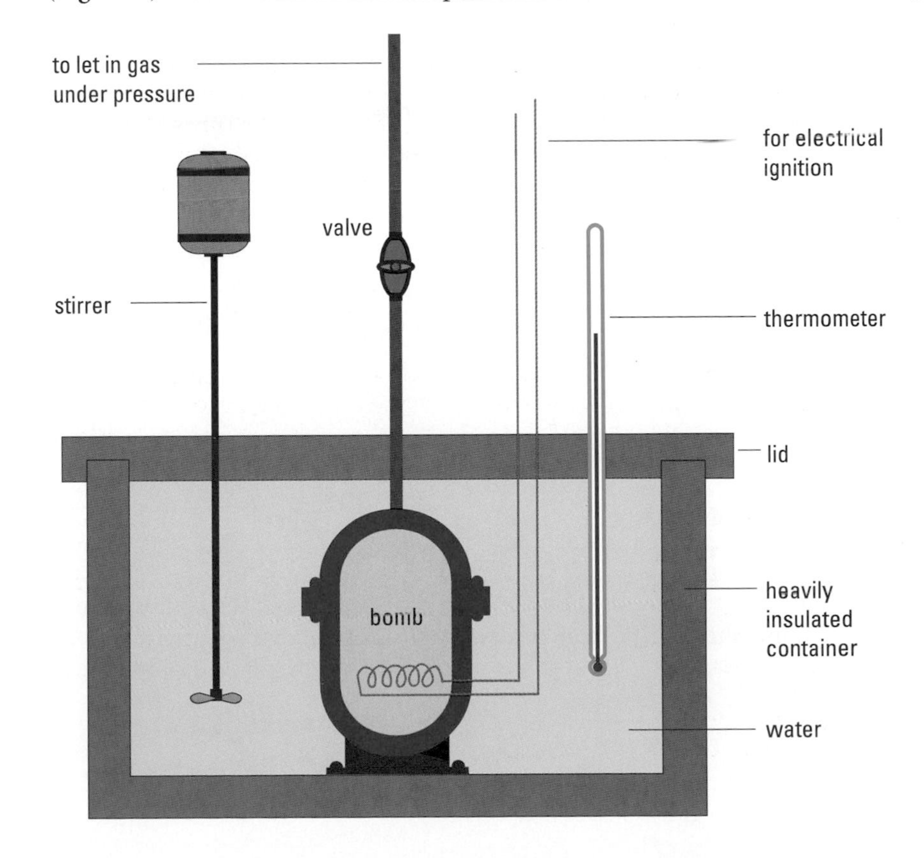

Figure 3
The reactants are placed inside the calorimeter's bomb, which is surrounded by the calorimeter water. Once the calorimeter is sealed and the initial temperature measured, the combustion reaction is initiated by an electric heater or spark. Stirring is essential in order to obtain a uniform final temperature for the water.

models it often exploded. Modern bomb calorimeters are strong enough to withstand explosive reactions. The energy change of the reaction is determined from the temperature change of the calorimeter.

Researchers use bomb calorimeters to measure energy changes for the combustion of fuels, oil, foodstuffs, certain crops, and explosives. But calorimeters that are larger and more sophisticated than polystyrene cups usually have a noticeable heat transfer to or from the components of the calorimeter. It is no longer possible to ignore the calorimeter materials as we did with the simple polystyrene calorimeter. The total energy change of the calorimeter is the sum of the energy changes of all of the components of the calorimeter:

$$\underset{\text{(calorimeter)}}{\Delta E_{\text{total}}} = \underset{\text{(water)}}{m_1c_1\Delta t} + \underset{\text{(containers)}}{m_2c_2\Delta t} + \underset{\text{(stirrer)}}{m_3c_3\Delta t} + \underset{\text{(thermometer)}}{m_4c_4\Delta t}$$

Because the temperature change, Δt, is identical for all components and the same components are used over and over again, this total energy calculation can be simplified. The different constants in the equation can be replaced by a single constant, C, the **heat capacity** of the particular calorimeter:

heat capacity: the energy absorbed or released by a system per degree Celsius (in units of, for example, kJ/°C; represented by C)

$$\Delta E_{\text{total}} = (m_1c_1 + m_2c_2 + m_3c_3 + m_4c_4)\,\Delta t$$

$$\Delta E_{\text{total}} = C\Delta t, \text{ where } C \text{ is the heat capacity of the calorimeter}$$

The total energy absorbed by the calorimeter is the energy transferred from the chemical system as heat, q. Therefore,

$$q = C\Delta t \text{ for a bomb calorimeter}$$

DID YOU KNOW ?

Heat Quantity Names and Units

The names and units for different heat quantities can be confusing. You have to be careful to distinguish among the following:

quantity of heat	q	J
specific heat capacity	c	J/(g·°C)
heat capacity	C	J/°C
specific heat	h	J/g

Manufacturers may provide a value for the heat capacity of the calorimeter, or the calorimeter may be calibrated by the user with a well-known standard before it is used for calorimetric analysis. Suppose a 1.50-g sample of sucrose, together with excess oxygen gas, is placed in the bomb of a calorimeter whose heat capacity is specified by the manufacturer as 8.57 kJ/°C. The temperature changes from 25.00°C to 27.88°C. What is the specific heat of combustion for sucrose? According to the law of conservation of energy, the energy released from the combustion (ΔE_c) equals the total heat transferred (q) to the calorimeter:

$$\underset{\text{(sucrose)}}{\Delta E_c} = \underset{\text{(calorimeter)}}{q}$$

The energy change for the chemical system is calculated, as before, using the mass and specific heat, and the total energy change of the calorimeter as $C\Delta t$:

$$mh_c = C\Delta t$$

$$1.50\text{ g} \times h_c = \frac{8.57\text{ kJ}}{\cancel{°C}} \times (27.88 - 25.00)\cancel{°C}$$

$$\underset{\text{sucrose}}{h_c} = \frac{16.5\text{ kJ}}{\text{g}}$$

Since the temperature of the calorimeter has increased, the combustion is exothermic and the specific heat of combustion of sucrose is reported as an exothermic 16.5 kJ/g.

Practice

Understanding Concepts

11. Canadian inventors have developed zeolite, a natural aluminum silicate mineral, as a storage medium for solar heat. Zeolite releases heat when hydrated (mixed with water to form a hydrate). In a test, zeolite is used to heat water in a tank that has a heat capacity of 157 kJ/°C. What is the energy change of hydration (ΔE_h) for zeolite if the temperature of the water increases from 27°C to 73°C?
12. An oxygen bomb calorimeter has a heat capacity of 6.49 kJ/°C. The complete combustion of 1.12 g of acetylene (ethyne) produces a temperature change from 18.60°C to 27.15°C. Calculate the specific heat of combustion, h_c, for acetylene, $C_2H_{2(g)}$.

Applying Inquiry Skills

13. Before energy changes of a reaction can be determined, a bomb calorimeter must be calibrated using a primary standard of precisely known energy content. Complete the Analysis section of the investigation report.

Question

What is the heat capacity of a newly assembled oxygen bomb calorimeter?

Experimental Design

An oxygen bomb calorimeter is assembled, and several samples of the primary standard, benzoic acid, are burned using a constant pressure of excess oxygen. The evidence that is collected (**Table 2**) determines the heat capacity of the calorimeter for future experiments.

Evidence

In *The CRC Handbook of Chemistry and Physics*, the specific heat of combustion for benzoic acid is reported as an exothermic 26.46 kJ/g.

Analysis

(a) Calculate the heat capacity of the bomb calorimeter for each trial.
(b) Answer the Question, using the average experimental value.

Making Connections

14. A Subaru Outback has a cooling system capacity of 6.0 L. The recommended coolant is a 1:1 V/V mixture of water and commercial automotive coolant. Canadians often call the coolant "antifreeze." Mixing coolant with water lowers the freezing point and also adds anti-corrosion compounds; both are features that protect the engine. The main ingredient of commercial coolant is the organic liquid 1,2-ethanediol (ethylene glycol), which has a lower specific heat capacity than water and is also notably more dense. The specific heat capacity of the coolant mixture is about 3.8 J/(g·°C), and its density is 1.06 g/mL. The car's cooling system regulates flow through the engine and radiator, maintaining coolant temperature (and thus engine temperature) at about 185°C, as the coolant flows out of the engine to the radiator.
(a) Find the heat capacity for the coolant in an Outback engine.
(b) What energy is removed from the total coolant volume of the engine by the radiator in driving conditions where the coolant reenters the engine at 150°C?
15. Besides the specific heat of combustion as determined in a bomb calorimeter, what other properties or factors are involved in evaluating alternative automobile fuels such as propane, ethanol, and hydrogen?

Answers

11. 7.2 MJ
12. exothermic 49.5 kJ/g
13. (a) 8.94 kJ/°C, 8.96 kJ/°C, 8.95 kJ/°C
    (b) 8.95 kJ/°C
14. (a) 24 kJ/°C
    (b) 0.85 MJ

Table 2: Calorimetric Evidence for the Combustion of Benzoic Acid

Trial	1	2	3
mass of $HC_7H_5O_{2(s)}$ (g)	1.024	1.043	1.035
initial temperature (°C)	24.96	25.02	25.00
final temperature (°C)	27.99	28.10	28.06

INQUIRY SKILLS

- ○ Questioning
- ○ Hypothesizing
- ○ Predicting
- ● Planning
- ● Conducting
- ● Recording
- ● Analyzing
- ● Evaluating
- ● Communicating

Investigation 12.2.2

Specific Heat of Combustion

Specific heats of combustion are usually determined in a bomb calorimeter. However, a bomb calorimeter is a relatively expensive piece of equipment that is common in university and industrial laboratories but rare in high schools. The purpose of this investigation is to create an apparatus to measure energy changes in a combustion reaction. Use a metal can or, if you did Investigation 12.1.1, modify the water heater you designed to obtain an approximate value for the specific heat of combustion of a hydrocarbon. Complete the **Experimental Design**, **Materials**, **Procedure, Analysis**, and **Evaluation** sections in the investigation report.

Question

What is the specific heat of combustion of paraffin from an ordinary paraffin candle?

Experimental Design

The burning of a paraffin candle is used to heat water in a calorimeter. Measurements of the mass of candle, mass (or volume) of water, and temperature change of the water are used to determine the specific heat of combustion of paraffin wax.

(a) Write a Procedure, including any safety and disposal instructions. Have your Procedure approved by your teacher before you start.

Materials

(b) List all supplies and equipment you will need to create or modify your calorimeter and to carry out the investigation, with approximate quantities and sizes. Include a diagram of your calorimeter.

Procedure

1. Carry out your Procedure.

Analysis

(c) Answer the Question.

Evaluation

(d) Evaluate your calorimeter design, plan, Materials, and Procedure, noting any flaws. Suggest some improvements, if necessary.

(e) Identify sources of experimental error or uncertainty and try to classify them as minor or major factors in this experiment.

(f) How confident are you about the answer you obtained?

Section 12.2 Questions

Understanding Concepts

1. For each of the following energy terms, state the symbol, typical units, and the situation in which this term may be used:
   (a) specific heat capacity (b) heat capacity
   (c) specific heat

2. In a calorimetry experiment, a student recorded the following evidence:

   volume of water = 125 mL

   initial temperature of the water = 19.6°C

   final temperature of the water = 30.8°C

   (a) What quantity of heat was transferred to the water?
   (b) Was the change in the chemical system endothermic or exothermic? Justify your answer.

Applying Inquiry Skills

3. Bomb calorimeters can be used to determine the energy content of foods by combustion analysis. Complete the **Analysis** section of the investigation report.

   Question

   Which substance, fat or sugar, has the higher specific heat of combustion?

   Experimental Design

   A sample of one component of fat (stearic acid, $HC_{18}H_{35}O_2$) is completely burned in a bomb calorimeter. The specific heat of combustion is determined and compared with the previously determined value for sucrose, 16.5 kJ/g.

   Evidence

   mass of stearic acid = 1.14 g

   heat capacity of calorimeter = 8.57 kJ/°C

   initial temperature = 25.00°C

   final temperature = 30.28°C

   Analysis

   (a) Calculate the specific heat of combustion of fat.
   (b) Compare your value for fat with the value for sucrose and answer the Question.

Making Connections

4. The energy content of foods is usually obtained from calorimetry experiments. Collect labels, or record energy information, from various components of different foods. Analyze this information and determine which types of foods contain the most energy per gram and which the least.
5. Oil companies are constantly developing and testing new gasolines to try to obtain an edge in the competitive consumer gasoline market.
   (a) What energy characteristic of gasoline is important in its use? Why?
   (b) What other characteristics of gasoline might also be investigated?

12.3 Heats of Reaction

All chemical reactions involve energy changes. Some reactions like combustion (burning) are obviously exothermic. You can feel the heat and see the light emitted from a burning campfire or fireplace. In an exothermic reaction, energy is a product just like the chemical products of the reaction. Exothermic reactions always release energy into the surroundings:

$$\text{reactants} \rightarrow \text{products} + \text{energy}$$

Endothermic reactions are less familiar in everyday life, but not unknown. If you put an Alka-Seltzer tablet in a glass of water, a chemical reaction takes place, as evidenced by the gas bubbles produced. If you feel the glass, you will notice that it feels cold. The reaction is absorbing heat from the surroundings (water, glass, air). In an endothermic reaction, energy is consumed like a reactant. Endothermic reactions always absorb energy from the surroundings:

$$\text{reactants} + \text{energy} \rightarrow \text{products}$$

You have used calorimetry to determine the specific heat of reaction for a particular substance. Now we can use this experimental value to write a chemical equation that includes the heat transferred. This chemical equation is called a **thermochemical equation**, and the quantity of heat transferred is called the **heat of reaction.**

thermochemical equation: a balanced chemical equation that includes the heat transferred to or from the surroundings

heat of reaction: the heat transferred in a reaction based on the amounts given by the coefficients of the balanced chemical equation (in units of, for example, kJ)

molar heat of reaction: the quantity of heat transferred in a reaction per mole of a specified substance (in units of, for example, kJ/mol); represented by ΔH_r

A chemical equation communicates the relative amounts, in moles, of all reactants and products. To be able to relate experimentally measured specific heats to chemical equations, the energy must be expressed as a quantity of heat per mole, such as kilojoules per mole; this quantity is called the **molar heat of reaction,** ΔH_r, for a particular substance. For example, the specific heat of combustion of octane, from a calorimetry experiment, is an exothermic 44.41 kJ/g. How can this be converted into a molar heat of combustion?

First, a conversion factor to convert kilojoules per gram into kilojoules per mole must be found. The mass in grams must be converted into an amount in moles. As you will recall from Unit 2, the conversion factor that was created by chemists to complete this conversion is the molar mass of the chemical. Using the molar mass as a conversion factor, we can convert the specific heat of combustion of octane, h_c, into the molar heat of combustion of octane, ΔH_c:

$$\underset{C_8H_{18(l)}}{\Delta H_c} = \frac{44.41\ \text{kJ}}{1\ \text{g}} \times \frac{114.26\ \text{g}}{1\ \text{mol}}$$

$$\underset{C_8H_{18(l)}}{\Delta H_c} = \frac{5074\ \text{kJ}}{\text{mol}}$$

This means that the combustion of one mole of octane releases 5074 kJ of heat. Notice how the units cancel in the above calculation. Cancelling units is always a good procedure to check to see if you are calculating correctly.

Molar heats of reaction must always specify the type of reaction (in this case the subscript used is "c" for combustion) and the substance reacted (in this case octane, $C_8H_{18(l)}$).

Sample Problem 1

Ethane is the second largest component of natural gas. If its specific heat of combustion is 51.85 kJ/g, what is the molar heat of combustion of ethane?

Solution

$h_{c_{C_2H_{6(g)}}} = 51.85 \text{ kJ/g}$

$$\Delta H_{c_{C_2H_{6(g)}}} = \frac{51.85 \text{ kJ}}{1 \text{ g}} \times \frac{30.08 \text{ g}}{1 \text{ mol}}$$

$$\Delta H_{c_{C_2H_{6(g)}}} = \frac{1560 \text{ kJ}}{\text{mol}}$$

The molar heat of combustion of ethane is an exothermic 1560 kJ/mol.

Thermochemical Equations

To combine a calculated molar heat of reaction with a chemical equation, you need to pay particular attention to the balancing of the equation. For example, the molar heat of combustion of octane (a major component of gasoline) is 5074 kJ/mol. Because combustion reactions are always exothermic, this means that 5074 kJ of heat is transferred to the surroundings during the combustion of one mole of octane. If we balance the chemical equation using "1" as the coefficient for octane, we can write the heat of reaction immediately without any calculations:

$$1\ C_8H_{18(l)} + \frac{25}{2}\ O_{2(g)} \rightarrow 8\ CO_{2(g)} + 9\ H_2O_{(g)} + 5074 \text{ kJ}$$

This equation is read as "one mole of octane reacts with twelve and a half (25/2) moles of oxygen to produce eight moles of carbon dioxide, nine moles of water, and five thousand and seventy-four kilojoules of heat." Of course, you may still write the balanced chemical equation with whole-number coefficients, but the heat of reaction must be doubled because 2 mol of octane are shown:

$$2\ C_8H_{18(l)} + 25\ O_{2(g)} \rightarrow 16\ CO_{2(g)} + 18\ H_2O_{(g)} + 10\ 148 \text{ kJ}$$

The heat term is part of the balanced equation, and if the coefficients are changed, then the heat of reaction is changed in the same way. In the previous example, if the coefficients are doubled, then the energy is doubled. This is necessary so that the equation agrees with the empirical molar heat of reaction, as shown below:

$$\Delta H_{c_{C_8H_{18(l)}}} = \frac{10\ 148 \text{ kJ}}{2 \text{ mol}}$$

$$\Delta H_{c_{C_8H_{18(l)}}} = \frac{5074 \text{ kJ}}{1 \text{ mol}}$$

To write a thermochemical equation, it is necessary to know the balanced chemical equation, the molar heat of reaction for one substance in the equation, and whether the reaction is endothermic (takes in energy) or exothermic (gives out energy). If a reaction is endothermic, remember to put the heat of reaction on the reactant side of the chemical equation. This heat is consumed, or used up, like the reactants and is listed along with the reactants. For example, the decomposition of water is endothermic and has a molar heat of reaction of 285.8 kJ/mol of water. The thermochemical equations may be written as

$$H_2O_{(l)} + 285.8 \text{ kJ} \rightarrow H_{2(g)} + \frac{1}{2} O_{2(g)}$$

or $2\ H_2O_{(l)} + 571.6 \text{ kJ} \rightarrow 2\ H_{2(g)} + O_{2(g)}$

Whether a particular reaction is endothermic or exothermic is not easy to recognize, so you will be given this information as part of any question. The only exception is a combustion reaction, which you are expected to know is exothermic.

Sample Problem 2

A student is experimenting with different substances to make a cold pack. From a calorimetry experiment she determines that the specific heat of solution of potassium bromate, $KBrO_{3(s)}$, is an endothermic 0.25 kJ/g. Calculate the molar heat of solution and write the thermochemical equation for the dissolving process.

Solution

$h_{s\,KBrO_3} = 0.25 \text{ kJ/g}$

$$\Delta H_{s\,KBrO_3} = \frac{0.25 \text{ kJ}}{\cancel{g}} \times \frac{167.00\ \cancel{g}}{1 \text{ mol}}$$

$$\Delta H_{s\,KBrO_3} = \frac{42 \text{ kJ}}{\text{mol}}$$

The molar heat of solution is an endothermic 42 kJ/mol and the thermochemical equation is written as

$$KBrO_{3(s)} + 42 \text{ kJ} \rightarrow KBrO_{3(aq)}$$

Sample Problem 3

Magnesium is commonly used in flares and fireworks. The combustion of magnesium (**Figure 1**) releases 24.7 kJ/g of magnesium. Calculate the molar heat of combustion of magnesium and write the thermochemical equation for this reaction.

Solution

$h_{c\,Mg} = 24.7 \text{ kJ/g}$

$$\Delta H_{c\,Mg} = \frac{24.7 \text{ kJ}}{\cancel{g}} \times \frac{24.31\ \cancel{g}}{1 \text{ mol}}$$

$$\Delta H_{c\,Mg} = \frac{600 \text{ kJ}}{\text{mol}}$$

The molar heat of combustion is an exothermic 600 kJ/mol and the thermochemical equation is written as

$$Mg_{(s)} + \frac{1}{2} O_{2(g)} \rightarrow MgO_{(s)} + 600 \text{ kJ}$$

or $2\ Mg_{(s)} + O_{2(g)} \rightarrow 2\ MgO_{(s)} + 1200 \text{ kJ}$

Figure 1
The energy released by the combustion or burning of magnesium is obvious from the bright light produced by this sparkler. A considerable quantity of heat is also produced in this exothermic reaction along with the white solid, magnesium oxide.

Finally, you may be asked to use a given thermochemical equation to determine a molar heat of reaction for a particular substance. The next sample problem shows an example of this kind of question.

Sample Problem 4

Hexane is a component of some naphtha fuels used in camping stoves. Use the following thermochemical equation to calculate the molar heat of combustion of hexane:

$$2\ C_6H_{14(l)} + 19\ O_{2(g)} \rightarrow 12\ CO_{2(g)} + 14\ H_2O_{(g)} + 7086\ kJ$$

Solution

$$\underset{C_6H_{14}}{\Delta H_c} = \frac{7086\ kJ}{2\ mol}$$

$$\underset{C_6H_{14}}{\Delta H_c} = \frac{3543\ kJ}{mol}$$

The molar heat of combustion of hexane is an exothermic 3543 kJ/mol.

Practice

Understanding Concepts

1. How are heats of reaction determined?
2. What does the term *specific* mean as part of a scientific quantity?
3. What are the units for specific heat of reaction and molar heat of reaction?
4. What conversion factor is used to convert between specific and molar heats of reaction?
5. Are combustion reactions endothermic or exothermic? State some common examples.
6. The specific heat of combustion of acetylene is determined calorimetrically to be 49.90 kJ/g.
   (a) What is the molar heat of combustion of acetylene?
   (b) Use the molar heat of combustion of acetylene to write a thermochemical equation.
7. Coal gasification converts coal (assume pure carbon) and water into a combustible mixture of carbon monoxide and hydrogen. The product mixture is called "coal gas" and is a cleaner fuel than coal for electric power-generating stations. From calorimetry, the specific heat of reaction for carbon is an endothermic 10.9 kJ/g.
   (a) What is the molar heat of reaction of carbon?
   (b) Write a thermochemical equation for the gasification reaction.
8. If carbon monoxide is not recovered and recycled in an industrial process, it can be burned to form carbon dioxide. This same conversion occurs in the catalytic converter of a car:

   $$2\ CO_{(g)} + O_{2(g)} \rightarrow 2\ CO_{2(g)} + 566\ kJ$$

   (a) Rewrite this chemical equation using one mole as the coefficient for carbon monoxide.
   (b) What is the molar heat of combustion of carbon monoxide?

Answers

6. (a) 1299 kJ/mol
7. (a) 131 kJ/mol
8. (b) 283 kJ/mol

Figure 2
In the human body, exothermic chemical reactions occur as fats and carbohydrates are metabolized.

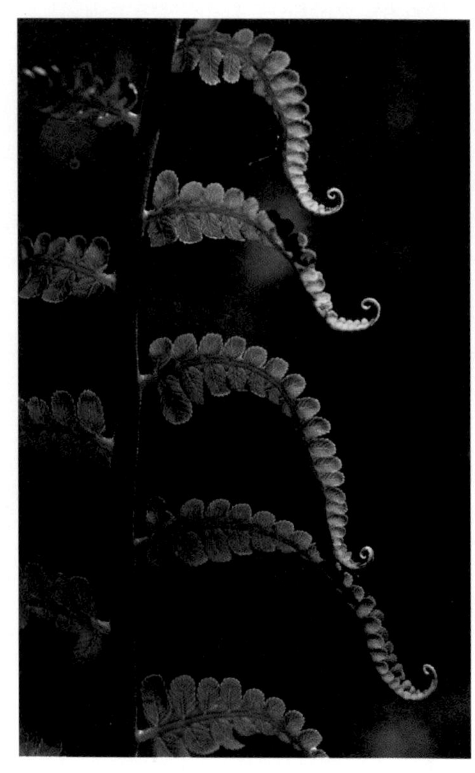

Figure 3
Plants use energy from the Sun in a series of endothermic reactions called photosynthesis.

bond energy: the energy required to break a chemical bond; the energy released when a bond is formed

9. Translate the molar heats of reaction given below into a balanced chemical equation, including the energy change as a term in the equation.
   (a) The molar heat of combustion for methanol is an exothermic 638.0 kJ/mol.
   (b) The molar heat of formation for liquid carbon disulfide is an endothermic 89.0 kJ/mol. The carbon disulfide is formed from its elements in a synthesis reaction.
   (c) The molar heat of roasting (complete combustion) for zinc sulfide is an exothermic 441.3 kJ/mol.
   (d) The molar heat of simple decomposition for iron(III) oxide to its elements is an endothermic 824.2 kJ/mol.
10. Pentane is a volatile component of gasoline. During the winter, more pentane is added to gasoline so that automobiles start more easily. The specific heat of combustion of pentane is determined calorimetrically to be 48.70 kJ/g. Use this information to write a thermochemical equation representing the complete combustion of pentane.

Reflecting

11. Do endothermic and exothermic only apply to chemical reactions? What are some other changes that can also be endothermic or exothermic?

A Theoretical Perspective on Energy Change

Energy transfer is an important factor in all chemical changes. Exothermic reactions, such as the combustion of gasoline in a car engine or the metabolism of fats and carbohydrates in a human body (**Figure 2**), release energy into the surroundings. Endothermic reactions, such as photosynthesis (**Figure 3**) or the decomposition of water into hydrogen and oxygen, remove energy from the surroundings. Knowledge of energy and energy changes is important to society and to industry, and the study of energy changes provides chemists with important information about chemical bonds. How do we explain the energy changes measured in calorimeters in terms of molecules, atoms, and bonds?

Just as glue holds objects together, electrical forces hold atoms together. In order to pull apart objects that are glued together, you have to supply some energy. Similarly, if atoms or ions are bonded together, energy is required to separate them. Separated atoms or ions release energy when they bond together again:

bonded particles + energy → separated particles
separated particles → bonded particles + energy

The stronger the bond holding the particles together, the greater the energy required to separate them. **Bond energy** is the energy required to break a chemical bond. It is also the energy released when a bond is formed. Even the simplest of chemical reactions may involve the breaking and forming of several individual bonds. The terms *exothermic* and *endothermic* are empirical descriptions of overall changes that can be explained by knowledge of bond changes. Consider the decomposition of water, for example:

$$2\,H_2O_{(l)} + \text{energy} \rightarrow 2\,H_{2(g)} + O_{2(g)}$$

The easiest way to supply the energy for this decomposition is electrically. This is known as the electrolysis of water. In this endothermic reaction, hydrogen–oxygen bonds in the water molecules must be broken before the hydrogen–hydrogen and oxygen–oxygen bonds in the products can be formed.

Breaking bonds requires energy and forming bonds releases energy. We know experimentally that the decomposition of water is endothermic, which means that it requires a constant input of energy to keep reacting. The theoretical interpretation is that it must take more energy to break the bonds in the reactants than is released when the new bonds in the products form (**Figure 4**).

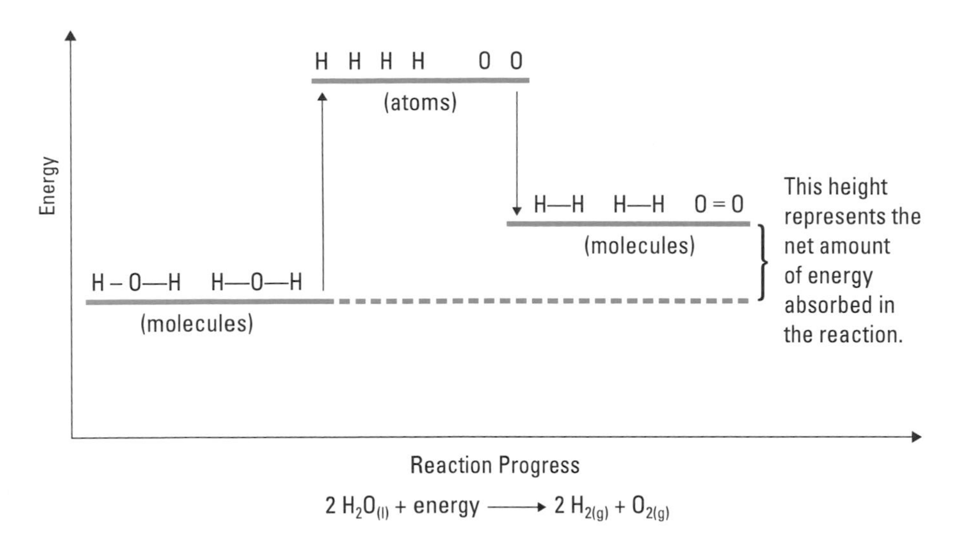

Figure 4
Since the overall change is endothermic, the energy required to break the O—H bonds must be greater than the energy released when the H—H and O═O bonds form.

For exothermic reactions, such as the formation of hydrogen chloride, the opposite is true (**Figure 5**). Exothermic reactions produce energy that transfers to the surroundings, usually as heat. In this case, more energy is believed to be released in forming new bonds than is absorbed in breaking bonds in the reactants. This is true for all exothermic reactions, including the combustion of fuels.

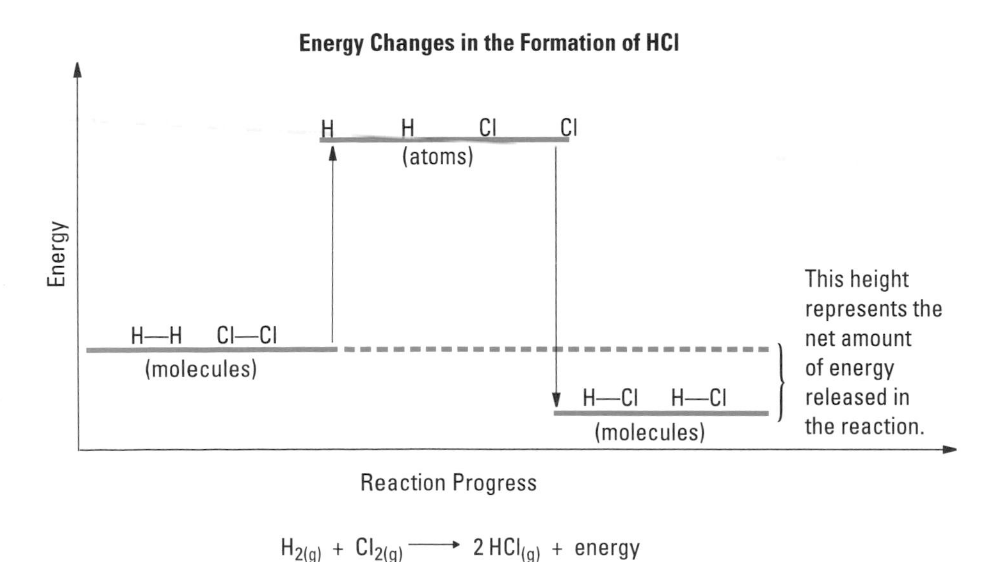

Figure 5
Energy is absorbed in order to break the H—H and Cl—Cl bonds, but more energy is released when the H—Cl bonds form. The overall result is an exothermic reaction.

Appendixes

APPENDIX A

Skills Handbook 606

A1: Scientific Inquiry 606
  Planning an Investigation 606
A2: Decision Making 612
  A Risk–Benefit Analysis Model 615
A3: Technological Problem Solving 616
A4: Lab Reports 618
A5: Math Skills 621
  Scientific Notation 621
  Uncertainty in Measurements 621
  Logarithms 623

APPENDIX B

Safety Skills 624

B1: Safety Conventions and Symbols 624
B2: Safety in the Laboratory 626

APPENDIX C

Reference 630

C1: Units, Symbols, and Prefixes 630
C2: The Elements 632
C3: Common Chemicals 634
C4: Cations and Anions 635
C5: Acids and Bases 636
C6: Line Spectra of the Elements 637

APPENDIX D

Answers 638

Scientific Inquiry

Planning an Investigation

In our attempts to further our understanding of the natural world, we encounter questions, mysteries, or events that are not readily explainable. We can use controlled experiments, correlational studies, or observational studies to attempt to answer these questions or explain the events. The methods used in scientific inquiry depend, to a large degree, on the purpose of the inquiry.

Controlled Experiments

A controlled experiment is an example of scientific inquiry in which an independent variable is purposefully and steadily changed to determine its effect on a second dependent variable. All other variables are controlled or kept constant. Controlled experiments are performed when the purpose of the inquiry is to create, test, or use a scientific concept.

The common components of controlled experiments are outlined in the flow chart below. *Even though the sequence is presented as linear, there are normally many cycles through the steps during the actual experiment.*

Process Description

Stating the purpose

Choose a topic that interests you. Determine whether you are going to create, test, or use a scientific concept and whether you are going to carry out a given procedure or develop a new experimental design. Indicate your decision in a statement of the purpose.

Asking the question

Your question forms the basis for your investigation. Controlled experiments are about relationships, so the question could be about the effects on variable A when variable B is changed. The question may also be about what causes the change in variable A. In this case, you might speculate about possible variables and determine which variable causes the change.

Hypothesizing/predicting

A hypothesis is a tentative explanation. You must be able to test your hypothesis, which can range in certainty from an educated guess to a concept that is widely accepted in the scientific community. A prediction is based upon a hypothesis or a more established scientific explanation, such as a theory. In the prediction you state what outcome you expect from your experiment.

Designing the investigation

The design of a controlled experiment identifies how you plan to manipulate the independent variable, measure the response of the dependent variable, and control all the other variables.

Example: A Test of the Collision–Reaction Theory

Stating the purpose: The collision–reaction theory sounds logical. The purpose of this investigation is to provide some concrete evidence to support or refute the collision–reaction theory.

Asking the question: How does changing the concentration of hydrochloric acid affect the time required for the acid to completely react with a fixed quantity of zinc?

Hypothesizing/predicting: According to the collision–reaction theory, if the concentration of hydrochloric acid is increased, then the time required for the reaction with zinc will decrease. The reasoning that supports this prediction is that a higher concentration of $HCl_{(aq)}$ produces more collisions per second between the aqueous ions in hydrochloric acid and the zinc atoms. More collisions per second would produce more reactions per second and, therefore, a shorter time would be required to consume the zinc.

Designing the investigation: The same amount of zinc metal is made to react with different known concentrations of excess hydrochloric acid. The time for the zinc to completely react is measured for each concentration of acid solution. The independent variable is the concentration of hydrochloric acid. The dependent variable is the time for the zinc to be consumed. The temperature of the solution, the quantity of zinc, the surface area of the zinc in contact with the acid, and the volume of the acid are all controlled variables.

There are many ways to gather and record your observations during your investigation. It is helpful to plan ahead and think about what data you will need and how best to record them. This helps to clarify your thinking about the question posed at the beginning, the variables, the number of trials, the procedure, the materials, and your skills. It will also help you organize your evidence for easier analysis.

Gathering, recording, and organizing observations

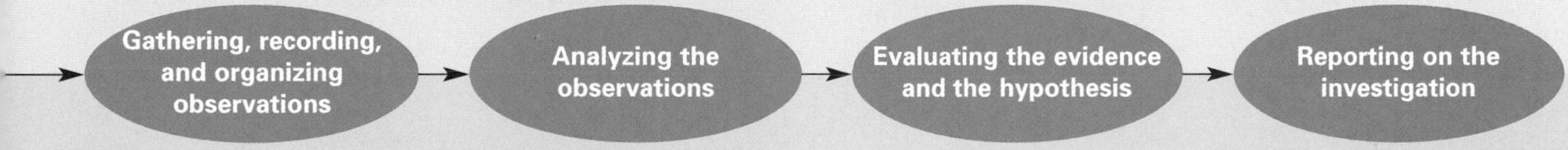

Time to completion for the reaction will be measured using a stop-watch and recorded in a table like **Table 1.**

After thoroughly analyzing your observations, you may have sufficient and appropriate evidence to enable you to answer the question posed at the beginning of the investigation.

Analyzing the observations

The observations will be presented in graphical format, with time on the *x*-axis and concentration of $HCl_{(aq)}$ on the *y*-axis. In this format any trends or patterns will be easier to see.

At this stage of the investigation, you will evaluate the processes that you followed to plan and perform the investigation. Evaluating the processes includes evaluating the materials, design, the procedure, and your skills. You will also evaluate the outcome of the investigation, which involves evaluating the hypothesis, i.e., whether the evidence supports the hypothesis or not. You must identify and take into account any sources of error and uncertainty in your measurements. Compare the answer created in the hypothesis/prediction with the answer generated by analyzing the evidence. Is the hypothesis acceptable or not?

Evaluating the evidence and the hypothesis

For a sample evaluation, see the Lab Report in Appendix A4.

In preparing your report, your objectives should be to describe your design and procedure accurately, and to report your observations accurately and honestly.

Reporting on the investigation

For the format of a typical lab report, see the sample Lab Report in Appendix A4.

Table 1: Reaction Time for Zinc with $HCl_{(aq)}$

Concentration of $HCl_{(aq)}$ (mol/L)	Time for reaction (s)
2.5	
2.0	
1.5	
1.0	
0.5	

Correlational Studies

When the purpose of scientific inquiry is to test a suspected relationship (hypothesis) between two different variables, but a controlled experiment is not possible, a correlational inquiry is conducted. In a correlational study, the investigator tries to determine whether one variable is affecting another without purposefully changing or controlling any of the variables. Instead, variables are allowed to change naturally. It is often difficult to isolate cause and effect in correlational studies. A correlational inquiry requires very large sample numbers and many replications to increase the certainty of the results.

The flow chart below outlines the components/processes that are important in designing a correlational study. The investigator can conduct the study without doing experiments or fieldwork, for example, by using databases prepared

Process Description

Choose a topic that interests you. Determine whether you are going to replicate or revise a previous study, or create a new one. Indicate your decision in a statement of the purpose.

In planning a correlational study, it is important to pose a question about a possible statistical relationship between variable A and variable B.

A hypothesis or prediction would not be useful. Correlational studies are not intended to establish cause–and–effect relationships.

The design of a correlational study identifies how you will gather data on the variables under study and also identifies potential sources. There are two possible sources—observation made by the investigator and existing data.

Stating the purpose → Asking the question → Hypothesizing/predicting → Designing the investigation

Example: Fish Death and Acidity

The purpose of this investigation is to determine whether there is a relationship between aquarium acidity and fish death and if so, what the relationship is.

Is there a statistical relationship between the acidity of the water in an aquarium and the death of fish, and if so, what is the relationship?

Does not apply.

It is generally considered to be unethical and inappropriate to conduct a controlled experiment in which the pH is manipulated in order to determine the acidity at which most fish die. However, we can obtain valuable data by involving a large sample (N=100) of home aquarium owners in a pH-monitoring program. Aquarium owners will measure the pH of their aquariums at regular intervals (every three days) over a period of three months. They will note and record the deaths of all fish. Participants will be expected to maintain their aquariums as they normally would and will also be asked to make other observations (e.g., the species of dead fish, symptoms leading up to the death) during the study period.

by other researchers to find relationships between two or more variables. The investigator can also make his or her own observations and measurements through fieldwork, interviews, and surveys.

Even though the sequence is presented as linear, there are normally many cycles through the steps during the actual study.

There are many ways to gather and record your observations during your investigation. It is helpful to plan ahead and think about what data you will need and how best to record them. This is an important step because it helps to clarify your thinking about the question posed at the beginning, the variables, the number of trials, the procedure, the materials, and your skills. It will also help you organize your observations for easier analysis.

Gathering, recording, and organizing observations

The results of the pH-monitoring study will be recorded in a table like **Table 2**. The pH will be measured and recorded every third calendar day for three months.

Table 2: Observations of Aquarium Acidity

Day and date	pH	Number of dead fish	Species of dead fish	Other observations
1 – 28/01/02				
2 – 31/01/02				
3 – 03/02/02				
4 – 06/02/02				

After thoroughly analyzing your observations, you may have sufficient and appropriate evidence to enable you to answer the question posed at the beginning of the investigation.

Analyzing the observations

We will analyze the data to determine if there is a relationship between the pH and the number of dead fish. We will also attempt to determine whether the relationship is statistically significant or is due simply to chance.

The additional data will be analyzed to determine if some species of fish are more affected by the acidity level than others. The additional data will also be analyzed for any warning signs that might alert the owner to a potential problem.

At this stage of the investigation, you will evaluate the processes used to plan and perform the investigation. Evaluating the processes includes evaluating the materials, design, the procedure, and your skills.

Evaluating the evidence

In order to determine whether our investigation provided valid evidence to answer our question we will need to ask several questions. Is the sample size large enough to enable us to generalize to the larger population? Are we reasonably confident that the participants measured the pH accurately and on schedule?

If we are confident that the evidence is reliable and valid, we may be able to make recommendations to aquarium owners to help them prevent the death of their fish. The investigation might lead us to conduct further studies or controlled experiments, to answer such questions, as "Are there other factors that are related to the death of aquarium fish?"

In preparing your report, your objectives should be to describe your design and procedure accurately and to report your observations accurately and honestly.

Reporting on the investigation

For the format of a typical lab report, see the sample Lab Report in Appendix A4.

Observational Studies

Often the purpose of inquiry is simply to study a natural phenomenon with the intention of gaining scientifically significant information to answer a question. Observational studies involve observing a subject or phenomenon in an unobtrusive or unstructured manner, often with no specific hypothesis. A hypothesis to describe or explain the observations may, however, be generated over time, and modified as new information is collected.

The flow chart below summarizes the stages and processes of scientific inquiry through observational studies.

Even though the sequence is presented as linear, there are normally many cycles through the steps during the actual study.

	Stating the purpose	Asking the question	Hypothesizing/predicting	Designing the investigation
Process Description	**Choose a topic that interests you. Determine whether you are going to replicate or revise a previous study, or create a new one. Indicate your decision in a statement of the purpose.**	**In planning an observational study, it is important to pose a general question about the natural world. You may or may not follow the question with the creation of a hypothesis.**	**A hypothesis is a tentative explanation. In an observational study, a hypothesis can be formed after observations have been made and information has been gathered on a topic. A hypothesis may be created in the analysis.**	**The design of an observational study describes how you will make observations relevant to the question.**
Example: Water Quality in Public Swimming Area	Although the bacterial quality of public swimming areas is normally tested by the municipal or provincial Department of Health, no chemical analysis is done unless a problem arises. The purpose of this investigation is to carry out an environmental assessment to determine the chemical quality of the local public swimming area.	What common chemicals are found in the local public swimming area and in what concentrations are they present?	At this point, we have no indication of which chemicals are present in the area, what their concentrations may be, and if there are any threats to swimmers. We have no hypothesis and can make no predictions.	We will take a sample of water from five different locations within the local swimming area each week for a month. Note will be made of other significant conditions (e.g., heavy rain, wind) that are present during the course of the study. The water samples will be tested for organic and inorganic chemicals that may pose a health hazard. The testing facilities at the Department of Health and the Chemistry Department at the local university will be used to determine the presence and concentrations of chemicals.

Table 3: Presence and Concentration of Chemicals in Public Swimming Area

Chemicals	Week 1					Week 2					Week 3					Week 4				
	Sample area					Sample area					Sample area					Sample area				
	1	2	3	4	5	1	2	3	4	5	1	2	3	4	5	1	2	3	4	5

There are many ways to gather and record your observations during your investigation. During your observational study, you should quantify your observations where possible. All observations should be objective and unambiguous as possible. Consider ways to organize your information for easier analysis.

Gathering, recording, and organizing observations

The data will be recorded in a table like **Table 3**. The chemicals to be tested for include the following:
—lead
—mercury
—cadmium
—nitrates/nitrites
—volatile organic compounds such as benzene, toluene, carbon tetrachloride
—petroleum products
—chlorine

After thoroughly analyzing your observations, you may have sufficient and appropriate evidence to enable you to answer the question posed at the beginning of the investigation. You may also have enough observations to form a hypothesis.

Analyzing the observations

The concentrations of the chemicals found in the swimming area are determined.

At this stage of the investigation, you will evaluate the processes used to plan and perform the investigation. Evaluating the processes includes evaluating the materials, the design, the procedure, and your skills. The results of most such investigations will suggest further studies, perhaps correlational studies or controlled experiments to explore tentative hypotheses you may have developed.

Evaluating the evidence and the hypothesis

We must determine if our sampling and testing procedures are appropriate. Is the number of samples sufficient? Were they taken at the proper sites? Was the testing of the samples carried out with care and precision?

The presence of chemicals in concentrations higher than the acceptable levels will alert us to potential problems with the swimming site. This might suggest further investigations to determine the possible source(s) of the chemical(s).

In preparing your report, your objectives should be to describe your design and procedure accurately, and to report your observations accurately and honestly.

Reporting on the investigation

For the format of a typical lab report, see the sample Lab Report in Appendix A4.

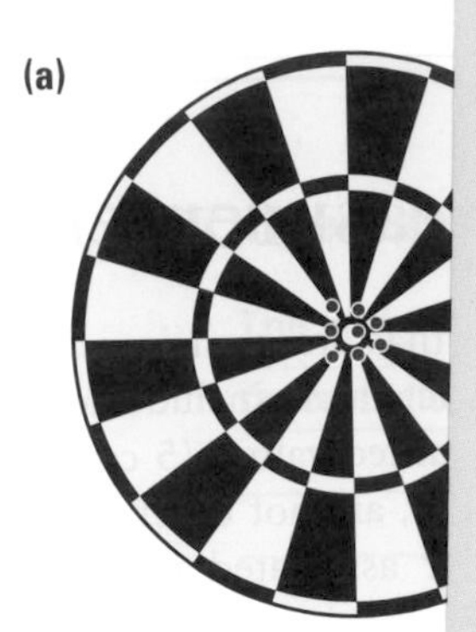

Common Chemicals

You live in a chemical world. As one bumper sticker asks, "What in the world isn't chemistry?" Every natural and technologically produced substance around you is composed of chemicals. Many of these chemicals are used to make your life easier or safer, and some of them have life-saving properties. Following is a list of selected common chemicals.

Common name	Recommended name	Formula	Common use/source
acetic acid	ethanoic acid	$HC_2H_3O_2$	vinegar
acetone	propanone	$(CH_3)_2CO_{(l)}$	nail polish remover
acetylene	ethyne	$C_2H_{2(g)}$	cutting/welding torch
ASA (Aspirin®)	acetylsalicylic acid	$C_6H_4COOCH_3COOH_{(s)}$	for pain-relief medication
baking soda	sodium hydrogen carbonate	$NaHCO_{3(s)}$	leavening agent
battery acid	sulfuric acid	$H_2SO_{4(aq)}$	car batteries
bleach	sodium hypochlorite	$NaClO_{(s)}$	bleach for clothing
bluestone	copper(II) sulfate pentahydrate	$CuSO_4 \cdot 5H_2O_{(s)}$	algicide, fungicide
brine	aqueous sodium chloride	$NaCl_{(aq)}$	water-softening agent
CFC	chlorofluorocarbon	$C_xCl_yF_{z(l)}$; e.g., $C_2Cl_2F_{4(l)}$	refrigerant
charcoal/graphite	carbon	$C_{(s)}$	fuel, lead pencils
citric acid	2-hydroxy-1,2,3-propanetricarboxylic acid	$C_3H_4OH(COOH)_3$	in fruit and beverages
carbon dioxide	carbon dioxide	$CO_{2(g)}$	dry ice, carbonated beverages
ethylene	ethene	$C_2H_{4(g)}$	for polymerization
ethylene glycol	1,2-ethanediol	$C_2H_4(OH)_{2(l)}$	radiator antifreeze
freon-12	dichlorodifluoromethane	$CCl_2F_{2(l)}$	refrigerant
Glauber's salt	sodium sulfate decahydrate	$Na_2SO_4 \cdot 10H_2O_{(s)}$	solar heat storage
glucose	D-glucose; dextrose	$C_6H_{12}O_{6(s)}$	in plants and blood
grain alcohol	ethanol (ethyl alcohol)	$C_2H_5OH_{(l)}$	beverage alcohol
gypsum	calcium sulfate dihydrate	$CaSO_4 \cdot 2H_2O_{(s)}$	wallboard
lime (quicklime)	calcium oxide	$CaO_{(s)}$	masonry
limestone	calcium carbonate	$CaCO_{3(s)}$	chalk and building materials
lye (caustic soda)	sodium hydroxide	$NaOH_{(s)}$	oven/drain cleaner
malachite	copper(II) hydroxide carbonate	$Cu(OH)_2 \cdot CuCO_{3(s)}$	copper mineral
methyl hydrate	methanol (methyl alcohol)	$CH_3OH_{(l)}$	gas line antifreeze
milk of magnesia	magnesium hydroxide	$Mg(OH)_{2(s)}$	antacid (for indigestion)
MSG	monosodium glutamate	$NaC_5H_8NO_{4(s)}$	flavour enhancer
muriatic acid	hydrochloric acid	$HCl_{(aq)}$	concrete etching
natural gas	methane	$CH_{4(g)}$	fuel
PCBs	polychlorinated biphenyls	$(C_6H_xCl_y)_2$; e.g., $(C_6H_4Cl_2)_{2(l)}$	in transformers
potash	potassium chloride	$KCl_{(s)}$	fertilizer
road salt	calcium chloride or sodium chloride	$CaCl_{2(s)}$ or $NaCl_{(s)}$	melts ice
rotten-egg gas	hydrogen sulfide	$H_2S_{(g)}$	in natural gas
rubbing alcohol	2-propanol (also isopropanol)	$CH_3CHOHCH_{3(l)}$	for massage
sand (silica)	silicon dioxide	$SiO_{2(s)}$	in glassmaking
slaked lime	calcium hydroxide	$Ca(OH)_{2(s)}$	limewater
soda ash	sodium carbonate	$Na_2CO_{3(s)}$	in laundry detergents
sugar	sucrose	$C_{12}H_{22}O_{11(s)}$	sweetener
table salt	sodium chloride	$NaCl_{(s)}$	seasoning
vitamin C	ascorbic acid	$H_2C_6H_6O_{6(s)}$	vitamin supplement
washing soda	sodium carbonate decahydrate	$Na_2CO_3 \cdot 10H_2O_{(s)}$	water softener

through the stopper until it shows from the other end. Ease the tubing or thermometer out of the borer.

- Protect your hands with heavy gloves or several layers of cloth before inserting glass into rubber stoppers.
- Be very careful while cleaning glassware. There is an increased risk of breakage from dropping when the glassware is wet and slippery.

Using Sharp Instruments Safely

- Make sure your instruments are sharp. Surprisingly, one of the main causes of accidents with cutting instruments is the use of a dull instrument. Dull cutting instruments require more pressure than sharp instruments and are therefore much more likely to slip.
- Select the appropriate instrument for the task. Never use a knife when scissors would work better.
- Always cut away from yourself and others.
- If you cut yourself, inform your teacher immediately and get appropriate first aid.
- Be careful when working with wire cutters or wood saws. Use a cutting board where needed.

Heat and Fire Safety

- In a laboratory where burners or hot plates are being used, never pick up a glass object without first checking the temperature by lightly and quickly touching the item, or by placing your hand near, but not on, the item. Glass items that have been heated stay hot for a long time, but do not appear to be hot. Metal items such as ring stands and hot plates can also cause burns; take care when touching them.
- Do not use a laboratory burner near wooden shelves, flammable liquids, or any other item that is combustible.
- Before using a laboratory burner, make sure that long hair is tied back. Do not wear loose clothing (wide long sleeves should be tied back or rolled up).
- Never look down the barrel of a laboratory burner.
- Always pick up a burner by the base, never by the barrel.
- Never leave a lighted laboratory burner unattended.
- If you burn yourself, *immediately* run cold water gently over the burned area or immerse the burned area in cold water and inform your teacher.
- Make sure that heating equipment, such as a burner, hot plate, or electrical equipment, is secure on the bench and clamped in place when necessary.
- Always assume that hot plates and electric heaters are hot and use protective gloves when handling.
- Keep a clear workplace when performing experiments with heat.
- When heating a test tube over a laboratory burner, use a test-tube holder and a spurt cap. Holding the test tube at an angle, facing away from you and others, gently move the test tube backwards and forwards through the flame.
- Remember to include a "cooling" time in your experiment plan; do not put away hot equipment.
- Very small fires in a container may be extinguished by covering the container with a wet paper towel or ceramic square.
- For larger fires, inform the teacher and follow the teacher's instructions for using fire extinguishers, blankets, alarms, and for evacuation. Do not attempt to deal with a fire by yourself.
- If anyone's clothes or hair catch fire, tell the person to drop to the floor and roll. Then use a fire blanket to help smother the flames.

Electrical Safety

- Water or wet hands should never be used near electrical equipment.
- Do not operate electrical equipment near running water or any large containers of water.
- Check the condition of electrical equipment. Do not use if wires or plugs are damaged, or if the ground pin has been removed.
- Make sure that electrical cords are not placed where someone could trip over them.
- When unplugging equipment, remove the plug gently from the socket. Do not pull on the cord.
- When using variable power supplies, start at low voltage and increase slowly.

Handling Chemicals Safely

Many chemicals are hazardous to some degree. When using chemicals, operate under the following principles:

1. Never underestimate the risks associated with chemicals. Assume that any unknown chemicals are hazardous.
2. If you can substitute, use a less hazardous chemical wherever possible.
3. Reduce exposure to chemicals to the absolute minimum. Avoid direct skin contact if possible.
4. When using chemicals, ensure that there is adequate ventilation.

The following guidelines do not address every possible situation but, used with common sense, are appropriate for situations in the high-school laboratory.

- Consult the MSDS before you use a chemical.
- Appropriate eye protection must be worn at all times where chemicals are used or stored. Wear a lab coat and/or other protective clothing (e.g., aprons, gloves).
- When carrying chemicals, hold containers carefully using two hands, one around the container and one underneath.
- Read all labels to ensure that the chemicals you have selected are the intended ones. Never use the contents of a container that has no label or has an illegible label. Give any such containers to your teacher.
- Label all chemical containers correctly to avoid confusion about contents.
- Never pipet or start a siphon by mouth. Use a pipet bulb or equivalent device.
- Pour liquid chemicals carefully (down the side of the receiving container or down a stirring rod) to ensure that they do not splash. Always pour from the side opposite the label—if everyone follows this rule, drips will always form on the same side, away from your hand.
- Always pour volatile chemicals in a fume hood or in a well-ventilated area.
- Never smell or taste chemicals.
- Return chemicals to their correct storage place. Chemicals are stored by hazard class.
- If you spill a chemical, use a chemical spill kit to clean up.
- Do not return surplus chemicals to stock bottles. Dispose of excess chemicals in the appropriate manner.
- Clean up your work area, the fume hood, and any other area where chemicals were used.
- Wash hands immediately after handling chemicals and before leaving the lab, even if you wore gloves.

Waste Disposal

Waste disposal at school, at home, or at work is a social and environmental issue. To protect the environment, federal and provincial governments have regulations to control wastes, especially chemical wastes. For example, the WHMIS program applies to controlled products that are being handled. (When being transported, they are regulated under the *Transport of Dangerous Goods Act*, and for disposal they are subject to federal, provincial, and municipal regulations.) Most laboratory waste can be washed down the drain, or, if it is in solid form, placed in ordinary garbage containers. However, some waste must be treated more carefully. It is your responsibility to follow procedures and dispose of waste in the safest possible manner according to the teacher's instructions.

Flammable Substances

Flammable liquids should not be washed down the drain. Special fire-resistant containers are used to store flammable liquid waste. Waste solids that pose a fire hazard should be stored in fireproof containers. Care must be taken not to allow flammable waste to come into contact with any sparks, flames, other ignition sources, or oxidizing materials. The method of disposal depends on the nature of the substance.

Corrosive Solutions

Solutions that are corrosive but not toxic, such as acids and bases, can usually be washed down the drain, but care should be taken to ensure that they are first either neutralized or diluted to low concentration. While disposing of such substances, use large quantities of water and continue to pour water down the drain for a few minutes after all the substance has been washed away.

Heavy Metal Solutions

Heavy metal compounds (for example, lead, mercury, and cadmium compounds) should not be flushed down the drain. These substances are cumulative poisons and should be kept out of the environment. Pour any heavy metal waste into the special container marked "Heavy Metal Waste." Remember that paper towels used to wipe up solutions of heavy metals, as well as filter papers with heavy metal compounds embedded in them, should be treated as solid toxic waste.

Disposal of heavy metal solutions is usually accomplished by precipitating the metal ion (for example, as lead (II) silicate) and disposing of the solid. Heavy metal compounds should not be placed in school garbage containers. Usually, waste disposal companies collect materials that require special disposal and dispose of them as required by law.

Toxic Substances

Toxic chemicals and solutions of toxic substances should not be poured down the drain. They should be retained for disposal by a licensed waste disposal company.

Organic Material

Remains of plants and animals can generally be disposed of in the normal school garbage containers. Animal dissection specimens should be rinsed thoroughly to rid them of any excess preservative and sealed in plastic bags.

Fungi and bacterial cultures should be autoclaved or treated with a fungicide or antibacterial soap before disposal.

First Aid

The following guidelines apply if an injury, such as a burn, cut, chemical spill, ingestion, inhalation, or splash in eyes, happens to yourself or to one of your classmates.

- If an injury occurs, inform your teacher immediately.
- Know the location of the first-aid kit, fire blanket, eyewash station, and shower, and be familiar with the contents/operation.
- If the injury is the result of chemicals, drench the affected area with a continuous flow of water for 30 min. Clothing should be removed as necessary. Inform your teacher. Retrieve the Material Safety Data Sheet (MSDS) for the chemical; this sheet provides information about the first-aid requirements for the chemical. If the chemicals are splashed in your eyes, have another student assist you in getting to the eyewash station immediately. Rinse with the eyes open for at least 15 min.
- If you have ingested or inhaled a hazardous substance, inform your teacher immediately. The MSDS will give information about the first-aid requirements for the substance in question. Contact the Poison Control Centre in your area.
- If the injury is from a burn, immediately immerse the affected area in cold water. This will reduce the temperature and prevent further tissue damage.
- In the event of electrical shock, do not touch the affected person or the equipment the person was using. Break contact by switching off the source of electricity or by removing the plug.
- If a classmate's injury has rendered him/her unconscious, notify the teacher immediately. The teacher will perform CPR if necessary. Do not administer CPR unless under specific instructions from the teacher. You can assist by keeping the person warm and reassured.

Units, Symbols, and Prefixes

Throughout *Nelson Chemistry 11* and in this reference section, we have attempted to be consistent in the presentation and usage of quantities, units, and their symbols. As far as possible, the text uses the Système international d'unités (SI). However, some other units have been included because of their practical importance, wide usage, or use in specialized fields. In our interpretations and usage, *Nelson Chemistry 11* has followed the most recent *Canadian Metric Practice Guide* (CAN/CSA–Z234.1–89), published in 1989 and reaffirmed in 1995 by the Canadian Standards Association.

SI Base Units

Quantity	Symbol	Unit name	Symbol
amount of substance	*n*	mole	mol
electric current	*I*	ampere	A
length	*L, l, h, d, w*	metre	m
luminous intensity	I_v	candela	cd
mass	*m*	kilogram	kg
temperature	*T*	kelvin	K
time	*t*	second	s

Some SI Derived Units

Quantity	Symbol	Unit	Unit symbol	Expression in SI base units
acceleration	$\vec{a}$	metre per square second	m/s^2	m/s^2
area	*A*	square metre	m^2	m^2
density	ρ, *D* *	kilogram per cubic metre	kg/m^3	kg/m^3
displacement	$\vec{d}$	metre	m	m
electric charge	*Q, q,* *e* *	coulomb	C	A·s
electric potential	*V*	volt	V	$kg{\cdot}m^2/(A{\cdot}s^3)$
electric field	*E*	volt per metre newton per coulomb	V/m N/C	$kg{\cdot}m/(A{\cdot}s^3)$
electric resistance	*R*	ohm	Ω	$kg{\cdot}m^2/(A^2{\cdot}s^3)$
energy	E, E_k, E_p	joule	J	$kg{\cdot}m^2/s^2$
force	*F*	newton	N	$kg{\cdot}m/s^2$
frequency	*f*	hertz	Hz	s^{-1}
heat	*Q*	joule	J	$kg{\cdot}m^2/s^2$
magnetic flux	Φ	weber	Wb	$kg{\cdot}m^2/(A{\cdot}s^2)$
magnetic field	*B*	Tesla weber per square metre	 Wb/m^2	T $kg/(A{\cdot}s^2)$
momentum	*P, p* *	kilogram metre per second	kg·m/s	kg·m/s
period	*T*	second	s	s
power	*P*	watt	W	$kg{\cdot}m^2/s^3$
pressure	*P* *p*	pascal newton per square metre	Pa N/m^2	$kg/(m{\cdot}s^2)$
speed	*v*	metre per second	m/s	m/s
velocity	$\vec{v}$	metre per second	m/s	m/s
volume	*V*	cubic metre	m^3	m^3
wavelength	λ	metre	m	m
weight	*W, w* *	newton	N	$kg{\cdot}m/s^2$
work	*W*	joule	J	$kg{\cdot}m^2/s^2$

* preferred

Defined (Exact) Quantities		
1 mL*	=	1 cm^3*
1 kL†	=	1 m^3†
1000 kg	=	1 t
1 Mg	=	1 t
1 atm	=	101.325 kPa
0°C	=	273.15 K
STP	=	0°C and 101.325 kPa
SATP	=	25°C and 100 kPa

*† assume that these are equivalent

REFERENCE

Numerical Prefixes

Prefix	Power	Symbol
deca-	10^{1}	da
hecto-	10^{2}	h
kilo-	10^{3}	k*
mega-	10^{6}	M*
giga-	10^{9}	G*
tera-	10^{12}	T
peta-	10^{15}	P
exa-	10^{18}	E
deci-	10^{-1}	d
centi-	10^{-2}	c*
milli-	10^{-3}	m*
micro-	10^{-6}	μ*
nano-	10^{-9}	n*
pico-	10^{-12}	p
femto-	10^{-15}	f
atto-	10^{-18}	a

* commonly used

Some Examples of Prefix Use

0.0034 mol = 3.4×10^{-3} mol = 3.4 **milli**moles or 3.4 mmol

1530 L = 1.53×10^{3} L = 1.53 **kilo**litres or 1.53 kL

Common Multiples

Multiple	Prefix
0.5	hemi–
1	mono–
1.5	sesqui–
2	bi–, di–
2.5	hemipenta–
3	tri–
4	tetra–
5	penta
6	hexa
7	hepta–
8	octa
9	nona–
10	deca–

Greek and Latin Prefixes

Prefix	Meaning	Prefix	Meaning
a–	not, without	hydro–	water
ab–	away from	hyper–	above
abd–	led away	hypo–	below
acro–	end, tip	infra–	under
aer–, aero–	air	inter–	between
agg–	to clump	intra–	inside of, within
agro–	land	intro–	inward
alb–	white	iso–	equal
allo–	other	lact–, lacti–, lacto–	milk
ameb–	change	leuc–, leuco–	white
amphi–	around, both	lys–, lyso–	break up
amyl–	starch	macro–	large
an–	without	meg–, mega–	great
ana–	up	melan–	black
ant–, anti–	opposite	mes–, meso–	middle
anth–	flower	micr–, micro–	small
aut–, auto–	self	mono–	one
baro–	weight (pressure)	morpho–	form, shape
bi–	twice	multi–	many
bio–	life	neo–	new
carcin–	cancer	oligo–	few
chlor–, chloro–	green	patho–	disease
chrom–, chromo–	colour	peri–	around
co–	with	pharmaco–	drug
cyan–, cyano–	blue	photo–	light
di–	two	pneum–	air
dors–	back	poly–	many
ec–, ecto–	outside	pseud–, pseudo–	false
em–	inside	pyr–, pyro–	fire
en–	in	radio–	ray
end–, endo–	within	sacchar–, saccharo–	sugar
epi–	at, on, over	sub–	beneath
equi–	equal	super–, supra–	above
erythro–	red	sym–, syn–	with, together
ex–, exo–	away, out	therm–, thermo–	temperature, heat
gastr–	stomach	tox–	poison
glyc–	sweet	trans–	across
halo–	salt	ultra–	beyond
hemi–	half	vitro	glass
hetero–	different	xanth–, xantho–	yellow
holo–	whole	xer–, xero–	dry
homo–	the same		

The Elements

Element	Symbol	Atomic number	Ionization energy (kJ/mol)	Electro-negativity	Electron affinity (kJ/mol)	Ionic radius (pm)	Common ion charge
actinium	Ac	89	509	1.1		111	3+
aluminum	Al	13	578	1.5	42.5	50	3+
americium	Am	95	578	1.3		97.5	3+
antimony	Sb	51	834	1.9	100.9	76	3+
argon	Ar	18	1521				
arsenic	As	33	947	2.0	78	222	
astatine	At	85		2.2	[270]	227	1–
barium	Ba	56	503	0.89	[14]	135	2+
berkelium	Bk	97	601	1.3		98	3+
beryllium	Be	4	899	1.5		31	2+
bismuth	Bi	83	703	1.9	91.3	96	3+
boron	B	5	801	2.04	26.7		
bromine	Br	35	1140	2.8	324.54	196	1–
cadmium	Cd	48	868	1.69		97	2+
calcium	Ca	20	590	1.00	1.78	99	2+
californium	Cf	98	608	1.3		95	3+
carbon	C	6	1086	2.55	121.85		
cerium	Ce	58	528	1.12		102	3+
cesium	Cs	55	376	0.7	45.50	169	1+
chlorine	Cl	17	1251	3.0	348.57	181	1–
chromium	Cr	24	653	1.6	64.3	64	3+
cobalt	Co	27	758	1.8	63.9	74.5	2+
copper	Cu	29	745	1.90	119.2	72	2+
curium	Cm	96	581	1.3		97	3+
dysprosium	Dy	66	572	1.22		91.2	3+
einsteinium	Es	99	619	1.3		98	3+
erbium	Er	68	589	1.24		89.0	3+
europium	Eu	63	547	1.2		94.7	3+
fermium	Fm	100	627	1.3		97	3+
fluorine	F	9	1681	4.0	328.16	136	1–
francium	Fr	87		0.7	[44]	180	1+
gadolinium	Gd	64	592	1.1		93.8	3+
gallium	Ga	31	579	1.6	29	62.0	3+
germanium	Ge	32	762	1.8	119.0	53.0	4+
gold	Au	79	890	2.4	222.75	91	3+
hafnium	Hf	72	680	1.3	[≈0]	78	4+
helium	He	2	2372				
holmium	Ho	67	581	1.23		90.1	3+
hydrogen	H	1	1312	2.1	72.55	10^{-3}/154	1+/1–
indium	In	49	558	1.7	29	81	3+
iodine	I	53	1008	2.5	295.15	216	1–
iridium	Ir	77	880	2.2	151.0	64	4+
iron	Fe	26	759	1.8	14.6	64.5	3+
krypton	Kr	36	1351				
lanthanum	La	57	538	1.10	[48]	106	3+
lawrencium	Lr	103				94	3+
lead	Pb	82	716	1.8	35.1	120	2+
lithium	Li	3	520	0.98	59.63	68	1+
lutetium	Lu	71	524	1.2		86.1	3+
magnesium	Mg	12	738	1.2		65	2+

Element	Symbol	Atomic number	Ionization energy (kJ/mol)	Electro-negativity	Electron affinity (kJ/mol)	Ionic radius (pm)	Common ion charge
manganese	Mn	25	717	1.5		80	2+
mendelevium	Md	101	635	1.3		114	2+
mercury	Hg	80	1007	1.9		110	2+
molybdenum	Mo	42	685	1.8	72.2	62	6+
neodymium	Nd	60	530	1.2		98.3	3+
neon	Ne	10	2081				
neptunium	Np	93	605	1.3		75	5+
nickel	Ni	28	737	1.8	111.5	72	2+
niobium	Nb	41	664	1.6	86.2	72	5+
nitrogen	N	7	1402	3.0			
nobelium	No	102	642	1.3		110	2+
osmium	Os	76	840	2.2	[19]	65	4+
oxygen	O	8	1314	3.50	140.98	140	
palladium	Pd	46	805	2.2	54.2	86	2+
phosphorus	P	15	1012	2.1	72.03	212	
platinum	Pt	78	870	2.2	205.3	70	4+
plutonium	Pu	94	585	1.3		86	4+
polonium	Po	84	812	2.0	[183]	65	4+
potassium	K	19	419	0.8	48.38	138	1+
praseodymium	Pr	59	523	1.13		99	3+
promethium	Pm	61	535	1.2		97	3+
protactinium	Pa	91	568	1.5		78	5+
radium	Ra	88	509	0.9		148	2+
radon	Rn	86	1037				
rhenium	Re	75	760	1.9	[14]	60	7+
rhodium	Rh	45	720	2.2		75	3+
rubidium	Rb	37	403	0.82	46.88	148	1+
ruthenium	Ru	44	711	2.2	[101]	77	3+
samarium	Sm	62	543	1.17		95.8	3+
scandium	Sc	21	631	1.3	18.1	81	3+
selenium	Se	34	941	2.4	194.96	198	
silicon	Si	14	786	1.8			
silver	Ag	47	731	1.93	125.6	126	1+
sodium	Na	11	496	0.93	52.87	95	1+
strontium	Sr	38	549	0.95	4.6	113	2+
sulfur	S	16	1000	2.5	200.41	184	
tantalum	Ta	73	761	1.5	31.1	68	5+
technetium	Tc	43	702	1.9	[53]	58	
tellurium	Te	52	869	2.1	190.15	221	2–
terbium	Tb	65	564	1.2	(–48)	92.3	3+
thallium	Tl	81	589	1.8	–9	144	1+
thorium	Th	90	587	1.3		94	4+
thulium	Tm	69	596	1.25		88.0	3+
tin	Sn	50	709	1.8	107.3	71	4+
titanium	Ti	22	658	1.54	7.6	68	4+
tungsten	W	74	770	1.7	78.6	65	6+
uranium	U	92	598	1.7		73	6+
vanadium	V	23	650	1.63	50.7	59	5+
xenon	Xe	54	1170				
ytterbium	Yb	70	603	1.1		86.8	3+
yttrium	Y	39	616	1.3	29.6	93	3+
zinc	Zn	30	906	1.65		74.0	2+
zirconium	Zr	40	660	1.4	41.1	79	4+

Bracketed values are calculated.
Values in parentheses are estimated.
Values in this table are taken from Lange's *Handbook of Chemistry*.

Common Chemicals

You live in a chemical world. As one bumper sticker asks, "What in the world isn't chemistry?" Every natural and technologically produced substance around you is composed of chemicals. Many of these chemicals are used to make your life easier or safer, and some of them have life-saving properties. Following is a list of selected common chemicals.

Common name	Recommended name	Formula	Common use/source
acetic acid	ethanoic acid	$HC_2H_3O_2$	vinegar
acetone	propanone	$(CH_3)_2CO_{(l)}$	nail polish remover
acetylene	ethyne	$C_2H_{2(g)}$	cutting/welding torch
ASA (Aspirin®)	acetylsalicylic acid	$C_6H_4COOCH_3COOH_{(s)}$	for pain-relief medication
baking soda	sodium hydrogen carbonate	$NaHCO_{3(s)}$	leavening agent
battery acid	sulfuric acid	$H_2SO_{4(aq)}$	car batteries
bleach	sodium hypochlorite	$NaClO_{(s)}$	bleach for clothing
bluestone	copper(II) sulfate pentahydrate	$CuSO_4 \cdot 5H_2O_{(s)}$	algicide, fungicide
brine	aqueous sodium chloride	$NaCl_{(aq)}$	water-softening agent
CFC	chlorofluorocarbon	$C_xCl_yF_{z(l)}$; e.g., $C_2Cl_2F_{4(l)}$	refrigerant
charcoal/graphite	carbon	$C_{(s)}$	fuel, lead pencils
citric acid	2-hydroxy-1,2,3-propanetricarboxylic acid	$C_3H_4OH(COOH)_3$	in fruit and beverages
carbon dioxide	carbon dioxide	$CO_{2(g)}$	dry ice, carbonated beverages
ethylene	ethene	$C_2H_{4(g)}$	for polymerization
ethylene glycol	1,2-ethanediol	$C_2H_4(OH)_{2(l)}$	radiator antifreeze
freon-12	dichlorodifluoromethane	$CCl_2F_{2(l)}$	refrigerant
Glauber's salt	sodium sulfate decahydrate	$Na_2SO_4 \cdot 10H_2O_{(s)}$	solar heat storage
glucose	D-glucose; dextrose	$C_6H_{12}O_{6(s)}$	in plants and blood
grain alcohol	ethanol (ethyl alcohol)	$C_2H_5OH_{(l)}$	beverage alcohol
gypsum	calcium sulfate dihydrate	$CaSO_4 \cdot 2H_2O_{(s)}$	wallboard
lime (quicklime)	calcium oxide	$CaO_{(s)}$	masonry
limestone	calcium carbonate	$CaCO_{3(s)}$	chalk and building materials
lye (caustic soda)	sodium hydroxide	$NaOH_{(s)}$	oven/drain cleaner
malachite	copper(II) hydroxide carbonate	$Cu(OH)_2 \cdot CuCO_{3(s)}$	copper mineral
methyl hydrate	methanol (methyl alcohol)	$CH_3OH_{(l)}$	gas line antifreeze
milk of magnesia	magnesium hydroxide	$Mg(OH)_{2(s)}$	antacid (for indigestion)
MSG	monosodium glutamate	$NaC_5H_8NO_{4(s)}$	flavour enhancer
muriatic acid	hydrochloric acid	$HCl_{(aq)}$	concrete etching
natural gas	methane	$CH_{4(g)}$	fuel
PCBs	polychlorinated biphenyls	$(C_6H_xCl_y)_2$; e.g., $(C_6H_4Cl_2)_{2(l)}$	in transformers
potash	potassium chloride	$KCl_{(s)}$	fertilizer
road salt	calcium chloride or sodium chloride	$CaCl_{2(s)}$ or $NaCl_{(s)}$	melts ice
rotten-egg gas	hydrogen sulfide	$H_2S_{(g)}$	in natural gas
rubbing alcohol	2-propanol (also isopropanol)	$CH_3CHOHCH_{3(l)}$	for massage
sand (silica)	silicon dioxide	$SiO_{2(s)}$	in glassmaking
slaked lime	calcium hydroxide	$Ca(OH)_{2(s)}$	limewater
soda ash	sodium carbonate	$Na_2CO_{3(s)}$	in laundry detergents
sugar	sucrose	$C_{12}H_{22}O_{11(s)}$	sweetener
table salt	sodium chloride	$NaCl_{(s)}$	seasoning
vitamin C	ascorbic acid	$H_2C_6H_6O_{6(s)}$	vitamin supplement
washing soda	sodium carbonate decahydrate	$Na_2CO_3 \cdot 10H_2O_{(s)}$	water softener

Cations and Anions

Common Cations

Ion	Name
H^+	hydrogen
Li^+	lithium
Na^+	sodium
K^+	potassium
Cs^+	cesium
Be^{2+}	beryllium
Mg^{2+}	magnesium
Ca^{2+}	calcium
Ba^{2+}	barium
Al^{3+}	aluminum
Ag^+	silver

Common Anions

Ion	Name
H^-	hydride
F^-	fluoride
Cl^-	chloride
Br^-	bromide
I^-	iodide
O^{2-}	oxide
S^{2-}	sulfide
N^{3-}	nitride
P^{3-}	phosphide

Ion Colours

Ion	Solution colour
Groups 1, 2, 17	colourless
$Cr^{2+}_{(aq)}$	blue
$Cr^{3+}_{(aq)}$	green
$Co^{2+}_{(aq)}$	pink
$Cu^{+}_{(aq)}$	green
$Cu^{2+}_{(aq)}$	blue
$Fe^{2+}_{(aq)}$	pale green
$Fe^{3+}_{(aq)}$	yellow-brown
$Mn^{2+}_{(aq)}$	pale pink
$Ni^{2+}_{(aq)}$	green
$CrO^{2-}_{4(aq)}$	yellow
$Cr_2O^{2-}_{7(aq)}$	orange
$MnO^{-}_{4(aq)}$	purple

Ion	Flame
Li^+	bright red
Na^+	yellow
K^+	violet
Ca^{2+}	yellow-red
Sr^{2+}	bright red
Ba^{2+}	yellow-green
Cu^{2+}	blue (halides) green (others)
Pb^{2+}	light blue-grey
Zn^{2+}	whitish green

Common Polyatomic Ions

Ion	Name	Ion	Name
$C_2H_3O_2^-$	acetate	CO_3^{2-}	carbonate
ClO_3^-	chlorate*	CrO_4^{2-}	chromate
ClO_2^-	chlorite	$Cr_2O_7^{2-}$	dichromate
CN^-	cyanide	HPO_4^{2-}	hydrogen phosphate
$H_2PO_4^-$	dihydrogen phosphate	$C_2O_4^{2-}$	oxalate
HCO_3^-	hydrogen carbonate (bicarbonate)	O_2^{2-}	peroxide
HSO_4^-	hydrogen sulfate (bisulfate)	SiO_3^{2-}	silicate
HS^-	hydrogen sulfide (bisulfide)	SO_4^{2-}	sulfate
HSO_3^-	hydrogen sulfite (bisulfite)	SO_3^{2-}	sulfite
ClO^-, OCl^-	hypochlorite	$S_2O_3^{2-}$	thiosulfate
OH^-	hydroxide	BO_3^{3-}	borate
NO_2^-	nitrite	PO_4^{3-}	phosphate
NO_3^-	nitrate	$P_3O_{10}^{5-}$	tripolyphosphate
ClO_4^-	perchlorate	NH_4^+	ammonium
MnO_4^-	permanganate	H_3O^+	hydronium
SCN^-	thiocyanate	Hg_2^{2+}	mercury(I)

*There are also corresponding ions containing Br and I instead of Cl.

Solubility of Ionic Compounds at SATP

		Anions						
		Cl^-, Br^-, I^-	S^{2-}	OH^-	SO_4^{2-}	CO_3^{2-}, PO_4^{3-}, SO_3^{2-}	$C_2H_3O_2^-$	NO_3^-
Cations	High solubility (aq) ≥0.1 mol/L (at SATP)	most	Group 1, NH_4^+ Group 2	Group 1, NH_4^+ Sr^{2+}, Ba^{2+}, Tl^+	most	Group 1, NH_4^+	most	all
		All Group 1 compounds, including acids, and all ammonium compounds are assumed to have high solubility in water.						
Cations	Low Solubility (s) <0.1 mol/L (at SATP)	Ag^+, Pb^{2+}, Tl^+, Hg_2^{2+} (Hg^+), Cu^+	most	most	Ag^+, Pb^{2+}, Ca^{2+}, Ba^{2+}, Sr^{2+}, Ra^{2+}	most	Ag^+	none

Acids and Bases

Oxyacids

Acid	Name
$HNO_{3(aq)}$	nitric acid
$HNO_{2(aq)}$	nitrous acid
$H_2SO_{4(aq)}$	sulfuric acid
$H_2SO_{3(aq)}$	sulfurous acid
$H_3PO_{4(aq)}$	phosphoric acid
$HC_2H_3O_{2(aq)}$	acetic acid
$HClO_{4(aq)}$	perchloric acid
$HBrO_{4(aq)}$	perbromic acid
$HIO_{4(aq)}$	periodic acid
$HClO_{3(aq)}$	chloric acid
$HBrO_{3(aq)}$	bromic acid
$HIO_{3(aq)}$	iodic acid
$HClO_{2(aq)}$	chlorous acid
$HClO_{(aq)}$	hypochlorous acid
$HBrO_{(aq)}$	hypobromous acid
$HIO_{(aq)}$	hypoiodous acid
$HFO_{(aq)}$	hypofluorous acid

Concentrated Reagents*

Reagent	Formula	Molar mass (g/mol)	Concentration (mol/L)	Concentration (mass %)
acetic acid	$HC_2H_3O_{2(aq)}$	60.05	17.45	99.8
carbonic acid	$H_2CO_{3(aq)}$	62.03	0.039	0.17
formic acid	$HCOOH_{(aq)}$	46.03	23.6	90.5
hydrobromic acid	$HBr_{(aq)}$	80.91	8.84	48.0
hydrochloric acid	$HCl_{(aq)}$	36.46	12.1	37.2
hydrofluoric acid	$HF_{(aq)}$	20.01	28.9	49.0
nitric acid	$HNO_{3(aq)}$	63.02	15.9	70.4
perchloric acid	$HClO_{4(aq)}$	100.46	11.7	70.5
phosphoric acid	$H_3PO_{4(aq)}$	98.00	14.8	85.5
sulfurous acid	$H_2SO_{3(aq)}$	82.08	0.73	6.0
sulfuric acid	$H_2SO_{4(aq)}$	98.08	18.0	96.0
ammonia	$NH_{3(aq)}$	17.04	14.8	28.0
potassium hydroxide	$KOH_{(aq)}$	56.11	11.7	45.0
sodium hydroxide	$NaOH_{(aq)}$	40.00	19.4	50.5

*Typical concentrations of commercial concentrated reagents

Acid–Base Indicators

Common Name	Colour of $HIn_{(aq)}$	pH range	Colour of $In^-_{(aq)}$	Common name	Colour of $HIn_{(aq)}$	pH range	Colour of $In^-_{(aq)}$
methyl violet	yellow	0.0 – 1.6	blue	*p*-nitrophenol	colourless	5.3 – 7.6	yellow
cresol red (acid range)	red	0.2 – 1.8	yellow	litmus	red	6.0 – 8.0	blue
cresol purple (acid range)	red	1.2 – 2.8	yellow	bromothymol blue	yellow	6.2 – 7.6	blue
thymol blue (acid range)	red	1.2 – 2.8	yellow	neutral red	red	6.8 – 8.0	yellow
tropeolin oo	red	1.3 – 3.2	yellow	phenol red	yellow	6.4 – 8.0	red
orange iv	red	1.4 – 2.8	yellow	*m*-nitrophenol	colourless	6.4 – 8.8	yellow
benzopurpurine-48	violet	2.2 – 4.2	red	cresol red	yellow	7.2 – 8.8	red
2,6-dinotrophenol	colourless	2.4 – 4.0	yellow	*m*-cresol purple	yellow	7.6 – 9.2	purple
2,4-dinotrophenol	colourless	2.5 – 4.3	yellow	thymol blue	yellow	8.0 – 9.6	blue
methyl yellow	red	2.9 – 4.0	yellow	phenolphthalein	colourless	8.0 – 10.0	red
congo red	blue	3.0 – 5.0	red	α-naphtholbenzein	yellow	9.0 – 11.0	blue
methyl orange	red	3.1 – 4.4	orange	thymolphthalein	colourless	9.4 – 10.6	blue
bromophenol blue	yellow	3.0 – 4.6	blue-violet	alizarin yellow r	yellow	10.0 – 12.0	violet
bromocresol green	yellow	4.0 – 5.6	blue	tropeolin o	yellow	11.0 – 13.0	orange-brown
methyl red	red	4.4 – 6.2	yellow	nitramine	colourless	10.8 – 13.0	orange-brown
chlorophenol red	yellow	5.4 – 6.8	red	indigo carmine	blue	11.4 – 13.0	yellow
bromocresol purple	yellow	5.2 – 6.8	purple	1,3,5-trinitrobenzene	colourless	12.0 – 14.0	orange
bromophenol red	yellow	5.2 – 6.8	red				

Line Spectra of the Elements

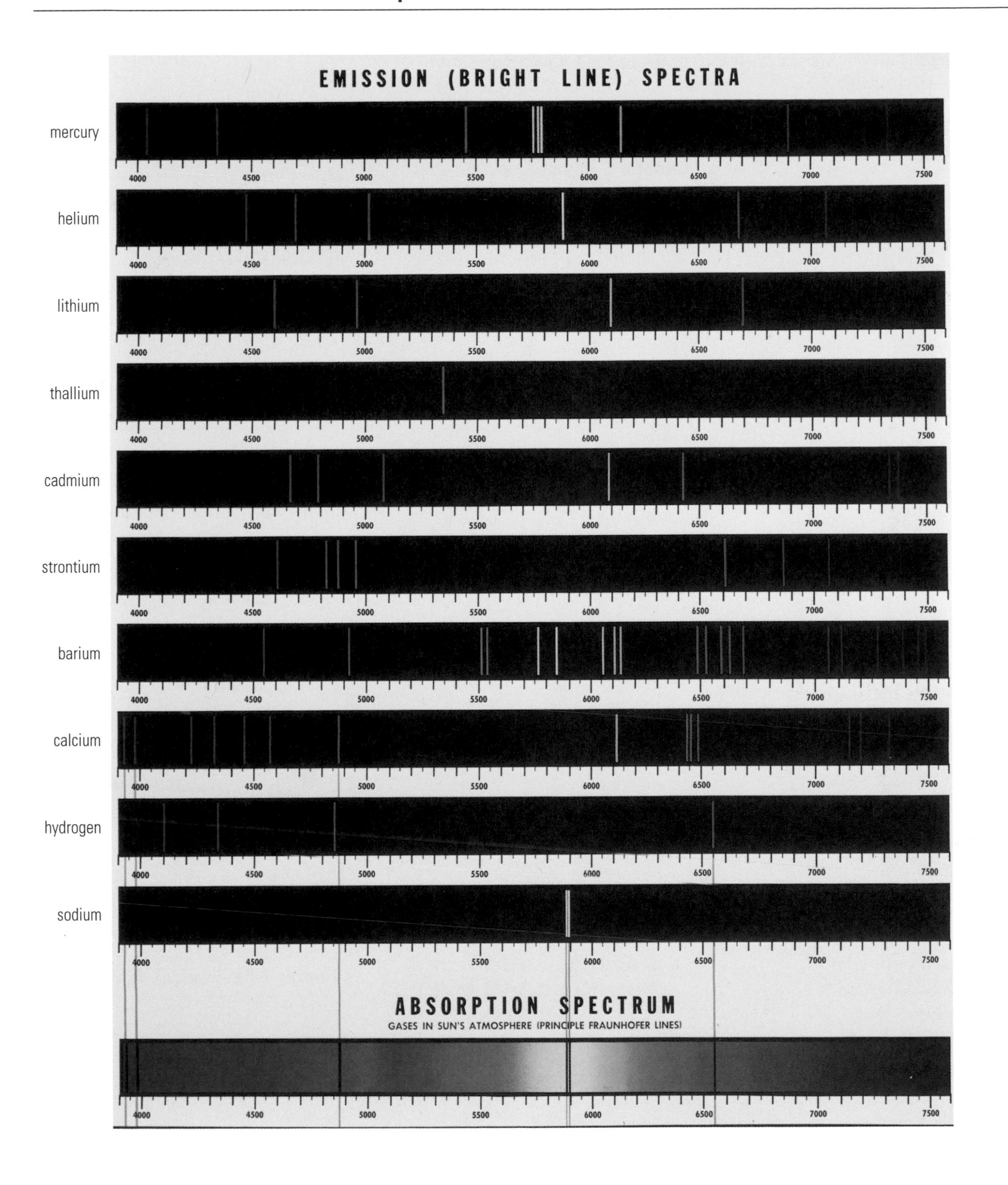

Answers

This section includes numerical answers to questions in Chapter and Unit Reviews that require calculation.

Chapter 1 Review, pp. 62–63

17. (a) 20, 20, 22
  (b) 38, 738, 52
  (c) 55, 55, 82
  (d) 26, 26, 33
  (e) 11, 11, 13

24. (c) <48.4 KJ/mol

Unit 1 Review, pp. 150–153

1. (a) 53, 74, 53
  (b) 15, 17, 15
  (c) 29, 35, 29
  (d) 80, 123, 80

3. 62.5 g

Chapter 4 Review, pp. 199–201

2. 24 u

3. 1.5 u

6. 28.11 u

8. (a) 100.09 g/mol
  (b) 92.02 g/mol
  (c) 286.19 g/mol

9. (a) 17.11 mol
  (b) 22.72 mol
  (c) 55.49 mol

10. (a) 48.0 g
  (b) 0.301 g
  (c) 240 kg

11. (a) 2.5×10^{23}
  (b) 4.6×10^{20}
  (c) 3.91×10^{23}
  (d) 1.76×10^{24}

12. (a) 294.34
  (b) 1.2×10^{24}
  (c) 1.3×10^{21}

13. (a) 35.4%
  (b) 52.9%
  (c) 9.1%

15. (a) $C_4H_5N_2O$
  (b) $C_8H_{10}N_4O_2$

16. (a) $C_4H_{14}O$
  (b) $C_8H_8O_2$

17. $C_8H_8O_3$

18. Na 42.1%; P 18.9%; O 39.0%

19. Na 33.2%; As 36.0%; O 30.8%

20. (a) $ZnCl_2$

Chapter 5 Review, pp. 252–253

8. 2.13×10^3 g

9. (a) 6.30 g
  (b) 1.47 g

10. (b) 39.4 g

11. 168 g

12. (c) 23.1 g
  (d) 97.8%

14. (a) 11.8 g
  (b) 70.9 g
  (c) 6.36 g

15. (b) 859.6 g
  (c) 527.9 g
  (d) 937.5 g

16. 3.77 kg

Unit 2 Review, pp. 256–259

1. (a) 2.87 kg
  (b) 73.8 g
  (c) 3.58 g
  (d) 0.115 g
  (e) 66.5 kg
  (f) 4.75 kg

2. (a) 9.73×10^{-2} mol
  (b) 0.186 mol
  (c) 84.2 mol
  (d) 0.832 mol
  (e) 0.100 mol
  (f) 2.98×10^{-4} mol

3. 5.00 g $O_{2(g)}$

4. 3.34 mmol

6. (a) 54.21% Ba, 20.53% Cr, 25.26% O
  (b) 49.54% Co, 10.10% C, 40.36% O
  (c) 20.66% Fe, 39.34% Cl, 4.48% H, 35.51% O

7. (a) 2:3:2:2
  (b) 2:3:2:3
  (c) 2:2:1

9. (a) 8:1:8
  (b) 2:15:12:6
  (c) 2:2:2:1

13. $C_{10}H_{14}N_2$

19. 95.9%

20. (b) 2.32 g
  (c) 8.38 g

21. (b) 3.48 g
  (c) 83.5%

22. $FePO_4{\cdot}4H_2O$

23. (a) 0.138 g
  (b) 0.0190 g

24. (a) 8.42 g
  (b) 11.6 g

25. (a) 146.6 g
  (b) 106.6 g

26. 2.93 g

27. (a) 4.67 g
  (d) 4.29 g

Chapter 6 Review, pp. 309–311

5. 5.0 g

6. 5.9% MF, 51 g; 2.0% MF, 0.15 kg; 1.2% MF, 0.25 kg

7. 0.30 L

8. 11 mg

9. (a) 0.70 mol/L
  (b) 0.125 mol/L
  (c) 2.0 mol/L
  (d) 0.66 mmol/L

10. (c) Na^+, 125 ppm; K^+, 138 ppm

11. 12.6 g

12. 42.8 mL

13. 61 mL

17. (a) 2.82 g

18. (a) 25.0 mL

Chapter 7 Review, pp. 358–359

9. (a) 0.138 L
  (b) about 0.35 L

10. 0.143 mol/L

14. (b) 0.528 mol/L

15. (b) 0.799 mol/L

Chapter 8 Review, pp. 403–405

6. (a) 2.1
  (b) 2.60

7. (a) 2.8×10^{-12}
  (b) 3.2×10^{-4}

10. 10:1

24. (b) 0.140 mol/L

25. (b) 7.31 mol/L

Unit 3 Review, pp. 408–411

13. 89 mg/L; 444 mg/L

14. 8.64 g/100 mL

15. 0.41 g

16. 20.0 mmol/L

17. (a) 75.0 ppm
    (b) 1.87 mmol/L

18. 1.2 L

19. (a) 0.573 mol/L
    (b) 649 g

21. (a) 41.3 mL
    (b) 1.85 L
    (c) 24 mmol/L
    (d) 767 mL

23. 3.5×10^{-11}; 7.2×10^{-4} mol/L; 4.506; 3.144; 1.4×10^{-8}

25. 6.90 L

26. 45.0 g

27. (a) 14 mg
    (b) 48 ppb

28. 0.32 mmol/L

34. (a) 8.1 mL

37. (b) 189 g
    (c) 14.7 times

Chapter 9 Review, pp. 456–457

6. (a) 99.3 kPa
   (b) 0.150 kPa
   (c) 253 kPa

7. (a) 273 K
   (b) 294 K
   (c) 310 K
   (d) 0 K

8. 8.23 L

9. (a) 150 mL
   (b) 135 kPa

10. 317°C

11. 302 kPa

12. (a) 2×10^3 m^3
    (b) 0.11 L

13. 2.53 kL, or 2.53 m^3

14. 6.4 kg

15. 1.03 kmol

Chapter 10 Review, pp. 492–493

3. 250 kPa

4. 96.83 kPa

5. (b) 6.67 L $CO_{2(g)}$, 1.43 L $N_{2(g)}$, 2.38 L $H_2O_{(g)}$

7. 24.8 L $N_{2(g)}$, 49.6 L $H_2O_{(g)}$, 12.4 L $O_{2(g)}$; total = 86.8 L

8. (a) 5.15 mmol
   (b) 162 kPa
   (c) 3.32×10^4 tubes

9. 254 kL, or 254 m^3

10. (a) 5.00×10^2 kL
    (b) 2.23 t

11. 0.11 GL

12. (a) 25.0 mL
    (b) 46.9 mL
    (c) 37.7 mg

13. 80 kPa $N_{2(g)}$, 120 kPa $H_{2(g)}$

15. (b) $V_{C_3H_{8(g)}} = 22.0$ L/mol
    (c) 1.78%

18. (a) 205 L

19. (a) 798 mL $CO_{2(g)}$, 1.20 L $H_2O_{(g)}$, 399 mL $SO_{2(g)}$

Unit 4 Review, pp. 496–499

5. (a) 0.41 MPa
   (b) 102 kPa
   (c) 45.6 MPa

6. (a) 0.21 mol
   (b) 0.924 mmol
   (c) 3.6 kmol

7. (a) 12.4 kL, or 12.4 m^3
   (b) 1.4 ML
   (c) 1.13 L

8. 170 kL, or 170 m^3

9. 0.33 mol

10. 2.6 kL

11. 8.96 g/L

14. 173 atm

15. 4.73 L

16. 1.2 m^3, or 1.2 kL

18. 97.0 kPa

19. $p_{N_2} = 78$ kPa, $p_{Ar} = 1$ kPa

20. (a) 1.00 L $NH_{3(g)}$, 1.25 L $O_{2(g)}$

21. (a) 32.8 L $CO_{2(g)}$, 16.4 L $N_{2(g)}$, 27.3 L $H_2O_{(g)}$, 2.73 L $O_{2(g)}$
    (c) 17.1 MPa

22. 2.7×10^2 kg

23. $V = 1.19$ L/mol

24. (a) 50 mL
    (b) reaction 1, 6 mol; reaction 2, 4 mol
    (c) 0.56 L
    (d) 2.1 L

25. (c) 9.90 mmol
    (d) 162 mL

26. $M = 101$ g/mol

27. (a) 29.1 mmol
    (b) Yield $CO_{2(g)} = 96.4\%$

28. (a) Yield $H_{2(g)} = 84\%$

30. (a) 5.54 GL $CO_{2(g)}$; 1.39 GL $H_2O_{(g)}$; 2.77 GL $NO_{2(g)}$

Chapter 12 Review, pp. 596–597

5. 988 kJ

6. 38.5°C

11. 414 g

12. 939 g

13. 37.7 kJ/g

15. (a) 373 kJ/mol
    (b) 6.21 MJ

17. (a) average 47.3 kJ/g
    (c) 6.74 MJ/mol

18. (a) 276 J/g; 171 J/g; 95 J/g

19. (a) 39.0 kJ/mol

21. (a) 2.0 MJ

Unit 5 Review, pp. 600–603

20. (a) 1.429 MJ/mol; 1.323 MJ/mol; 1.258 MJ/mol

21. (a) 400 kJ/mol
    (b) 133 kJ/mol
    (c) 33.3 kJ

22. 64.5 kJ/mol

23. (c) 1.41 MJ

24. 21 g

25. 286 kJ/mol

26. 25.6°C

29. (a) 19.9 kJ/g; 27.7 kJ/g
    (b) 638 kJ/mol; 1.28 MJ/mol

31. (a) 5.7 GJ
    (b) $50

Glossary

A

absolute zero: believed to be the lowest possible temperature

acid: a compound that ionizes in water to form hydrogen ions (according to Arrhenius' theory); a proton donor (according to the Brønsted-Lowry concept); a substance that, in aqueous solution, turns blue litmus red (empirical definition)

acid rain: any form of natural precipitation that has an abnormally high acidity

actinide: actinium and the 13 elements that follow it in the seventh row of the periodic table, elements 89 to 102

activity series: a list of elements arranged in order of their reactivity, based on empirical evidence gathered from single displacement reactions

actual yield: the amount of product that is actually obtained at the end of a procedure

acyclic hydrocarbons: open-chain hydrocarbons without any rings of carbon atoms

addition polymerization: a reaction in which unsaturated monomers combine with each other to form a polymer

agricultural runoff: the surface water, with its load of pollutants in solution and in suspension, that drains off farmland

alchemy: a medieval chemical philosophy or practice, the principal goals of which were to transmute elements (e.g., lead to gold), to cure all illness, and to manufacture an essence that would allow long life

alicyclic hydrocarbons: hydrocarbons that have a structure based on rings of carbon atoms

aliphatic compound: one that has a structure based on straight or branched bond chains or rings of carbon atoms; not including aromatic compounds, for example, benzene

alkali metal: an element in Group 1 of the periodic table

alkaline earth metal: an element in Group 2 of the periodic table

alkane: a hydrocarbon with only single bonds between carbon atoms

alkenes: a hydrocarbon family of molecules that contain at least one carbon–carbon double bond; general formula, C_nH_{2n}

alkyl group: a hydrocarbon group derived from an alkane by the removal of a hydrogen atom; often a substitution group or branch on an organic molecule

alkynes: a hydrocarbon family of molecules that contain at least one carbon–carbon triple bond; general formula, C_nH_{2n-2}

alloy: a homogeneous mixture (a solution) of two or more metals

alpha decay: the process in which alpha particles are emitted

alpha particle: the nucleus of a helium atom

alpha ray: a stream of alpha particles

amphiprotic: a substance capable of acting as an acid or a base in different chemical reactions; an entity that can gain or lose a proton (sometimes called amphoteric)

analogy: a comparison of a situation, object, or event with more familiar ideas, objects, or events

aqueous solution: a solute dissolved in water

aquifer: an underground formation of permeable rock or loose material that can produce useful quantities of water when tapped by a well

artificial transmutation: the artificial bombardment of a nucleus by a small entity such as a helium nucleus, a proton, or a neutron

atmospheric pressure: the force per unit area exerted by air on all objects

atom: the smallest particle of an element that has all the properties of that element (theoretical definition)

atomic mass: the relative mass of an atom on a scale on which the mass of one atom of carbon-12 is exactly 12 u; represented by A_r

atomic mass unit: a unit of mass defined as 1/12 the mass of a carbon-12 atom; represented by u

atomic number: the number of protons present in the nucleus of an atom of a given element; represented by Z

atomic radius: a measurement of the size of an atom in picometres (1 pm = 10^{-12} m)

Avogadro's constant: the number of entities in one mole: 6.02×10^{23}; represented by N_A

Avogadro's theory: equal volumes of gases at the same temperature and pressure contain equal numbers of molecules

B

base: an ionic hydroxide that dissociates in water to produce hydroxide ions (according to Arrhenius' theory); a proton acceptor (according to the Brønsted-Lowry concept); a substance that, in aqueous solution, turns red litmus blue (empirical definition)

beta decay: the process in which beta particles are emitted

beta particle: an electron emitted by certain radioactive atoms

beta ray: a stream of beta particles

binary compound: a compound composed of two kinds of atoms or two kinds of monatomic ions

bond energy: the energy required to break a chemical bond; the energy released when a bond is formed

bonding capacity: the number of electrons lost, gained, or shared by an atom when it bonds chemically

Boyle's law: as the pressure on a gas increases, the volume of the gas decreases proportionally, provided that the temperature and amount of gas remain constant; the volume and pressure of a gas are inversely proportional

C

calorimeter: an isolated system consisting of some chemical system surrounded by a measured quantity of water and other components

calorimetry: the technological process of measuring energy changes in chemical systems

carbon-14 dating: a technique that uses radioactive carbon-14 to identify the date of death of once-living material

catalyst: a substance that speeds up the rate of a reaction without undergoing permanent change itself

catalytic converter: a device that uses catalysts to convert pollutant molecules in vehicle exhaust to less harmful molecules

Charles's law: the volume of a gas varies directly with its temperature in kelvin, if the pressure and amount of gas are constant

chemical bond: the forces of attraction holding atoms or ions together

chemical change: any change in which a new substance is formed

chemical energy: energy derived directly from a chemical reaction; for example, burning of hydrocarbons in a furnace

chemical equation: a representation of a chemical reaction that indicates the chemical formulas, relative number of entities, and states of matter of the reactants and products

chemical nomenclature: the system, such as the one approved by IUPAC, of names used in chemistry

coefficient: a whole number indicating the ratio of molecules or formula units of each substance involved in a chemical reaction

collision–reaction theory: a theory stating that chemical reactions involve collisions and rearrangements of atoms or groups of atoms and that the outcome of collisions depends on the energy and orientation of the collisions

combined gas law: the product of the pressure and volume of a gas sample is proportional to its absolute temperature in kelvin; $pv = kT$

combustion reaction: the reaction of a substance with oxygen, producing oxides and energy

complete combustion: the reaction of an element or compound with oxygen to produce the most common oxides, for example, carbon dioxide, sulfur dioxide, nitrogen dioxide, and water

compound: a pure substance that can be broken down by chemical means to produce two or more pure substances (empirical definition); a substance containing atoms of more than one element combined in fixed proportions (theoretical definition)

concentrated: having a relatively large quantity of solute per unit volume of solution

concentration: the amount of a given solute in a solution

conjugate acid: the acid formed by adding a proton (H^+) to a base

conjugate acid–base pair: an acid–base pair that differs by one proton (H^+)

conjugate base: the base formed by removing a proton (H^+) from an acid

continuous spectrum: the pattern of colours observed when a narrow beam of white light is passed through a prism or spectroscope (empirical definition)

cooling rate: the heat flowing out of a substance per unit time (in units of, e.g., J/min)

coordinate covalent bond: a covalent bond in which both of the shared electrons come from the same atom

corrosion: the unwanted reaction of metals with chemicals in the environment

covalent bond: the attractive force, between two atoms of nonmetallic elements, that results when electrons are shared by the atoms

cracking: reaction of a larger hydrocarbon molecule that breaks it into two or more smaller hydrocarbon molecules

crude oil: the fraction of petroleum that is liquid at normal pressure, consisting of higher molar mass hydrocarbons than natural gas; it is refined to separate it into gas, liquid, and solid hydrocarbon components

crystal lattice: a regular, ordered arrangement of atoms, ions, or molecules

cycloalkane: a class of alicyclic hydrocarbon with only single bonds between carbon atoms that are bonded to form a ring structure

Dalton's law of partial pressures: the total pressure of a mixture of nonreacting gases is equal to the sum of the partial pressures of the individual gases

decomposition reaction: a chemical reaction in which a compound is broken down into two or more simpler substances

diagnostic test: an empirical test to detect the presence of a chemical

diatomic: a molecule composed of two atoms of the same or different elements

dilute: having a relatively small quantity of solute per unit volume of solution

dilution: the process of decreasing the concentration of a solution, usually by adding more solvent

dipole–dipole force: an attractive force acting between polar molecules

dissociation: the separation of ions that occurs when an ionic compound dissolves in water

double displacement reaction: a reaction in which aqueous ionic compounds rearrange cations and anions, resulting in the formation of new compounds

E

electrical conductivity: the ability of a material to allow electricity to flow through it

electrolyte: a compound that, in aqueous solution, conducts electricity (empirical definition)

electron: a negatively charged subatomic particle

electron affinity: the energy change that occurs when an electron is accepted by a neutral atom in the gaseous state

electron cloud: the region of an atom in which electrons are most probably located

electron dot diagram: a representation of an atom or ion, made up of the chemical symbol and dots indicating the number of electrons in the valence energy level; also called Lewis symbol

electronegativity: a number that describes the relative ability of an atom, when bonded, to attract electrons

element: a pure substance that cannot be broken down into simpler substances by chemical means (empirical definition); a substance composed entirely of one kind of atom (theoretical definition)

empirical definition: a statement that defines an object or process in terms of observable properties

empirical formula: simplest whole-number ratio of atoms or ions in a compound

empirical knowledge: knowledge coming directly from observations

endothermic: changes that usually involve a decrease in the temperature of the surroundings; energy is transferred as heat from the surroundings to a chemical system

endpoint: the point in a titration at which a sharp change in a property occurs (e.g., a colour change)

energy level: a state with definite and fixed energy in which an electron is allowed to move (theoretical definition)

excess reagent: the reactant that is present in more than the required amount for complete reaction

exothermic: changes that usually involve an increase in the temperature of the surroundings; energy is transferred as heat from a chemical system to the surroundings

F

flame test: a diagnostic technique in which a metallic compound is placed in a flame and the colour produced is used to identify the metal in the compound

formula unit: the simplest whole-number ratio of atoms or ions of the elements in an ionic compound

fractional distillation: the separation of components of petroleum by distillation, using differences in boiling points; also called fractionation

fusion: a nuclear change in which small nuclei combine to form a larger nucleus accompanied by the release of very large quantities of energy

G

galvanizing: coating iron or steel with zinc to prevent rusting

gamma rays: high-energy (short wavelength) electromagnetic radiation emitted during radioactive decay

gas constant: the constant of variation that relates the pressure in kilopascals, volume in litres, amount in moles, and temperature in kelvin of an ideal gas; represented by R

geometric (cis-trans) isomers: organic molecules that differ in structure only by the position of groups attached on either side of a carbon–carbon double bond; a cis isomer has both groups on the same side of the molecular structure; a trans isomer has the groups on opposite sides of the molecular structure

geothermal energy: energy from the heat of the Earth; for example, water pumped into and out of hot rock

global warming: the increase in the average temperature of Earth's atmosphere

gravimetric stoichiometry: the procedure for calculating the masses of reactants or products in a chemical reaction

greenhouse effect: a theory stating that heat is trapped near Earth's surface by carbon dioxide gas, atmospheric water vapour, and some other gases

ground state: the lowest energy level that an electron can occupy (theoretical definition)

group: a column of elements in the periodic table; sometimes referred to as a family

H

half-life: the time it takes for one-half the nuclei in a radioactive sample to decay

halogen: an element in Group 17 of the periodic table

hard water: water containing an appreciable concentration of calcium and magnesium ions

heat: the energy being transferred between substances (in units, e.g., of kilojoules and represented by the quantity symbol q)

heat capacity: the energy absorbed or released by a system per degree Celsius (in units of, e.g., kJ/°C; represented by C)

heat of reaction: the heat transferred in a reaction based on the amounts given by the coefficients of the balanced chemical equation (in units of, e.g., kJ)

high solubility: with a maximum concentration at SATP of greater than or equal to 0.1 mol/L

homogeneous mixture: a uniform mixture of only one phase

hydrate: a compound that decomposes to an ionic compound and water vapour when heated (empirical definition); a compound that contains water as part of its crystal structure (theoretical definition)

hydrocarbons: organic compounds that contain only carbon and hydrogen atoms in their molecular structure

hydrogen bond: a relatively strong dipole–dipole force between a positive hydrogen atom of one molecule and a highly electronegative atom (F, O, or N) in another molecule

hydrogen polyatomic ion: a bi-ion; a polyatomic ion with an available hydrogen ion (e.g., hydrogen carbonate (bicarbonate) ion, hydrogen sulfite (bisulfite) ion)

hydronium ion: a hydrated hydrogen ion (proton), conventionally represented as $H_3O^+_{(aq)}$

ideal gas: a hypothetical gas composed of hypothetical particles that would have zero size, travel in exactly straight lines, and have no attraction to each other (zero intermolecular force)

ideal gas law: the product of the pressure and volume of a gas is directly proportional to the amount and the kelvin temperature of the gas; $pv = nRT$

immiscible: two liquids that form separate layers instead of dissolving

incomplete combustion: the reaction of an element or compound with oxygen to produce some oxides with less oxygen content than the most common oxides, for example, carbon monoxide and nitrogen monoxide

insoluble: having a negligible solubility at SATP

intermolecular force: the attractive force between molecules

intramolecular force: the attractive force between atoms and ions within a compound

ion: a charged entity formed by the addition or removal of one or more electron(s) from a neutral atom (theoretical definition)

ionic bond: the electrostatic attraction between positive and negative ions in a compound; a type of chemical bond

ionic compound: a pure substance formed from a metal and a nonmetal

ionization: any process by which a neutral atom or molecule is converted into an ion

ionization energy: the amount of energy required to remove an electron from an atom or ion in the gaseous state

isotope: atoms of an element that have the same number of protons and neutrons; there may be several isotopes of the same element that differ from each other only in the number of neutrons in the nuclei (theoretical definition)

isotopic abundance: the percentage of an isotope in a sample of an element

IUPAC: the International Union of Pure and Applied Chemistry; the international body that approves chemical names, symbols, and units

Kelvin temperature scale: a temperature scale with zero kelvin (0 K) at absolute zero and the same size divisions as the Celsius temperature scale

kinetic molecular theory: a theory stating that all matter is made up of particles in continuous random motion; temperature is a measure of the average speed of the particles

landfill leachate: water that has filtered through or under a landfill site, picking up pollutants during its passage

lanthanide: lanthanum and the 13 elements that follow it in the sixth row of the periodic table, elements 57 to 70

law of combining volumes: the law stating that when measured at the same temperature and pressure, volumes of gaseous reactants and products of chemical reactions are always in simple, whole-number ratios; also called Gay-Lussac's law of combining volumes

law of conservation of mass: the law stating that during a chemical reaction matter is neither created nor destroyed

law of constant composition: the law stating that compounds always have the same percentage composition by mass

law of definite proportions: the law stating that a specific compound always contains the same elements in definite proportions by mass

Lewis structure: a representation of covalent bonding based on Lewis symbols; shared electron pairs are shown as lines and lone pairs as dots

Lewis symbol: a representation of an atom or ion, made up of the chemical symbol and dots indicating the number of electrons in the valence energy level; an electron dot diagram

limiting reagent: the reactant that is completely consumed in a chemical reaction

line spectrum: a pattern of distinct lines, each of which corresponds to light of a single wavelength, produced when light consisting of only a few distinct wavelengths passes through a prism or spectroscope (empirical definition)

London dispersion force: an attractive force acting between all molecules, including nonpolar molecules

lone pair: a pair of valence electrons not involved in bonding

low solubility: with a maximum concentration at SATP of less than 0.1 mol/L

mass number: the sum of the number of protons and neutrons present in the nucleus of an atom; represented by *A*

mass spectrometer: a sophisticated instrument used for studying the structures of elements and compounds; one application is to precisely determine the mass and abundance of isotopes

metal: an element that is a good conductor of electricity, malleable, ductile, and lustrous

metalloid: an element located near the "staircase line" on the periodic table; having some metallic and some nonmetallic properties

miscible: liquids that mix in all proportions and have no maximum concentration

model: a mental or physical representation of a theoretical concept

molar concentration: the amount of solute, in moles, dissolved in one litre of solution

molar heat of reaction: the quantity of heat transferred in a reaction per mole of a specified substance (in units of, e.g., kJ/mol); represented by ΔH_r

molar mass: the mass, in grams, of one mole of a substance; the SI unit for molar mass is g/mol

molar volume: the volume that one mole of anything occupies at a specified temperature and pressure

mole: the amount of a substance; the number of entities equivalent to Avogadro's number (6.02×10^{23}); the number of carbon atoms in exactly 12 g of a carbon-12 sample; the unit of stoichiometry

molecular compound: a pure substance formed from two or more nonmetals

molecular formula: a group of chemical symbols representing the number and kind of atoms covalently bonded to form a single molecule

mole ratio: the ratio of the amount in moles of reactants and/or products in a chemical reaction

monomers: the smallest repeating unit of a polymer

multivalent: the property of having more than one possible valence

N

natural gas: the fraction of petroleum that vaporizes at normal temperature and pressure, consisting of low molar mass hydrocarbons; also primarily methane sold as heating fuel

net ionic equation: a way of representing a reaction by writing only those ions or neutral substances specifically involved in an overall chemical reaction

neutral: having no effect on either red or blue litmus paper; neither acidic nor basic

neutralization: a competition for protons that results in a proton transfer from the acid to the base (according to the Brønsted-Lowry concept); a reaction between an acid and a base that results in a pH closer to 7 (empirical definition)

neutralization reaction: a double displacement reaction between an acid and a base to produce an ionic compound (a salt) and usually water

neutron: an uncharged subatomic particle in the nucleus of an atom

noble gas: an element in Group 18 of the periodic table

nonelectrolyte: a compound that, in aqueous solution, does not conduct electricity (empirical definition)

nonmetal: an element that is generally a nonconductor of electricity and is brittle

nuclear energy: energy derived directly from a nuclear reaction; for example, a nuclear reactor for production of electricity

nucleon: any particle in the nucleus of an atom

nucleus: the small positively charged centre of the atom

octet rule: a generalization stating that when atoms combine, the covalent bonds between them are formed in such a way that each atom achieves eight valence electrons (two in the case of hydrogen)

orbit: a circular (spherical) path in which an electron can move around the nucleus (theoretical definition)

organic chemistry: the study of organic compounds

organic compounds: compounds that contain carbon, except $CO_{(g)}$, $CO_{2(g)}$, and ionic compounds with carbon

oxyacid: an acid containing oxygen, hydrogen, and a third element

oxyanion: a polyatomic ion containing oxygen

partial pressure: the pressure a gas in a mixture would exert if it were the only gas present in the same volume and at the same temperature; represented by *p*

parts per million: unit used for very low concentrations; represented by ppm

percentage ionization: the percentage of molecules that form ions in solution

percentage yield: the ratio, expressed as a percentage, of the actual or experimental quantity of product obtained (actual yield) to the maximum possible quantity of product (theoretical yield) derived from a gravimetric stoichiometry calculation

period: a row in the periodic table

periodic law (according to Mendeleev): Mendeleev's statement that the properties of the elements are a periodic (regularly repeating) function of their atomic masses

periodic law: a rule, developed from many observations, stating that when the elements are arranged in order of increasing atomic number, their properties show a periodic recurrence and gradual change (modern definition)

periodic trend: a gradual and consistent change in properties within periods or groups of the periodic table

petroleum: a complex gas and/or liquid mixture composed mostly of hydrocarbons, obtained by drilling into underground deposits

pH: a measure of the acidity of a solution; the negative logarithm, to the base 10, of the molar concentration of hydrogen ions, $[H^+]$

polar covalent bond: a covalent bond formed between atoms with significantly different electronegativities; a bond with some ionic characteristics

polar molecule: a molecule that is slightly positively charged at one end and slightly negatively charged at the other because of electronegativity differences

polyatomic ion: a covalently bonded group of atoms with an overall charge

polyatomic molecule: a molecule consisting of more than two atoms of the same or different elements

polymerization: a type of chemical reaction involving the formation of very large molecules (polymers) from many small molecules (monomers)

polymers: a long-chain molecule made up of many small identical units (monomers)

precipitate: form a low-solubility solid from a solution (verb); the solid formed in a chemical reaction or by decreased solubility (noun)

pressure: force per unit area

pressure and temperature law (Gay-Lussac's law): the law stating that the pressure exerted by a gas varies directly with the absolute temperature if the volume and amount of gas remain constant

principal quantum number: a number specifying the theoretical energy level of an electron in an atom; represented by n

proton: a positively charged subatomic particle in the nucleus of an atom

pure water: deionized or distilled water

qualitative: describes a quality or change in matter that has no numerical value expressed

qualitative chemical analysis: the identification of substances present in a sample; may involve several diagnostic tests

quantitative: describes a quantity of matter or degree of change of matter; involving measurements related to number or quantity

quantitative analysis: measuring the quantity of a substance

quantum mechanics: a theory of the atom in which electrons are described in terms of their energies and probability patterns

radioactive: capable of spontaneously emitting radiation in the form of particles and/or gamma rays

radioactive decay: the spontaneous decomposition of a nucleus

radioisotope: a radioactive isotope of an element, occurring naturally or produced artificially

reactant: a substance that participates in a chemical reaction and that is consumed during the reaction

refining: the physical and/or chemical process that converts complex organic mixtures into simpler mixtures or purified substances

reforming: reaction of two or more smaller hydrocarbon molecules that combines them into a larger or more branched hydrocarbon molecule

relative atomic mass: the mass of an element that would react with a fixed mass of a standard element, currently carbon-12

representative element: an element in any of Groups 1, 2, and 13 through 18

SATP (standard ambient temperature and pressure): exactly 25°C and 100 kPa

saturated hydrocarbons: hydrocarbons with only single bonds in their molecules; containing a maximum number of hydrogen atoms

saturated solution: a solution containing the maximum quantity of a solute at specific temperature and pressure conditions

single displacement reaction: the reaction of an element with a compound to produce a new element and a new compound

soda-lime process: a water-softening process involving sodium carbonate and calcium hydroxide, in which calcium carbonate and magnesium carbonate are precipitated out

solar energy: energy directly and indirectly from solar radiation; for example, directly through a window or indirectly through the water cycle or winds

solubility: a property of a solute; the concentration of a saturated solution of a solute in a solvent at a specific temperature and pressure

solute: a substance that is dissolved in a solvent to form a solution (e.g., salt, NaCl)

solution: a homogeneous mixture of substances composed of at least one solute and one solvent

solution stoichiometry: a method of calculating the concentration of substances in a chemical reaction by measuring the volumes of solutions that react completely; sometimes called volumetric stoichiometry

solvent: the medium in which a solute is dissolved; often the liquid component of a solution (e.g., water)

specific heat: the heat transferred per unit mass (in units of, e.g., J/kg); represented by h

specific heat capacity: the quantity of heat required to change the temperature of a unit mass of a system by one degree Celsius (in units of, e.g., J/(g·°C)); represented by c

spectator: an entity or substance, such as an ion, molecule, or ionic solid, that does not change or take part in a chemical reaction

stable octet: a full shell of eight electrons in the outer energy level of an atom

standard curve: a graph used for reference, plotted with empirical data obtained using known standards

standard solution: a solution for which the precise concentration is known

stock solution: a solution that is in stock or on the shelf (i.e., available); usually a concentrated (possibly even saturated) solution

stoichiometric: involving a single, whole-number ratio of ions, as in a balanced chemical equation

stoichiometry: the study of the relationships between the quantities of reactants and products involved in chemical reactions

STP (standard temperature and pressure): exactly 0°C and 101.325 kPa

strong acid: an acid that ionizes almost completely (>99%) in water to form aqueous hydrogen ions (theoretical definition)

strong base: an ionic hydroxide that dissociates 100% in water to produce hydroxide ions (according to Arrhenius' theory and the "reaction-with-water" theory)

structural formula: a representation of the number, types, and arrangement of atoms in a molecule, with dashes representing covalent bonds

structural isomers: chemicals with the same molecular formula, but with different structures and different names

synthesis reaction: a chemical reaction in which two or more simple substances combine to form a more complex substance; also known as a combination reaction

temperature: a measure of the average kinetic energy of a substance's particles; represented by *t*, when using degrees Celsius

tertiary compound: a compound composed of three different elements

theoretical knowledge: knowledge based on abstract ideas created to explain observations

theoretical yield: the amount of product that we predict will be obtained, calculated from the equation

theory: a comprehensive set of ideas that explains a law or a large number of related observations

thermal decomposition: a decomposition reaction that occurs when the reactant is heated

thermochemical equation: a balanced chemical equation that includes the heat transferred to or from the surroundings

titrant: the solution in the buret during a titration

titration: a laboratory procedure involving the carefully measured and controlled addition of a solution from a buret into a measured volume of a sample solution

total ionic equation: a chemical equation that shows all high-solubility ionic compounds in their dissociated form

transition: movement of an electron from one energy level to another (theoretical definition)

transition metal: an element in Groups 3 through 12 of the periodic table

transmutation: the changing of one element into another as a result of radioactive decay

transuranic elements: elements that follow uranium in the periodic table, elements 93+

trend: a gradual and consistent change in properties within periods or groups of the periodic table

triad: a group of three elements with similar properties

unified atomic mass unit: a unit of mass for atoms; 1/12 of the mass of a carbon-12 atom (theoretical definition); represented by u

unsaturated hydrocarbon: a reactive hydrocarbon whose molecules contain double and triple covalent bonds between carbon atoms; for example, alkenes and alkynes

valence: the charge on an ion

valence electrons: those electrons that occupy the highest shell of an atom and are used by the atom to form chemical bonds (theoretical definition)

van der Waals forces: weak intermolecular attractions, including London dispersion forces and dipole–dipole forces

weak acid: an acid with characteristic properties less than those of a strong acid (empirical definition); an acid that ionizes only partially (<50%) in water to form hydrogen ions, so exists primarily in the form of molecules (theoretical definition)

weak base: a chemical that reacts less than 50% with water to produce hydroxide ions (according to the "reaction-with-water" theory)

word equation: a representation of a chemical reaction using only the names of the chemicals involved

yield: the amount of product that is obtained in a chemical reaction

Index

A

Absolute zero, 430
Acetylene, 545, 554
Acid–base indicators, 636
Acid–base reactions, 393–97
Acid–base theories, 378–91
Acid–base titration, 394–95
Acid rain, 115
Acids. *See also* Acid–base
  defined, 363, 387
  formulas of, 363
  naming, 98–100
  properties of, 378
  and solutions, 268
  strong, 365, 378–79, 385
  weak, 365, 378–79, 385
Actinides, 19
Activity series, 125
  and single displacement reactions, 126–27
Actual yield, 238
Acyclic hydrocarbons, 535
Addition polymerization, 555
Addition reaction, 545
Agricultural runoff, 295
Air, components of, 449
Air bags, 416
Air quality, 449–52
Alchemy, 24
Alcohol, 267, 348
Alicyclic hydrocarbons, 535
Aliphatic compounds, 508
Alkali metals, 19
Alkaline earth metals, 19
Alkanes, 509–10
  branch–chain, 530–32
  properties of, 537–39
  straight-chain, 529
  structural diagrams, 533–34
Alkenes, 543–45
  naming, 545–46
  properties of, 548–50
  structural diagrams, 546–47
Alkyl group, 529
Alkynes, 543–45
  naming, 545–46
  properties of, 548–50
  structural diagrams, 546–47
Alloys, 129, 132
  common, 134
Alpha decay, 216, 217
Alpha particles, 29–30, 217
Alpha rays, 217
Aluminum, 132
Ammonia, 118, 212, 287, 480
Amphiprotic, 387
Anaesthesia, 487–88
Anaesthetic technician, 490
Analogies, 22
Anhydrous, 95
Anions, 69
  list of, 635
Aqueous solutions, 267–68
  colour of, 341–42
Aquifer, 291
Archimedes, 429
Aristotle, 23–24
Aromatic compounds, 508
Arrhenius, Svante, 277–78, 362–64, 390–91
  acid–base definitions, 381, 382–84, 389
Art conservationist, 208
Artificial transmutation, 220
Atmosphere, 419, 458
Atmospheric pressure, 424, 487, 488
Atomic mass, 14–15, 26–31
  defined, 28
  relative, 163–64
Atomic mass unit (u), 164
Atomic number *(Z)*, 26
Atomic radius, 51
  and first ionization energy, 56
Atomic theory
  development of, 24–25
  modern, 37–47
Atoms, 23
  developing model of, 22–26
  predicting common ions of, 70–71
Avogadro, Amedeo, 168, 466
Avogadro's constant, 168
Avogadro's theory, 466, 469

B

Baking soda, 387
Ball-and-spring model, 535
Ball-and-stick model, 509
Bartlett, Neil, 18
Bases, 19. *See also* Acid–base
  defined, 363, 387
  naming, 100
  properties of, 378
  and solutions, 268
  strong, 384–85
  weak, 384–85
Beckman, Arnold, 371
Becquerel, Henri, 216
Bends, the, 442
Benzene, 508
Berzelius, Jπns Jakob, 10, 506
Beta decay, 216, 218
Beta particles, 30, 218
Beta rays, 218
Binary compounds, 90–92
Bioamplification, 372
Biological oxygen demand (BOD), 337
Biosphere, 464
Blood plasma, 347
Bohr, Niels, 37–40
Bomb calorimeter, 577–78
Bond energy, 586
Bonding capacity, 76
Boyle, H.E.G., 489
Boyle, Robert, 427, 428
Boyle's law, 427
Boyle's machine, 488, 489
Bradley, Susan, 261
Branch-chain alkanes, 530–32
Breathalyzer, 204, 348
Bromine water test, 549
Brønsted, Johannes, 386, 390
Brønsted-Lowry concept, of acids/bases, 386–88, 390
Bronze, 10
Brooks, Harriet, 30
Brown, Robert, 419
Brownian motion, 419
Bruce Nuclear Generating Station, 221
Buckminsterfullerene, 555
Bunsen, Robert Wilhelm, 342

C

Calcium carbide, 554
Calorimeter, 573, 575
  bomb, 577–78
Calorimetry, 573–78
CANDU reactor, 34–35
Carbon cycle, 504–505
Carbon dioxide, 161, 458
  atmospheric, 117, 520
  and global warming, 521–23
Carbon-14 dating, 31
Carbon monoxide, 160–61, 526–27
Carbon-12, 164
Catalysts, 112
Catalytic converters, 112, 452
Cathode ray tube, 24–25
Cations, 69
  list of, 635
Cavendish, Henry, 429
Cellular respiration, 115
Chadwick, James, 25
Charles, Jacques, 429, 432
Charles's law, 432
Chemical bonds, 69
Chemical changes, 108
  mechanism for, 109–10
  representing, 111
Chemical energy, 566
Chemical equations, 111
  balancing, 210–14
Chemical formulas, calculating, 187–93
Chemical laboratory technician, 208
Chemical nomenclature, 89
Chemical reactions
  efficiency of, 238
  evidence of, 108
  industrial, 242–43
  types of, 115
  yield of, 238–39

Chemicals, list of common, 634
Chemistry, in technology, 245–49
Chemistry teacher, 377
Chernobyl, 35
Chlorofluorocarbons (CFCs), 476–79
Cis-trans isomers, 551
Cochineal, 240
Coefficients, 111
Cold packs, 574
Coleridge, Samuel Taylor, 264
Collision–reaction theory, 109–10, 480
Colour, qualitative analysis by, 341–42
Combination reactions, 118
Combined gas law, 436–38
Combustion
  complete, 519–20
  incomplete, 526–27
Combustion analyzer, 185
Combustion reactions, 115
  and the atmosphere, 117
  energy changes and, 577–78
Commercial diver, 490
Common logarithms, 623
Complete combustion, 519–20
Compounds, 10
  aliphatic, 508
  aromatic, 508
  classifying, 66
  hazardous, 537
  names and formulas of, 89–100
  organic, 506–10
  with polyatomic ions, 94
  proportions in, 160–62
  tertiary, 94
Compressed gases, 441–42
Concentrated reagents, list of, 636
Concentrated solution, 281
Concentrations
  calculations, 284–87, 288–89
  defined, 281
  low, 282–83
  percentage, 281
Condensed structural diagrams, 535
Conjugate acid, 388
Conjugate acid–base pair, 388
Conjugate base, 388
Contaminants, 293
Continuous spectrum, 39
Controlled experiments, 606
Cooley, Jean, 501
Cooling rate, 572
Coordinate covalent bonds, 78
Correlational studies, 608
Corrosion, 130
Corrosion engineer, 135
Covalent bonds
  coordinate, 78
  formation of, 75–77
  polar, 82–84
  strength of, 79
Cozens, Frances L., 3
Cracking reactions, 517
Crisscross rule, 90–91
Crude oil, 512
Curie, Marie, 216
Crystal lattice, 69–70
Crystallization, 321
Cycloalkanes, 535–36
  properties of, 537–39

D

Dacron, 173
Dalton, John, 24, 163, 390–91
Dalton's law of partial pressure, 460–63
Davy, Sir Humphry, 389
Decision-making skills, 612–15
Decomposition reactions, 119–20
Deep-sea diving, 442
Democritus, 23
Destructive distillation, 518
Diagnostic radiographer, 33
Diagnostic test, 12
Diatomic, 170
Diatomic molecules, 75
Dilute solution, 281
Dilution, 302–304
  and pH, 372
Dipole–dipole forces, 86
Dissociation, 278, 333, 363
Döbereiner, Johann, 15
Dosimeter, 219
Double displacement reactions, 136–39
Drinking water, 291–98
Dumas, Jean-Baptiste-André, 245

E. coli, 298
Efficiency, of chemical reaction, 238
Einstein, Albert, 22, 38, 390, 391
Electrical conductivity, 66
Electrolytes, 66, 267, 277–78
Electron, 25
Electron affinity, 56–57
Electron cloud, 46
Electron dot diagrams, 72
Electronegativity, 57–58
  in chemical bonds, 83–84
  of selected elements, 127
Elements
  line spectra of, 637
  list of, 632–33
  naming, 10–11
  organizing, 14–15
Empedocles, 24
Empirical definition, 10
Empirical formula, 185
  calculating, 187–88
Empirical knowledge, 22
Endothermic, 571
Endpoint, 395
Energy
  chemical, 566
  geothermal, 566
  nuclear, 566
  solar, 566
  thermal, 567
  use of, 591–93
Energy changes
  classifying, 566–71
  theoretical perspective on, 586–87
Energy level, 38
Energy resource, 566
Environmental chemist, 377
Equations
  chemical, 111, 210–14
  net ionic, 334
  nuclear, 216–22
  thermochemical, 582, 583–85
  total ionic, 333
  word, 111
Erlenmeyer flask, 394
Estimating, 622
Ethene (ethylene), 544, 592–93
Ethyne (acetylene), 545
Excess reagent, 204, 344
  calculating, 230–31
  in quantitative analysis, 205–206
Exothermic, 571

Faraday, Michael, 277, 390
Fertilizers, 213–14, 351
Filtration, 138
First ionization energy, 54–56
Fission, 34, 221
Flame emission spectroscopy, 42
Flame test, 42, 342
Forensic anthropologist, 33
Forensic chemist, 162, 208
Formula unit, 69
Fossil fuels, 507, 520
  alternatives to, 592
  uses of, 591–93
Fourney, Ron, 155
Fractional distillation, 512–14
Freon, 476
Fusion, 221–22

G

Galvanizing, 130
Gamma radiation, 30
Gamma rays, 216, 218
Gas constant, 443
Gases
  applications of, 487–88
  compressed, 441–42
  ideal, 443
  mixtures of, 460–64
  molar mass of, 446–48
  molar volume of, 469–70
  reactions of, 466–70
  reactions producing, 139
  solubility of, 317
Gas laws, 423–38
Gasoline, 266, 598

Gas state, explaining, 420
Gas stoichiometry, 480–82
Gatorade, 270
Gay-Lussac, Joseph, 435, 466
Gay-Lussac's law, 435, 466, 467
Geometric isomers, 551
Geothermal energy, 566
Germanium, 16
Gesner, Abraham, 518
Geysers, 441
Giguère, Paul, 383, 390
Glass making, 245
Global warming, 521–23
Gold
- extracting, 129
- in jewellery, 134
- properties of, 12

Gravimetric stoichiometry, 224
Greenfreeze technology, 479
Greenhouse effect, 117, 521–23
Ground-level ozone, 449, 450–53
Ground state, 38
Group, 18

H

Haber, Fritz, 212
Haber process, 118, 212
Half-life, 30
Halogens, 19
Hard water, 328
Heat, 567–70
Heat capacity, 578
Heat of reaction, 582–87
Heisenberg, Werner, 46
Helium, 442, 443, 470
Heterogeneous mixtures, 266
High solubility, 323
Hoffman apparatus, 119
Homogeneous mixtures, 266
Homologous series, 509
Honey, 321
Hooke, Robert, 428
Hot-air balloon, 446
Household Hazardous Product Symbols (HHPS), 624
Hydrates, 95
Hydrocarbons, 479, 507
- acyclic, 535
- alicyclic, 535
- classification of, 508
- complete combustion of, 519–20
- and global warming, 521–23
- incomplete combustion of, 526–27
- saturated, 538
- unsaturated, 548

Hydrochloric acid, 267
Hydrofluoroether (HFE), 479
Hydrogen, 28
- line spectrum of, 39

Hydrogen bonds, 86–87, 275
Hydrogen compounds, properties of, 363
Hydrogen peroxide, 158
Hydrogen polyatomic ions, 383
Hydrologic cycle, 292
Hydrometallurgist, 135
Hydronium ion, 383

Ibuprofen, 243
Ice, 87
Ideal gas, 443
Ideal gas law, 443–45
Immiscible, 319
Incomplete combustion, 526–27
Industrial chemical reactions, yield in, 242–43
Insoluble, 323
Intermolecular forces, 80, 86, 273
International Union of Pure and Applied Chemistry (IUPAC), 10, 89
Intramolecular force, 80, 273
Iodine, tincture of, 267
Ionic bonding, 68–73
Ionic bonds, 69
- representing, 72–73

Ionic compounds, 66
- binary, 90–92
- explaining properties of, 69–70
- formation of, 70
- liquid, 362
- solubility of, 635
- in water, 277–78

Ionic halides, 70
Ionic radius, 51–54
Ionization, 363
Ionization energy, 54–56
Ions, 51
- colours of, 635
- and the human body, 71

Isolated system, 573
Isomers, structural, 529
Isotopes, 27–28
Isotopic abundance, 165

Kelvin, Lord, 431, 432
Kelvin temperature scale, 431
Kinetic energy, 420
Kinetic molecular theory, 109–10, 419–20
- and Dalton's law of partial pressures, 462–63

Kirchhoff, Gustav Robert, 342
Knowledge, changes in, 390–91
Kolbe, Adolph, 506
Kyoto Protocol, 522

Lab reports, 618–20
Landfill leachate, 293
Lanthanides, 19
Lasers, 3
Lavoisier, Antoine, 389, 460
Law of combining volumes, 466–68
Law of conservation of energy, 573, 574
Law of conservation of mass, 24, 460
Law of constant composition, 24
Law of definite proportions, 162
Laws, 462
Laws of thermodynamics, 573
Lead, 112
Leavening, 438
LeBlanc process, 202, 245–46
Lemieux, Raymond, 506
Lewis, G.N., 72
Lewis structure, 75
Lewis symbols, 72
Limiting reagent, 204, 344
- calculating, 230–31
- determining, 233–34

Line spectra of the elements, 637
Line spectrum, 39
Line structural diagrams, 535
Litmus test, 268
Logarithmic graphs, 623
Logarithms, 623
London dispersion forces, 86, 279
Lone pair, 76
Lothar Meyer, Julius, 15
Lowry, Thomas, 386, 390
Low solubility, 323

M

Magnesium, 180, 584
Mass, calculations, 288–89
Mass number *(A)*, 27
Mass spectrometer, 28, 190
Material Safety Data Sheets (MSDS), 624
Math skills, 621–23
Matter
- early Greek theories of, 23–24
- states of, 418–20

McLennan, John, 443
Measurements, uncertainty in, 621–22
Measuring, 622
Mendeleev, Dmitri, 15–17
Meniscus, 87
Metalloids, 10, 11
Metallurgical engineer, 135
Metals, 10
- extracting, 129

Meteorologist, 490
Meteorology, 487
Methane, 507
Michelangeli, Diane, 413
Miscible, 319
Models, 22
Molar concentration, 283, 351
Molar heat of reaction, 582
Molar mass, 168–70
- of a gas, 446–48
- of new compounds, determining, 190–93

Molar volume, of gases, 469–70
Mole, 168, 283
Mole concept, calculations involving, 171–75

Molecular compounds, 66
  explaining properties of, 80
  naming, 97–98
Molecular formula, 186
  calculating, 189–90
Molecular motion, 421
Molecules, distances between, 421
Mole Day, 171
Mole ratio, 212
Monomers, 555
Montreal Protocol, 478–79
Morveau, Guyton de, 89
Moseley, H.G.J., 26
Multivalent, 91
Multivalent metals, 91–92

N

Nagaoka, Hantaro, 25
Nanticoke electric power generating station, 566
Natural gas, 512
Net ionic equations, 334
Neutral, 268
Neutralization, 139, 370, 387
Neutron, 25
Neutron number *(N)*, 27
Newlands, Alexander, 15
Nichrome, 43
Nitrates, 213–14
Nitrogen cycle, 449–50
Nitrogen fixation, 449
Nitroglycerine, 474
Noble gases, 18, 19
Nonaqueous mixtures, explaining, 279
Nonelectrolytes, 66, 267
Nonmetals, 10
Northern lights, 347
Nuclear energy, 566
Nuclear equations, balancing, 216–22
Nuclear power and waste, 34–35
Nuclear radiation, characteristics of, 29
Nuclear reactor technician, 33
Nucleons, 216
Nucleus, 25

O

Observational studies, 610
Octane, 517
Octet rule, 76
Orbit, 38
Organic chemistry, 506
Organic compounds, 506–10
Organic molecules, diversity of, 555
Oxalic acid, 401
Oxides, resulting from combustion reactions, 115
Oxyacids, 99, 636
Oxyanions, 94
Oxygen
  natural levels of, 337
  and respiration, 463–64
Ozone, ground-level, 449, 450–53
Ozone-destroying substances (ODSs), 478–79
Ozone layer, 458, 475–79

P

Partial pressure
  applications of, 463–64
  Dalton's law of, 461–63
Particle accelerators, 220
Parts per million, 282
Pauling, Linus, 57
Percentage composition, calculating, 178–85
Percentage concentration, 281–82
Percentage ionization, 365
Percentage yield, calculating, 238–39
Period, 18
Periodic law, 15, 17
Periodic table, 15–19
  trends in, 48–58
Periodic trends, 16
Petrochemical engineer, 557
Petrochemicals, from fossil fuels, 592–93
Petroleum, refining, 512–14
Petroleum geologist, 557
pH, 368–71
pH meter, 371
Photochemical smog, 108
Photosynthesis, 119, 586
pH scale, 370
Pickering nuclear power station, 567
Pipet, 304
Planck, Max, 38
Plastics technologist, 557
Polar covalent bonds, 82–84
Polarity, and solubility, 275
Polar molecules, 85
Polyatomic ions, 78
  compounds with, 94–95
  list of, 635
Polyatomic molecules, 75
Polyethylene, 555
Polymerization, 555
Polymers, 555
Polypropylene, 556
Popcorn, 439
Precipitate, 136, 324
Precipitation reactions, 138, 331–32
Precision, 621
Prefixes, 631
Pressure
  defined, 424
  and volume, 424–28, 466–68, 469
Pressure and temperature law, 434–35
Principal quantum number, 46
Products, calculating masses of, 223–27
Propane, 480, 590
Proportions, in compounds, 160–62
Proton, 25
Proust, Joseph, 162
Pure water, 301

Q

Qualitative, 341
Qualitative chemical analysis, 341–45
Quantitative, 24, 341
Quantitative analysis, 204–206, 347–53
Quantum mechanical theory, 45–47
Quantum mechanics, 45

R

Radioactive, 29
Radioactive decay, 30, 216–19
Radioactivity, 216
Radioisotopes, 29
Radiologist, 33
Random error, 621
Reactants, 110
  calculating masses of, 223–27
Reactions
  of gases, 466–70
  in solution, 331–34
Refinery laboratory technician, 557
Refining, 507
  petroleum, 512–14
Reforming reactions, 517
Relative atomic mass, 163–64
Representative elements, 19
Respiration, 463–64
Risk–benefit analysis, 615
Rose hips, 194
Rutherford, Ernest, 25

S

Safety, in the laboratory, 626–29
Safety conventions and symbols, 624–25
Salt of Ammon, 89
Saturated hydrocarbons, 538
Saturated solution, 314
Scientific inquiry skills, 606–11
Scientific notation, 621
Sequential qualitative chemical analysis, 343–44
Sewage treatment, 337–40
Significant digits, 622
Single displacement reactions, 123–34
SI units, 630
Smoke detectors, 217
Soda-lime process, 328
Soddy, Frederick, 27
Sodium carbonate, 202, 245–47
Sodium hydroxide, 302
Solar energy, 566
Solubility, 136–37
  of gases, 317
  of solids, 314
  and temperature, 316, 319
  in water, 319
Solubility categories, 323–25
Solubility curve, 314
Solubility table, 137
Solute, 136, 266

Solutions
 concentration of, 281–89
 defining, 266–68
 explaining, 272–79
 preparation of, 300–304
 reactions in, 331–34
 saturated, 314
 standard, 300, 395
 working with, 376
Solution stoichiometry, 348
Solvay, Ernest, 246–47, 325
Solvay process, 246–47, 325–26
Solvent, 136, 266
Sørenson, Søren, 369
Space-filling model, 509
Specific heat, 572
 of combustion, 578
Specific heat capacity, 568
Spectator, 334
Spectrophotometer, 342
Spectroscopy, 342
Spruce budworm, 242
Stainless steel, 134
Stable octet, 70
Stalactites/stalagmites, 64, 65, 321
Standard ambient temperature and pressure (SATP), 10, 11, 424
Standard atmospheric pressure, 424
Standard curve, 205–206
Standard pressure, 424
Standard solutions, 300, 395
Standard temperature and pressure (STP), 10, 424
States of matter, 418–20
Stock solutions, 302
Stoichiometry, 162
 gas, 480–82
 gravimetric, 224
 solution, 348
Straight-chain alkanes, naming, 529
Strong acids, 365, 378–79, 385
Strong bases, 384–85
Structural diagrams, 509, 535
Structural formula, 76
Structural isomers, 529
Substitution reactions, 539
Sucrose, 506
Sulfuric acid, 283
Surroundings, 573
Symbols, 630–31
Synthesis reactions, 118–19
Systematic error, 621

T

Technological problem-solving skills, 616–17
Technology
 chemistry in, 245–49
 environmental issues, 248–49
Teflon, 555
Temperature
 defined, 571
 and pressure, 434–35, 466–68, 469
 and solubility, 316, 319
 and volume, 429–33, 466–68, 469
Tertiary compounds, 94
Theoretical knowledge, 22, 23
Theoretical yield, 238
Theories, 22, 390–91, 462
Thermal decomposition, 120
Thermal energy, 567
Thermochemical equations, 582, 583–85
Thomson, J.J., 24, 391
Thomson, Sir William (Lord Kelvin), 431
Titrant, 394
Titration, 348, 394–95
Torricelli, Evangelista, 424
Total ionic equations, 333
Transition, 38
Transition metals, 19
Transmutation, 217
 artificial, 220
Transuranic elements, 19
Trends, periodic, 16
Triad, 15
Tripp, Charles, 507
TRIUMF cyclotron, 220

Ultraviolet (UV) radiation, 475–78
Uncertainty in measurements, 621–22
Unified atomic mass unit (u), 28
Units, 630–31
Ununbium, 148
Unsaturated hydrocarbons, 548

Valence, 90
Valence electrons, 46, 69
Van den Broek, A., 17
Van der Waals, Johannes, 86, 444
Van der Waals forces, 86, 444
Vapour pressure, 464
Vapour trails, 520
Ventilation, 487–88
Vitalism, 506
Vitamins
 natural, 194
 solubility of, 278
Volatile organic compounds (VOCs), 449
Volume
 calculations, 288–89
 molar, 469–70
 and pressure, 424–28
 and reactions of gases, 466–70
 and temperature, 429–33
Von Liebig, Justus, 389

Walkteron, Ontario, 298
Waste water treatment, 337–40
Water
 contamination of, 292–94
 displacement of, 464
 drinking, 291–98
 hard, 328
 ionic compounds in, 277–78
 molecular substances in, 273
 pure, 301
 softening, 328–29
 solubility in, 319
 as universal solvent, 264, 266, 279
 waste, 337–40
Water of hydration, 95
Water mixtures, explaining, 273
Water-quality analyst, 377
Water treatment, 295–98
Weak acids, 365, 378–79, 385
Weak bases, 384–85
Williams, James, 507
Willson, Thomas Leopold, 554
Wöhler, Friedrich, 506
Word equations, 111
Workplace Hazardous Materials Information System (WHMIS), 624–25
Wrought iron, 225

Yield
 calculating, 238–39
 in industrial chemical reactions, 242–43

Credits

Unit 1: Matter and Chemical Bonding
Unit opener: p. vi, p. 2 © Richard Megna/Fundamental Photographs; p. 3 right Frances Cozens; p. 5 Dave Starrett; p. 6 Anne Bradley

Chapter 1: p. 9 © Duomo/CORBIS; p. 10 © Royal Ontario Museum/CORBIS; p. 11 © Glen Allison/PhotoDisc; p. 15 © Bettmann/CORBIS; p. 22 Dave Starrett; p. 24 © Christel Gerstenberg/CORBIS; p. 28 © TEK Image/Science Photo Library; p. 33 clockwise from top left © Roger Ressmeyer/CORBIS, © Karen Huntt Mason/CORBIS, © Tim Beddow/Science Photo Library, © Roger Ressmeyer/CORBIS; p. 35 top to bottom © CP Picture Archive (Wide World Inc.), © CP Picture Archive (Wide World Inc.); p. 37 © Bettmann/CORBIS; p. 39 © Wabash Instrument Corp./Fundamental Photographs; p. 44 all © Lester V. Bergman/CORBIS; p. 46 Dave Starrett; p. 57 © Bettmann/CORBIS.

Chapter 2: p. 65 © David Muench/CORBIS; p. 66 Richard Siemens; p. 70 © Margaret Oechsli/Fundamental Photographs; p. 87 Richard Siemens.

Chapter 3: p. 107 © CP Picture Archive (Paul Hourigan); p. 108 left © Hamish Robertson, right © Stephanie Maze/CORBIS; p. 109 top Dave Starrett, bottom © Richard Megna/Fundamental Photographs; p. 114 (a) © Richard Megna/Fundamental Photographs, (b) Dave Starrett, (c) Dave Starrett, (d) Anne Bradley; p. 117 Dave Starrett; p. 118 © Richard Megna/Fundamental Photographs; p. 124 Dave Starrett; p. 129 © Museum of History & Industry/CORBIS; p. 130 top © Paul Almasy/CORBIS, bottom © CORBIS/Magma; p. 134 top © Hamish Robertson, bottom Anne Bradley; p. 135 top & bottom left © Charles O'Rear/CORBIS, right © Dave Thompson/Life File/PhotoDisc; p. 136 Dave Starrett.

Unit 2: Quantities in Chemical Reactions
Unit opener: p. vii, p. 154 © TEK Image/Science Photo Library; p. 155 Ron Fourney.

Chapter 4: p. 159 © CORBIS/Magma; p. 160 © CP Picture Archive (Tom Hanson); p. 161 top Anne Bradley, bottom Richard Siemens; p. 163 © National Research Council Canada; p. 168 Richard Siemens; p. 169 Richard Siemens; p. 173 top © James King-Holmes/Science Photo Library, bottom © Science Photo Library; p. 174 Dave Starrett; p. 177 Dave Starrett; p. 178 © Martin Bond/Science Photo Library; p. 194 © S. Solum/PhotoLink/PhotoDisc; p. 195 © The State Russian Museum/CORBIS; p. 196 Dave Starrett; p. 197 © CORBIS/Magma; p. 201 left © Robin Adshead; The Military Picture Library/CORBIS, right © CORBIS/Magma.

Chapter 5: p. 203 © Gail Mooney/Coribs; p. 204 top © Michael S.Yamashita/CORBIS, bottom © CORBIS/Magma; p. 205 Dave Starrett; p. 208 top © George W. Wright/CORBIS, bottom left © CORBIS/Magma, bottom right © Photodisc/Ryan McVay; p. 209 Dave Starrett; p. 212 © CORBIS/Magma; p. 216 © Bettmann/CORBIS; p. 219 © Bob Rowan; Progressive Image/CORBIS; p. 220 © TRIUMF/G. Roy; p. 221 top photo © Robert Del Tredici, bottom © Photodisc Imaging; p. 224 © Dr. E.R. Degginger; p. 225 © Lowell Georgia/CORBIS; p. 227 Richard Siemens; p. 238 Dave Starrett; p. 240 (a) (b) © Bob Krist/CORBIS, (c) © Alison Wright/CORBIS; p. 242 Queens University, Kingston; p. 243 Anne Bradley; p. 245 top © Charles O'Rear/CORBIS, bottom © James L. Amos/CORBIS; p. 247 © Science Photo Library; p. 254 © Photodisc/C Squared Studios; p. 256 © CORBIS/Magma; p. 257 Lucille Davies; p. 258 Dave Starrett; p. 259 © Paul A.Souders/CORBIS.

Unit 3: Solutions and Solubility
Unit opener: p. viii, p. 260 © WorldSat International Inc., 2001 www.worldsat.ca All Rights Reserved; p. 261 Susan Bradley.

Chapter 6: p. 265 © CORBIS/Magma; p. 266 © CORBIS/Magma; p. 267 top Anne Bradley, bottom left & right Richard Siemens; p. 270 Anne Bradley; p. 271 Anne Bradley; p. 272 top & bottom Dave Starrett; p. 274 Dave Starrett; p. 281 Dave Starrett; p. 283 Dave Starrett; p. 285 Anne Bradley; p. 287 Dave Starrett; p. 293 © Galen Rowell/CORBIS; p. 297 © Eva Omes, City of Toronto Works and Emergency Services; p. 298 © CP Picture Archive (Kevin Frayer); p. 301 left Richard Siemens, right © Ohaus Corporation; p. 302 top © Culligan of Canada Ltd., bottom Richard Siemens; p. 303 Anne Bradley; p. 304 Richard Siemens; p. 309 left, right Anne Bradley; p. 311 © Natalie Fobes/CORBIS.

Chapter 7: p. 313 Anne Bradley; p. 314 © Richard Megna/Fundamental Photographs; p. 321 top Dave Starrett, bottom left © Richard Hamilton Smith/CORBIS, bottom right Dave Starrett; p. 323 Dave Starrett; p. 324 all Dave Starrett; p. 328 left Water Soft DS-32P Softener courtesy of Shelson Crawford, Water Softening and Purification Specialist, right © Francoise Sauze/Science Photo Library; p. 333 Jeremy Jones; p. 337 © Eva Omes, City of Toronto Works and Emergency Services; p. 341 Dave Starrett; p. 342 Richard Siemens; p. 343 © Galen Rowell/CORBIS; p. 347 top © Katsunori Nagase/Economic Tourism and Development, bottom © Photodisc/Keith Brofsky; p. 350 Richard Siemens;p. 351 © Potash & Phosphate Institute; p. 354 Richard Siemens; p. 358 Dave Starrett.

Chapter 8: p. 361 © Ted Spiegel/CORBIS; p. 365 Richard Siemens; p. 369 Richard Siemens; p. 371 Larry Stepanowicz/Fundamental Photographs; p. 372 top © W.Perry Conway/CORBIS, bottom CORBIS/Magma; p. 376 Dave Starrett; p. 377 left Dave Starrett, top right © Jerry Mason/Science Photo Library, bottom right © Colin Cuthbert/Newcastle University/Science Photo Library; p. 379 Richard Siemens; p. 383 © Université Laval, Quebec; p. 386 © Science Photo Library; p. 387 top Dave Starrett, bottom Anne Bradley; p. 389 top © Leonard de Selva/CORBIS, middle left © Michael Nicholson/CORBIS, middle right © Austrian Archives/CORBIS, bottom © Bettmann/CORBIS; p. 393 © CP Picture Archive (Paul Darrow); p. 394 Richard Siemens; p. 406 Anne Bradley.

Unit 4: Gases and Atmospheric Chemistry
Unit opener: p. viii, p. 412 © Bob Rowan; Progressive Image/CORBIS; p. 413 Diane MIchelangeli.

Chapter 9: p. 417 © Richard Olivier/CORBIS; p. 418 top © Alfred Pasieka/Science Photo Library, middle © Sheila Terry/Science Photo Library, bottom © Hamish Robertson; p. 421 Photo courtesy of Boreal Laboratories 1-800-387-9393; p. 428 © Bettmann/CORBIS; p. 429 © Bettmann/CORBIS; p. 438 © CORBIS/Magma; p. 440 Photo courtesy of Boreal Laboratories 1-800-387-9393; p. 441 top ©

CORBIS/Magma, bottom © Photodisc/Steve Cole; p. 442 top © CORBIS/Magma, bottom © Lowell Georgia/CORBIS; p. 446 © Photodisc/Russell Illig; p. 451 © Yann Arthus-Bertrand/CORBIS; p. 456 © David Vaughan/Science Photo Library.

Chapter 10: p. 458 © David Muench/CORBIS; p. 459 © NASA/Science Photo Library; p. 464 © Roger Ressmeyer/CORBIS; p. 468 © Martin Bond/Science Photo Library; p. 469 Richard Siemens; p. 470 © Photodisc/John A. Rizzo; p. 471 © Todd Gipstein/CORBIS; p. 472 Anne Bradley; p. 474 © Paul A. Souders/CORBIS; p. 475 Honda Canada, Inc.; p. 477 © NASA/Science Photo Library; p. 479 photo © Greenpeace Canada; p. 480 Richard Siemens; p. 490 left © Staffan Widstrand/CORBIS, top right © John Greihm/Science Photo Library, bottom right © CORBIS; p. 492 © Alan Towse/CORBIS; p. 494 CORBIS/Magma; p. 495 left © Ralph White/CORBIS, right © Roger Ressmeyer/CORBIS; p. 497 © Jonathan Blair/CORBIS; p. 499 © Bettmann/CORBIS.

Unit 5: Hydrocarbons and Energy
Unit opener: p. 1, p. 500 © Joseph Sohm; ChromoSohm Inc./ CORBIS; p. 501 Jean Cooley; p. 503 © Jamie Bloomquist/oi2.

Chapter 11: p. 506 top © Adam Woolfitt/CORBIS, inset Anne Bradley; p. 507 C 280-0-0-0-14/Archives of Ontario; p. 513 CORBIS/Magma; p. 517 © Hamish Robertson; p. 518 Imperial Oil Archives; p. 527 © Sheila Terry/Science Photo Library; p. 545 © Michael Coyne/The Image Bank; p. 549 Richard Siemens; p. 555 Richard Siemens; p. 557 clockwise from left © Maximilian Stock Ltd./Science Photo Library, © Stevie Grand/Science Photo Library, © Roger Ressmeyer/CORBIS, © Jack Fields/CORBIS.

Chapter 12: p. 565 CORBIS/Magma; p. 566 provided courtesy of Ontario Power Generation; p. 566 top provided courtesy of Ontario Power Generation, bottom © Dr. Ray Clark (FRPS) & Mervyn de Calcinagoff (FRPS)/Science Photo Library; p. 569 © CP Picture Archive (Jacques Boissinot); p. 570 © Bettmann/CORBIS; p. 584 © Photodisc/Don Farrall; p. 586 top © david Turnley/CORBIS, bottom © Photodisc/Adalberto Rios Szalay/Sexto Sol; p. 590 top © Sergio Dorantes/CORBIS, bottom © Frank Jenkins Enterprises; p. 592 © Trans-Canada Pipeline; p. 598 top © Hamish Robertson, bottom © and courtesy of Petro-Canada; p. 600 © and courtesy of Petro-Canada.

Appendix C: p. 637 © Wabash Instrument Corp./Fundamental Photographs.

Illustrators: Andrew Breithaupt, Deborah Crowle, Irma Ikonen, Dave McKay, Linda Neale, Peter Papayanakis

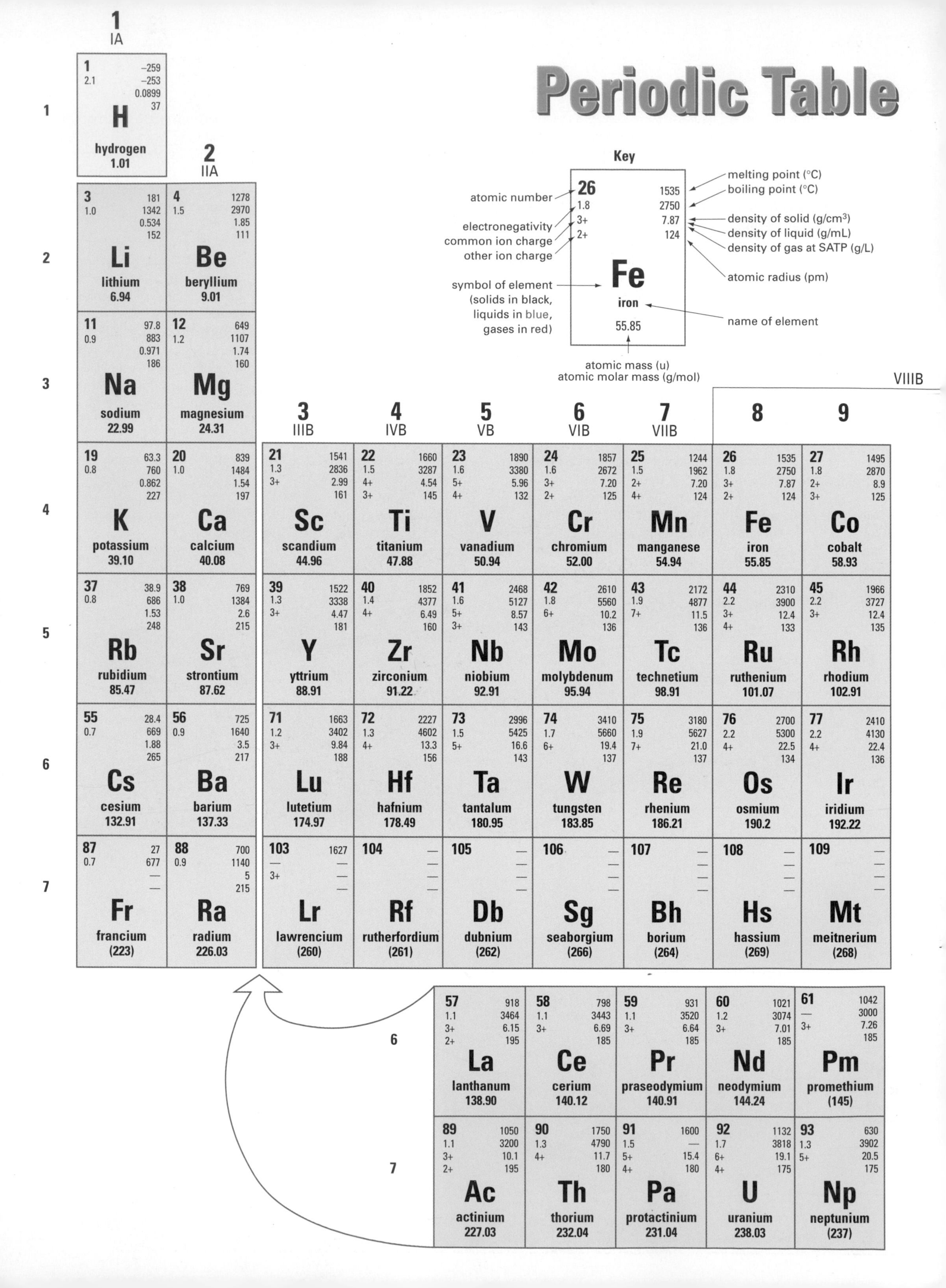

Periodic Table
Key
26 1.8 3+ 2+ 1535 2750 7.87 124 Fe iron 55.85
atomic number
electronegativity
common ion charge
other ion charge
symbol of element (solids in black, liquids in blue, gases in red)
melting point (°C)
boiling point (°C)
density of solid (g/cm3)
density of liquid (g/mL)
density of gas at SATP (g/L)
atomic radius (pm)
name of element
atomic mass (u)
atomic molar mass (g/mol)
1 IA
2 IIA
3 IIIB
4 IVB
5 VB
6 VIB
7 VIIB
8
9
VIIIB
1
2
3
4
5
6
7
1 2.1 −259 −253 0.0899 37 H hydrogen 1.01
3 1.0 181 1342 0.534 152 Li lithium 6.94
4 1.5 1278 2970 1.85 111 Be beryllium 9.01
11 0.9 97.8 883 0.971 186 Na sodium 22.99
12 1.2 649 1107 1.74 160 Mg magnesium 24.31
19 0.8 63.3 760 0.862 227 K potassium 39.10
20 1.0 839 1484 1.54 197 Ca calcium 40.08
21 1.3 3+ 1541 2836 2.99 161 Sc scandium 44.96
22 1.5 4+ 3+ 1660 3287 4.54 145 Ti titanium 47.88
23 1.6 5+ 4+ 1890 3380 5.96 132 V vanadium 50.94
24 1.6 3+ 2+ 1857 2672 7.20 125 Cr chromium 52.00
25 1.5 2+ 4+ 1244 1962 7.20 124 Mn manganese 54.94
26 1.8 3+ 2+ 1535 2750 7.87 124 Fe iron 55.85
27 1.8 2+ 3+ 1495 2870 8.9 125 Co cobalt 58.93
37 0.8 38.9 686 1.53 248 Rb rubidium 85.47
38 1.0 769 1384 2.6 215 Sr strontium 87.62
39 1.3 3+ 1522 3338 4.47 181 Y yttrium 88.91
40 1.4 4+ 1852 4377 6.49 160 Zr zirconium 91.22
41 1.6 5+ 3+ 2468 5127 8.57 143 Nb niobium 92.91
42 1.8 6+ 2610 5560 10.2 136 Mo molybdenum 95.94
43 1.9 7+ 2172 4877 11.5 136 Tc technetium 98.91
44 2.2 3+ 4+ 2310 3900 12.4 133 Ru ruthenium 101.07
45 2.2 3+ 1966 3727 12.4 135 Rh rhodium 102.91
55 0.7 28.4 669 1.88 265 Cs cesium 132.91
56 0.9 725 1640 3.5 217 Ba barium 137.33
71 1.2 3+ 1663 3402 9.84 188 Lu lutetium 174.97
72 1.3 4+ 2227 4602 13.3 156 Hf hafnium 178.49
73 1.5 5+ 2996 5425 16.6 143 Ta tantalum 180.95
74 1.7 6+ 3410 5660 19.4 137 W tungsten 183.85
75 1.9 7+ 3180 5627 21.0 137 Re rhenium 186.21
76 2.2 4+ 2700 5300 22.5 134 Os osmium 190.2
77 2.2 4+ 2410 4130 22.4 136 Ir iridium 192.22
87 0.7 27 677 — — Fr francium (223)
88 0.9 700 1140 5 215 Ra radium 226.03
103 — 3+ 1627 — — — Lr lawrencium (260)
104 — — — — Rf rutherfordium (261)
105 — — — — Db dubnium (262)
106 — — — — Sg seaborgium (266)
107 — — — — Bh borium (264)
108 — — — — Hs hassium (269)
109 — — — — Mt meitnerium (268)
6
7
57 1.1 3+ 2+ 918 3464 6.15 195 La lanthanum 138.90
58 1.1 3+ 798 3443 6.69 185 Ce cerium 140.12
59 1.1 3+ 931 3520 6.64 185 Pr praseodymium 140.91
60 1.2 3+ 1021 3074 7.01 185 Nd neodymium 144.24
61 — 3+ 1042 3000 7.26 185 Pm promethium (145)
89 1.1 3+ 2+ 1050 3200 10.1 195 Ac actinium 227.03
90 1.3 4+ 1750 4790 11.7 180 Th thorium 232.04
91 1.5 5+ 4+ 1600 — 15.4 180 Pa protactinium 231.04
92 1.7 6+ 4+ 1132 3818 19.1 175 U uranium 238.03
93 1.3 5+ 630 3902 20.5 175 Np neptunium (237)

s p d f

of the Elements

Period	10	11 IB	12 IIB	13 IIIA	14 IVA	15 VA	16 VIA	17 VIIA	18 VIIIA
1									2 — / X −272 / −269 / 0.179 / 50 **He** helium 4.00
2				5 2.0 / X 2300 / 2550 / 2.34 / 88 **B** boron 10.81	6 2.5 / X 3550 / 4827 / 2.26 / 77 **C** carbon 12.01	7 3.0 −210 / −196 / 1.25 / 70 **N** nitrogen 14.01	8 3.5 −218 / −183 / 1.43 / 66 **O** oxygen 16.00	9 4.0 −220 / −188 / 1.70 / 64 **F** fluorine 19.00	10 — / X −249 / −246 / 0.900 / 62 **Ne** neon 20.18
3				13 1.5 660 / 2467 / 2.70 / 143 **Al** aluminum 26.98	14 1.8 / X 1410 / 2355 / 2.33 / 117 **Si** silicon 28.09	15 2.1 44.1 / 280 / 1.82 / 110 **P** phosphorus 30.97	16 2.5 113 / 445 / 2.07 / 104 **S** sulfur 32.06	17 3.0 −101 / −34.6 / 3.21 / 99 **Cl** chlorine 35.45	18 — / X −189 / −186 / 1.78 / 95 **Ar** argon 39.95
4	28 1.8 / 2+ / 3+ 1455 / 2730 / 8.90 / 124 **Ni** nickel 58.69	29 1.9 / 2+ / 1+ 1083 / 2567 / 8.92 / 128 **Cu** copper 63.55	30 1.6 / 2+ 420 / 907 / 7.14 / 133 **Zn** zinc 65.38	31 1.6 / 3+ 29.8 / 2403 / 5.90 / 122 **Ga** gallium 69.72	32 1.8 / 4+ 937 / 2830 / 5.35 / 123 **Ge** germanium 72.61	33 2.0 817 / 613 / 5.73 / 121 **As** arsenic 74.92	34 2.4 217 / 684 / 4.81 / 117 **Se** selenium 78.96	35 2.8 −7.2 / 58.8 / 3.12 / 114 **Br** bromine 79.90	36 — / X −157 / −152 / 3.74 / 112 **Kr** krypton 83.80
5	46 2.2 / 2+ / 4+ 1554 / 2970 / 12.0 / 138 **Pd** palladium 106.42	47 1.9 / 1+ 962 / 2212 / 10.5 / 144 **Ag** silver 107.87	48 1.7 / 2+ 321 / 765 / 8.64 / 149 **Cd** cadmium 112.41	49 1.7 / 3+ 157 / 2080 / 7.30 / 163 **In** indium 114.82	50 1.8 / 4+ / 2+ 232 / 2270 / 7.31 / 140 **Sn** tin 118.69	51 1.9 / 3+ / 5+ 631 / 1750 / 6.68 / 141 **Sb** antimony 121.75	52 2.1 450 / 990 / 6.2 / 137 **Te** tellurium 127.60	53 2.5 114 / 184 / 4.93 / 133 **I** iodine 126.90	54 — / X −112 / −107 / 5.89 / 130 **Xe** xenon 131.29
6	78 2.2 / 4+ / 2+ 1772 / 3827 / 21.5 / 138 **Pt** platinum 195.08	79 2.4 / 3+ / 1+ 1064 / 2808 / 19.3 / 144 **Au** gold 196.97	80 1.9 / 2+ / 1+ −39.0 / 357 / 13.5 / 160 **Hg** mercury 200.59	81 1.8 / 1+ / 3+ 304 / 1457 / 11.85 / 170 **Tl** thallium 204.38	82 1.8 / 2+ / 4+ 328 / 1740 / 11.3 / 175 **Pb** lead 207.20	83 1.9 / 3+ / 5+ 271 / 1560 / 9.80 / 155 **Bi** bismuth 209.98	84 2.0 / 2+ / 4+ 254 / 962 / 9.40 / 167 **Po** polonium (209)	85 2.2 302 / 337 / — / 142 **At** astatine (210)	86 — / X −71 / −61.8 / 9.73 / 140 **Rn** radon (222)
7	110 — / — / — / — **Uun** ununnilium (269, 271)	111 — / — / — / — **Uuu** unununium (272)	112 — / — / — / — **Uub** ununbium (277)	113	114 — / — / — / — **Uuq** ununquadium (285)	115	116 — / — / — / — **Uuh** ununhexium (289)	117	118 — / — / — / — **Uuo** ununoctium (293)

Period									
6	62 1.2 / 3+ / 2+ 1074 / 1794 / 7.52 / 185 **Sm** samarium 150.36	63 — / 3+ / 2+ 822 / 1527 / 5.24 / 185 **Eu** europium 151.96	64 1.1 / 3+ 1313 / 3273 / 7.90 / 180 **Gd** gadolinium 157.25	65 1.2 / 3+ 1356 / 3230 / 8.23 / 175 **Tb** terbium 158.92	66 — / 3+ 1412 / 2567 / 8.55 / 175 **Dy** dysprosium 162.50	67 1.2 / 3+ 1474 / 2700 / 8.80 / 175 **Ho** holmium 164.93	68 1.2 / 3+ 1529 / 2868 / 9.07 / 175 **Er** erbium 167.26	69 1.2 / 3+ 1545 / 1950 / 9.32 / 175 **Tm** thulium 168.93	70 1.1 / 3+ / 2+ 819 / 1196 / 6.97 / 175 **Yb** ytterbium 173.04
7	94 1.3 / 4+ / 6+ 641 / 3232 / 19.8 / 175 **Pu** plutonium (244)	95 1.3 / 3+ / 4+ 994 / 2607 / 13.7 / 175 **Am** americium (243)	96 — / 3+ 1340 / 3110 / 13.5 / — **Cm** curium (247)	97 — / 3+ / 4+ 986 / — / 14 / — **Bk** berkelium (247)	98 — / 3+ 900 / — / — / — **Cf** californium (251)	99 — / 3+ 860 / — / — / — **Es** einsteinium (252)	100 — / 3+ 1527 / — / — / — **Fm** fermium (257)	101 — / 2+ / 3+ 1021 / 3074 / — / — **Md** mendelevium (258)	102 — / 2+ / 3+ 863 / — / — / — **No** nobelium (259)